P9-BHV-262

THE CELL

Volume II

THE CELL

Biochemistry, Physiology, Morphology

THE CELL

Biochemistry, Physiology, Morphology

Edited by

JEAN BRACHET

Faculté des Sciences, Université libre de Bruxelles
Bruxelles, Belgique

ALFRED E. MIRSKY

The Rockefeller Institute
New York, New York

VOLUME II

CELLS AND THEIR COMPONENT PARTS

1961

ACADEMIC PRESS, New York and London

ACADEMIC PRESS INC.
111 FIFTH AVENUE
NEW YORK 3, N. Y.

United Kingdom Edition
Published by
ACADEMIC PRESS INC. (LONDON) LTD.
17 OLD QUEEN STREET, LONDON SW 1

Library of Congress Catalog Card Number 59-7677

PRINTED IN THE UNITED STATES OF AMERICA

LIST OF CONTRIBUTORS

ROBERT D. ALLEN, *Department of Biology, Princeton University, Princeton, New Jersey*

JEAN BRACHET, *Faculté des Sciences, Université libre de Bruxelles, Bruxelles, Belgique*

A. J. DALTON, *National Cancer Institute, National Institutes of Health, Public Health Service, Department of Health, Education and Welfare, Bethesda, Maryland*

DON W. FAWCETT, *Department of Anatomy, Harvard Medical School, Boston, Massachusetts*

S. GRANICK, *The Rockefeller Institute, New York, New York*

ALFRED E. MIRSKY, *The Rockefeller Institute, New York, New York*

KURT MÜHLETHALER, *Department of General Botany, Laboratory for Electron Microscopy, Swiss Federal Institute of Technology, Zurich, Switzerland*

ALEX B. NOVIKOFF, *Department of Pathology, Albert Einstein College of Medicine, Yeshiva University, New York, New York*

SYOZO OSAWA, *Biological Institute, Faculty of Science, Nagoya University, Nagoya, Japan*

ERIC PONDER, *Nassau Hospital, Mineola, New York*

KEITH R. PORTER, *The Rockefeller Institute, New York, New York**

* After September 1, 1961: The Biological Laboratories, Harvard University, Cambridge, Massachusetts.

PREFACE

For the past hundred years—from some time before 1855, the year of Virchow's celebrated aphorism *Omnis Cellula e Cellula*—investigation of the cell has been marked by a sequence of revolutionary movements. The most recent of these movements began some fifteen years ago. In each revolutionary change the cell has been viewed dynamically; structure and function have been correlated. From the present movement in cytology there is emerging a picture of the cell that is even more dynamic than heretofore. An important reason for the dynamic quality of the new cytology is that biochemistry is playing a far more significant role in cell studies than it ever did before. The greater role of biochemistry in cytology has been made possible by developments in cytochemistry, by the new methods for isolation and analysis of cell organelles, by techniques for exploiting the potentialities of radioisotopes. At the same time innovations in microscopy, notably in electron microscopy, are bringing structure down to the molecular level. The prospect before us is of a fusion of all cell biology at the molecular level.

The present volume, Volume II, is devoted to the study of the cell constituents: the cell membrane, plant cell walls, ameboid movement, cilia and flagella, mitochondria, lysosomes and related particles, chloroplasts, Golgi apparatus, the ground substance, the interphase nucleus and its interaction with the cytoplasm. The cell constituents have not been considered from a mere descriptive viewpoint: their biochemical activities and their interactions with other cell organelles have been emphasized as well.

J. Brachet

April, 1961 A. E. Mirsky

CONTENTS

THE CELL: *Biochemistry, Physiology, Morphology*

COMPLETE IN 5 VOLUMES

VOLUME I

PART I: METHODS

PART II: PROBLEMS OF CELL BIOLOGY

Volume III: MEIOSIS AND MITOSIS

Contents

Volume IV: SPECIALIZED CELLS, PART 1

Contents

Volume V: SPECIALIZED CELLS, PART 2

Contents

CHAPTER 1

The Cell Membrane and Its Properties

By ERIC PONDER

I. Difficulties Inherent in the Treatment of the Subject

The idea of the existence of a cell membrane is familiar to all biologists and can be illustrated by quoting what some of the best-known investigators have to say about it: "The distinction between the interior of a cell and the medium which surrounds it is maintained by a membrane. . . . Continued existence of the cell is dependent on the ability of this membrane to permit passage of some substances and to prevent that of others" (Davson and Danielli, 1943). "We shall therefore assume, in the absence of evidence to the contrary, that the interfaces at the outer boundary of the cytoplasm . . . are the seats of characteristic permeability and may be thought of as semi-permeable membranes" (Brooks and Brooks, 1941). "There is relatively little dispute at present over the existence of a boundary to free diffusion at the cell surface. This part of the cell, the *plasma membrane*, though not the only barrier to free diffusion within the cell, is nevertheless the important one with respect to the exchanges of molecules and ions between a cell and its environment" (Parpart and Ballentine, 1952). "All permeability theories have in common that the resistance to diffusion is located in the so-called plasmalemma or plasma membrane, which is the outer boundary of the cytoplasm" (Frey-Wyssling, 1948).

Although these statements do not all say quite the same thing, they say or imply that there is a barrier to free diffusion, at least of some substances, between the cell and the medium surrounding it, that this barrier is situated at the cell surface, and that it is a special structure called the cell membrane. Leaving out of account for the moment the fact that a selective barrier to free diffusion may result when a system contains two phases, no third phase or "membrane" between the two phases being required, we meet with some ambiguity when we inquire into the characteristics of the special structure situated at the cell surface. This is because "one has never succeeded in making the lemma [membrane] visible as an individual layer" (Frey-Wyssling, 1948), with the result that we have no way at the present time of distinguishing between the "cell wall" or "surface ultrastructure" and the "membrane." Quite contrary to Frey-Wyssling's remark that "like all physiological problems, the problem of physiological permeability is founded on morphological considerations," we have here a case in which observations of permeability properties have led to tentative conclusions about the morphology, and particularly about the minimum requirements for the morphology of a surface membrane; as time has passed, the qualifying terms "tentative" and "minimum" have been lost sight of, to such an extent that, as late as 1952, we come across the statement: "On the whole, then, the ghost of the mammalian red cell, when prepared by osmotic and even most lytic means, represents as closely as possible a plasma membrane" (Parpart and Ballentine, 1952), which can profitably be compared with Höber's (1945) more easily defensible description of the red cell ghost as being "only partly pieces of the plasma membrane and partly remnants of an [underlying] skeleton or network."

The essential point is that when we are asked to describe the cell membrane, situated, if it is situated at all, in the surface layers of the cell, we are asked to describe something the properties of which are inferred (some of the inferences being open to challenge) "via physiological experiments or reasoning" [Frey-Wyssling (1948), quoting Davson and Danielli], and something which has not yet been separated from the adjacent cell wall or surface ultrastructure, the properties of which are almost entirely based on direct observation. On the one hand, these direct observations of the properties of the surface ultrastructure do not lead one to a description of the permeability properties of the membrane, nor, on the other hand, do the permeability observations from which the existence and properties of the membrane are inferred lead one to a description of the properties of the surface ultrastructure. There are no definite answers to such questions as: Is the membrane a continuous outer layer of the surface ultrastructure, or a discontinuous layer throughout the surface structure (as, by analogy, in the case of a "defense in depth")? The resulting ambiguity is undesirable, for if we measure a property of the surface layers, however prepared, it is always possible for someone whose conception of the membrane does not include this property to object that it belongs to the surface ultrastructure or the cell wall, and is not a property of the membrane; in this way, the measurable properties of the membrane can be indefinitely reduced in number until there is nothing left but a conception, usually based on the interpretation of permeability observations.

It must be admitted that most biologists have not been much concerned with this dilemma. Most of the conclusions about the structure of the "membrane" are based, in reality, on observations of the properties of the red cell ghost or stroma; these conclusions will be critically examined in the sections which follow (particularly II), but the general practice is to accept them at their face value, particularly if they are supported by conclusions (usually inferred) regarding the membrane properties of some other kind of cell. The present situation can be illustrated by two relevant quotations. The first is from a book entitled "Permeability of Natural Membranes" written by two investigators (Davson and Danielli, 1943), who had the courage to say in their preface: "As J. Loeb complained many years ago, obscure or inexplicable phenomena in biology are fashionably brought into the currency of 'knowledge' by way of the Philosopher's Stone, 'a change in permeability.' When to this the more modern elixir of 'surface action' is added, night unto night sheweth knowledge." They also say, in their concluding chapter: "Of the details of the structure of the plasma membrane we still know extremely little." It may be objected that this book was published in 1943, before the electron microscope came into general use. But a second quotation can be taken from "Exposés actuels: Problèmes de structures, d'ultrastructures et de fonctions cellulaires," dated 1955: "We have analyzed the arguments put forward (regarding the membrane or surface ultrastructure, particularly of the mammalian red cell and its ghost) so as to show how unconvincing is the existing evidence set forth in recent reviews. To read these, one is led to believe that the structure of the membrane has been demonstrated, even if only partially, whereas all that has existed in the minds of the original investigators has been a number of purely suggestive ideas" (Dervichian, 1955).

As for the prototype of the membrane, the mammalian red cell ghost, the usual procedure is to hemolyze red cells and then to free them, by repeated washing or similar processes, from as much material as can be removed. What remains is a thin, rugged structure which does not become thinner with further washing. This is identified with the membrane, although there is nothing to indicate that the membrane of permeability theory is the most rugged part of the surface ultrastructure and the most resistant to the processes used to prepare it; indeed, there are several reasons to think that it is not. The only thing that can be said for this approach is that it results in a material which can be analyzed chemically and which has measurable physical properties, largely dependent, however, on the process by which it is obtained.

In the descriptions that follow, it will be convenient to retain the usual loose terminology, which does not distinguish clearly between the cell wall, the surface ultrastructure, and the cell membrane, for otherwise it would be impossible to give descriptions at all. The diversity of the experimental material and the diversity of the properties that have led to conclusions regarding both the surface ultrastructure and the cell membrane is very great; it is rare, indeed, for two investigators to use the same material or the same methods. It is essential, accordingly, to scrutinize both the methods and the material carefully, and to remember that if *A* represents *B*, *A* and *B* must be similar not only when compared in some particular way, but when compared in other ways as well. If, for example, the electron microscope were to reveal a structure in the red cell ghost, prepared in a particular manner, which is just what the membrane of permeability theory calls for, it would be fatal to the conclusion that this is the "real" structure of the membrane if some independent kind of observation were to show that the same red cell ghost does not behave in some way which permeability theory requires, e.g., as regards volume changes in media of different tonicity, penetrability to ions, or some other essential property of the membrane-invested cell.

II. The Red Cell Membrane

A. *Properties and Structure of Red Cell Ghosts: Variation in Their Properties According to Methods Used to Prepare Them*

So many conclusions regarding the properties of the cell membrane are based on studies of the properties of mammalian red cell ghosts, or on deductions from permeability studies of mammalian red cells and their ghosts, that it is necessary to scrutinize this material closely. This is all the more important because the mammalian ghost has been uncritically assumed to be an object of constant composition and as having constant properties, regardless of the way in which it is derived from the corresponding red cell by the process of hemolysis. Some investigators, ignoring the complexity of the material, have even gone the length of saying that the mammalian red cell ghost represents, as nearly as possible, a plasma membrane.

Ghosts can be made from red cells by any process that produces hemolysis, e.g., exposure to hypotonic media, to heat, to low temperatures, to ultrasonics, or to a large variety of hemolysins. As examined with ordinary optics, they may all look alike or at least similar, but it is easy to show, by using phase optics, that there are almost as many kinds of ghost as there are methods by which they are prepared. Ponder (1952) has described some of the properties of human red cell ghosts prepared by hemolysis in hypotonic media, without and with subsequent "reversal of hemolysis" in phosphate, Veronal, and CO_2-saturated systems, as well as those ghosts produced by freezing and thawing and of ghosts produced by the action of lysins. The differences are tabulated, and it is apparent that these different varieties of ghost are far from identical. This is only an extension of such well-known observations as that the electrophoretic mobility of ghosts produced by saturating a hypotonic system with CO_2 is less than that of ghosts prepared by some other method, or that the surface appearance of the ghost, whether observed with phase contrast or with the electron microscope, depends on the nature of the lysin used to hemolyze the red cell (Ponder *et al.*, 1953). There are other more obvious differences, particularly in the amount of residual hemoglobin which the ghost contains when hemolysis takes place in media of different tonicities (Ponder and Barreto, 1957). For certain purposes, such as the measurement of the thickness of the ghost conceived as "the membrane," it may be desirable that the quantity of residual hemoglobin which the ghost contains be as small as possible, but the removal of hemoglobin from a surface ultrastructure by special methods such as Hillier and Hoffman's (1953) method of "successive hemolysis" may produce a situation as artificial as that produced when lipids are removed from naturally occurring lipoproteins by organic solvents.

Possibilities such as these could be debated endlessly. In the meantime, the three propositions which are not worth considering are (1) that all kinds of ghosts, however prepared, are substantially identical, (2) that ghosts prepared by some particular procedure such as saturation of a hemolyzed system with CO_2 (Parpart, 1942) or the procedure known as successive hemolysis (Hillier and Hoffman, 1953) are in some sense "more real" than ghosts prepared by other methods, and (3) that ghosts, however prepared, can be safely regarded as representations of the cell membrane. At best, they may be representations of a surface ultrastructure, probably modified by the method of preparation.

B. Chemical Composition: Lipids, Proteins, and Lipoproteins

The surface ultrastructure of the mammalian red cell and its ghost contains lipids and proteins which, in their natural state, are probably lipoproteins or lipid-protein complexes. The ratio of lipid to protein varies with species, but the general picture as presented by leading investigators in the field (Parpart and Dziemian, 1940; Parpart and Ballentine, 1952) can be taken as representative.

The ratio of total lipid to protein (by weight) lies between 1:1.60 and 1:1.82. Extraction with organic solvents shows that some of the lipid

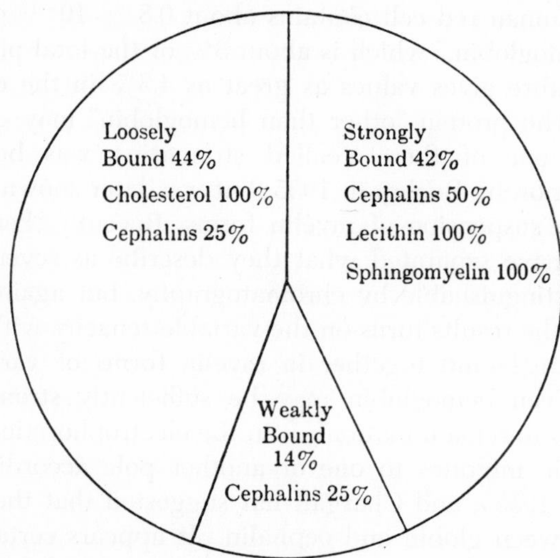

FIG. 1. Binding of lipids in the plasma membrane of beef red cells expressed as per cent of the total lipid. Lipid fraction per cent is that of each fraction. From Parpart and Ballentine (1952).

is strongly bound, some loosely bound, and some weakly bound; this is illustrated by Fig. 1, taken from Parpart and Ballentine's review (1952). The points that seem to be well established are that all or almost all (Williams *et al.*, 1941) of the lipids of the intact red cell are found in the ghost, at least when it is prepared by Parpart's method in which cold hypotonic hemolyzates are saturated with CO_2, and that the cholesterol of the ghost is "bound" to protein. Physical chemists are agreed, at least at the present time, that lipids do not exist except in combination, of varying degrees of stability, with proteins, and that extraction procedures are apt to result in laboratory creations. The lipid:protein ratios mean that each protein molecule has some seventy lipid molecules associated with it, more or less firmly, but at the same time it is recognized that the lipids and proteins are in a state of dynamic equilibrium, rapidly exchanged, and dependent in their proportion on the physiological state of the animal (Bodansky, 1925, 1931; Altman *et al.*, 1951; Muir *et al.*, 1951). It is of some interest, particularly in connection with the views of Dervichian (see below), that the *molar* concentration of the total lipid per red cell is nearly the same as that of the cell hemoglobin (5.7×10^{-16} *M*).

The protein composition of the red cell ghost is as yet unsettled, at least so far as detail is concerned. Assuming a molecular weight of 200,000, each human red cell contains about 0.8×10^{-12} gm. of protein "other than hemoglobin," which is about 3% of the total protein present; the older literature gives values as great as 4.3% in the dog and 7.8% in the sheep. The protein "other than hemoglobin" may consist of several fractions; one of these (called stromatin) was believed to be isolated in solution by Boehm in 1935, but was later shown by Furchgott (1940) to be a suspension of myelin forms. Recently, Perosa and Raccuglia (1952) have separated what they describe as several lipoprotein components distinguishable by chromatography, but again the difficulty in interpreting the results turns on the variable tenacity with which lipids and proteins are bound together in myelin forms of varying size and composition. Even hemoglobin may be sufficiently strongly bound to myelin forms as to remain motionless in the electrophoretic field, whereas free hemoglobin migrates to one or another pole according to the pH (Ponder *et al.*, 1955), and Chargaff has suggested that there may be an ionic bond between globin and cephalin. It appears certain that in the case of the myelin forms that arise from the surface of the red cell or of the ghost and that contain hemoglobin, the latter is not a contaminant but a complex (Bessis *et al.*, 1954; Ponder, 1956).

Since the study of the proteins of the red cell and ghost ultrastructure presents so many difficulties when they are investigated in something approaching their natural state, it is understandable that attempts have been made to approach the problem by methods that involve extraction with organic solvents. Calvin and his associates (see Dandliker *et al.*, 1950; Moscovitch and Calvin, 1952) hemolyze red cells with water, dry the ghosts, and suspend them in water. After standing at pH 7.3, the material is centrifuged at high speed, and the sediment is washed several times at pH 9.0. It becomes cream white in color, and is called *reticulin* or *stromin.* When dried, it is still white, and when extracted with ether, it yields *elenin.* Electron microscope examination of reticulin shows rod-shaped particles measuring from 5 to 11 μ long and from 0.3 to 1.3 μ wide. Hillier and Hoffman tentatively identify their "plaques," or alignment of plaques, with elenin. Extraction of elenin with alcohol gives a protein supposedly similar to the *stromatin* of Jorpes (1932) and of Boehm (1935), the molecules of which are elongated and show birefringence of flow, but it should be remembered that the stromatin of Boehm was shown by Furchgott (1940) to be, not an isolated protein, but a suspension of myelin forms.

In the preparation of elenin there is a step, after the extraction of ether-soluble lipids from reticulin with ether, in which elenin is thrown down. The supernatant fluid at this stage gives a precipitate if brought to pH 6.4 in the absence of salt, or to pH 5.0 in its presence. The precipitate is another protein fraction, called S *protein*. The *a* protein separated by electrophoresis from ghosts made by freezing and thawing (Stern *et al.*, 1945) has the mobility of reticulin, elenin, or stromatin, while their *b* protein has the mobility of S *protein*.

To this multiplicity of protein fractions must be added hemoglobin, often complexed with lipoprotein or lipids. Two other fractions have been described (1) a globulin which remains in the supernatant fluid at the first stage of the preparation of reticulin, and (2) a carbohydrate-poor albumin, called the *antisphering factor*, which is easily removed from the red cell surface. There is now doubt as to whether it is really a component of the surface ultrastructure of the red cell.

Considering the number of physicochemical and immunological procedures which can be used for establishing the homogeneity of proteins, and that none of them except a few determinations of electrophoretic mobility has been systematically applied to the protein "fractions" of the ghost, the only safe conclusion is that terms such as stromin, elenin, stromatin, etc., are labels with which to mark the product of a particular treatment of the material and are not to be taken, at least in the meantime, as designations of protein fractions in the usual sense.

C. *Thickness and Degree of Hydration*

An estimate of the thickness of the red cell or ghost surface ultrastructure can be arrived at by a variety of methods; these yield values ranging from 50 A. to at least five times that figure. The methods fall into several classes.

1. *Measurements in "Hemoglobin-free" Ghosts*

There are methods that measure the dry weight, or the quantity of lipids and proteins separately, contained in the ghost and expressed as a percentage of that in the red cell, the ghosts being prepared by methods such as those of Parpart (1942) or of Hillier and Hoffman (1953), both of which reduce the hemoglobin content of the ghost to a minimum. These methods may be suspected of giving minimum values, because determined attempts to remove hemoglobin may also remove other substances (proteins as well as lipids) by a leaching-out process (see below, in connection with the analytical leptoscope).

Parpart and Ballentine find that the dried material of hemoglobin-free rabbit red cell ghosts comprises 1.0 ± 0.2% of the dry weight of the red cell. The volume of the latter being 57 μ^3 and its area 110 A.^{2}, the thickness of the dry surface ultrastructure would be about 52 ± 10 A., without corrections being made for density. A double layer of all the available lipids would give a thickness of about 30 A., leaving some 22 A. for the thickness of the protein components in their dry state. Hydration would increase this thickness, as also would the inclusion of any hemoglobin in the ultrastructure. Like Waugh and Schmitt (1940), Parpart and Ballentine think that the degree of hydration is small, and remark that a comparison of the permeability constant for rabbit red cells to ethylene glycol with the free diffusion constant for the same substance suggests that the water in the ultrastructure is continuous with the water of the environment only over about 0.1% of the surface of the red cell. A very different result, however, is obtained if the permeability constant and the free diffusion constant for O_2 are compared (Hartridge and Roughton, 1927; Dirken and Mook, 1931); here the ratio is 10% instead of 0.1%. Arguments of this kind are not convincing, the nature of the penetrating substance probably having much to do with the result. Further, Parpart and Ballentine think that if the lipid is present in a three-dimensional arrangement rather than in a double layer, the degree of hydration would be decreased rather than increased, but whether this conclusion is right or wrong depends entirely on what form the lipoprotein complexes of the surface are supposed to take. Some investigators, who regard intercalated water as an indispensable structural component of lipoprotein complexes (e.g., Dervichian, 1949), would arrive at an opposite conclusion. Parpart and Ballentine reject, quite properly, the conclusion of Mitchison (1950, 1953) that the surface ultrastructure is 5000 A. thick and more than 90% hydrated. On the other hand, they misinterpret Fricke and Curtis' observation (1935) that both the red cell and the ghost have a high electrical resistance as meaning that the degree of hydration is necessarily small; a "high" resistance, in Fricke and Curtis' experiments, means a resistance high relative to the resistance of 1% NaCl (about 50–100 times greater) and has little bearing on the point in question.

Disregarding these small and indecisive points, the conclusion to be drawn from Parpart and Ballentine's observations on hemoglobin-free ghosts is essentially that the dry thickness is about 52 A., and that the degree of hydration, which would determine the wet thickness, is uncertain. Finally, they quote Hillier and Hoffman's electron microscope observations (1953) as being in accordance with their views.

2. *Measurements in Ghosts That Contain Hemoglobin*

There are methods that prepare ghosts by hemolyzing red cells with water, restoring the hemolyzate to isotonicity, and measuring the dry weight of the ghosts so as to give a value for that part of the cell which does not diffuse away into a hypotonic medium (Fricke *et al.*, 1939). The ghosts prepared by this process contain some hemoglobin, and hemoglobin is also contained in the medium surrounding them. The quantity of hemoglobin, both in the ghosts and in the supernatant fluid can be estimated either as hemoglobin or as iron, and the quantity contained in the ghosts can be subtracted from the dry weight. The result ought to be comparable with the dry weight of hemoglobin-free ghosts, but the latter is 1.0 ± 0.2% of the red cell volume, whereas the method with which we are now concerned gives the higher value of between 1.8 and 2.8%, with a "best value" of 2.1%. This would give a dry thickness for the surface ultrastructure of about 120 A., or about twice the thickness given by the methods which start with hemoglobin-free ghosts. This difference is not due to the higher content of hemoglobin, since this is subtracted; it could, however, be due to the ghosts being prepared in a more "gentle" manner (i.e., with fewer washings) than the ghosts prepared by Parpart's method.

3. *Ghosts Prepared by Intermediate Methods*

There are methods intermediate between those described in Sections 1 and 2 above. The inclusion of this category is tantamount to repeating that the value obtained for the thickness of the surface ultrastructure of the ghost depends on the way in which it is prepared.

Williams *et al.* (1941) prepare ghosts or "posthemolytic residues" by washing red cells, hemolyzing them in cold water, and washing the ghosts repeatedly in a citrate buffer at pH 5.5. The product obtained by this process differs in color, i.e., in hemoglobin content, from animal to animal. The values found for the volume of the dry ghost as a percentage of the cell volume varies from 1.3 in the ox, 2.2 in the rabbit, 2.3 in the sheep, and 3.3 in man, from each of which something must be subtracted if hemoglobin is not a component of the ultrastructure. In the case of human ghosts, 25% of the dry weight is made up of hemoglobin; the volume of the human cell being 86 μ^3 and its area 167 μ^2, the thickness of the surface ultrastructure, with hemoglobin subtracted, works out at about 125 A., i.e., about the same as Fricke and his associates' value.

The question of the usefulness of many washings certainly becomes a serious one when the results of the methods of Tishkoff and co-workers (1953) are examined. They hemolyze dog red cells in cold water and then precipitate the ghosts by a modification of Ballentine's method, in which sodium acetate is added to bring the pH of the hemolyzate to 4.6. The precipitated ghosts are washed with 1:10,000 acetic acid, either by a "short wash" procedure which involves two washings, or by a "long wash" procedure which involves two additional washings during the course of 2 days. Washing four times instead of twice results in a loss of about 25% of the lipids and 25% of the proteins, and the amount of hemoglobin, estimated as iron, is larger in the "long wash" procedure than in the "short wash" procedure.

Apart from the evidence provided by these experiments that substantial amounts of lipids and proteins are lost from ghosts that have been washed repeatedly, there is no evading the possibility that hemoglobin often may be a component of the surface ultrastructure of the ghost, and not a mere contaminant. That hemoglobin is actually complexed with the myelin forms derived from the surface of the ghost has been proved, as nearly as is possible to prove a point in the little-investigated domain of complex formation, both by the observations of Bessis and associates (1954) and by the electrophoretic behaviors of hemoglobin, ghosts, and myelin forms, which show that stromata and even myelin forms retain an appreciable amount of hemoglobin, presumably because the latter is complexed to the lipoproteins of their ultrastructures (Ponder *et al.*, 1955). Prolonged washing with selected media, resulting in hemoglobin-free ghosts, may therefore produce a superficially desirable result, which is obtained, in reality, by breaking down hemoglobin-lipoprotein complexes, liberating lipid and protein along with hemoglobin, and thus leading to a spuriously small value for the thickness of the surface ultrastructure.

4. *Leptoscopic Measurements*

Measurement of the thickness of dried ghosts can be made with Waugh and Schmitt's analytical leptoscope (Waugh and Schmitt, 1940; Waugh, 1950; personal communication, Waugh, 1953). There is little to be added about the results of this elegant method (see Ponder, 1955), which depends on a comparison of the intensity of reflected light from dried ghosts superimposed on steps of barium stearate of known thickness. When the ghost reflects light of the same intensity as that reflected from the step on which it lies, the ghost thickness and the known thickness of the step are, to a first approximation, equal. Waugh and his students, Gray and Taub, have improved the original leptoscope by making it completely objective (a "refraction micrometer") and have used it to carry out analyses of the thickness of red cell ghosts. These have confirmed some of the earlier observations of Waugh and Schmitt and have also allowed for the entrapment of air in the dried specimen. When this is done, the maximum lipid thickness turns out to be about 40 A. (rabbit ghosts), but difficulties still remain regarding the total thickness, which depends in a remarkable way on the time required for hemolysis, i.e., on the particular conditions under which the ghost is formed. The maximum thickness, however, is at pH 6.0 as in the original work of Waugh and Schmitt.

There are two points in which the leptoscopic measurements are still at variance with results obtained by other methods. The first concerns the degree of hydration of the ghost, which, when examined with the leptoscope, is air-dried. Heating to 100°C. reduces the thickness by less than 20%; this would lead to the conclusion that the surface is quite anhydrous, containing less water, indeed, than would be expected to be associated with the proteins which make up more than half of it (bound water of protein, about 15%). Measurements made in the opposite direction, i.e., of the quantity of water which dry ghosts can take up from isotonic saline, point to the rehydrated structures containing about 60% water (Ponder, 1954). Both types of experiment, however, are open to criticism, and an opinion as to which is the more likely must rest, in the meantime, on what one's conception of red cell and ghost structure may be. The importance of the point will be appreciated when it is remembered that a dry "membrane" with a thickness of 200 A. would be 240 A. thick when wet if the degree of hydration were 20%, but between 500 and 600 A. if the degree of hydration were 60%.

The other point on which leptoscopic measurements are questionable is that they show the thickness of the ghost ultrastructure to be greater in the regions of the biconcavities of the ghost, the increased thickness being largely due to protein. This does not agree with results obtained with the electron microscope, which shows that the ghost is thicker in the region of its rim, probably because of the greater inclusion of hemoglobin there (Ponder *et al.*, 1952a). This difference in result may again be due to the method used to prepare the ghost, and there is always the question whether the "membrane" or "surface ultrastructure" of the ghost corresponds to anything that exists at the surface of the red cell (Dervichian, 1955).

D. Observations Made with the Electron Microscope

The electron microscope provides somewhat conflicting evidence about the thickness of the red cell ghost, less convincing evidence about the fine structures of the surface, and quite a confused body of information bearing on the way in which the surface is changed or injured by

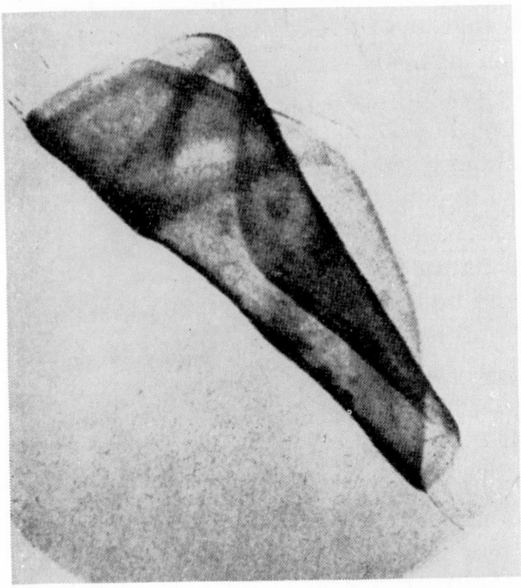

FIG. 2. Electron microscope photograph of a red cell ghost, one of the first pictures of its kind to be published. Kindly sent to me by Dr. Hans Zwickau in 1941.

various agents or by methods of preparation. Thickness is best investigated by the method of thin sectioning, while changes in the surface, such as the changes which occur during storage (Ponder *et al.*, 1954), are better investigated by the methods of moulage and shadowing; the conclusions which bear on fine structure are reached by a combination of both methods with their many variations. Unfortunately, we have here a situation that was once thought to apply to "the" red cell ghost (Fig. 2). As in the case of the staining methods of fifty or sixty years ago, investigators have been more eager to obtain spectacular results than to explore, in a systematic manner, the effects of pH, ionic strength, and, in general, the ways in which their material has been prepared. There is therefore nothing new in each investigator believing that his method is not only the best but the only reliable one, and that his interpretation of an electron microscope image, scarcely likely to be easily interpreted, is both unique and true.

1. *Thickness*

As regards thickness, the values found depend both on the method used and on the kind of ghost. Sectioning gives smaller values than moulage, which does not distinguish with certainty between the surface ultrastructure and the material underlying it. Inescapably, the more the ghosts are washed and the less hemoglobin the final product contains, the assumption that the "real" ultrastructure or membrane is free of hemoglobin is an assumption, no doubt convenient, but serving little purpose except that of simplifying an unknown situation. Hillier and Hoffman's (1953) thickness for the structure which they regard as the "membrane" is about 60 A. (dry). Wolpers (thin sectioning) (1956) gives 150–300 A. as the thickness of his ghosts, and the shadowed moulages of Bessis and his collaborators (1950) give at least 500 A. Few investigators familiar with the nature of the material and with the ways in which it can be prepared would be impressed with the difference of thickness reported (50–500 A.) even although an order of 10 is involved, and it makes the matter worse that the degree of hydration is subject to an uncertainty of almost the same order.

An important point is that most investigators who are rarely in agreement as to the thickness of the ghost surface ultrastructure agree that there are local variations in the thickness. Waugh and Schmitt (1940), Wolpers (1956), Latta (1952), and Bernhard (1952) describe the surface layers as being either variable in thickness or as being less sharply delimited on their inner aspect than on their outer. In some cases the greatest thickness is described as twice the smallest. These differences are quite local, i.e., 1 μ or less apart, and are not related to the increased thickness over the biconcavities (leptoscopic observations) or to the increased thickness at the rim of the ghost (electron microscope observations) (Ponder *et al.*, 1952a). The latter apparent contradiction is probably due to two different kinds of ghost, prepared by two different methods, containing different amounts of hemoglobin differently distributed. "When two competent investigators obtain two entirely different results, it is rarely due to error; they are almost always performing two different experiments" (Otto Rahn).

2. *The Fine-Structure Problem*

The status of the "fine-structure problem" as it applies to the surface of the red cell ghost has certainly been advanced by the use of the electron microscope. As early as 1942, Wolpers and Zwickau (see also Wolpers, 1956) described a fibrillar structure not unlike the fibers in Bernhard's photographs, which are arranged in a herringbone fashion (Fig. 3). Two of the most recent electron microscope pictures are those of Hoffman *et al.* (1956) (Fig. 4) and of Hillier and Hoffman (1953) (Fig. 5). A special method, open in some respects to criticism and called the method of successive hemolysis, frees the ghost of hemoglobin,

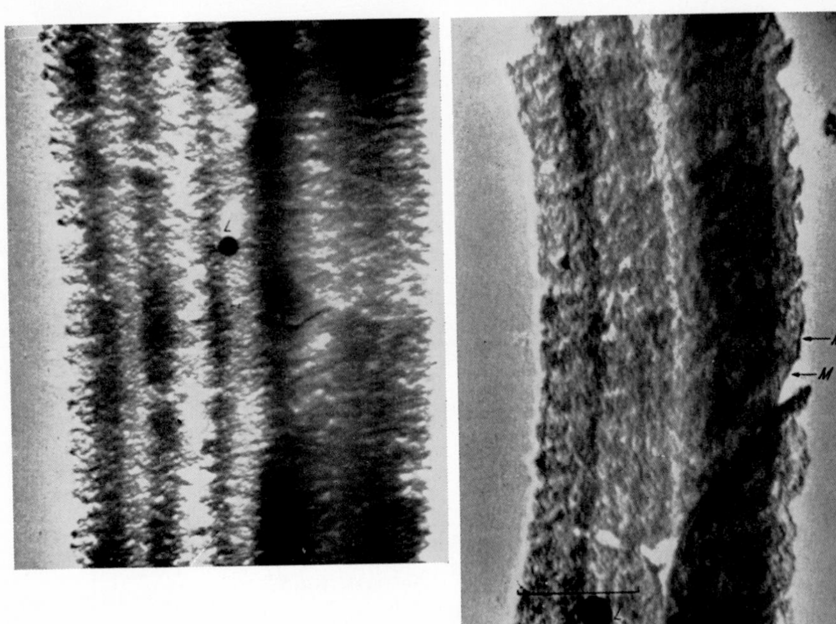

FIG. 3. Slices across the ultrastructure of an unhemolyzed red cell. Electron microscope preparation shadowed with gold. Magnification about × 20,000 in the case of the figure on the right; somewhat less in the case of the figure on the left. The round object is a latex particle of known size. From an original photograph kindly sent to me by Dr. W. Bernhard.

and the electron microscope then shows fibrils with superimposed "plaques," some 30 A. high and about 200 A. in diameter (Figs. 4 and 5). Hillier and Hoffman's pictures are quite like those of Bernhard, and both structures (a combination of fibrils and plaques) can be interpreted in terms of classic permeability theory; it is, indeed, a kind of restatement of Bechold's (1921) description of a hypothetical fibrous structure with the spaces between the fibrils filled with lipid. Hillier and Hoffman suggest that the spaces between the plaques are actually pores. It is curious that no variation in the fine structure has yet been reported, considering the modern tendency to regard the surface structure as a mosaic rather than a uniform sheet. Some competent electron microscopists regard Hillier and Hoffman's fine structure as largely artifactive because of the way in which the ghosts are prepared (particularly because of denaturation and precipitation of proteins and splitting-off of naturally occurring lipoproteins, which occurs when ghosts are washed repeatedly in hypotonic media) [see Danon *et al.* (1956) and also Lovelock (1955) who give clear evidence of the loss of lipids and lipoproteins at low ionic strengths and after repeated washing].

It ought to be pointed out here that (until recently) there has been an essential difference between the methods of the French and American investigators. The former have tended to use relatively low magnification, whereas in America the tendency has been to use high magnification, sometimes empty. This difference has to be reckoned with when the effects of injurious agents on the ghost surface structure is considered. Thus Latta (1952) finds no effects when ghosts are treated with most hemolytic agents (Fig. 6); Hillier and Hoffman find only a uniform loss of hemoglobin; but Bessis and his collaborators (cf. also Bundham and Wright, 1953) (Fig. 7) have observed a variety of changes in the surfaces of ghosts exposed to lysins. Irregular depressions ("craters"), spikes containing hemoglobin ("spicules") and, above all, myelin forms arising from the craters are regularly found at low magnifications (Ponder *et al.*, 1952b; Ponder *et al.*, 1953) (Figs. 8, 9, and 10). These myelin forms often coalesce to form a "glue" surrounding the ghost; it is not difficult to imagine that the glue could be removed by washing the ghost, leaving something more tenuous behind. Finally, some of the structures shown by moulages at relatively low magnification, e.g., the lacy system of "microsomes," about 1000 A. in diameter and described by Bessis and Bricka (1950), are now known to be artifacts (Fig. 11).

A point that escapes most electron microscopists who describe fine structure or the absence of it in the red cell ghost is that it is essential that the *particular* variety of ghost in which the structure is found should be shown to have at least a reasonable number of the properties attributed to red cell ghosts in general, e.g., they should be shown to be able to shrink and swell in varying tonicities, to be relatively impermeable to certain substances, to show something resembling the resistance and capacity associated with ghosts (and with the intact red cell), and so on. If the fine structure of the ghost is found to be abnormal (Hoffman *et al.*, 1956; see also Braunsteiner *et al.*, 1956), some of the other abnormalities of the ghost or of the red cell should be demonstrated as well.

Electron microscope photographs are more easily understood when seen than when they are described. Since there is so much variation of opinion among various microscopists, the photographs (Figs. 12–17) will give the reader a good idea of the present state of affairs.

FIG. 4. A human red cell ghost enlarged about 19,500 times. The whitish areas are the folded surface of the ghost. From an electron microscope photograph by Joseph F. Hoffman *et al.* (1956).

E. Polarization Optics

This complex subject can be simplified by saying that the red cell ghost shows a small birefringence which varies from one region to another. This result has been obtained with a unique polarization microscope.

All that remains is to trace briefly the history of the investigation of birefringence in the red cell ghost. This was first observed by Schmitt and associates (1936, 1938) and is of a low order of magnitude. That it

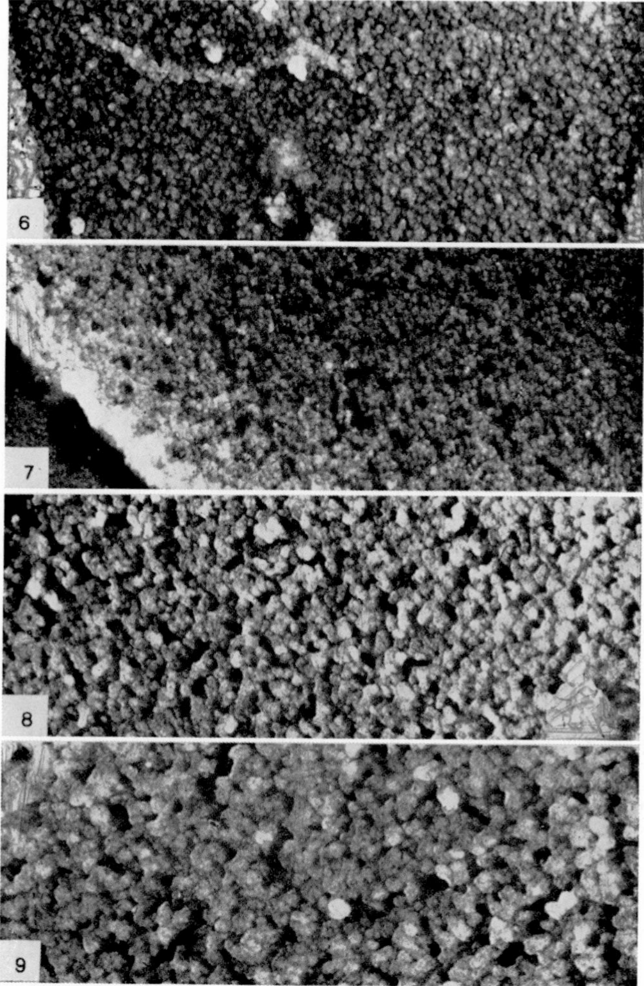

FIG. 5. Electron microscope photographs at four different magnifications (the highest × 82,000) to show the plaquelike structure of the surface. From an original photograph kindly sent to me by Dr. J. Hillier; see Hillier and Hoffman (1953).

occurs at all means either that the surface ultrastructure of the ghost has birefringent molecules in it (*intrinsic birefringence*) or that its molecules are arranged in one of several special ways which give rise to *bire-*

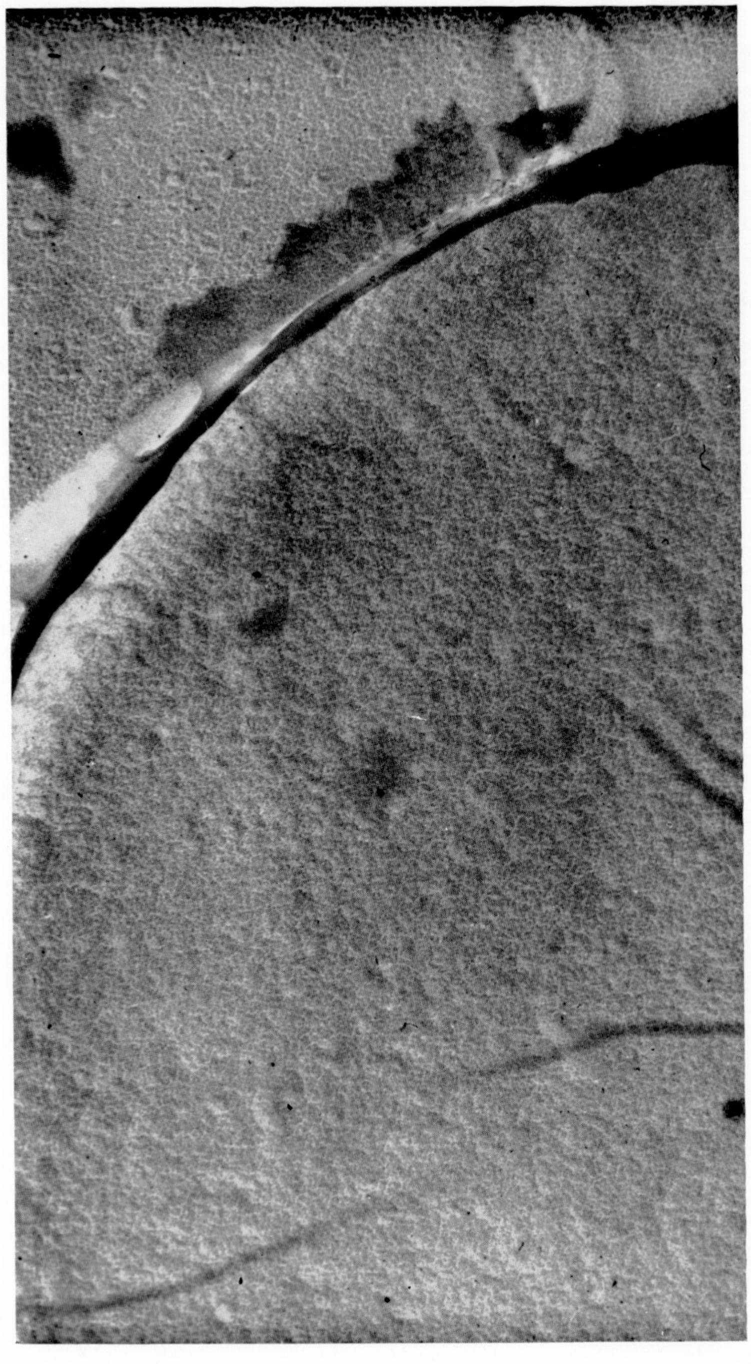

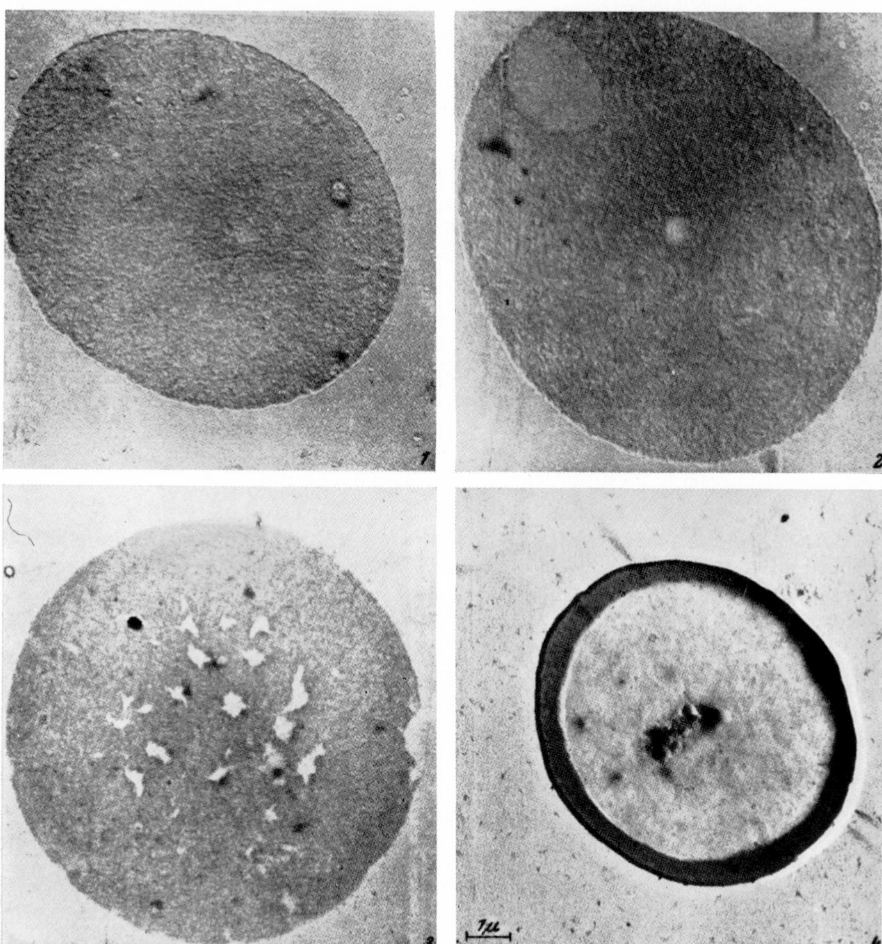

FIG. 7. (1) Normal erythrocyte membrane, gold shadowing; (2) normal erythrocyte membrane, gold shadowing; (3) normal erythrocyte membrane without shadowing; (4) partially hemoglobinized erythrocyte with residual hemoglobin at the periphery, gold shadowing. Electron microscope photograph from a paper by Bundham and Wright (1953) (the original photographs were kindly sent to me by Dr. Claude-Starr Wright).

FIG. 6. Part of the surface and edge of a normal unfixed rabbit red cell. After Latta (1952). The author does not think that any real structure is shown in the photograph.

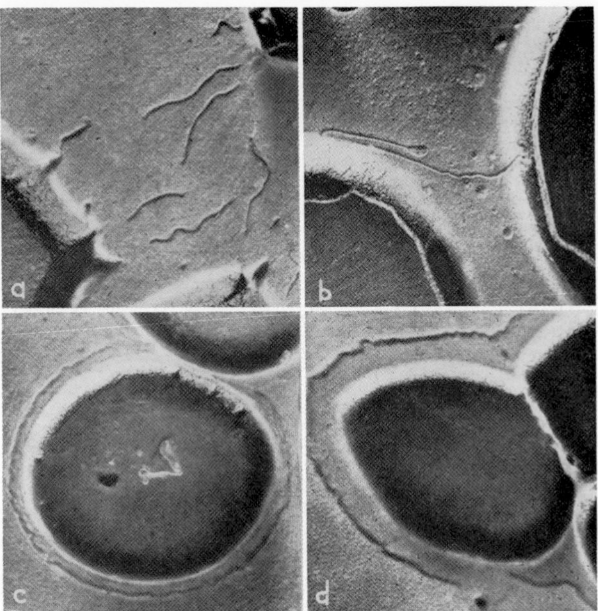

FIG. 8. Myelin forms and peripheral rings. a and b. Myelin forms, some still attached to the surface of the cell and some detached. c and d. Peripheral rings. Notice in (c) the myelin form arising from a crater and, in (d), a fusion of two peripheral rings. From Ponder *et al.* (1952b).

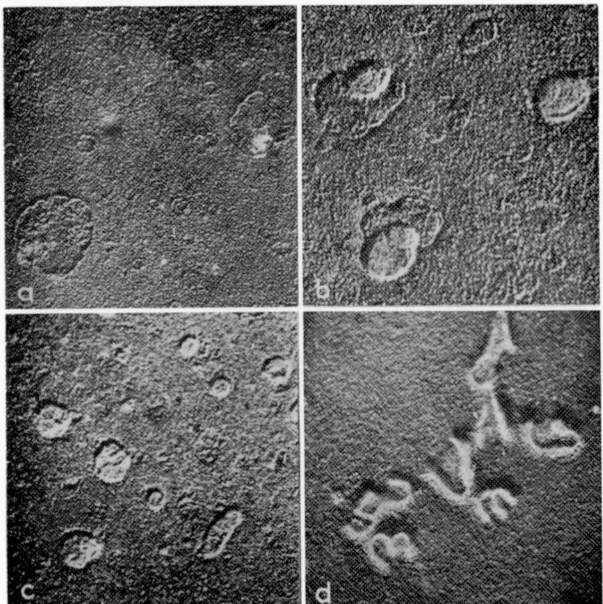

FIG. 9. Irregularities observed on the surface of red cells. a. Craters of type I. b. Craters of type II. c. Craters of type III. d. Spicules of type II. From Ponder *et al.* (1952b).

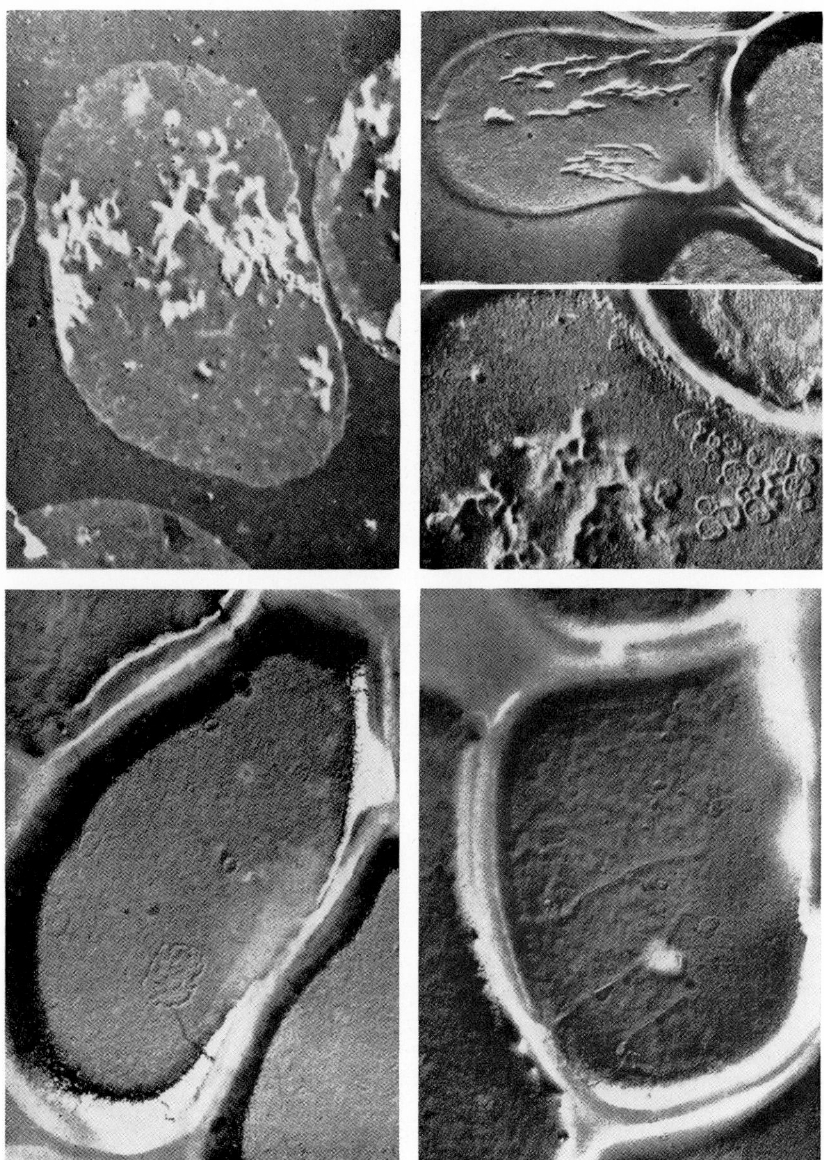

FIG. 10. Changes in the red cell surface as a result of the action of lysins. Upper left, sodium taurocholate. Upper right, sodium lauryl sulfate. Lower left, digitonin. Lower right, saponin. From Ponder *et al.* (1953).

fringence of form, or both. When rods or layers of one refractive index are arranged in a medium of another refractive index, birefringence of form can be calculated, on reasonable assumptions, by a formula due to

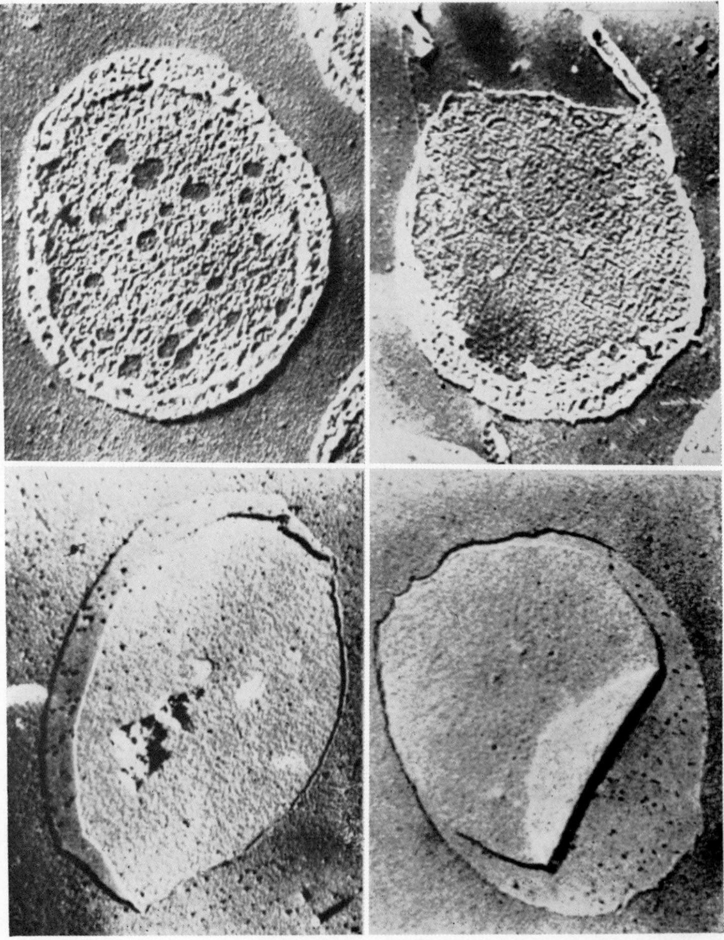

FIG. 11. Four electron microscope photographs of the red cell showing (above) a system of microsomes which may be artifactive. Below and to the left, craters, and below and to the right, the upper and lower surfaces of the ghost ultrastructure clearly separated. From an original photograph kindly sent me by Dr. Marcel Bessis; see Bessis and Bricka (1950).

Wiener. When the refractive index of the medium is the same as that of the rods or layers, Wiener's relation gives a value of zero. Such a value can be sought experimentally by immersing the ghost in media of different refractive indexes, and although a value of zero may not be found, it is usually possible to find a minimum value; the difference between this and zero then gives the intrinsic birefringence of the material which makes up the rods or layers. Intrinsic birefringence is usually positive in sign, and intrinsic birefringence and birefringence of form are usually additive.

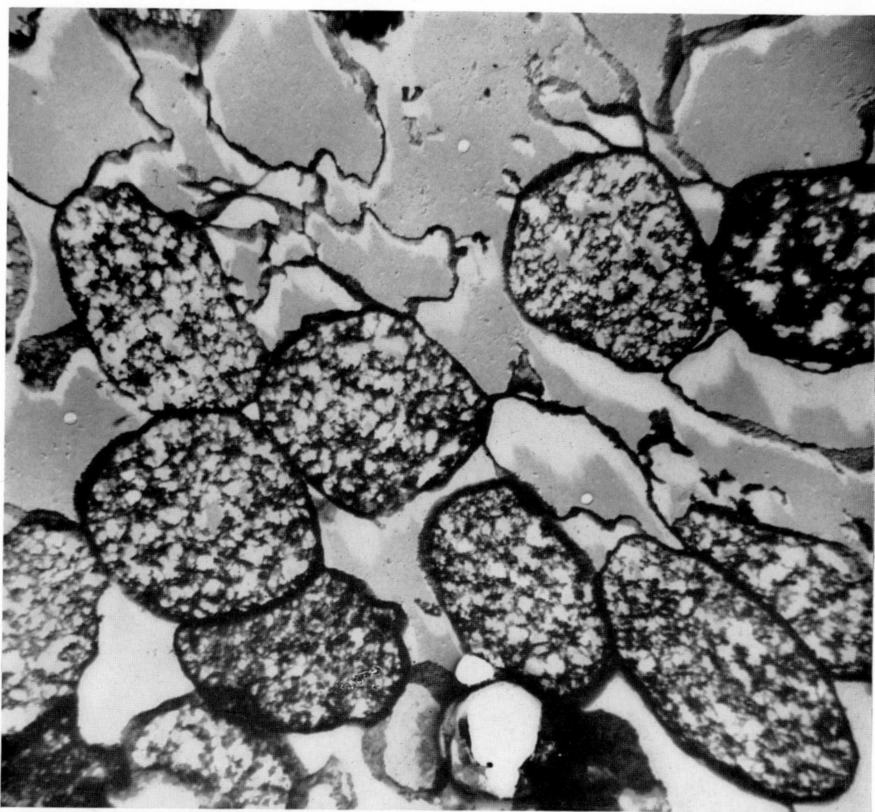

FIG. 12. Electron microscope photograph of cells, some of which are empty and some of which still contain hemoglobin. Magnification: × 4800. From an original photograph kindly sent me by Dr. W. Bernhard.

Schmitt and his associates found a very small birefringence in the rabbit red cell ghost prepared by freezing and thawing, and they concluded that the smallness of the birefringence is due to a birefringence of form being off-set by a positive intrinsic birefringence. The latter can be reduced by treating the ghost with lipid solvents, and so the form birefringence of the proteins ought to be negative in sign. As might be

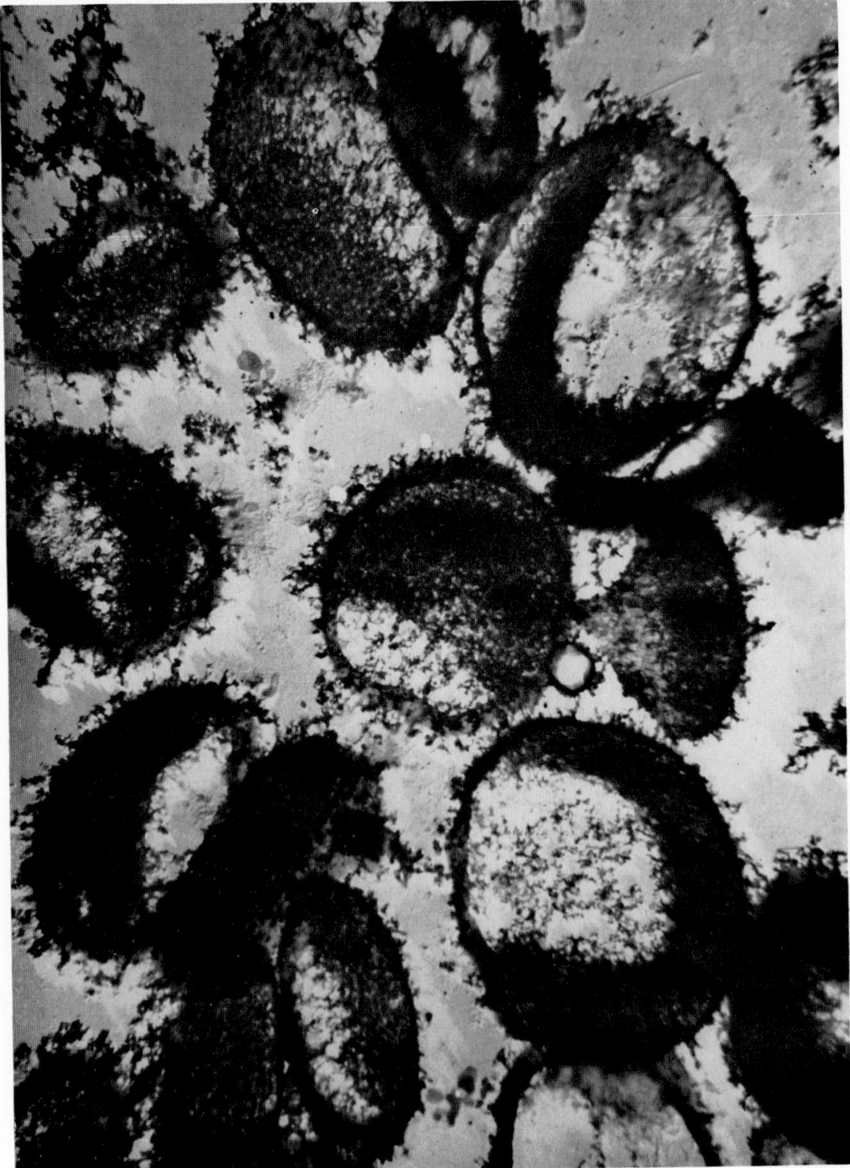

Fig. 13. Partial hemoglobinization at an earlier stage than that shown in Fig. 12. From an original photograph kindly sent to me by Dr. W. Bernhard.

expected, this negative birefringence is reduced by placing the ghosts in media of high refractive index. In view of these results, the most likely arrangement is that the lipid molecules are intrinsically birefringent and arranged radially like a barricade, while the proteins lie in lamellae with

FIG. 14. Partial hemolysis: electron microscope photograph of the surface. Cells fixed in 30% formol. From an original photograph kindly sent to me by Dr. W. Bernhard.

their long axes tangential to the surface of the ghost. On a series of reasonable but not entirely reliable assumptions, the observed birefringence of the whole ultrastructure (a retardation of about 5 A.) would result from a structure about 500 A. thick (when hydrated), but it ought to be made clear that these early observations were only semiquantitative and that polarization optics cannot be safely used to support or refute ideas regarding the localization of proteins, lipids, etc., particularly as the two may show strong interaction. The situation is further complicated by the possibility that the residual hemoglobin of the ghost is not a contaminant, but a strongly interacting component of the surface structure.

More recently Swann and Mitchison (1950) have made great im-

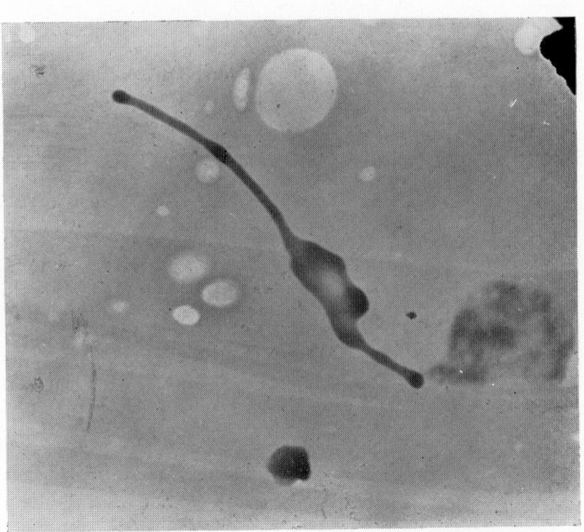

FIG. 15. A myelin form showing dense areas, probably corresponding to the presence of hemoglobin. From an original photograph kindly sent to me by Dr. Fritz Jung.

FIG. 16. Electron microscope photograph of part of the surface of the red cell in which the dense hemoglobin has contracted away from what is construed to be a thin surface ultrastructure. The thickness of the surface ultrastructure is estimated as being about 500 A. From an original photograph kindly sent to me by Dr. John Rebuck.

provements in the measurement of birefringence, and by applying these methods to ghosts prepared by placing red cells in glycerol, Mitchison (1953) has arrived at the conclusion that the surface ultrastructure of the ghost is about 5000 A. thick, highly hydrated, and occupying about

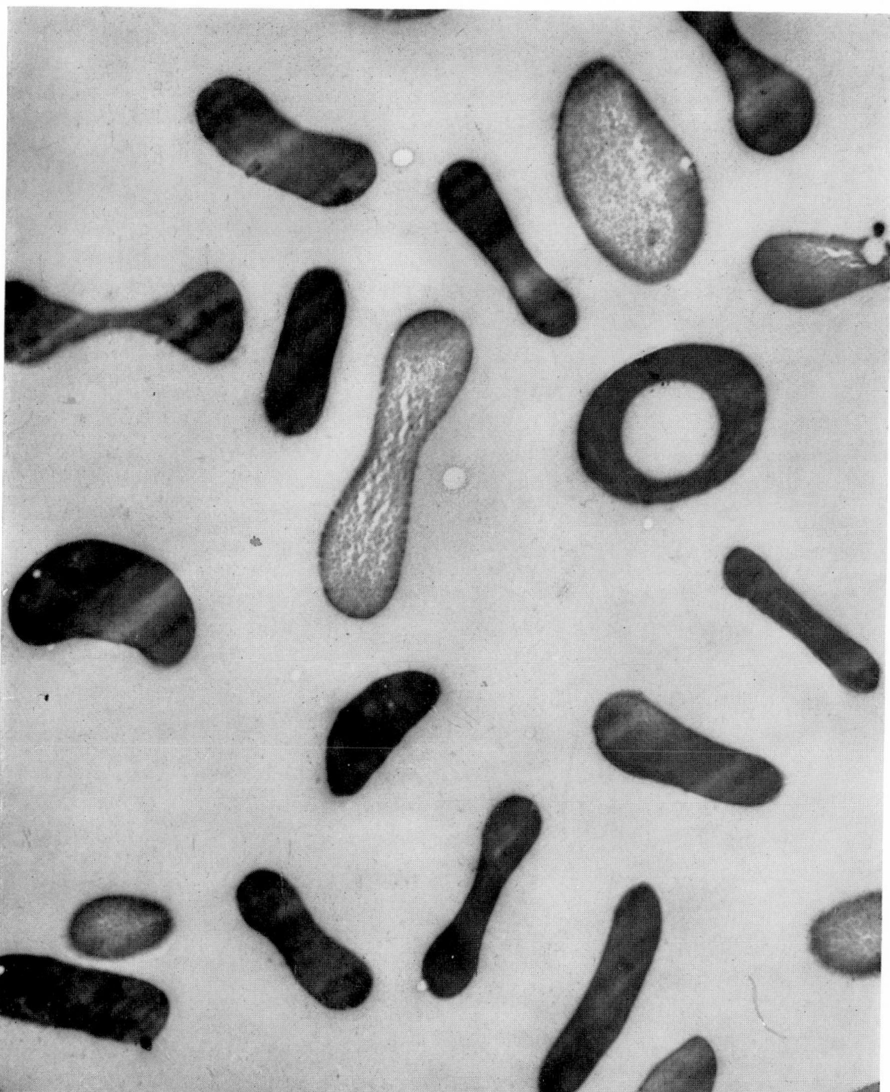

FIG. 17. Thin sections of red cells. Note the dark peripheral region of the red cell seen on edge in the middle of the picture. This may be an artifact. From an original photograph kindly sent to me by Dr. Marcel Bessis.

half the volume of the ghost. This estimate, which is at variance with all others, seems to be due to an unfortunate selection of material, for the ghosts formed by placing red cells in glycerol are visible with the light microscope, and show birefringence, only if they are partially hemoglobinized (Ponder and Barreto, 1954); when completely free of hemoglobin, they are visible with phase optics but are invisible with the light microscope and do not show the birefringence which Mitchison describes and which leads to the estimated thickness of 5000 A. As a result, his proposed model of the arrangement of the molecules in a very thick ghost ultrastructure need not be discussed. At first sight it would appear that a similar criticism could be directed at the early results of Schmitt and co-workers, for their rabbit ghosts were certainly not entirely hemoglobin-free; their conclusions rest, however, not so much on the magnitude of the observed birefringence and on its spatial relation to the edge of the ghost as on the way in which it can be changed experimentally by the use of lipid solvents and of media of different refractive index. Finally, in the background of all the observations, there lurks the question whether a "natural" red cell ghost is ever hemoglobin-free. There is always the possibility that all hemoglobin-free ghosts are laboratory creations in which interactions between lipids, hemoglobin, and other proteins have been modified or destroyed. For the moment, this question is unanswered, and is so unwelcome that it is usually ignored.

F. Other Properties Bearing on Structure

There are such a large number of observations bearing on the structure of the red cell and its ghost that it will be possible to refer to a few of them only, and even then in the briefest fashion.

(1) One of the most important observations is that ghosts prepared from human blood by the technique of "reversal of hemolysis" (hemolysis by a very hypotonic medium, followed by restoration of isotonicity by the addition of NaCl) fall into at least three distinct populations: large spherical ghosts with a volume of about 150 μ^3, biconcave discoidal ghosts with a volume of about 90 μ^3, and small crenated ghosts with a volume as small as 15 μ^3. These three kinds of ghost, with very different volumes, coexist in the same tonicity, which may be varied from 0.3 to 1.7. This behavior is clearly incompatible with the van't Hoff-Mariotte law, which states that the volume of an object that can be described as an "osmometer" (i.e., an object that changes its volume in media of different tonicities by the exchange of water alone) is a linear function of the reciprocal of the tonicity of the medium. That this law applies, except in a very general kind of way, to the mammalian red cell has been repeat-

edly questioned (Ponder, 1950b; Sidel and Solomon, 1957); that it applies at all to the type of ghost resulting from "reversal of hemolysis" is an untenable proposition. This quite unexpected observation of the existence of three types of ghost in the same tonicity probably means that the structure of the ghost interferes with its "osmotic behavior" and that the structure of some ghosts is different from that of others (Ponder and Barreto, 1957). Such differences should appear in electron microscope images, if these really reflect the ultrastructure of the ghost.

(2) The heterogeneity of volume and shape of ghosts should not be confused with the *local* heterogeneity of the red cell surface required by the mosaic theory of red cell structure. The idea of a mosaic rests, for the most part, on deductions from permeability theory, but the scintillation phenomenon observed in red cells under suitable conditions suggests that local, although temporary, states of heterogeneity exist. "Scintillation," "flicker," "rhythmic movement," or "vibratory movement" was first described by Browicz in 1890, and his observations were confirmed by Cabot in 1901; Pulvertaft rediscovered the phenomenon in 1949, using phase optics, and a good description of it is that the red cell surface shows a fine waviness that has been likened to the movements produced by wind blowing over a field of wheat. The three principal explanations that have been advanced to account for the phenomenon are that the flicker is due to slight and temporary differences in osmotic pressure affecting the position of surface molecules or groups of molecules that are not firmly bound to their neighbors; that the flicker is due to local metabolic changes, particularly glycolysis, with which its activity goes hand in hand (Blowers *et al.*, 1951); and that the underlying cause is a change in the state of the surface molecules, and perhaps the molecules of the interior as well, from a state in which they can move freely, as in a sol, to a state in which their movement is more restrained, as in a gel (Tompkins, 1954). Whatever the explanation, the essential point is that all the red cell surface is not in the same state at any instant of time. Scintillation has not yet been described in ghosts.

(3) An important property of the red cell ghost is that, when prepared by a variety of methods, it retains some hemoglobin tenaciously. There are two views regarding this "residual" hemoglobin, the one being that it is a contaminant (upheld by the school of Parpart and his collaborators), and the other being that hemoglobin is complexed with other surface proteins and with lipids. The first idea rests on there being methods that give hemoglobin-free ghosts (e.g., Hillier and Hoffman's method of "successive hemolysis"), whereas the second idea is supported by experiments by Bessis and his collaborators (1954), who have shown

that the myelin forms derived from sickle cells contain the abnormal hemoglobin S, and by electrophoresis experiments (Ponder *et al.*, 1955) which show that some of the hemoglobin in a system containing either ghosts or myelin forms remains attached to the almost immobile ghosts or myelin forms, while the remainder of the hemoglobin moves freely either as an anion or as a cation, according to the pH. The observations of Bernstein *et al.* (1938) and of Tishkoff and associates (1953) favor

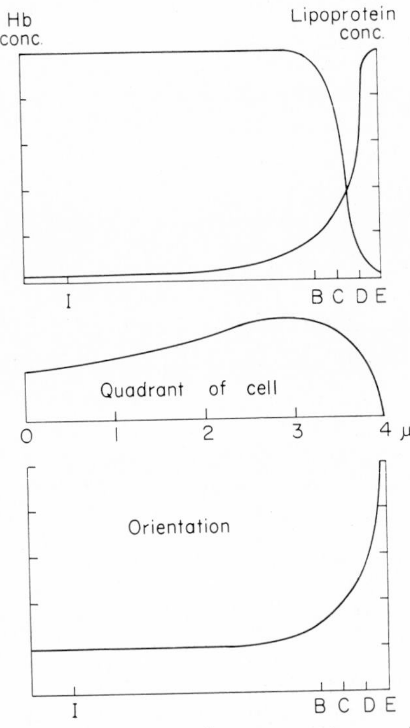

FIG. 18. Representation of the supposed concentrations and orientations of hemoglobin (Hb) and lipoproteins in a quadrant of the red cell. For further explanation, see text.

the latter conclusion, but meantime it is a matter of weighing one body of evidence against another. Ponder's conception of structure (1951a) in which lipids, proteins, and a small amount of associated hemoglobin are found in the surface layers, while the interior of the cell contains hemoglobin principally, perhaps with a small amount of supporting structure, is at least worth considering (Fig. 18). There are a number of observations, all relating to the problem of the thickness of the red cell ghost, which suggest that washing so completely as to remove all the hemo-

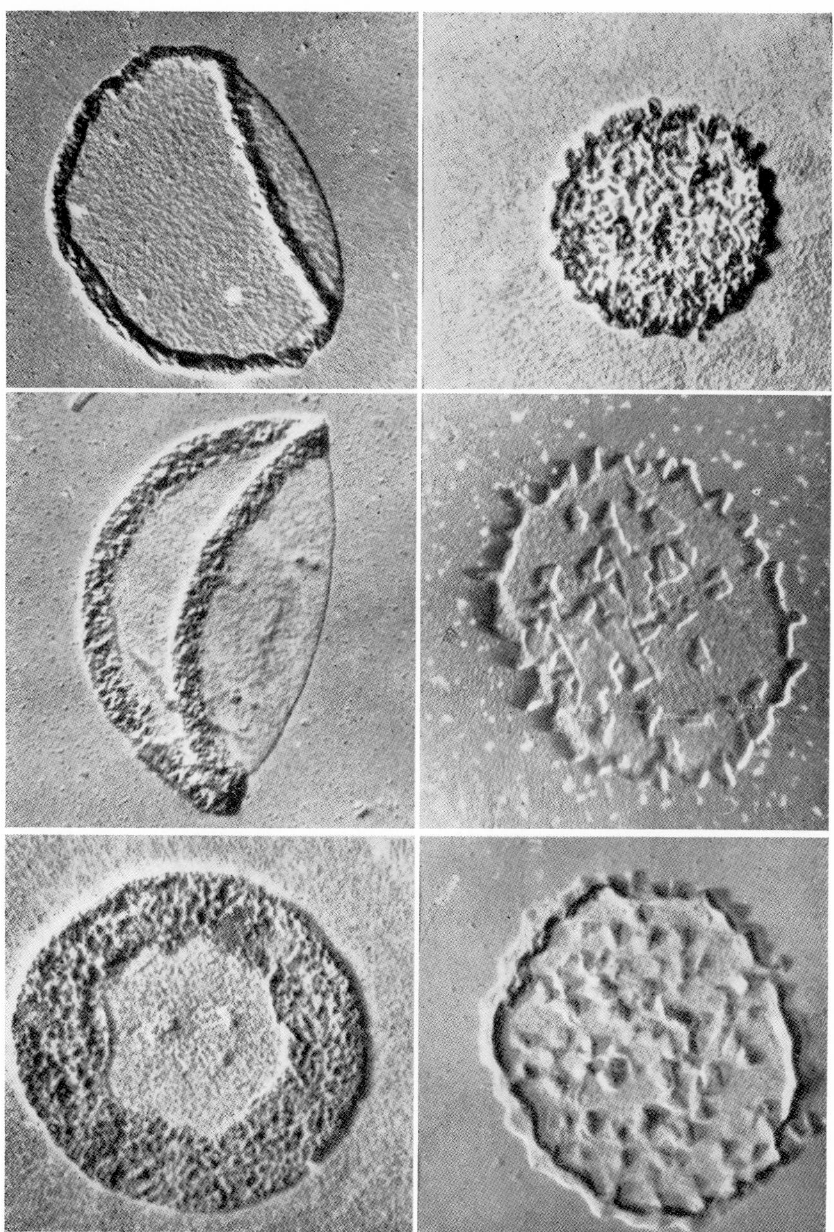

FIG. 19. Electron microscope photographs of discoidal and spherical ghosts containing different amounts of hemoglobin. At the left, discoidal ghosts. At the right, crenated ghosts. The hemoglobin forms a peripheral ring in the discoidal ghosts and is concentrated in the spicules of the crenated ghosts. In this figure, the concentration of hemoglobin increases from above downward. From Ponder *et al.* (1952a).

globin results in the removal of other substances as well, but it would be unfair to represent these as conclusions.

(4) Red cell shape and the disk-sphere transformation are as yet not related to the problem of surface structure in any definite way. With the exception of the observations of Ponder and associates (1952a), which

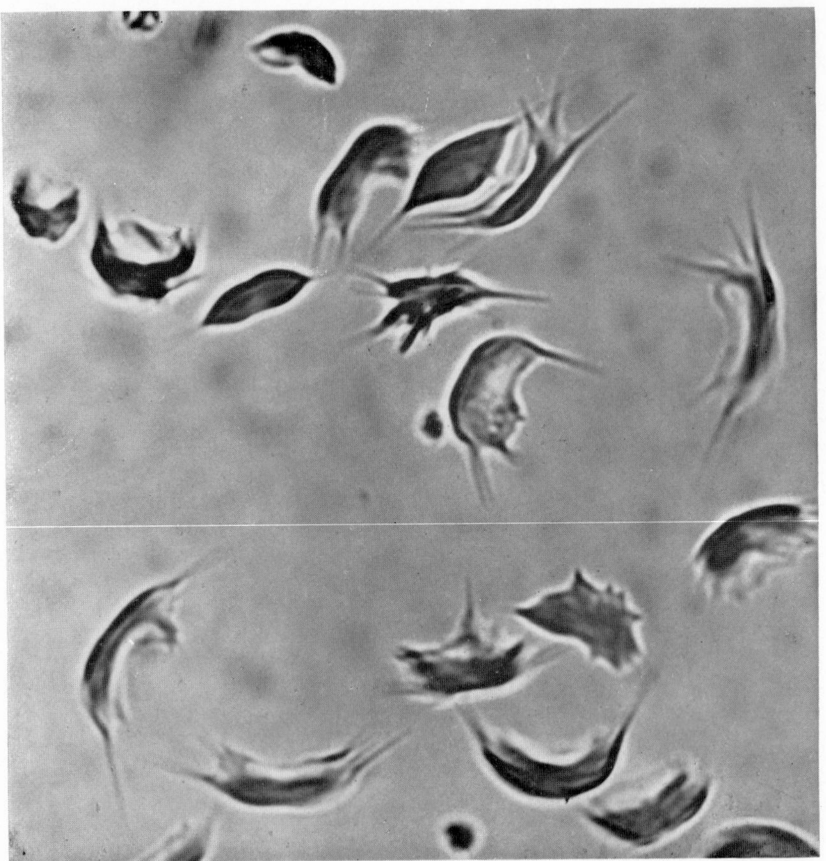

FIG. 20. Sickle cells. From an original photograph kindly sent to me by Dr. J. W. Harris.

are concerned with the distribution of hemoglobin in discoidal and spherical ghosts (Fig. 19), together with leptoscopic determinations of ghost thickness, there is virtually no information on the question. The transformation from disk to sphere, which is *reversible* in many cases, must involve a decrease of surface area of about 30%; this is usually attributed to a folding of the surface of the disk (as in crenation), but little is known about it. Sphering by lecithin and cephalin is at least accompanied by the uptake, presumably at the red cell surface, of lecithin or cephalin.

(5) Observations on the volume of red cells and ghosts fragmented by heat have led to the idea that both the cell and its ghost fragment as

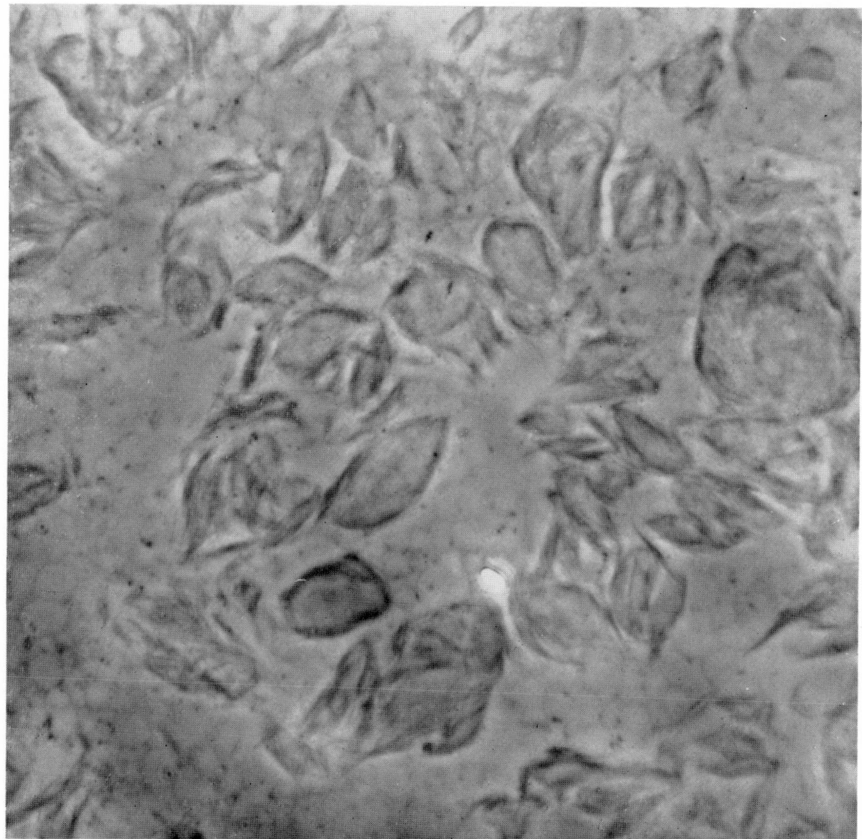

FIG. 21. Tactoid formation in hemoglobin derived from sickle cells. From an original photograph kindly sent me by Dr. J. W. Harris.

solid bodies rather than as shells containing hemoglobin in solution (Ponder, 1951a). This would imply the existence of a more or less ordered structure, and there is X-ray evidence in support of this idea (Dervichian *et al.*, 1947).

(6) The sickling phenomenon (Figs. 20 and 21) occurs when there is only a small proportion of hemoglobin remaining in the ghost (Ponder,

1951b), and since it is the abnormal hemoglobin S that causes sickling, it would be a simple explanation of the sickling of ghosts if hemoglobin S were a component of the surface ultrastructure (see Bessis *et al.*, 1954) (Fig. 22).

(7) Under certain conditions, ghosts have the property of contracting (Fig. 23) until they reach a volume of about 15 μ^3, which is about the volume that would be occupied by their "fixed framework" (Fricke *et al.*, 1939; Ponder and Barreto, 1953). The extent of this contraction can be used, somewhat roughly, as a measure of the thickness of the surface ultrastructure (250–1000 A., wet).

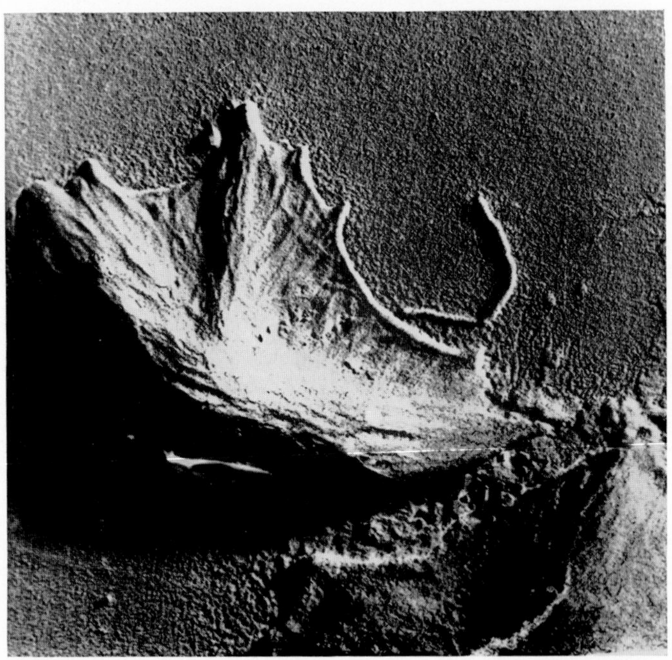

FIG. 22. A sickled red cell with rigid myelin forms. From an original photograph kindly sent me by Dr. Marcel Bessis.

G. The Membrane of the Cell and Particularly the Surface of the Ghost Regarded as an Artifact

This point of view has been taken by Dervichian (1955) in a review on the structure and permeability of cell membranes in general. The membranes discussed are the plasma membranes (plasmalemmas) which are in immediate contact with the cytoplasm and which are supposed to prevent diffusion of the cytoplasm outward into the surrounding medium and to regulate the penetration of dissolved substances inward from the surrounding medium. The existence of the membrane is always invoked as an interpretation for what is called the "permeability" of the cell. By "permeability" is usually meant the greater or lesser facility with which a substance dissolved in the external medium can penetrate into the interior of the cell, and it is a common opinion that the magnitude

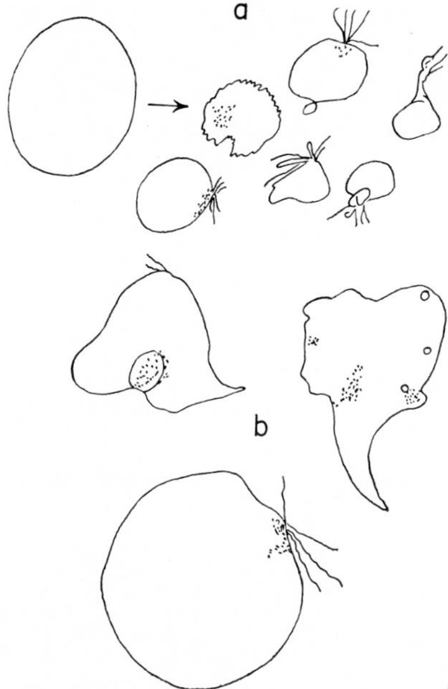

FIG. 23. Spontaneous shrinking of human red cell ghosts. From Ponder and Barreto (1954).

of the penetration of various substances is regulated by the presence of a surface membrane acting like a sieve or a layer (lipid) in which the penetrating substance dissolves. Starting from this point of view, we are forced to admit special properties of the membrane of certain cells so as to be able to interpret the facility with which some ions penetrate preferentially into the cell interior and accumulate there. At the same time, to account for different degrees of penetration of two kinds of substances (as, for example, urea or the sugars on the one hand, and inorganic substances on the other), the existence of a mosaic membrane formed by the juxtaposition or the superposition of protein and lipid has been imagined. Thus, in a general way, the ideas of the membrane and of permeability are bolstered up, the one by the other, to such an extent that in some cases the existence of the membrane is based entirely on permeability considerations and is, in reality, an arbitrary explanation of the way in which penetration and accumulation of different substances occur in different parts of the cell. The rate of penetration, as well as the rate of accumulation of various substances, however, is not necessarily related to the existence of a membrane.

The problem of the existence of a surface membrane ought to be separated from the problem of selective permeability. Further, the question ought to be asked whether it is necessary to postulate a membrane with special properties so as to explain the facts of penetration and accumulation. The question really is whether or not it is possible to account for biological phenomena in the simple terms of what is known about physicochemical properties and their effects, matters that are certainly better established than the purely hypothetical properties of a model of a membrane, the physicochemical nature of which is largely based on assumptions.

1. *Distinction between "Membrane" and Ghost*

It is first of all necessary to distinguish between what might be a membrane or, more precisely, a superficial region of an intact red cell, and the ghost which one obtains after hemolysis. It is a gratuitous assumption to suppose that the ghost constitutes a sac with a more or less spongy structure enclosing hemoglobin which is lost at the moment of hemolysis. Nothing proves that the ghost represents a membrane of a pre-existing nature situated at the surface of the intact red cell, and there is no way of rigorously deciding from the study of the structure of the ghost what the structure of the intact red cell may be. Chemical analysis shows that practically all the lipids (phospholipids, glycerides, cholesterol, etc.) present in the red cell can be recovered from stromata after hemolysis; one cannot, however, say that these constituents were already in the superficial regions before hemolysis, and it is possible that the appearance of the ghost is nothing but the consequence of a precipitation occurring in a mixture of protein and lipids. Such a precipitation obviously can occur only where the constituents of the cell and the external medium come into contact with each other, and surface precipitates obtained with droplets of artificially formed coacervates actually give rise to sacs similar to ghosts.

2. *Analogous Physicochemical Observations*

It is known that lipoprotein associations can separate into two fluid phases at equilibrium and that this depends not only on the proportion of the different constituents but also on the pH and the concentration of different electrolytes in the system (Dervichian and Magnant, 1947; Dervichian, 1949). Displacements of the equilibrium, whether due to the variation of the proportions, the changes in pH, or changes in electrolyte concentration, can result in the solubilization (where the two originally separate fluid phases become a single phase) or to a loss of solubility resulting in precipitation.

Dervichian and Magnant (1947) working with droplets of a coacervate made essentially of hemoglobin, a nucleonate, and a lipid component have observed phenomena that recall in an astonishing manner the hemolysis of red cells and the appearance of "ghosts" as the result of the action of various agents known to be hemolytic. Although the concentration of hemoglobin in these droplets of coacervate are as high as those in the red cell, there exists a fundamental difference between the behavior of red cells and that of the droplets. This difference is that the latter fuse when they come in contact with each other and that, in most cases at least, it is certain that there is no semipermeable membrane preventing the diffusion of hemoglobin toward the exterior, i.e., the droplets result from a simple equilibrium of phases. On introducing a solution containing a hemolysin into a preparation of droplets between slide and coverslip, one can see the droplets of the coacervate rapidly developing vacuoles. These vacuoles increase in size, often fusing so as to occupy almost all the interior of the droplet and leaving nothing except a thin membrane-like layer separating them from the surrounding solution. In this way, droplets may become sacs, which often remain spherical. In the course of this development, one can see continual rearrangements of the internal vacuoles, which shows that their walls are fluid. On the other hand, the final spherical sac has a rigid appearance. By the addition of hemolyzing agents to the droplets of the coacervate, new equilibria are established that result in the appearance of a more concentrated phase, enclosing less concentrated components in a new form and making up the walls of the vacuoles and of the "ghost" which is the final result. These interactions take place throughout the mass of the droplet, but precipitation probably occurs only at its periphery, i.e., at the place where the hemolytic material comes into contact with the material of the droplet. Because of this, the final sac, the walls of which are made up of precipitated material, takes on the initial form of the droplet.

If these artificial stromata are dried, it would be reasonable to expect that they would give rise to little crystals of hemoglobin which would be more or less perfect according to their admixture with lipids, other proteins, etc. Supposing further that we extract the lipids with organic solvents, the precipitate would then be expected to become more or less spongy because of the removal of the lipid portion, thus leaving the appearance of holes.

We can now ask in a tentative manner the following questions: Can we deduce from the fact that each droplet of the coacervate can give rise to a spherical sac that this "ghost" pre-exists in the surface of the drop and that it has the same structure as that observed after drying? Can we pretend, after having extracted with an organic solvent the lipids contained in these sacs of precipitated material, that the holes which we see correspond to definite "sites" where these lipids are localized and that there exists a mosaic at the surface of the droplet made up of protein and lipid areas? We know that in the present case, all the lipid and protein constituents were in the solution contained at equilibrium in a separate phase by reason of their interaction, but once a precipitate has formed, particularly after the loss of water by drying, the resulting material may be very different according to the pH, dilution, hemolytic agents employed, etc., and, further, according to the different proportions of the constituents.

The observations made by Waugh and Schmitt (1940) on the variation of the thickness of stromata with pH can now be otherwise interpreted. In the neighborhood of pH 6, the proteins that contribute to the composition of the ghost are probably in the neighborhood of their isoelectric point, i.e., they are present under conditions in which precipitation is at an optimum. As precipitation occurs, however, by association of the proteins with other constituents (lipids included) which are present in the cell, or at least in its superficial region, it is easy to understand that the thickness will be greatest at pH 6. The extraction by an organic solvent removes the lipids and cannot fail to diminish the thickness of the ghost. Nevertheless, the maximum thickness remains at pH 6, showing clearly the influence of pH on the quantity of protein precipitated in the coacervate.

A critique of this kind shows how dangerous it is to assume that the composition or physicochemical structure of the thin wall of the sac which we refer to as a ghost is the same as that of the surface layers of the red cell. Many of the properties of the surface layers of red cell ghosts (e.g., thickness and degree of hydration, electron microscope appearances, polarization optics, etc.) cannot properly be thought of as throwing much light on the properties and structure of the intact red cell membrane.

III. Permeability Observations

A. *Permeability to Water: "Osmotic Laws"*

1. *Application of the Van't Hoff-Mariotte Law*

Almost since the beginning of permeability theory, the simplifying assumption has been made that the mammalian red cell can be represented by a model in which a thin, nonrigid, nonelastic, and nonextensible surface membrane permeable to water and to anions, but not to cations, surrounds a quantity of hemoglobin and salts in solution. Such a model would shrink or swell by the transfer of water alone when placed in solutions of higher or lower osmotic pressure than that in its interior, and would be expected to behave as a "perfect osmometer" in accordance with the van't Hoff-Mariotte law. This law relates the cell volume V to the tonicity T and can be expressed in several ways, one of the most convenient of which is

$$V = W \frac{(a - aT)}{aT + 1} + 1 \tag{1}$$

Here V is the volume of the cell expressed as a fraction of V_0, the initial volume in an isotonic medium, where V_0 is put equal to unity. W is the fraction of the cell volume occupied by water. An isotonic solution is one that has a tonicity of unity and in which the cell neither swells nor shrinks. It is usual to take the animal's own plasma as the ideal isotonic solution of tonicity $T = 1.0$ and to assign the tonicity 1.0 to solutions of electrolytes, sugars, etc., in which the red cell volume remains unchanged.

If a is very large, the above expression becomes

$$V = W\,(1/T - 1) + 1. \tag{2}$$

and this can be written as

$$V - 1 = W\,(1/T - 1) = f(T,a). \tag{3}$$

i.e., if V is plotted against $1/T$, a straight line with a slope of W results. If a is relatively small, the more exact expression for the van't Hoff-Mariotte law can be rearranged in a similar way, but it is necessary to compute functions $f(T,a)$ which show the value of the fraction on the right side of the equation for a series of values of T and a series of values of a. The function $f(T,a)$ is equal to

$$\frac{1}{T + 1/a} - \frac{T}{T + 1/a} \tag{4}$$

and if V is plotted against $f(T,a)$, computed for the particular value of a in the experiment under consideration, a straight line should again result if the van't Hoff-Mariotte law applies and if the water exchange alone is involved. The slope of this line is again W, the fraction of the cell volume occupied by water.

Since shrinking and swelling cannot go on indefinitely, a further assumption has been made that the swelling increases as the tonicity decreases until a critical volume, V_h, is reached in a critical tonicity T_h. Any volume increase beyond this results in hemolysis. When the critical volume V_h is reached, the surface area of the cell, which has by this time become spherical, is A_h, and the simplifying hypothesis also states that A_h is the same as the initial surface area of the discoidal red cell of volume V_0 (Jacobs, 1926; Castle and Daland, 1937; Ponder, 1937). This hypothesis has been useful in experiments on permeability since it has simplified matters to assume that the lysis of a red cell corresponds to its having reached its critical volume because of the entry of water, *and that the critical volume is constant* for the various types of system under consideration. It further explains why the tonicity in which red cells hemolyze is a function of their shape. If all the red cells were to obey the van't Hoff-Mariotte law and if all contained the same fractional amount of water, each would swell to the same multiple of its initial volume. Further, if the surface:volume ratio were the same for all, i.e., if all the red cells had the same shape, all of them would swell to their critical volume in the same tonicity, T_h, and all would hemolyze simultaneously. In reality, some red cells hemolyze in tonicities higher than others, and so it can be concluded that the water content, the shape, or both, of the cells in the initial population are different. The tonicity, T_h, in which 50% of the red cells hemolyze is that in which the cell of average shape and average water content reaches its critical volume, the exact relation being, in the case of a "perfect osmometer"

$$T_h = \frac{W}{(V_h/V_0 - 1) + W} \tag{5}$$

V_h/V_0 being a complex function of red cell shape because V_0 and V_h are enclosed within the surfaces of the same area but of different shapes. Using this expression as an approximation, it can be shown that the radius r of the sphere which the average mammalian red cell assumes when its critical volume is reached is 0.28 A, A being the semiaxis major of the cell in its initial discoidal form. In the case of the human and rabbit red cells, a sphere with this radius r has a volume of about 2.1 V_0. A critical volume of 1.76 V_0 has been found experimentally for the rabbit red cell in hypotonic serum (Ponder, 1937), and Castle and Daland (1937) have found a critical volume of 1.6 V_0 for human red cells in hypotonic oxalated plasma. Both these values are less than the calculated 2.1 V_0; the simplifying hypothesis that an increase in red cell volume can occur until the area enclosing the volume begins to exceed A_0 is accordingly not a very good one. Hemolysis certainly takes place before the volume increases enough to require a stretching of the surface ultrastructure.

The critical volumes at which red cells hemolyze when the medium surrounding them is other than hypotonic plasma are often much smaller than the volumes that could be contained within a surface of area A_0. Critical volumes tend to be less in systems containing washed red cells than they are in hypotonic plasma, and the effect of hemolysins in reducing the critical volume is often great. For example, if the system contains saponin, the critical volume may be as small as 1.2 V_0. Presumably this is because the lysin weakens the structure of the cell.

There are also limits to the extent to which the red cell can shrink in the hypertonic media. These have not been thoroughly studied, but both Gough and Krevisky have found that only about half of the cell water can be removed by placing it in a medium as hypertonic as 20% NaCl. The remaining half of the water is not osmotically transferable under these conditions. Removal of half the cell water raises the concentration of hemoglobin to about 60%. Nearly all hemoglobins become crystalline, paracrystalline, or gelated in concentrations as great as this, and water may be very difficult to remove by osmotic means from such crystals or gels.

2. *Values of R*

When measurements of the volumes of red cells in hypotonic media are made, the values of V usually fall on a fairly good straight line when plotted against $1/T$ or $f(T,a)$, at least until T becomes so small that the critical volume of the most osmotically fragile cells is reached. The line should have a slope equal to W, the fraction of the cell volume occupied by water. Usually, however, its slope is less than this, so that RW has to be substituted for W in the expressions which describe the swelling of a "perfect osmometer" (Ponder, 1950b; Sidel and Solomon, 1957).

Different explanations have been advanced to account for the necessity of the introduction of the constant R into the equations for swelling so as to reconcile the results of experiments with the van't Hoff-Mariotte law. It should be recognized that although the values of R found experimentally are unquestionably influenced by the details of the measurements, R values less than 1.0 have been found by many independent investigators using a large variety of methods. No method, indeed, gives

$R = 1.0$ consistently. Taken in the aggregate, the many determinations that have been made lead to the conclusion that although the mammalian red cell rarely behaves as a "perfect osmometer," it nevertheless displays a regularity in the way in which it swells in hypotonic media.

The first explanation to be considered for the value of R usually being less than 1.0 is that some of the water in the red cell is "bound." Drabkin's (1950) studies of the amount of water that is not solvent water has led him to suggest that about 15% of the red-cell water does not exhibit the normal properties of the remainder, but this does not account for R values smaller than about 0.85, and many such values have been observed. One would expect that the amount of bound water in normal red cells would presumably be a reasonably constant factor; it is possible, on the other hand, that the fraction of bound water is different in the different metastable states which the red cell can assume. Crenated red cells, for example, tend to be poor osmometers, i.e., to have a low value of R, and the paracrystalline red cells of the rat in cold citrate scarcely change in volume at all when placed in hypotonic media, i.e., the value of R is substantially zero. Sickled red cells, although not crystalline, have either an interior or a thick surface ultrastructure sufficiently well oriented to be birefringent, and these also are poor osmometers.

An alternative explanation is that the activity of the salts in the red cell decreases as a more orderly state of orientation develops in its interior, the red cell becoming more rigid and developing a resistance to deformation which is similar to elasticity. Teorell (1952), indeed, starts with the expression

$$TV = \text{constant} \tag{6}$$

as in the gas law, and then introduces two correction terms, a and b, so as to derive a relation that fits the experimental data. These data are for ghosts, but the reasoning would follow for intact red cells. He rewrites the expression as

$$(T + a)\ (V - b) = \text{constant} \tag{7}$$

where b is the volume $(V_0 - RW)$ which does not enter into the water exchange, i.e., into the volume occupied by hemoglobin and by bound water. The constant on the right-hand side of the equation is equal to RW. The meaning of this expression is obscure, but Teorell regards it as indicating that the ghost, and perhaps the intact red cell also, has some degree of elasticity or rigidity.

The red cell is sometimes called a "perfect osmometer" when it obeys the van't Hoff-Mariotte law after the addition of correction terms such as R, a, or b. This is justifiable only if it can be shown independently that the fraction of the cell occupied by substances other than water is equal to b, that the elasticity or rigidity has the magnitude represented by a, and that the fraction of water bound or constrained by internal orientations is properly represented by R. It is not justifiable to call a red cell a "simple" or "perfect" osmometer because its volume changes follow the van't Hoff-Mariotte law with any convenient values of b, a, or R inserted, although this is often done. The history of the subject leaves one astonished at the extent to which the idea, originally introduced as a simplifying assumption, that the red cell is a "perfect osmometer" has been accepted in spite of the clearest experimental evidence to the contrary.

A few papers have recently appeared in which the osmotic behavior of the red cell has been studied, not from the standpoint of volume determinations, but by comparing the freezing points of intact or hemolyzed red cells with those of the media in which they are suspended (Williams *et al.*, 1959; Olmstead, 1960). The conclusion is that, except for small departures, the freezing point of the red cell interior is the same as that of the surrounding medium, which may be serum or plasma, or hypotonic or hypertonic serum or plasma ranging from a tonicity of about 0.5 to 1.4. The small departures observed can be accounted for by an observation of Adair (1928), who found that the osmotic pressure of hemoglobin varies with its concentration in a curvilinear fashion. However, the systems used by Williams *et al.* contained the cells and serum of defibrinated blood, and accordingly, even if volume measurements were used, would obey the van't Hoff-Mariotte law with a correction factor R of 0.9 or more, and the systems in which R is much smaller could scarcely be converted into systems containing "perfect osmometers" by applying Adair's data. In these letter systems, moreover, R is very variable, and to this can be added that Olmstead concludes that water efflux and water influx in the rabbit erythrocyte are probably each under different physicochemical control.

3. *Volume Changes of Ghosts*

When we come to consider the volume changes in hypotonic solutions, not of intact red cells, but of ghosts, we find an entirely new situation.

The relatively straight line obtained by plotting $V/(1\text{-}p)$ against $f(T,a)$, p being the fractional number of red cells which have hemolyzed and become ghosts, becomes subject to upward and downward departures (Fig. 24). The upward departures from linearity are best accounted for by assuming that the ghost has a certain rigidity and occupies an appreciable volume in the column of cells plus ghosts. The reason for this is probably that the ghost usually retains some hemoglobin. To account for the curve for $V/(1\text{-}p)$ plotted against $f(T,a)$ passing through a maximum and departing in a downward direction from linearity at low tonicities which produce 50% hemolysis or more, a relation can be assumed between the critical volume V_h and the value RW for the cells of a heterogeneous population. This is an attractive idea, for it is not unlikely that compactly arranged molecules of red cells that show low values of R (crenated, paracrystalline, and sickled red cells) form a structure which breaks down relatively easily when the molecules are separated, as they must be when the cell swells. Alternatively, we

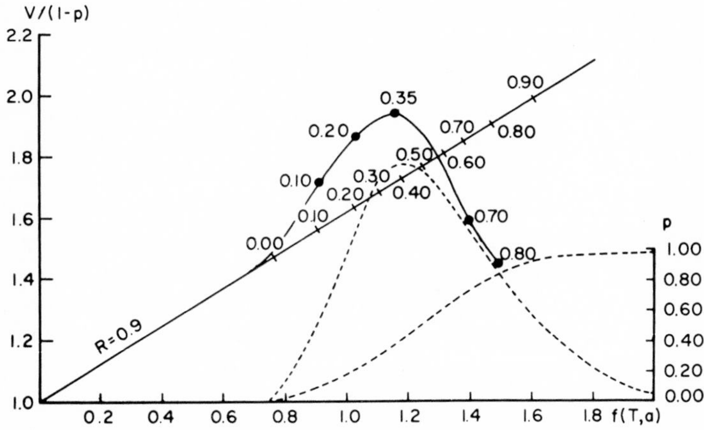

FIG. 24. A linear tonicity-volume relation and the relation observed experimentally (solid line with a slope of $R = 0.9$). The values of p are marked off on both relations. The percentage hemolysis curve is the dotted sigmoid curve referred to the ordinate at the right of the figure. The other dotted curve is its differential.

can suppose that as the tonicity decreases the rigidity of the ghost decreases, so that it contributes less to the volume of the packed cells plus packed ghosts than it does when the tonicity is higher.

These conclusions regarding the rigidity of ghosts can be extended to systems in which the ghosts are produced by the action of hemolysins (Ponder, 1950a). The results obtained with either saponin or the bile salts indicate that the rigidity of the ghost is a function of the concentration of the lysin which produces it.

None of these points raise any particular difficulty regarding the red cell obeying the van't Hoff-Mariotte law, provided that sufficient correction terms are added. When we come to consider the osmotic behavior of the ghost itself [in this case, the human red cell ghost (Ponder and Barreto, 1957)], we find a state of affairs which is altogether unexpected. When the ghosts are made by hemolyzing them in very dilute NaCl and then restoring the tonicity of the surrounding medium to values such as 1.7, 1.0, 0.5, and 0.3, it is found that the volume occupied by the ghosts is linear with $1/T$ although it is only about half that of the red cells from which the ghosts were derived. This looks as if the ghosts were osmometers and as if the van't Hoff-Mariotte law accounted for their behavior, at least in a general way. When the individual ghosts in a system of selected tonicity are examined, however, it is found that there are at least three types making up the entire population. The first is a large spherical ghost of volume about 160 μ^3. The second is a discoidal, and often biconcave, ghost of volume about 90 μ^3, and the third is a crenated ghost which may have a volume as small as 15 μ^3. The volumes can be measured quite well by the use of phase microscopy together with electronic flash, and the proportion of large spherical ghosts, discoidal ghosts, and crenated ghosts in any population can be found by counting the percentage number of each. We have here a situation in which ghosts of volume 16 μ^3, 90 μ^3, and 150 μ^3 (or a little more) coexist in the same tonicity, from which it is obvious that the ghosts do not behave as osmometers. The linear relation between the *average* volume of the populations of ghosts when plotted against $1/T$ seems to be fortuitous, and makes it look as if the van't Hoff-Mariotte law were obeyed, at least in a modified form, although inspection of the individual ghosts of the population shows that this is an impossible conclusion. One possibility is that the ghost has intrinsic factors which are concerned with its shape, and that these are as important as the factors which are concerned with its osmotic behavior.

4. *Penetration of Water*

Assuming that cells are actually invested with membranes, somewhat as defined by Davson and Danielli (1943), the amount of information on the penetration of water is neither great nor reliable. Lucké and coworkers (1939) give a table which contains information regarding the comparative permeability to water in the case of the sea urchin, sand dollar, starfish, eggs of algae, pulp cells of onion, leaf cells of ferns, marine annelids, marine mollusks, fibroblasts, leucocytes, and erythrocytes. The permeabilities vary from 0.1 to 3.0 and are expressed as the number of cubic microns of water which pass through 1 μ^2 of cell surface per minute per atmosphere difference in osmotic pressure at a temperature of about 20°C. McCutcheon and Lucké (1928) found that the permeability constant of *Arbacia* eggs was increased when they were allowed to swell in pure hypotonic dextrose solution. They have also described a number of antagonistic actions between salts. Ionic antagonisms that affect the permeability of marine eggs to water have also been studied by Fukuda (1936). Jacobs (1932) and Jacobs and Parpart (1932) have studied the influence of salts on the rate of penetration of water into the erythrocyte, and the permeability of leucocytes to water has been studied by Shapiro and Parpart (1937), the permeability constant being found to lie between the high value for the erythrocyte and the low value for *Arbacia* eggs. Other published works are referred to in Chapter X of Davson and Danielli's monograph (1943). The reader should also refer to a very scholarly review by Dick (1959), which calls attention to the many assumptions, some of them wholly unjustified, which have been introduced into the treatment of osmotic equilibria of cells and the rates of penetration of substances into them.

5. *Criticisms of "Perfect Osmometer" Concept*

Dervichian (1955), after referring to an article of Lapicque (1925) which contains remarks on the possibly illusory role of membranes and their permeability, proceeds to criticize the idea that a cell can be considered as a "perfect osmometer" even although the swelling (or shrinking) obeys the simplified relation

$$(T + a)\ (V + b) = \text{constant} \tag{8}$$

where b is a correcting constant which represents the volume of the "inert substance" of the cell. He calls attention to the fact that b, far from being a simple corrective term, may take on values that are astonishingly great. For example, if b is no more than 11% of the value of the cell volume for unfertilized sea urchin eggs, it is equal to 40% of the total volume in the case of the red cells of mammals and 64% in the case of certain yeasts. He asks how one can possibly say that the volume of the cell varies as a function of the osmotic pressure and as a "perfect osmometer" when one has to discount 64% of the cell volume. In criticizing the work of Lucké (1932) on the fragmented sea urchin egg, he points out that the increase in volume by the dilution of the sea water in which the egg is immersed depends on the *composition* of each of the internal phases of the cell, and not on the semipermeability of a hypothetical membrane.

The speed of penetration of water into *Amoeba* has been studied by Løvtrup and Pigón (1951). These investigators have suggested that the "permeability constants" are nothing except constants of diffusion applicable to a substance dissolved in water, although when the matter under consideration is the determination of the penetration of the solvent itself, e.g., water, it is more rational to speak of a "constant of filtration." They have proposed a relation which assumes that the quantity of water filtered is expressed by this "constant of filtration" and is related to the osmotic pressure. The demonstration of the mathematical relation may be rigorous enough in itself, but one must not lose sight of the fact that it differs from the hypothesis which states that there is really a filtering membrane at the surface of *Amoeba* and that the penetration of water is regulated by this hypothetical filter.

As Dervichian remarks in the same review, the term permeability, and still more, the term semipermeability, belongs to a terminology applied to artificial membranes and has unfortunately evoked in the minds of those who use it the idea of a porosity which prevents or slows down the passage of molecules. Similarly, the phenomena of swelling and of osmosis have been wrongly associated with permeability since these phenomena are, in reality, independent of the presence or the absence of a membrane. The vagueness of the definitions and the conditions for measuring them has been vigorously criticized by Rashevsky and Landahl (1940) and, in general agreement with their conclusions, it is difficult to separate experimental facts from interpretations when one tries to analyze the various papers on the subject. Too often, the results reported have no significance unless one admits, *a priori,* the hypotheses put forward by the authors.

6. *Colloid-osmotic Hemolysis*

The concentration of protein in an intact cell immersed in plasma or in an isotonic solution is always greater than it is outside. When in solution, the protein acts as a large anion at a pH above its isoelectric point, and if the inside were to be separated from the outside by a membrane freely permeable to both anions and cations, there would be an unequal distribution of ions inside and outside and a corresponding difference in osmotic pressure. Because of this osmotic pressure difference, water would enter the cell; in the absence of cohesive and elastic forces, the cell would swell to the point of cytolyzing.

If the membrane were permeable to both anions and cations, only water could move across it and would do so until any difference in osmotic pressure was abolished. If the membrane were permeable to anions but not to cations, the former but not the latter would be unequally distributed, but osmotic equilibrium, once established, would be maintained. If the membrane were permeable to cations but not to anions, the former would be distributed in the Donnan ratio, but again, osmotic equilibrium would be maintained.

The consequences of a free permeability to *both* anions and cations in the presence of high concentrations of protein inside the cell was pointed out by Jacobs, restated by Davson, and was used as an argument that cation exchanges could not occur between cells and hypertonic or isotonic or hypotonic media, the reason being that if they did, the cells would swell indefinitely until they cytolyzed because of the colloid-osmotic pressure of the protein (dissociated or not) inside them. By 1940, we find permeability to cations no longer appearing as something to be denied, but as a step in a proposed mechanism of hemolysis or cytolysis. Suppose that a cell is normally impermeable to cations. No osmotic effects are produced by the unequal distribution of the anions in the presence of ionized protein, but if a lytic or injurious agent were to render the cells permeable to cations also, swelling would begin and continue. This hypothesis (the "dual hypothesis") was used to account for hemolysis by certain lysins (Davson and Danielli, 1938), for some of the behavior of ghosts (Davson and Ponder, 1938), and for photodynamic hemolysis (Davson and Ponder, 1940). In 1941, it was described by Wilbrandt under the name of colloid-osmotic hemolysis and was considered in much greater detail in a later and important paper by Wilbrandt (1948). The reasoning is still essentially the same: injury leads to a permeability to cations that is not normally present, and permeability to cations leads to a swelling that stops only when the cell cytolyzes.

What is unsatisfactory about the "dual hypothesis" is that it assumes a normal impermeability to cations that does not exist, and that it is indefinite in a quantitative sense. What occurs is better stated without using the terms permeable and impermeable. Granted that the protein inside the cell exerts a colloid-osmotic pressure, when the cell is in its normal environment this is presumably balanced by the concentrations and activities of the solutes in the environment being equal to the concentrations and activities of the solutes in the cell, including the protein. Under such circumstances, there can be no unidirectional movement of water and no change in the cell volume. There can be ion exchange without a volume change, however, provided that the loss of sodium ions, carrying some water with them, is balanced from the standpoint of concentration and activity by the gain of potassium ions carrying the same amount of water with them. This is usually what occurs, and the net effect of such an exchange has been called a "physiological impermeability to cations." When the transport mechanisms of the cell are impaired or when the structure of the cell is modified by the action of injurious substances, the restriction that potassium and sodium must be exchanged millimole for millimole may be removed, and the two ions may move in different directions at different rates. If the ion inside the cell moves out more quickly than the ion outside moves in, the cell shrinks. If the reverse, it swells. To simplify the picture, suppose that an exchange of only one cation for another cation is involved and that the internal ion is called *I* and the external ion *E*. Let the rate of loss of *I* be large while the rate of entry of *E* is small. The cell will then shrink and the concentration of *I* becomes equal inside and outside. *E* will nevertheless enter the shrunken cell, however slowly, because the cell still contains protein with its colloid-osmotic pressure. The entry of *E*, which takes water with it, produces swelling that continues until the critical volume of the cell is reached and the cell cytolyzes. The obvious conditions that could prevent this happening would be a change in the state of the protein, a restoration of the restriction that the exchange must be millimole for millimole, or that there shall be no exchange at all.

The unsatisfactory feature of the hypothesis becomes clear when stated in this way. What happens to the cell volume depends essentially on the concentrations of *I* and *E* and on their rates of penetration across the surface of the cell. Although an approximate expression can be developed (Ponder, 1951c), what will happen under a special set of conditions involves so many constants, few of which can be determined independently, that it is possible to account for almost any volume change by making suitable assumptions. As might be expected, ion exchanges and red cell volume changes are poorly related in experiment, and there are some ion exchanges in which the result is a shrinking of the cell to its original volume after a stage of swelling and in which there is no evidence of the effect of the colloid-osmotic pressure of the contained protein. Further, the colloid-osmotic hypothesis indirectly carries with it an implication that the average cell hemolyzes at a particular critical volume. In the case of the red cell and the lysins digitonin and saponin, no such volumes are ever attained, and the possibility that lysins produce the condition for ion exchange, for colloid-osmotic hemolysis, and also affect the critical volume complicates the situation greatly.

The utility of the colloid-osmotic hypothesis, moreover, depends on the form of the question which it is expected to answer. Can the cation exchange observed be the cause of the hemolysis or cytolysis that occurs? The answer will generally be yes. We observe a cation exchange, sodium for potassium; will it necessarily result in hemolysis or cytolysis? The answer will often be no. Since we are not given values for all the constants, does the hypothesis allow us to make quantitative predictions? Generally, it does not. The "dual hypothesis" or the colloid-osmotic hypothesis is accordingly one which provides insight into many situations, but which frequently is not sufficiently specific.

B. *Methods for Studying Permeability*

A variety of methods are used for studying permeability, and these can be applied to erythrocytes, marine and other eggs, leucocytes, yeasts, bacteria, plant cells, muscle cells, nerve axons, etc. The methods fall into several classes: these comprise chemical methods; methods that measure volume changes; methods (used almost exclusively in the case of the red cell) that measure hemolysis or fragility, and, exceptionally, changes in the pigment that the cells contain; and methods that measure changes in the conductivity of the medium surrounding the cell due to the leakage of electrolytes from the latter. There are also methods that measure changes in the conductivity of cell suspensions due to a change in the volume of the cells, and changes in the capacity or impedance of the cells themselves.

1. *Chemical Methods*

The substance to be studied may be added directly to whole blood and samples of the blood removed after measured times, centrifuged, and either the cells or the supernatant fluid analyzed. Alternatively, samples of the cells may be added directly to a solution of a penetrating substance, the cells being centrifuged down and either the cells or the supernatant fluid analyzed. This method is not applicable to rapidly penetrating substances such as potassium. More recently, instead of estimating potassium or sodium chemically, it has become common practice to estimate them by flame photometry. Radioactive penetrating substances may also be used.

2. *Volume Changes*

When cells are placed in an isotonic solution to which a penetrating substance is added, they first shrink because of the loss of water and subsequently swell because of the gain of water as the substance enters. The rate of the shrinkage and swelling enables the rate of penetration of the substance to be determined. If the substance penetrates slowly, the volume changes may be measured by the hematocrit, by diffraction if the cells can be converted into spheres, or by the extent to which light, measured with a photoelectric cell, passes through the suspension. Parpart (1935) has described a method for recording and analyzing the changes in light transmission, continuous records being obtained with a photocell, a galvanometer, and a camera. The analysis is made by means of calibration curves, and volume changes are measured in relative rather than in absolute units. Over a large range of NaCl concentrations, the relation between the volume V and the opacity (expressed as $E =$

— log transmission) is EV = constant, but the value of the constant may vary from one suspension to another (Wilbur and Collier, 1943). Unfortunately, the method suffers from the disadvantage that the opacity of a suspension is a function of several variables.

3. *Methods Depending on Hemolysis, Fragility Changes, and Changes in the Hemoglobin in Red Cells*

The method that depends on hemolysis is essentially the same as the opacimetric method already referred to. First used by Griyns in 1896, the method has been used by Jacobs (1930), who has devised a simple method in which one observes the minimum depth of a suspension of red cells necessary to prevent the perception of a glowing filament. As hemolysis proceeds, the opacity of the suspension becomes less, and so there is an increase in the minimum depth through which the glowing filament can be observed. The way in which the minimum depth varies with time can be recorded on a kymograph. Alternatively, the instrument may be set so that when the filament is just visible, a known degree of hemolysis has occurred, and the time required for this to be reached is used as an indication of the rate of penetration of the substance (usually nonelectrolytes).

Fragility is the reciprocal of the dilution of the isotonic medium surrounding the cells which produces a given degree of hemolysis. By measuring the dilution of the initial isotonic concentration required to produce a given degree of hemolysis, approximations as to the amount of the substance which has entered the cell can be made.

Hartridge and Roughton (1927) have measured the rate of penetration of CO and O_2 into red cells, these gases causing changes in the absorption spectrum. Keilin and Mann (1941) have used the difference between the absorption spectra of acid and alkaline hemoglobin to study penetration by anions.

A disadvantage attached to the hemolysis method which is rarely remarked upon is that the added substance which permeates into the cell and causes it to swell may also change the critical volume at which hemolysis occurs.

4. *Electrical Methods*

The conductivity method consists in measuring the resistance r_1 of a suspension of red cells at a frequency of 2000–4000 cycles, throwing down the cells, measuring the resistance r_2 of the supernatant fluid, and then calculating the volume concentration ϱ from the expression

$$\varrho = \frac{r_1/r_2 - 1}{r_1/r_2 + 1/X} \tag{9}$$

where X is a form factor which depends upon the shape of the cells; if the cells are spheres, $X = 2$, but for the discoidal cell of the rabbit or man, $X =$ about 1.1. It is frequently stated that the resistance of red cells is practically infinite, i.e., the cells appear to be nearly perfect non-conductors. This is a misconception. Since ϱ cannot be found by any independent method with a precision of greater than about 1%, it is possible that the red cell has a conductance of about 1% of that of the medium in which it is suspended.

The difficulty in applying this method is to ascertain the value for the form factor X. Velick and Gorin (1939) use a method in which r_1 and r_2 are measured for the suspension, the suspension diluted with an equal volume of isotonic plasma or saline, the suspension diluted with 2 volumes of isotonic plasma or saline, and so on. In this way, the term X can be found either by solving equations simultaneously in pairs or by a graphical method. This method would be highly satisfactory if it were not for the fact that the measured resistances r_1 and r_2 become subject to large errors as ϱ decreases to 0.2, and if it were generally practical to use large values of ϱ. The method is also subject to stirring errors which, indeed, are its greatest drawback. The stirring errors are analogous to the changes in light transmission which occur when a red cell suspension is stirred, although they are opposite in direction. They can be eliminated only by converting the cells into spheres, but even if this is done by adding an agent such as distearyl lecithin, it is possible that new effects are introduced.

There are two other electrical methods which can be used under certain circumstances. The first applies to single giant cells such as *Valonia* or *Nitella* or the axon of the squid. A direct current is passed across the cell wall, which in the case of the plant cells is made up of the protoplasm, the plasma membrane, the vacuolar membrane, and the cell sap. In the case of the squid axon, the direct current passes through the axoplasm and then through the enveloping membrane. When direct current is applied, it must not be allowed to flow for more than a few seconds, otherwise polarization of the membranes occurs. The current strength must also be small enough not to release an excitation wave. The direct current with the squid nerve has been found to be about 1000 ohms/cm.^{2}, in *Valonia* to be about ten times this amount, and in *Nitella* about 100–200 times this amount. The second method consists in applying high-frequency alternating current either to single cells or to cell suspensions. The general effect is like that of charging a condenser, the static capacity of which is $C = Q/V$, Q being the charge and V the potential difference. This static capacity has been found to be about 1 microfarad/cm.[2] for a variety of cells, e.g., red cells, sea urchin eggs, muscle, nerve, and plant cells. This complex method is beyond the scope of this chapter, but will be found considered in detail in a paper by Cole (1940). Under conditions which are thought to be correlated with variations in ion permeability, such as excitation and current flow, injury, and death, the membrane resistance is found to be considerably altered, but the capacity is usually relatively unaffected. It may be concluded that the cell membrane has a permeability to ions represented by a resistance, and an ion-impermeable structure represented by a capacity and a dielectric loss. Furthermore, only a small fraction of the membrane is normally permeable to ions. The mechanism of the ion permeability is not obvious from resistance measurements, but the capacity and dielectric loss suggest that the impermeable structure may be highly organized and several molecules thick.

C. *Equations Used in the Study of Permeability*

These equations are thoroughly studied in Davson and Danielli's monograph (1943, see Chapter IV) and in the monograph of Brooks and Brooks (1941, see Chapter I). They arise in their most understandable form in papers by Jacobs. The fundamental equation for a change in cell volume with time is

$$dV/dt = kA(p - P) \tag{10}$$

where k is a constant depending upon the properties of water and of the cell membrane, the area A of the latter, p the variable osmotic pressure within the cell, and P the fixed osmotic pressure in the external hypotonic solution. By making reasonable assumptions, this equation may be integrated to give one of two equations

$$k = \frac{p_0V_0}{P^2At} \log \frac{p_0V_0 - PV_0}{p_0V_0 - PV} - \frac{V - V_0}{PAt} \tag{11}$$

or

$$k = \frac{V^2 - V_0^2}{2p_0V_0At} \tag{12}$$

according to whether P is or is not equal to zero. The cell will stop swelling or shrinking when $dV/dt = 0$ or when $p = P$, at which time its volume will be V; if this volume is less than the critical volume V_h, the cell will remain intact and in equilibrium with its surroundings, but if V exceeds V_h, the cell will hemolyze. The relation between the original differential equation and the two integrated forms involves a number of subsidiary assumptions; one of these is that the cell possesses a membrane or surface ultrastructure so thin or so weak that it cannot support an excess of osmotic pressure in either direction; while others are that the volume changes are due to the transfer of water alone, that the interior of the cell is substantially the same as in an ideal solution, and that the interior itself is structureless and devoid of elasticity.

The second of these integrated expressions can be used to determine the relation between the time t, the cell volume V, and the osmotic pressure of the medium in which the cells are placed. A series of such curves is illustrated in Fig. 25, one curve corresponding to each value

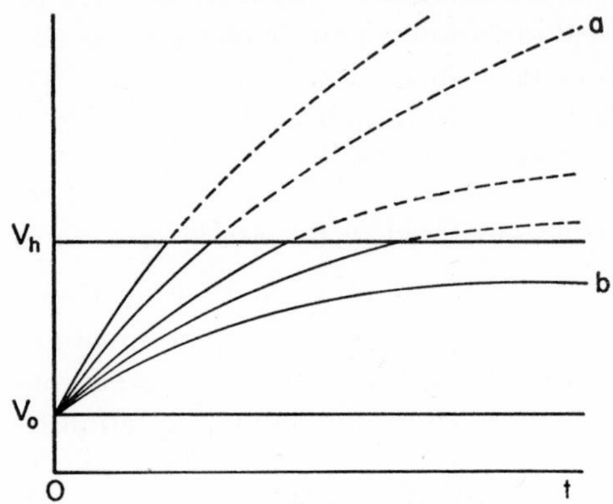

FIG. 25. Increase in volume of red cells in a hypotonic medium as a function of time t. V_0 is the original volume of the cell. V_h is the critical volume. If the medium is too hypotonic, the value of V_h is exceeded (as in curve a) and lysis takes place. If the hypotonicity of the medium is not very great, the value V_h is not reached at all, and a curve such as that marked b results. After Jacobs (1927).

of P. The ordinate shows increasing values of V between V_0 and V_h, and the abscissa shows the time required for these volumes to be reached. Above the value V_h, the cells have no real existence because they have hemolyzed. In the case of the curve marked b, a finite volume of the cell is reached at equilibrium.

When water enters along with dissolved substances, the rate of hemolysis depends primarily on the permeability of the cell to the penetrating substance, e.g., urea or glycerol. There are an extraordinary number of species variations connected with the permeability of red cells to penetrating substances. These have been studied, in particular, by Jacobs (1931), and in connection with them the curious effect of copper on the inhibition of the penetration of glycerol into the red cells of man and the rat by Jacobs and Corson (1934), Jacobs (1946), and also LeFèvre (1947).

For further information regarding the equations used in analyzing the permeability of cells to dissolved substances, reference is made to papers by Jacobs and Stewart (1932), by Jacobs (1933), and to the extensive review by Jacobs published in 1932.

D. Conditions at Equilibrium

If we confine ourselves to the particular condition at which the rate of volume increase dV/dt is zero, and if we express all volumes in relative rather than in absolute units, we can make use of a beautiful method of computation introduced by Jacobs and Stewart (1947). A series of symbols are needed for the mathematical statement; they are:

Initial components of the cell:

Water	$w_0 = 1$
Nonpenetrating anions	A
Nonpenetrating cations	B
Nonpenetrating neutral molecules	M
Hb in all forms (osmolar)	P

External concentrations of equilibrium:

Nonpenetrating anions	C_A
Nonpenetrating cations	C_B
Penetrating anions	C_a
Penetrating cations	C_b
Nonpenetrating neutral molecules	C_M
Penetrating neutral molecules	C_m
Proteins in all forms (osmolar)	C_P
External pH at equilibrium	pH_e

Unknown quantities:

Cell water at equilibrium	w
Internal pH at equilibrium	pH_i
Ionized hemoglobin at equilibrium	p

Equilibrium distribution ratios:

Penetrating anions	r_a
Penetrating cations	r_b
Penetrating neutral molecules	r_m

Abbreviations:

$C = C_A + C_a + C_B + C_b + C_M + C_P$
$D = A + B + M + P$
$E = B - A - f(P, pH_i)$
$Q_a = C_b/C_a,\ Q_b = C_a/C_b,\ R_a = C/2C_a,\ R_b = C/2C_b$

A series of assumptions must now be made. These are that all the salts are completely ionized, that all ionic valencies are unity, that all activity coefficients are unity, and that the osmotic pressure is linear with the sum of the concentrations of ions and molecules. The principal assumptions which are doubtful are that all activity coefficients are approximately equal to 1.0, that there are no elastic forces involved, that the osmotic pressure of the intracellular hemoglobin (Hb) is substantially the same as the osmotic pressure as measured in solution, that cations do not normally pass across the surface, and that a specified percentage of complete hemolysis corresponds to a definite amount of swelling of the cells and to the entry of a definite amount of water. By a process of simplification and elimination, entirely reasonable in relation to the model under consideration once the assumptions are admitted, one gets a series of general equations:

$$w = \frac{D + E}{C - 2r_bC_b} = \frac{D - E}{C - 2r_aC_a} \tag{13}$$

$$r_a = 10^{\mathrm{pH}_i - \mathrm{pH}_e} = \frac{R_aE \pm \sqrt{Q_aD^2 + (R_a^2 - Q_a)E^2}}{D + E} \tag{14}$$

$$r_b = 10^{\mathrm{pH}_e - \mathrm{pH}_i} = \frac{R_bE \pm \sqrt{Q_bD^2 + (R_b^2 - Q_b)E^2}}{D - E} \tag{15}$$

The application of the method can be illustrated by using it in connection with a variety of special situations, of which Jacobs and Stewart (1947) consider seven. The first is that of the normal red cell in saline, where we have some terms present and other terms absent, e.g., cations do not permeate (by assumption), i.e., C_b is zero or absent, and has not to be considered.

Present: $P\ B\ C_a\ C_B$ Absent: $A\ M\ C_A\ C_b\ C_M\ C_P$

So $C = 2C_a$
$D = B + P$
$E = B - f(P, \mathrm{pH}_i)$
$Q = 0$
$R = 1$

$$\text{Therefore } r_a = \frac{2E}{D + E}$$

$$w = \frac{D + E}{2C_a}$$

This is essentially the same expression as that derived by Jacobs and Parpart (1931) by a much longer procedure. It leads directly to their equation

$$\frac{W_1}{W_2} = \frac{2B - 1 - F_1}{2B - 1 - F_2} \tag{16}$$

in which B is the ratio of base in the cells to cell hemoglobin, where F_1 and F_2 are the amounts of base bound per unit quantity of hemoglobin under two conditions which correspond to the presence of W_1 and W_2 in the cell (e.g., two different temperatures or two different pH's), and where the concentration of salt in the suspension medium is constant ($C_1 = C_2$).

A second important case is that of a cation-permeable red cell in saline, where we have

Present: $P\ C_a\ C_b$ Absent: $A\ B\ M\ C_A\ C_B\ C_M\ C_P$

$$\begin{aligned} \text{So } C &= 2C_a \\ D &= P \\ E &= -f(P,\text{pH}_i) \\ Q_a &= 1 \\ R_a &= 1 \end{aligned}$$

$$\text{Therefore } r_a = \frac{D+E}{D+E} = 1$$

$$w = \frac{D+E}{2C_a - 2C_a} = \infty$$

The result of a red cell's becoming cation-permeable ought therefore to be an indefinite swelling and eventual hemolysis. This kind of hemolysis is called "colloid-osmotic hemolysis" (Wilbrandt, 1941), and the process by which a lysin first causes the red cell to become cation-permeable and then to swell and hemolyze as a consequence of its cation permeability is referred to as the "dual mechanism for hemolysis" (Davson and Ponder, 1938, 1940).

Jacobs and Stewart (1947) consider five other special situations: the normal erythrocyte in plasma, the effect of nonpenetrating molecules on colloid-osmotic hemolysis, anomalous osmotic swelling, unequal distribution of an ammonium salt, hemolysis in an ammonium salt, and the swelling produced by NH_4OH. The general equations are able to describe a large number of actual occurrences (some of them, like anomalous osmotic swelling, unexpected) in a satisfactory way, although the description is often more qualitative than quantitative because numerical values for volumes are not sought. This treatment of the subject bears much the same relation to reality as the formal description of the kinetics of hemolysis bears to what really happens in hemolytic systems. In each case, some of the assumptions are demonstrably wrong, but taken as a whole, each set of equations acts as a background of reference against which experimental results can be observed.

E. Permeability to Nonelectrolytes

1. Permeability of the Red Cell

Investigations of this subject began with the work of Griyns (1896) and of Hedin (1897). Griyns used the hemolysis method and was interested in whether substances penetrated the red cell or not, paying little attention to the rates involved. He was, however, able to draw the important conclusion that "substances with the same chemical groupings behave similarly" which is true today.

Hedin, using the freezing point-depression method, found that cane sugar, dextrose, and galactose do not penetrate; that arabinose, mannitol, and adonitol penetrate only slightly within 24 hours; that ethylene glycol, glycerol, and erythritol penetrate in the order given in times which vary from a few minutes up to 28 hours; and that amides, urea, aldehydes, ketones, and ethers penetrate too rapidly for rates to be compared. He discusses his results in relation to Overton's theory of lipid solubility, the more rapidly penetrating substances being the more lipid soluble, with the exception of urea and ethylene glycol.

Kozawa (1914), using a hemolysis technique, made it clear that there are real differences in permeability in different red cells. With human red cells, he found the results shown below, and similar results were obtained with the red cells of the ape. The cells were permeable to arabinose, xylose, galactose, mannose, sorbose, glucose, and levulose; slightly permeable to adonitol; and impermeable to sucrose, maltose, lactose, glycine, alanine, mannitol, and dulcitol. The red cells of the ox, pig, rabbit, guinea pig, goat, horse, camel, and cat were all impermeable to these substances, whereas dog cells were permeable to glucose. Following this early work, many investigators have made more detailed studies of the species variation in permeability, and these are summarized in Davson and Danielli's monograph (1943, see Chapter VIII). Jacobs and Glassman (1937) have given a brief account of results on species characteristics, the permeabilities to glycerol, urea, and thiourea of the red cells of a number of species of fishes, birds, reptiles, and mammals being found to be different. In fishes, ethylene glycol penetrates fastest, while urea and thiourea penetrate at rates which are variable from species to species. In birds, ethylene glycol and glycerol penetrate rapidly, whereas thiourea enters the cells much more slowly, although more rapidly than urea. In reptiles, urea penetrates relatively rapidly, ethylene glycol next, and thiourea much more slowly. In mammals, urea penetrates rapidly, ethylene glycol enters the cells more slowly, and thiourea still more slowly.

The question of the relation of the rate of penetration of a substance with a low lipid solubility to its molecular volume has been investigated by Mond and Hoffmann (1928), Mond and Gertz (1929), and by Höber and Orskov (1933). The latter conclude that the absolute value of the "molecular refraction," i.e., the volume of the molecule, is the determining factor only when a given series of homologous substances is considered. On the other hand, there is no correlation for a given species between the rates of hemolysis and the molecular volumes when substances belonging to different homologous series are used. Their results make it clear that besides the molecular volume of the molecule and its lipid solubility, there is at least one other factor which determines the permeability of a membrane to a given substance. This is the importance of specific chemical groupings, e.g., Hedin's finding that OH groups tend to slow the rate of penetration of many molecules. Quantitative work on the effect of introducing OH groups into the propane molecule has been published by Jacobs and collaborators (1935), and it is evident that the position of the OH group is important, the compound with its OH groups on adjacent carbon atoms penetrating more rapidly than that with the OH groups at the ends of the molecule. The rate of penetration of a nonelectrolyte accordingly depends on at least five important variables: species characteristics, lipid solubility, molecular volume, specific chemical grouping, and position of the groups on the molecule. All these factors are of fundamental importance with the exception of the third, the molecular volume. In any one homologous series, penetration decreases as molecular volume increases, but so many other properties in such a series vary in the same way as molecular volume that it is impossible to tell at the present time whether the rate of penetration depends essentially on it.

From the permeability data, it would appear that the membrane of ox red cells behaves to a first approximation as a homogeneous lipid layer. On the other hand, in the case of the red cells of man, rat, and rabbit, only a small part of the cell is specially adapted to allow passage of glycerol. It is doubtful, however, whether statements such as this signify anything more than that different kinds of red cells possess different species characteristics, and so little is known about the structure of the red cell membrane that a statement to the effect that the membrane of ox red cells behaves as a homogeneous lipid layer whereas the membranes of the red cells of man, rat, and rabbit have a mosaic structure is liable to be misleading.

2. *Plant Cells*

A large amount of work has been done on the permeability of plant cells, but very little has been published in a form suitable for quantitative treatment. This work is of interest because it raises the question of the plasma membrane acting both as a selective solvent and a molecular sieve.

Collander and Barlund (see Collander, 1937) have measured the permeability of *Chara* to nonelectrolytes and plotted the results against olive oil-water partition coefficients. The graph is, in a general fashion, linear, and Collander concludes that there is a relatively close relation between the oil-water solubility of substances and their permeability constants. Nevertheless, whereas medium and large molecules penetrate the plasma membrane only when dissolved in the lipids, smaller molecules can also penetrate in some other way, and so the plasma membrane seems to act both as a selective solvent and as a molecular sieve. Danielli, however, (in Davson and Danielli's monograph, 1943, Appendix A) concludes that there is no justification for plotting the permeability against partition coefficients; instead, one should plot $PM^{1/2}$ against partition coefficients. Here, P is the permeability of the membrane in gram moles crossing unit area per second per gram mole liter concentration difference, and M is the molecular weight of the penetrating substance. If this is done, the situation is improved, and Danielli concludes that the plasma membrane of *Chara* cells may be a homogeneous lipid layer. The arguments, however, are largely qualitative, and the results available are hardly sufficient for positive conclusions.

3. *Permeability of Bacteria*

The information that has been obtained on the permeability of bacteria is confusing. Collander (1937) gives a few values for *Bacterium paracoli* which are not greatly dissimilar from those of cells known to have a lipid membrane. On the other hand, a large amount of work has been done on *Beggiatoa mirabilis,* and this indicates that the cell membrane behaves as a molecular sieve, although most of the results are qualitative. It has been said that the properties of the *Beggiatoa* membrane are those of a nonlipid molecular sieve, diffusion taking place through water-filled pores. There is no evidence of any correlation with partition coefficients. Nevertheless, it can be calculated that the total cross-sectional area of the "pores" is only about 500 A.^{2}, and since a pore large enough to admit a substance such as saccharose must have an area of about 30–40 A.^{2}, there can only be between 10 and 20 pores in the membrane of this bacterial cell. A similar calculation for *Bacterium paracoli* is still more puzzling, for the area occupied by the "pores" is less than 0.1 A.^{2}, which is less than the area of any molecule. Its membrane cannot be considered an inert molecular sieve unless it is enormously thicker than most plasma membranes. Consideration of the space available inside a small cell, indeed, shows that there is hardly enough space for a plasma membrane of the type found in large cells. A plasma membrane 50 A. thick would occupy 0.1% of the volume of *Beggiatoa mirabilis*, 3% of *Escherichia* (*Bacterium*) *coli*, and 100% of the virus of foot-and-mouth disease. Thus, when a size is reached at which a plasma membrane of 50 A. becomes a large part of the cell, we do not find cells, but only viruses.

4. *Marine Eggs*

The permeability of these has been studied by R. S. Lillie [1916 to 1918; see Lillie (1918)]. His work was mainly confined to permeability to water, but work on nonelectrolyte solutes is more recent and is mainly due to Jacobs, Lucké, McCutcheon, and their colleagues. Lucké and his colleagues (1939) have compared *Chaetopterus* and *Cumingia* eggs with *Arbacia* eggs and have found that the former are more permeable to ethylene glycol and glycerol than are *Arbacia* eggs. The permeability of *Arbacia* eggs to ethylene glycol and to water is approximately doubled on fertilization. The ions sodium, potassium, and calcium have a marked effect on the permeability to water but little effect on the permeability to ethylene glycol.

5. *Experiments on Bacteria*

Some experiments have been interpreted as showing that the surface membrane is a molecular sieve (for *Beggiatoa mirabilis*) and that it is a lipid layer through which substances pass with greater or less ease according to their solubility in lipids. This double point of view leaves one at a loss to explain how other cells, such as the human red cell, can present the two types of permeability simultaneously. Collander and Barlund have revived an idea suggested by Nathanson to the effect that there is a mosaic membrane which is a sieve and made up of lipids at one and the same time. According to the relative importance of these regions, penetration of different substances in different cells is supposed to be regulated either by their lipid solubility or by the size of their molecules. As Dervichian (1955) remarks, membranes with the strangest properties have been invented out of pure imagination in order to reconcile hypotheses which can account for restricted observations only. While it is impossible to reject these theories, it is necessary to remember that their reality or even their likelihood is only speculative.

Whatever may be the complexity of the structure of the cell, it can be considered from a standpoint of physical chemistry as a heterogeneous system made up by the juxtaposition of phases in dynamic equilibrium.

The penetration of substances from the external medium into the cell leads to the problem of the partition of a substance between two or more different phases and to the speed of penetration as related to the rapidity of the partition. It is known that a partition coefficient of a substance between two nonmiscible solvents is related to the solubility of the substance in each solvent. As a first approximation, the concentrations at equilibrium in the different phases are related to the solubilities of the substance undergoing partition. In general, a substance that penetrates into the interior of a cell is more soluble there than it is in the surrounding medium, and, further, some structures fix substances in the cell interior more than do others. Considering the importance of lipids in penetration phenomena, it is useful to recall what we know about the mechanism of the solubility of organic substances in lipids in aqueous solution. Fatty substances, such as glycerides, cholesterol and its esters, are held in an aqueous phase because of their association with lipids, such as the phosphatides, which are much more soluble in water. This association occurs in micelles in which the long-chain molecules lie side by side, oriented with their polar groups turned outward toward the water while their paraffin chains are turned inward. Just as water can be considered to be dissolved in the mass of polar groups, similarly, nonpolar substances, such as benzene or chloroform, can penetrate into the interior of the micelles and become dissolved between the paraffin chains attached to the lipid molecules. Organic substances which are normally insoluble in water thus pass into solution because of the presence of lipids which have an affinity for water. Further, although one thinks of the interior of the cell as being an aqueous solution, one can think of the sum of all the interior of the lipid molecules as being a lipid phase itself dispersed in water. It has been shown, for example (Dervichian and Magnant, 1946) that dyes that are insoluble in water but soluble in organic solvents can be maintained in an apparently intracellular aqueous phase because of the presence of swollen myelin forms. It has been shown also that dyes that are water soluble accumulate in the aqueous phases of lipids by penetrating and orienting themselves in the mixed layers lying side by side with layers of lipid molecules.

This example of solubility in lipids which possess a considerable aqueous phase leads to generalizations about solubility when the partition of a substance between the different phases is considered. It will be seen that the term "solubility" expresses the affinity of a substance in terms of chemical properties possessed by one of the media. One cannot account for the penetration of a substance into the interior of the cell without considering the idea of diffusion, even if one wishes to retain the point of view that penetration is governed by the presence of a sieve-like membrane. Think of the diffusion of a substance in a single medium. Here we have the relation (Fick's law)

$$\frac{1}{A} \cdot \frac{dn}{dt} = D \cdot \frac{dc}{dx} \tag{17}$$

which expresses the idea that the rate of diffusion of the dn/dt (number of molecules dissolved displaced in time) is proportional to the concentration gradient dc/dx (the derivative of a concentration c in relation to the displacement dx in the direction in which the diffusion is occurring). The number of molecules displaced per second can be divided by the area A to give the number of molecules crossing the surface of 1 cm.^{2} per second. Obviously, diffusion of dissolved molecules is impeded by contacts with the molecules of the solvents, diffusion thus occurring more rapidly when the viscosity of the medium is small, and for the same viscosity, diffusion being slower as the size of the diffusing molecules increases. The viscosity effect and that of the size of the diffusing molecule is included in the diffusion coefficient D, which is smaller as the viscosity of the medium increases or as the particle size becomes greater.

In the case of the penetration of a substance into a cell, the substance passes from an initial phase into a different phase. Since the rate of diffusion is subject to a discontinuity, the illusion of the existence of a sieve-like membrane can result. Even if the phase boundaries are diffuse, Fick's law cannot be held to apply unless the movements of the diffusing molecules are disorderly and unless their displacements occur in a medium under isotropic conditions, both from the point of view of chemical composition and physical properties. The conditions are the opposite, however, in the case of crossing an interface which separates two phases, and a molecule is attracted as the affinity of one of the media for the molecule is greater or smaller than the affinity of the other medium. Because of this, the probability of a molecule passing in one direction may be very different from the probability of its passage in another. Not only is the partition of the substance between the two media conditioned by its solubility in each, but also the rate at which the molecules moving at random will penetrate in either direction is governed similarly. This leads to the following conclusions. As in all diffusion phenomena, the rate of penetration is necessarily dependent on the size of the particles and the viscosity of the medium, and this rate is strongly influenced by the "solubility" in the two media; because of this, the rate of movement in one direction and in the other is, in general, different.

F. Electrical Phenomena and "Membrane" Potentials

The presence of large ionized molecules, such as those of the proteins, result in there being an inequality in the number of small ions, such as sodium and chloride in the exterior and in the interior of the cell. This is the Donnan effect. There is, however, a further selective accumulation of some ions which does not depend on their charge but rather on their nature.

The inequality between the concentrations of various ions inside the cell and outside it may be remarkable. In the muscle cell, for example, there is a hundred times as much potassium inside as outside (Ringer's solution), eighty times as much sodium ion inside as outside, while the concentration of the chloride ion inside the cell is only about a hundredth of that outside. Because of the difference between the mobilities u and v of ions and cations of the same valency, it is apparent that there will be a difference in electrical potential between two media in which the respective concentrations of the same ionic valency are c_1 and c_2. For ideal solutions, this difference is

$$\Delta E = -\frac{u - v}{u + v}\frac{RT}{F}\log\frac{c_1}{c_2} \tag{18}$$

where F is the Faraday. This difference of potential is often called a membrane potential because one may have to use a membrane to separate the two phases, but it should be understood that it is the difference in concentration which is the cause of the potential difference and not the presence of a membrane.

The difference of potential between the interior and the exterior of cells was measured more than twenty years ago and verifies the relation given above. Such experiments have been carried out, for example, with *Valonia* by Damon and Osterhout (1930), the concentration c_2 in the external medium being varied by placing the cell in dilute sea water. These investigators found that the difference in potential is actually described by the expression given above.

For the electrical behavior of large plant cells, the student is referred to Osterhout (1933) and to Blinks (1940). Both Höber (1945) and Davson and Danielli (1943, see Chapter XV) have interesting material on the nature of selective permeability and bioelectric potentials, but most of the information refers either to the large plant cells already described or to muscle and nerve, both of which present special features.

G. *Permeability to Ions*

1. *Anions*

The literature contains many studies of the anion permeability of the red cell, but they are nearly always semiquantitative. This branch of our knowledge still remains unsatisfactory. In the case of the red cell, however, it is generally true that anions permeate much more rapidly than cations.

No change in red cell volume is more often referred to than that which occurs on the addition of CO_2 to blood. As a matter of fact, the early experimental evidence is unsatisfactory (Ponder, 1934). Nasse, Hamburger, von Limbeck, Gurber, and a number of other workers were involved in this question of the chloride shift (water shift) on the addition of CO_2, but as time went on, the extent of the water shift became smaller and smaller until Van Slyke and colleagues (1923) arrived at a volume increase of 0.4% for a change in CO_2 tension of about 10 mm. An extreme case is that of Joffe and Poulton (1920), who could detect no evidence of increase in cell volume accompanying increases of CO_2 tension (20 mm. to about 600 mm.). The whole subject illustrates the position of one who is confronted with a deduction which follows from theoretical principles, possibly sound in themselves, and at the same time with a failure to demonstrate the validity of the conclusions by available experimental methods. It is often said that a water shift *must* take place when CO_2 is added to blood unless the Second Law of Thermodynamics does not hold. It is also possible that some of the initial assumptions are wrong.

The reader is referred to papers by Maizels (1934), by Höber (1936), by Dziemian (1939), whose results are opposed to those of Höber, and to a paper by Parpart (1940), this dealing with Cl-SO_4 exchange. The subject of anion penetration is certainly in an unsatisfactory state from a quantitative point of view, and this may partly be due to the way in which the red cells have been treated; according to Keilin and Mann (1941), the permeability to anions is lost when the cells are washed with acid phosphate buffer.

2. *Cations*

No aspect of a penetration of ions into or out of the red cell has undergone such a change as the behavior of cations. In 1931, when Jacobs wrote his review on the permeability of the erythrocyte, impermeability to cations was regarded as a necessity, the argument being that if both anions and cations could penetrate, the red cell would hemolyze by the process referred to as "colloid-osmotic hemolysis." Nevertheless, Ashby (1924) and Kerr (1929) had shown that the red cell membrane is not cation impermeable. Their results were usually referred to and then ignored or, at best, explained away. By 1941, the results of Ashby and Kerr had been confirmed in one manner or another by Scudder and Smith (1940), Harris (1941), Dean and associates (1941), and others. In 1940, Davson found a slow cation exchange (loss of sodium, gain in potassium) in the cat red cell, which contains an excess of Na/K. In the dog (Kerr, 1926; Yannet *et al.*, 1936) and cat (Robinson and Hegnauer, 1936; Hegnauer and Robinson, 1936; Cohn and Cohn, 1939), there is unmistakable evidence of a movement of cations. The *in vivo* permeability to cations is smaller in the rabbit and in man (Eisenman *et al.*, 1937). In recovery from diabetic acidosis, 10–15% increase in red cell potassium accompanies increases in organic acid soluble phosphates (Guest, 1942), and in the monkey with malaria, the potassium lost and the sodium gained by red cells may be 30% of the initial potassium lost and over 80% of the initial sodium gained (Overman, 1947).

Nevertheless, we find a paper by Davson and Danielli (1938) which barely concedes the permeability of the red cell to cations. They put forward the point of view that "abnormal permeabilities" can be produced by centrifuging, exposure to hypotonicity, to heavy metals, and to a variety of hemolysins in low concentration; even substances such as guaiacol, which does not produce hemolysis in the concentrations used, may cause a 16% loss of cell potassium. Among the conditions which produce these "abnormal permeabilities" are storage at 4°C., which produces a considerable potassium loss increasing with time, changes in pH, and exposure to narcotics. It is a question as to when the penetration or "transfer" of cations, such as sodium and potassium, became generally accepted; it was a slow process, the idea of cation "transfer" beginning with the paper by Wilbrandt in 1937, which showed that the addition of fluoride resulted in a marked increase in the rate of loss of potassium, and ending about ten years later with the studies of Ponder on prolytic loss of potassium from human red cells. These prolytic losses (i.e., losses of potassium that occur when the red cells are exposed to low concentrations of lysins) are only conspicuous cases of the cation exchanges ("transfers") that occur in the absence of hemolysins. It is important to note that the potassium losses, either in the presence of lysins or in their absence, are usually accompanied by a mole for mole sodium uptake. As a result a considerable ion exchange may take place without the red cell showing conspicuous volume changes. By 1951, the subject of ion transfer had reached a state of being the subject of a review (Sheppard, 1951).

3. *Penetration of Ions into Other Cells*

The early work on the permeability to ions in *plant cells* is principally due to Overton (1895) and Osterhout (1911), both of whom depended upon production of plasmolysis by a hypertonic solution. Overton, who used both nonelectrolytes and salts, concluded that plant cells are impermeable to salts because the plasmolysis is permanent. Osterhout, however, found that plasmolyzed *Spirogyra* cells will, under certain conditions, deplasmolyze and later plasmolyze again, the deplasmolysis being apparently due to the penetration of the salt (NaCl). Osterhout emphasized that in *Spirogyra* cells there is no plasmolysis in 0.375 *M* NaCl nor in 0.195 *M* $CaCl_2$, whereas in a mixture of 100 ml. of the NaCl solution and 10 ml. of the $CaCl_2$ solution, plasmolysis occurs, presumably because the permeability in the mixture of the two salts is sufficiently low to allow of plasmolysis. Similar results were obtained by Brooks (1916) with *Laminaria* leaves, by Fitting (1919) with *Tradescantia*, and by Kaho (1921) with lupine roots, and so it would seem that pure solutions of NaCl increase the permeability to salts, whereas $CaCl_2$ antagonizes this effect.

Osterhout (1922) made a very detailed study of the influence of the ionic composition of the medium surrounding the cell by compressing disks cut from *Laminaria* between two conductivity electrodes and measuring the electrical resistance at low frequencies. The results are summarized in his well-known monograph "Injury, Recovery, and Death in Relation to Conductivity and Permeability."

Apart from Osterhout's results on marine algae (particularly *Valonia*, *Nitella*, and *Chara*), the interested reader is referred to papers by M. M. Brooks (1925), Jacques (1937), Cooper and his collaborators (1929), Hoagland and Davis (1923), and S. C. Brooks (1938), who was one of the first to use radioactive potassium. The reasoning employed in some of these papers is doubtful, particularly because at the time the possibility of a metabolic ion transfer was not considered.

The small amount of work on the permeability of yeasts, the larger amount of work on the permeability of aquatic animals, and the still larger amount of work on the permeability of muscle and nerve to ions is covered in Chapter XIII of Davson and Danielli's monograph (1943). As regards the latter, the penetration of ions seems to be as much related to ion transfer as it is to the simple theories of the penetration of a membrane.

H. Permeability to Weak Electrolytes

Strong bases and strong acids resemble strong neutral salts in being completely ionized and in being unable to pass into cells unless they produce injury. The lipid components of the surface membrane have little ability to dissolve ions, and if neutral salts penetrate at all, they must do so because the membrane has an adequate pore size. Undissociated molecules, on the other hand, such as are contained in solutions of weak bases and weak acids, may penetrate the lipid layer at a rate which depends on their solubility in it. The classic experiment on the penetrating power of undissociated molecules in a solution of a weak electrolyte is that of Osterhout (1925), who investigated the distribution of H_2S between sea water and the sap of *Valonia* cells. The normal pH

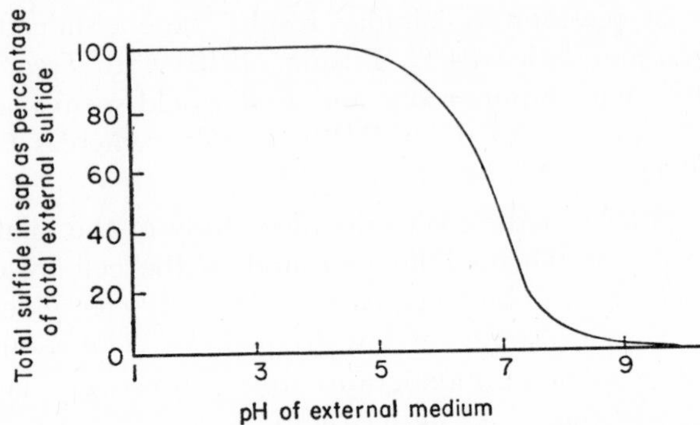

FIG. 26. The relation of the total sulfide in sap as a percentage of the total external sulfide at pH 9. After Osterhout (1925).

of this sap is 5.8, and the pH of sea water was adjusted to a range of values from 5 to 10 by addition of HCl or NaOH. H_2S was then introduced into the sea water, and its distribution at equilibrium observed (Fig. 26). It can be calculated that at pH 5, H_2S is almost completely present as a free acid, whereas at pH 10 it is dissociated into the ions HS and S. Analysis showed that at pH 10, the inside concentration of H_2S was nearly zero, but at pH 5, it was close to the outside concentration.

Dervichian (1955) calls attention, however, to a series of experiments performed by McCutcheon and Lucké (1924) in which the penetration of brilliant cresyl blue, a weak base, was studied in three different kinds of organisms, a vegetable cell (*Nitella*), an egg (*Asterias*), and an entire animalcule (*Gonionema*). In these experiments the pH was modified both in the external medium and in the interior of the cells. It was found that when the pH of the cell interior remained constant, the penetration of the dye increased with increasing alkalinity in the external medium; on the other hand, if the pH of the external medium was kept constant, the penetration of the dye was inhibited as the interior of the cell became more alkaline. These experiments, generally ignored, are in contradiction to the hypothesis of a relation between permeability and the state of ionization. Similarly, experiments by Albert and his collaborators (1945) have led to a conclusion which is almost exactly the opposite of the opinion that only the nonionized forms of a material (in this case, derivatives of acridine) are able to penetrate.

In this connection, the behavior of a droplet of a coacervate in an electric field should be considered, because the opinion that substances which are ionized do not penetrate is usually supported by an argument which supposes that the ions of the substance are repelled by the surface charge of the cell. Studies of the behavior in an electrical field of droplets of a coacervate formed by mixing gelatin and gum arabic have been made by Bungenberg de Jong and Dekher (1935), who have shown that the charge on the droplet varies according to the proportion of the two constituents, becoming zero for a certain composition. The resultant charge therefore depends on the sum of the charges of the two components in the mass of the droplet. These charges neutralize each other when the components are in certain proportions, but for other proportions the resultant charge is positive or negative according to whether one or the other of the two substances is in greater concentration. Droplets of coacervates do not possess a surface membrane, and it should be recalled that a sphere filled uniformly with an electric charge behaves like a sphere which is charged only superficially. If we were to suppose that it is the surface of a cell which, by its charge, repels substances in the ionic state and prevents them from penetrating, it would be necessary to suppose that the surface charge of the cell could change sign according to the situation, since repulsion would have to be as much the cause of the nonpenetrability of a positive ion as it would be of a negative ion.

I. Active Transfer

The most complete account of transport and accumulation in biological systems is that of Harris (1956). While concessions are made to matters as uncertain as "the membrane," "the ghost," "membrane permeability," etc., this book is extremely instructive. Active ion transport will be considered only with respect to red blood cells in order to avoid reviewing transport in muscle, nerve, kidney, etc., all of which present special problems beyond the scope of this chapter.

1. Sodium Movement

Sodium moves in a complex manner, and there appear to be at least three fractions of unequal motility. About 2 mmoles/liter is very easily exchanged and may be associated with the cell surface. About 10 meq./liter exchanges easily at 37°C., and about 2 meq./liter cells of the sodium is very difficult to exchange (Sheppard *et al.*, 1951; Gold and Solomon, 1955).

This situation is similar to one that arose much earlier (Ponder, 1946) in connection with prolytic losses. The question may be put in this form: is the loss of potassium a loss of all the potassium from some of the cells, a loss of some of the potassium from all of the cells, or a combination of these two extreme situations? This point is undecided, but is important because if anything except the second extreme situation (loss of some of the potassium from all the cells) were the actual situation, we would have to challenge the validity of the conclusions drawn from nearly all permeability experiments except those based on single cells. There are certainly cases in which the heterogeneity of certain properties of cells is so great that anything approaching an average value is almost meaningless (Ponder, 1956).

2. Potassium Transfer

Again, the potassium transfer in red cells is not uniform. There is a labile fraction of about 5% of the cell potassium which is much more readily exchanged than the remainder (Ponder, 1951c; Solomon and Gold, 1955). Many observers believe that the transfer of potassium in an inward direction and the transfer of sodium in an outward direction are controlled by the same mechanism. Maizels (1949) thinks that the sodium excretion is the primary process, potassium accumulation being secondary to it and perhaps passive, this view resting principally on the apparent necessity of a cation-permeable red cell swelling and hemolyzing in a sodium-rich medium unless there is some mechanism for expelling the sodium which diffuses into it from outside. Nevertheless, potassium accumulates in systems in which red cells are bathed with LiCl and CsCl instead of NaCl (Ponder, 1950c), and it may be remarked that when the transfer of sugars and amino acids is taken into account, the number of "pumping mechanisms" becomes improbably great. It is probable that some important factor is being overlooked in the present treatment of the transfer of these substances and that we are still at that stage of complexity which usually precedes the clear understanding of a subject.

Many substances, including heavy metal ions, surface-active agents, and narcotics produce changes in the ion transfers. Irradiation with X-rays has a similar effect (Sheppard and Beyl, 1951). Carbon monoxide, cyanide, azide, and dinitrophenol do not affect the uptake of potassium by cells initially low in potassium, but substances that affect anaerobic glycolytic reactions, such as fluoride and iodoacetate, have an effect. It is generally agreed (Ponder, 1950c; Solomon, 1952) that the glucose consumption per liter of red cells is about 2.3 mmoles per hour at 37°C. The energy equivalent to this glycolysis corresponds to about 110 calories per hour, but that required to bring about the potassium and sodium movements against the concentration gradients requires only about 10% of this (Ponder, 1950c; Harris and Maizels, 1952; Solomon, 1952).

The uptake of phosphate by red cells has been followed by Anderson (1942), Prankerd and Altman (1955), and others. Sulfate is not contained in red cells but can be exchanged for chloride (Parpart, 1940). Comparatively little work has been done on the transport of other ions.

3. *Transfer of Sugars*

Both Wilbrandt and Rosenberg (1950) and LeFèvre (1948) explain sugar movement by the sugars forming a temporary combination with some part of the cell surface, the compound being both formed and decomposed by enzymatic reactions at the outer and inner surface of the membrane. The sugar is then transported by a "carrier." This idea of "carrier molecules" is in accordance with Osterhout's picture of the accumulation mechanism in which potassium enters the cell by combining with carrier molecules at the surface, the combination of carrier molecules plus ions diffusing inward and the ions being split off in regions of the ultrastructure remote from the surface or even in more deeply seated regions. This passage requires energy derived from metabolism, both for the building of new carrier molecules and for the destruction of those already engaged in carrying ions. In the case of the transport of potassium in preference to sodium, the carrier molecules would select the former in preference to the latter. In the absence of metabolism, there is no inward diffusing carrier or exchanger capable of binding potassium preferentially. Under these circumstances, the only directions in which ions can move are those of their concentration gradients. The fraction of the cell surface involved may be small, as in the case of copper-inhibited transport of glycerol (Jacobs and Corson, 1934; LeFèvre, 1948). The reader is directed to two short reviews, one by Sheppard (1951) and the other by Ponder (1951c). The latter concerns potassium-sodium exchanges both in the direction of their diffusion gradients and against their diffusion gradients, referring particularly to prolytic ion exchanges. If the active transfer of inorganic salts, sugars, and amino acids occurring against their concentration gradients is linked to metabolism and is brought about by the action of enzymes, the enzymology of the red cell must be quite complex; in fact, as has been remarked above, the number of enzymes involved would be improbably large. The reader is referred to an excellent review by Denstedt in "Blood Cells and Plasma Proteins; Their State in Nature," edited by James L. Tullis in 1953. Denstedt gives an account of some of the enzymatic pathways that may be involved both aerobically and anaerobically, and his concept of the cell surface is that it is a dynamic part of the cell and that its behavior is controlled by metabolic activity. He thinks that the "semipermeability of the membrane" is not a fixed property but a manifestation of the metabolic state of the cell subject to continuous adjustment according to requirements. It must be emphasized that some of these views are vague, and Denstedt himself says that it may be that the metabolic approach to cell preservation is being overemphasized at the present time. As regards the position of the many enzymes that must be involved in a system such as this, Maizels (1949) has suggested that the enzyme systems are situated on the inside of the surface ultrastructure.

IV. Theories Regarding Membrane Structure

A. *Submicroscopic Structure of Cell Membranes*

As Dervichian has remarked, there exists an extraordinarily large number of reports on what is called the "permeability of the cell," but direct studies on the superficial region of the cell are comparatively rare, and the few facts we have are those that concern the surface of red cells. It is almost always these that are referred to when the details of the "structure of the membrane" are under discussion. A brief summary of the microscopic structure of the red cell membrane will summarize much of what has been said in other parts of this chapter.

As we have seen, a lipid layer theory, a sieve or ultrafilter theory, and a mosaic theory of the typical membrane have all been advanced, and there is no doubt that the surface ultrastructure involves both proteins and lipids. Almost certainly, these lipids and proteins are complexed, and if we allow a *dry* average thickness of the surface ultrastructure of 50–100 A., it can be easily imagined that complexes occur in the surface ultrastructure throughout a depth in the wet state of the cell of 300 A. or more.

Winkler and Bungenberg de Jong (1941) regard the ultrastructure as a complex system consisting of phosphatide-calcium ions, together with serum proteins, the surface layer being covered with a phosphatide film, stabilized by cholesterol, and the positive choline groups of the phosphatide being linked to a more deeply seated stromatin (Fig. 27). Their model of the cell also contains a very superficial layer in which an incomplete film of polar lipids turn their lipophilic ends inward toward the phosphatide layer and their hydrophilic ends outward. The oriented lipid molecules are considered as just sufficient to cover the surface of the cells in the manner indicated, the layer of stromatin below it being 120 A. thick. The thickness of the total surface ultrastructure is about

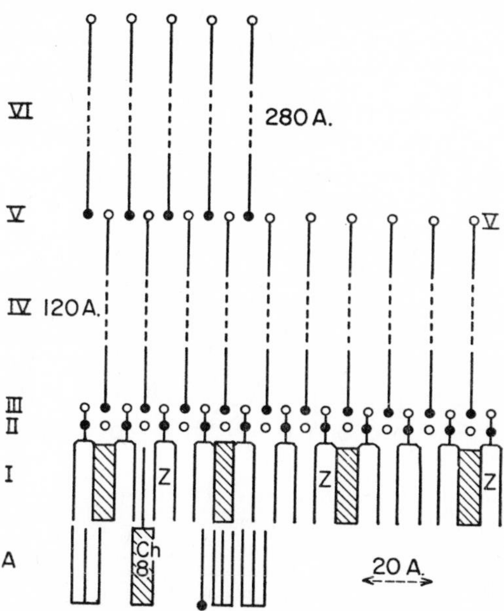

FIG. 27. Molecular structure of the surface of the red cell according to Winkler and Bungenberg de Jong (1941). Dots represent ionic groups; circles, cationic groups; shaded areas, cholesterol; Z, phosphoric acid; and dots with lines attached, fatty acids.

150 A., which is not greatly different from the 250 A. obtained by Wolpers (1956) (electron microscope measurements). This complex scheme explains many properties of red cells; e.g., it makes allowance both for the lipid layer theory, and the molecular sieve theory, there being a lipid film with molecular pores where the cholesterol covering is lacking. Unfortunately, there are a number of difficulties associated with this model, one of which is that the analysis of red cells does not show the presence of calcium. It is true that the amount of calcium present is so small that it might escape detection in analysis, but its inclusion in the scheme is based on nothing more than the fact that calcium is present in blood

plasma, together with convenience. Frey-Wyssling (1948) points out that stromatin is more likely to be parallel to the surface than radially arranged, and inclines rather to the orientations based on birefringence and described by Schmitt *et al.* (1936, 1938).

Schmitt and his colleagues came to the conclusion that the surface ultrastructure is a composite body with alternate protein and lipid lamellae, the latter not necessarily continuous (see Fig. 28). In this model the lipids are oriented radially and the proteins tangentially. Whether a complex ultrastructure such as this should be thought of as supporting the lipid layer theory, the sieve theory, or the mosaic theory is doubtful.

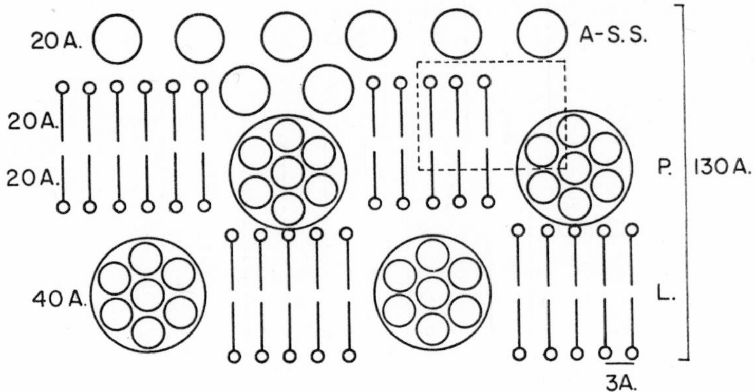

FIG. 28. Diagrammatic pattern for the red cell surface proposed by Ponder (1948). *A-S.S.*, antisphering substance; *L.*, lipid palisades; *P.*, protein components, seen in cross section.

Considering what is now known about the interaction between lipids and proteins (hemoglobin, incidentally, being included since it is held by the ghost tenaciously), it is questionable whether the ideas of permeability as being determined by lipid layers, by molecular sieves, or by mosaic ultrastructures have any clear meaning. This is particularly true if substances enter and leave cells by being transported by "carrier molecules," and if the properties of the surface layers are dependent on metabolic processes.

Figure 29 shows the conception of Parpart and Ballentine (1952). The figures to the left and to the right show the proposed structure as seen tangential and as seen perpendicular to the surface. Protein is indicated by crosslines, water by stippling, nonaqueous phase by clear areas. Phospholipid is represented by rectangles with circular heads, and cholesterol by rectangles. The figure in the center shows the change in orientation of lipid molecules during hemolysis, which is conceived to be reversible. No comment need be made regarding this figure; it is simply a conception of surface structure which, so far as the evidence goes, is neither better nor worse than the structures proposed by Ponder (1948), and by Winkler and Bungenberg de Jong (1941).

B. Theories of Meyer and of Teorell

The starting point of these new ideas, both of which originated in 1935, is that a molecular framework or surface ultrastructure represents a gigantic, polyvalent, and immobile cation or anion. The same ideas may be applicable to the cytoplasm also, which, because of its ampho-

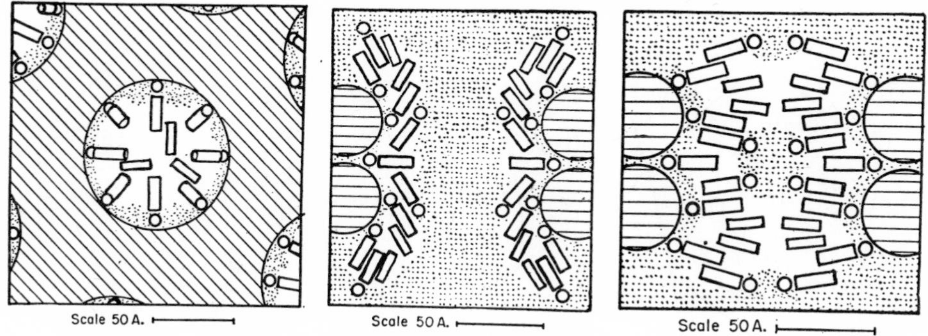

FIG. 29. Suggested structure of the red cell surface shown in a section tangential to the surface (left). Protein is designated by crosslines, water by stippling, and nonaqueous phase by clear areas. Phospholipid is represented by rectangles with circular heads, and cholesterol by rectangles. The figure at the right shows the same arrangement in a section perpendicular to the surface. The central figure shows the change of orientation of lipid molecules during hemolysis. From Parpart and Ballentine (1952).

teric character, can act either as a cation or as an anion according to the pH. (See Teorell, 1935.)

It can be imagined that in the meshes of the surface ultrastructure, carboxyl groups or amino groups or both are fixed as immobile members of the main valency chains (see Fig. 30). The surface of the cell may accordingly be either anionic, cationic, or amphoteric. This theory of submicroscopic structure of the surface layers and the cytoplasm may seem to take into account only a sievelike action, but lipid solubility is also included since the framework, especially in its outer regions, contains lipids and phosphatide molecules. Wilbrandt (1941) rightly concludes that no sharp distinction can be made between the effects of filter action and solubility.

A framework in the form of a polyvalent immobile ion in contact with a true solution represents a Donnan system even although no membrane is present. This is obvious from the fact that the migration of the colloid framework into the surrounding solution is impossible, whereas mobile ions can migrate freely. This is the condition required for a Donnan equilibrium, and no semipermeable membranes are necessary.

Frey-Wyssling (1948) gives a short account of the theory of Meyer, who combines the above results with the velocity of ion migration in a membrane possessing framework structure in order to arrive at a quantitative expression for the penetration. The interested reader should certainly refer to this material since, to a certain extent, it leads to a synthesis of the theories of permeability in biology. Each of the quan-

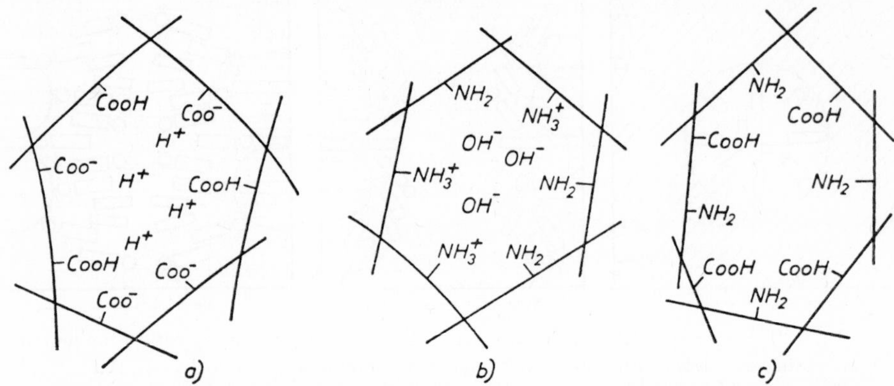

FIG. 30. A diagram to illustrate Meyer's and Teorell's permeability theory. From left to right, an ionic molecular framework, a cationic molecular framework, and an amphoteric molecular framework.

tities occurring in the final expression refers to a principle different from that of the usual permeability theories. There is a quantity U which refers to the ion mobility and which measures the resistance of a sieve-like filter. In a hydrophilic framework with wide meshes U for cations and U for anions would be equal to the ion migration velocities in water. By narrowing the meshes, however, larger organic ions are impeded, and the filter effect will influence the quantities U. The effect of the solubility in lipids is accounted for by the distribution coefficients I. The concentration gradient is expressed by c, and the cell activity constant A is related to the electrical phenomena accompanying the penetration. If the framework has a negative charge, i.e., if it behaves like an anion, A is positive, whereas in the reverse case of a positively charged framework, A is negative. For the amphoteric case, A may be positive or negative according to the pH of the environment. Meyer's theory is based on potentiometry and so allows only the study of ion permeability, which is of greater importance for metabolism than the penetration of nonelectrolytes. Meantime, its application to cytoplasmic permeability is difficult, since so many quantities have to be accounted for and since so few are known. Nevertheless, it is a theory which will bear watching in the future, particularly as membranes of the classic type are not necessarily involved.

V. CONCLUSION

If the reader of this chapter has concluded that the author is not convinced about the structure, or even the necessary existence, of the cell membrane as it is generally described, he will not be far wrong. Without going so far as to say that all Dervichian's criticisms are valid, it is certainly true that many of the conclusions about the cell membrane and its permeability are based on pre-existing ideas, on unallowable simplifications, as well as on a disregard of both physical chemistry and of the results of experiments on the cells themselves. The conclusion that the mammalian red cell is a "perfect" or even a good osmometer is, for example, incredible, yet the papers that describe it as such are made known to every student of general physiology. Starting with such a rejection of confirmed experimental data, it is easy to go backward and to question a variety of findings that were once thought to be true, e.g., the assumed impermeability of red cells to cations (not true), the overlooking of the possibility of active ion transfer (now the subject of a book), the lipid, the sieve, and the mosaic theories of the membrane (convenient), and finally the idea that cells are invested with a membrane at all. Long ago, Griesbach thought that a cell need not have a surface semipermeable membrane and that it was sufficient for it to have an interface between it and its environment, i.e., to be a collection of coacervates, although he did not know what a coacervate was at the time he wrote. This idea has been ignored for at least the last sixty years, more because of the popularity of the semipermeable membrane than because of evidence against it. Even the controversy between Rollett and Norris (about 1890), which divided physiologists into two camps, and which was decided in favor of Norris and of a surface membrane, was not, as is generally supposed, concerned with a surface membrane at all; neither Rollett nor Norris thought this to be necessary except insofar as the surface layers might be "modified to some slight distance inward by contact with the fluid surrounding them," i.e. that they might have properties which we would now describe as properties of an interface (Ponder, 1948). If this author were to be asked whether he believes that a cell membrane, lipid, sievelike, or mosaic in structure, perhaps with enzyme systems incorporated in it, is *solely* responsible for the entrance and egress of substances in the case of the typical cell, he would have to reply that he does not know, and that, on the basis of the existing evidence, he cannot know.

Acknowledgments

This review has been supported by a grant from the Eli Lilly Company, Special Grants Committee.

I have to thank Mr. Don Allen for the reproductions of the photographs illustrating the review and Miss Ruth J. Mandelbaum for the line drawings.

References

Adair, G. (1928). *Proc. Royal Soc.* **A120**, 573.

Albert, A., Rubbo, S. D., Goldacre, R. J., Davey, M. E., and Stone, J. D. (1945). *Brit. J. Exptl. Pathol.* **26**, 160.

Altman, K. I., Whatman, R. N., and Salomon, K. (1951). *Arch. Biochem. Biophys.* **33**, 168.

Anderson, R. S. (1942). *Am. J. Physiol.* **137**, 539.

Ashby, W. (1924). *Am. J. Physiol.* **68**, 585.

Bechold, H. (1921). *Münch. med. Wochschr.* **68**, 127.

Bernhard, W. (1952). *Nature* **170**, 359.

Bernstein, S. S., Jones, R. L., Erickson, B. N., Williams, H. H., Avrin, I., and Macy, I. G. (1938). *J. Biol. Chem.* **122**, 507.

Bessis, M., and Bricka, M. (1950). *Rev. hématol.* **5**, 396.

Bessis, M., Bricka, M., Breton-Gorius, J., and Tabuis, J. (1954). *Blood* **9**, 39.

Blinks, L. R. (1940). *Cold Spring Harbor Symposia Quant. Biol.* **8**, 204.

Blowers, R. E., Clarkson, M., and Maizels, M. (1951). *J. Physiol. (London)* **113**, 228.

Bodansky, M. (1925). *Biochem. J.* **63**, 239.

Bodansky, M. (1931). *Proc. Soc. Exptl. Biol. Med.* **28**, 628.

Boehm, G. (1935). *Biochem. Z.* **282**, 22.

Braunsteiner, H., Gisinger, E., and Pakesch, F. (1956). *Blood* **11**, 753.

Brooks, M. M. (1925). *Am. J. Physiol.* **72**, 222.

Brooks, S. C. (1916). *Am. J. Botany* **9**, 483.

Brooks, S. C. (1938). *J. Cellular Comp. Physiol.* **11**, 247.

Brooks, S. C., and Brooks, M. M. (1941). "The Permeability of Living Cells," Vol. 19 of Protoplasma Monographien. Borntraeger, Berlin.

Bundham, S., and Wright, C. S. (1953). *J. Clin. Invest.* **32**, 979.

Bungenberg de Jong, H. G., and Dekher, W. A. L. (1935). *Kolloid-Beih.* **43**, 143.

Castle, W. B., and Daland, G. A. (1937). *A.M.A. Arch. Internal Med.* **60**, 949.

Cohn, W. E., and Cohn, E. T. (1939). *Proc. Soc. Exptl. Biol. Med.* **41**, 445.

Cole, K. (1940). *Cold Spring Harbor Symposia Quant. Biol.* **8**, 110.

Collander, R. (1937). *Trans. Faraday Soc.* **33**, 985.

Cooper, W. C., Dorcas, M. J., and Osterhout, W. J. V. (1929). *J. Gen. Physiol.* **12**, 427.

Damon, E. B., and Osterhout, W. J. V. (1930). *J. Gen. Physiol.* **13**, 445.

Dandliker, W. B., Moscovitch, M., Zimm, B., and Calvin, M. (1950). *J. Am. Chem. Soc.* **72**, 5587.

Danon, D., Nevo, A., and Marikovsky, Y. (1956). *Bull. Research Council Israel* **6E**, 36.

Davson, H., and Danielli, J. F. (1938). *Biochem. J.* **32**, 991.

Davson, H., and Danielli, J. F. (1943). "Permeability of Natural Membranes." Macmillan, New York.

Davson, H., and Ponder, E. (1938). *Biochem. J.* **32**, 756.

Davson, H., and Ponder, E. (1940). *J. Cellular Comp. Physiol.* **15**, 67.

Dean, R., Noonan, T. R., Haege, L., and Fenn, W. O. (1941). *J. Gen. Physiol.* **24**, 353.

Dervichian, D. G. (1949). *Discussions Faraday Soc. No.* **6**, 7.

Dervichian, D. G. (1955). *In* "Exposés actuels: Problèmes de structures, d'ultrastructures et de fonctions cellulaires," Chapter IV, p. 103. Masson, Paris.

Dervichian, D. G., and Magnant, C. (1946). *Bull. soc. chim. biol.* **28**, 426.

Dervichian, D. G., and Magnant, C. (1947). *Ann. inst. Pasteur* **73**, 841.

Dervichian, D. G., Fournet, G., and Guinier, A. (1947). *Compt. rend. acad. sci.* **224**, 1848.

Dick, D. A. T. (1959). *Intern. Rev. Cytol.* **8**, 387.

Dirken, M. N., and Mook, H. W. (1931). *J. Physiol.* (*London*) **73**, 349.

Drabkin, D. L. (1950). *J. Biol. Chem.* **185**, 231.

Dziemian, A. J. (1939). *J. Cellular Comp. Physiol.* **14**, 103.

Eisenman, A. J., Hald, P. M., and Peters, J. P. (1937). *J. Biol. Chem.* **118**, 289.

Fitting, H. (1919). *Jahrb. wiss. Botan.* **59**, 1.

Frey-Wyssling, A. (1948). "Submicroscopic Morphology of Protoplasm and Its Derivatives." Elsevier, New York.

Fricke, H., and Curtis, H. J. (1935). *J. Gen. Physiol.* **18**, 821.

Fricke, H., Parker, E., and Ponder, E. (1939). *J. Cellular Comp. Physiol.* **13**, 69.

Fukuda, T. R. (1936). *J. Cellular Comp. Physiol.* **7**, 301.

Furchgott, R. F. (1940). *Cold Spring Harbor Symposia Quant. Biol.* **8**, 224.

Gold, G. L., and Solomon, A. K. (1955). *J. Gen. Physiol.* **38**, 389.

Griyns, G. (1896). *Arch. ges. Physiol. Pflüger's* **63**, 86.

Guest, G. M. (1942). *A.M.A. Am. J. Diseases Children* **64**, 401.

Harris, E. J. (1941). *J. Biol. Chem.* **141**, 579.

Harris, E. J. (1956). "Transport and Accumulation in Biological Systems." Academic Press, New York.

Harris, E. J., and Maizels, M. (1952). *J. Physiol.* (*London*) **118**, 40.

Hartridge, H., and Roughton, F. W. (1927). *J. Physiol.* (*London*) **62**, 232.

Hedin, S. G. (1897). *Arch. ges. Physiol. Pflüger's* **68**, 229.

Hegnauer, G. H., and Robinson, E. J. (1936). *J. Biol. Chem.* **116**, 769.

Hillier, J., and Hoffman, J. F. (1953). *J. Cellular Comp. Physiol.* **42**, 203.

Hoagland, H., and Davis, H. (1923). *J. Gen. Physiol.* **5**, 629.

Höber, R. (1936). *J. Cellular Comp. Physiol.* **7**, 367.

Höber, R. (1945). "Physical Chemistry of Cells and Tissues," Section 5, 17. Blakiston, Philadelphia, Pennsylvania.

Höber, R., and Orskov, S. L. (1933). *Arch. ges. Physiol. Pflüger's* **231**, 599.

Hoffman, J. F., Wolman, I. J., Hillier, J., and Parpart, A. K. (1956). *Blood* **11**, 946.

Jacobs, M. H. (1926). *Harvey Lectures Ser.* **22**, 146.

Jacobs, M. H. (1930). *Biol. Bull.* **58**, 104.

Jacobs, M. H. (1931). *Proc. Am. Phil. Soc.* **70**, 363.

Jacobs, M. H. (1932). *Biol. Bull.* **62**, 178.
Jacobs, M. H. (1933). *J. Cellular Comp. Physiol.* **3**, 121.
Jacobs, M. H. (1946). *Biol. Bull.* **91**, 237.
Jacobs, M. H., and Corson, S. A. (1934). *Biol. Bull.* **67**, 325.
Jacobs, M. H., and Glassman, H. N. (1937). *Biol. Bull.* **73**, 387.
Jacobs, M. H., and Parpart, A. K. (1931). *Biol. Bull.* **60**, 95.
Jacobs, M. H., and Parpart, A. K. (1932). *Biol. Bull.* **63**, 224.
Jacobs, M. H., and Stewart, D. R. (1932). *J. Cellular Comp. Physiol.* **1**, 71.
Jacobs, M. H., and Stewart, D. R. (1947). *J. Cellular Comp. Physiol.* **30**, 79.
Jacobs, M. H., Glassman, H. N., and Parpart, A. K. (1935). *J. Cellular Comp. Physiol.* **7**, 197.
Jacques, A. G. (1937). *J. Gen. Physiol.* **20**, 737.
Joffe, J., and Poulton, E. P. (1920). *J. Physiol. (London)* **54**, 129.
Jorpes, E. (1932). *Biochem. J.* **26**, 1488.
Kaho, H. (1921). *Biochem. Z.* **123**, 284.
Keilin, D., and Mann, T. (1941). *Nature* **148**, 493.
Kerr, S. E. (1926). *J. Biol. Chem.* **67**, 271.
Kerr, S. E. (1929). *J. Biol. Chem.* **85**, 47.
Kozawa, S. (1914). *Biochem. Z.* **60**, 231.
Lapicque, L. (1925). *Ann. physiol. physicochim. biol.* **1**, 85.
Latta, H. (1952). *Blood* **7**, 508.
LeFèvre, P. G. (1947). *Biol. Bull.* **93**, 224.
LeFèvre, P. G. (1948). *J. Gen. Physiol.* **31**, 505.
Lillie, R. S. (1918). *Am. J. Physiol.* **45**, 406.
Lovelock, J. E. (1955). *Biochem. J.* **60**, 692.
Løvtrup, S., and Pigón, A. (1951). *Compt. rend. trav. lab. Carlsberg sér. chim* **28**, 1.
Lucké, B. (1932). *J. Cellular Comp. Physiol.* **2**, 193.
Lucké, B., Hartline, H. K., and Ricca, R. A. (1939). *J. Cellular Comp. Physiol.* **14**, 237.
McCutcheon, M., and Lucké, B. (1924). *J. Gen. Physiol.* **6**, 501.
McCutcheon, M., and Lucké, B. (1928). *J. Gen. Physiol.* **12**, 129.
Maizels, M. (1934). *Biochem. J.* **28**, 337.
Maizels, M. (1949). *J. Physiol. (London)* **108**, 247.
Mitchison, J. M. (1950). *Nature* **166**, 347.
Mitchison, J. M. (1953). *J. Exptl. Biol.* **30**, 1.
Mond, R., and Gertz, H. (1929). *Arch. ges. Physiol. Pflüger's* **221**, 623.
Mond, R., and Hoffmann, F. (1928). *Arch. ges. Physiol. Pflüger's* **219**, 467.
Moscovitch, M., and Calvin, M. (1952). *Exptl. Cell Research* **3**, 33.
Muir, H. M., Perrone, J. C., and Popjak, G. (1951). *Biochem. J.* **48**, IV.
Olmstead, E. G. (1960). *J. Gen. Physiol.* **43**, 707.
Osterhout, W. J. V. (1911). *Science* **34**, 187.
Osterhout, W. J. V. (1922). "Injury, Recovery, and Death in Relation to Conductivity and Permeability." Lippincott, Philadelphia, Pennsylvania.
Osterhout, W. J. V. (1925). *J. Gen. Physiol.* **8**, 131.
Osterhout, W. J. V. (1933). *Cold Spring Harbor Symposia Quant. Biol.* **1**, 166.
Overman, R. R. (1947). *Federation Proc.* **6**, 147.
Overton, E. (1895). *Vjschr. naturforsch. Ges. Zurich.* **40**, 1.
Parpart, A. K. (1935). *J. Cellular Comp. Physiol.* **7**, 153.
Parpart, A. K. (1940). *Cold Spring Harbor Symposia Quant. Biol.* **8**, 25.

Parpart, A. K. (1942). *J. Cellular Comp. Physiol.* **19**, 248.
Parpart, A. K., and Ballentine, R. (1952). *In* "Modern Trends in Physiology and Biochemistry" (E. S. G. Barron, ed.), p. 135. Academic Press, New York.
Parpart, A. K., and Dziemian, A. J. (1940). *Cold Spring Harbor Symposia Quant. Biol.* **8**, 17.
Perosa, L., and Raccuglia, G. (1952). *Experentia* **7**, 382.
Ponder, E. (1934). "The Mammalian Red Cell and the Properties of Hemolytic Systems," No. 6 of Protoplasma Monographien. Borntraeger, Berlin.
Ponder, E. (1937). *J. Exptl. Biol.* **14**, 267.
Ponder, E. (1946). *J. Gen. Physiol.* **30**, 235.
Ponder, E. (1948). "Hemolysis and Related Phenomena." Grune & Stratton, New York.
Ponder, E. (1950a). *Rev. hématol.* **5**, 580.
Ponder, E. (1950b). *J. Gen. Physiol.* **33**, 177.
Ponder, E. (1950c). *J. Gen. Physiol.* **33**, 745.
Ponder, E. (1951a). *J. Exptl. Biol.* **28**, 567.
Ponder, E. (1951b). *Compt. rend. soc. biol.* **145**, 1665.
Ponder, E. (1951c). *J. Gen. Physiol.* **34**, 359.
Ponder, E. (1952). *J. Exptl. Biol.* **29**, 605.
Ponder, E. (1954). *Nature* **173**, 1139.
Ponder, E. (1955). "Red Cell Structure and Its Breakdown," Vol. 10 of Protoplasmatologia. Springer, Vienna.
Ponder, E. (1956). *Rev. hématol.* **11**, 123.
Ponder, E., and Barreto, D. (1953). *Acta Hematol.* **12**, 393.
Ponder, E., and Barreto, D. (1954). *J. Gen. Physiol.* **39**, 319.
Ponder, E., and Barreto, D. (1957). *Blood* **12**, 1016.
Ponder, E., Bessis, M., and Bricka, M. (1952a). *Compt. rend. acad. sci.* **235**, 96.
Ponder, E., Bessis, M., Bricka, M., and Breton-Gorius, J. (1952b). *Rev. hématol.* **7**, 550.
Ponder, E., Bessis, M., and Breton-Gorius, J. (1953). *Rev. hématol.* **8**, 276.
Ponder, E., Bessis, M., Breton-Gorius, J., Guinier, A., Antzenberg, P., and Dervichian, D. G. (1954). *Rev. hématol.* **9**, 123.
Ponder, E., Ponder, R., and Barreto, D. (1955). *Rev. hématol.* **10**, 531.
Prankerd, T. A. J., and Altman, K. I. (1955). *Biochem. J.* **58**, 622.
Pulvertaft, R. J. V. (1949). *J. Clin. Pathol.* **2**, 281.
Rashevsky, N., and Landahl, H. D. (1940). *Cold Spring Harbor Symposia Quant. Biol.* **8**, 9.
Robinson, E. J., and Hegnauer, G. H. (1936). *J. Biol. Chem.* **116**, 779.
Schmitt, F. O., Bear, R. S., and Ponder, E. (1936). *J. Cellular Comp. Physiol.* **9**, 89.
Schmitt, F. O., Bear, R. S., and Ponder, E. (1938). *J. Cellular Comp. Physiol.* **11**, 309.
Scudder, J., and Smith, M. (1940). *Cold Spring Harbor Symposia Quant. Biol.* **8**, 269.
Shapiro, H., and Parpart, A. K. (1937). *J. Cellular Comp. Physiol.* **10**, 160.
Sheppard, C. W. (1951). *Science* **114**, 85.
Sheppard, C. W., and Beyl, G. (1951). *J. Gen. Physiol.* **34**, 691.
Sheppard, C. W., Martin, W. R., and Beyl, G. (1951). *J. Gen. Physiol.* **34**, 411.
Sidel, V. W., and Solomon, A. K. (1957). *J. Gen. Physiol.* **41**, 243.
Solomon, A. K. (1952). *J. Gen. Physiol.* **36**, 57.
Solomon, A. K., and Gold, G. L. (1955). *J. Gen. Physiol.* **38**, 371.

Stern, K. G., Reiner, M., and Silber, R. H. (1945). *J. Biol. Chem.* **161**, 731.
Swann, M. M., and Mitchison, J. M. (1950). *J. Exptl. Biol.* **27**, 226.
Teorell, T. (1935). *Proc. Soc. Exptl. Biol. Med.* **33**, 282.
Teorell, T. (1952). *J. Gen. Physiol.* **35**, 669.
Tishkoff, G. H., Robscheit-Robbins, F. S., and Whipple, G. H. (1953). *Blood* **8**, 459.
Tompkins, E. H. (1954). *J. Lab. Clin. Med.* **43**, 527.
Van Slyke, D. D., Wu, H., and McLean, F. C. (1923). *J. Biol. Chem.* **56**, 765.
Velick, S., and Gorin, M. (1939). *J. Gen. Physiol.* **23**, 753.
Waugh, D. (1950). *Ann. N.Y. Acad. Sci.* **50**, 835.
Waugh, D., and Schmitt, F. O. (1940). *Cold Spring Harbor Symposia Quant. Biol.* **8**, 233.
Wilbrandt, W. (1937). *Trans. Faraday Soc.* (Properties of Membranes), p. 956.
Wilbrandt, W. (1941). *Arch. ges. Physiol. Pflüger's* **245**, 22.
Wilbrandt, W. (1948). *Arch. ges. Physiol. Pflüger's* **250**, 569.
Wilbrandt, W., and Rosenberg, T. (1950). *Helv. Physiol. et Pharmacol. Acta* **9**, 86.
Wilbur, K. M., and Collier, H. B. (1943). *J. Cellular Comp. Physiol.* **22**, 233.
Williams, H. H., Erickson, B. N., and Macy, I. G. (1941). *Quart. Rev. Biol.* **16**, 80.
Williams, T. F., Fordham, C. C., Hollander, W., and Welt, L. G. (1959). *J. Clin. Invest.* **38**, 1587.
Winkler, K. C., and Bungenberg de Jong, H. G. (1941). *Arch. néerl. physiol.* **25**, 431.
Wolpers, C. (1956). *Klin. Wochschr.* **34**, 61.
Wolpers, C., and Zwickau, K. (1942). *Folia Haematol.* **66**, 211.
Yannet, H., Darrow, D. C., and Carey, M. K. (1936). *J. Biol. Chem.* **112**, 477.

CHAPTER 2

Plant Cell Walls

By KURT MÜHLETHALER

I. The Chemical Nature and Submicroscopic Structure of the Constituents of the Cell Wall

A. *Cellulose*

The most important constituent of the plant cell wall is cellulose. It forms the structural framework within which other wall substances, such as pectin, lignin, hemicellulose, etc., are embedded. For this reason, as well as its technical importance, cellulose has been intensively investigated, and many of its chemical, physical, and morphological properties are known.

For more than a century, it has been known that this substance hydrolyzes in strong acid to give glucose, and, in fact, the yield is almost quantitative. The elementary structural unit, glucose, has been shown by Haworth (1925) to be a six-membered ring structure with an oxygen bridge connecting carbon atoms numbers 1 and 5. When two such heterocyclic molecules fuse through the removal of a molecule of water, the disaccharide cellobiose is formed (this compound is stereochemically different from maltose, the disaccharide formed in starch hydrolysis). Further polymerization, involving the end hydroxyl groups of the cellobiose molecule, yields high-polymer chains the length of which may vary greatly from one molecule to another. Generally, the average value for the number of glucose residues per chain is estimated to be about 3000.

In the cell wall these chain molecules are not isolated, but occur in bundles formed of parallel chains (Fig. 1). These bundles are called *microfibrils* and consist of about 2000 cellulose chain molecules giving a diameter of the order of 100–250 A. (Rånby, 1949; Frey-Wyssling *et al.*, 1948; Mühlethaler, 1949a; Preston, 1951). When observed in the electron microscope, these minute bundles appear as fibrils of uniform diameter with a length exceeding several microns. X-Ray diffraction studies showed that the cellulose molecules are ordered, to a considerable extent, in a crystalline pattern. The earliest investigations were made by Sponsler (1922), Sponsler and Dore (1926), and Meyer and Mark (1928). From these investigations came the fact that the repeating period of 10.3 A. along the fiber axis is identical with the length of a cellobiose molecule, thus establishing the cellulose molecule as a chain of cellobiose units. The other dimensions of the unit cell are: $a = 8.34$ A., $c = 7.9$ A., $\beta = 84°$ (Meyer and Mark, 1928). The bonds responsible for crystallinity are different in different directions: in the direction of the chain, the glucose units are connected by primary covalent bonds; the lateral aggregation of chains is brought about by weaker cohesion bonds (hydrogen bonds).

The X-ray diagram interpretation permits certain conclusions about the extent of the crystalline regions. The greater the number of crystalline planes reflecting the X-rays, the sharper the interference spots appear on the film. Small crystalline regions give more diffuse spots. According to measurements by Hengstenberg (1928), the breadth of the crystalline regions or *micelles* is 50–60 A. in the direction at right angles to the 101 plane; micelle length is about 600 A., giving a length to breadth ratio of 10:1. These are only average values and, within the microfibrils, micelle

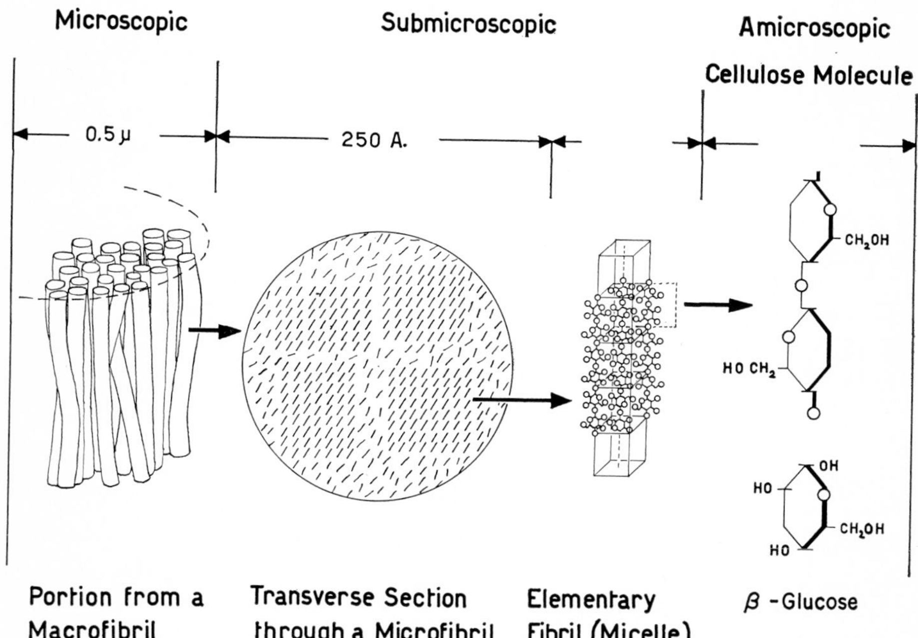

FIG. 1. Structural elements of cellulose at various degrees of magnification.

dimensions may vary considerably. The crystalline regions involve about 100 cellulose chain molecules held together in crystalline array for a distance equivalent to about 150 consecutive glucose residues. Because, as has been mentioned above, the polymerization number of the cellulose chain molecules is far greater, averaging 3000 residues, a given chain molecule may be a component of several different micelles (Frey-Wyssling, 1936). In the regions between micelles, the cellulose chains are not ordered. As is shown in Fig. 1, a cross section through a microfibril may pass through about 2–4 micelles.

The cell wall fibrils visible in the light microscope are called *macro-*

fibrils and are composed of many parallel microfibrils seen as an aggregate; the macrofibril may contain as many as 400 microfibrils. The capillary spaces between the microfibrils in the cell wall are filled by incrusting substances such as pectin, lignin, etc. The orientation of microfibrils in walls of various cell types will be considered in Section VII.

B. *Hemicelluloses*

The hemicelluloses form a complex group of carbohydrate polymers found in cell walls, and the exact character of the polymerization is not yet known. The common characteristic of the hemicelluloses is their solubility, which is greater than that of cellulose. Hydrolysis in dilute warm acid yields pentoses, primarily arabinose and xylose, as well as hexoses, chiefly mannose and galactose. The extent to which these substances contribute to cell wall structure is quite variable. In the cotton hair, they compose only a few per cent of the total wall mass; in collenchymatous cell walls, up to 50% (Frey-Wyssling, 1935). As is the case with cellulose, the pentose and hexose units polymerize to form long-chain molecules. The hemicellulose xylan, such as that found in wood and straw by Hampton *et al.* (1929), is formed of primary valence chains that can be distinguished from cellulose only by the absence of CH_2OH side groups. According to Astbury *et al.* (1935), such hemicellulose chains could replace some of the cellulose chains within the microfibril. The hemicelluloses termed hexosans form not only structural materials in the cell wall, but also serve as reserve foodstuffs, in the manner of starch. One finds, for example, large amounts of hemicellulose in the date seed, where these compounds comprise the thick central layer of the secondary cell wall. In these palm seeds, the hemicellulose is mannan, composed of mannose chains. In the seeds of legumes, however, one finds galactans as foodstuff and wall material. Because the mannan of the ivory nut and the galactomannan of the date palm yield X-ray diffraction diagrams, it may be concluded that these hemicellulose molecules aggregate to form crystalline micelles in the manner described for cellulose (Herzog and Gonell, 1924).

C. *Pectic Substances*

Pectic substances are generally localized in the middle lamella, the adjacent primary wall, or the outermost layers of the secondary wall. The elementary structural unit of this polymer has been shown by Ehrlich (1917) to be galacturonic acid. Additional constituent molecules are glucuronic acid and arabinose. The most important chemical difference from cellulose is the presence of a COOH group in the position of the CH_2OH of cellulose. This group gives considerable hydrophily to the pectic substances and also gives the capacity to form salts with calcium and magnesium. This can be demonstrated microchemically by treating plant tissue with sulfuric acid, whereupon numerous crystals of calcium sulfate precipitate out in the region of the middle lamella. Because of their hydrophily, pectic substances are usually found in a highly swollen, water-saturated state. Cell walls contain pectic substances as incrustations or, as in the middle lamella and corner thickenings of collenchyma, in layers between cellulose lamellae. The removal of pectic substances from the cell wall can be brought about through treatment with oxidizing agents such as hydrogen peroxide; this removal leaves the tissue macerated. Pectic substances appear structureless in the electron microscope, which indicates that the molecules are not ordered and occur at random.

D. Lignin

In contrast to the above-mentioned wall substances, lignin does not occur alone in the wall but is found only in association with other wall constituents. Like the other wall constituents, it is a high polymer; but the elementary units are not as homogeneous as is the case with the polysaccharides. The very fragmentary knowledge of lignin structure makes it difficult to characterize. The term "lignin" cannot be considered to refer to a specific chemical substance; it is rather a designation for a group of high molecular weight amorphous compounds that are chemically very similar. As early as 1897, Klason (cited in Schubert, 1954) advanced the view that lignin was a condensation, or polymerization, product of coniferyl alcohol. This compound is structurally closely related to such derivatives of lignin as vanillin, and syringaldehyde. Recent investigations with radioactive glucosides (Freudenberger *et al.*, 1955) have shown that tissues in the process of lignification contain a cell-bound β-glucosidase. This enzyme would split the glucoside *d*-coniferin which was present into glucose and coniferyl alcohol; the latter being immediately converted into lignin by the reductases present. The polymerization does not occur in linear fashion as with cellulose, but takes place in all directions of space. Microchemically, lignin can be detected with basic dyes such as gentian violet and chrysoidine. The yellow color formed by treatment with aniline sulfate and the cherry red appearing upon staining with phloroglucinol and hydrochloric acid are characteristic for lignin. Chlor-zinc-iodine gives a yellow color to a lignified cell wall.

Lignification usually begins in the region of the middle lamella and is most prominent where several cell boundaries come together; the process then spreads gradually toward the cell lumen. After lignification, the cell wall generally appears thicker than before. This is clearly seen in the stone cells of young pears (*Pyrus communis*), where lignified cells appear scattered among unlignified ones (Frey-Wyssling, 1935). Here the lignified cell walls are two to four times as thick as the walls without lignin. Furthermore, Alexandrov and Djaparidze (1927) found that the removal of lignin from the stone cells of quince (*Cydonia oblonga*) reduced the thickness of the walls to half their previous value. Lignification is thus always associated with a swelling of the wall; this stems from the fact that the lignin is deposited within the original cellulose framework. Only in rare cases does delignification occur in nature. According to Frey (1928), the lignification process could be stimulated as part of a plant's reaction to the application of pressure. Thus water plants, which have a minimum of mechanical pressure to overcome because of the buoyant effect of water, have much less lignification than land plants. Secondary xylem elements, which withstand mechanical pressure, lignify more strongly than those that are subject to tensile stress (stretching).

E. *Cuticular Substances*

All above-ground organs of the archigoniates and angiosperms are coated with a thin cuticular layer which serves to protect the underlying cells from desiccation. All cuticular substances, such as cutin and the closely related suberin, are very hydrophobic compounds. They consist of various high polymers of saturated and unsaturated fatty acids and oxygen-containing fatty acids, for example phloionolic acid, $C_{21}H_{42}(OH)$-COOH, and the acid of cork $COOH(CH_2)_6COOH$ (Frey-Wyssling, 1935). Cuticular substances are detected microchemically by the reagents Sudan III and corallin, but it is not possible to distinguish between the above-mentioned substances with staining reactions. The electron microscopical investigations of Sitte (1955) on bottle cork and potato skins showed the suberin present in the form of lamellae. Suberin itself is amorphous. The negative birefringence of the cuticle, first observed by Ambronn (1888), stems from the easily melted and easily extracted waxes that are embedded in the suberin.

In addition to the above-mentioned *framework substances* (cellulose and hemicellulose), one may group wall constituents as *incrustations* (pectin, lignin) and *adcrustations* (cuticular substances). These last are merely deposited upon the wall proper and contain no cellulose.

F. *Mineral Deposits*

In addition to the described high-polymer incrustations one finds in many cell walls incrustations of inorganic nature, primarily compounds containing calcium carbonates and silicates. The same substances appear also in the protoplasm or vacuole of the living cell in the form of crystals.

Heavy deposits of calcium compounds are found especially in the hairs of many borages, crucifers, and cucurbits as well as in the cell walls of many green algae, particularly the Conjugales and Charales (Küster, 1951). The deposition of calcium and magnesium carbonates in cell walls apparently is amorphous.

Silicious cell walls are widespread throughout the plant kingdom, especially among the Equisetaceae, Gramineae, and Cyperaceae. Because the mineralization of the wall is stronger with increasing age, the depositing process is regarded physiologically as the cell's mechanism for ridding itself of excess minerals.

In summary, the plant cell wall can be shown to consist of the substances listed in Table I.

II. Microscopic Structure of Cell Walls

The plant cell wall is a product of protoplasmic activity, and, in the higher plants, its development begins in the formation of the cell plate immediately after nuclear division. In the microscope, one observes initially many fine droplets which then fuse to form a thin isotropic layer, the middle lamella. The observation of the origin and development of the wall is most beautifully observed in the zoospores of certain algae. The just-formed zoospores are at first naked, but form an initial cell wall, the *primary wall,* within a few hours after fertilization. According to Whitaker (1931), the cell wall formation in freshly shed eggs of *Fucus* can occur within an hour. Fifteen minutes after fertilization, the surface of the egg is still viscid; after 45 minutes it is semisolid, and, after 60 minutes it is firm. A positive reaction for cellulose with iodine and sulfuric acid is obtained as early as 90 minutes after fertilization. The primary wall is very thin and shows no internal structure. Upon the inner surface of the primary wall, which consists of cellulose and a high percentage of pectin, the *secondary wall* is deposited. This secondary wall can become very thick and in the case of mature fibers almost fills the entire cell lumen. In the bast fibers of *Linum,* the Urticaceae, and the Asclepiadaceae, this wall is almost pure cellulose. In the light microscope, an inner, middle, and outer layer of the secondary wall can be distinguished. The polarized light microscope shows the inner and outer layers to be very highly birefringent when a cross section of a secondary wall is observed. The central layer displays weaker birefringence, because the cellulose chains here are running generally in the direction of the cell axis. After strong swelling of the wall with alkali, all three layers

TABLE I
PLANT WALL SUBSTANCES

Substance	Elementary unit	Optical properties	Staining reaction
Cellulose	Glucose	Positive birefringence	Chlor-zinc-iodine (stains violet)
Hemicellulose	Arabinose, xylose, mannose, galactose	Partially birefringent	None specific
Pectic substances	Glucuronic and galacturonic acids	Isotropic	Ruthenium red
Lignin	Coniferyl alcohol	Isotropic	Phloroglucinol hydrochloride (stains rose); chlor-zinc-iodine (stains yellow)
Cuticular substances	Fatty acids	Negative birefringence	Sudan III (stains orange)
Mineral deposits	Calcium and magnesium in the form of carbonates or silicates	Isotropic	

are seen to be composed of fine lamellae whose thickness lies at the limits of resolution of the light microscope. Balls (1919) and Kerr (1937), observing these lamellae in the cotton fiber, considered them to be daily depositions of cellulose. Cotton hairs produced under constant conditions of light and temperature did not show these lamellae (Anderson and Kerr, 1938). More recent investigations by Barrows (1940) have put a different interpretation on this matter. Barrows was able to show that the number of lamellae in various plants could appreciably exceed the number of growing days and thus lamella formation could not be the result of a simple daily rhythm. The exact number of secondary wall lamellae is very difficult to estimate because they lie very close together in fibrous cells. According to counts by van Wisselingh (1925), over 100 lamellae could be present in the epidermal cell walls of cruciferous seeds.

When one observes the surface of mature cell walls, it is often possible to detect fine striations winding in a helical fashion around the cylindrical wall. These are seen most clearly in the algae (*Valonia, Cladophora*, etc.). Each wall layer has only one striation direction, the striations being parallel. The striation direction in adjacent lamellae may be quite different. From these observations it was proposed that the cell wall was composed of very fine fibrils, and Balls (1922) estimated that these fibrils would have a diameter of about 0.4 μ. Investigation with the electron microscope has shown that these striations, barely visible in the light microscope, actually are inhomogeneities in lamellae of microfibrils (Fig. 6). It will be shown in a later section that the cellulose chains run parallel to each other within a given lamella. The growth of the cell tends to produce linear tears in the lamella; the tears appear between parallel microfibrils and accentuate the striated appearance. Since these fine tears always run parallel to the orientation of the cellulose, they provide the basis of a reliable method of determining the direction of cellulose chains with the light microscope. The striations in *Valonia* are particularly well seen under side illumination (Wilson, 1955). This principle is demonstrated on a larger scale with the striation-like grooves on a phonograph record. The grooves are best seen not when the disk is illuminated directly from above, but when the light beam makes a small angle with the disk. Using the proper oblique illumination, Wilson was able to clarify the complex fibrillar pattern within the lamellar wall of *Valonia*. The striations in wood tracheids are especially clear (Strasburger, 1882); here the fibrils in the outer layers run in a less steep helix than do the fibrils of the inner layers.

Young parenchymatous cells do not have a uniform cell wall. In the planes of contact with neighboring cells, the wall contains many pores.

When the pores are very small and are traversed by strands of protoplasm running from cell to cell, they are called *plasmodesmata*. Larger pores are termed *pits*. They may occur singly or in groups (pit fields). Some pits develop into rather complicated structures which, after the death of the protoplast, can act as valves with respect to the passage of fluids through the cell. These pits will be the subject of Section VI.

III. Indirect Methods of Investigation

A. *Swelling Characteristics*

The wall of the living cell always contains water, which is absorbed until a specific saturation point is reached. The changes in dimension with the uptake and loss of water of both living and dead cell walls were extensively investigated by Naegeli in 1864 and formed part of the basis for his micellar theory. Von Mohl (1859) had discovered that cell walls were birefringent when viewed in a polarized light microscope and thus were optically anisotropic. Birefringence had previously been encountered only in crystals, and the optical anisotropy of cell walls was difficult to explain. The fact that these same cell walls could swell under appropriate treatment and then return to their original dimensions made it clear that the walls were not single homogeneous crystals. Naegeli proposed an inhomogeneous structure for cell walls and maintained that they were composed of many small individual optically anisotropic crystallites, each crystallite being an elongated polyhedron, shaped somewhat like a brick (Naegeli, 1928). The small crystallite was termed a *micelle*. The micelles were thus optically anisotropic and homogeneous, the wall as a whole was optically anisotropic and inhomogeneous, because the swelling properties showed that it consisted of two or more phases. When one lets dried flax fibers swell in water there is an increase in breadth of 20% compared to an increase in length of only 0.5% (von Höhnel, 1905). According to Naegeli's hypothesis, the water taken up would be adsorbed between the crystalline micelles. The water taken up is thus a measure of the total surface area of the crystallites within the cell wall. From the strong anisotropy of swelling, it can only be concluded that the micelles are ordered and are not isodiametric. Katz (1924, 1925) made a great contribution to botany when he was able to prove this hypothesis. He used methods of X-ray diffraction to determine the approximate size of the micelles and their crystallinity. When he studied swollen cell walls he found that both the size and crystallinity of the micelles remained constant despite the fact that they were spread apart by the water taken up. With the theory on a firm basis, it became possible to use exact measurements of the anisotropy of swelling as an indication of the orientation of the micelles.

B. Polarized Light Microscopy

The cell wall can be placed in such a position that it appears bright when viewed between crossed Nicol prisms in a polarized light microscope. Thus the wall is birefringent and has two different indexes of refraction at right angles to each other in the plane observed. In fiber cells (Fig. 2), the largest index of refraction lies within the plane of the wall parallel to the fiber axis and is designated $n\gamma$. The smallest index of refraction, $n\alpha$, lies within the plane of the wall and at right angles to the fiber axis (transverse). In the case of cellulose, the intermediate index of refraction, $n\beta$ is equal to $n\alpha$ (Ambronn, 1892). The indicatrix of the cellulose crystallite is, like all crystals, a triaxial index ellipsoid. When one knows the orientation of the micelles within the wall, by X-ray diffraction methods, for example, one can determine with which crystallographic direction of the cellulose lattice the three optical axes ($n\alpha$, $n\beta$, $n\gamma$) correspond. Investigation shows that the direction of the largest index of refraction ($n\gamma$) is identical with the long axis of the micelles and cellulose chains (Frey-Wyssling, 1935). The determination of the main index of refraction is made with the immersion method, the method used on mineralogical specimens, which involves finding a solution that causes the contours of the fiber to disappear completely when examined in monochromatic light (sodium, 5890 A. emission band). The index of refraction of the solution (measured directly in a refractometer) then corresponds with the largest index of refraction of the wall. The values found for various cell walls are generally similar, $n\gamma$ being about 1.596 and $n\alpha$ being about 1.525 (Frey, 1926). From these values the birefringence of native cellulose ($n\gamma$-$n\alpha$) is calculated as 0.071. This value is about eight times as large as those for quartz and gypsum.

Because, as already mentioned, the morphological long axis of the micelles corresponds to the direction of the largest index of refraction, one can determine the direction of the crystallites using the polarized light microscope (Fig. 2). It is done in the following manner: the object is placed between the crossed Nicol prisms of the microscope and rotated on the stage until one of the four positions is reached when the object appears dark (extinction position). This shows that the crystallite axis is parallel to the plane of vibration of transmitted light of one or the other Nicol prism (Ambronn and Frey, 1926). The prisms are at 90° to each other, and a second step is required to eliminate the ambiguity. The object is rotated 45° in the plane of the stage to the "bright position," and a comparison crystal, whose largest index of refraction is 45° from the plane of transmitted light of either prism, is inserted into the microscope tube. This comparison crystal, typically a sheet of gypsum, is of precise thickness, and polarized light passing through it acquires a pink color. If the large index of refraction of the object and that of the comparison crystal are in the same direction, the object's effect on the beam of polarized light will be added to the effect already produced by the comparison crystal; this makes the object appear blue in contrast to the background pink. When the large index of refraction of the object is at right angles to that of the comparison crystal, the object will appear yellower than the background color. This is because the interference effect of the object is subtracted from the effect of the comparison crystal. In this manner, one can quickly determine the direction of the highest index of refraction of the object.

The most important cell wall textures are assembled in Fig. 2. The most perfect orientation of cellulose is realized in the *fiber texture*. Here the micelles and the microfibrils (the latter as observed in the electron microscope) run exactly parallel to a given direction. The texture of the elongate prosenchyma cells is, however, almost always wound in a helical fashion, and this is termed a *helical texture*. The angle of helical winding varies both with cell type and with the age of the cell. In hemp, the angle is about 2°; in ramie, 3–5°; in cotton fibers, 25–35°; and in wood fibers it may reach 50°. When the helical texture becomes less and less steep, it may reach 90° from the cell axis, and this special case is called *ring texture* (Fig. 2). These transversely oriented micelles are found most commonly in the strengthening rings of scalariform vessel cells.

In contrast to these cells, which all show a parallel texture, stand the cell walls with a *scattered texture*. If the crystalline regions lie scattered about the direction of the cell axis, this texture is termed *fiberlike*. The value of the largest index of refraction of the cellulose is equal to that of the cell only when the micelles are perfectly parallel; when scattering is present, the value for the wall's largest index of refraction falls, while that of the crystallites, of course, remains the same. Thus the measured $n\gamma^*$ of the cell is smaller than $n\gamma$ of cellulose; the greater the scatter, the relatively smaller $n\gamma^*$ becomes. Cell walls with scattered texture display all possible intermediates between axial and transverse orientation, as did cells with parallel texture. If no fibrillar or crystallite direction predominates, then the texture is termed *foliate* or *isotropic*. This texture is most common in the walls of parenchyma cells which are isodiametric. When the surface of such cells is viewed in the polarized light microscope, it does not appear bright at any position of rotation, being isotropic. The beam passes through crystallites in all possible orientations, and one crystallite cancels the effect of another, so there is no net effect on the beam and no birefringence is observed. Walls with a favored

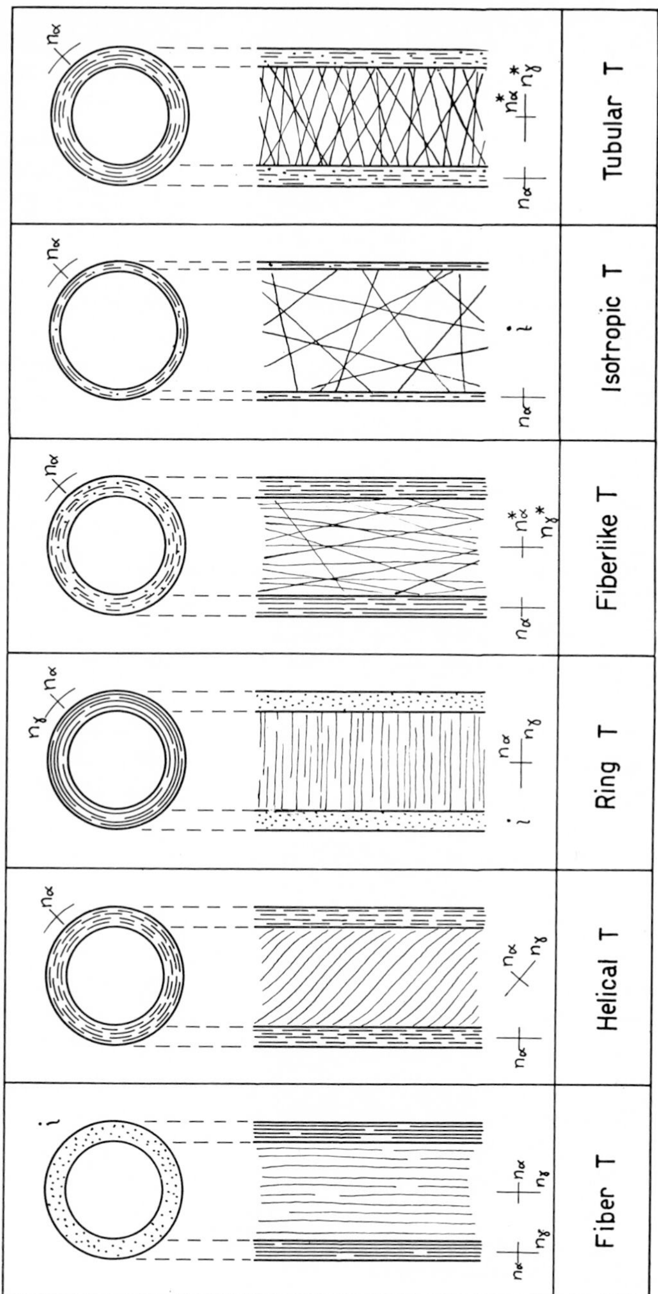

FIG. 2. Diagrams of the various types of cell wall textures T., as originally proposed from studies with the polarized light microscope. After Frey-Wyssling (1935).

orientation of texture in the transverse direction have a *tube texture.* In surface view, such a wall appears optically negative (largest index of refraction, $n\gamma^*$ is at right angles to the long axis of the cell); a radial section through the wall appears optically positive. [Of the two indexes of refraction in view ($n\beta^*$ and $n\alpha^*$) the larger, ($n\beta^*$), is parallel to the longitudinal axis of the cell.] This structure is found in tracheids of broadleaf trees, vessels, sieve tubes, latex tubes, etc. (Frey-Wyssling, 1935).

Recent investigations with the electron microscope have confirmed the presence of these textures proposed on the basis of indirect methods. Because the axis of the micelles coincides with the axis of the microfibril, the polarized light microscope remains an effective instrument for the determination of cell wall texture. For the study of mean microfibrillar orientation within a cell wall, these optical methods are unsurpassed.

C. X-Ray Diffraction Methods

In contrast to the techniques of polarized light investigation, X-ray diffraction methods are not easily applicable to the study of single-cell wall layers or single cells. To obtain a clear X-ray diagram, a considerable number of cells, such as a bundle of parallel fiber cells, must be placed in the X-ray beam. The origin of the spots on an X-ray diagram is based upon the beam's regular reflection from the heavy atomic centers arranged in the periodic fashion characteristic of crystals. The X-ray diagram is thus direct evidence of the presence of a crystalline structure. As mentioned above, the proper interpretation of a diagram can yield the dimensions of the unit cell of cellulose as well as the orientation and size of the crystallites within the cell wall. The values for the unit cell of cellulose have already been given. The orientation of the cellulose chains can be determined from the arrangement of the interference spots on the diagram. When a diagram is made of cellulose fibers with well-oriented cellulose chains running in the direction of the cell axis, the interference spots are quite sharp and form a regular, symmetrical pattern. If the crystallites are merely scattered around a preferred direction, the spots broaden into arcs. When the scattering effect is augmented by the presence of helical windings (chains in the "front" and "rear" walls having different effects on the beam), the scattering is so great that adjacent arcs fuse to form large arcs (sickle diagram). Finally, the crystallites may be completely scattered at random, in which case the arcs broaden completely to form a diagram of concentric circles. The long dimension of the "sickle" arcs formed by helical arrangements of cellulose in cell walls subtends an angle directly related to the angle between the course of the helical windings and the cell axis. The breadth of the spot or arc is a function of the size of the crystallites present; very small crystallites give diffuse spots. Sharp spots are obtained only when the crystallites present have at least one dimension greater than 1000 A. The presence of noncrystalline material creates a diffuse background fog which is evently distributed over the diagram. A thorough discussion of X-ray diffraction methods can be found in the book by Preston (1952).

More exact information about the orientation of microfibrils in the cell wall can now be obtained with the electron microscope as will be seen in the next section.

IV. The Cell Wall in the Electron Microscope

A. *The Structure of Microfibrils*

When plant cell walls are treated with hydrolyzing solutions to remove all noncellulosic constituents, the cellulose remaining appears always in the form of microfibrils of a diameter of 100–200 A. (Fig. 3). Study of a wide range of celluloses found in nature, including cellulose in fungi (slime molds), algae, tunicates, archigoniate and phanerogamic plants, has always revealed this fibrillar form. The length of the microfibril can exceed several microns, and length appears to vary much more than the diameter. Estimation of microfibrillar length is very difficult because the microfibrils are always present in great numbers and rarely can a single microfibril be distinguished from its neighbors for a long distance.

The origin of microfibrils has not been extensively studied. Investigations with the extracellular formation of cellulose by the bacterium *Acetobacter xylinum* (Mühlethaler, 1949b; Colvin *et al.*, 1957) (Fig. 4) have shown that cellulose formation does not require direct contact with protoplasm. Rather it appears that the living cell provides only the elementary units and the enzymes necessary for polymerization, cellulose formation taking place at some distance from the cell. This formation can be very rapid. Colvin *et al.* (1957) have published photographs which show that cellulose-free bacteria can produce microfibrils several microns long within a minute. After longer periods of time, longer microfibrils are observed, microfibrillar diameter remaining constant. Thus the microfibrils apparently display growth at the tips. Why the diameter does not increase during growth is not clear. It appears that the crystallization of high polymers follows certain thermodynamic laws and for some physical reason stops after a diameter of 100–200 A. is attained. Comparison of such varied high polymers as vanadium pentoxide (Fig. 20), actomyosin, nucleic acid, and silk fibrils shows that a fibrillar breadth in the range 100–200 A. is widespread (Balashov and Preston, 1955).

The arrangement of the crystallites within the microfibril is still the object of investigation. According to Rånby and Ribi (1950), a cross section through a microfibril should reveal either a completely crystalline or noncrystalline arrangement. In other words, micellar diameter would equal microfibrillar diameter and crystalline sections of the microfibril

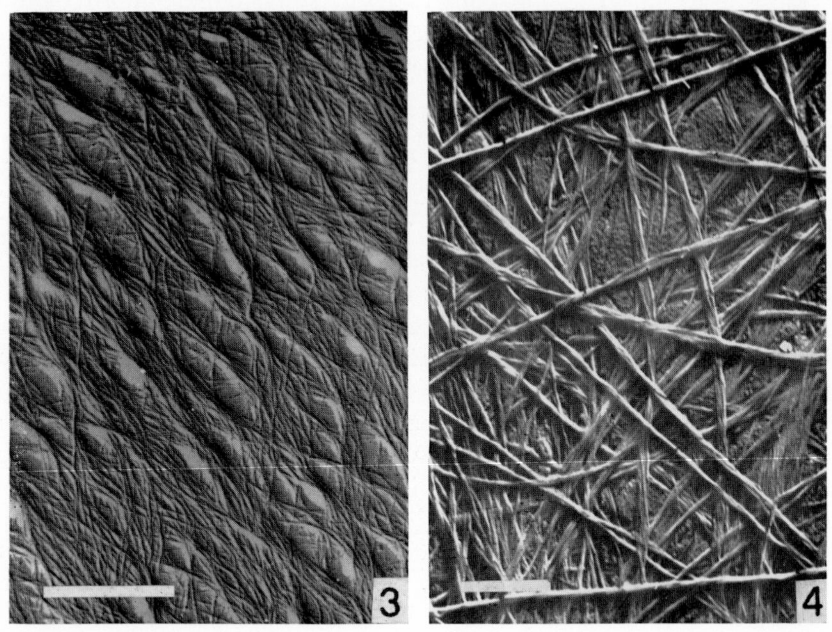

Fig. 3. Microfibrils in the cell wall of root cells of *Allium cepa* after the extraction of noncellulosic substances. Magnification: × 18,000.

Fig. 4. Fibrillar sheet from the extracellular cellulose produced by *Acetobacter xylinum*. After Mühlethaler (1949b). Magnification: × 12,000.

would be separated by noncrystalline sections. These interpretations are based on the finding that a brief and gentle hydrolysis of cellulose gives a cellulose sol which appears, in the electron microscope, to be composed of rodlets about 500 A. long and 100 A. in diameter. The original diameter of the microfibril has not been decreased. This experiment indicates that the noncrystalline regions of the microfibril, which are most easily attacked by the hydrolyzing acid, are located intermittently along the microfibril axis. The remaining rodlets, representing the more crystalline cellulose, can be dissolved into their constituent cellulose molecules only after further and stronger hydrolysis.

In thicker fibrils (200–250 A.), this simple crystal association is unlikely. According to Vogel (1954), the microfibrils of ramie (diameter 173–203 A.) can be broken up into smaller slatlike units with a cross section of 30 × 80 A. when subjected to hydrolysis or ultrasonic disintegration. These little slatlike segments have a length of 300–1000 A. and are termed *elementary fibrils* (Fig. 1). In contrast to the crystallites found by Rånby and Ribi (1950), these elementary fibrils could be associated side by side as well as end to end within the microfibril. According to the diameter of the microfibril, from two to four such crystallites could be found in a cross section, the spaces between them being filled with noncrystalline cellulose, as shown in Fig. 1 (Frey-Wyssling and Mühlethaler, 1951; Frey-Wyssling, 1954). In higher plants, the microfibrils may be considered as being composed of pure cellulose (with crystalline and noncrystalline regions) because they show the same diameter before and after the extraction of noncellulosic substances (Mühlethaler, 1949a). In the algae the situation is less uniform, and it is possible that molecules of hemicellulose could be incorporated between cellulose molecules of the microfibril (Preston, 1957).

The microfibrils are, as shown in Fig. 3, unbranched and individualized; they show, however, some tendency to aggregate. This is seen especially well in old cultures of the bacterium *Acetobacter xylinum* (Fig. 4), where the single microfibrils are apparently united into multiple-strand cables. According to Colvin (personal communication), the fibrils here are not perfectly straight but are twisted along a very long helix. Similar associations of microfibrils have been observed in cell walls (Frey-Wyssling, 1951). Their origin is brought about by hydrogen bonds, and they are formed only when the concentration of fibrils is so great that they touch each other. As soon as the microfibrils are separated by the deposition of incrusting substances, the association is dispersed and the microfibrils are again individualized.

B. Structure of the Primary Wall

The first-formed cell wall is termed the primary wall. As is shown by a portion of such a membrane in *Valonia* (Fig. 5), the microfibrils run in all directions within the plane of the wall and, because the cell is essentially spherical, there is no preferred direction of scattering. If the cell is elongated, as is the case with fiber initials, one observes a preferred scattering in the transverse direction. During elongation this texture is converted to one with scattering about the longitudinal direction because the stretching has reoriented the original microfibrils. Thus wall structure is not constant during cell elongation but responds to the deforming effects of polar cell growth. The explanation of this reorienting effect requires precise knowledge of the arrangement of the microfibrils within the primary wall, and this will now be discussed.

According to the earlier view of primary wall structure, the growth of the wall proper involved the insertion of new microfibrils within the interstices of the loosely built cellulose framework already present. This wall growth process, taking place throughout the volume of the cell wall,

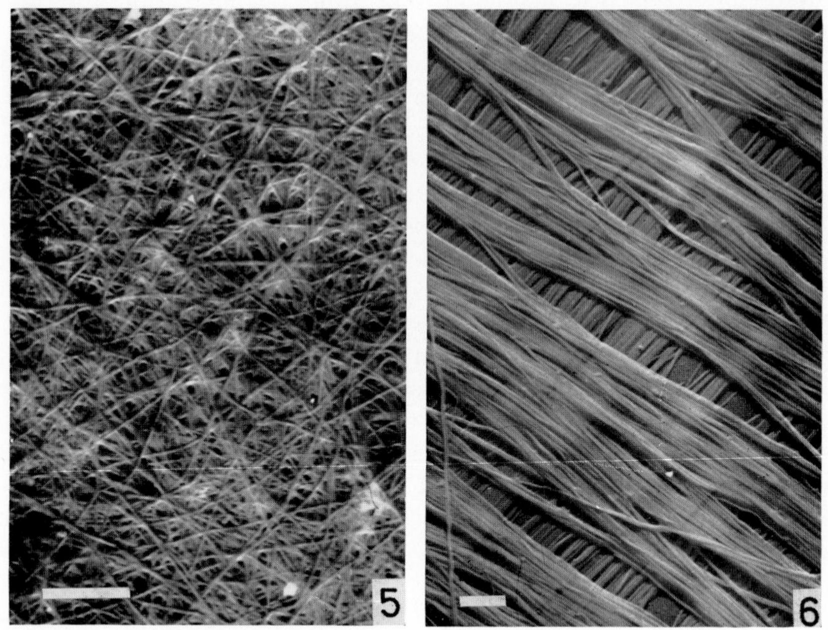

FIGS. 5 and 6. Cell wall of a *Valonia* sporeling.

FIG. 5. Primary wall with scattered texture.

FIG. 6. Secondary wall with parallel texture. After Steward and Mühlethaler (1953). Magnification: × 12,000.

was termed growth by intussusception. No matter whether microfibrils were pushed into spaces in wall, grew into such spaces by tip growth, or condensed there *in situ* the theory of intussusception involved the key assumption that new microfibrils were intermingled with ones previously present. When a primary wall is observed in surface view in the electron microscope, its structure appears as if it could have been built up by the depositing of microfibrils one on top of the other, without interweaving. The microfibrillar texture seen in the electron microscope is the projection, into a plane, of a very loosely built framework of microfibrils (Fig. 5). In the natural condition, the wall is swollen by the hydrophilic pectic substances and the microfibrils are generally far apart from each other. A chemical analysis of the growing *Avena* coleoptile cell wall reveals a dry mass of only 7.5% (Frey-Wyssling, 1952). Thus 92.5% of the wall consists of water. In corn coleoptiles 32 mm. long, about two-thirds of the dry substance of the wall is noncellulosic. Thus the growing wall consists of only about 2.5% cellulose by volume. In cambium cells, Preston (1952) found a cellulose volume of about 8%. The separation of microfibrils in the natural state is about 30–35% greater than that observed in electron microscope pictures (Mühlethaler, 1950a).

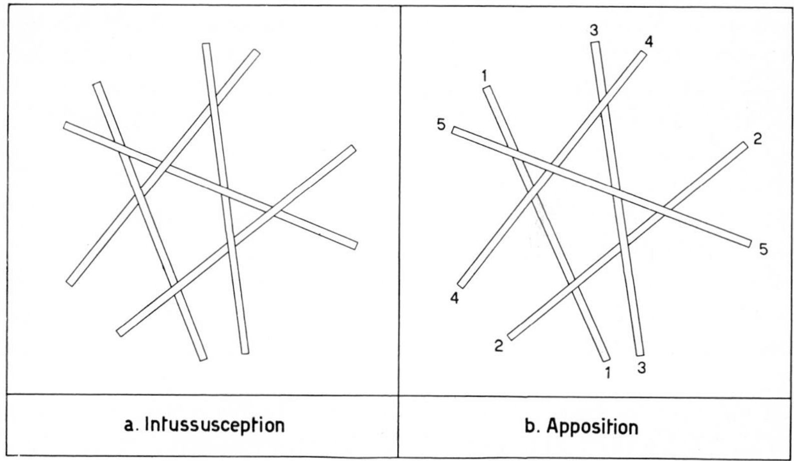

Fig. 7. Configuration of microfibrils in a primary cell wall. a. Intussusception-growth model. b. Apposition-growth model. First-formed fibril is number *1*, the second, *2*, etc.

Two apparently similar primary wall configurations which show different textures are presented side by side in Fig. 7. If the formation of the primary wall involves the placement of new microfibrils within the pre-existing microfibrillar framework, as was postulated under the theory of intussusception, then the texture should correspond to the *interwoven* one shown in Fig. 7, a. Successively formed microfibrils are numbered, the first formed being number one. In this figure, some of the new microfibrils run underneath those previously present. If the formation of new microfibrils takes place by apposition with successive microfibrils being deposited with varying directions, then the texture given in Fig. 7, b is obtained. Here young fibrils (higher numbers) always lie on top of older ones (lower numbers). All microfibrils seen in primary walls appear to correspond with the apposition texture shown in Fig. 7, b—lying one on top of another. The configuration shown in Fig. 7, b could also arise through the formation of fibrils within the pre-existing framework, provided growth of the new microfibril did not involve any interweaving. This possibility is considered unlikely. It is clear that the observed apposition texture would be easier to stretch than the intussusception texture, where the intertwined fibrils could not easily glide past one another. Since the apposition texture (Fig. 7) is supported by observations of many electron micrographs, it may be concluded that formation of the primary wall and the deposition of the layered secondary wall both proceed by the same mechanism, namely that of apposition. The main difference between the two types of wall formation lies in the much looser and more scattered apposition of new wall microfibrils in the primary wall. Wall layers deposited in the secondary wall are composed of densely packed essentially parallel microfibrils (Fig. 6).

Evidence that the addition of new wall substance occurs at, or near, the inner surface of the cell wall is found in the study by Green (1958) on the elongation of the internode cells of *Nitella*. Growth in these cells is evenly distributed along the cell axis. When elongating cells are transferred to medium containing the radioisotope tritium, the isotope is incorporated into the wall as a substitute for hydrogen. From this moment on, newly made wall will be radioactive and, if growth is by apposition, the wall initially present will remain as an isotope-free outer layer, which will be thinned out as it covers an ever larger surface. Assuming apposition, this layer will absorb relatively less and less of the radiation coming from the interior of the wall as the cell surface expands. Knowing that about 50% of the radiation is absorbed by an isotope-free cell wall, one can calculate the expected relation between the relative amount of radiation absorbed by the outer layer and the amount of surface expansion that has taken place in isotopic medium. The decrease in the amount of absorption by the outer layer corresponds quantitatively to that predicted on the assumption of growth by apposition.

C. *Secondary Cell Wall*

The transition from the primary wall structure to the parallel orientation of microfibrils in the secondary cell wall is not sudden. The transitional lamellae (Fig. 8) show successively less scatter of microfibrils, and finally the parallel texture of the secondary wall is attained. The lamellar appearance of secondary walls, first noticed in wood cells after swelling, is especially well seen in the algae where these lamellae remain separated from each other by the persistence of pectic layers. The lamellae of fiber cells are not visible in cross sections viewed in the electron microscope because the layers are too closely adpressed (Figs. 32 and 33). Fig. 6 shows a surface view of the secondary lamellae of *Valonia,* and it may be seen that the microfibrillar direction, while constant within a lamella, varies from one layer to the next by a certain angle. This consistent change of direction is about 120°. Thus lamellae numbers 1, 4, 7, $(1 + 3n)$, etc. have the same direction as do numbers 2, 5, 8, $(2 + 3n)$,

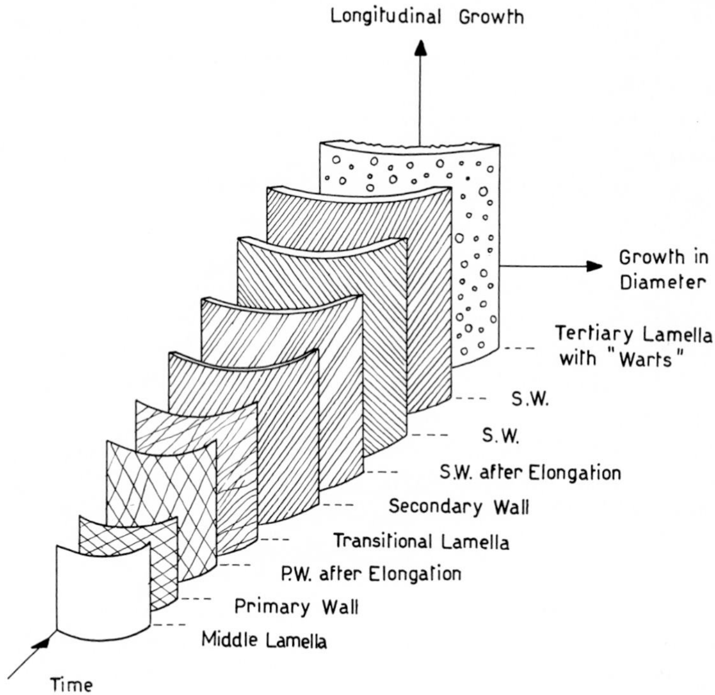

FIG. 8. Time sequence of the formation of the various types of cell wall layers in a tracheid. *P.W.*, primary wall; *S.W.*, secondary wall.

etc. (Steward and Mühlethaler, 1953). The forces responsible for this regular reorientation are unknown. The crossed lamellar structure apparently gives added strength to the cell wall. Ordinarily, secondary wall formation begins when the cell has attained its final size. When, as is the case with *Valonia,* area increase continues after the formation of secondary wall, the lamellae tear along lines parallel to the microfibrils and the linear breaks in the lamellae appear as striations when the wall is viewed in the light microscope. Microfibrils connecting two successive lamellae have not yet been observed. The formation of lamellae appears to follow a certain rhythm and the making of one layer seems to involve the entire cell surface.

The succession of types of secondary cell wall lamellae in higher plants is given schematically in Fig. 8. The outermost layers often have a helical arrangement steeper than that found in the inner, more recently formed, layers. This is explained by the fact that some extension can take place after secondary wall formation has started, the stretching drawing out of the helixes.

The last-formed wall layer, the *tertiary lamella* differs both in chemical composition and morphological structure from the primary and secondary wall.

D. The Tertiary Cell Wall

The presence of a tertiary cell wall and its staining properties have been recently well established by Bucher (1953). First mention of this innermost wall layer can be traced, however, to Hartig (1855), who observed that the mature cell is enclosed by an innermost wall layer, which differs in properties and structure from the secondary wall. The innermost wall also coats the pit canals and was given the name "Ptychode" (Hartig, 1855). A few years later, Sanio (1860) made some further observations and was the first to call this structure the tertiary wall. Since that time, no important study was made of this structure until it was rediscovered by Bucher (1953). Its morphological structure was determined in the electron microscope. As is shown in Fig. 9, the tracheids of *Abies* possess an innermost layer covered with many "warts" whose size can vary in different plants. In *Callitris* (Fig. 10), this layer is covered with many droplike protrusions, while, in *Abies pectinata,* only numerous little warts are present. It is not possible to separate this layer from the rest of the wall by mechanical methods. This is best done, according to Meier (1955), by fungal decomposition. The fungus *Chaetomium globosum* ("soft rot") frees this layer from the rest of the wall but does not destroy the microfibrils or incrustations within it. The microfibrils are also resistant to the attack of other fungi, such as *Merulius,* which quickly attacks cellulose. It is therefore unlikely that these microfibrils are made of cellulose. Experiments with paper chromatography have led to the proposal that they are made of xylan (Meier, 1955). According to Bucher (1953) the tertiary cell wall can be stained with Victoria Blue B (Ciba). Macerated wood is stained with this dye, then swollen with copper ethylenediamine, and finally washed. The tertiary wall and the primary wall appear blue, the secondary wall is red. Tertiary walls prepared in this way show, in the conifers, a helical texture or appear homogeneous. Nothing is known of the chemical composition of the warts. They are almost always present in the genus *Pinus;* they are absent in *Picea.*

V. The Mechanism of Orientation and Growth

A. *General*

When one considers the diversity of wall structures just described one naturally asks the question: What could be the nature of the physical forces responsible for arranging submicroscopic fibrils in so many different ways? The most obvious solution is to attribute the proper forces for wall formation to the protoplasm. Thus the early botanists often discussed the point of whether the cell wall was thoroughly penetrated by the protoplast and was thus a living structure or whether it was merely a nonliving excretion product such as starch.

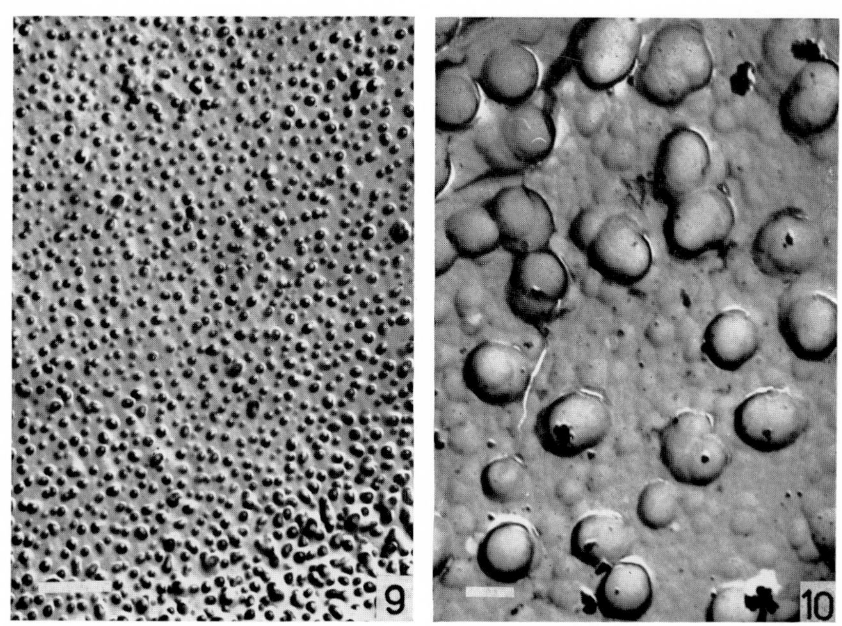

Figs. 9 and 10. Electron micrographs of two tertiary lamellae.

Fig. 9. Inner wall of a tracheid of *Abies pectinata* with many "warts." Magnification: × 10,000.

Fig. 10. Tertiary lamella of a *Callitris* tracheid. Magnification: × 5000.

It was important first to settle the question whether the cell wall could grow without being in contact with the protoplast. The usual condition in plant tissues (meristems, fiber cells, conducting cells, etc.) shows the cell wall in direct contact with the living cell. Fitting (1900) was very interested in this question and found some exceptions to the general rule in the spore walls of *Selaginella* and *Isoetes,* where the cell wall is capable of growth despite the absence of direct contact with the protoplasm. As is shown in Fig. 11, the protoplasm is contracted into a ball-like structure in the interior of the macrospore. It is in contact with the cell wall only in a restricted area at the tip of the cell; the rest of the wall is separated from the protoplast by a vacuole. Despite the separation, the two cell walls, endospore and exospore, display an observable growth. The investigations by Fitting (1900) were later confirmed by many other authors (Lyon, 1901; Denke, 1902; Campbell, 1902; Strasburger, 1907). Wentzel (1929) found a similar condition in liverworts, and comparable observations have been made for the cell

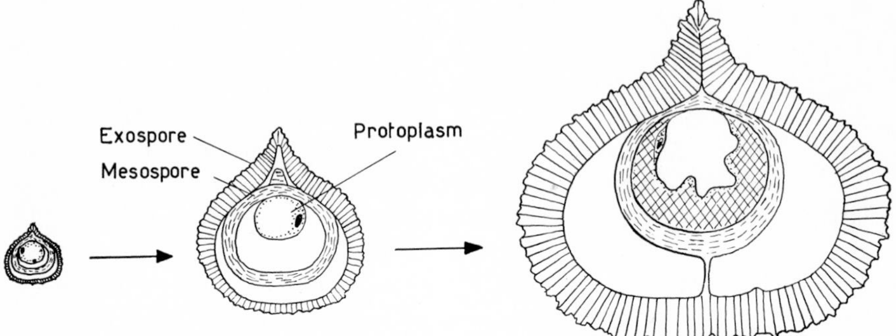

FIG. 11. Growth of the macrospore of *Selaginella helvetica* showing wall growth without direct contact with the protoplast. After Fitting (1900).

wall of the pollen grain (Beer, 1911, Geitler, 1937). The growth in these spore walls was explained by the presence of protein in the wall proper, this protein being capable of wall synthesis as is the protoplast. Investigations of the protein content of cell walls showed, however, that only small amounts were present. According to Wood (1926) it amounts to only 0.001%, but Thimann and Bonner (1933) found up to 12% protein. The present view is to regard the protein found in cell walls as coming from the protoplasm contained in channels through the wall, such as pits and plasmodesmata. Another possible source of protein in the secondary wall was suggested by the observations of Krabbe (1887), who noted, in *Nerium* that during the formation of the lamellae of the secondary wall some protein was trapped between wall layers and thus became incorporated into the wall.

Some further evidence for the presence of living substance in the cell wall was based on the fact that certain chemicals which killed protoplasm also caused changes in the wall (Lloyd and Uhléla, 1926). Dead rhizoids stain immediately with Congo red, whereas living ones remain colorless. Brauner (1933) explained the reason for differential staining of living and dead cell walls when he showed that death of the cell rendered the tonoplast fully permeable and that constituents of the vacuolar fluid were then free to impregnate the wall. For example, death could liberate tannic compounds from the vacuole and these could be adsorbed by the wall, rendering it stainable. This "mordanting" effect of the vacuolar contents after death had actually been mentioned by Pfeffer (1886) long ago.

According to the contemporary view, the cell wall is not considered to be a living system capable of extending its own structure through synthesis. Once the protoplasm is removed, the wall must be considered to be a dead shell that functions solely as a supporting structure.

There is a general opinion that the growth of plant cells is a slow process and that the increase in surface area is relatively small. Measurements on the filaments of grass anthers have shown that elongation can proceed at the rate of 2.5 mm. per minute (Schoch-Bodmer and Huber, 1945). The growth period is naturally short, lasting only a few minutes. Epidermis cells of the moss *Pellia* increase in length sixty- to eightyfold in a few days (Overbeck, 1934).

Apparently a most important force for the extension of the wall is that of turgor. In some cases plasmolysis of turgid cells can lead to a shrinkage of 30% (Martens, 1931). Recent views on the growth process stem from Went (1929), who proposed that the plant growth substance auxin increased the plasticity of the wall so that it would be especially easily extended by turgor pressure and irreversibly elongated. Next came the question whether the growth substance acted directly on the wall or whether it produced effects in the cytoplasm that in turn altered the wall. The opinion that the effect is direct (Ruge, 1937) is no longer widely held. Thimann (1951) showed that the growth substance acted in some way on cell respiration, since respiratory inhibitors also inhibit growth. It was concluded that auxin affected first the cell metabolism by more or less activating it. More recent observations by Hackett and Thimann (1953) on the auxin-induced growth of potato tissue disks, failed to show a stimulating effect of auxin on respiration. According to Bonner and Wildman (1947), the growth substance plays no specific role in metabolism but has instead a general stimulating effect. This could come through an action on a central pool of energy-rich substance. (The increased metabolism leads then to an intensive water uptake, which would stimulate the extension of the cell.)

Growth is invariably accompanied by an increase in both cell volume (water uptake) and cell wall area. Many investigators have chosen one or the other increase as the primary (and most important) one. If the properties of the wall remain constant while auxin stimulates the entry of water by some metabolic pump mechanism (Thimann, 1951), this could lead to irreversible extension of the wall. If, at the other extreme position, the tendency of water to enter the cell remains constant, an auxin-induced increase in wall plasticity could lead to growth stimulation. Choice between these two alternatives is still open, although several things point to the correctness of the second view—that auxin action involves a stimulation of wall expansion and that water uptake follows passively. To demonstrate a direct relation between auxin and water uptake, the other physical method of water uptake of the cell, that of turgor (diffusion-pressure deficit) must be eliminated. If auxin application could lead to growth in cells reduced to zero diffusion-pressure deficit (incipient plasmolysis), then the presence of auxin-stimulated water uptake would be demonstrated. To the author's knowledge this has not been convincingly done. It is, however, possible to explain failures of this type of experiment by saying that conditions of incipient plasmolysis so dehydrate the cytoplasm that it is incapable of functioning in water uptake and growth. Evidence for the opposite role of auxin —that of an action on the primary wall—has come from Ordin *et al.* (1955), who showed that auxin application leads to increased metabolic turnover of methyl groups in the growing wall. From the point of view of an investigator of wall structure, this action of auxin on the cell wall is quite reasonable. It has been mentioned that new wall is formed only at the inner surface of the wall and that the rest of the wall is stretched and deformed during growth. Since the wall is composed of fibrils in a matrix, a decrease in viscosity of the matrix could well lead to the stimulation of wall extension. This decrease in viscosity could be directly related to the auxin-induced methyl-group metabolism. Morphological details of the expansion of cell wall will be treated in the next section.

B. "Tearing" Growth

The first investigations concerning cell wall changes during elongation were made with algal cells. The marine siphonaceous algae are especially favorable material for such studies since the cells are of macroscopic size. In the nineteenth century Noll (1887) studied growing thallus tips of *Derbesia* and *Caulerpa.* He succeeded in staining the cell wall with Berlin blue while not causing great harm to the cytoplasm. Continued culture of these plants gave growth with the newly built primary wall appearing colorless. He was able to observe that the old blue-stained wall was stretched by the growth of the cell until finally it tore. The addition of new wall substance occurred through apposition at the inner surface of the wall. In the mature cell, the fragments of the original blue-stained wall remained as "islands" on the surface of the colorless wall. Similar observations on algal cells were made by Klebs (1888), using the dye Congo red on *Vaucheria.*

It was objected by Reinhardt (1899) and Zacharias (1891) that this tearing could be the result of some harmful effects of the staining process. Schmitz (1880) and Strasburger (1882) were, however, able to observe

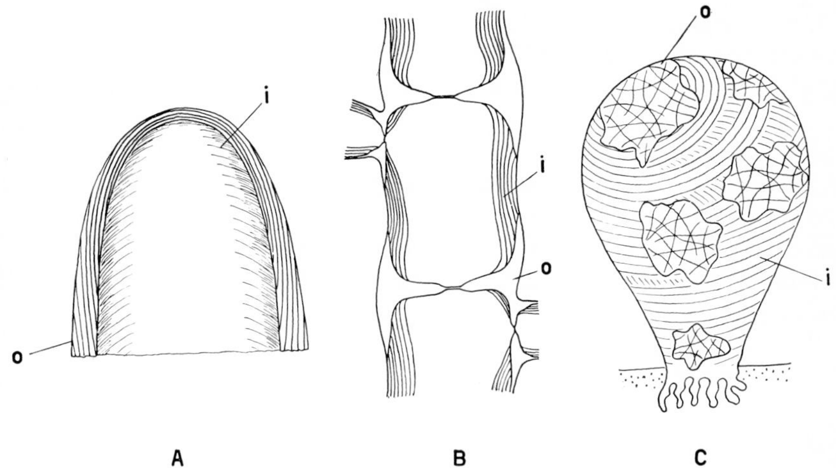

FIG. 12. Enlargement of the cell wall through the tearing of the outer layers, *o;* inner layers, *i.* A. *Bornetia* (after Strasburger, 1882). B. *Callithamnion* (after Berthold, 1886). C. *Valonia* sporeling.

this effect of the growth of inner wall layers rupturing older ones in *Bornetia* (Fig. 12, A), where this may be seen without staining. In many other cases, the bursting of the primary wall is undoubtedly a part of the normal growth process and the presence of the torn layers leaves no question. Tearing of the outer lamellae is especially well seen during the development of endogenic hairs of many algae. Berthold (1886) studied filaments of *Callithamnion* and noted how the rapid growth of the inner layers of the wall regularly broke previously present outer lamellae in the region of the center of the cylindrical cell (Fig. 12, B).

Recent investigation (Fig. 12) on growing sporelings of *Valonia* (Steward and Mühlethaler, 1953) have confirmed these earlier observations. Sporelings only 16 hours old show an entire primary wall (Fig. 5).

With further cell growth, the wall is stretched and finally torn. The fragments of the primary wall may then be recognized as patches on the outer surface of the secondary wall (Fig. 12, C). Lamellae of the secondary wall tear only along lines parallel to the microfibrils and the spaces that open up through this tearing are not filled with new microfibrils.

There is the question whether this tearing process is a normal part of wall growth in the cells of higher plants as well as those of algae. As the following section will reveal, in higher plants the stretching process rarely results in tearing, but instead has the effect of loosening the wall already present and reorienting the fibrils in the direction of elongation. Tears in higher plant walls have been observed in collenchyma cells with the electron microscope (Fig. 34). The corner thickenings are pushed apart by the growth of the cell and the middle lamella and primary wall are torn. The sequence corresponds to that in *Callithamnion* filaments (Fig. 12, B). The tearing process is apparently enhanced by the presence of inextensible incrustations around the microfibrils, like cutin or lignin. These substances would prevent the smooth slipping of fibrils past each other. Most primary walls of higher plants have pectic substances as the incrusting material, and this may make relative movement of the microfibrils reasonably easy.

C. *Multinet Growth*

In the cells of higher plants the course of membrane extension proceeds in the manner of multinet growth as first described by Roelofsen and Houwink (1953). Surface expansion involves the opening up of the existing microfibrillar texture, as is sketched in Fig. 13. The original primary wall, with microfibrils oriented more or less in the transverse

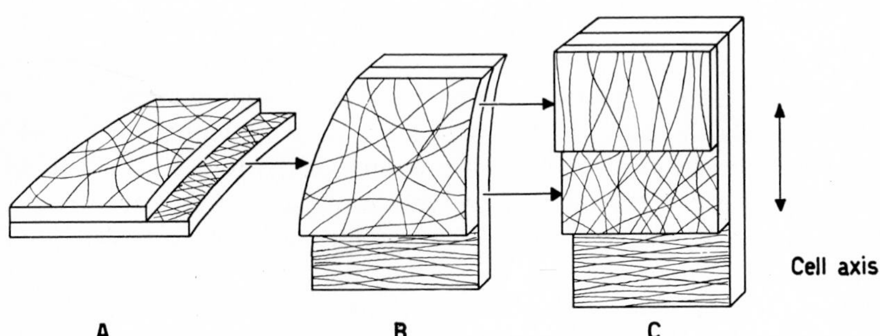

FIG. 13. Multinet growth in the growing cotton hair: A, near the tip; B, where the tip merges into the tubular part; C, in the tubular part. Layer boundaries are drawn for illustrative purposes only; the transition is gradual. After Houwink and Roelofsen (1954).

direction, is stretched in the direction of growth and its constituent microfibrils are reoriented in this same direction. At the same time, new microfibrils are deposited at the inner surface of the wall. These are laid down with the characteristic transverse orientation, and, after further growth, all but the last wall layers eventually undergo the reorientation to the longitudinal direction. In the mature cell wall, the first-formed microfibrils are seen running longitudinally scattered on the outer wall surface (Fig. 16). The whole process is apparently passive, corresponding to the behavior expected from an inert framework. Multinet growth is especially well seen in the latex tubes of *Euphorbia splendens*. Three pictures by Moor (1956) of cells of different sizes are assembled in Figs. 14–16.

From the orientation (Figs. 14–16) of the outermost fibrils of the mature wall one can make conclusions about the main direction of previous growth. Fiber cells, growing primarily in length, have the outermost microfibrils running longitudinally; vessel cells, growing mostly in diameter, have the outermost fibrils in the transverse direction.

D. *Tip Growth*

The cells of pollen tubes, fibers, root hairs, tracheids, etc., which are extremely elongated in the axial direction, do not grow equally over the entire surface but only at their tips. Depending on whether growth is at one or both ends of the cell, the growth is termed uni- or bipolar tip growth. Surface extension is usually restricted to a relatively small zone at the tip (Fig. 17). The greater the distance from the tip, the slower the growth rate. In Fig. 17, growth represented by *cd–c′d′* is greater than that shown as *ab–a′b′*, and regions lying on the cylindrical surface have already ceased to grow. At the tips of these cells, one sees mostly loose fibrils in the electron microscope (Mühlethaler, 1950a). The protoplasm must exert a strong pressure in the direction of growth and continuously push the structure at the tip over to the sides. Tip growth is regarded as localized multinet growth. Below the tip, the wall is not changed because no surface increase takes place. In fiber cells, the formation of the secondary wall can begin before the end of cell elongation. This additional thickening starts in the center of the cell and advances toward the two growing tips. Thus a single cell can form primary and secondary wall simultaneously.

E. *Mosaic Growth*

Mature cell walls can, under certain conditions, resume growth through the localized change of wall structure. Typical examples are root hair formation and the formation of the sieve plate. Such localized regions of structure changes appear spontaneously, and this type of wall-forming activity was termed "mosaic growth" by Frey-Wyssling and Stecher (1951). The microfibrils of the existing wall are pushed aside and are later seen forming a wrinkle around the edge of the opening. The pattern of mosaic growth in sieve-plate formation in *Cucurbita pepo* (Frey-Wyssling and Müller, 1957) will serve as an example. The end walls of columns of sieve cells are perforated by groups of large pores (sieve plates) through which protoplasmic strands connect adjacent cells. Thus adjacent cells in the column must form openings that coincide exactly. One stage of development is shown in Fig. 18. One of the adjacent sieve plates is already completely formed, while the other is still in the process of construction. During the dissolution of the middle lamella in the maceration of the specimen, the two adjacent sieve plates shifted a bit so the pores in the two plates do not exactly coincide. One sees that the pores of the immature plate are still crossed by a great many microfibrils. In the mature plate, these fibrils have been pushed to the edge of the pore where they reinforce the sieve structure. It appears that the protoplasm has pushed these fibrils aside in forming the pore. In order to ensure that the new pores correspond to those of the mature plate, it is most likely that the formation of the pores is directed by the mature cell. A process similar to this is found in the formation of the torus membrane in the development of bordered pits between adjacent cells (Figs. 29 and 31).

F. Extracellular Wall Formation

In the types of wall formation discussed above, the cell wall was directly built by the protoplast and the wall surrounded the protoplast in more or less intimate contact. A few cases will now be discussed in which several naked protoplasts may secrete a cell wall that later becomes quite distant from the cells that produced it. In contrast to the usual intracellular wall that directly surrounds the protoplast, these walls secreted by naked cells which do not remain fixed to the wall are termed *extracellular walls.* It has already been mentioned that the cellulose fibrils produced by the bacterium *Acetobacter xylinum* are extracellular. The masses of cellulose produced by these cultures, tangled mats up to a centimeter thick, are not produced according to any morphological principle and cannot be considered walls (Fig. 4). There is no apparent

---

FIGS. 14–16. Three growth stages of the latex-tube wall of *Euphorbia splendens.*

FIG. 14. Primary wall before elongation. Magnification: × 30,000.

FIG. 15. Wall during elongation. Magnification: × 30,000.

FIG. 16. Mature cell wall. Magnification: × 20,000. After Moor (1956).

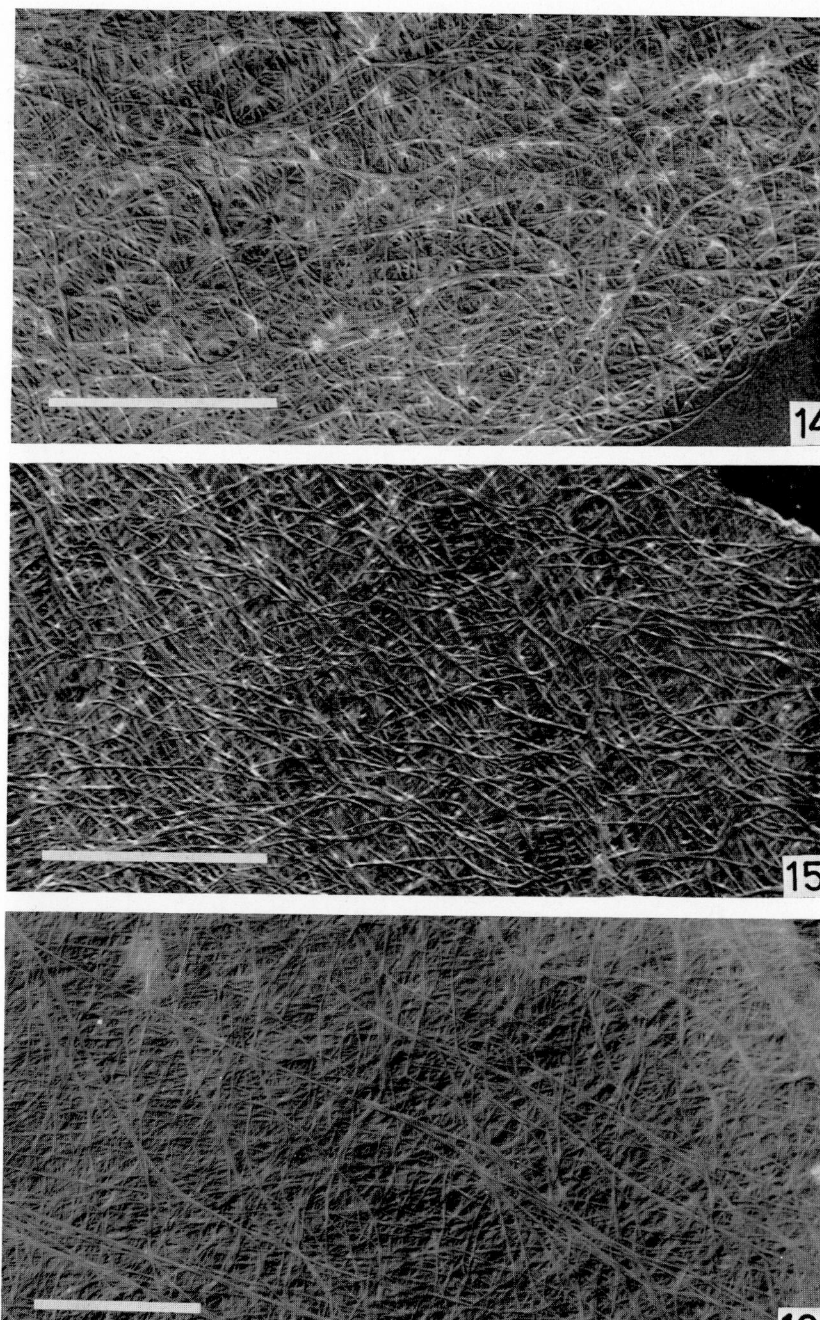
14
15
16

coordination among the cells of the culture in producing these essentially amorphous clumps of cellulose.

Careful observations on the development of the slime mold *Dictyostelium discoideum* by Raper and Fennell (1952), Bonner (1944, 1952), and Bonner *et al.* (1955) have shown that the formation of the fruiting

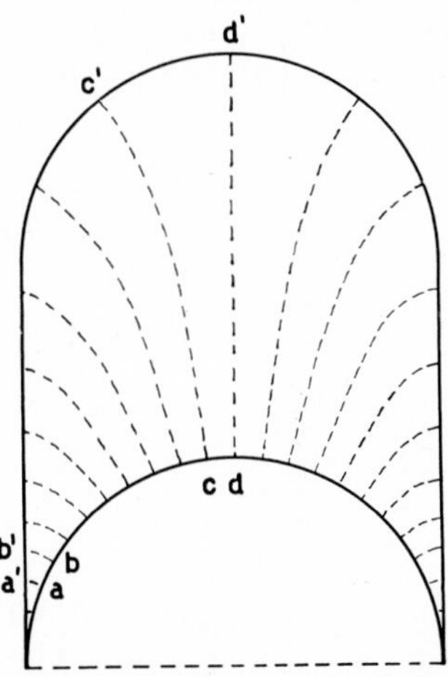

FIG. 17. Changes at the root-hair tip during elongation. Surface expansion is relatively greater as the very tip is approached. After Reinhardt (1899).

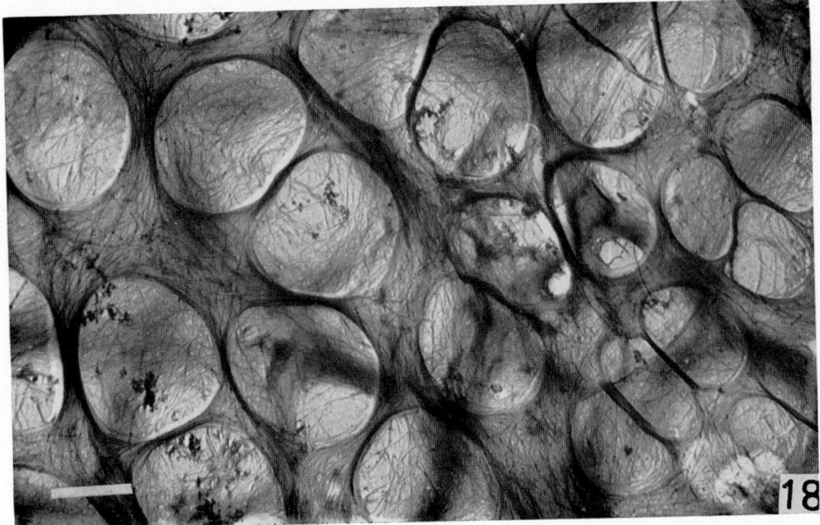

FIG. 18. Mosaic growth. Developmental stages of a sieve plate in the stem of *Cucurbita pepo.* Magnification: × 11,000. After Frey-Wyssling and Müller (1957).

body, the sorophore, involves the production of both extra and intracellular cellulose membranes. Individual amebae aggregate to form a "pseudoplasmodium" and this assemblage of amebae is actually differentiated; the apical portion of the pseudoplasmodium is composed of amebae that will form the stalk (pre-stalk cells) and the other amebae

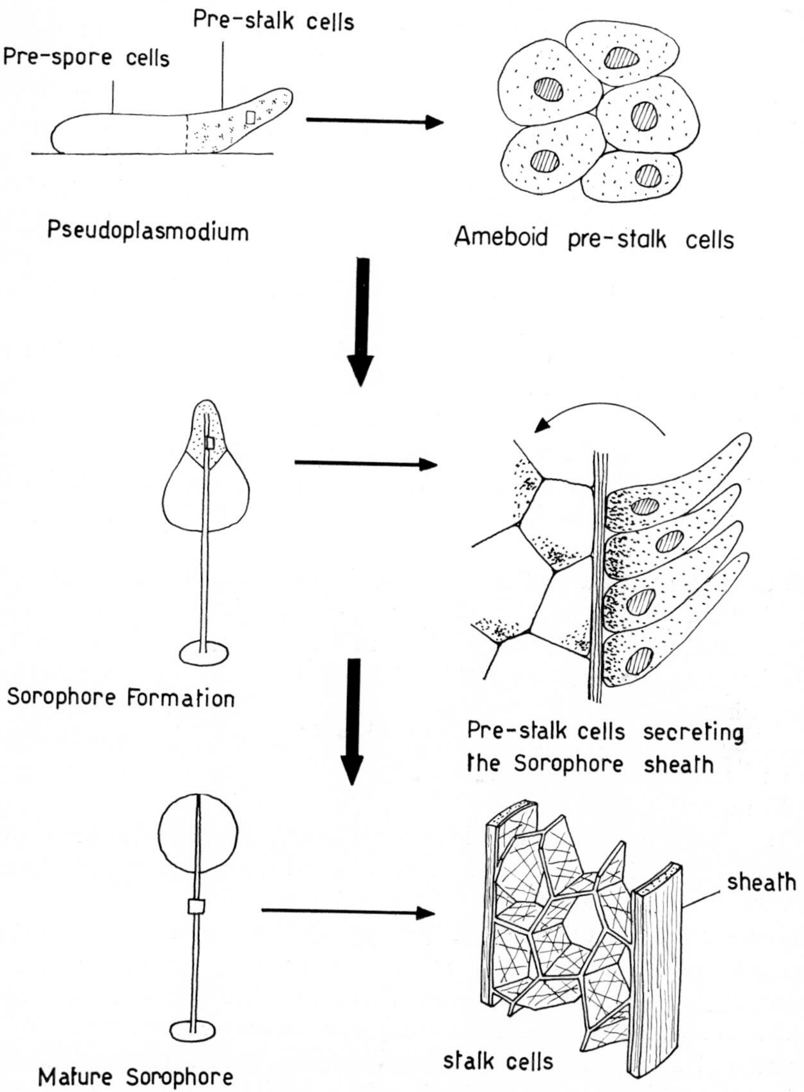

FIG. 19. Development of the sorophore of *Dictyostelium discoideum* involving both extra- and intracellular wall formation. After Bonner *et al.* (1955).

to the rear will form spores and are called pre-spore cells (Fig. 19). After the pseudoplasmodium has stopped migrating, the amebae form a drop-like structure with the pre-stalk cells on top. A vertical tube-shaped cellulosic structure is secreted within the amebal mass and this forms the lowest part of the stalk. Pre-stalk cells migrate up the outside of this tube and, while moving, secrete cellulose upon it. When the cells reach the top of the tube, they extend the tube's length a bit before migrating over the rim of the tube and then down into the interior of the tube. Stalk-cell movement ends inside the tube, and these packed cells now each form a more or less typical primary wall. Addition of cellulose to the top of the tube lengthens it, and this elongation of the sheath, combined with the filling of the tube with stalk cells, gives the fruiting body height. As the stalk elongates the spore cells are drawn up into the air. When the stalk-forming process is over, they become spores ready for dispersal (Fig. 19). After first secreting the stalk sheath and then forming a typical cell wall, the stalk cells die.

The whole process of fruiting-body formation in this organism has been investigated by Mühlethaler (1956) and Gezelius and Rånby (1957) with the electron microscope. As is universally the case, the cellulose is present in the form of microfibrils. In the first-formed stalk sheath, these fibrils are parallel and in the axial direction (fiber texture); the wall later formed around each individual ameba has a scattered texture. This is the only known case where a cell produces a well-oriented wall first, and then one with a scattered texture. When these two walls are compared in the electron microscope, it is seen that the first (secreted) wall has a dense packing of well-ordered microfibrils, while the second cell wall has a loose, scattered texture. The same observations are made in comparing primary and secondary walls. One might draw the conclusion that the degree of alignment of the microfibrils could be in part a function of the concentration of microfibrils. In higher concentrations, the fibrils mutually interfere and the available space is most efficiently filled by the parallel arrangement. Model experiments can be made with organic or inorganic substances that form fibrils: these fibrils can be formed at various concentrations and the orientation observed. Typical model-substances are vanadium pentoxide and the tobacco mosaic virus. The former can form fibrils several microns long with a uniform diameter of about 100 A. (Mühlethaler, 1948; Suito and Takiyama, 1954). As is shown in Fig. 20, vanadium pentoxide can assume a scattered as well as a parallel texture, depending on concentration. Similar observations have been made with tobacco mosaic virus (Wyckoff, 1947). At low concentration, the fibrils run at random among each other; at high concentration,

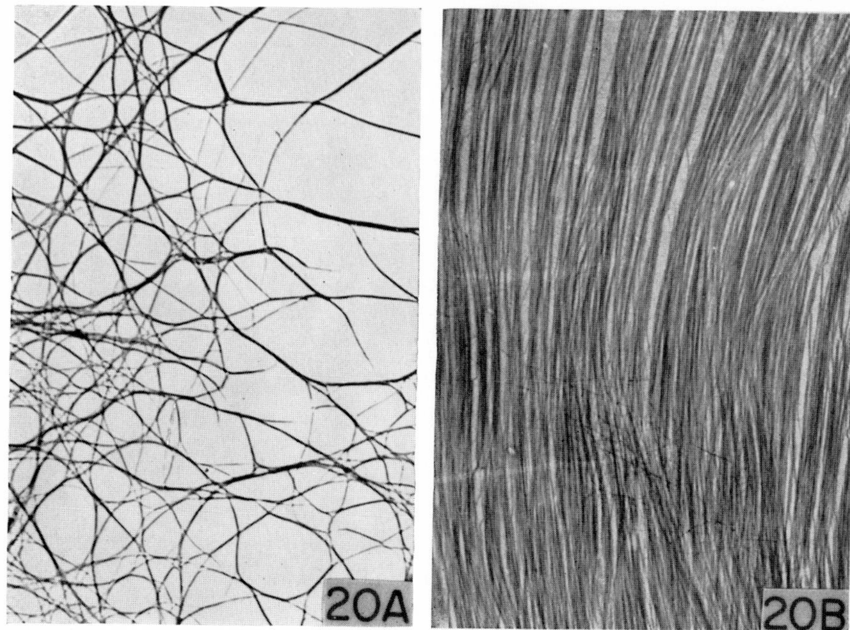

FIG. 20. Scattered (A) and parallel (B) textures of a vanadium pentoxide gel as a function of concentration. Magnification: × 15,000. After Suito and Takiyama (1954).

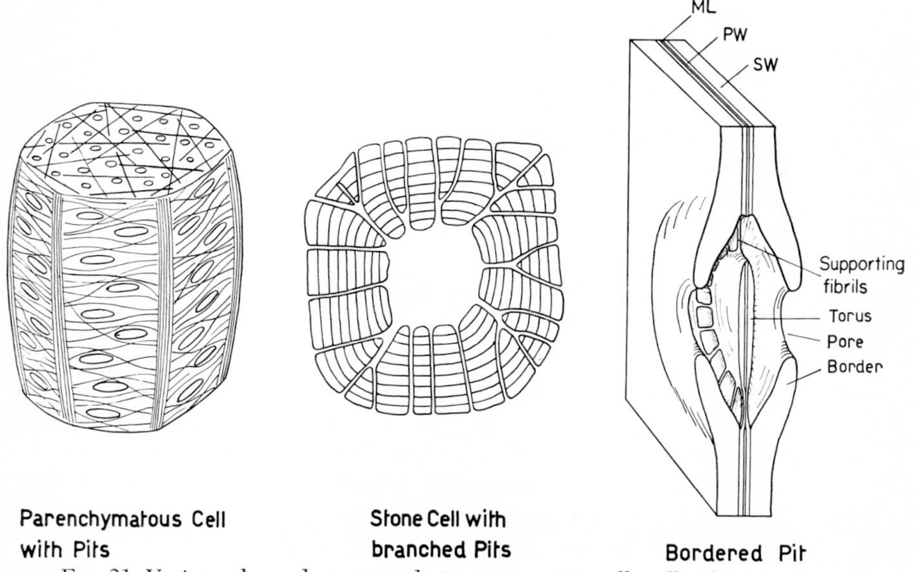

FIG. 21. Various plasmodesmata and pit structures in cell walls of parenchymatous and woody cells. *ML*, middle lamella; *PW*, primary wall; *SW*, secondary wall.

they are parallel to each other. These model experiments indicate that the differences in structure between primary and secondary cell walls could be based on a physicochemical mechanism.

VI. The Structure of Cell Connections

When one observes a cell wall in the light or the electron microscope, it is easily seen that neighboring cells are connected by pores through their contacting cell walls (Fig. 21). These pores, filled with connecting strands of protoplasm when the adjacent cells are living, are called *plasmodesmata* when the pores are small and lie at the limits of resolution of the light microscope. Larger pores are termed *pits* (Figs. 22–27). The work of Strasburger (1898), A. Meyer (1920), Mühldorf (1937), and Küster (1951) indicates that these structures are present in almost all tissues of higher plants. Their frequency can be very high in certain tissues. Using the electron microscope, Strugger (1957) found about six or seven pits per square micron in the root cells of *Allium cepa,* and thus a meristematic cell 20 μ long on each side could contain of the order of 20,000 protoplasmic connections. An earlier estimate came from Kuhla (1900), who found 10–38 pits on 100 μ^2 of cambiform cell surface. As is shown in Fig. 22, the plasmodesmata and pits are not evenly distributed over the entire surface. In most plant cells, the end walls are perforated by small round plasmodesmata and the radial and tangential walls show many large, elliptical pores (Figs. 26 and 27).

During cell elongation, these pores experience displacement from one another. In the parenchyma cells of the *Avena* coleoptile, the number of pits present per cell is constant throughout elongation (Wardrop, 1955). During elongation, the growth of the wall between the pits pushes them farther apart and the number of pits per unit surface area falls. Scott *et al.* (1956) found a similar situation in growing onion root cells. Studying this problem in *Elodea canadensis,* Wilson (1957) found a different relationship. There was an increase in the number of pits per cell until the stem internode length reached 12 mm.; thereafter the number per

---

FIGS. 22–27. Electron micrographs of various pit structures.

FIGS. 22 and 23. Surface view of a parenchyma cell in the root of *Brassica napus.* Pits are present only in the planes of contact with neighboring cells.

FIG. 24. Cross-wall under stronger magnification (× 10,000).

FIG. 25. Section through a cell wall showing plasmodesmata, *Lilium martagon* (× 10,000).

FIG. 26. Tangential wall of the root meristem of *Allium cepa* showing elliptical pits (× 10,000).

FIG. 27. Pits in a tracheid of *Agathis aranchiaceae* (× 2000).

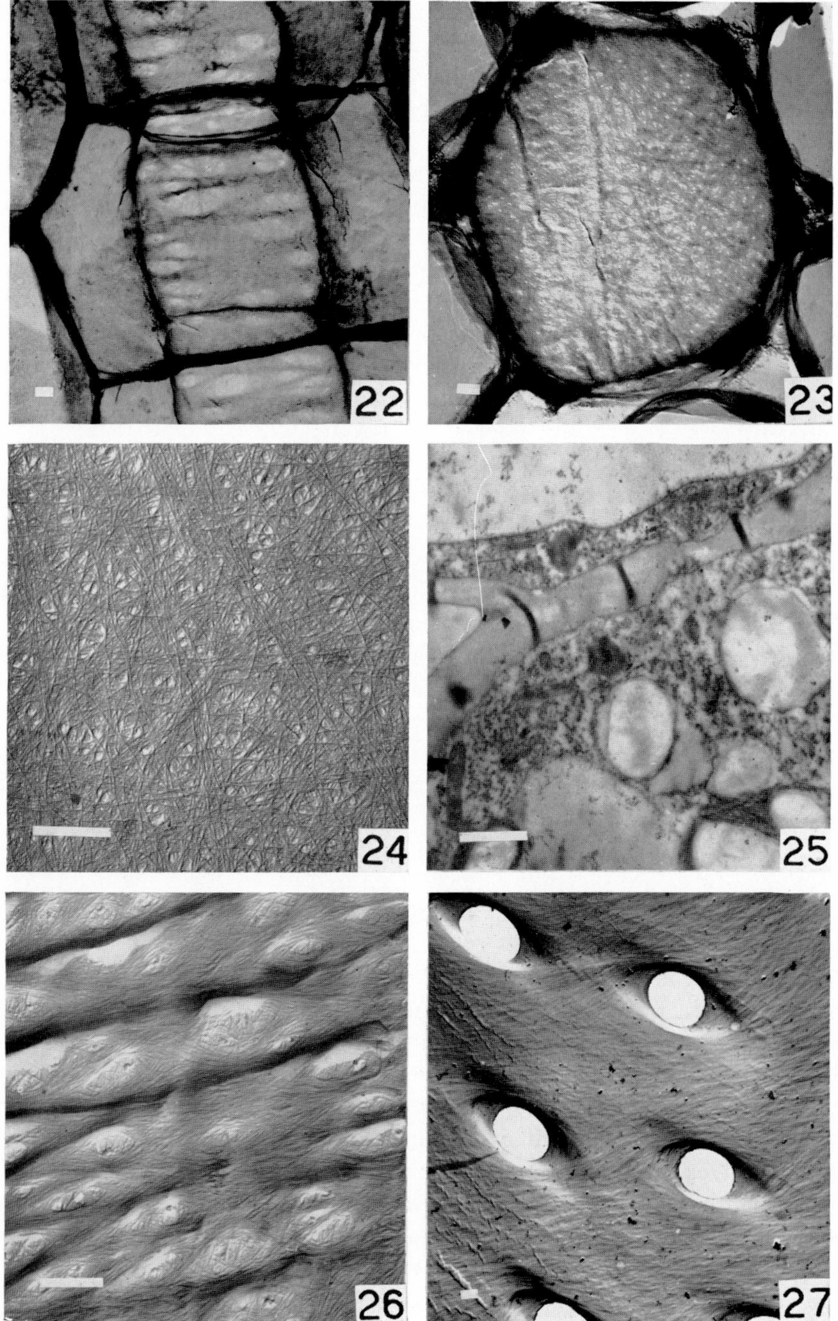
22
23
24
25
26
27

cell was constant. During later secondary cell wall growth, many pits can be covered over, isolating the protoplast (Mühlethaler, 1950b).

The tracheids of evergreens have many bordered pits on the radial and tangential walls, and these openings can act as valves governing the flow of water through the cell (Figs. 28–31). The size, number, and structure of the pits not only controls water flow in the intact tree, but is also important in wood technology, since the penetration of wood-preserving agents is also affected by the character of the pits.

These structures have been microscopically investigated by Russow (1883), Bailey (1913, 1916, 1957), and Buvat (1954). The secondary wall partially overgrows the pore opening and forms a raised border to the pit, the *margo* (Figs. 21 and 28). The elevation of the border from the plane of the primary wall leaves an open space. The space between adjacent pits is partially blocked by the presence, in the plane of the primary wall, of a disk-shaped body, the *torus,* which is suspended in position by a large number of fibrils that radiate from its margin (Figs. 29 and 31). In the living plant the torus remains in position and fluids moving through the pair of apposed pits flow in between these strands that radiate from the torus. If a fluid begins to pass through the pit pair too rapidly, the torus is pushed from its central position until it rests against the pit border, blocking the further flow through the border. Thus the torus acts as a "stopper" in a simple valve.

The replica technique of specimen preparation for the electron microscope has been frequently applied to wood pit structure (Liese and Fahnenbrock, 1952; Liese and Hartmann-Fahnenbrock, 1953; Liese and Johann, 1954a, b; Eicke, 1954; Harada and Miyazaki, 1952; Frey-Wyssling *et al.,* 1953, 1955, 1956). Electron micrographs of various views and sections of pits are given in Figs. 28–31. In Fig. 28, the view of the pore with its raised border is especially clear. If the border is sectioned away, the torus and its suspending fibrils are clearly seen (Figs. 29 and 31). When the valve action of the torus is ineffective, structural investigation has shown that the inner surface of the border is covered with projecting warts of the tertiary wall (Fig. 30). Thus the contact between the displaced torus and the border is incomplete. In pine, these warts are always present, but they are absent in spruce, fir, larch, and Douglas fir. Technical difficulties in impregnating various woods with preservatives (telephone poles, for example) can be traced to the absence of these warts on the inner surface of the border and the consequent effectiveness of the valve mechanism.

The formation of bordered pits can be followed through the study of tracheids of varying age (Frey-Wyssling *et al.,* 1956). Primary tracheids, still in the stage of bipolar tip growth, show no pits at the ends of the cell. The start of differentiation occurs only after surface area increase has ceased. The microfibrils of the primary wall appear to be aggregated into compound strands that radiate out from the center of the pit region. In the early stages, many criss-crossing single microfibrils are still connected to the larger radiating strands. The central area, which will be occupied by the torus, is bounded by microfibrils that run in a circular direction. The original scattered microfibrils of the torus represent the

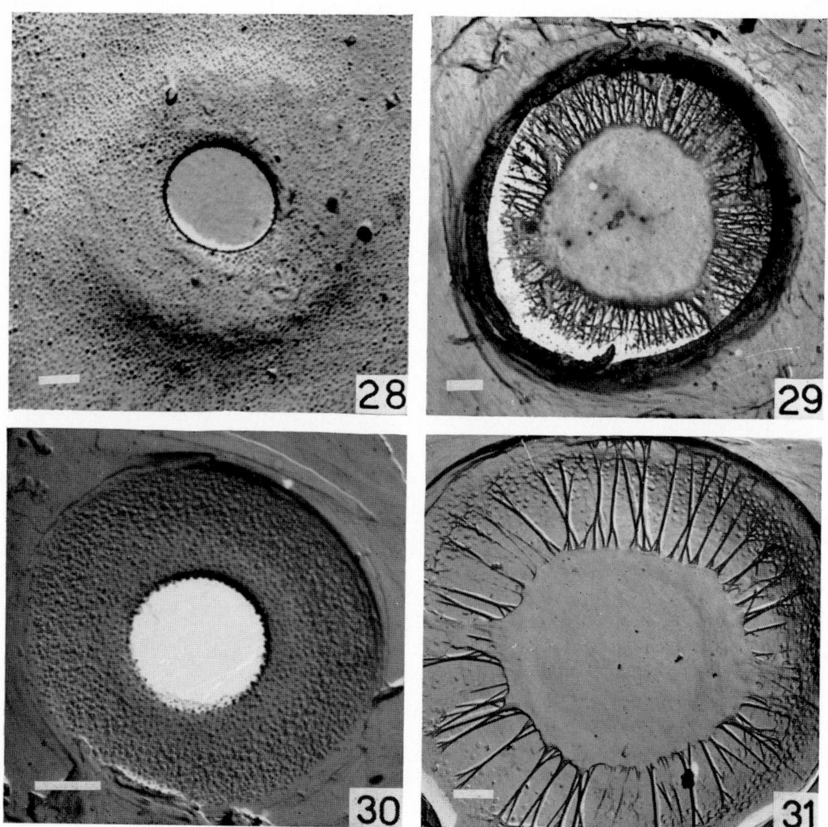

FIGS. 28–31. Structure of the bordered pit.

FIG. 28. View of a bordered pit showing the raised border and the pore. Magnification: × 4000.

FIG. 29. Similar, but with the upper border removed by sectioning; the torus is seen suspended by the radiating fibrils (× 4000).

FIG. 30. Inner surface of a pit border (the other border and the torus have been removed). Note the "warts" of the tertiary wall (× 5000).

FIG. 31. Bordered pit of *Pinus sylvestris* (× 4000).

primary wall and the well-ordered circular microfibrils may be considered a secondary wall. The radiating strands already described will serve to suspend the torus in its central position in the mature pit. At the same time, the border is forming and, while raised above the plane of the torus, it continues to grow and form an ever smaller pore, like the closing of an iris diaphragm. The pore is finally smaller than the torus. Apart from the region of the pit, the fibrils of the secondary wall run in the typical helical fashion.

As well as these complicated bordered pits, there are also (in *Pinus*, for example), the so-called window pits. These appear in contact with the medullary rays and are large structures 20 μ or more wide, closed by the thin membrane.

VII. Membrane Structure in Various Cell Types

A. *Fibers and Tracheids*

The optical properties and microfibrillar orientation of these cells have already been described. In Fig. 32 is a cross section through a bundle of flax fibers; a section of the secondary wall is shown in greater magnification (Fig. 33). Lamellation is not visible here because of the close packing; it is also difficult to determine the helical angle in the various regions of the wall. Optical investigations on the helical angle have shown that it need not be the same on both the radial and tangential walls of the same cell. This is most striking in wood tracheids. According to Preston (1952), 2-year-old cedar tracheids have an angle of ascent of the helix up to 56° on the radial walls, while the tangential walls have an angle of only 36°. With increasing age of the cell, this relation changes. In 6-year-old tracheids, the angle in the radial wall is 50° and that in the tangential wall is 46°. Despite the decrease in difference with age, a small difference always remains. The explanation of these differences is not clear; it may be predetermined in the cambial initials.

In fibers subject to great stress, it is often noticed that certain portions of the wall buckle (Frey-Wyssling, 1953). In the light microscope, these lines of damage appear as little, cracked regions along the fiber axis. The angle between the cracks and the cell axis may vary between 40° and 60° according to the force exerted. These cracks do not represent breakages, but rather openings in the secondary wall texture. Small mechanical pressures are able to open up wall textures without disrupting cell associations. This shows that the side-to-side cohesion of the cellulose microfibrils is weaker than the binding forces of the middle lamella. Despite the opening up of the texture, the original resistance to stretch of the fiber remains the same since the microfibrils are unharmed.

B. *Collenchyma Cells*

Another cell type associated with strengthening plant tissues is that of collenchyma. Unlike other cell wall types, the secondary wall here is not evenly thickened; secondary wall is produced primarily in the

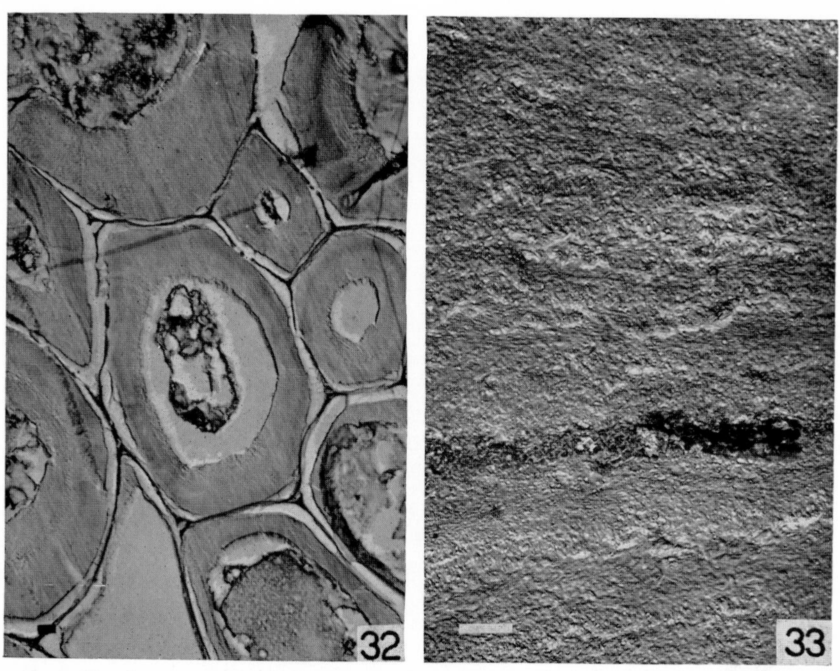

FIG. 32. Cross section through a fiber bundle of flax (*Linum usitatissimum*). Magnification: × 2000.

FIG. 33. Longitudinal section through a very thick secondary cell wall (× 8500).

corners of the cell (Fig. 34). In cross section, the three or four corner thickenings of typical collenchyma cells are seen to be associated with the coming together of longitudinal walls of neighboring cells. This so-called corner collenchyma is often found in stems of the Labiatae and Umbelliferae. This strengthening tissue is present in the elongating parts of the stem, as well as in the mature regions. The corner thickenings appear, then, long before the end of cell elongation. As is shown in Fig. 34, the cellulose lamellae are separated by broad amorphous layers of pectic substances. As was mentioned previously, the old wall layers become torn by cell extension and are replaced by new ones. Roelofsen and Kreger (1951) showed that, in the collenchyma of *Petasites vulgaris*, cell extension has the effect of orienting the usually amorphous pectic substances in the direction of elongation.

C. *Water-conducting Elements*

Tracheids and xylem elements are classified according to the nature of the bands of secondary wall that give these conducting elements

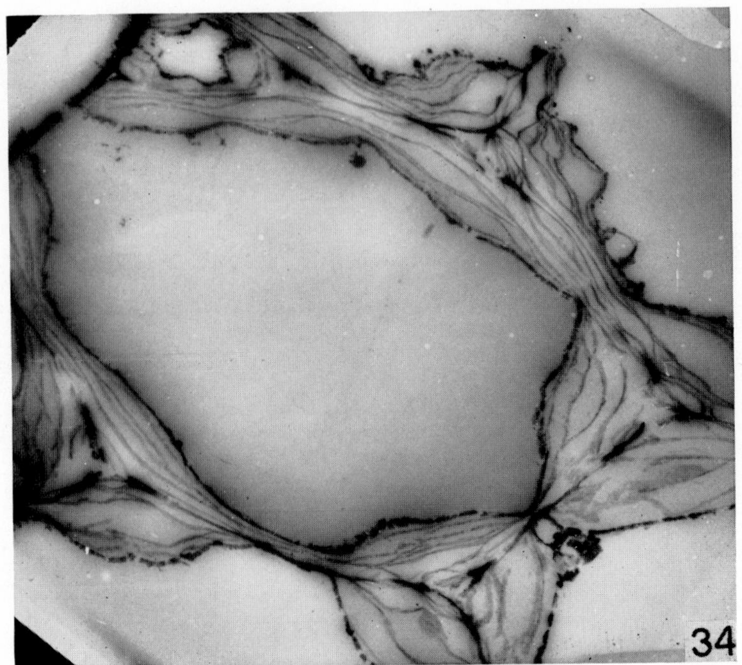

FIG. 34. Collenchyma cell from the stem of *Ricinus communis;* the middle lamella is torn. Magnification: × 1200.

added strength. These thickenings may take a ring, helical, or net form. In tracheids, thickening may be reticulate or scalariform; in xylem elements, helical or annular (ring). Without these special wall thickenings, these cells, which function after death of the protoplasm, would be crushed by the surrounding cells, interrupting the flow of water in the cell column. The simplest structure is that of xylem elements with annular bands of thickening apposed on the inner surface of the primary wall (Fig. 35). The first-formed microfibrils of the primary wall are reoriented to the axial direction by the effects of multinet growth. The ring bands, or tightly wound helical bands, have a parallel texture of microfibrils and these structures, formed interior to the primary wall, may be considered the result of localized secondary wall formation. These bands are often very broad in monocotyledons, whereas in dicotyledons they are generally very narrow. These thickenings are only loosely connected to the primary wall and can easily be pulled away from it. The cross section of the thickenings is usually round, but other shapes can occur (Baecker, 1922). Lignification is restricted to these bands and gives them added strength.

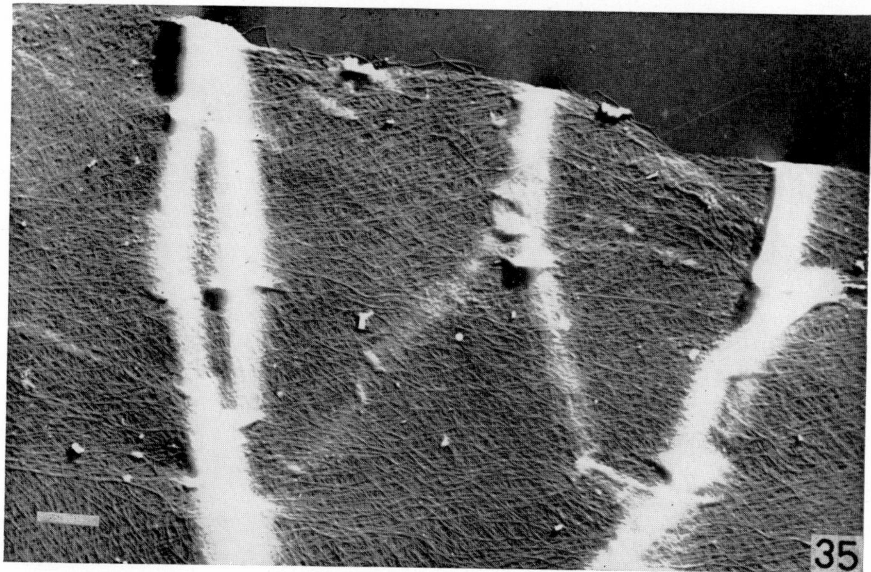

FIG. 35. Xylem vessel element from the root of *Hyacinthus* with helical secondary thickening. Magnification: × 8000.

The formation of the helical bands was first investigated by Dippel (1898). The protoplasm was observed to form narrow streams along the primary wall and then form thickenings along the streams. These reports, however, have never been satisfactorily confirmed.

D. Epidermal Cells

Epidermal cells are usually characterized by the presence of a thick, cutinized wall on that side of the cell which forms the outer surface of the organ (Fig. 36). The most thoroughly investigated epidermal cells are those from *Clivia nobilis* (Ambronn, 1888; Frey-Wyssling, 1935; Anderson, 1928; M. Meyer, 1938). In the light microscope, three layers are detectable in the outer epidermal wall: a cellulose wall adjacent to the cell lumen, then a compact layer of cutin, and finally, on the outside, a thin cuticle (Fig. 36). The optical properties of these three layers are different. The cellulose appears optically positive, the compact cutinized layer appears negative, and the cuticle is isotropic. When the cuticular layer is warmed to about 100° C., the negative birefringence disappears and returns only after the preparation is cooled (Ambronn, 1888). A similar effect is apparent in cork and seems to be based on the presence

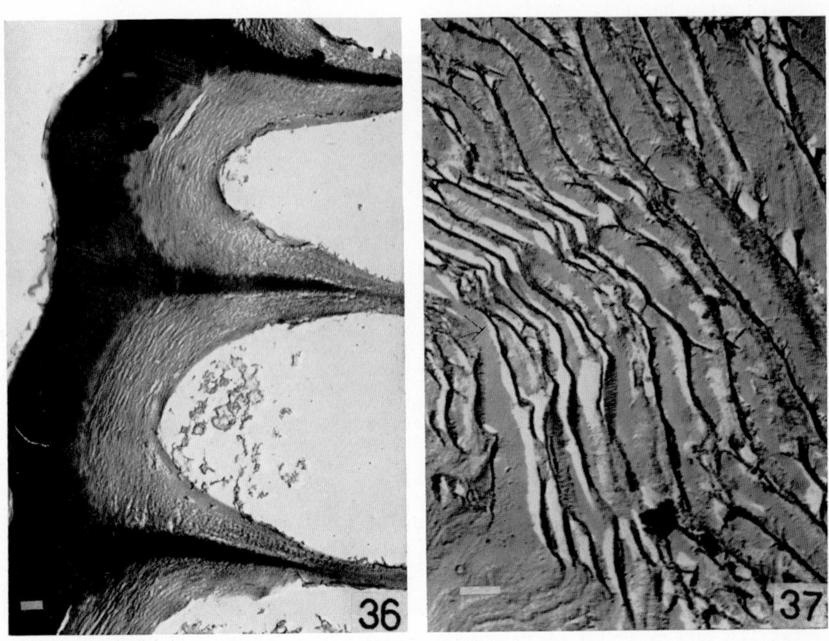

FIGS. 36 and 37. Epidermal cell of *Clivia nobilis*.

FIG. 36. Section through the cellulose and cutinized layers (dark zone). Magnification: × 2500.

FIG. 37. Lamellar structure of the cellulose-containing layer (× 4000).

of an oriented substance of low melting point within the cutin. According to the investigations of M. Meyer (1938), this substance is a kind of wax. These melting experiments show that cutin itself is isotropic.

As is seen in the electron micrograph (Fig. 37), the cellulose layer shows a typical lamellation. It is especially well seen here, and also in collenchyma cells, because the cellulose layers are separated by less dense layers of pectic substances. The amount of pectin per layer increases with distance from the cell lumen. The boundary between the cellulose layer and the cutin is sharply defined by a pectin sheet. This pectin layer would correspond to the middle lamellae found in tissue cells. The cutin layer is an adcrustation and increases gradually in thickness. In a manner similar to lignin formation, formation of the cutin layer appears to involve the transport of low molecular weight precursors through the cellulose to the cutin layer where polymerization takes place *in situ,* distant from the protoplasm. In the electron microscope (Fig. 36), the cutinized substances are contrast-rich and can be easily distinguished from cellulose. No internal structure has been found within the cutinized layer.

Many plasmodesmata were observed in the outer epidermal wall of various plants by Schumacher and Halbsguth (1939), as well as by Lambertz (1954). They lead from the protoplasm out to the cuticle and can be branched. The investigations of Lambertz (1954) showed that these channels were apparent only at certain times of the day. Material fixed between 9 A. M. and 2 P. M. seldom showed the plasmodesmata, but material fixed after 5 P. M. almost always had these canals. Light and temperature, as well as time of day, influenced the appearance of plasmodesmata. The number of plasmodesmata in *Primula pulverulenta* was about 1000 to 1500 per 50 μ^2. That means that the outer wall of an epidermal cell can contain about eight or nine thousand plasmodesmata. In angiosperms, these structures are most numerous in leaves and young stems and are less common in petals, nectaries, anthers, etc. With respect to such physiological problems as gas exchange and respiration, these findings may be very important.

E. Pollen-Grain Cell Walls

The most complicated cell wall morphology is encountered among the pollen grains. In the older literature, only the inner layer (*intine*) and the structured outer layer (*exine*) are distinguished. On the basis of exacting microscopical observation, the pollen analysts have worked out finer subdivisions of the wall (Erdtman, 1952). As is shown in Fig. 38,

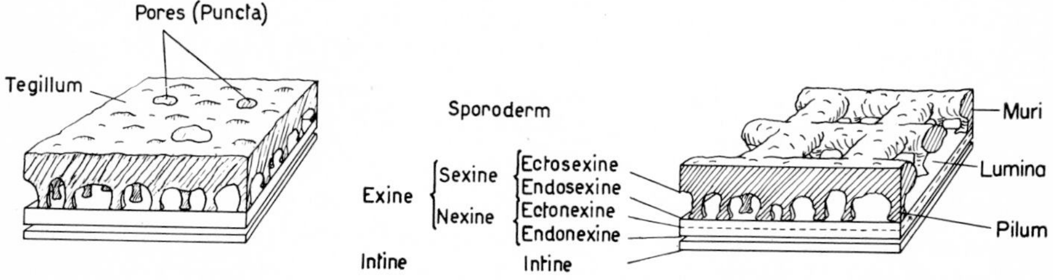

FIG. 38. Explanatory sketch of the sporoderm of the pollen grain.

the intine and exine are subdivided into two layers. The outer part of the exine, the most highly sculptured part of the wall, is called the sexine (*s*culptured *exine*), the underlying *n*onsculptured part was called the *nexine*. Further subdivisions are shown in Fig. 38.

The sexine can exist as a continuous layer (tegillate sexine), in the form of striations (striate sexine), or in a netlike configuration (reticulate sexine). This outermost layer is connected with the underlying basal

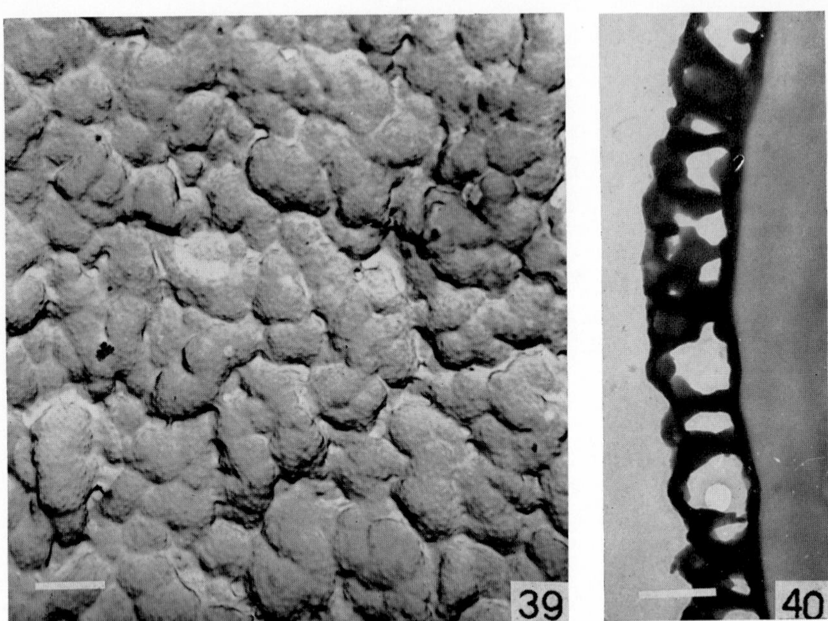

FIGS. 39 and 40. Pollen grain walls from *Pinus laricio*.
FIG. 39. Replica of the upper surface. Magnification: × 8000.
FIG. 40. Section through the sporoderm (× 10,000).

layer, the nexine, by little rods (pila, bacula) as shown in Figs. 39 and 40 (Morán and Dahl, 1952; Mühlethaler, 1953, 1955; Afzelius *et al.*, 1954; Sitte, 1953). As an example of a sporoderm cell wall, a replica and a cross section of a pollen grain (*Pinus laricio*) are reproduced in Figs. 39 and 40.

The sexine is formed of a continuous membrane with irregularly formed knoblike projections. The *pila* connects this covering membrane with the nexine. The intine appears structureless and can vary greatly in thickness among various pollen grains.

Chemically, the exine is made of sporopollenin, and the intine consists of a cellulose framework incrusted with pectin. We must thus consider the intine as a cell wall and the exine as an additional adcrustation corresponding to the cuticular layer of the outer epidermal cell wall. The origin of these complicated pollen membrane structures is not yet known.

VIII. Summary

The plant cell wall consists primarily of cellulose, hemicellulose, pectin, and lignin. The cellulose, present in the form of fibrils of diameter 100–200 A., serves as the framework that supports the other constituents. In growing (primary) walls, these fibrils are scattered in the plane of the wall, whereas in the later-formed secondary wall they are in a closely packed parallel array. Former investigations have shown that new cellulose is added to the wall by apposition at the inner surface in both primary and secondary walls. A lamellar structure is generally found in secondary walls, and the direction taken by the fibrils in successive lamellae may vary by a fixed angle. Within the cellulose skeleton noncellulosic substances are deposited. The final wall-building process involves the formation of a tertiary wall layer interior to the secondary wall. This layer differs in chemical and morphological properties from the others.

During growth the outer portions of the cell wall are subjected to severe stretching, and these regions either tear or are drawn out in the direction of growth. This latter distortion of the outer wall regions usually reorients the outermost microfibrils of the wall from their original transverse position into the direction of growth. The various types of cell wall growth were classified into tearing growth, multinet growth, tip growth, and mosaic growth. The structural changes associated with these various types of growth were described. Finally, several very highly specialized cell walls were discussed; these included fibers, collenchyma, xylem elements, epidermal cells, and pollen grains.

Acknowledgments

I wish most heartily to thank Dr. Paul Green of the Department of Botany, University of Pennsylvania, for the translation of this manuscript into English.

To Professor A. Frey-Wyssling I am especially grateful for many suggestions and references.

References

Afzelius, B. M., Erdtman, G., and Sjöstrand, F. S. (1954). *Svensk Botan, Tidskr.* **48**, 155.

Alexandrov, W. G., and Djaparidze, L. I. (1927). *Planta* **4**, 467.

Ambronn, H. (1888). *Ber. deut. botan. Ges.* **6**, 226.

Ambronn, H. (1892). "Anleitung zur Benützung des Polarisationsmikroskopes," Leipzig.

Ambronn, H., and Frey, A. (1926). "Das Polarisationsmikroskop." Akademie Verlagsgesellschaft, Leipzig.

Anderson, D. B. (1928). *Jahrb. wiss. Botan.* **69**, 501.

Anderson, D. B., and Kerr, Th. (1938). *Industr. Eng. Chem.* **30**, 48.
Astbury, W. T., Preston, R. D., and Norman, A. G. (1935). *Nature* **136**, 391.
Baecker, R. (1922). *Sitzber. Akad. Wiss. Wien. Math.-naturw. Kl.* **131**, 139.
Bailey, I. W. (1913). *Forestry Quart.* **11**, 12.
Bailey, I. W. (1916). *Botan. Gaz.* **62**, 133.
Bailey, I. W. (1957). *Holz Roh u. Werkstoff* **15**, 210.
Balashov, V., and Preston, R. D. (1955). *Nature* **176**, 64.
Balls, W. L. (1919). *Proc. Roy. Soc.* **B90**, 542.
Balls, W. L. (1922). *Proc. Roy. Soc.* **B93**, 426.
Barrows, F. L. (1940). *Contribs. Boyce Thompson Inst.* **11**, 161.
Beer, R. (1911). *Ann. Botany (London)* **25**, 199.
Berthold, G. (1886). "Studien über Protoplasmamechanik," Leipzig.
Bonner, J. T. (1944). *Am. J. Botany* **31**, 175.
Bonner, J. T. (1952). *Am. Naturalist* **86**, 79.
Bonner, J., and Wildman, S. G. (1947). *Growth* **10**, 51.
Bonner, J. T., Chiquoine, A. D., and Koldrie, M. Q. (1955). *J. Exptl. Zool.* **130**, 133.
Brauner, L. (1933). *Flora* **127**, 190.
Bucher, H. (1953). "Die Tertiärlamelle von Holzfasern und ihre Erscheinungsformen bei Coniferen." Unters. Cellulosefabrik Attisholz, Solothurn, Switzerland.
Buvat, R. (1954). *Compt. rend. acad. Sci.* **239**, 1667.
Campbell, D. H. (1902). *Ann. Botany (London)* **16**, 419.
Colvin, J. R., Bayley, S. T., and Beer, M. (1957). *Biochim. et Biophys. Acta* **23**, 652.
Denke, P. (1902). *Botan. Centr. Beih.* **12**, 182.
Dippel, L. (1898). "Das Mikroskop," Braunschweig.
Ehrlich, F. (1917). *Chemiker- Ztg.* **41**, 197.
Eicke, R. (1954). *Ber. deut. botan. Ges.* **67**, 213.
Erdtman, G. (1952). "Pollen Morphology and Plant Taxonomy of Angiosperms." Almquist & Wiksells, Stockholm.
Fitting, H. (1900). *Botan. Ztg.* **58**, 107.
Freudenberger, K., Reznik, H., Fuchs, W., and Reichert, M. (1955). *Naturwissenschaften* **42**, 29.
Frey, A. (1926). *Kolloidchem. Beih.* (Ambronn-Festschrift) p. 40.
Frey, A. (1928). *Ber. deut. botan. Ges.* **46**, 444.
Frey-Wyssling, A. (1935). "Die Stoffausscheidung der höheren Pflanzen." Springer, Berlin.
Frey-Wyssling, A. (1936). *Protoplasma* **25**, 261.
Frey-Wyssling, A. (1951). *Holz Roh u. Werkstoff* **9**, 333.
Frey-Wyssling, A. (1952). *Symposia Soc. Exptl. Biol.* **6**, 320.
Frey-Wyssling, A. (1953). *Holz Roh u. Werkstoff* **11**, 283.
Frey-Wyssling, A. (1954). *Science* **119**, 80.
Frey-Wyssling, A., and Bosshard, H. H. (1953). *Holz Roh u. Werkstoff* **11**, 417.
Frey-Wyssling, A., and Mühlethaler, K. (1951). *Fortschr. Chem. org. Naturstoffe* **8**, 1.
Frey-Wyssling, A., and Müller, H. R. (1957). *J. Ultrastruct. Research* **1**, 17.
Frey-Wyssling, A., and Stecher, H. (1951). *Experientia* **7**, 420.
Frey-Wyssling, A., Mühlethaler, K., and Wyckoff, R. W. G. (1948). *Experientia* **4**, 475.
Frey-Wyssling, A., Mühlethaler, K., and Bosshard, H. H. (1955). *Holz Roh u. Werkstoff* **13**, 245.
Frey-Wyssling, A., Bosshard, H. H., and Mühlethaler, K. (1956). *Planta* **47**, 115.

Geitler, L. (1937). *Planta* **27**, 426.
Gezelius, K., and Rånby, B. G. (1957). *Exptl. Cell Research* **12**, 265.
Green, P. (1958). *Am. J. Botany*, **45**, 111.
Hackett, D. P., and Thimann, K. V. (1953). *Am. J. Botany* **40**, 183.
Hampton, H. A., Haworth, W. N., and Hirst, E. L. (1929). *J. Chem. Soc.* **31**, 1739.
Harada, H., and Miyazaki, Y. (1952). *J. Japan. Forestry Soc.* **34**, 350.
Hartig, T. (1855). *Botan. Ztg.* **13**, 161, 185, 393, 409, 433, 461.
Haworth, W. N. (1925). *Nature* **116**, 430.
Hengstenberg, J. (1928). *Z. Krist.* **69**, 271.
Herzog, R. O., and Gonell, H. W. (1924). *Naturwissenschaften* **12**, 1153.
Houwink, A. L., and Roelofsen, P. A. (1954). *Acta Botan. Néerl.* **3**, 385.
Katz, J. R. (1924). *Ergeb. exakt. Naturw.* **3**, 332.
Katz, J. R. (1925). *Ergeb. exakt. Naturw.* **4**, 171.
Kerr, Th. (1937). *Protoplasma* **27**, 229.
Klebs, G. (1888). *Untersuch. Botan. Inst. Tübingen* **2**, 489.
Krabbe, G. (1887). *Jahrb. wiss. Botan.* **18**, 436.
Kuhla, F. (1900). *Botan. Ztg.* **58**, 29.
Küster, E. (1951). "Die Pflanzenzelle," Fischer, Jena.
Lambertz, P. (1954). *Planta* **44**, 147.
Liese, W., and Fahnenbrock, M. (1952). *Holz Roh u. Werkstoff* **10**, 197.
Liese, W., and Hartmann-Fahnenbrock, M. (1953). *Biochim. et Biophys. Acta* **11**, 190.
Liese, W., and Johann, I. (1954a). *Planta* **44**, 269.
Liese, W., and Johann, I. (1954b). *Naturwissenschaften* **41**, 579.
Lloyd, F. E., and Uhléla, V. (1926). *Trans. Roy. Soc. Can. III* **20**, 45.
Lyon, F. M. (1901). *Botan. Gaz.* **32**, 124.
Martens, P. (1931). *Cellule rec. cytol. histol.* **41**, 15.
Meier, H. (1955). *Holz Roh u. Werkstoff* **13**, 323.
Meyer, A. (1920). "Morphologische und physiologische Analyse der Zelle der Pflanzen und Tiere." Jena.
Meyer, K. H., and Mark, H. (1928). *Ber. deut. chem. Ges.* **61**, 593.
Meyer, M. (1938). *Protoplasma* **29**, 552.
Moor, H. (1956). Diplomarbeit ETH. Zürich.
Moran, F. H., and Dahl, A. O. (1952). *Science* **116**, 465.
Mühldorf, A. (1937). *Botan. Centr. Beih.* **A56**, 171.
Mühlethaler, K. (1948). *Makromol. Chem.* **2**, 143.
Mühlethaler, K. (1949a). *Biochim. et Biophys. Acta* **3**, 15.
Mühlethaler, K. (1949b). *Biochim. et Biophys. Acta* **3**, 527.
Mühlethaler, K. (1950a). *Ber. schweiz. botan. Ges.* **60**, 614.
Mühlethaler, K. (1950b). *Biochim. et Biophys. Acta* **5**, 1.
Mühlethaler, K. (1953). *Mikroskopie* **8**, 103.
Mühlethaler, K. (1955). *Planta* **46**, 1.
Mühlethaler, K. (1956). *Am. J. Botany* **43**, 673.
Naegeli, C. (1864). *Sitzber. math. naturw. Abt. bayer. Akad. Wiss. München* **1**, 282; **2**, 114.
Naegeli, C. (1928). "Die Micellartheorie." Ostwald's Klassiker, No. 227. Engelmann, Leipzig.
Noll, F. (1887). *Abhandl. senckenberg. naturforsch. Ges. No.* **15**, 101.
Ordin, L., Cleland, R., and Bonner, J. (1955). *Proc. Natl. Acad. Sci. U.S.* **41**, 1023.
Overbeck, F. (1934). *Z. Botan.* **27**, 129.

Pfeffer, W. (1886). *Untersuch. Botan. Inst. Tübingen* **2**, 179.
Preston, R. D. (1951). *Discussions Faraday Soc.* **11**, 165.
Preston, R. D. (1952). "The Molecular Architecture of Plant Cell Walls." Wiley, New York.
Preston, R. D. (1957). *Meeting 9th Intern. Congr. Cell Biol. St. Andrews, 1957.*
Rånby, B. G. (1949). *Acta Chem. Scand.* **3**, 649.
Rånby, B. G., and Ribi, E. (1950). *Experientia* **6**, 12.
Raper, K. B., and Fennell, D. (1952). *Bull. Torrey Botan. Club* **79**, 25.
Reinhardt, M. O. (1899). *Schwendener-Festschrift* p. 425.
Roelofsen, P. A., and Houwink, A. L. (1953). *Acta Botan. Néerl.* **2**, 218.
Roelofsen, P. A., and Kreger, D. R. (1951). *J. Exptl. Botany* **2**, 332.
Ruge, U. (1937). *J. Botany* **31**, 1.
Russow, E. (1883). *Botan. Zentr.* **13**, 171.
Sanio, C. (1860). *Botan. Ztg.* **18**, 201, 209.
Schmitz, F. (1880). *Sitzber. Niederrhein. Ges. Natur- u. Heilkunde Bonn.*
Schoch-Bodmer, H., and Huber, P. (1945). *Verhandl. naturforsch. Ges. Basel* **56**, 343.
Schubert, W. (1954). *Holz Roh u. Werkstoff* **12**, 373.
Schumacher, W., and Halbsguth, W. (1939). *Jahrb. wiss. Botan.* **87**, 324.
Scott, F. M., Hammer, K. C., Baker, E., and Bowler, E. (1956). *Am. J. Botany* **43**, 313.
Sitte, P. (1953). *Mikroskopie* **8**, 290.
Sitte, P. (1955). *Mikroskopie* **10**, 178.
Sponsler, O. L. (1922). *Am. J. Botany* **9**, 471.
Sponsler, O. L., and Dore, W. H. (1926). *Colloid Symp. Monogr.* **4**, 174.
Steward, F. C., and Mühlethaler, K. (1953). *Ann. Botany (London)* **17**, 295.
Strasburger, E. (1882). "Ueber den Bau und das Wachstum der Zellhäute." Fischer, Jena.
Strasburger, E. (1898). *Jahrb. wiss. Botan.* **31**, 511.
Strasburger, E. (1907). *Flora* **97**, 123.
Strugger, S. (1957). *Protoplasma* **48**, 231.
Suito, E., and Takiyama, K. (1954). *Proc. Japan. Acad.* **30**, 752.
Thimann, K. V. (1951). *Growth* **10**, 5.
Thimann, K. V., and Bonner, J. (1933). *Proc. Roy. Soc.* **B113**, 126.
van Wisselingh, C. (1925). "Die Zellmembran," Linsbauer Handbuch der Pflanzenanatomie, Lieferung 11. Borntraeger, Berlin.
Vogel, A. (1954). *Makromol. Chem.* **11**, 111.
von Höhnel, F. (1905). "Die Mikroskopie der technisch verwendeten Faserstoffe," 2. Aufl., S. 21. Hartleben, Wien, Leipzig.
von Mohl, H. (1859). Cited after Frey-Wyssling (1935).
Wardrop, A. B. (1955). *Australian J. Botany* **3**, 137.
Went, F. W. (1929). *Rec. trav. botan. néerl.* **25**, 1.
Wentzel, R. (1929). Dissertation. Marburg.
Whitaker, D. M. (1931). *Biol. Bull.* **61**, 294.
Wilson, K. (1955). *Ann. Botany (London)* **19**, 289.
Wilson, K. (1957). *Ann. Botany (London)* **21**, 1.
Wood, F. M. (1926). *Ann. Botany (London)* **40**, 547.
Wyckoff, R. W. G. (1947). *Biochim. et Biophys. Acta* **2**, 139.
Zacharias, E. (1891). *Flora* **74**, 466.

CHAPTER 3

Ameboid Movement

By ROBERT D. ALLEN

I. Introduction

A. *Characteristics and Occurrence*

Cell movements that involve form changes brought about through cytoplasmic streaming are usually designated as ameboid movement. This process generally results in locomotion if the cells so engaged are attached to some substratum.

In its simplest form, the cytoplasmic streaming pattern of an ameboid cell consists of an axial stream continuously displaced through a tubular body. There are many variations on this theme, however, which are still referred to as ameboid movement. While some amebae are predominantly monopodial, others send out many pseudopodia and are either temporarily or permanently polypodial. The shapes of pseudopodia vary in the extremes from the stout, almost cylindrical *lobopodia* of the larger species to the fine, straight, filamentous *filopodia* or the branching, anastomosing *reticulopodia* of the foraminifera.

Ameboid movement occurs in some form throughout Class Sarcodina, not only in the Order Amoebaea, where it is most commonly studied (free-living and parasitic amebae and testacea such as *Difflugia* and *Arcella*), but in the Orders Foraminifera and Rhizomastigina (ameboflagellates) as well. The other orders in this class exhibit locomotion and protoplasmic streaming, but their movement probably should not be called ameboid according to the above definition. In the radiolaria and heliozoa, for example, protoplasmic streaming takes place along rigid skeletal spines. The proteomyxida (e.g., *Labyrinthula;* see Watson and Raper, 1957) are not strictly ameboid; they appear to slide along a slime track by an unknown mechanism. The slime molds are classified by the protozoologists as the Order Mycetozoa and by the botanists as Myxomycetes. During part of their life cycle they exist as myxamebae, and later fuse to form a plasmodium (e.g., *Physarum polycephalum*). Despite the fact that slime molds form from the fusion of single amebae and even in the plasmodial stage somewhat resemble huge amebae, the movements and streaming of plasmodia have been studied almost exclusively by plant physiologists. In view of the fact that an excellent review of protoplasmic streaming in plants has just been published (Kamiya, 1959), plasmodial streaming will not be considered in detail in the present chapter.

Ameboid movement is a process of prime importance in the development of perhaps all multicellular organisms. Evidence of this comes from accounts of the aggregation of dissociated sponge cells (Galtsoff, 1923) and of the aggregation of myxamebae of the cellular slime molds (Bonner, 1950). Ameboid movements can be seen in embryos that are sufficiently transparent and probably occur in others as well, for many embryonic and some adult tissue cells move like small amebae when growing in tissue culture. Several excellent motion pictures of such processes have been made in recent years.

Even the unfertilized ova of marine invertebrates, which are often considered to be examples of "undifferentiated protoplasm," may exhibit striking streaming and sometimes ameboid movements. Some eggs, such as those of sponges, move about naturally by means of ameboid movement. Others show streaming movements within restraining membranes; when these membranes are removed, ameboid locomotion occurs. Some eggs in which these phenomena have been observed and studied are those of *Arbacia* (Kitching and Moser, 1940), *Chaetopterus* (Lillie, 1902), and *Spisula* (Rebhun, 1961).

B. Relationship to Other Forms of Protoplasmic Movement

Some form of protoplasmic streaming occurs in almost all organisms. In a recent review of protoplasmic streaming in plants, Kamiya (1959) has classified types of streaming into the following categories: agitation, circulation, rotation, fountain streaming, streaming along definite tracks, tidal streaming, and shuttle streaming. He has also pointed out several cases in which one type of streaming changes to another. Many of these types of streaming occur also in animal cells. Rotatory and circulatory streaming occur in animal cells; *Paramecium, Stentor,* and many other ciliates exhibit circulatory streaming which is apparently important in ingestion and assimilation of food substances. Some primitive slime molds demonstrate the "overlap" between the protoplasmic streaming in plants and that associated with ameboid movement. *Reticulomyxa filosa* (Nauss, 1949) ordinarily exhibits constant "sleeve" type movement (fountain streaming); when changing location in culture, attachment of a fountain-streaming pseudopod to the substratum brings about locomotion. There are many other streaming phenomena in animal and plant cells which are rarely studied because their mechanisms seem so obscure (cf. Andrews, 1955).

Because streaming is very much the same whether it occurs in animal or in plant cells, its study should be pursued wherever it occurs. The division of streaming phenomena into two categories, plant and animal, is superficial and is a reflection of biologists' categorizing of themselves as botanists and zoologists and the consequent restriction of their studies exclusively to either plant or animal cells.

The hope has often been entertained that all streaming phenomena would eventually prove to have a common mechanism. Most authors agree nowadays that if such a common mechanism exists, it is contractility. Contractility in its simplest form seems to be the most likely explanation for the "peristaltic" contractile waves that spread along *Euglena* and many other flagellates (cf. Kamiya, 1939). In ameboid cells and in other protoplasmic streaming phenomena, neither the fact nor the localization of contractility has been very clear. Whether the basic mechanism of protoplasmic streaming is similar in diverse cells remains to be demonstrated; for the time being it appears prudent to investigate each streaming phenomenon separately with the hope that similarities will become apparent.

C. *Historical Sketch*

1. *Discovery*

It is difficult to believe that more than a century passed after the invention of the microscope until the first ameba was described. The first observation has usually been credited to Rösel von Rosenhof (1755). Kudo (1959) among others has questioned this and has reproduced Rösel's original plate showing a large ameboid organism, the "kleine Proteus," to which Linnaeus later gave the name *Chaos.* If Rösel von Rosenhof's observations were not of an ameba but of a slime mold, as some think, it would then seem to be Müller (1786) who discovered the ameba.

2. *Early Contractility Theories*

Dujardin (1835, 1838) was evidently the first to suggest a mechanism for ameboid movement. He proposed that all living things were made of the same kind of substance, "sarcode," which was endowed with what he considered to be the inherent properties of life; among these were contractility and extensibility, which he postulated to be the basis for all movement. This view, or some variation of it, was held by nearly every biologist for the following half century. While Dujardin did not differentiate between the inner and outer layers of an ameba, considering them both to have the properties of contractility and extensibility, Ecker (1849) expressed disagreement with this view and pointed to the fact that the flow of granular cytoplasm originated at the posterior end of the organism, suggesting that pressure by the contraction of the outer layer only was the cause of streaming. Later, Wallich (1863) stated quite definitely that contraction of the outer layer of the cell was the cause of flow. The idea of ectoplasmic contraction was developed still further by

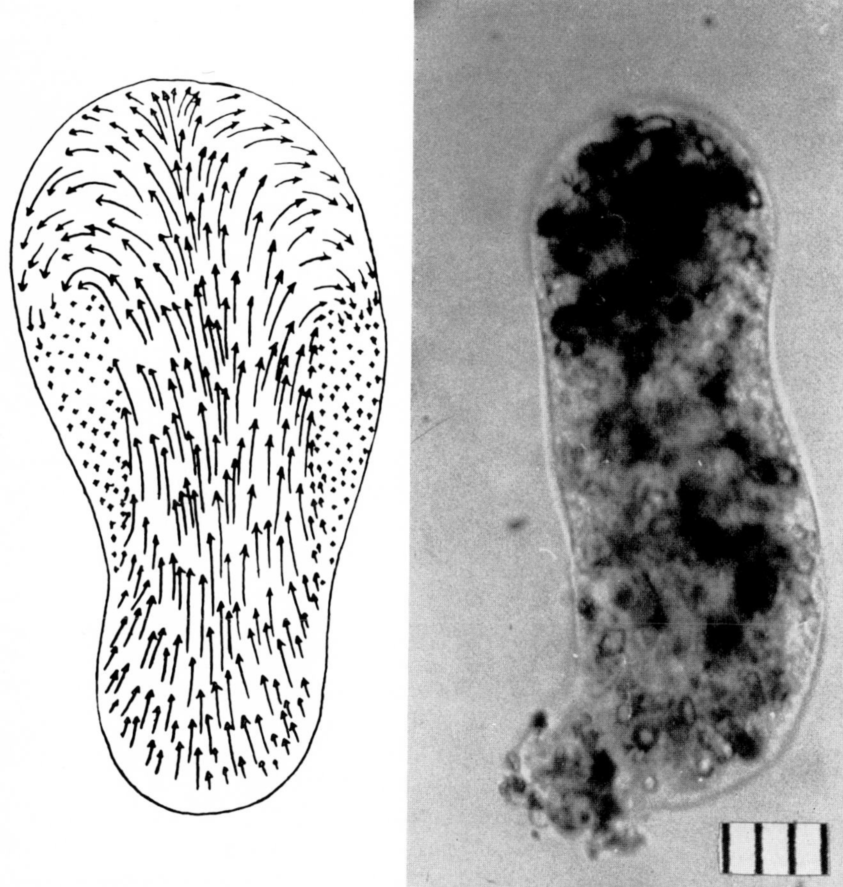

FIG. 1. Left: the pattern of cytoplasmic streaming in *Pelomyxa palustris* after Schulze (1875). A later account by Mast (1934) differed in a few details. Right: a photograph by Dr. J. L. Griffin of a *P. palustris* collected during the summer of 1959 in the same pond (in the center of Falmouth, Massachusetts) in which Mast (1934) obtained his specimens.

F. E. Schulze (1875) on the basis of a careful analysis of the pattern of streaming movements in *Pelomyxa* (cf. Fig. 1). Schulze also believed that the inner cytoplasm was fluid. In many ways his views were quite similar to those adopted fifty years later by Pantin (1923) and Mast

(1926). A more complete account of the early contractility theories of ameboid movement is given in a review by De Bruyn (1947).

3. *Surface Tension Theories*

The rise of physical chemistry during the middle part of the nineteenth century, coupled with the realization among biologists that descriptions and explanations of biological processes were most meaningful when expressed in physical and chemical terms, led to an understandable yet unfortunate digression in the theory of ameboid movement of nearly forty years' duration. By the 1880's surface tension phenomena were well understood, and the demonstration that the spreading of oil on water was a surface tension phenomenon led Berthold (1886) to propose a surface tension theory of ameboid movement. He suggested that the formation of pseudopodia was due to an interplay among the surface tensions between substrate and protoplasm, protoplasm and water, and water and substrate. A more extreme view was adopted by Bütschli (1892) and later by Rhumbler (1898). They assumed that merely local reduction of surface tension could cause pseudopodial extension; and some observers even suggested that the ameba carried forward in its endoplasm soaplike substances, the function of which would be to reduce surface tension at the anterior part of the cell. It was well known at the time that ameboid movement could be "closely imitated" by model systems of liquid droplets of known properties (e.g., water, oil, ether, glycerol, mercury). Perhaps the most striking model experiment was that devised by Bernstein (1900), who showed that a drop of mercury would exhibit "ameboid locomotion" in nitric acid in response to the presence of a crystal of potassium bichromate. The sight of such a phenomenon must have greatly excited biologists of the time, for here indeed was a vivid demonstration of how simple physics and chemistry could be invoked to "explain" a biological phenomenon.

The various model experiments demanded that the surface layer of an ameba should move backward. Schulze (1875) reported slight backward movement of this kind in the granular ectoplasm. Others (e.g., Rhumbler) confirmed this observation, and the pattern of streaming of the outermost cytoplasmic inclusions seemed to be the "fountain streaming" that the surface tension theory required.

The surface tension theory became less attractive when it was shown that the surface of *Amoeba verrucosa,* a rolling form, moves *forward* instead of backward (Jennings, 1904). Dellinger (1906) confirmed this for *A. verrucosa,* but showed by ingeniously viewing amebae from the side that *A. proteus* "walks" on its pseudopodia (Fig. 2) in a way incompatible with the idea that the surface of an ameba was sufficiently fluid to be moved by the feeble forces of surface tension. When Mast and Root (1916) showed that an ameba could cut a paramecium in two with its food cups, it became perfectly clear that the surface layer of the ameba was, indeed, far too rigid to be affected by surface tension in the

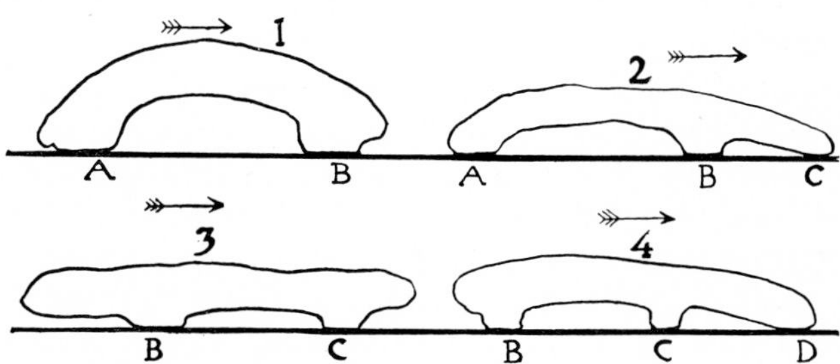

FIG. 2. Drawing of an ameba as observed from the side to show the steps in locomotion. From Dellinger (1906).

way proposed by Berthold, Bütschli, Rhumbler, and others. A more complete account of the surface tension theory of ameboid movement can be found in reviews by Tiegs (1928), Mast (1926), and De Bruyn (1947).

4. *The Return to Contractility Theories*

The surface tension theories were gradually replaced by contractility theories surprisingly similar in concept to, but expressed in different terms from, those which had been proposed in the previous century. The principal reason that biologists had earlier been willing to relinquish contraction as a mechanism for ameboid movement was that the mere word contractility offered no ultimate explanation of ameboid movement in physicochemical terms, for the structure of the contractile material was below the limit of visibility with the microscope. When the newer contractility theories appeared, they were expressed in terms of colloid chemistry, the methods of which offered the hope, which microscopy had not, that the "structure" of protoplasm and the mechanism of contractility might be understood. As De Bruyn (1947) pointed out in his review, there has always been a close dependence of theories of ameboid movement on the prevailing conception of protoplasmic structure (cf. also Section II, C, 1 and 2).

When Dellinger (1906) rejected surface tension, he returned to a contractility theory which was little different from that proposed by Dujardin. He gained the impression from watching both *Amoeba* and *Difflugia* from above and from the side that the endoplasm contracted in order to pull the animal along. From the fact that rolling an ameba along with a needle did not cause its contents to flow, and from the fact that the inclusions were ". . . held in definite positions under conditions that would indicate that the endosarc (endoplasm) has definite structure," he concluded that "a coarse reticulum of contractile substance distributed through the endosarc would account for the phenomena as we have observed them."

Subsequent theories of ameboid movement are usually referred to cumulatively as "the sol-gel theory" because they assume a marked difference in consistency between the endoplasm and ectoplasm. Actually, the idea of a consistency difference goes back at least to Schulze (1875). Although Hyman (1917) was the first to apply the terms sol and gel with reference to ameba cytoplasm, Pantin (1923) and Mast (1926) organized a coherent theory of ameboid locomotion based on the idea of gelation of the endoplasm at the anterior end of an ameba, and solation of the ectoplasm at the posterior. The motive force for streaming was thought to be pressure either from contraction of the ectoplasm involving syneresis (Pantin, 1923) or from "elastic recoil" of the ectoplasm (Mast, 1926, 1931a).

The more recent development of these ideas and an evaluation of the evidence on which they are based will be discussed later (Section III).

D. Remarks on Methods and General Approach

The preceding section gives some idea of the diversity of the many theories that have been formulated over the past 120 years concerning ameboid movement. Nearly all of these have been of a simple nature. This diversity suggests strongly that the phenomenon itself may not be so simple as has been generally assumed. Not only the theories but also the observations on which they have been based have been often contradictory. In some cases some observers may possibly have been swayed by their preconceptions, but more often such contradictory observations can be attributed to important but subtle differences in experimental conditions.

Difficulties in interpreting data and observations in the literature are caused most frequently by taxonomic uncertainties and by lack of control by the observer of the nutritional condition and physiological state of the living material. There is no question that such factors as culture conditions, nutritional state (whether fed recently or starved, type of food, etc.), and temperature profoundly affect the rate and pattern of streaming in any species. More important still, observations with the microscope have not always been carried out with care to avoid compression, local heating, or exposure to excessively intense light. All these factors have contributed to observations that were therefore misleading, for, had these factors been considered and controlled, the observations would have been different.

Nothing retards the understanding of a phenomenon so much as a lack of experimental data. Ameboid movement has been one of the most difficult biological problems to attack experimentally because of the extreme lability of the process itself; nearly every modification of the environment causes cessation of streaming. Largely for this reason, there have been relatively few *decisive experiments* which give insight into the mechanism of the process. Instead, conclusions have usually had to be drawn from observations that were subject to a choice of interpretations.

II. Observational Basis for Theories of Ameboid Movement

A. *Structure and Cytology of the Large, Free-living Amebae*

The most complete descriptions of the form, cytology, and movement of amebae have been of the larger species, particularly of the "proteus group" of the Genus *Amoeba,* and of the giant, multinucleate ameba, *Chaos chaos* (*Pelomyxa carolinensis, Amoeba carolinensis, Chaos carolinensis*).[1] Many smaller species have been described in detail, but their movements have not been studied extensively.

1. *General Form of Body and of Pseudopodia*

The oft-repeated textbook description of an ameba as a "shapeless sac of protoplasm" is quite incorrect. Each species has a characteristic range of shapes and average number of pseudopodia (which may depend to some extent upon environmental conditions). Schaeffer in 1920 listed several distinguishing characteristics for each member of the "proteus group," for until that time no distinction had been made in descriptions of form and movement of *A. proteus, A. dubia,* and *A. discoides,* all of which are now standard organisms for many kinds of investigations.

Figure 3 shows photographs of these three species and of *Chaos*

[1] Some protozoologists (see especially Kudo, 1959) have referred to the giant ameba as *Pelomyxa.* Although the historical reasoning for this may be correct, it is very misleading to place in the same genus two amebae which, while both multinucleate, differ so radically in form, manner of locomotion, cytoplasmic inclusions, and enzyme content. We agree with King and Jahn (1948), who favor a separate genus, *Chaos,* for the giant ameba.

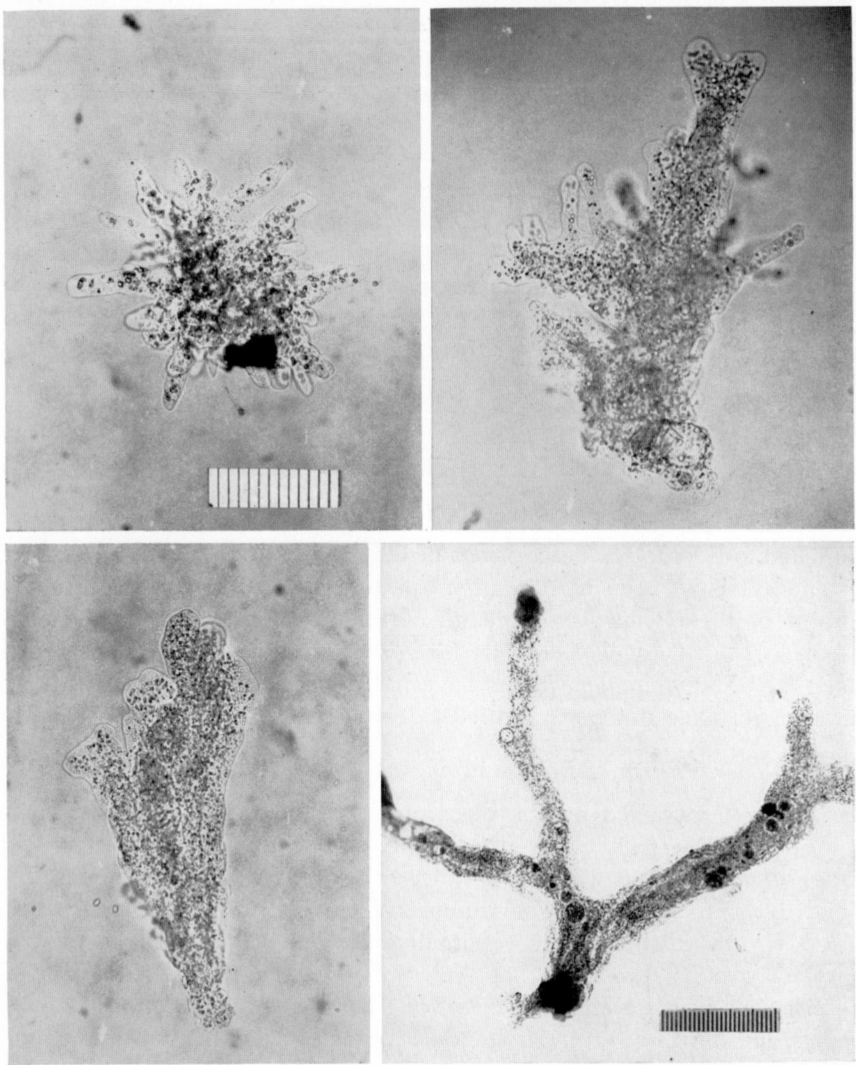

FIG. 3. Photographs of four species of amebae referred to frequently in the text. Upper left, *Amoeba dubia*; upper right, *A. discoides*; lower left, *A. proteus*; all photographed at the same magnification. Scale = 10 μ. Lower right, *Chaos chaos* (*Chaos carolinensis, Amoeba carolinensis, Pelomyxa carolinensis*) at a lower magnification. The specimen of *A. dubia* is shown characteristically carrying debris on its uroid (above the focal plane).

chaos. A. proteus has longitudinal ectoplasmic ridges, finely granular cytoplasm, and relatively few pseudopodia when in active locomotion. *A. discoides* is quite similar except for less prominent ectoplasmic ridges, a discoid nucleus, and a tendency toward branching pseudopodia. *A. dubia* ordinarily exhibits many more pseudopodia than either *A. proteus* or *A. discoides,* and its cytoplasmic crystals are much larger than those of the other species. It apparently lacks heavy spherical bodies. *Chaos chaos* is remarkably similar in form and in the number and size of cytoplasmic inclusions to *Amoeba proteus,* but is multinucleate and much larger than the uninucleate species. Small specimens of *Chaos* could easily be mistaken for *A. proteus* if the nuclear differences were disregarded; even the ectoplasmic ridges of the two species are quite similar.

The pseudopodia of the larger amebae are usually roughly cylindrical (the broader ones are somewhat flattened), with hemispherical tips. Among the smaller species, long pointed pseudopodia are also found. Under the simplest conditions for the study of ameboid movement, an ameba may progress by means of a single, cylindrical pseudopod. This type of movement usually results in the formation of a wrinkled "tail piece" or *uroid.* In predominantly monopodial specimens, temporary lateral pseudopoda retract into stumps which retreat posteriorly and eventually merge with the uroid. In polypodial amebae, for example *A. dubia* or some specimens of *Chaos chaos,* each retracting pseudopod resembles a uroid for a short time until it is absorbed into the main body mass.

2. *The Cell Surface*

Most early workers regarded the ameba surface as an interface between two immiscible fluids, water and protoplasm (cf. Schaeffer, 1920). Mast (1926), however, presented evidence that the surface of the ameba was covered by a definite sheath, which he called the *plasmalemma.* The evidence was as follows: (1) a double line was visible at the surface, instead of the single line observed at interfaces; (2) particles of lampblack adhering to the surface moved forward as if on a membrane sliding over the ectoplasmic layer; and (3) the plasmalemmas of broken cells remained intact as structures showing wrinkles and folds. Recent electron micrographs of thin sections have confirmed Mast's contention and have shown that the plasmalemma has a definite and interesting ultrastructure (Fig. 4): in *A. proteus* a double membrane about 200 A. thick, with filaments about 80 A. in diameter extending 1100 to 1700 A. out into the medium (Pappas, 1956, 1959; Mercer, 1959).

The surface of *Chaos chaos* was found to be similar except that the filaments were only 40–60 A. thick. Evidently these observations cannot be generalized to all amebae, for Pappas (1959) was unable to find similar structures in the surface of *Hartmanella rysodes,* which instead had only a single membrane 60 A. thick. Several authors have claimed that the plasmalemma is covered by a layer of slime (Lewis, 1951); in any case, Bairati and Lehmann (1953) believe that the plasmalemma, or

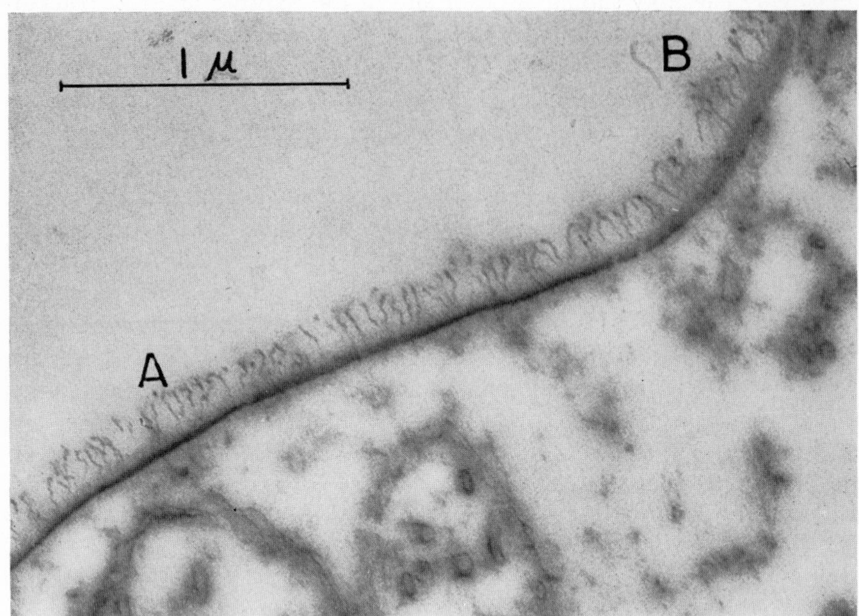

FIG. 4. Electron micrograph of the plasmalemma of *Amoeba proteus.* The thickness of the plasmalemma at *A* is about 200 A.; the greater thickness at *B* is due to tangential section. The outer filaments are about 80 A. in diameter and 1100 A. long. Courtesy of Dr. G. Pappas (1959).

at least the surface layer of *Amoeba proteus,* contains mucoprotein; the evidence is (1) metachromasy, (2) positive Hotchkiss reaction, and (3) apparently specific action of hyaluronidase on the surface of fixed cells.

There are several indications that amebae can form new plasmalemma rather rapidly under certain conditions. When an ameba has been feeding, some of the original plasmalemma becomes the lining of the food vacuoles and must be replaced. A specimen of *Chaos chaos* confined to a narrow capillary and forced to move monopodially often develops an enormously attenuated uroid. This must represent a significant increase in surface area. On the other hand, there is no evidence that the ameba renews its surface membrane every time it moves its own length as Goldacre and Lorch (1950) and Goldacre (1952a) have stated. In fact, the evidence cited in Section II, B, indicates that almost no formation of membrane takes place in the anterior region of the ameba.

The permeability of the surface layer of *Chaos chaos* has been studied by means of the cartesian diver balance with isotopic water by Løvtrup and Pigón (1951). The amount of water entering a cell (and therefore the amount being actively transported out at equilibrium) was found to correspond to 2–4% of the cell's total volume per hour.

3. *Cytoplasmic Inclusions*

a. Food vacuoles. Amebae in wild cultures (rice or hay infusions) or in mass cultures fed on *Tetrahymena* or other food organisms exhibit food vacuoles containing food in various stages of digestion. The microscopic changes undergone by these food vacuoles during digestion have been reviewed by Andresen (1942, 1956) and by Andresen and Holter (1945). Recent electron microphotographs of Pappas (1959) and of Borysko and Roslansky (1959) add new details concerning the fine structure of the vacuoles and their contents.

It is desirable for many studies of ameboid movement, and necessary for many biochemical investigations, to dispose of food vacuoles. Although depriving the cells of food for 24–36 hours effectively reduces the size and perhaps number of food vacuoles without causing damage or abnormalities of movement, it is doubtful whether it is possible or feasible to rid amebae entirely of the contents of their food vacuoles. Cohen (1957) has found that traces of the contents of food vacuoles remain for as long as 3 weeks after the initiation of fasting. Muggleton and Danielli (1958) have shown, moreover, that an inadequate diet can cause reproductive damage to *A. proteus* and *A. discoides.* The difficulties involved in disposing of the contents of food vacuoles have not always been considered in biochemical work with amebae. The usefulness of these cells in biochemical work will be limited, to a certain extent, until a satisfactory defined liquid medium is devised.

b. Crystals. The cytoplasm of many of the larger amebae, for example the proteus group and *Chaos chaos,* contains numerous crystalline inclusions. Some of the smaller species also contain crystals, but their occurrence is by no means universal.

According to most descriptions, the crystals of *A. proteus* and *Chaos chaos* are enclosed in vacuoles (Mast, 1926; Mast and Doyle, 1935; Andresen, 1942, 1956; Andresen and Holter, 1945; Torch, 1959; Pappas, 1959); however, Singh (1938) found crystals to be free in the cytoplasm and thought the vacuoles, where found, to be induced. In *A. dubia* the crystals are much larger than those of the other species, and vacuoles, when present, can be seen easily with either phase contrast, interference contrast, or bright field microscopes (Fig. 5). In carefully handled, uncompressed specimens of this species, the crystals lie free in the cytoplasm (Griffin, 1959; Allen and Roslansky, 1959; Torch, personal com-

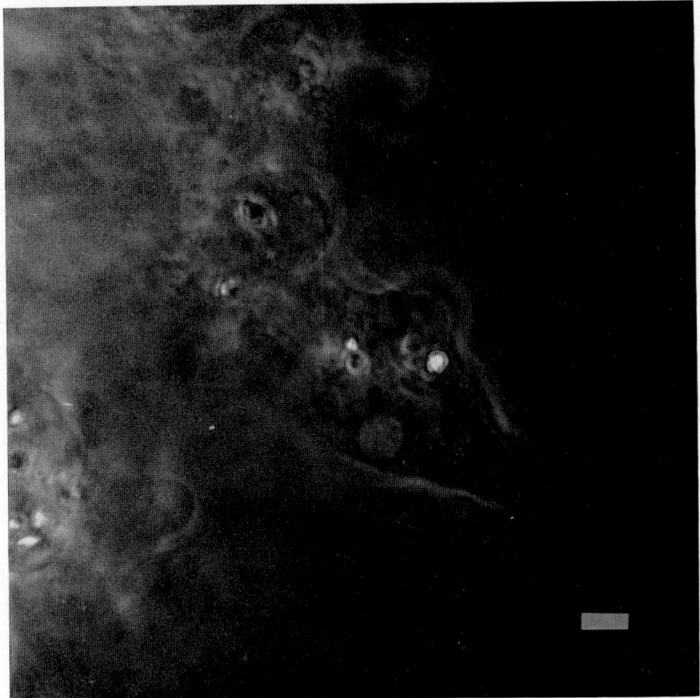

FIG. 5. A thin pseudopod of *A. dubia* showing a crystal free in the cytoplasm. Nearby are two clear vacuoles such as have frequently been observed around crystals after compression or other injury.

munication). On the other hand, vacuoles can be induced to form around crystals by compression under a coverglass, exposure to heat or intense light, fixation, other kinds of rough treatment, and, under some conditions, centrifugation. It is much more difficult to be certain whether the crystals of *A. proteus, A. discoides,* or *Chaos chaos* are contained in vacuoles, for the crystals themselves are smaller (see Fig. 6) and more difficult to see in the intact ameba without compressing the cell. However, in slender pseudopodia of *A. proteus* the majority of crystals appear to be perfectly free in the cytoplasm. In the case of *Chaos chaos*, there is more variability among cells, but at least some crystals are free in all cells that are handled with the precautions mentioned above. Gradual centrifugation brings the crystals of the smaller species into a tight pocket at the centrifugal pole of the cell instead of into a suspended layer as has been found in most previous studies. In *Chaos*, however, different proportions of the crystals collect at the bottom and in the suspended layer in different cells.

It seems likely, therefore, that most of the crystal vacuoles which have been described in the literature have been induced by some form of moderately rough handling. The solid surfaces of crystals might be expected to be excellent foci for the formation of vacuoles in the cytoplasm.

The crystals themselves tend to be bipyramidal or thin, flat plates; rods have also been described. Many authors have considered them to be excretory products or stored food reserves. Until recently there were no decisive data; now, however, Griffin (1959, 1960) in the author's laboratory and Grunbaum *et al.* (1959) at the Carlsberg Laboratory have produced data that show conclusively that the crystals are new nitrogen excretion products. Griffin (1959, 1960) has positively identified the platelike crystals as carbonyldiurea on the basis of evidence from microanalysis, X-ray diffraction pattern, infrared absorption spectra, and petrographic analysis. The bipyramidal crystals can be "recrystallized" from water as platelike crystals of carbonyldiurea but do not recrystallize as bipyramids; this fact led Griffin (1959) to the view that the bipyramids also were probably carbonyldiurea, or at least a closely related compound. The data of Grunbaum *et al.* (1959) suggest that the bipyramids are not carbonyldiurea, but a related substance. Probably both substances are breakdown products of purine metabolism through allantoic acid.

The size and shape of crystals depends to some extent on species (see Fig. 6). Those of *A. dubia* average about 8.5 μ (up to 12 μ) for the bipyramids and as large as 30 μ in longest dimensions for the plates. In *A. proteus* and *Chaos chaos* the crystals are smaller, averaging a little less than 3 μ in each species for the bipyramids; plates are much smaller than in *A. dubia*, usually 5 μ or less on a side. Some of the bipyramidal crystals are truncated on one end, others on both ends. The bipyramids are not birefringent, but the plates are markedly so; some show themselves to be interpenetration twins under the polarizing microscope (Griffin, 1959).

c. Heavy spherical bodies. Mast (1926) described the presence in *Amoeba proteus* of "refractile bodies" and believed that they originated in food vacuoles (cf. also Mast and Doyle, 1935; Singh, 1938). Later, Andresen (1942) described similar structures in the cytoplasm of *Chaos chaos* and renamed them "heavy spherical bodies." Either name would appear to be appropriate, for almost all that is known about them is that they are heavy and that they have a high refractive index. These structures have a microscopically visible structure, but so far they have not been sufficiently well preserved to be identified in electron micrographs (e.g., Cohen, 1957; Pappas, 1956, 1959; Greider *et al.*, 1956; Mercer, 1959).

Despite the fact that the heavy spherical bodies collect in the most centrifugal portion of cells subjected to centrifugation in a density gradient, there is some doubt that these bodies are actually more dense than the crystals (cf. preceding section). It has been pointed out above that the crystals of gradually centrifuged amebae share the bottom layer with the heavy bodies; it is, therefore, not possible to determine which is more dense without isolating both from the cell. This does not appear to have been done with the heavy spherical bodies.

d. Lipid droplets. Most well-fed amebae contain lipid droplets, but these make up only a very small portion of the cytoplasmic volume; in *Chaos chaos* they range in size from 0.5 to 3 μ and occupy about 0.5% of the volume of the cell (Andresen, 1956).

e. Mitochondria and alpha granules. Mast (1926) distinguished two kinds of small granules: alpha granules, which were about 0.25 μ in diameter, and beta granules, or mitochondria. Too little is known at present about the structure and chemistry of alpha granules to be certain whether or not they are a discrete class of particulates. It is not certain whether alpha granules have been seen in electron micrographs; those of Pappas (1959) do not show any, but Cohen (1957) has pointed to particles in *A. proteus* which are at least of the proper size to be alpha granules. Ameba mitochondria, however, have been observed in electron microscopical studies of several species (Cohen, 1957; Pappas, 1956, 1959; Pappas and Brandt, 1958; Borysko and Roslansky, 1959). The mitochondria of *A. proteus* are about 1–1.5 μ long and quite similar to those reported for other protozoa (cf. Figs. 7 and 8); those of *Chaos chaos* are somewhat larger and more complex in internal structure.

f. Contractile vacuole. The single contractile vacuoles of uninucleate amebae and the numerous contractile vacuoles of *Chaos chaos* are sur-

---

FIG. 6. The crystalline cytoplasmic inclusions of three species of amebae isolated from frozen-dried cells and mounted in immersion oil (Griffin, 1959). Upper right, *Chaos chaos*; upper left, *Amoeba proteus*; lower left, *A. dubia*; lower right, crystals from *A. dubia* between crossed polars to show the birefringence and interpenetration twinning pattern characteristic of some of the plates. The bipyramids are isotropic. Scale = 10 μ. Courtesy of Dr. J. L. Griffin.

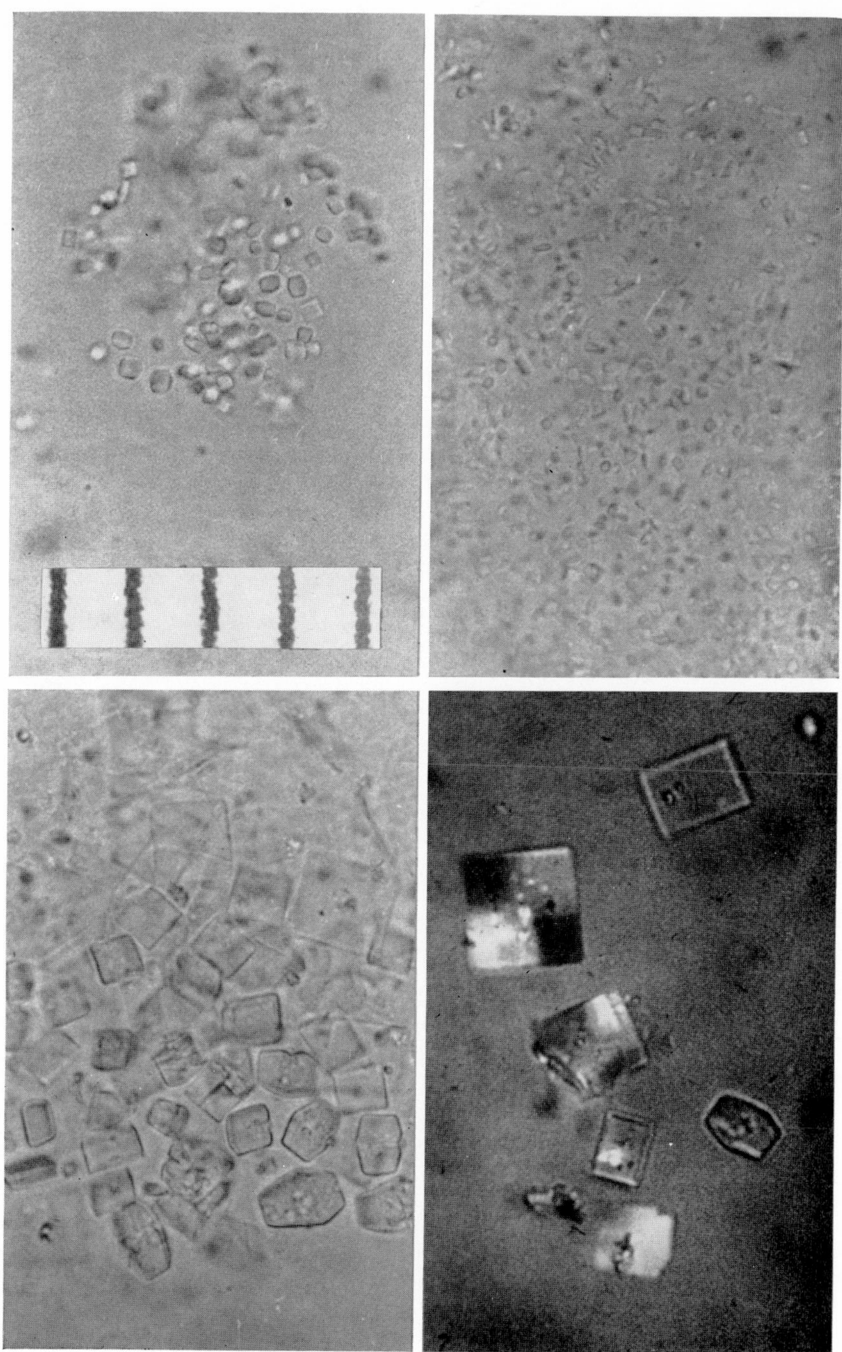

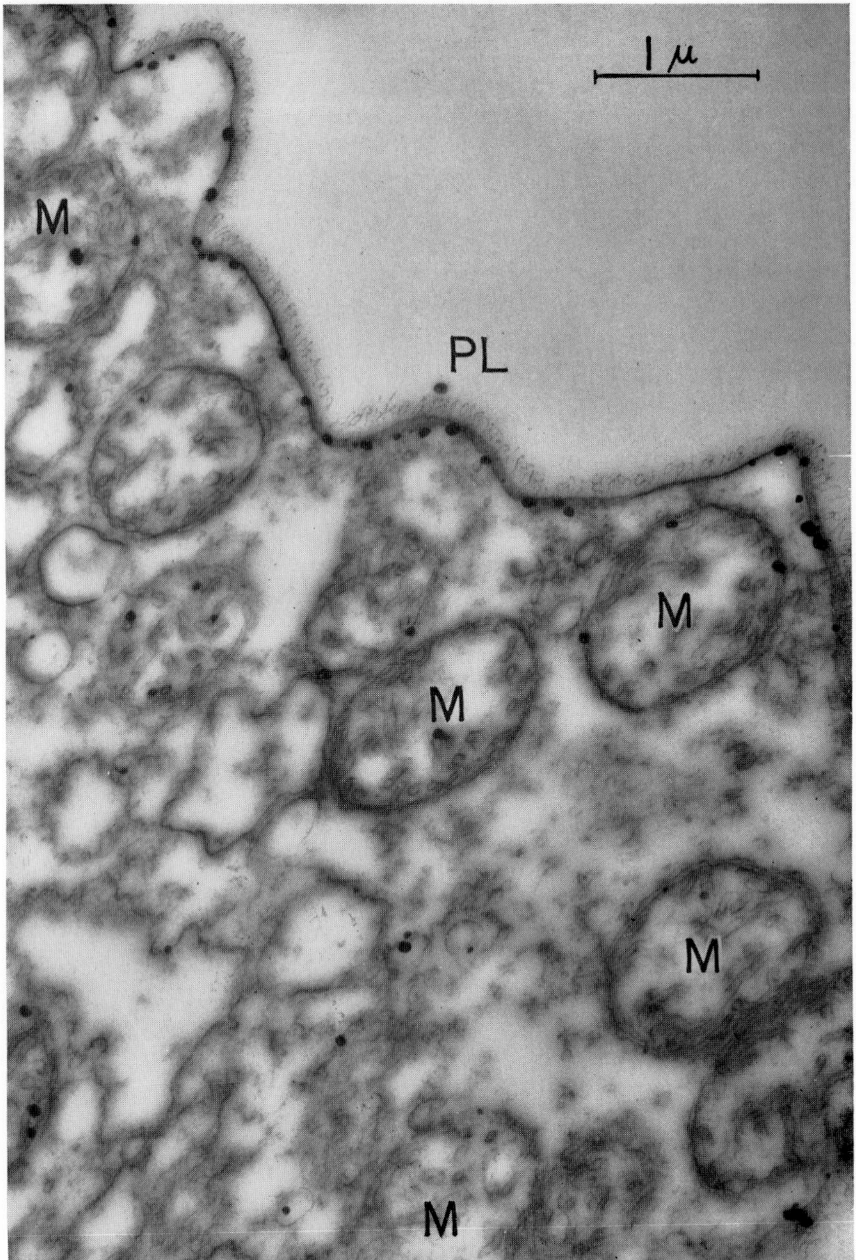

FIG. 7. Electron micrograph of a section of *Amoeba proteus* showing the plasmalemma (*PL*) and several mitochondria (*M*). Note also the alveolar structures of various shapes and sizes. Courtesy of Dr. G. Pappas (1959).

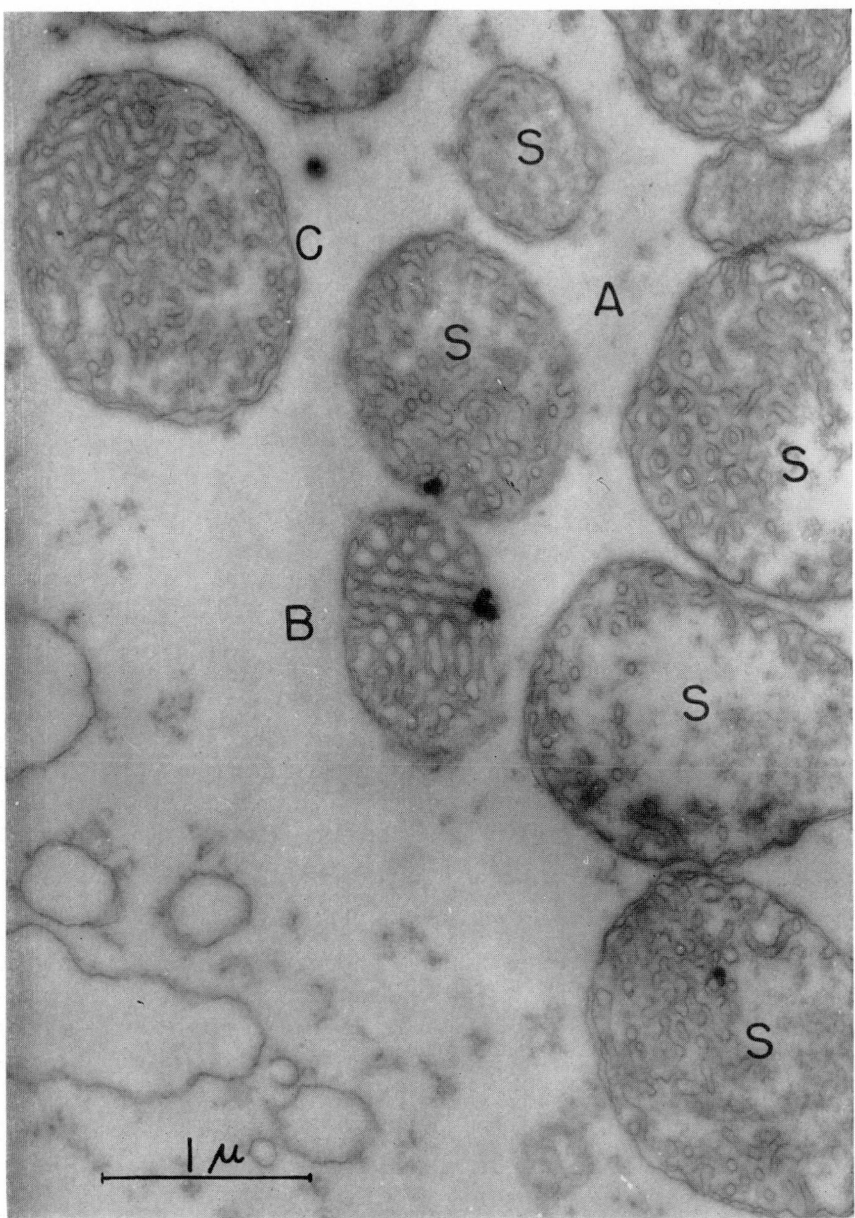

FIG. 8. Electron micrograph of a group of mitochondria (*A*, *B*, and *C*) of *Chaos chaos,* various ones of which show different profiles of their inner membrane patterns. Most mitochondria show an area of the stroma (*S*) in which no projections of the inner mitochondrial membrane are seen. Courtesy of Dr. G. Pappas (1959).

rounded by a layer of mitochondria which can be seen easily with the light microscope (cf. Andresen, 1942, 1956). The electron micrographs of Pappas and Brandt (1958) and of Pappas (1959) show a layer 2 μ thick of small vesicles, which in turn are surrounded by a layer of mitochondria (Fig. 9). Pappas and Brandt believe the vesicles represent the vehicles for the transport of water from the mitochondrial layer to the contractile vacuole. It is likely that the walls of at least some of the vesicles merge with the expanding wall of the contractile vacuole. In a way, this is almost the exact reverse of the course of events in pinocytosis (cf. Pappas and Brandt, 1958).

g. *Other vacuoles, the ground substance, and the endoplasmic reticulum.* The cytoplasm of amebae contains a number of vacuoles the significance of which is unknown. Clear vacuoles are most often seen in compressed, centrifuged, starved, injured, vitally stained, or fixed cells. In view of the fact that almost any experimental treatment of an ameba (even sucking it into a pipette) may disturb it to a certain extent, it is extremely difficult to be certain whether the vacuoles "normally present" are natural or induced. Pappas (1959) has commented on the particularly large number of vacuoles that are present in *A. proteus* and *Chaos chaos* in comparison to *Hartmanella.* The latter also exhibited a well-defined endoplasmic reticulum similar to that found in vertebrate cells, whereas the cytoplasm of *Amoeba* and *Chaos* showed little evidence of an extensive endoplasmic reticulum. The explanation for these facts may lie in the fact that the larger amebae are not sufficiently rapidly fixed by osmic acid. Goldacre (1952b) has described the response of amebae to osmic acid. It would appear that considerable alteration of cytoplasmic structure may occur before fixation is accomplished.

4. *The Nucleus*

The nuclei of the larger uninucleate amebae possess birefringent nuclear membranes (Schmidt, 1939) which present a honeycomb appearance in the electron microscope (Pappas, 1959) (cf. Fig. 10). Sometimes evaginations are seen in the nucleus, suggesting that the latter may contribute materials to the cytoplasm.

Just inside the nuclear membrane are numerous "dense bodies" which

---

FIG. 9. A high magnification electron micrograph of the region immediately surrounding a contractile vacuole of *Chaos chaos.* Vesicles (*V*) nearest to the contractile vacuole may contribute their contents to the contractile vacuole (*CV*) during diastole. A golgi complex (*G*) is seen in the lower right. Courtesy of Dr. G. Pappas (1959).

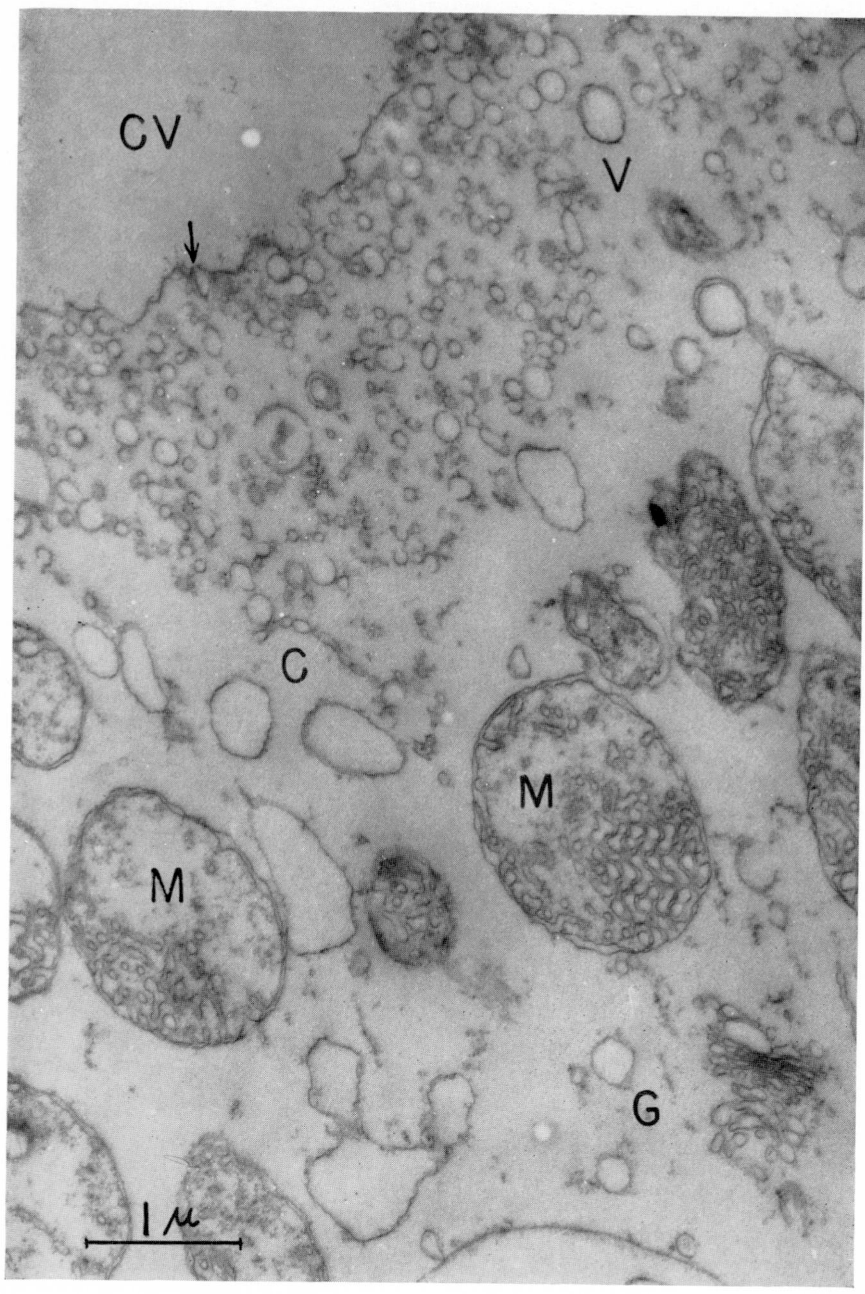
CV
V
C
M
M
G
1 μ

have often been called "nucleoli"; these measure from 0.4 to 1.7 μ in diameter. According to Chalkley (1936), these bodies are basophilic but contain no deoxyribonucleic acid (DNA). The more internal regions of the nucleus show a diffuse Feulgen reaction and contain some interesting helical structures which are too large to be DNA molecules and too small to be chromosomes (Pappas, 1956, 1959; Mercer, 1959) (Fig. 10). The longest helixes are 2500–3000 A. in length and 250–280 A. in diameter. The distance between repeating turns is 140 A.

B. Movement of the Surface Layer

Jennings (1904) published the first extensive observations of movements of the surface layer of amebae; he found that particles of soot attached to the top surface of *A. verrucosa* were carried forward along the top surface and over the advancing tip, and came to rest in contact with the substratum. This observation was made during the high point in the popularity of the surface tension theory, which demanded just the opposite movement of the ameba surface. Consequently, Jennings' work is considered to have been instrumental in forcing the abandonment of the surface tension theory. Jennings believed that other species besides *A. verrucosa* exhibited "rolling motion" of this type, but later workers, especially Dellinger (1906) and Mast (1926), showed this not to be the case. Particles of soot or carbon clinging to the plasmalemma (or perhaps to slime covering it) in *A. proteus* move forward, keeping a more or less constant relationship to the tip of the advancing pseudopod. When a pseudopod branches anterior to an attached carbon particle, the behavior of the particle is compatible with the view that the plasmalemma is a pliable, relatively permanent sheath which is pulled forward by continuing extension of the ectoplasmic tube. The almost unbelievable extent of the pliability of the ameba surface is illustrated by the fact that when an ameba is impaled by a microneedle, attached particles flow around the needle (Griffin and Allen, 1960).

During normal progression, the plasmalemma adheres strongly at restricted points of attachment to both the ectoplasmic tube and the substratum. It is only in this way that the streaming mechanism can bring about locomotion, for, in the absence of attachment to the substratum, "fountain streaming" occurs instead of locomotion (Section II, C, 2). The

---

FIG. 10. Electron micrograph of a portion of the nucleus of *Chaos chaos*. The nuclear membrane made up of two membranes (*NM*) and an inner fibrillar network (*F*) is seen at the top. (*N*) nucleolus; clusters of helixes are seen at *A* and *B*. Courtesy of Dr. G. Pappas (1959).

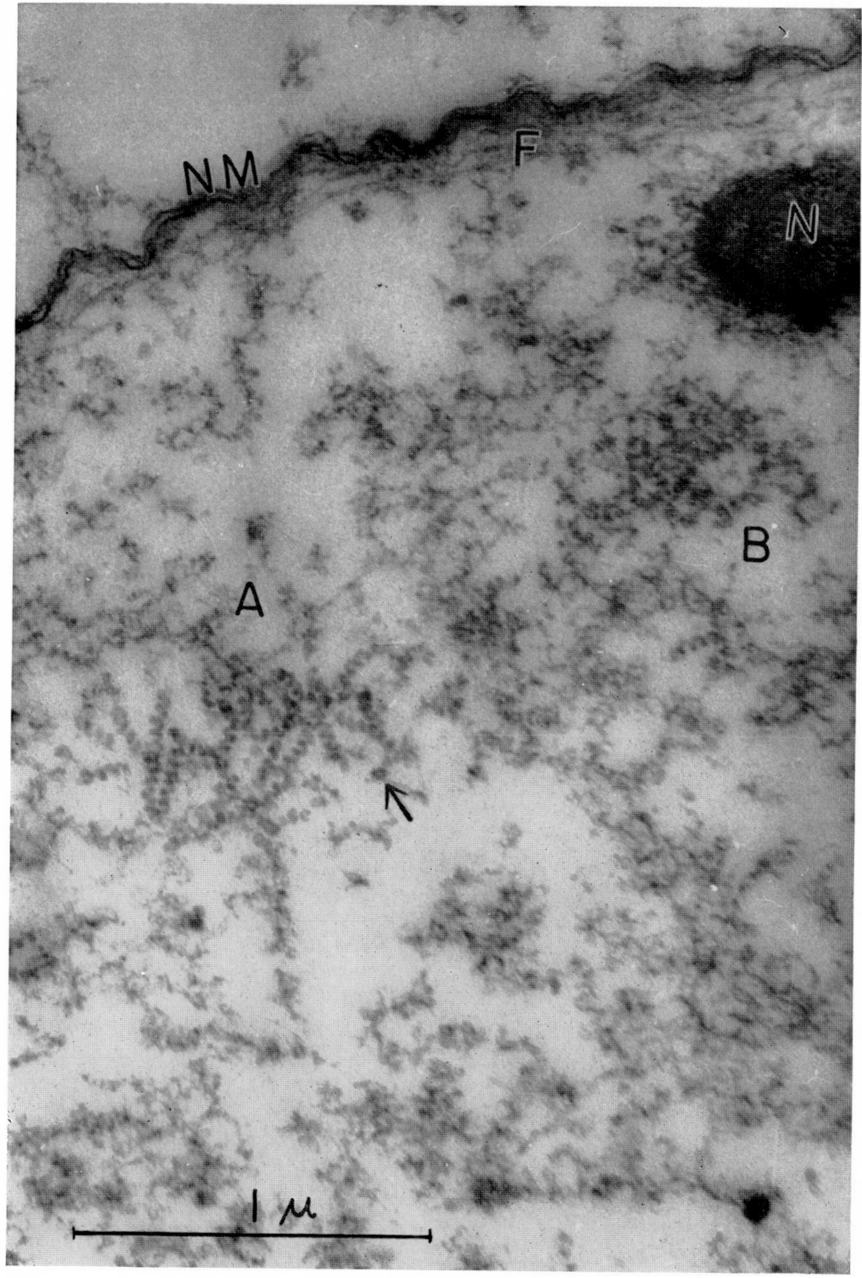
NM
F
N
B
A
1 μ

space between the plasmalemma and the granular ectoplasmic tube has been called the hyaline ectoplasm. It can be inferred from the surface movements described above that this layer, where present, is highly fluid. The hyaline ectoplasmic layer is probably absent in the tail region and at points of attachment to the substratum.

C. *Cytoplasmic Streaming Movements and Their Interpretation*

1. *General Remarks on Fluids: Their Consistency and Flow*

As De Bruyn (1947, p. 2) has pointed out in his review: "The various theories of amoeboid movement have been greatly influenced by the prevailing concepts of protoplasmic structure. This is not surprising; it has always been evident that the final explanation of the mechanism of protoplasmic movement must ultimately be based on the structure or structural elements of protoplasm. Such other functional cellular phenomena as metabolism, respiration and secretion may or may not depend on protoplasmic structure, but the phenomenon of protoplasmic movement *must* be so based. For this reason concepts of protoplasmic movement must take into account protoplasmic structure, just as theories of protoplasmic structure cannot ignore protoplasmic movement."

The word "structure" implies something visible; unfortunately, present techniques are not sufficiently refined or sensitive to reveal more than fragmentary details of the structure of the more labile biocolloids. On the other hand, estimation of cytoplasmic consistency is a more easily attainable goal, and some notion of cytoplasmic structure can be deduced from consistency.

An observer can obtain considerable information about a fluid, particularly of its consistency, from its manner of flow. Since consistency and flow are two inseparable properties of ameba cytoplasm, it is important to know as much as possible about their relationship, both in model systems and in the ameba.

"Flow" of a fluid (as observed in a capillary, for example) involves both *displacement* of the fluid mass as a whole and its *deformation*. Only deformation of a fluid provides clear information regarding its consistency. In the following discussion, we shall consider only permanent deformations; that is, we shall disregard flow elasticity, which is well known to occur in a number of fluids.

The definition of *consistency* proposed by the Society of Rheology is "... that property of a material by which it resists permanent change of shape, and is defined by the complete force-flow relation" (Scott Blair, 1938, p. 24). The force-flow relation is perhaps most easily understood by considering a hypothetical model of two plates separated by a fluid the consistency of which is under investigation (Fig. 11). If one plate is set in motion by a force, the fluid is deformed as adjacent layers or lamellae of fluid slide over one another. The rate of deformation or the *velocity gradient* of the fluid is the velocity of the fluid, divided by the distance between the plates. The velocity gradient has the dimensions of 1/seconds. The velocity gradient, then, is the "flow" part of the force-flow relation. The "force" part is the *shear stress,* or force required to maintain the velocity gradient per unit area of the moving plate.

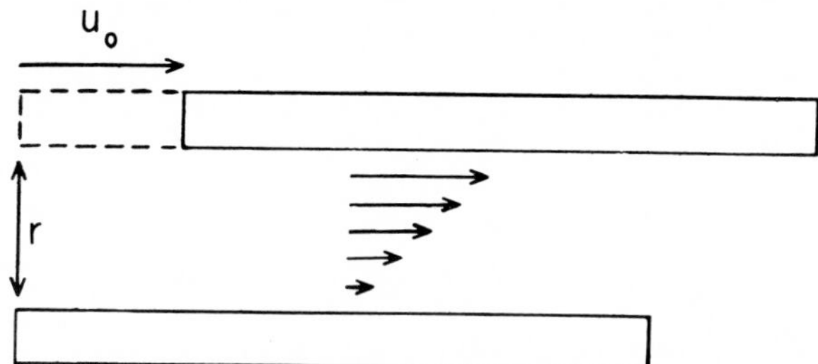

FIG. 11. A diagram showing the laminar flow of a Newtonian fluid separating two plates, one of which is in motion with velocity u (see text). The velocity gradient is du/dr.

For simple fluids at moderate stress, the force-flow relation is linear; such fluids are commonly referred to as Newtonian, because they obey Newton's hypothesis that the velocity gradient is directly proportional to shear stress. The proportionality constant is the *viscosity.* For many fluids, the force-flow curve is nonlinear; these are referred to as non-Newtonian and the reciprocal of the slope of the force-flow curve at some point in the stress axis represents the (variable) "*apparent viscosity.*" Figure 12 shows two hypothetical force-flow curves, one (*A*) for a Newtonian fluid, the other (*B*) for a non-Newtonian (pseudoplastic) one; the apparent viscosities of the two fluids might be equal at high shear stress in spite of differences in shape of the force-flow curve as a whole (consistency). It is for this reason that it is particularly important to make no assumptions regarding the shape of the force-flow curve for cytoplasm, for a low viscosity measurement can convey an entirely false impression concerning the consistency and therefore the "structure" (or especially the absence thereof) in cytoplasm.

One of the most favorable situations in which to study the consistency of a fluid is to observe its flow in a capillary. It is possible to construct a force-flow relation either from the rate of outflow from a capillary under different pressure gradients (by the Poiseuille-Hagen equation) or from the profile of the velocities exhibited by adjacent layers of fluid along the radius of the capillary. We shall consider here only the latter case, which has direct applications to cytoplasmic streaming in the ameba.

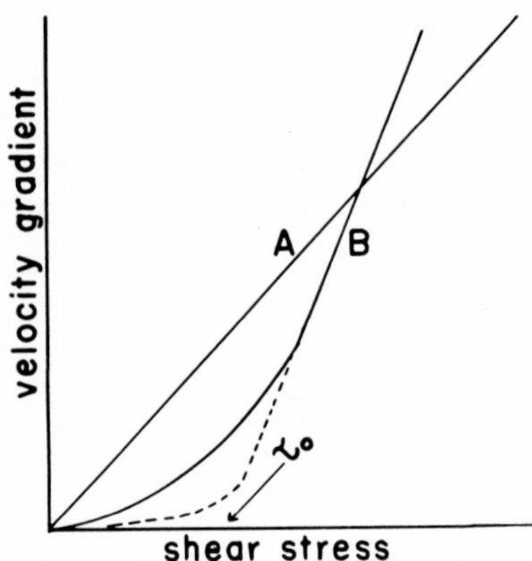

FIG. 12. Schematic force-flow curves for (*A*) Newtonian and (*B*) non-Newtonian fluids (see text).

When a fluid exhibits equilibrium flow, the pressure gradient applied to the cross-section area of the capillary balances the shear stress at the perimeter, so that

$$\tau 2\pi r = \left(-\frac{dp}{dx}\right)\pi r^2 \tag{1}$$

where τ is the shear stress (dynes/cm.^{2}), r is the radius, and dp/dx is the pressure gradient (dynes/cm.^{3}). In a Newtonian fluid, the shear stress is proportional to the product of the viscosity and the velocity gradient; that is, the viscosity is a proportionality constant. However, if we let μ represent a variable, the apparent viscosity, then the following expression of Newton's law applies to any fluid:

$$\tau = \mu \frac{du}{dr} \tag{2}$$

where du/dr is the velocity gradient (1/sec.). By substituting Eq. (2) into Eq. (1)

$$\frac{du}{dr} = \left(-\tfrac{1}{2}\frac{dp}{dx}\right)\frac{r}{\mu} \tag{3}$$

The integrated form of this equation provides the basis for interpreting the rheological significance of velocity profiles:

$$\int du = u_0 - u = \left(-\tfrac{1}{2}\,\frac{dp}{dx}\right)\frac{rdr}{\mu} \tag{4}$$

From Eq. (1) it can be seen that the shear stress in a capillary increases in direct proportion to the radius, from a minimum at the center to a maximum at the walls. Since the viscosity of a Newtonian fluid is independent of stress, its velocity profile in capillary flow is always a paraboloid of revolution as is shown by integrating Eq. (4) for the case in which the viscosity is taken to be constant ($\mu = K$):

$$u_0 - u = \left(-\tfrac{1}{2}\,\frac{dp}{dx}\right)\frac{r^2}{K} \tag{5}$$

On the other hand, a non-Newtonian fluid showing an inverse relation between stress and viscosity ($\mu = a/r$) would exhibit a flattened "plug-flow" velocity profile (Fig. 13):

$$u_0 - u = \left(-\tfrac{1}{2}\,\frac{dp}{dx}\right)\frac{r^3}{3a} \tag{6}$$

Such would be the case for a fluid exhibiting pseudoplasticity or structural viscosity, whether from false body (as in paints) or from thixotropy. Fluids with a definite yield point show a velocity profile which is not only flattened but truncated as well. This was pointed out by Kamiya and Kuroda (1958) in their analysis of the velocity profiles of slime mold

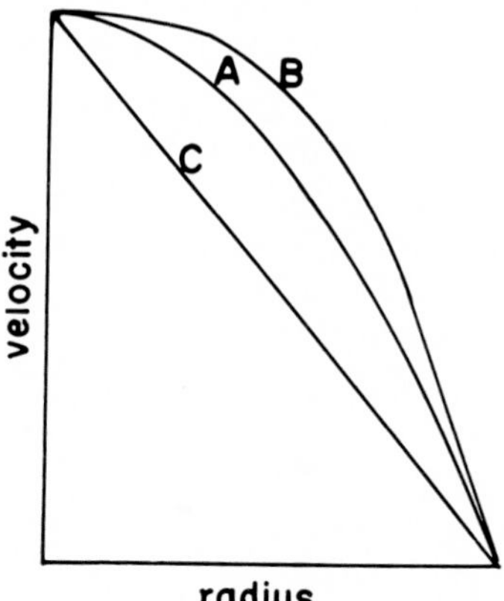

FIG. 13. Velocity profiles for capillary flow: *A*, Newtonian fluid (constant viscosity throughout the stream); *B*, a gradient of apparent viscosity from high at the center to low at the walls; and *C*, a gradient from a high viscosity at the walls to low at the center (see text).

endoplasm, which they correctly interpreted as indicating non-Newtonian behavior.

Even more specific rheological information about a fluid can be obtained from the velocity profile alone if the slope at each point in the radius is plotted as a function of the radius. Since the stress is a function of the radius (Eq. 1), and the slope of the velocity profile is the velocity gradient, such a plot is the force-flow relation of the fluid. From Eq. (1) it can also be seen that evaluation of the pressure gradient (if possible) permits the abscissa of the force-flow relation to be specified in terms of shear stress (dynes/cm.^{2}). This is an important step, for then the slope of the force-flow relation is the apparent fluidity, and its inverse is the apparent viscosity expressed in poise, the c.g.s. unit (water has a viscosity of 0.01 poise). It is only from the range of apparent viscosities of a fluid that we can gain any notion of its consistency when compared to that of common fluids such as water and glycerol.

2. *The Consistency of Ameba Cytoplasm*

a. Evidence from streaming. The "sol-gel" concept of ameba structure, which has played such a large part in theories of ameboid movement not only during the recent past but also in the last century (Section I, C, 2), seems to have been based largely on the observed fact that the endoplasm "flows" through a stationary tube of ectoplasm. Mast (1926) realized that solids and liquids are differentiated by the ease with which their spatial relations are changed and watched for signs that cytoplasmic inclusions "tumbled over one another" as evidence of fluidity. While Mast thus recognized the importance of deformation as a criterion of fluid consistency, he did not rigorously distinguish between deformation and displacement of the endoplasm, for he lacked present knowledge of the rheological significance of velocity profiles (Section II, C, 1). Mast (1926) believed on the basis of his observations that the entire endoplasmic stream had a fluid consistency, while the ectoplasm was more rigid, and he therefore assigned the somewhat prejudicial terms "plasmasol" and plasmagel" to the endoplasm and ectoplasm, respectively.

From Section II, C, 1, it is apparent that the velocity profile across a cylindrical pseudopod would provide the most meaningful data from which to gain information on cytoplasmic consistency. With this in mind, Allen and Roslansky (1959) have measured velocity profiles across relatively thin pseudopodia of *Chaos chaos*. These pseudopodia are practically circular in cross section; furthermore, the endoplasmic stream contains sufficient inclusions to serve as markers for the velocities of adjacent layers of fluid (Fig. 14). The velocity profile for the anterior half of the endoplasmic stream was found to be essentially plug flow; i.e., a flattened and often asymmetrically truncated parabaloid somewhat similar to that found by Kamiya (1950) and Kamiya and Kuroda (1958) for endoplasmic flow in a myxomycete filament. On the other hand, velocity profiles in the tail region were pointed rather than flattened; this is what would be expected if there were a gradient of consistency from the more fluid central region to the less fluid peripheral region. There are, however, several other possible interpretations for the pattern of flow in the tail. Allen and Roslansky constructed force-flow curves (Fig. 15) from their velocity profile data on the assumption that the motive force was a constant pressure gradient. The motive force (nature or site of action unspecified) was measured by applying an external pressure gradient by means of a capillary attached to a pressure screw and manometer. The range of pressures required to halt an advancing pseudopod of *Chaos chaos* 0.194 cm. in length was found to be 1.10–3.2 cm. of water. Thus the minimum average "pressure gradient" in the ameba was 5.6×10^3 dynes/cm.^{3}. This minimum value sets the lower limits of the apparent viscosities calculable from the force-flow relation. The apparent viscosities in the axial endoplasm were found to vary between 0.74 and 8.9 poise, with the highest values at the front and lowest in the tail. On the other hand, the apparent viscosity at the edge of the stream, where the shear stress is highest, was calculated as 0.44–1.5 poise in the anterior half, but as high as 9.3 in the tail. Although these values are approximate and are based on the assumption of a constant pressure gradient along the specimen, they serve as an indicator of the degree of "structure" present in the cytoplasm. The truncated velocity profile which is found on at least one side of each stream indicates a yield point, such as is characteristic of gels.

The exact interpretation of endoplasmic velocity profiles will have to await further information regarding the nature and site of action of the motive force (see Section III, A). One conclusion can be drawn from these data which is not dependent upon the correctness of the pressure gradient hypothesis; there can be no question that the endoplasm has a non-Newtonian consistency. In other words, *the endoplasm is not a structureless sol, and therefore it need not be entirely passive in ameboid movement.* In this connection, Allen and Roslansky (1959) made two observations that were not readily explained by the pressure hypothesis. First, the velocity profiles were not always symmetrical on the two sides of the stream visible under the microscope; this could mean either unequal consistency on the two sides or unequal application of the motive force, whatever its nature. It is difficult to imagine how pressure could

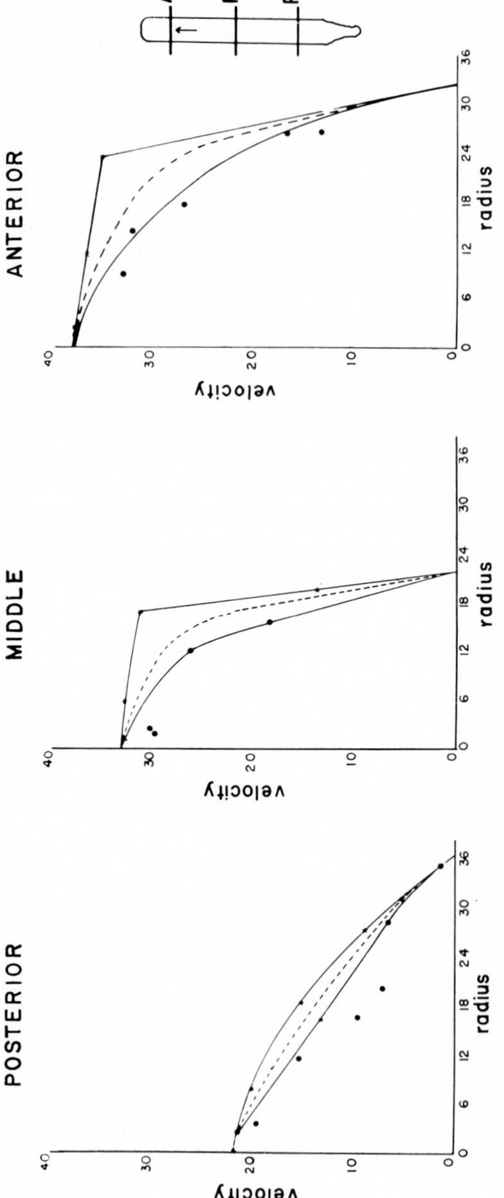

FIG. 14. Endoplasmic velocity profiles for the anterior, posterior, and middle regions of the giant ameba *Chaos chaos*. From Allen and Roslansky (1959).

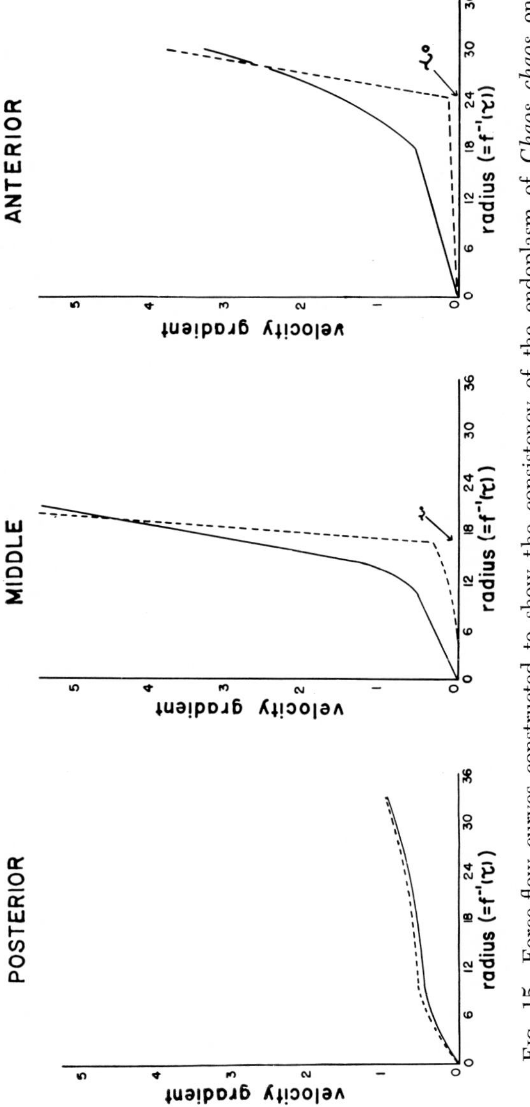

FIG. 15. Force-flow curves constructed to show the consistency of the endoplasm of *Chaos chaos* on the assumption of pressure-induced flow (see text). From Allen and Roslansky (1959).

be applied unequally. The second observation was that an externally applied pressure gradient did not exert so great an influence on endoplasmic streams shared by pairs of nearby advancing and retracting pseudopodia as would be expected from the assumption of a fluid endoplasm flowing under pressure. Instead, the endoplasmic stream behaved somewhat as a structural unit, for it could be diverted into a pseudopod perpendicular to it only by pressure gradients sufficiently high to cause the cell to burst.

The error in assuming that endoplasmic streaming is itself evidence of fluidity can be seen if the patterns of streaming in attached and unattached specimens are compared. In attached amebae, the endoplasmic stream diverges in all directions just posterior to the hyaline cap, and becomes motionless with respect to the environment; in becoming stationary, the cytoplasm of this region *appears* to become gelated. In amebae which are unable to become attached to the substratum, the streaming events inside the cell are the same as in an attached specimen, except that now the ectoplasm *appears* to "flow" backward and the forward movement of the endoplasm seems correspondingly slower. The pattern of flow corresponds exactly to the "fountain streaming" which played such an important role in the arguments regarding the surface tension theory. In both attached and unattached specimens, ectoplasm is continually formed from the endoplasm; the only difference is that attachment causes the ectoplasmic tube to be extended forward. Since only the velocity gradients in the ameba give any information regarding consistency in different regions, whatever is true regarding the consistency of attached amebae must also be true for unattached ones, provided the motive force and the pattern of streaming remain exactly the same in the two cases, as appears to be the case. Figure 16 has been constructed schematically to show how the normal pattern of flow in an attached ameba would be expected to be modified if attachment were prevented. Judging from the account of Mast (1926, p. 400-401) of streaming without locomotion and from similar observations of our own, it seems nearly certain that the rate and pattern of streaming are identical in the two cases. It is difficult at present to make quantitative comparisons because no satisfactory way has yet been found to prevent attachment. Schaeffer (1920, p. 11-12) placed amebae in gelatin solutions; we have had somewhat more success with quartz capillaries and silicone-treated slides, but none of these procedures gives uniform results. There appears to be no doubt that poor attachment to the substratum was the cause of so many early observations of "fountain streaming" (e.g., Schulze, 1875; Rhumbler, 1898; Rand and Hsu, 1927). Movement of this kind is often a sign of poor culture conditions; amebae which have been deprived of food are, as a rule, weakly attached to the substratum (Griffin, 1959).

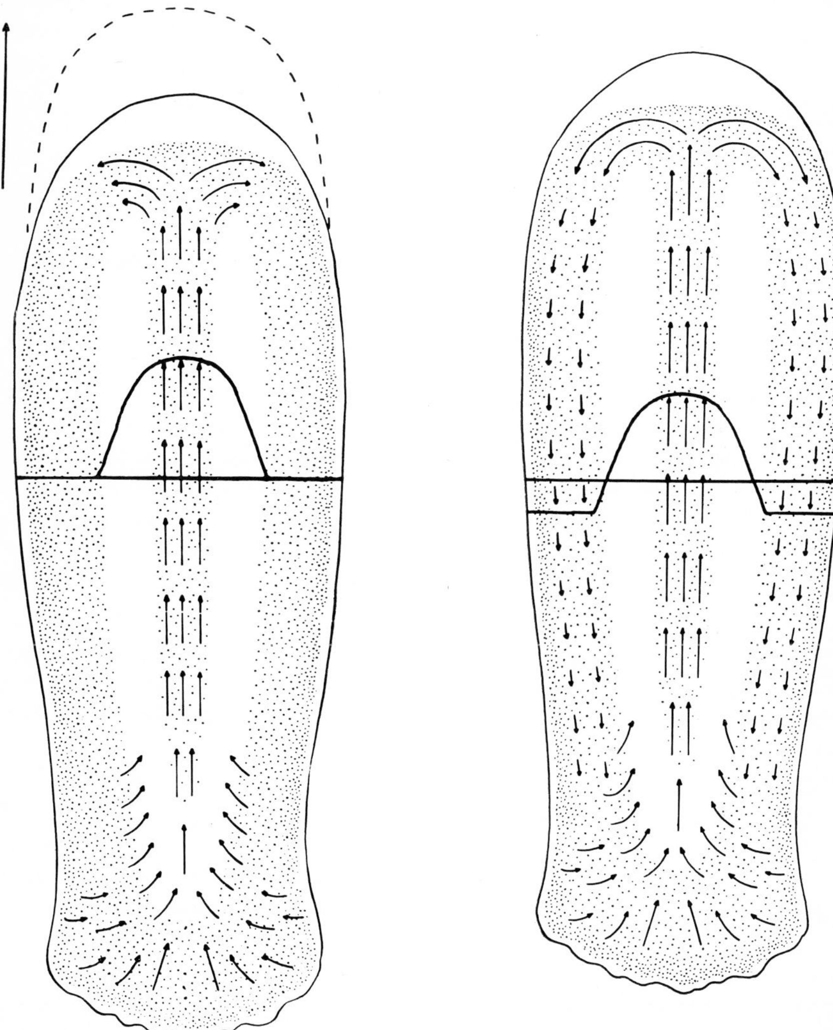

FIG. 16. A schematic diagram to show the relation between streaming in attached (progressing) and unattached (fountain-streaming) amebae (see text).

b. Evidence from Brownian motion. Dellinger (1906) was apparently the first to point out that the form assumed by an ameba was incompatible with the view that the cytoplasm was sufficiently fluid to be affected by surface tension. Dellinger's contention was later elegantly supported by the hydrostatic pressure experiments of Marsland and Brown (1936), who showed that liquefaction of the cytoplasm in an elongated pseudopod caused the ameba to become spherical or even break up into many small droplets under the influence of surface tension.

Mast (1926) was also of the opinion that the ability of an ameba to send out an elongated pseudopod was ample evidence that at least part of the cytoplasm, the ectoplasm (plasmagel), was a gel. He was surprised, then, to find active Brownian motion in all parts of the ectoplasm. According to Mast's (1926, p. 379) account: "If a given granule in the plasmagel is carefully observed, it is found that, while it may move continuously for an apparently indefinite period of time, it does not progress beyond the boundaries of a very small area. Similar granules suspended in water are not thus limited in their movements." Mast recognized that two interpretations were possible. (1) The ectoplasm might consist of a fibrous network with fluid interstices; that is, a system of two continuous phases. (2) The ectoplasm might be a two-phase system with one of the phases enclosed in vacuoles; that is, an alveolar structure. Because Mast had seen crystals and some other inclusions in vacuoles (see Section II, A, 4, b), he was inclined to believe that the cytoplasmic structure was basically alveolar, and his figures were drawn interpretively according to this view (cf. Fig. 17). While the interpretation of Mast's observations of restricted Brownian motion is still an open question, there seems now to be more justification for the view that two continuous phases exist. The crystal vacuoles and some (but not all) other vacuoles that Mast saw were probably induced by compression. It is now well known that the hyaline fluid, which has a lower refractive index than the rest of the cytoplasm, can be filtered through gelated regions of the cytoplasm (Landau *et al.*, 1954; Allen and Roslansky, 1958; Goldacre, 1952b); for this reason, the ectoplasm may be more like a sponge than a gel packed with alveoli. The phase contrast microscope makes it possible to see small vacuoles very easily in pseudopodia (Fig. 5); no such vacuoles can be seen around individual mitochondria or alpha granules which exhibit restricted movement. According to Arena (1943), the ground cytoplasm of the ameba is not alveolor after removal of the larger alveoli by centrifugation.

Mast also made an attempt to watch Brownian motion in the endoplasm. The smallest particles, which contribute the most interesting information regarding cytoplasmic consistency, would be impossible to observe because of their rapid displacement and because of optical interference from light scattering above and below the stream. Mast was able to watch individual crystals in the stream and noted that some were in pronounced motion within their vacuoles and others appeared to be held fast. Under the conditions in which we have observed crystals in the endoplasmic stream, they have not been surrounded by vacuoles.

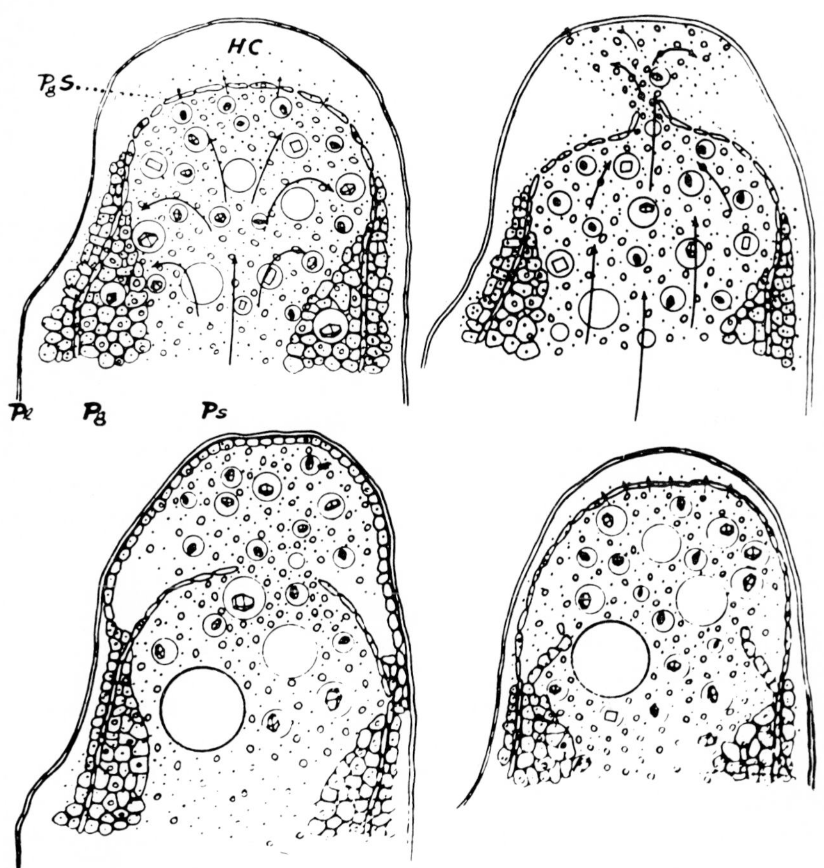

FIG. 17. Mast's (1926) interpretation of events in the anterior portion of an advancing pseudopod. The plasmagel (*Pg*) drawn with an alveolar structure to account for restricted Brownian movement (see text). The "plasmagel sheet" (*PgS*) was also postulated to account for the failure of "plasmasol" (*Ps*) to mix with the fluid of the hyaline cap (*HC*).

There is a marked difference in their Brownian motion in the axial endoplasm and when isolated in the medium; in the former it is negligible.

It is doubtful whether a quantitative study of Brownian motion in ameba endoplasm is technically feasible, or whether the results of such a study could be interpreted in rheological terms. One reason is that the heterogeneity which Mast's observations indicated, might be expected to be present in the endoplasm as well as the ectoplasm. The careful study of Brownian motion in *Spirogyra* cytoplasm by Baas Becking *et al.* (1928) showed quite clearly that heterogeneity and nonrandom movements of particles can give rise to a wide range of spurious "viscosity" values. Since the various equations for determining viscosity from quantitative measurements of Brownian motion require the fluid to be both homogeneous and Newtonian, it is necessary to know first whether these requirements are satisfied. The observations reported in this and the preceding sections indicate that in cytoplasm they are not. A further complication for the study of Brownian motion in streaming cytoplasm is that particles sometimes exhibit sudden, unexplained translatory movements as if pulled for a short distance by contractile fibers. This is a common observation in amebae, especially during a change in the direction of movement. Jarosch (1956) has devised a simple method for determining whether the movements of a particle are truly random (Brownian) or directed (agitation streaming), and has applied this method to the early steps in the organization of cytoplasmic streaming in cells of the inner epidermis of *Allium cepa.* Stewart and Stewart (1959) also have described directed movements of single inclusions in quiescent slime mold cytoplasm. They calculated the "viscosity" of the endoplasm and of the channel wall protoplasm of the slime mold from observations of Brownian motion and found values of "0.02–0.06 poise" for both regions despite the fact that the channel wall cytoplasm is a gel. Since these values conflict with the information available on the consistency of slime mold endoplasm from velocity profiles and since it does not seem likely that the ectoplasm and endoplasm of slime molds should have such low and identical viscosity values, it seems more reasonable to interpret the rapid Brownian motion as occurring in local regions of low viscosity in the interstices of a spongy gel. There is also another possibility: that the movements which were thought to be Brownian were less obvious examples of the directed movement which they also described.

Apparently only one serious attempt has been made to study quantitatively the consistency of ameba cytoplasm by Brownian motion. Pekarek (1930) measured Brownian motion of inclusions in the cytoplasm of compressed, immobilized, vitally stained cells. In view of the fact the neutral red is somewhat toxic and induces vacuoles and artifacts (the "neutral red bodies") (Andresen, 1944; Torch, 1959), it is not surprising that low viscosity values were obtained. It is not completely certain whether the measurements themselves apply to immobilized cytoplasm or to the contents of induced vacuoles.

c. Evidence from centrifugation. Some of the earliest attempts to gain rheological information regarding ameba cytoplasm were made by centrifugation. Heilbrunn (1929a, b) centrifuged *A. dubia* and found that three seconds' exposure to an acceleration of 128 *g* was sufficient to sediment the cytoplasmic crystals into the centrifugal half of the cell. Assuming the cytoplasm to be Newtonian, Heilbrunn (1929b) calculated the "protoplasmic viscosity" from a modified form of Stokes' law and obtained a value about twice that of water. While the value calculated was not claimed to be more accurate than the second digit (0.019 poise), several assumptions were involved in the selection of material, in the use of the falling sphere method in a heterogeneous fluid, and in the estimate of densities, all of which proved to be incorrect (cf. Allen, 1960).

Until the invention of the centrifuge microscope by Harvey and Loomis (1930), only after the rotor had stopped was it possible to observe the effects of centrifugation. By building the objective and tube (with two prisms) of a microscope into a centrifuge rotor, Harvey and Loomis made it possible for later workers to make clear observations on cells subjected to a wide range of centrifugal accelerations.

Harvey and Marsland (1932, p. 85) used this instrument to observe the displacement of inclusions in amebae (*A. dubia* and *A. proteus*) under centrifugal acceleration: "The heavy crystals of *Amoeba dubia* always fall in 'jerks' even when moving through a visibly clear field. They move and stop and move and stop as if they met invisible obstructions. The same is true of *A. proteus.* This behavior is marked and apparent in all parts of the animal and suggests that the discontinuous movement must be due to a structure in the cytoplasm rather than to adherence of the crystals as they move along a sticky external surface. Frequently, a stream of heavy granules will flow through what must be a small channel."

We have recently (Allen, 1960) confirmed and extended the results of Harvey and Marsland (1932). Three species of amebae (*A. dubia, A. proteus,* and *Chaos chaos*) were subjected to accelerations applied both suddenly and gradually. When the acceleration was applied gradually, it was found that individual inclusions broke away suddenly and exhibited discontinuous displacement. The range of accelerations in which this occurred differed slightly for each of the three species because of differences in the size of inclusions. Those inclusions in the posterior half were more readily displaced than those in the anterior half, and the endoplasmic stream showed evidence of presenting a barrier to the displacement of crystals. *A. proteus* showed the astonishing capability of resuspending crystals in its axial endoplasm under centrifugal fields of up to 170 *g*. At this acceleration the reduced weight of each crystal is six to seven times that of an equal volume of gold at gravity.

Sudden acceleration was applied by means of a centrifuge microscope rotor driven by a heavy flywheel through a magnetic clutch arrangement. With this centrifuge microscope it was possible to achieve a constant terminal acceleration of as much as 400 *g* within a fraction of a second. When a sudden acceleration of 225 *g* was applied to a centrifugally moving specimen of *Chaos chaos*, it was found that the displacement of nuclei and cytoplasmic inclusions was most rapid in the shear zone and hardly noticeable in the axial endoplasm or in the ectoplasm. Specimens moving perpendicularly in relation to the axis of acceleration exhibited a greater degree of stratification in the posterior part than in the anterior.

In view of these observations, it appears doubtful whether centrifuge data can be used to express the consistency of the cytoplasm in numerical terms (even as a range of apparent viscosities). Each inclusion has a slightly different size, and different classes of inclusions have different densities; thus the shear stress on each inclusion reaches the yield point (or critical shear stress) of the cytoplasm at a slightly different time. The result is that the interior of the cell is exposed to a more or less continuous "hailstorm" of falling inclusions which break away over a period of several minutes through the range of accelerations in which the shear stress exerted by each inclusion exceeds the cytoplasmic yield value. The densities of the various cytoplasmic inclusions are not known except for the cytoplasmic crystals. These crystals have a density of 1.745 ± 0.005 in the platelike crystalline form, which Griffin (1959, 1960) has shown to be carbonyldiurea. If the ground substance is assumed to have a density of 1.01 (4% protein), then the shear stress of each crystal can be estimated by treating it as a sphere in a homogeneous fluid:

$$\tau = \frac{r\,(0.73)cg}{3}$$

where r is the radius, c the centrifugal acceleration (in gravities), and g is the gravitational constant (980 dynes/sec.^{2}). If the fluid were not homogeneous, but a gel of some sort, the stress would be applied more nearly to the cross-sectional area of the sphere instead of over its entire surface, and would be perhaps four times as large. Since the crystals are not spherical, the shear stress they exert would depend also on their orientation.

Because the crystals of the various species studied vary in size, the acceleration at which they break away and fall through the cytoplasm is slightly different. The stress at which crystals begin to fall was calculated to be of the same order of magnitude as the yield stress observed in velocity profiles by Allen and Roslansky (1959), or about 5 dynes/cm.^{2}. This is approximately the minimum stress exerted by a crystal corresponding to a sphere with a radius of 2 μ at 100 *g*.

The results of studies with the centrifuge microscope (Harvey and Marsland, 1932; Allen, 1960) leave little doubt that the greater part of the cytoplasm of the ameba possesses a rheological consistency compatible with the view that tenuous and labile "structure" is present, and incompatible with the view recently reiterated by Heilbrunn (1958) that the "viscosity" of ameba cytoplasm is in the range of a few centipoise. There is no evidence to support the latter view.

d. Evidence from heavy spheres in the cytoplasm. Allen and Griffin (1960) have observed that *A. dubia* (and sometimes *A. proteus* or *Chaos chaos*) will ingest spheres of polystyrene, glass, oils, iron, gold, and even mercury (Fig. 18). Of these, gold spheres are heaviest and are nontoxic and therefore offer the greatest opportunity of studying the fall of particles exhibiting a range of rather low shear stresses. Many amebae can ingest 50–100 or even more gold spheres of all sizes from 1 to 26 μ in diameter without much change in their normal pattern of locomotion. Those amebae that were studied contained at most only a few spheres, usually of different sizes; these were originally ingested in food cups, but the fluid was soon taken into the cytoplasm leaving the particles presumably surrounded by a tight membrane.

The behavior of gold spheres in ameba cytoplasm provides a clear test for the correctness of the results of centrifugation and velocity profile analysis, for if the cytoplasm possessed a low, Newtonian viscosity such a sphere would be heavy enough to "rattle around" in the cell whenever the latter was turned on the rotating stage of a horizontal microscope. On the other hand, a non-Newtonian fluid exhibiting pseudoplastic flow would support the smaller spheres, allow some discontinuous movement of the somewhat larger spheres that exert a shear stress close to the yield value of the fluid, and would offer much less resistance to the fall of the largest spheres.

The behavior of gold spheres in *A. dubia* was found to be exactly according to expectation on the basis of the data from velocity profile analysis and centrifugation (see previous sections). Spheres smaller than 7 μ in diameter rarely exhibited any fall, and then only at the edge of the endoplasmic stream (shear zone). Those between 7 and 8 μ in diameter exerted sufficient shear stress to exhibit occasional and limited discontinuous displacement in the tail region or in the shear zone, but were clearly prevented from falling across the axial endoplasm or into the ectoplasm. It has frequently been observed that even considerably

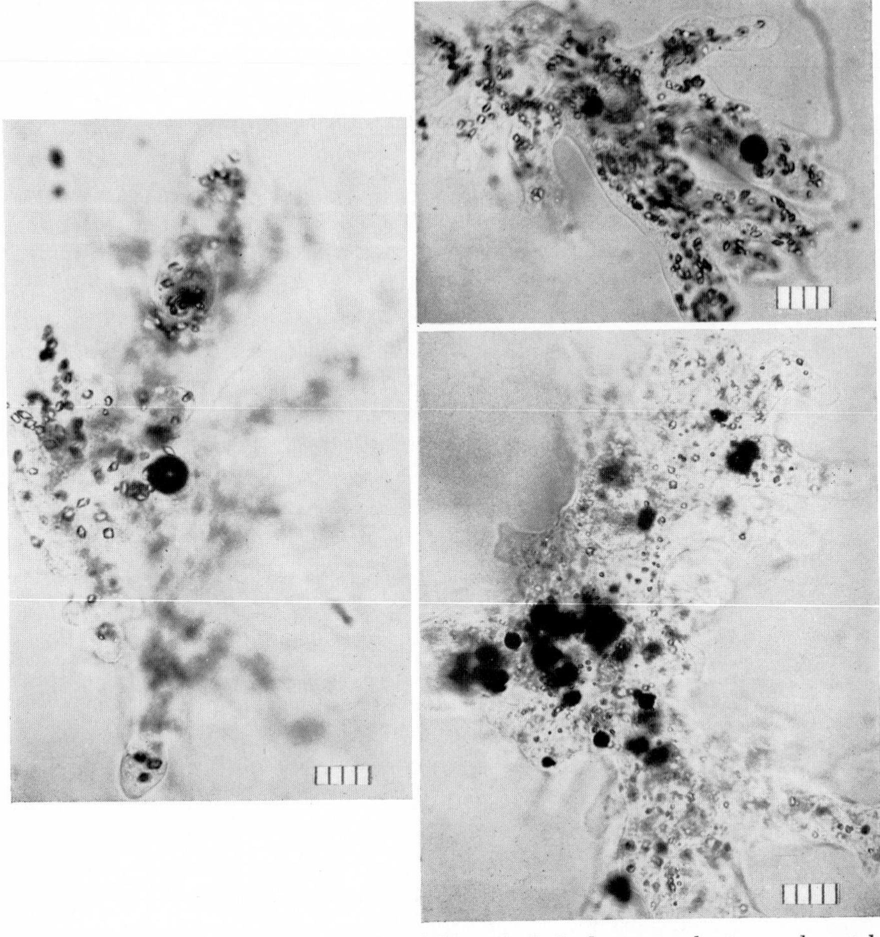

FIG. 18. Photographs of *Amoeba dubia* which had previously ingested metal spheres. Specimen shown in the left and upper right frames contained gold spheres; that in the lower right frame contained spheres of carbonyl iron. Scale = 10 μ.

larger (10–16 μ) spheres may ride along on top of the axial endoplasm, apparently somehow trapped in the shear zone and unable to fall, even obliquely, through the axial endoplasm. When such fall does occur, it is difficult always to be certain whether the particle has passed through the axial endoplasm or around it in the shear zone. In most cases it appeared to be the latter.

One situation in which the relation of cytoplasmic structure to ameboid movement was revealed might be mentioned. The following event was filmed and has been observed and analyzed several times; it is typical of several observations that were made on spheres considerably larger than 7–8 μ in diameter. In this instance, a sphere 24 μ in diameter exhibited discontinuous fall in the endoplasm of a cell that was being slowly rotated in order to keep the sphere in motion. During one of its discontinuous passages across the cell, it evidently brought about some form of stimulation, for the typical rounding-up response of the cell was elicited. During the 20 seconds when the cell was rounded and motionless, the sphere fell back and forth freely and rapidly. However, just *before* streaming resumed, the cytoplasm rather quickly developed an increased resistance to the movement of the sphere and the sphere soon resumed the slower, discontinuous fall it had exhibited before stimulation had occurred. The almost inescapable conclusion from this observation is that the "structure," which can be inferred to develop in the cytoplasm, is closely associated with the ability of the cell to move.

There is other evidence to suggest a close relationship between "structure" and movement. *A. dubia* rendered immobile by 0.02 *M* KCl become spherical, and their crystals sink to the bottom of the cell within a few minutes (Heilbrunn and Daugherty, 1931). Griffin (unpublished) has observed the gradual settling of crystals in the tail region of amebae which were forced to assume a monopodial form in a small capillary; such amebae gradually lose their ability to move, and streaming comes almost to a halt. The fact that crystals fall under these conditions indicates that the consistency of the cytoplasm has become altered and exhibits a rather low apparent viscosity even at the low shear stress exerted by the crystals. The apparent viscosity is still considerably higher than Heilbrunn (1929b) estimated, for the crystals are much more dense than he assumed (1.74 as opposed to 1.1).

Yagi (1959) has recently measured the (apparent) viscosity of ameba endoplasm by moving nickel particles magnetically and found a general inverse relationship between streaming rate and apparent viscosity, with values between 12.0 and 0.09 poise. Considering the difficulties inherent in applying Stokes' law to intracellular particle displacements (see Allen, 1960), these data are in rather good numerical agreement with the analyses of endoplasmic velocity profiles by Allen and Roslansky (1959). It is probable that some of the variability in Yagi's data were due not only to differences in streaming rate, but to variations in the velocity gradients in the various regions in which the nickel particles were moved. A student in the author's laboratory, Mr. W. Wallin (1960), succeeded in demonstrating elastic recoil on the part of most of the endoplasm of *A. dubia* when ingested iron particles were displaced by means of an electromagnet. Recoil was not due to the stretching of the vacuolar membrane surrounding the particle.

e. Evidence from hydrostatic pressure experiments. Hydrostatic pressure has been shown to be an extremely useful tool in the study of the relationships between structure and function in cells. It follows from Le Chatelier's principle that gels which increase in volume and absorb heat when setting should be solated (liquefied) by hydrostatic pressure. Protoplasmic gels are of this type (Marsland, 1942). It should be emphasized, however, that if ameba cytoplasm is contractile, it should be expected to behave similarly to muscle, which is thrown into a contracted state by hydrostatic pressure (Brown, 1957). Since the contractile elements in ameba cytoplasm are certainly less concentrated and presumably less organized than those of muscle, it might be expected that hydrostatic pressure would first cause contraction and then solation. Marsland and Brown (1936) showed that cytoplasmic streaming both in amebae and in plant cells can be arrested by hydrostatic pressures of about 5000 lb./in.^{2} Probably the effect of pressures of this magnitude is general solation as is indicated by the centrifugation data of Landau *et al.* (1954). Small pressure increments cause effects which are subject to alternative interpretations: at first pseudopodia become thinner, but then develop bulbous tips. Assuming this to be the action of surface tension on solated cytoplasm, Marsland and Brown (1936) pointed out that the "liquefying action" most readily affected newly formed pseudopodia; this interpretation was in line with Mast's (1926) impression that there was a gradient of ectoplasmic rigidity from the more rigid tail region to the less rigid front. However, as was pointed out above, recent centrifuge-microscope data indicate a rigidity gradient opposite in direction to that proposed by Mast (Allen, 1960). Therefore, the hypothesis that the bulbous tips are the result of cytoplasmic contraction seems at least plausible.

f. A generalized view of ameba cytoplasmic structure: a proposed terminology. While the information available on the cytoplasmic consistency in various regions of the ameba is still incomplete, it appears quite certain that the "gel tube filled with sol" concept of ameba structure is incorrect. Therefore, it is suggested that the "plasmagel, plasmasol" terminology of Mast be abandoned in favor of terms that denote only position and function and carry no connotations regarding consistency.

It is proposed that the older terms *ectoplasm* and *endoplasm* be used instead (Rhumbler, 1898; Schaeffer, 1920). The latter should be subdivided into *axial endoplasm,* where little if any velocity gradient is developed, and the *shear zone,* where a steeper velocity gradient is found. It is further proposed that the regions of interconversion of ectoplasm and endoplasm be given the names *fountain zone* for the anterior portion where the endoplasm splits and becomes everted to form the ectoplasm, and the *zone of recruitment* for the region in the posterior part of the cell where the endoplasmic stream is recruited from stationary ectoplasm.

The terms hyaline cap, hyaline ectoplasm, and plasmalemma should be retained, but there appears to be no further need for, or significance to, the "plasmagel sheet," which Mast (1926, 1931a) postulated to account for the failure of the granular endoplasm of the shear zone to mix with the hyaline cap. Both materials were assumed to be fluid.

The concept of ameba cytoplasmic structure which emerges from the previous sections is that represented in Fig. 19. It appears that not only the ectoplasmic layer, but also the axial endoplasm possesses gel structure. While the data for comparing the relative strengths of the gel in these regions are not extensive, it would appear that the axial endoplasmic "gel" is quite tenuous, but nevertheless continuous with the probably stronger ectoplasmic gel through the granular cytoplasm of the fountain zone. There are some indications that the tail region recruitment zone is more fluid than the axial endoplasm; it is for that reason that the shear zone and recruitment zone have been drawn as continuous in Fig. 19; this may be an oversimplification. Figure 20 shows the actual appearance of the uroid, hyaline cap, endo- and ectoplasm of *Amoeba discoides.*

3. *Optical Properties of the Cytoplasm*

a. Refractive index. Amebae are not suitable objects for study with immersion refractometry because the usual immersion media induce either osmotic changes or pinocytosis. Interference microscopy, however, has been used to measure refractive index differences in different parts of actively moving amebae (Allen and Roslansky, 1958). The average refractive index in the posterior portion of *A. proteus* was found to be 1.3401, corresponding in dry mass to a protein concentration of about 3.95%; for the anterior region the lower values of 1.3391, and 3.39% were found. Although considerable scatter was found in the data, it appears certain that a gradient in refractive index arises as a result of movement, and not as a result of contractile vacuole function. The fringe-in-field photograph in Fig. 21 clearly shows that the measurements were in the first order, so that the recent data of James (1959) must be reinterpreted.

Ameba ground cytoplasm appears perfectly homogeneous in phase and interference microscopes, except for obvious vacuoles and inclusions: there are no indications of alveolar structure as postulated by

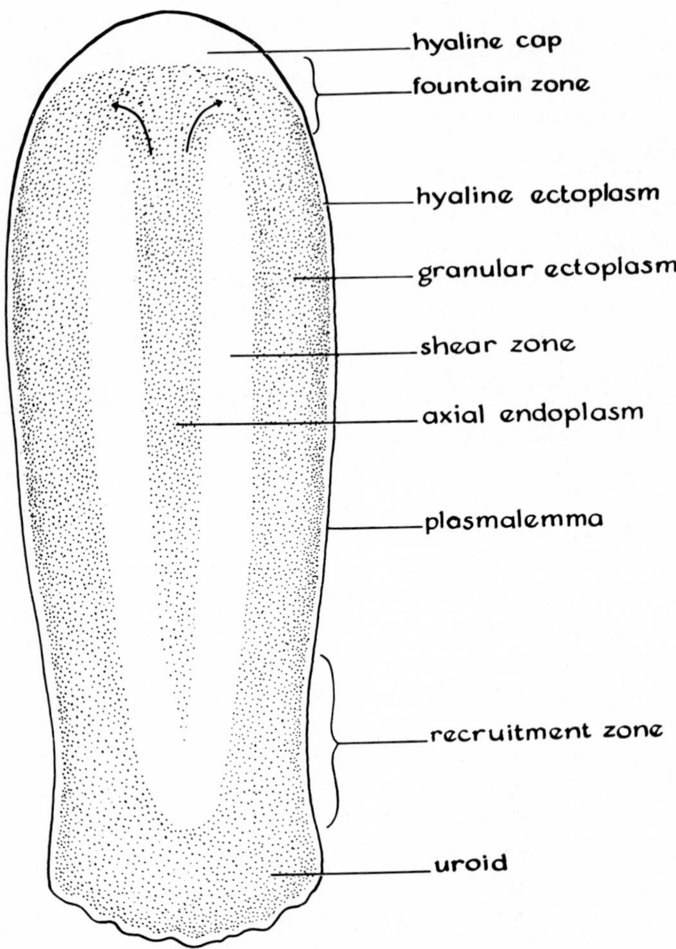

FIG. 19. A proposed terminology for different regions of the cytoplasm of the ameba (see text).

Mast (1926). Vacuoles can be induced easily by rapid compression or other kinds of injury; in one favorable case the refractive index of two such injury vacuoles was found to be equivalent to a 1.15% protein solution. The refractive index of the hyaline cap fluid is probably still lower, equivalent to about 1% protein or a little less (Allen and Roslansky, 1958). The mechanism by which injury vacuoles form is not known; a likely possibility is that a tear in the structural phase results in the collection of interstitial fluid, which is somehow partitioned off and later resorbed. The ultracentrifugation data of Marshall *et al.* (1959)

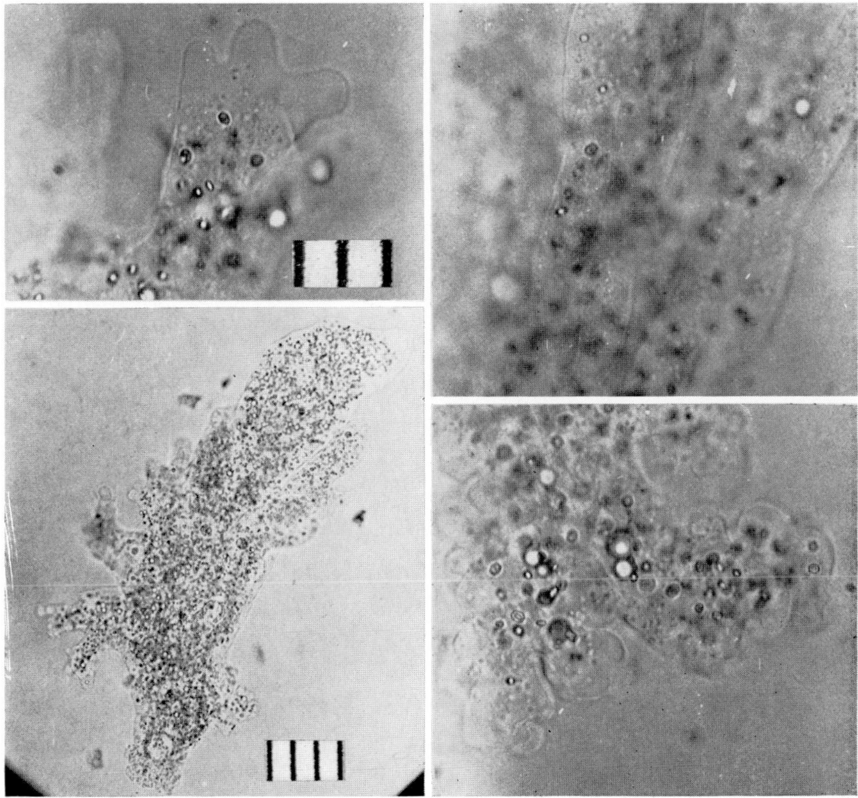

FIG. 20. Photographs of *Amoeba discoides* showing: (lower left) the general shape of the pseudopodia, the nucleus, and the uroid; (upper left) the hyaline cap region of an advancing pseudopod; (upper right) ectoplasmic ridges; and (lower right) some of the many "villous projections" on the uroid. Crystals and heavy spherical bodies are clearly visible in all frames. The smaller and less distinct inclusions are mitochondria. Scale = 10 μ.

suggest that the "expressed plasmasol," which contained only about 2% protein, may have lacked at least some of the structural proteins responsible for movement. However, this approach can be expected to yield much valuable information, for little is known at present regarding the proteins of the ground substance of amebae.

The ability of ameba cytoplasm to form a hyaline cap during movement, or a large hyaline zone on decompression from hydrostatic pressure (Landau *et al.*, 1954) or on treatment with anesthetics (Goldacre, 1952b), together with its propensity for forming hyaline injury vacuoles, leave us

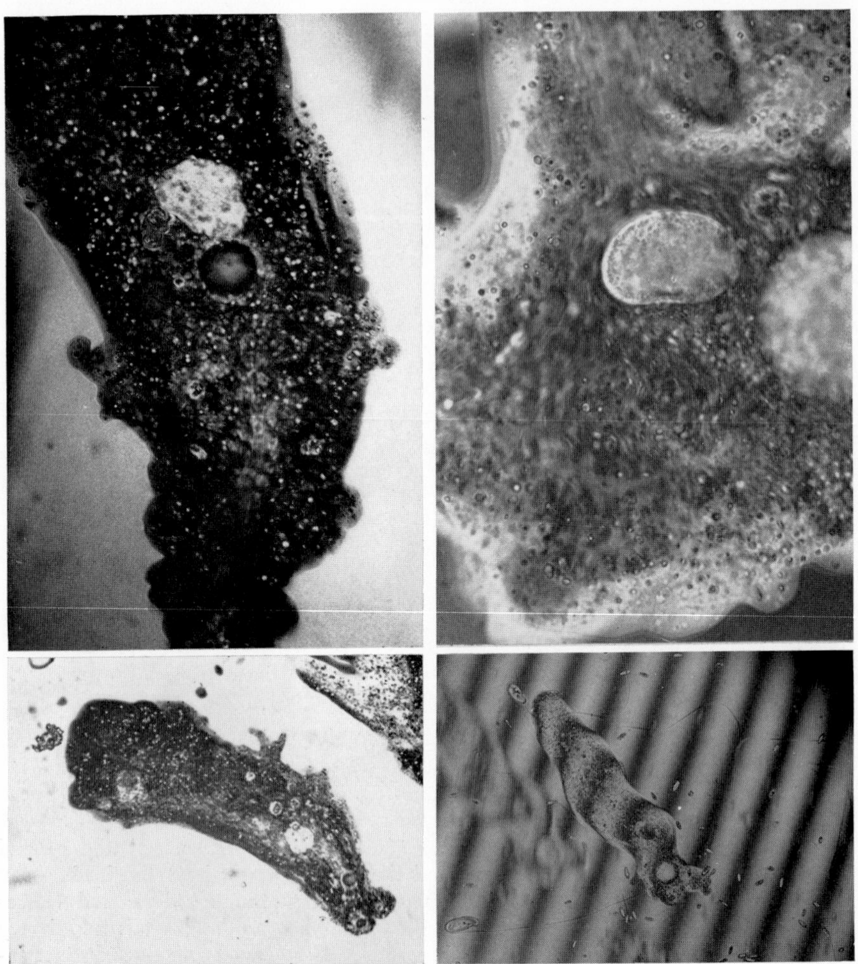

FIG. 21. *Amoeba proteus* photographed with an American-Optical-Baker Interference Microscope. Upper and lower left: flattened specimens showing the gradient in refractive index reported by Allen and Roslansky (1958). Upper right: a large uncompressed specimen showing the nucleus with its peripherally located "nucleoli." Lower right: a fringe-in-field photograph of a small *A. proteus* to show that the retardation is in the first order. The fringes were brought into the field by placing a drop of water on top of the coverglass.

to conclude that the cytoplasm consists of at least two phases which are probably continuous (i.e., not separated into alveoli). The mobility of these phases with respect to one another may be an important factor in the mechanics of ameboid movement.

b. Birefringence. Birefringence is clearly visible in the hyaline ectoplasm and nuclear membrane of *Ameoba verrucosa* (Schmidt, 1939) and in the surface of other amebae (Mitchison, 1950). While the pseudopodia of tissue culture cells and snail amebocytes and the filopodia of foraminifera show positive axial birefringence, the only birefringence visible in the larger amebae is that apparently due to the membrane (Mitchison, 1950). Failure until recently to detect birefringence along the axis of an ameba pseudopod can be ascribed to the light scattering of numerous inclusions (crystals, refractile bodies, etc.), which prevents the eye from distinguishing the subtle intensity differences between background and object, even with a sensitive compensator. Because the total concentration of proteins in the ground substance is only about 4%, very little birefringence could be expected in the ground cytoplasm even if the proteins were highly oriented. Consequently, the organism in which to look for birefringence would be *Chaos chaos,* the pseudopodia of which may be from 100 to 175 μ thick. Allen (1958) found evidence of weak pseudopodial birefringence by a simple photometric method employing a rotating compensator to feed symmetrical birefringent signals to a recording photomultiplier circuit. Asymmetry of signals caused by the introduction of an ameba pseudopod into the field could be compensated by retardation of from 1 to 10 A. Taken as a whole, then, the *average* birefringence of a pseudopod does not exceed 2×10^{-5}. More recent investigations, however, have succeeded in localizing the birefringent material and establishing the probable nature of birefringence. (Allen and Maddux, 1960). *Amoeba proteus* oriented at 45° from the plane of the polarizer were driven back and forth on a hydraulic scanning stage so that a short, narrow slit focussed in the object plane and transmitting elliptically polarized light scanned across successive regions of the cell from front to back. The retardation at each point scanned was recorded as a ± D.C. voltage by a null-point synchronous detector analyzing the symmetry of signals from an electron-optical light modulating crystal driven at 10 kc. Sensitivity of the instrument was 1 volt/A. By comparing the position, within the pseudopodial cross section, of the maximum retardation with the maximum thickness of the ectoplasmic tube, shear zone, and axial endoplasm, it was possible to show that birefringence in the anterior part of a pseudopod was primarily localized in the shear zone. Since significant velocity gradients are developed in this region

(Section II, C, 2 and Fig. 14), this is probably flow birefringence. The most highly birefringent region in the cell, however, was found to be the tail endoplasm. This was not expected, for the velocity gradients in this region are much lower than those in the shear zone (Fig. 14). However, birefringence in this region is explained satisfactorily as due to strain by the fountain zone contraction theory (Section III, E, 2).

4. Details of Cytoplasmic Streaming in the Ameba

a. The path of single inclusions. If an inclusion in the axial endoplasm is followed, it is found to pass rapidly to the fountain zone, where it is deflected toward the periphery and becomes stationary in relation to the environment; there it remains until the ameba has moved roughly its own length (about 2 minutes for *A. proteus* and *A. discoides*), whereupon it re-enters the endoplasmic stream (Fig. 16). Inclusions first observed in the shear zone behave somewhat differently; they may tumble about and remain there, or may become more or less closely associated with either the ectoplasm or the axial endoplasm. Similarly, Mast's (1926) observations that single inclusions or even pieces of "plasmagel" may break off from the ectoplasm and join the endoplasmic stream (i.e., in the shear zone) indicate that the inclusions in the shear zone are probably derived from the edges of both the axial endoplasm and the ectoplasmic tube.

b. The fountain zone and hyaline cap. The pattern of streaming in the fountain zone is complex and variable and therefore difficult to analyze even in monopodial specimens. Of several fragmentary accounts, Mast's (1926 and 1931a) description was the most extensive, but expressed in terms of an erroneous concept of ameba cytoplasmic structure. Because it was then believed that the entire endoplasm had the consistency of a sol, the existence of a "plasmagel sheet" had to be postulated in order to account for the failure of the "plasmasol" and hyaline cap fluids to mix. Mast and Prosser (1932) interpreted spurts of endoplasmic flow through the fountain zone as a periodic bursting of the plasmagel sheet, as shown in Fig. 17.

At the time that Mast and his co-workers studied ameboid movement, it seemed quite straightforward that the endoplasm merely poured out from the mouth of the ectoplasmic tube where it became gelated. It was noted that the thickness of the "sol" and "gel" layers changed relative to one another under various experimental conditions. The "gel/sol ratio" was found to be about 1.3 at the anterior portion of the cell at 22–23°C. (Mast and Prosser, 1932).

The presence of plug flow and structure in the axial endoplasm forces us to re-examine the events occurring in the fountain zone and to reinterpret some of the observations of Mast and his co-workers. There seems to be no clear evidence that a structure such as the "plasmagel sheet" really exists. Instead, it may be more simply assumed that the granular cytoplasm, when it contracts, sends out a fluid by the well-known process of syneresis (see Section III, E, 1).

The pattern of cytoplasmic movement within the ameba appears to be identical in attached (progressing) and unattached (fountain streaming) specimens (see Section II, C, 2, a, and Fig. 16). From this streaming pattern it is quite clear that the endoplasm becomes everted in the fountain zone. There is no indication of a layering process, i.e., radial movement in the ectoplasmic tube, after the eversion has taken place.

If simple eversion of the endoplasmic stream were all that occurred in the fountain zone, then the cross-sectional areas of the endoplasmic stream (A_s) and ectoplasmic tube (A_t) would be equal, and the "gel/sol ratio," A_t/A_s, would be equal to unity. In fact, however, the A_t/A_s ratio is always greater than unity, and varies in an interesting manner when the physical and chemical environment is altered. For example, the A_t/A_s ratio varies inversely with temperature over the range in which ameboid movement is possible. At the upper temperature limit (32–34°C.) the value of A_t/A_s approaches unity as movement ceases (Mast and Prosser, 1932). Chemical factors such as pH, relative cation concentrations, and light intensity also affect the A_t/A_s ratio (Pitts and Mast, 1933, 1934; Mast and Stahler, 1937). It will be pointed out in Section III, E, 2 that changes in A_t/A_s ratio are readily interpreted as a measure of cytoplasmic contraction in the fountain zone.

One characteristic feature of cytoplasmic streaming through the fountain zone is its occurrence in spurts. It was probably this fact which gave Mast (1926) the impression that the hypothetical "plasmagel sheet" burst periodically when the pressure of the fluid behind it became sufficiently high. These spurts of the granular endoplasm flowing through the fountain zone are always synchronized with blisterlike formations of the hyaline cap. This is particularly evident when the spurts occur first on one side of the advancing pseudopod and then on the other. Alternation of spurts of this kind results in the uneven build-up of the advancing ridge of the ectoplasmic tube; it also results in an uneven, "swaggering motion" in the endoplasm which has been noted in velocity profiles (Allen and Roslansky, 1958).

c. Shortening of the ectoplasm. A number of authors have cited the observation that the ectoplasm "contracts" during locomotion as evidence for the pressure theory of ameboid movement (e.g., Pantin, 1923; Mast,

1926; Goldacre and Lorch, 1950). It is certainly true that the tail ectoplasm *shortens,* but this does not tell us whether the shortening is active or passive, or whether, if it is active, sufficient tension is developed to constitute a motive force for streaming. Roslansky (unpublished) in the author's laboratory has analyzed several motion pictures to determine the degree and location of ectoplasmic shortening. While individual granules, which serve as markers, do not always move so regularly as might be expected, the data indicate that most of the shortening occurs in the posterior third or quarter of the ectoplasmic tube.

The shortening of the ectoplasmic tube is accompanied by an increase in width, as can be seen in the data of Mast and Prosser (1932), who found "gel/sol ratios" ranging from 1.3 at the anterior end to 2.4 at the posterior end. It is not clear why the thickness of the ectoplasm should increase in the anterior half of the cell where no apparent shortening occurs.

d. The recruitment zone and uroid. The endoplasmic stream arises from the inner surface of the posterior third of the ectoplasmic tube. Inclusions from this region leave the ectoplasm quite independently and unpredictably and join the endoplasmic stream. In monopodial specimens of *A. proteus* and *Chaos chaos,* where the recruitment zone is most easily studied, the axial endoplasm arises predominantly from the uroid and the posterior parts of the ectoplasmic tube.

An illuminating way to study events in this region of the cell is to analyze a motion picture of streaming in reverse. It is immediately apparent that one could never mistake reversed movements in the recruitment zone for those in the fountain zone; they are entirely different and not opposites of one another.

e. Initiation of streaming. Although investigation of the initiation of streaming in a previously quiescent cell would have considerable significance for the theory of ameboid movement, very little attention has been given to this matter in animal cells. A few observations might be mentioned which may point up the need for further study. In a rounded, nonmotile cell, organized streaming is often preceded by somewhat disorganized endoplasmic "churning movements." The first pseudopodia to appear often erupt almost explosively and exhibit pronounced fluctuations in streaming velocity. Most cells begin movement after quiescence with many pseudopodia and then gradually become monopodial. It is impossible to generalize, however, for some specimens exhibit exactly the reverse sequence of events.

f. Reversal of streaming. Goldacre and Lorch (1950) have stressed the fact that amebae rarely alter their direction and have proposed on this basis that the uroid may contain a "tail organizer" which plays a role in the streaming process. There is no doubt, however, that amebae can change their direction if acted upon by any appropriate stimulus. Polypodial specimens exhibit frequent reversals in individual pseudopodia; sometimes a pair of adjacent pseudopodia passes a stream of endoplasm back and forth several times.

When the direction changes in individual pseudopodia, the endoplasm usually first reverses in the old posterior part; a wave of reversal then travels toward the old front. The hyaline cap continues to form until the wave of reversal has nearly reached the fountain zone.

g. Oppositely directed streaming. Schaeffer (1920) and Mast (1926) both described situations in monopodial amebae in which the direction of the entire cross section of endoplasmic stream was opposite in two different regions of the same pseudopod. In one case both ends of an ameba became "tails" so that centrally directed streaming tended to cause the ameba to become spherical. In the other case, the middle of a pseudopod acted as the "tail" and directed flow toward both ends. Mast (1926) regarded both of these types of movement as explainable by the "sol-gel" theory.

Another type of oppositely directed streaming, a description of which appears not to have been published, is that in which two adjacent streams in the same cross section of pseudopod move in opposite directions. This kind of streaming is occasionally seen briefly in cells recovering from centrifugation or injury. A film published and distributed by Dr. W. D. Lewis shows a short sequence of oppositely directed streams in an unattached *Chaos chaos*. This kind of streaming usually occurs in the tail, and is often transient.

5. *Streaming in Dissociated Ameba Cytoplasm*

It was reported a few years ago in a brief note (Allen, 1955) that cytoplasm from broken amebae exhibits organized streaming movements when dissociated from the plasmalemma and hyaline ectoplasm. The phenomenon is best observed when a *Chaos chaos* (or *A. proteus*) is drawn with distilled water or culture medium into a quartz or glass capillary and the capillary is immersed in oil. The cytoplasm is liberated from the cell membranes by shattering the capillary with a sharp scalpel as close as possible to both ends of the ameba. Streaming either continues uninterrupted or resumes after a short pause regardless of whether the naked cytoplasm is in contact with distilled water, culture medium, or oil at the broken ends of the capillary. Within seconds, the remains of the plasmalemma are swept to one end of the capillary, and the dissociated cytoplasm comes into contact with the capillary wall.

If the cytoplasm has not been too severely disrupted by the shattering of the capillary, the pattern of streaming immediately after rupture is very similar to the fountain streaming of an intact, unattached cell. The endoplasm streams toward the previous anterior end and becomes everted just as in normal fountain streaming. The rate of streaming is often initially double that in the intact cell, as if by breaking the cell some kind of "restraint" had been removed.

Within a few minutes the pattern of streaming changes; the original fountain pattern breaks up into what corresponds roughly to radial sectors of the original pseudopodial structure. That is, the capillary is filled with from a few to many U-shaped, cylindrical bodies of cytoplasm. The cytoplasm "flows" along one arm of each U toward, through, and back away from the bend. The location of the bend apparently corresponds always with the former advancing front of the cell.

Recently, Allen *et al.* (1960) have analyzed this phenomenon in detail and have reported data which are in excellent agreement with the fountain-zone contraction theory (Section III, E, 2). In their analysis, the following points were emphasized: 1. The velocity of cytoplasmic movement is 2 to 3 times more rapid toward than away from the bend, indicating a shortening at the bend. 2. The arm retreating from the bend has a cross-sectional area 2-3 times greater than the arm advancing toward the bend. This indicates that a thickening takes place at the bend. 3. The velocity distribution across adjacent, but oppositely moving arms of the same U-shaped unit showed a velocity distribution clearly indicating a marked consistency difference between the two arms; material advancing toward the bend (the endoplasmic arm) had a truncated parabolic velocity profile similar to that exhibited by the endoplasm of the intact ameba (Allen and Roslansky, 1959), while the velocity profile across the retreating (ectoplasmic) arm was much flatter, indicating a more rigid gel consistency. 4. In the presence of traces of calcium, vacuoles believed to result from syneresis form suddenly at the bend. 5. The temporal sequence of events in sporadic streaming indicates that movement toward and away from the bend result from an initial contraction at the bend which simultaneously results in equal tension on the endoplasmic arm and compression on the ectoplasmic arm. Since the contraction remains localized at the bend, it must be propagated (like the contraction of smooth and cardiac muscle cytoplasm) at approximately the velocity of endoplasmic advance relative to the bend. Evidence will be presented later (Section III, E, 2) which strongly supports the hypothesis that a contraction analogous to the one demonstrated in dissociated cytoplasm occurs at the anterior end of the intact ameba (see also Allen, 1961).

The literature of protoplasmic streaming in plants contains several interesting reports of streaming activity, not in dissociated cytoplasm as described above, but in fragments of plant cells (Yotsuyanagi, 1953; Jarosch, 1957; see also the review by Kamiya, 1959).

D. *Locomotion and Streaming in Other "Ameboid" Cells*

1. *Small Amebae*

A rather large number of genera and species of small amebae have been described over the past several decades. Many of the descriptions have been based on very few specimens, which were observed under unspecified conditions; furthermore, taxonomic divisions have sometimes been based chiefly on manner of locomotion, body form, and shape of pseudopodia. It seems doubtful whether new species of amebae should be described unless they have been raised in culture and some attempt has been made to determine the extent of their structural and behavioral variation during their life cycles, in various nutritional states, etc. The careful study of *Hartmanella astronyxis* by Ray and Hayes (1954) showed very clearly that the types of pseudopodia formed are very dependent upon many factors, such as the medium, nutrition, substratum, stage of the life cycle, and probably various environmental factors as well. It was suggested on the basis of this study that the kind of pseudopodia formed and manner of locomotion were probably not satisfactory taxonomic criteria for the genus *Hartmanella.* The same may well prove true of the other genera and species of small amebae.

In general, the smaller amebae have been neglected as material for the study of ameboid movement. This may be due in part to taxonomic uncertainties, but it is also due in a considerable degree to the variability in types of ameboid movement found in the smaller cells and to the difficulties in observing details of the streaming process.

In their form, the smaller amebae resemble tissue cells of mammals much more closely than do the larger amebae. Some of the former are polarized and move quite like leucocytes. This is especially true of the parasitic species, *Endamoeba terrapinae,* which is sold by Turtox. Other species are less polarized and resemble macrophages in their movement. Among the less polarized smaller species, some, like *Vexillifera telmathalassa* (Bovee, 1956), have conical pseudopodia and no uroid. Some of the smaller amebae also show an anterior "ruffle" similar to that seen in some tissue cells in culture.

The descriptions of locomotion in smaller amebae have for the most part been too brief to add anything new to the more extensive descriptions in larger species. There seems to be little doubt that there is a need for further work on the details of movement in these cells.

2. *Tissue Cells*

It would be outside the scope of this paper to review the vast literature concerning ameboid movements of various tissue cells throughout the animal kingdom. For present purposes, it is sufficient to point out that cell movements which have been called ameboid are a general phenomenon which is probably of central importance in developmental mechanics. Even the eggs of marine invertebrates, which have often been considered to be the most "generalized" of cells, engage in striking ameboid movements: some under normal conditions, as is the case with the eggs of sponges; and others only when their membranes are removed. However, streaming movements go on in the cytoplasm of eggs and many other cells; it is probable that these movements cannot be used for locomotion in many because the cells possess relatively rigid outer layers. The importance of ameboid movement in development has been pointed out many times (see Section I, A).

The only studies on ameboid movement itself in tissue cells have been with mammalian material. Lewis (1939) concluded from his observations on neutrophile and eosinophile leucocytes that their movement was similar in most respects to that described by Mast (1926) in *A. proteus.* He therefore interpreted tissue cell movements in terms of the sol-gel theory, with the added feature that the ectoplasmic gel was supposed to be automatically contractile, for contractile tension was believed to be a property of all gels. Lewis noted that his leucocytes exhibited constrictions which remained stationary while the cell advanced; these he considered to be evidence of ectoplasmic contraction. De Bruyn (1944–1946) made careful studies of the manner of movement of several types of tissue cells and came to some important conclusions. First, he found that the constriction rings of Lewis were caused by the plasma clot environment and occurred in the same place in cells which followed the same route through the clot. De Bruyn found that lymphocytes on a flat coverglass were more polarized than heterophiles, which often changed their direction; the former were most often observed to move in the shape of a "handmirror." In a plasma clot, however, these cells became more nearly cylindrical and moved with a twisting, writhing, "wormlike" movement. Shortening of the ectoplasm was observed. De Bruyn's study (1944, 1945) included records of the changes in the outline of cells during the course of considerable movement. The tracings provide an excellent way of demonstrating the difference between "polarized" and "unpolarized" cells. It was shown that lymphocyte migration has two definite phases: (1) a locomotory phase in which the cells are polarized, and (2) a nonlocomotory phase in which the cells are very active, but move polypodially without much net progression. Lymphocytes were found to exhibit intermediate forms of movement while undergoing hypertrophy to form macrophages. De Bruyn's conclusions from his study were in agreement with Lewis' that the events in tissue cells were similar in many ways to those in *A. proteus* as described by Mast, but he was more cautious in his acceptance of the sol-gel theory.

More recent work, especially that of Robineaux (1954, 1959) and Robineaux and Nelson (1955), has demonstrated the exceptionally fine control of tissue cell pseudopodia during phagocytosis. The observation, that neutrophile polymorphonuclear leucocytes selectively ingest bacteria that are strongly attached to erythrocytes following immuno-adherence, indicates the sensitivity with which the engulfing pseudopodia are controlled, and, at the same time, the strength with which these pseudopodia act; the immuno-adherence of the bacteria to the red cells was so firm that the bacteria could not be disengaged by micromanipulation (Robineaux and Nelson, 1955). Leucocytes demonstrate perhaps an even more phenomenal control over pseudopodial movements in the selective ingestion of lysed nuclei during lupus erythematosus cell formation; the phagocytizing pseudopodia slip in between the nuclear membrane and the cytoplasm and leave the latter behind (Robineaux, 1959).

There does not seem to be any evidence to contradict the view of Lewis (1939) that the mechanism of tissue cell movement is probably identical with that of ameboid movement. The essential features of the two processes are quite similar. A possible exception to this statement is the role of the hyaloplasmic "veils" or "ruffles" on many tissue cells, which, while similar in some ways to the ameba's hyaline cap, are much more active in movement, phagocytosis, and pinocytosis than their counterpart in the ameba (Robineaux, 1954, 1959).

Although the embryological literature contains many accounts of the ameboid activity of dissociated embryonic cells and many cases of ameboid activity inferred from changes in cellular distribution, there have until recently been few direct observations to show the long suspected importance of ameboid movement in embryonic development. Recently, Dan and Okazaki (1956) and Gustafson and Kinnander (1956) have demonstrated the importance of contracting pseudopodia (some of which resemble filopodia) of both primary and secondary mesenchyme cells of the sea urchin embryo during gastrulation. The observations of Dan and Okazaki were demonstrated by physiological experiments and those of Gustafson and Kinnander by a remarkable time-lapse film. There will doubtless be found many other examples of the performance of physical work by pseudopodia in developmental processes.

Not only embryonic and some adult tissue cells, but also some neoplastic cells have been studied. Enterline and Coman (1950) have reviewed the observations, some of which were made almost a hundred years ago, on ameboid movements in neoplastic cells. It has been considered virtually certain for some time that the invasiveness of neoplastic cells is due largely to the migration of unattached cells into interstitial spaces by ameboid movement. While the rate of locomotion of cancer cells appears not to be very rapid, it has been calculated that movement from the mammary gland to the axillary lymph nodes would take only about a month by ameboid movement alone. Enterline and Coman noted an interesting phenomenon: the movement of clusters of three to five Shope rabbit papilloma cells as a coordinated unit, with the whole mass changing shape. This is quite reminiscent of the movement of aggregating amebae of various cellular slime molds (Bonner, 1950, 1959).

3. *Testaceae*

The shelled rhizopods present a group of ameboid cells in which considerable diversity has evolved in the type of shell construction but which move, according to most accounts, in a manner quite similar to amebae. *Difflugia* forms long, cylindrical pseudopodia which attach to the substratum and pull the shell along (Dellinger, 1906) (Fig. 22). These pseudopodia also can wave about with surprising rapidity, apparently by local contractions of the ectoplasm (Mast, 1931b). Local contractions also cause the formation of hyaline blebs, which Mast considered to be a sign of syneresis. Propagated contractions apparently occur as the result of stimulating a pseudopod. Dellinger (1906) and Mast (1926) both thought that *Amoeba* and *Difflugia* move by similar

---

FIG. 22. Phase contrast photographs taken at intervals of about 10 seconds to show the locomotion of *Difflugia*. (1) An almost perfectly cylindrical pseudopodium becomes narrow at the tip before stopping, then shortens (2) forming a lateral "hyaline cap." Note the difference in refractive index between the ectoplasmic tube and the endoplasm on one side, and hyaline material on the other. (3) A lateral pseudopod forms, quickly narrows, and stops (4) and is replaced by another pseudopod. Meanwhile the originally basal part of these pseudopodia (1–3) shortens (4–6) as the shell is displaced. Scale = 10 μ.

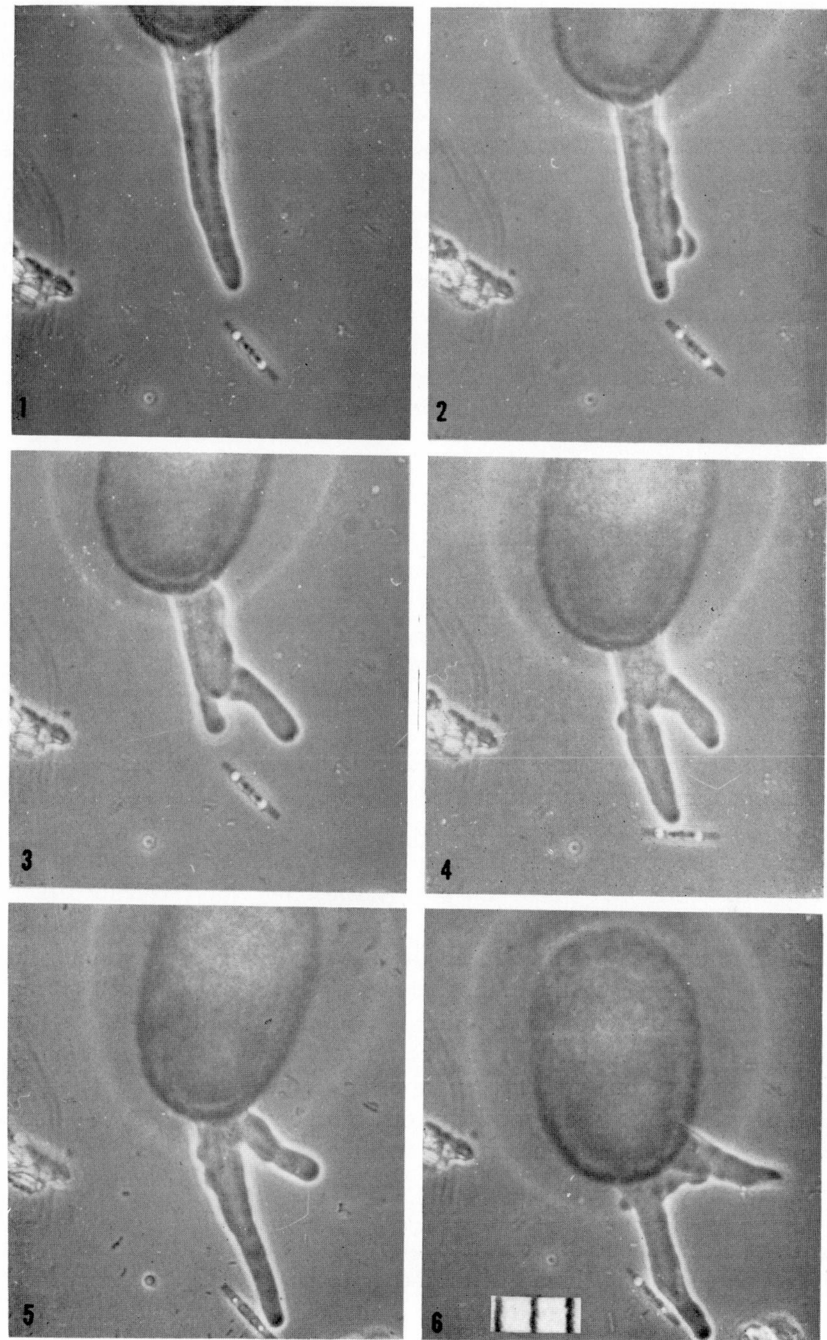
1
2
3
4
5
6

mechanisms, but they held widely divergent views as to what the mechanism was. Dellinger's view that both the endo- and ectoplasm were contractile was not very different from that of Dujardin (1938). Mast, on the other hand, interpreted movement according to the sol-gel theory. The evidence on which both of these views were based was inadequate. Details of endoplasmic flow are difficult to discern because of the scarcity of inclusions sufficiently large to serve as markers in cinematographic analyses. This difficulty has undoubtedly hampered studies not only of the endoplasm, but also of ectoplasmic contraction and conduction, which are apparently much more pronounced in the testaceans than in amebae. If this difficulty could be overcome, the testaceans might prove to be a valuable group of organisms in which to study ameboid movement because they form very regular pseudopodia.

4. Foraminifera

Protoplasmic streaming in foraminifera is so complex and unique a phenomenon that it is even doubtful from the descriptions whether the term "ameboid movement" should be applied to it. While it is true that locomotion of foraminifera is brought about through "protoplasmic streaming," the extension of fine, anastomosing pseudopodia (reticulopodia), and changes in cell form, these activities are so different from the corresponding ones in amebae that we can only hope that the mechanisms of the two processes might be similar.

One of the earliest and most concise descriptions of streaming movements in foraminifera is that of Leidy (1879, p. 279-280): "In the emission of the pseudopodial filaments of *Gromia terricola,* the protoplasm pours into the mouth of the shell in a slow manner, and gradually envelopes the body. . . . From the protoplasmic envelope delicate streams extend outwardly, at first emanating from the front; they more or less rapidly multiply and radiate in all directions. Gradually extending, they fork into branches of the utmost tenuity. Contiguous branches freely join or anastomose with one another, and thus establish an intricate net, which in its full extent covers an area upward of four times the diameter of the body of *Gromia.* The pseudopodal net incessantly changes—puts forth new branches in any position, while others are withdrawn—diminishing and disappearing in one spot, while it spreads and becomes more complex in another spot.

"*Gromia terricola,* with its pseudopodal net fully spread, like its near relatives, reminds one of a spider occupying the center of a circular web. If we imagine every thread of the latter to be a living extension of the animal under the same control as its limbs, the spider would be a nearer likeness to the *Gromia*. Over each and every thread of the pseudopodal net *Gromia* has as complete control as if the threads were permanently differentiated limbs acted on by particular muscles, and directed in their movements by nervous agency. Threads dissolve their connections and are withdrawn; new ones are formed and establish other connections: they bend; they contract into a spiral; they occasionally move like the lashing of a whip, and indeed produce almost every conceivable variety of motion. Not infrequently spindle-like accumulations of protoplasm occur in the course of pseudopodial threads. . . ." "The pseudopodal extensions of *Gromia* consist of pale granular protoplasm with coarser and more defined granules. The latter are observed to be in incessant motion along the threads, flowing in opposite directions in all but those of the greatest delicacy."

It is immediately apparent that a pressure gradient along a reticulopodium could not account for continuous two-way streaming such as Leidy (1879) and later workers have described. Sandon (1934) also emphasized that bidirectional streaming occurred in threads a micron or less in diameter, and that the rate of "flow" was independent of diameter. The latter observation showed that the "flow" along a reticulopodium is not comparable to flow of a fluid in a capillary. Sandon also found no evidence of sol-gel changes or of any contracting tube of ectoplasm. His observation that injurious stimuli cause the reticulum to break up into droplets or short rods may be readily interpreted in terms of the solation of the threads, in the same manner as hydrostatic pressure causes solation of an extended ameba with consequent break-up into droplets.

Recently, Jahn and Rinaldi (1959) have thoroughly investigated the streaming movements of *Allogromia laticollaris*. This marine foraminiferan has been successfully cultured for a number of years by Arnold (1955), who has described its life history and movement. Jahn and Rinaldi also found no evidence of a gel tube, and they regard "streaming" in the finest pseudopodia as the oppositely directed displacement of two continuous portions of a long filament which is folded at the tip of an advancing pseudopodium (Figs. 23, 24). They found bidirectional streaming not only in the finest pseudopodia, but throughout the entire reticulopodial net including the anastomoses, which migrate along the main strands to which they are connected, exhibiting bidirectional streaming at the same time.

Jahn and Rinaldi concluded from their study that much of the cytoplasm of *Allogromia* is organized into fibrils, not only in the narrowest reticulopodia, but also in the nodes and even in the large pseudopodia

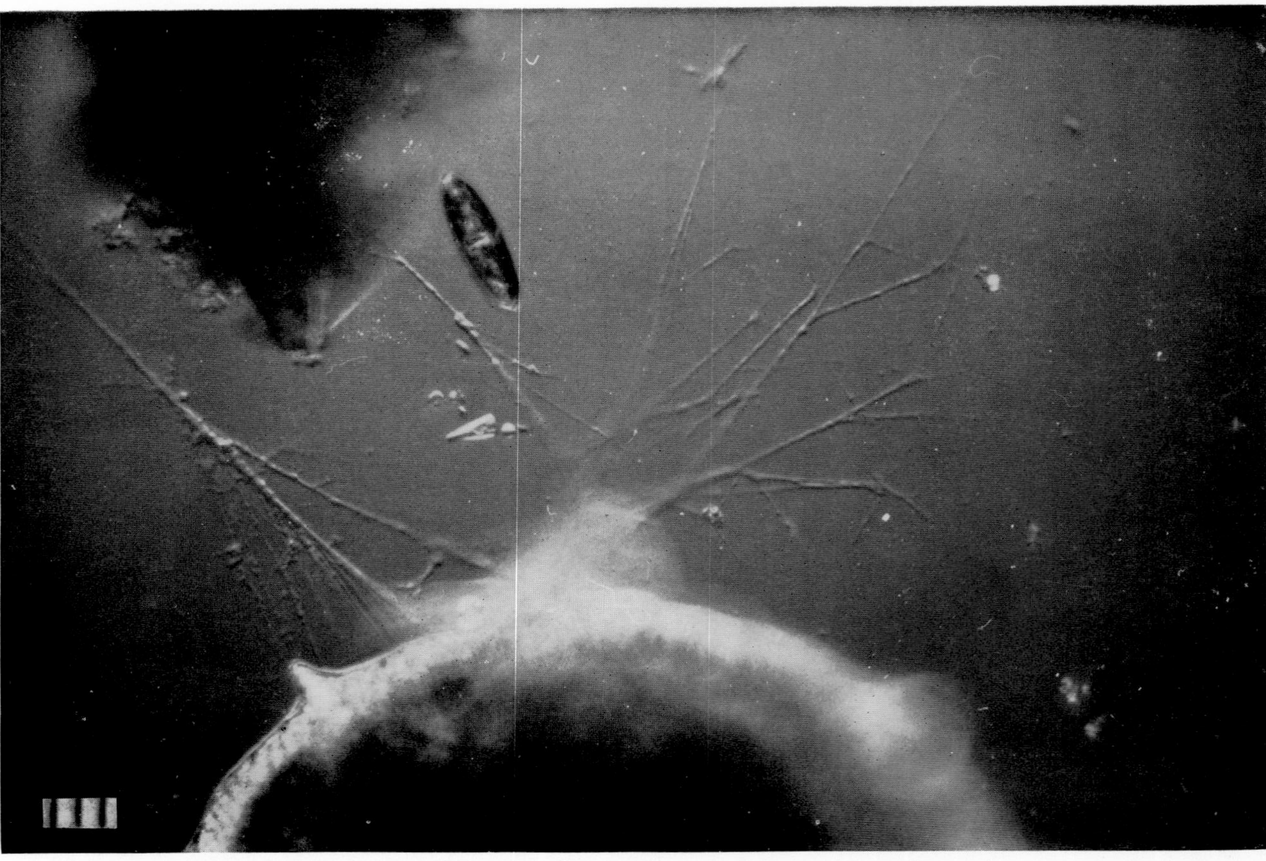

FIG. 23. A photograph of *Polystomella crispa* taken with the BBT/Krauss-Nomarski interference contrast microscope to show the appearance of a portion of the reticulopodial network. Scale = 10 μ.

issuing from the mouth of the shell. Despite Leidy's suggestion that the manner of streaming suggested central control (quotation above), Jahn and Rinaldi found that excised portions of the net could continue their typical bidirectional streaming for hours after separation from the main

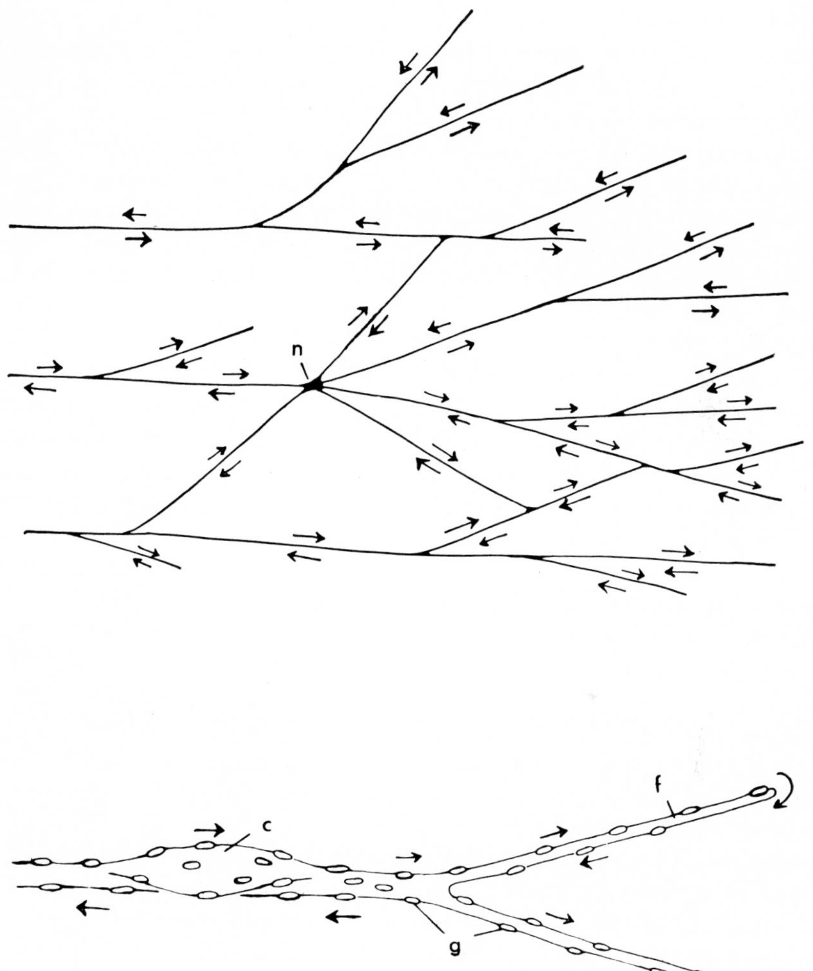

FIG. 24. Diagrams showing the general shape, structure, and streaming pattern in (above) three anastomosing reticulopodia, and (below) the distal portion of one of the finer pseudopodia with a single bifurcation into branches about 1μ in diameter. Arrows show the direction of streaming; (*n*) node, (*g*) granules, (*c*) small cytoplasmic masses, (*f*) actively moving filament. After Jahn and Rinaldi (1959).

protoplasmic mass. On the basis of this and other evidence, Jahn and Rinaldi proposed an "active shearing mechanism" to account for the opposite displacement of the two contiguous portions of a hyaline filament which seem to make up a single streaming unit. The mechanism is more descriptive than explanatory, but it does propose that the motive force is somehow applied at the "interface" between the contiguous portions of the filament, and not at one end or at both ends. However, in view of the fact that contractility has long been considered as a likely motive force for protoplasmic movement of many kinds, it is unfortunate that this analysis did not include data that would help to decide whether contractility might be involved in these movements. The fact that no gel-tube is present and no apparent sol-gel changes occur is no reason to exclude contractility, particularly in view of the descriptions of Leidy (1879), Jepps (1942), and others, which indicate that reticulopodia can bend, wave, contract, and oscillate.

III. Theories of Ameboid Movement

Superficially it would seem to be a modest goal to formulate a theory of ameboid movement in which the general nature and exact site of action of the motive force were delineated. There is as yet no common agreement on whether this goal has been achieved. One of the purposes of this paper is to outline a new hypothesis which seems at present to explain satisfactorily the observations that have been made on movement on amebae and related cells. Truly decisive experiments remain to be done. It can be hoped that a pair of alternative theories will suggest such experiments.

A. *Possible Sites for Application of the Motive Force*

In a monopodial ameba, there would seem to be four ways in which the motive force of ameboid movement could be applied in order to account for the forward displacement of the endoplasm. The force could be applied (1) *peripheral to the endoplasmic stream,* as for example by means of a contraction of the ectoplasmic tube to generate a pressure gradient; (2) *at the advancing end of the ectoplasmic tube* to pull the endoplasm toward the front by a contraction in the general region of the fountain zone; (3) *within the endoplasmic stream* by any kind of active expansion process bearing against the posterior region of the inner wall of the ectoplasm; or (4) *at the edge of the endoplasmic stream* by a unidirectional force (i.e., as opposed to pressure) applied between the endoplasm and ectoplasm to displace them in opposite directions.

B. Possible Sources of the Motive Force

A great many mechanisms known or suspected to be capable of performing mechanical work have been suggested to be the motive force for ameboid movement. Among these have been contractility, sol-gel changes, syneresis and imbibition, diffusion drag forces, surface tension, and elastic tension. Most modern theories have been based on one or more of the first three of these mechanisms.

The ultimate source of energy for ameboid movement can probably be assumed to be the same as that for the performance of work in all cells, that is, high-energy phosphate compounds such as adenosine triphosphate (ATP). Whether these compounds are involved in the final transition of chemical energy to mechanical energy which produces ameboid movement is not known. For this reason the results of experiments with ATP (injected or added to the medium) must be interpreted with caution.

C. The Decisive Importance of Endoplasmic Consistency

Just as ectoplasmic consistency proved to be decisive in the abandonment of the surface tension theory (cf. Section I, C, 3), endoplasmic consistency is of possibly decisive importance in localizing the motive force of ameboid movement.

1. If the endoplasm were a Newtonian sol, it would have to be moved passively by forces external to itself (Section III, A). Under these conditions, a pressure gradient would seem to be the only attractive hypothesis to explain ameboid movement.

2. If the endoplasm were markedly non-Newtonian, we could infer the presence of "structure" (Section II, C, 1). The motive force could then be applied externally as above, or in a number of other ways. The endoplasm could be pulled from the front; or it could also develop its own motive force by some kind of active expansion process (e.g., imbibition of fluid) bearing against the posterior region of the inner ectoplasmic wall. If the cytoplasm were organized into bundles or sheets of fibrils, it could perhaps develop a motive force between moving layers by an "active shearing mechanism" such as Jahn and Rinaldi (1959) have recently proposed for reticulopodial streaming in foraminiferans.

The earliest data on endoplasmic consistency (Mast, 1926; Heilbrunn, 1929a, b; Pekarek, 1930) seemed to indicate that the endoplasm was a structureless sol; therefore, the pressure theory of ameboid movement has been regarded as the most likely up to present time. In Section II, C, 2, it has been pointed out on the basis of (1) the evidence obtained from velocity profiles (Allen and Roslansky, 1959), (2) similar data for slime mold cytoplasm by Kamiya and Kuroda (1958), (3) observations in the centrifuge microscope (Harvey and Marsland, 1932; Allen, 1960), and (4) the behavior of gold spheres in the cytoplasm (Allen and Griffin, 1960), that an entirely different concept of ameba cytoplasmic structure has arisen. The ectoplasmic tube appears to be structurally continuous with the axial portion of the endoplasm, as has been represented diagrammatically in Fig. 19. Only the shear zone and tail endoplasm show signs of fluidity.

It must be recognized that present knowledge of cytoplasmic structure and consistency in the ameba is very incomplete. There is only limited information on elastic deformation, although data have been published for fibroblast cytoplasm (Crick and Hughes, 1949). Nevertheless, it appears certain that the cytoplasm of the ameba is not a Newtonian sol of low viscosity. Claims that it is have been based on application of shear stresses in excess of the physiological range or on observations on abnormal cells.

D. *Alterations in Cytoplasmic Consistency and the Sol-Gel Theory*

Present evidence seems quite clear in indicating that there are variations in apparent viscosity (or perhaps yield stress) in different regions of the endoplasm of a moving ameba (Section II, C, 2). However, it is not clear whether these differences reflect (1) variations in the shear stress to which different regions of the cytoplasm are subjected by the motive force (whatever it may be) or (2) changes in over-all consistency (see definition on page 158). A decision between these two possibilities could be reached only on the basis of complete force-flow curves for the cytoplasm in different regions of the cell. This kind of data cannot be obtained, for it is not possible to examine the origin of the force-flow relation in a fluid which is already flowing under stress. The closest approximation to a complete force-flow relation of ameba cytoplasm is obtained from the velocity profile data of Allen and Roslansky (1959), in which the force-flow relation is plotted on the assumption of pressure-induced flow (see Fig. 14).

The distinction between local apparent viscosity differences on the one hand and true consistency differences on the other may seem artificial, but it is only by making this distinction that a rigorous decision could be made as to whether true sol-gel changes such as occur in inanimate gels (gelatin, methylcellulose, or actomyosin) under changing conditions (e.g., temperature and pressure) might occur as an integral part of ameboid movement. It is important to realize that what are spoken of loosely as "sol-gel transformations" *in the cell* are differences in apparent viscosity, and may or may not represent true changes in consistency.

One of the most effective ways of investigating the relationship between cytoplasmic gel structure and ameboid movement is the application of hydrostatic pressure (Marsland and Brown, 1936; Marsland, 1942; Landau *et al.*, 1954; Marsland, 1956; Zimmerman *et al.*, 1958). There seems to be little doubt that high hydrostatic pressures exert a general liquefying effect on ameba cytoplasm in accordance with expectation based on the classification of cytoplasmic gels as Freundlich type II gel systems. The effects of sudden pressure increments on pseudopodial form is quite striking: pseudopodia bulge anteriorly to form "terminal spheres" (Fig. 25). This effect has been interpreted as pressure-induced solation followed by passive deformation of the anterior cytoplasm into a spheroidal shape by surface tension forces. In view of the fact that no information is available on local consistency changes in the region of newly formed terminal spheres, other interpretations are possible. For example, it is possible that pressure increases the extent of contraction in the fountain zone as predicted by the fountain zone contraction theory (Section III, E, 2). If small pressure increments initially increased the extent of contraction (before it caused solation), there would be an immediate thickening of the ectoplasmic tube which would closely resemble what has been described as the formation of terminal spheres.

There can scarcely be any doubt that different regions of ameboid cells exhibit either different apparent viscosities or different over-all consistencies. It also appears clear that temperature, hydrostatic pressure, and various chemical agents can bring about consistency changes in either the "solation" or "gelation" directions. What is lacking, however, is direct evidence that consistency changes of this type bear any "causative" relation to protoplasmic streaming. The idea that sol$\rightleftharpoons$gel changes themselves might provide the motive force for ameboid movement was abandoned when it became apparent that the volume change was in the wrong direction to account for streaming (cf. de Bruyn, 1947).

In recent years, there has been a greater tendency to regard the hydrostatic pressure and chemical data as indicating the importance of the integrity of the ectoplasmic gel, on the assumption that this was the portion of the cell most likely to undergo contraction (Landau *et al.*, 1954; Marsland, 1956; Zimmerman *et al.*, 1958). The observation that gold spheres fall in the cytoplasm of stimulated cells until just before movement resumes (page 175) suggests that the maintenance of axial endoplasmic structure may also be important. A more general conclusion would be that any gelated region in the cytoplasm may be important for the maintenance of movement. Recently Hirshfield *et al.* (1958) have shown that the nucleus plays a definite and important role in maintaining gel structure in the ameba. This may go a long way toward explaining the role of the nucleus in maintaining the organization of ameboid movement. Enucleated amebae can move for some hours or even days after enucleation, but their movements are uncoordinated and sporadic.

E. Contractility Theories

1. The Evidence for Contractility

Before considering contractility theories of cell movement, it is necessary to evaluate the various kinds of evidence on which these theories are based. In muscle, it is not difficult to establish the occurrence of a contraction, for tension and work are easily measured physical quantities. Contracting material also shortens and thickens unless prevented from performing work in an isometric contraction. While shortening and thickening of a material can mean that it has contracted actively, it is also possible that it has been passively deformed.

In some inanimate gels, contraction is accompanied by syneresis. Actomyosin gels in particular undergo marked syneresis when contraction is induced by the addition of ATP (Szent-Györgyi, 1947).

To establish that cytoplasm in some part of an ameboid cell contracts actively, it should be demonstrated that part of the cell develops tension or performs work. Shortening and thickening constitute evidence of the "compatible with" type which may be considered as strengthened by evidence of localized syneresis.

There is good reason to believe that cytoplasm of ameboid cells should be contractile. The isolation of an ATP-sensitive, actomyosin-like protein from ameboid slime molds (Loewy, 1952; Ts'o *et al.*, 1956) suggests the possible presence of macromolecules similar to those responsible for muscle contractility.

One of the strongest evidences of the ability of ameba cytoplasm to contract is the striking "decompression contraction" which occurs within 10–15 seconds after an ameba is released from prolonged exposure to hydrostatic pressure (Fig. 25). When the granular cytoplasm contracts away from the plasmalemma, it leaves behind a peripheral hyaline fluid (from syneresis) which is later reimbibed (Landau *et al.*, 1954). Goldacre (1952b) has described similar phenomena when amebae are exposed to various anesthetics.

Another evidence of contraction in the ameba is the formation of the hyaline cap, from which we infer syneresis of the granular cytoplasm

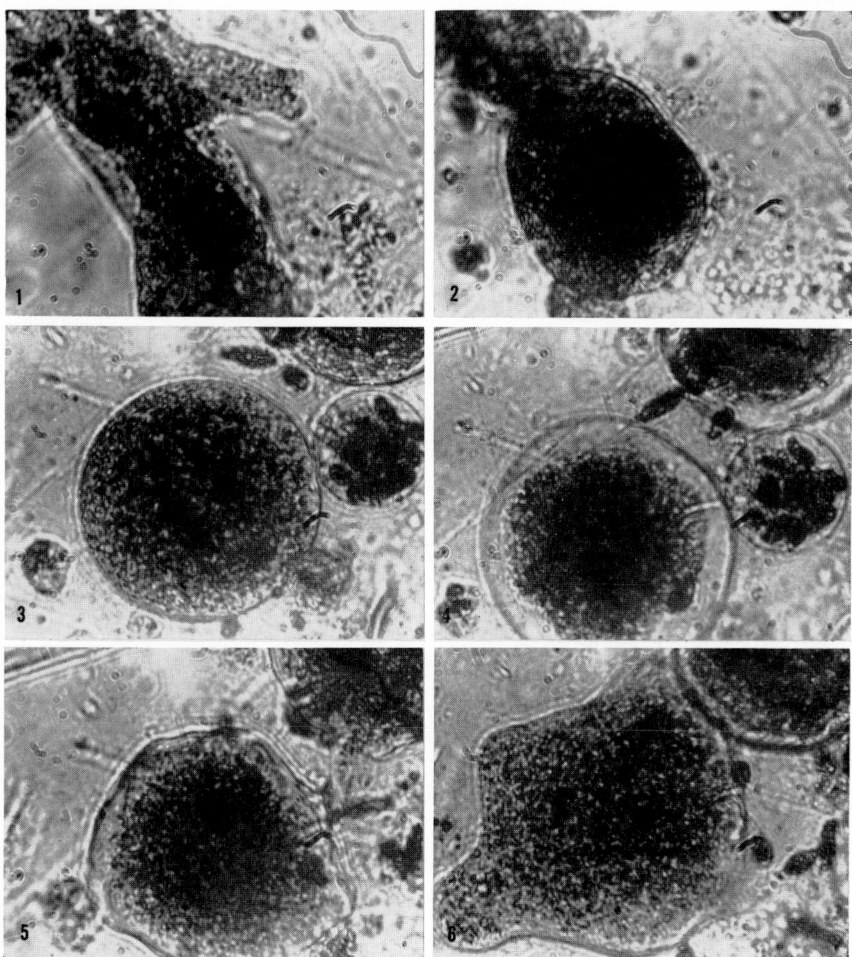

Fig. 25. Changes in the form and movement in the same specimen of *A. proteus* during and after exposure to a hydrostatic pressure of 6000 lb./in.^{2} at 25°C. (1) Normal form at atmospheric pressure. (2) After 5 minutes at 6000 lb./in.^{2}. Note the rounded form of the main cytoplasmic mass and the pinched-off part of the large pseudopodium. (3) Fifteen minutes later, pressure still maintained. Specimen has rotated 90°. (4) Fifteen *seconds* after pressure was reduced to atmospheric level. Note the marked contraction of the granular cytoplasm leaving a broad hyaline zone beneath the plasmalemma. (5) Ninety seconds after decompression. Note the first signs of ameboid activity. (6) Seven minutes after decompression. Ameboid activity now quite vigorous. Courtesy of Landau *et al.* (1954).

(Section III, E, 2, b). The hyaline cap could not be composed of water entering from the medium, for the rate of penetration of water is too low to account for the volume of fluid which appears in the hyaline cap (cf. data of Løvtrup and Pigón, 1951).

Dissociated ameba cytoplasm is very clearly contractile. It not only undergoes shortening and extension, but exhibits syneresis as well (page 186). As was pointed out previously, one of the most interesting features of contractile movements in dissociated cytoplasm is their occurrence in waves which are propagated along discrete sections of cytoplasm.

2. *Localization of Contraction in the Ameba*

a. The ectoplasmic tube. It has been stated above (Section II, C, 4, c) that the posterior region of the ectoplasm shortens as an ameba progresses. Most authors have interpreted this shortening as a contraction which produces tension and have cited it in support of the theory that ameboid movement is caused by pressure generated by ectoplasmic contraction (Pantin, 1923; Mast, 1926; de Bruyn, 1947; Goldacre and Lorch, 1950; Goldacre, 1952a, b; Marsland, 1956; Allen and Roslansky, 1958).

The ectoplasmic contraction hypothesis has until recently been the most satisfactory basis for a theory of ameboid movement, even though there has been no clear evidence that the shortening of the ectoplasm represents a true contraction. In fact, when the endoplasm was considered to be a Newtonian sol, the establishment of a pressure gradient by means of ectoplasmic contraction seemed to be the only tenable theory of ameboid movement. If the shortening of the ectoplasmic tube were accompanied by syneresis, this might be considered adequate evidence of contraction. The existence of the expected anterior-posterior gradient in average refractive index (Section II, C, 3, a) was formerly tentatively interpreted as evidence for a tide of hyaline fluid originating from contracting tail ectoplasm and flowing to the anterior tip of the cell where it is visible as the hyaline cap (Allen and Roslansky, 1958). This interpretation did not, however, account for the periodic nature of hyaline cap formation. The gradient of refractive index could also be explained by the fact that the ratio of ectoplasm to endoplasm is greater in the posterior region than in the anterior. If there were a refractive index difference between these two layers a gradient such as that which was found would result.

The ectoplasmic contraction theory has recently been extended by Goldacre and Lorch (1950) to the molecular level in a simple and attractive theory of ameboid movement and osmotic work. This theory combines the ectoplasmic contraction and sol-gel theories of Pantin (1923) and Mast (1926) and is expressed in terms of the folding and unfolding of protein molecules. Ectoplasmic contraction is explained as the folding of interlinked (gelated) protein molecules in the posterior ectoplasm; solation of the ectoplasm in the recruitment zone is considered to be the superfolding of these same molecules so that they relinquish the linkages which held them together in the gel form. When the folded molecules pass to the front of the cell, they unfold and reform linkages, according to the theory, and cause the gelation that extends the ectoplasmic tube.

It is interesting to note that a similar idea occurred to Engelmann (1879) as a purely hypothetical model long before the idea of conformation changes in protein molecules was proposed. Engelmann suggested the presence of hypothetical "inotagmen," or micelle-like particles, which could undergo changes in shape.

The idea of changes in conformation of proteins as a basis for muscle contraction and protoplasmic streaming came under consideration again as a result of demonstrations by X-ray diffraction that some proteins, especially some of the albuminoids could be made to change from extended to condensed forms (Meyer, 1929; Astbury, 1939). Until ten or fifteen years ago, it was taken for granted that protein folding occurred in muscle contraction. More recent X-ray diffraction patterns of muscle have, however, rendered this rather unlikely (Astbury, 1947; Huxley, 1953). The most recent advances in the ultrastructure of muscle indicate that adjacent actin and myosin filaments do not shorten, but instead appear to be somehow pulled past one another (cf. Hanson and Huxley, 1955; Szent-Györgyi, 1958).

The theory of Goldacre and Lorch is particularly useful, for it can be put to experimental test. Unfortunately, the supporting evidence obtained so far, while compatible with the theory, is also subject to other interpretations. The evidence is as follows:

1. From the well-known fact that unfolded (denatured) proteins adsorb more dye than folded (native) molecules, it was predicted that proteins of the entire ectoplasmic tube would, if unfolded, readily adsorb a dye such as neutral red from the medium, then desorb it in the tail region where folding was postulated to occur. According to the proposed model, some of the stained cytoplasm should then be passed forward into the endoplasmic stream for further accumulatory cycles. This prediction was in general fulfilled for *Pelomyxa* (Okada, 1930), *Amoeba discoides* (Goldacre and Lorch, 1950), and *Amoeba proteus* (Prescott, 1953). However, there are several difficulties in interpreting the phenom-

enon of dye accumulation in amebae. First, Goldacre and Lorch (1950) erroneously assumed that the plasmalemma of the ameba is continually formed at the front and destroyed at the rear of the cell; this assumption conflicts with the observations of many workers on the behavior of particles attached to the plasmalemma (Section II, B). Therefore the dye would have to be taken up almost entirely by the ectoplasm. It is also questionable, and of considerable importance, which component of the cytoplasm is stained by neutral red. This dye is well known to be somewhat toxic to amebae and to produce granular artifacts and neutral red vacuoles (Andresen, 1944). Goldacre (1952a) and Noland (1957) noted that the accumulation of dye in the tail appeared to be associated with local regions of coagulation. This and other factors besides conformation changes in proteins could account for the accumulation of dye in the tail. First, the entire outermost region of the ectoplasm, which would be expected to trap much of the dye, normally accumulates in the tail before joining the endoplasmic stream. Second, the surface area of the tail is greatly increased by the presence of folds and wrinkles; for this reason, the rate of penetration of dye should be expected to be highest in this region of the cell.

2. The results of injecting ATP into different parts of amebae also conformed to expectation according to Goldacre and Lorch's theory. Injection of 1–3% ATP into the tail was stated to cause more rapid streaming; into the front, reversal; and into the middle, cessation of organized streaming followed by "bubbling" movements. While these results are certainly compatible with the theory, they do not actually prove anything, for there is no evidence that the ATP injected had any influence on the contractile machinery of the cell. An ameba is capable of a variety of responses to mechanical or chemical stimulation such as would unavoidably occur on microinjection. The anterior portion is particularly sensitive, so that stimulation in this region would be expected to cause reversal. Acceleration of streaming by injection in the tail would be subject to different interpretations depending on the time interval before the cell responded to the injected ATP; a delay of even a few seconds might result in the transport of the injected ATP to the anterior portion of the cell to affect some process there. To interpret experiments of this sort, it would be valuable to know, in addition to the controls reported by Goldacre and Lorch, the number of experiments, the variation in results, and the duration of delay in the effects. The work of Hoffmann-Berling (1954, 1958) has shown that ATP can have the effect of causing either contraction or relaxation in cell model systems depending on concentration and on the type of contractile system involved. This

work illustrates the wide gap that exists between the observation of effects of injected ATP on ameboid movement and our understanding of the mechanism of those effects.

The concept of ameba cytoplasmic structure presented in Section II, C, is not in good agreement with the model proposed by Goldacre and Lorch. When more becomes known about the ultrastructure of different regions of ameba cytoplasm it may be worth while to modify their model. The idea that proteins fold and unfold in biological processes such as protoplasmic streaming or muscle contraction is an interesting speculation; unfortunately there is no clear evidence at the present time that any such changes occur in cytoplasmic contractility.

b. The fountain zone. In the description of events in the fountain zone (Section II, C, 4, b) it was pointed out that after a section of cytoplasm has passed through the fountain zone, its cross-sectional area becomes several times greater. As long as the endoplasm was considered to be a Newtonian sol, there was no reason to regard the events in the fountain zone as anything but the pouring out of the fluid endoplasm to extend the ectoplasmic tube. However, this view is no longer tenable, for the endoplasm is not fluid and does not "pour" out of the ectoplasmic tube; instead it becomes everted in regular stream-lines (Section II, C, 2, a). Therefore, the increase in cross-sectional area can be interpreted as a thickening of cytoplasm as it is displaced through the fountain zone into the ectoplasm. That this thickening is accompanied by a shortening is shown by the fact that granules decelerate as they enter the fountain zone (Allen *et al.*, 1960).

Since the axial endoplasm possesses structure (see Section II, C, 2) it can transmit tension between the relatively rigid advancing rim of the ectoplasmic tube and the uncontracted axial endoplasm. The ectoplasmic tube extends forward only if attached through the plasmalemma to the substratum (Section II, B).

The idea expressed above, perhaps best referred to as the "fountain zone contraction hypothesis," is illustrated in Fig. 26. This hypothesis has received strong positive support from the recent detailed analysis of streaming in cytoplasm dissociated from the giant ameba, *Chaos chaos* (Allen *et al.*, 1960). As was pointed out in Section II, C, 5, each streaming unit of dissociated cytoplasm shortens, thickens, loses water (syneresis), and shows indirect but unmistakable signs of developing tension in bringing about cytoplasmic streaming. Thus the fountain zone contraction hypothesis offers a satisfactory explanation for streaming in dissociated cytoplasm.

In the case of the intact cell there is as yet no conclusive evidence that tension is produced by a contraction in the fountain zone, but there is indirect evidence, referred to above, for the occurrence of such a contraction. The following observations appear to confirm or be compatible with the fountain zone contraction hypothesis.

1. The A_t/A_s ("sol-gel") ratio changes profoundly under various experimental conditions (Mast and Prosser, 1932; Pitts and Mast, 1933, 1934). Low temperature was found to increase this ratio, and high temperature to lower it until movement stopped before the value of A_t/A_s

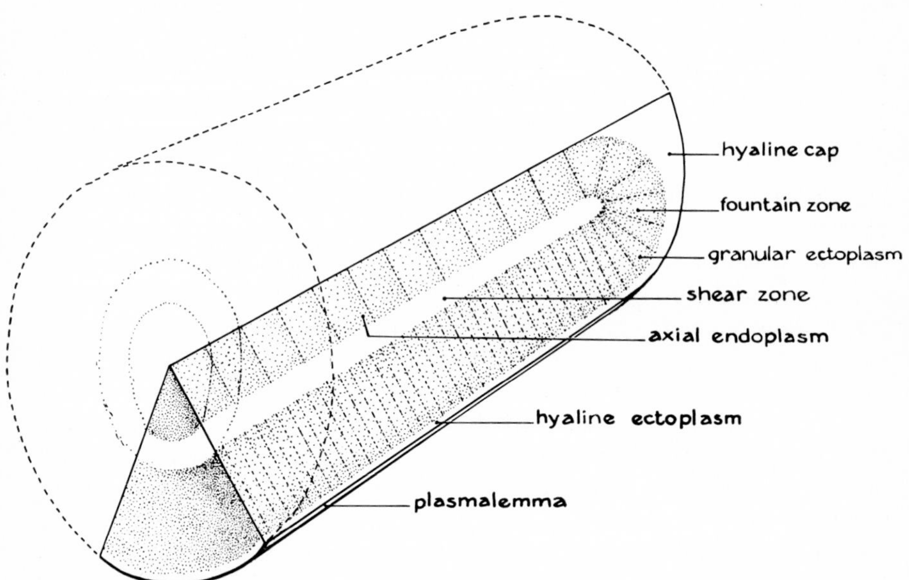

FIG. 26. A schematic wedge-shaped section of an advancing pseudopod to show how a contraction of cytoplasm in the fountain zone could bring about endoplasmic streaming.

reached unity (Mast and Prosser, 1932). At that time these data were interpreted as showing that temperature affected the sol$\rightleftharpoons$gel equilibrium. It is now clear, however, that the temperature data were in the *wrong direction* to be explained on this basis, for cytoplasmic gels solate at reduced temperatures and high hydrostatic pressure (Marsland, 1956; Landau *et al.*, 1954). The increase in A_t/A_s ratio at low temperatures (Mast and Prosser, 1932) and at somewhat elevated hydrostatic pressures (an alternative explanation for the "terminal sphere phenomenon" discussed in Section II, C, 2) can be interpreted as an increased *extent* of contraction in the fountain zone. Factors which independently affect the velocity of streaming may be explained as affecting the velocity of propagation of that contraction, for the contraction must be propagated posteriorly along the axial endoplasm at a rate equal to the velocity of streaming (measured relative to the pseudopodial tip) in order to keep the contraction localized in the fountain zone.

2. The consistency difference between the endoplasm and ectoplasm is compatible with the idea that the endoplasm and ectoplasm are uncontracted and contracted states of the same material. This would not be surprising since similar consistency changes are well known to occur in muscle contraction.

3. The syneresis represented by hyaline cap formation is best explained as fluid pressed out of the nearby fountain zone. During reversal of direction, the hyaline cap continues to form in an old pseudopod as long as streaming continues in the original direction. A new pseudopod begins immediately to form its own hyaline cap which often reappears at a different frequency from those on other pseudopodia. If our interpretation that the hyaline cap is formed during the fountain zone contraction is correct, then it would appear necessary that the hyaline cap fluid flow posteriorly through the hyaline ectoplasmic channel to the tail region, where the hyaline fluid must return to the ectoplasm as it undergoes extension in the recruitment zone.

4. The effects of stimulation by light are most pronounced at the anterior tip of the cell. Mast (1932) demonstrated that sudden illumination of the entire "plasmagel sheet" area (i.e., in the fountain zone) caused rapid cessation of movement. Streaming stopped first in the fountain zone and a wave of cessation traveled posteriorly through the endoplasm. Illumination of half of the fountain zone caused pseudopod formation to be diverted to the opposite side (Fig. 27). These observations could be readily explained as a direct effect of light on the contractile mechanism in the fountain zone.

The effects of light on the ameba are complicated and not easily understood on the basis of available data. Whereas sudden exposure to bright light causes cessation of streaming, Mast and Stahler (1937) showed that after a period of accommodation (Fig. 28) amebae began to move more rapidly at intensities up to 15,000 meter-candles (Fig. 29). The increased rate of movement could possibly be due to an enhancement of contraction after the accommodation period, for the "gel/sol ratio" was found to undergo a marked increase after prolonged exposure to light (Fig. 30).

5. The response of an ameba to touch is highly localized in the fountain zone. Goldacre (1952b) found that a blunt microneedle applied against the hyaline cap elicited no response until the needle, along with

the underlying region of plasmalemma, had come in contact with the granular cytoplasm. The ameba's response was to divert its pseudopod formation to one side in a manner similar to its response to light.

The ameba's response to touch is undoubtedly of primary importance in food getting. For this reason, a theory of movement which locates not only the generation, but also the delicate regulation of the motive force at the anterior portion of the cell would be advantageous in explaining the complex series of events involved in the formation of food cups. The ability of an ameba to "sense" food and move toward it (Schaeffer, 1916)

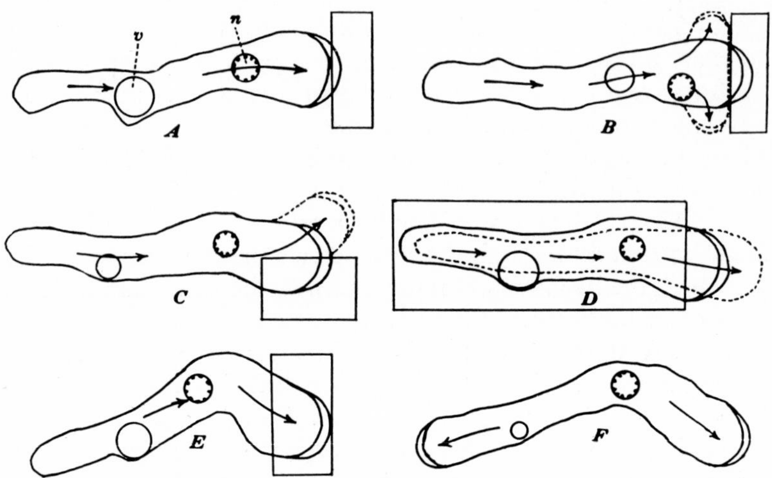

FIG. 27. Camera lucida drawings of "*Amoeba* x" illustrating the response to localized illumination. Rectangular areas, regions of high intensity illumination; arrows show direction of streaming. Dotted lines in B, C, and D show cell form shortly after illumination. E and F, before and after illumination for a few minutes. From Mast (1932).

and to form different kinds of food cups depending on the size, shape, and motility of the food organism (Kepner and Edwards, 1918), are not easily explained on the basis of theories assuming the motive force to be located in the posterior region.

There are a number of phenomena already referred to in Section II, D, 2 which show clearly the delicate and precise control of tissue cell pseudopodia during phagocytosis (Robineaux, 1954, 1959; Robineaux and Nelson, 1955). Such a degree of precision of control is more difficult to account for by the tail contraction theory than by a theory which places the control mechanism at the front of the pseudopod itself.

6. The fountain zone contraction hypothesis is the only one that

offers an explanation for streaming in opposite directions within the same cross section of pseudopod (Section II, C, 4, f) as well as the opposite streaming which occurs in broken cells (Section II, C, 4, g). The "streaming" exhibited by folded sections of dissociated cytoplasm may

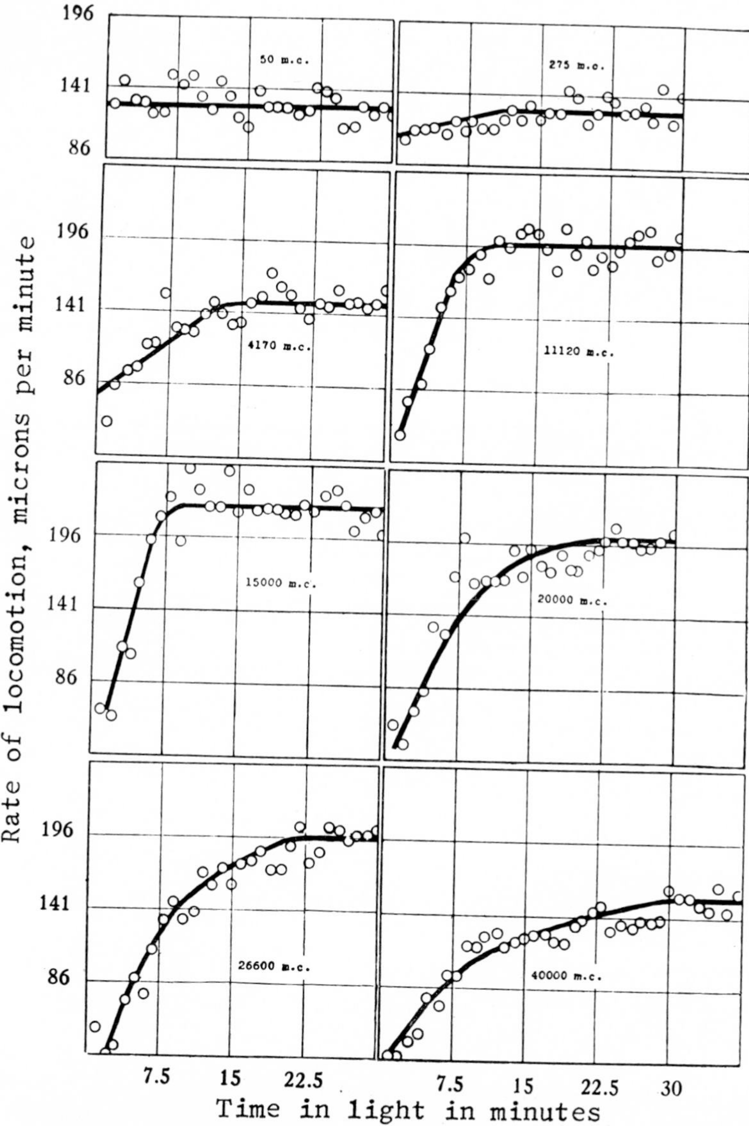

FIG. 28. The rate of locomotion of *Amoeba proteus* in relation to the duration of exposure to light. From Mast and Stahler (1937).

then be explained by considering the folded sections to be comparable to the wedge-shaped section of a pseudopod represented diagrammatically in Fig. 26. One of the arms of this folded section, comparable to the ectoplasmic arm in Fig. 26, would be contracted and therefore more rigid (as in a contracted muscle fiber). If either end of the section were anchored, propagation of the contraction along such a section

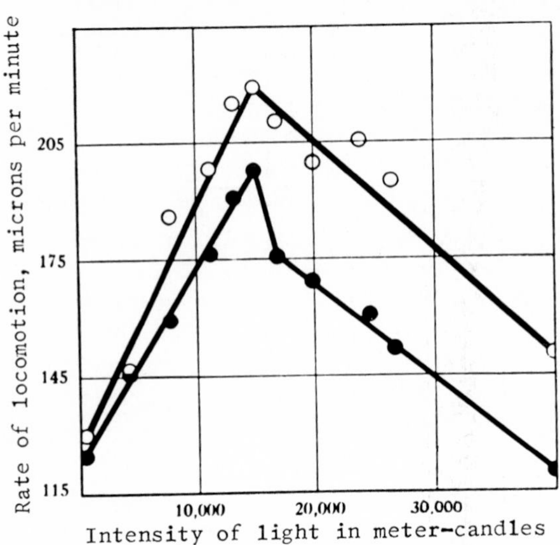

FIG. 29. The rate of locomotion of *A. proteus* in relation to the intensity of incident light. From Mast and Stahler (1937).

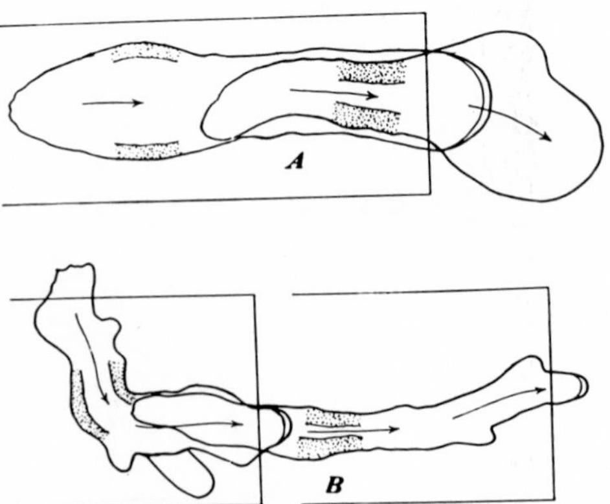

FIG. 30. Camera lucida drawings of *Amoeba proteus* showing the increase in gel/sol ratio as a result of illuminating all but the anterior end. Rectangular areas were illuminated; stippled areas show the thickness of the ectoplasm (plasmagel). After Mast (1932).

would cause the bend to travel ahead of the contracted region. However, if either the bend were anchored or the entire section were free, then the bend would remain relatively stationary and the cytoplasm would move in reference to the bend, appearing to "stream" through it.

F. Concluding Remarks

The general idea that a contraction of cytoplasm anchored either at a fountain zone or at some other region in which streaming cytoplasm makes a sharp bend is perhaps applicable to other kinds of protoplasmic streaming besides ameboid movement. The opposite streaming of adjacent bodies of cytoplasm in *Reticulomyxa filosa* (Nauss, 1949) are impossible to explain on the basis of any kind of pressure theory. Descriptions of movement in this slime mold are not sufficiently complete to let us be certain whether the explanation proposed here for ameboid movement might also be applicable to *Reticulomyxa.* However, this organism does exhibit fountain streaming, so that it does appear possible that the fountain zone contraction hypothesis might explain the pattern of adjacent opposite streams on the basis of pulling rather than pressure. In protoplasmic streaming generally, there has been great difficulty in localizing the site of action of the motive force with certainty (Kamiya, 1959). The possibility that the motive force could reside in such a position as to pull rather than push a body of streaming cytoplasm has been overlooked, probably because of assumptions regarding the Newtonian nature of the cytoplasm.

The fountain zone contraction hypothesis may also be used to explain the type of streaming which occurs in foraminifera. Reticulopodia have no fountain zone such as is found in the ameba. Instead, according to Jahn and Rinaldi (1959), they consist of a filament which is folded double with the bend at the tip. If Jahn and Rinaldi are correct in their assertion that the fundamental unit of streaming in these pseudopodia is this folded filament, it is then possible to make two alternative hypotheses. Either an "active shearing mechanism" such as Jahn and Rinaldi proposed displaces the arms of the filaments in opposite directions with the bend remaining passive; or a propagated contraction occurs at the bend, so that one of the arms is contracted cytoplasm (as in the ectoplasmic arm of the wedge-shaped section of an ameba pseudopod: Fig. 26) and the other extended cytoplasm. Propagation of the contraction toward the extended arm would result in the "streaming" of the substance of the filament through the bent region at the tip, as is seen to occur.

The advantage of the latter hypothesis is that it can be tested. It also offers a possible common ground for muscular contraction, ameboid movement, and reticulopodial streaming. The active shearing mechanism is one of several submicroscopic mechanisms (see also Loewy, 1949) that are of limited value because of the paucity of experiments they suggest.

The present situation with regard to theories of ameboid movement may be summarized as follows:

1. Surface tension theories are now only of historic interest.

2. The sol-gel theory is at best descriptive; as such, it is inaccurate on the basis of recently acquired information on cytoplasmic consistency. There is apparently no known way in which sol-gel transformations per se could perform the work involved in ameboid movement. This is not to say, however, that consistency changes do not accompany ameboid movement.

3. Much evidence is in favor of active contraction as the motive force for ameboid movement and probably at least some other kinds of protoplasmic streaming as well. Although the posterior region of the ectoplasmic tube has been regarded until the present as the most likely site for contraction to occur in the ameba, the weight of present evidence in the author's opinion favors the view that a propagated contraction occurs in the fountain zone at the anterior region of each pseudopod.

4. The ectoplasmic contraction theory is tenable only for the ameba and is inapplicable to some slime molds and to probably all foraminiferans. On the other hand, the fountain zone contraction hypothesis seems applicable not only to the ameba, but to many other kinds of protoplasmic streaming as well. It has the further advantage of calling attention to some possible parallels to muscle contraction.

There are very few biological problems in which so many theories have been proposed to explain so few data, as has been the case of ameboid movement. It is to be hoped that future research will provide decisive experiments with which we can choose among the several theories that now must be considered as plausible. It is also hoped that attention of many workers will be focused on some of the less frequently studied ameboid cells. There is such diversity among these cells that it seems virtually certain that important factors await discovery which have been overlooked in our preoccupation with the large, free-living amebae.

References

Allen, R. D. (1955). *Biol. Bull.* **109**, 339.

Allen, R. D. (1958). *Anat. Record* **132**, 403.

Allen, R. D. (1960). *J. Biophys. Biochem. Cytol.* **8**, 379.

Allen, R. D. (1961). *Exptl. Cell Research, Suppl.* In press.

Allen, R. D., and Griffin, J. L. (1960). In preparation.
Allen, R. D., and Maddux, W. (1960). In preparation.
Allen, R. D., and Roslansky, J. D. (1958). *J. Biophys. Biochem. Cytol.* **4**, 517.
Allen, R. D., and Roslansky, J. D. (1959). *J. Biophys. Biochem. Cytol.* **6**, 437.
Allen, R. D., Cooledge, J., and Hall, P. J. (1960). *Nature* **187**, 896.
Andresen, N. (1942). *Compt. rend. trav. lab. Carlsberg, Sér. chim.* **24**, 139.
Andresen, N. (1944). *Compt. rend. trav. lab. Carlsberg, Sér. chim.* **25**, 147.
Andresen, N. (1956). *Compt. rend. trav. lab. Carlsberg, Sér. chim.* **29**, 435.
Andresen, N., and Holter, H. (1945). *Compt. rend. trav. lab. Carlsberg, Sér. chim.* **25**, 107.
Andrews, E. A. (1955). *Biol. Bull.* **108**, 121.
Arena, U. F. de la. (1943). *Bol. Anat. (Habana)* **1**, 13.
Arnold, Z. M. (1955). *Univ. Calif. (Berkeley) Publ. Zool.* **61** *No. 4,* 167.
Astbury, W. T. (1939). *Ann. Rev. Biochem.* **8**, 113.
Astbury, W. T. (1947). *Proc. Roy. Soc.* **B134**, 303.
Baas Becking, L. C. M., Bakhuyzen, H. V. D. S., and Hotelling, H. (1928). *Verhandel. Koninkl. Akad. Wetenschap. Amsterdam Afdeel. Natuurk. (Tweede Sectie) Deel* **25**, 5, 1.
Bairati, A., and Lehmann, F. E. (1953). *Exptl. Cell Research* **5**, 220.
Bernstein, J. (1900). *Arch. ges. Physiol. Pflüger's* **80**, 628.
Berthold, G. (1886). "Studien über Protoplasmamekanik." A. Felix, Leipzig.
Bonner, J. T. (1950). *Biol. Bull.* **99**, 143.
Bonner, J. T. (1959). "The Cellular Slime Molds." Princeton Univ. Press, Princeton, New Jersey.
Borysko, E., and Roslansky, J. (1959). *Ann. N.Y. Acad. Sci.* **78**, 432.
Bovee, E. C. (1956). *J. Protozool.* **3**, 155.
Brown, D. E. S. (1957). In "Influence of Temperature on Biological Systems" (F. H. Johnson, ed.), pp. 83-110. Ronald Press, New York.
Bütschli, O. (1892). "Untersuchungen über mikrokopische Schäume und das Protoplasma." Leipzig.
Chalkley, H. W. (1936). *J. Morphol.* **60**, 13.
Cohen, A. I. (1957). *J. Biophys. Biochem. Cytol.* **3**, 859.
Crick, F. H., and Hughes, A. F. W. (1949). *Exptl. Cell Research* **1**, 37.
Dan, K., and Okazaki, K. (1956). *Biol. Bull.* **110**, 29.
De Bruyn, P. P. H. (1944). *Anat. Record* **89**, 43.
De Bruyn, P. P. H. (1945). *Anat. Record* **93**, 295.
De Bruyn, P. P. H. (1946). *Anat. Record* **95**, 177.
De Bruyn, P. P. H. (1947). *Quart. Rev. Biol.* **22**, 1.
Dellinger, O. P. (1906). *J. Exptl. Zool.* **3**, 337.
Dujardin, F. (1835). *Ann. Sci. nat. Zool.* **4**, 343.
Dujardin, F. (1838). *Ann. Sci. nat. Zool.* **10**, 230.
Ecker, A. (1849). *Z. wiss. Zoöl. Abt.* **A1**, 218.
Engelmann, T. W. (1879). *Hermann's Handb. Physiol.* **1**, 343.
Enterline, H. T., and Coman, D. R. (1950). *Cancer* **3**, 1033.
Galtsoff, P. S. (1923). *Biol. Bull.* **45**, 153.
Goldacre, R. J. (1952a). *Intern. Rev. Cytol.* **1**, 135.
Goldacre, R. J. (1952b). *Symposia Soc. Exptl. Biol. No.* **6**, 128.
Goldacre, R. J., and Lorch, I. J. (1950). *Nature* **166**, 497.
Greider, M. H., Koster, W. J., and Frajola, W. J. (1956). *J. Biophys. Biochem. Cytol.* **2**, *Suppl.,* 445.

Griffin, J. L. (1959). Ph.D. Thesis. Princeton Univ., Princeton, New Jersey.
Griffin, J. L. (1959). Unpublished observations.
Griffin, J. L. (1960). *J. Biophys. Biochem. Cytol.* (1960). **7**, 227.
Griffin, J. L., and Allen, R. D. (1960). *Exptl. Cell Research* **20**, 619.
Grumbaum, B. W., Møller, Max K., and Thomas, R. S. (1959). *Exptl. Cell Research* **18**, 385.
Gustafson, T., and Kinnander, N. (1956). *Exptl. Cell Research* **11**, 36.
Hanson, J., and Huxley, H. E. (1955). *Symposia Soc. Exptl. Biol. No.* **9**, 228.
Harvey, E. N., and Loomis, A. L. (1930). *Science* **72**, 42.
Harvey, E. N., and Marsland, D. A. (1932). *J. Cellular Comp. Physiol.* **2**, 75.
Heilbrunn, L. V. (1929a). *Protoplasma* **8**, 58.
Heilbrunn, L. V. (1929b). *Protoplasma* **8**, 65.
Heilbrunn, L. V. (1958). *Protoplasmatologia* **2**, Cl 1.
Heilbrunn, L. V., and Daugherty, K. (1931). *Physiol. Zoöl.* **4**, 635.
Hirshfield, H. I., Zimmerman, A. M., and Marsland, D. A. (1958). *J. Cellular Comp. Physiol.* **52**, 269.
Hoffmann-Berling, H. (1954). *Biochim. et Biophys. Acta* **15**, 226.
Hoffmann-Berling, H. (1958). *Biochim. et Biophys. Acta* **27**, 247.
Huxley, H. E. (1953). *Proc. Roy. Soc.* **B141**, 59.
Hyman, L. H. (1917). *J. Exptl. Zool.* **24**, 55.
Jahn, T. L., and Rinaldi, R. A. (1959). *Biol. Bull.* **117**, 100.
James, T. W. (1959). *Ann. N.Y. Acad. Sci.* **78**, 501.
Jarosch, R. (1956). *Protoplasma* **47**, 478.
Jarosch, R. (1957). *Biochim. et Biophys. Acta* **25**, 204.
Jennings, H. S. (1904). *Carnegie Inst. Wash. Publ. No.* **16**, 129.
Jepps, M. W. (1942). *J. Marine Biol. Assoc. United Kingdom* **25**, 607.
Kamiya, N. (1939). *Ber. deut. botan. Ges.* **57**, 231.
Kamiya, N. (1950). *Cytologia (Tokyo)* **15**, 183.
Kamiya, N. (1959). *Protoplasmatologia* **8**, *3a*, 1.
Kamiya, N., and Kuroda, K. (1958). *Protoplasma* **44**, 1.
Kepner, W. A., and Edwards, J. G. (1918). *J. Exptl. Zool.* **24**, 381.
King, R. L., and Jahn, T. L. (1948). *Science* **107**, 293.
Kitching, J. A., and Moser, F. (1940). *Biol. Bull.* **78**, 80.
Kudo, R. R. (1959). *Ann. N.Y. Acad. Sci.* **78**, 474.
Landau, J. V., Zimmerman, A. M., and Marsland, D. A. (1954). *J. Cellular Comp. Physiol.* **44**, 211.
Leidy, J. (1879). Fresh-water rhizopods of North America. *Rept. U.S. Geol. Survey No.* **12**.
Lewis, W. H. (1939). *Arch. exptl. Zellforsch. Gewebezücht.* **23**, 1.
Lewis, W. H. (1951). *Science* **113**, 473.
Lillie, F. R. (1902). *Arch. Entwicklungsmech. Organ.* **14**, 477.
Loewy, A. G. (1949). *Proc. Am. Phil. Soc.* **93**, 326.
Loewy, A. G. (1952). *J. Cellular Comp. Physiol.* **40**, 127.
Løvtrup, S., and Pigón, A. (1951). *Compt. rend. trav. lab. Carlsberg, Sér. Chim.* **28**, 1.
Marshall, J., Schumaker, V. N., and Brandt, P. W. (1959). *Ann. N.Y. Acad. Sci.* **78**, 515.
Marsland, D. A. (1942). *In* "The Structure of Protoplasm" (W. Seifriz, ed.), pp. 127-161. Iowa State College Press, Ames, Iowa.
Marsland, D. A. (1956). *Intern. Rev. Cytol.* **5**, 199.
Marsland, D. A., and Brown, D. E. S. (1936). *J. Cellular Comp. Physiol.* **8**, 167.

Mast, S. O. (1926). *J. Morphol. and Physiol.* **41**, 347.
Mast, S. O. (1931a). *Protoplasma* **14**, 321.
Mast, S. O. (1931b). *Biol. Bull.* **61**, 223.
Mast, S. O. (1932). *Physiol. Zoöl.* **5**, 1.
Mast, S. O. (1934). *Physiol. Zoöl* **7**, 470.
Mast, S. O., and Doyle, W. L. (1935). *Arch. Protistenk.* **86**, 278.
Mast, S. O., and Prosser, C. L. (1932). *J. Cellular Comp. Physiol.* **1**, 333.
Mast, S. O., and Root, F. M. (1916). *J. Exptl. Zool.* **21**, 33.
Mast, S. O., and Stahler, N. (1937). *Biol. Bull.* **73**, 126.
Mercer, E. H. (1959). *Proc. Roy. Soc.* **B150**, 216.
Meyer, K. H. (1929). *Biochem. Z.* **214**, 253.
Mitchison, J. M. (1950). *Nature* **166**, 313.
Muggleton, A., and Danielli, J. F. (1958). *Nature* **181**, 1738.
Müller, O. F. (1786). "Animalcula Infusoria Fluviatilia et Marina."
Nauss, R. N. (1949). *Bull. Torrey Botan. Club* **76**, 161.
Noland, L. E. (1957). *J. Protozool.* **4**, 1.
Okada, Y. K. (1930). *Arch. Protistenk.* **70**, 131.
Pantin, C. F. A. (1923). *J. Marine Biol. Assoc. United Kingdom* **13**, 24.
Pappas, G. D. (1956). *J. Biophys. Biochem. Cytol.* **2**, 221.
Pappas, G. D. (1959). *Ann. N.Y. Acad. Sci.* **78**, 448.
Pappas, G. D., and Brandt, P. W. (1958). *J. Biophys. Biochem. Cytol.* **4**, 485.
Pekarek, J. (1930). *Protoplasma* **11**, 19.
Pitts, R. F., and Mast, S. O. (1933). *J. Cellular Comp. Physiol.* **3**, 449.
Pitts, R. F. and Mast, S. O. (1934). *J. Cellular Comp. Physiol.* **4**, 237.
Prescott, D. M. (1953). *Nature* **172**, 593.
Rand, H. W., and Hsu, S. (1927). *Science* **65**, 261.
Ray, D. L., and Hayes, R. E. (1954). *J. Morphol.* **95**, 159.
Rebhun, L. I. (1961). In preparation.
Rhumbler, L. (1898). *Arch. Entwicklungsmech. Organ.* **7**, 103.
Robineaux, R. (1954). *Rev. hématol.* **9**, 364.
Robineaux, R. (1959). *In* "Mechanisms of Hypersensitivity," (Henry Ford Hospital Symposium), pp. 371-412. Little, Brown, Cambridge, Massachusetts.
Robineaux, R., and Nelson, R. A., Jr. (1955). *Ann. inst. Pasteur* **89**, 254.
Rösel von Rosenhof, A. J. (1755). *Monatlichherausgegebenen Insecten-Belustigung* **3**, 622.
Roslansky, J. D. (1959). Unpublished observations.
Sandon, H. (1934). *Nature* **133**, 761.
Schaeffer, A. A. (1916). *J. Exptl. Zool.* **20**, 529.
Schaeffer, A. A. (1920). "Amoeboid Movement." Princeton Univ. Press, Princeton, New Jersey.
Schmidt, W. J. (1939). *Protoplasma* **33**, 44.
Schulze, F. E. (1875). *Arch. mikroscop. Anat. u. Entwicklungsmech.* **11**, 329.
Scott Blair, G. W. (1938). "An Introduction to Industrial Rheology." Blakiston, Philadelphia, Pennsylvania.
Singh, B. N. (1938). *Quart. J. Microscop. Sci.* **80**, 601.
Stewart, P. A., and Stewart, B. T. (1959). *Exptl. Cell Research* **17**, 44.
Szent-Györgyi, A. (1947). "Muscular Contraction." Academic Press, New York.
Szent-Györgyi, A. (1958). *Science* **128**, 699.
Tiegs, O. W. (1928). *Protoplasma* **4**, 88.
Torch, R. (1959). *Ann. N.Y. Acad. Sci.* **78**, 407.

Torch, R. (1960). Personal communication.
Ts'o, P. O. P., Jr., Bonner, J., Eggman, L., and Vinograd, J. (1956). *J. Gen. Physiol.* **39**, 325.
Wallich, G. C. (1863). *Ann. and Mag. Nat. Hist.* [3] **11**, 365.
Wallin, W. (1960). Unpublished observation.
Watson, S. W., and Raper, K. S. (1957). *Gen. Microbiol.* **17**, 368.
Yagi, K. (1959). *Dobytsugaku Zasshi* **68**, 317.
Yotsuyanagi, Y. (1953). *Cytologia (Tokyo)* **18**, 146.
Zimmerman, A.M., Landau, J. V., and Marsland, D. A. (1958). *Exptl. Cell Research* **15**, 484.

CHAPTER 4

Cilia and Flagella

By DON FAWCETT

I. Introduction

Cilia and flagella are motile, hairlike appendages on the free surface of cells. When they are few in number and long, in proportion to the size of the cell, it is customary to call them *flagella;* when they are numerous and relatively short they are called *cilia.* Flagella usually have an undulant motion, and cilia have a pendular stroke; flagella generally beat independently whereas cilia are coordinated, but no clear-cut morphological or physiological distinction can be made between the two and the terms are often used interchangeably. In the plant kingdom flagella are confined to the zoospores and gametes of certain algae and aquatic fungi and to the spermatozoids of mosses, ferns, and two groups of primitive trees. In the animal kingdom, they are of more widespread occurrence. Cilia are characteristic of a major group of the Protozoa and are found in one or more epithelial tissues in animals of nearly every metazoan phylum. Many protozoans, and the spermatozoa of most metazoans, swim by means of flagella. They are also found on a number of fixed-cell types such as the choanocytes of sponges, the gastroderm cells of coelenterates, and certain cells of the vertebrate nephron.

Cilia play an important role in such diverse physiological processes as *locomotion, alimentation, circulation, respiration, reproduction,* and *sensory reception.* They are organs of locomotion, not only for unicellular organisms, but also for flatworms and even some gastroderms. Sessile protozoans, bryozoans, and mollusks use them to create currents in the surrounding water that sweep food into the oral apparatus. They transport food through the alimentary tract of mollusks and some echinoderms. In sipunculoids, echiuroids, and certain annelids, in which a true circulatory system is poorly developed or entirely lacking, cilia maintain a slow circulation of body fluid. In some vertebrates the circulation of the cerebrospinal fluid is attributed to the beating of cilia lining the ventricles. They have an important protective role in the respiratory tract of mammals, where they continuously eliminate from the body, bacteria and foreign particles inhaled and trapped in the layer of mucus that lines the airway. The ciliated epithelia of the reproductive ducts participate in the transport of the gametes. In addition to having functions that depend upon their motility, cilia may be transformed in the course of cell differentiation into nonmotile structures having a sensory function. For example, the cnidocils that trigger the discharge of nematocysts in cnidaria are modified cilia and the photosensitive elements in the retina of mollusks, amphibians, and mammals arise by modification of cilia.

The electron microscope in recent years has revealed in cilia and flagella a complex internal structure consisting of eleven longitudinal fibrils arranged in a precise pattern that is amazingly constant throughout the plant and animal kingdoms. This uniformity of structure has stimulated widespread interest in such problems as the evolution and functional significance of this master plan; the origin and chemical nature of the internal fibrils; the mode of contraction of cilia, and the mechanism of coordination of their beat. It is the purpose of this chapter to draw together the results of this renewed interest in the ciliary apparatus. In its preparation we have drawn freely upon the classic monograph by Gray (1928) and the reviews of Klein (1932) and Lucas (1932) to achieve broader coverage than is available in the recent literature and to present the newer developments in proper historical perspective.

II. Structural Organization of the Ciliary Apparatus

The essential components of the ciliary apparatus are: the *cilium,* the slender cylindrical process projecting from the free surface of the cell, and its *basal body,* the intracellular organelle which is its origin and kinetic center. In addition to these two components, which are both essential for motility, fibrous *rootlets* occur in some ciliated epithelia. One or more of these originate from the lower[1] pole of each basal body and run downward for variable distances into the cytoplasm. They are generally regarded as supporting elements. The corresponding fibrous structures in ciliates run along the rows of cilia parallel to the surface of the organism, and in addition to their supporting function, are believed to play a role in the integration of ciliary activity.

The basal bodies are embedded in a refractile, firmly gelated ectoplasmic layer of clear cytoplasm just beneath the cell surface and are uniformly spaced in straight, parallel rows. The number of cilia per cell varies from a few to three hundred in different epithelial tissues, and may range up to 14,000 on some Protozoa. In normal function the timing of the beat of all the cilia on a cell is under the control of an intracellular coordinating mechanism, but the individual cilia are usually free to move independently of one another. This is not invariably true in Protozoa, however, where groups of cilia may be bound together in a common matrix to form compound motile organelles. A row of cilia may thus be joined to form a *membranelle,* or a cluster of cilia may fuse into a pyri-

[1] As used here, *vertical* and *horizontal* refer, respectively, to the directions *perpendicular* and *parallel* to the cell surface. *Upper* and *lower* poles of the basal body are used synonymously with *distal* and *proximal* to distinguish the end nearer the cell surface from that more deeply situated in the cytoplasm. *Forward* and *backward* are used to designate directions along the rows of basal bodies corresponding to the directions of the *effective* and *recovery* stroke of the cilia.

form structure called a *cirrus*, resembling a watercolor brush. These compound protozoan organelles are made up of cilia that have the same fine structure as those of ciliated cells in general.

A. *The Cilium or Flagellum*

1. *Historical Considerations*

Although to the early cytologists using the light microscope, cilia appeared devoid of internal structure, Engelmann postulated as early as 1868 that their contractile protoplasm was composed of fine fibrillar elements of molecular dimensions. It was speculated that these submicroscopic fibrils were aligned parallel to the long axis of the cilium in its resting phase but shortened into globular form during contractions. The first visual evidence of fibrillar structure in a motile cell process was reported in 1887 by Jensen, who described the appearance of numerous minute fibrils when sperm flagella were disrupted by excessive pressure on the cover glass. This observation on the fraying of the tip of the sperm tails was confirmed and extended by Ballowitz (1888), and it was apparently he, who first interpreted these fibrils as contractile elements running the full length of the flagellum. This view was shared by few of his contemporaries.

The majority of investigators of that period reasoned that the contractile substance of the cilium would be most effective if combined with an elastic supporting element, and therefore they envisioned a supple rod or fiber in the core of the cilium surrounded by a layer of contractile protoplasm. In spite of the observations by Jensen and Ballowitz of multiple fibrils in sperm tails, this so-called *axial filament* of cilia was usually depicted as a single strand. Then in 1909, Dellinger demonstrated four or more longitudinal fibrils in the cilia and flagella of several different protozoans by teasing under the microscope, and Korschikov (1923) confirmed the presence of several fibrils in stained preparations of similar material. Corresponding observations were subsequently reported for epithelial cilia (Erhard, 1910; Grave and Schmitt, 1925) and for the cilia on the male gametes of certain algae (Mühldorf, 1930; Midduno, 1934).

With the development of the electron microscope in the 1940's, motile microorganisms were among the first biological materials examined, and the presence of internal fibrils in protozoan cilia and sperm flagella was soon reaffirmed (Harvey and Anderson, 1943; Schmitt *et al.*, 1943). Specimen preparation for electron microscopy at that time consisted of fixing intact or mechanically disrupted cells in osmium vapor and allowing them to dry onto the surface of a Formvar film, which could be transferred to the specimen grid of the microscope. To gain contrast and accentuate surface irregularities, metal was evaporated at a low angle onto the specimen. The image then recorded in electron micrographs had a "shadowed" three-dimensional appearance that facilitated interpretation.

Applying these methods to spermatozoa from two different species, Hodge (1949) found that the number of fibrils in the tails was consistently eleven, of which two differed slightly from the other nine (Fig. 1). Other authors, studying protozoans, variously estimated the number of fibrils per cilium to be from 8 to 12. At the time, there seemed to be no reason to doubt that the number varied to this extent from species to species. The first indication of uniformity of structure in a wide range of different forms came from the work of Irene Manton and her associates in the Botany Department at Leeds, who, from 1950 onward, carried out an extensive series of comparative investigations on plant cilia, making skillful use of the best techniques then available (Manton and Clarke, 1950, 1951a, 1952; Manton, 1952). In preparations of fern spermatozoids, the cilia often disintegrated so that their component fibrils were well spread upon the supporting film and could easily be counted (Fig. 2). Manton found in these plant cilia, as Hodge had earlier in animal sperm tails (Fig. 1), that the number was always eleven and that two of the fibrils differed from the others in being thinner and more closely adherent. It was suggested that, in the intact cilium, this odd pair was centrally placed with the other nine arranged around it (Manton and Clarke, 1951a). In some thirty plant species studied in the next three years, the number of ciliary fibrils was always found to be eleven, and it was concluded that this number was probably of universal occurrence in plant cilia. By 1952 enough evidence had accumulated for Manton and Clarke to publish the diagrammatic reconstruction of a cilium reproduced here as Fig. 3. Nine peripheral fibrils are shown forming a tube around a central pair. Each fiber is enclosed in a separate sheath and divided into two halves by a radially oriented septum. The two central fibers were presumed to be double, like the other nine, and are shown enclosed in a membrane called the *core sheath.* A regular periodic beading observed along one side of isolated fibers of dismembered cilia was interpreted as the residue of a helical component interposed between the core sheath and the outer row of longitudinal fibrils. This was represented in the diagram as the *spiral-tube lining.* This depiction of the internal organization of the cilium was based upon an extensive experience with plant material and included as much information as could be gained from the examination of dissociated cilia in whole mounts. Efforts to apply similar methods of study to epithelial cilia of animals

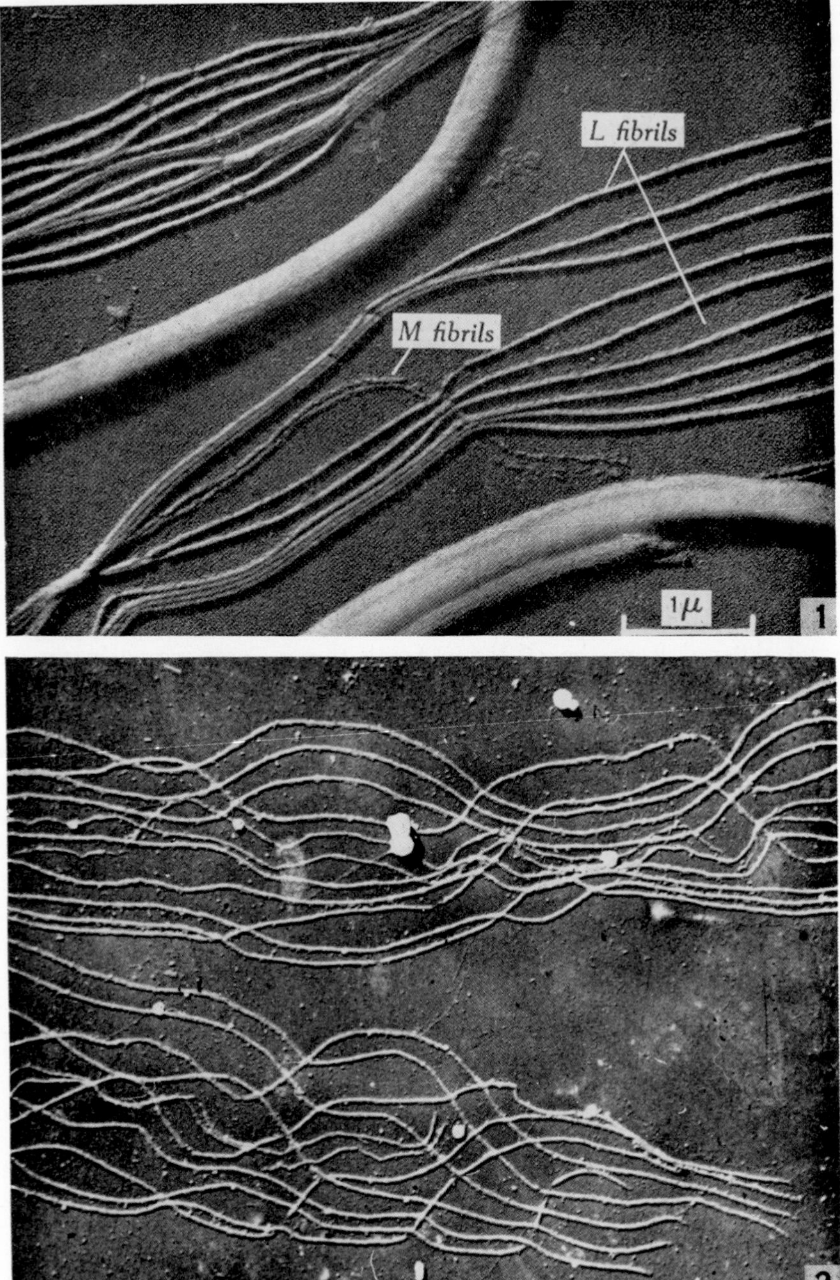
L fibrils
M fibrils
1μ
1
2

were disappointing and led only to a verification of the fact that these cilia also contained multiple fibrils (Engstrom, 1951). Further analysis of the number and disposition of the fibrils had to await the development of methods for the preparation of ultrathin tissue sections in which their normal spatial relations were preserved.

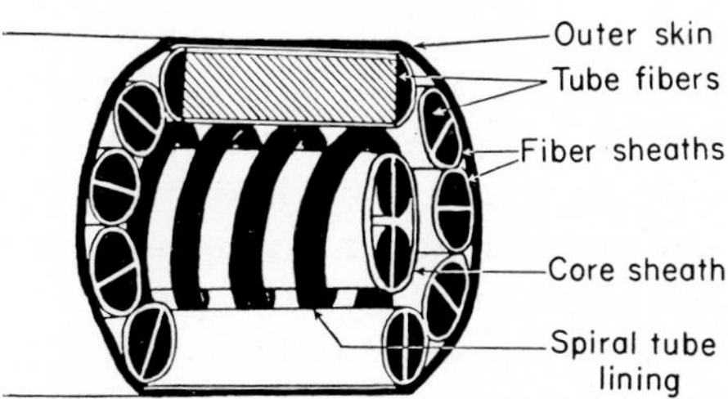

FIG. 3. Manton's diagrammatic reconstruction of the cilium of *Sphagnum*, based upon early electron microscopic studies of dismembered cilia (Manton and Clarke, 1952). This prediction of the three-dimensional arrangement of the components of the cilium proved to be remarkably accurate. Subsequent studies, employing thin sections, have confirmed most of the structural relations depicted in this diagram. The central fibrils have been shown to be single instead of double as shown here; the existence of a spiral tube lining is doubtful, and the presence of a core sheath is controversial.

By 1953 the necessary technical advances in fixation (Palade, 1952), embedding (Newman, 1950), and microtomy (Porter and Blum, 1953), had been achieved to permit Fawcett and Porter (1954) to examine epithelial cilia from a wide range of animal species in thin sections. From invertebrates to man the number of internal fibrils in the cilia was invariably eleven, and the study of cross sections provided a gratifying confirmation of the salient features of Manton's diagrammatic reconstruction of the cilium. The description of the structure of the ciliary apparatus

---

FIG. 1. Electron micrograph of intact and dismembered tails of cock spermatozoa. In the disrupted tails, nine similar fibrils of uniform size (L fibrils) are seen, and two odd fibrils which are thinner and apparently somewhat more fragile (M fibrils). It was believed that the M fibrils were normally in the axis of the tail with the other nine arranged around them. This assumption was later substantiated by electron micrographs of transverse sections of intact sperm tails. From Grigg and Hodge (1949).

FIG. 2. Parts of two dismembered cilia from a spermatozoid of the male fern. The number of strands is eleven and two of these differ slightly from the others. This and similar micrographs from a large variety of plant material first clearly established the constancy in the number of fibrils in plant cilia. Specimen shadowed with uranium. Magnification: × 12,500. From Manton and Clarke (1951a).

which follows is based upon electron micrographs of thin sections of the ciliated epithelia of Metazoa studied by Fawcett and Porter (1954) or Rhodin and Dalhamn (1956) and cilia of Protozoa investigated by Sedar and Porter (1955), Roth (1957), and Gibbons and Grimstone (1960).

2. *The Current Concept of the Structure of Cilia and Flagella*

Cilia and flagella have a cylindrical shaft approximately 0.2 μ in diameter tapering near the tip. The cilia of epithelia are 5–10 μ long whereas flagella range from these dimensions upward in length to 150 μ or more in some of the Protozoa. In the interior of each of these vibratile processes there are eleven longitudinal fibers—two fibers in the center and the other nine arranged in a cylinder around them (Fig. 4). This bundle of fibers evidently corresponds to the structure which light microscopists formerly called the *axial filament.* The term *axial filament complex* is now applied to this fasciculus consisting of two central and nine peripheral fibers. In cross sections the cut ends of the central fibers are circular, about 200 A. in diameter and 300 A. apart from center to center (Fig. 6, A). They have a dense surface layer (50–70 A.) and a less dense interior, which gives them a tubular appearance. The outer fibers are oblong in section with the short axis (180–220 A.) radial and the long axis (300–350 A.) roughly parallel to the ciliary membrane. Each is divided into two halves by a radially oriented septum, which is continuous with the dense surface layer of the fiber and is of the same thickness (Fig. 5). The peripheral fibers thus have a "double-barreled" or

---

Fig. 4. Transverse section through the cilia (*Cl*) of a cell in the epithelium of the mouse oviduct, from a paper by Fawcett and Porter (1954) which first demonstrated in thin sections the constancy in number and arrangement of the internal fibrils in cilia of a large variety of animal species. This work confirmed the main points of Manton's tentative reconstruction of the cilium (Fig. 3) and established that the uniformity of structure which she found in the plant kingdom held true throughout the animal kingdom as well. All the sectioned cilia show the same pattern of nine peripheral and two central fibrils, and the orientation of the central pair is the same in all the cross sections. Lines drawn through their centers are parallel, and it is believed that the direction of ciliary beat (see double arrow) is perpendicular to these lines. The smaller profiles devoid of internal structure are microvilli (*Mv*) of the brush border of the cell.

Fig. 5. Cross section through cilia of the epithelium lining the rat trachea. The double nature of the peripheral fibrils is especially clear in this micrograph (at the arrows). Each fibril has a dense exterior and a less dense interior, giving it a tubular appearance, and each appears to be divided into two compartments by a radially oriented septum. One of the cilia shown here (*X*) is damaged and its limiting membrane has broken down into small vesicles. Magnification: × 121,000. (Courtesy Dr. J. Rhodin.)

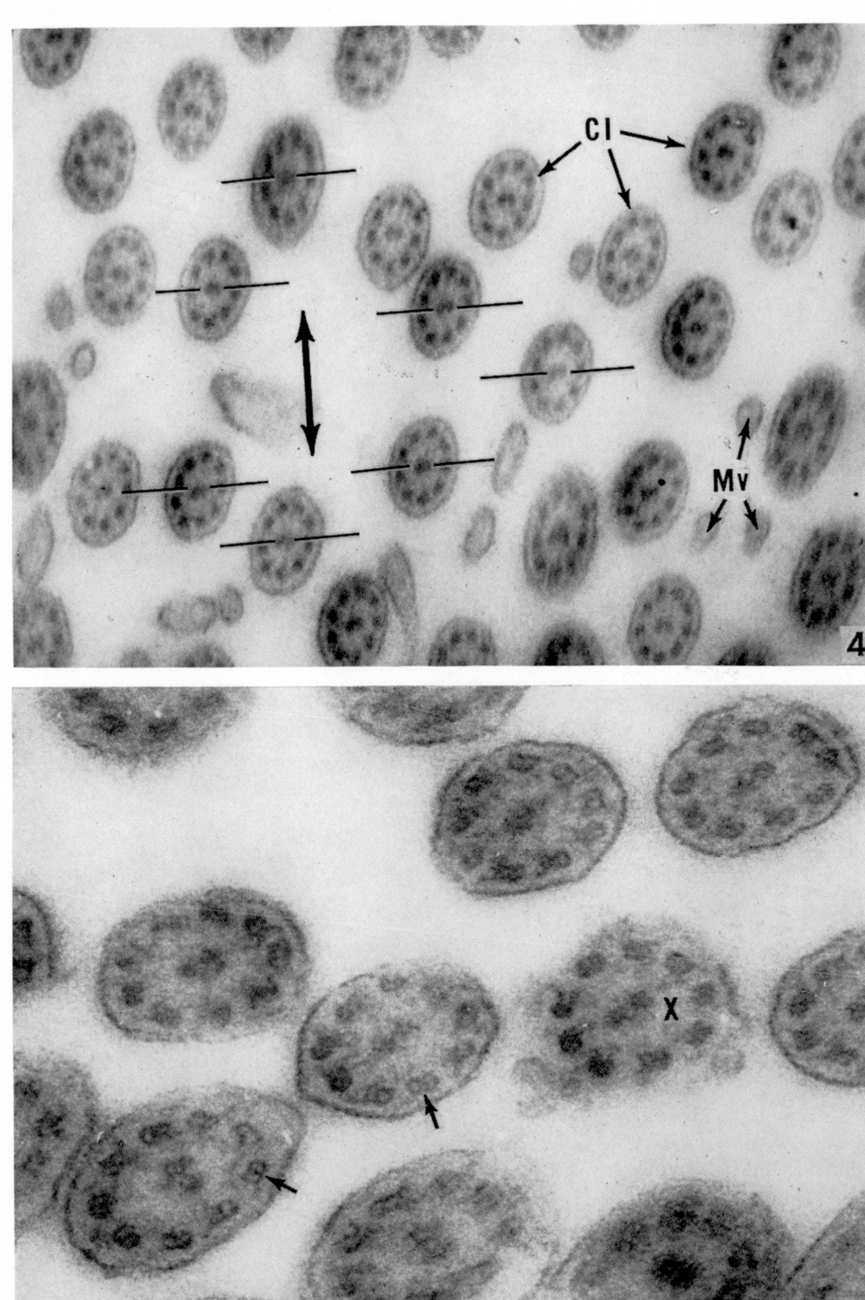
Cl
Mv
4
X
5

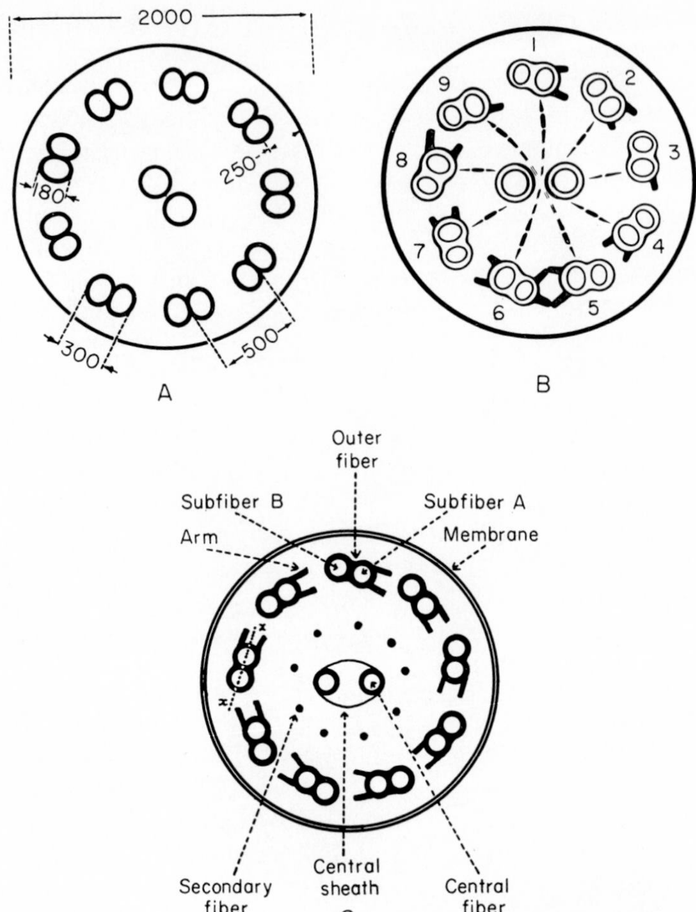

FIG. 6. A. Diagram illustrating the disposition of fibrils in a cross section of a typical cilium and giving the approximate dimensions in Angström units, based upon the work of Fawcett and Porter (1954) and of Rhodin and Dalhamn (1956).

B. Schematic representation of a sea urchin sperm flagellum, modified from Afzelius (1959). Some inequality in the size of the two halves of the peripheral double fibrils is suggested, and between the fibrils are short projections referred to as "arms." The interrupted lines represent a radial distribution of linear densities observed in the interfibrillar matrix (see also Fig. 17).

C. Diagram of a flagellum from Gibbons and Grimstone (1960). These authors find that the arms are consistent in direction and are always associated with the same subfiber. The pattern of densities in the matrix, interpreted by Afzelius as radial lamellae, are depicted here as cross sections of a set of secondary fibers interposed between the central and outer fibers.

"figure-of-eight" appearance in section whereas the central single fibers have a simple circular profile.

These components of cilia and flagella are routinely demonstrable in sections of a wide variety of plant and animal materials and are presumed to be of universal occurrence. With special staining methods and high resolution electron microscopy, a number of additional details of fine structure have recently been revealed in flagella of invertebrate spermatozoa (Afzelius, 1959) and in those of certain protozoans (Gibbons and Grimstone, 1960). These new structural features may also prove to be common to all motile processes when similar methods of specimen preparation have been more widely applied. Afzelius found that the outer double fibers are partitioned into two subfibers of slightly unequal size and that short projections, which he called *arms,* arise from one subfiber of each doublet and extend toward the next fiber (Figs. 6, B and 17). According to Gibbons and Grimstone (1960), there are usually two of these arms and, contrary to the condition shown in Afzelius' diagram (Fig. 6, B), these point in the same direction on all nine fibers (Figs. 6, C and 7). This direction is clockwise from the point of view of an observer looking along the flagellum from base to tip.

In longitudinal sections, the arms do not appear to be continuous flanges but are approximately rectangular structures, 150 A. long, 50 A. thick, and spaced 130 A. apart. The subfiber that bears the arms (subfiber A) is slightly smaller and its interior is somewhat more dense than that of the other (subfiber B). The subfibers of each doublet do not both lie on the circumference of a circle, as was formerly assumed to be the case. Instead, subfiber A lies slightly closer to the center of the flagellum than subfiber B, so that a line passing through the centers of the two (*x–x* in Fig. 6, C) would make an angle of 5°–10° to a line drawn tangent to the flagellum opposite the center of that fiber.

The internal fibers of the cilium maintain their relative positions with such a high degree of constancy that one must assume they are bound together by direct connections or by being embedded in a firm matrix. In routine preparations the matrix appears homogeneous, but in micrographs of exceptional contrast it is possible to discern a spokelike pattern of linear densities extending from the central pair toward subfiber A of each of the nine outer fibers (Figs. 6, B and 17). In depicting these radial markings diagrammatically, Afzelius showed them as lines with thickenings midway between the central pair and the outer fibers. Instead of lines, Gibbons and Grimstone find in this location nine dots of appreciable density, which they interpret as cross sections of an intermediate set of slender longitudinal fibers. The nine dots are very clearly shown in their beautiful micrographs of cross sections of protozoan flagella (Fig. 7) but, the delicate sinuous lines in longitudinal sections, which they interpret as the secondary *fibers,* are less convincing. At present it remains unsettled whether the faint pattern of densities seen in the flagellar matrix represents a second group of longitudinal fibers or a set of radial connections between the central and the outer fibers.

In the central region of the flagellum in transverse section, thin lines can be seen running out from one of the central fibers and curving round to join the other. These vague markings are tentatively interpreted by Gibbons and Grimstone as a *central sheath* enveloping the axial pair of fibers. Slanting lines seen in this region in longitudinal sections suggest to them the possibility that this central sheath may consist of minute filaments coiled round the two central fibers. It will be recalled that Manton and Clarke (1952) described a *core sheath* and postulated a *spiral-tube lining* (Fig. 3) on the basis of micrographs of disrupted cilia. When neither of these structures were found in the early thin sections of cilia, other workers dismissed them as artifactitious. It is a tribute to their keen observation of whole mounts that eight years later, improved methods of preservation and staining have made it possible to identify

---

FIG. 7. Cross sections of flagella and basal bodies of the flagellate *Pseudotrichonympha.* Comparison of this splendid electron micrograph by Gibbons and Grimstone with those of Figs. 4 and 5 provides an interesting measure of the rapid technical progress in electron microscopy between 1954 and 1960. At the left of the figure are several flagella; at the right, is a group of basal bodies. The double contour of the unit membrane limiting the flagella is clearly shown, as are also the two central and nine outer fibers. Each of the latter consists of two subfibers (A and B), of which one bears small arms that project clockwise toward the next fiber in the row. Between the central and outer fibers are nine dots, which are interpreted as sections of a set of secondary fibers. The wall of each basal body consists of nine groups of three aligned tubular elements (*a, b,* and *c*) set at an angle, with subunit *a* of each joined by a slender dense strand to subunit *c* of the next and so on around the ring. In addition, there is a central cylinder in the basal body connected to the triplets in the wall by nine radially disposed linear densities (see interpretive diagram Fig. 9, G).

FIG. 8. Sperm tail of the bat *Myotis* sectioned transversely through the principal piece. The spermatozoon illustrated here has an anomalous double tail with two flagella in a common membrane but, in other respects, its fine structure is entirely normal. Both the flagella shown here are surrounded by the dense fibrous sheath of the principal piece. In the outer row of longitudinal fibers of each axial filament complex, the two subfibers of each doublet differ in their density. Subfiber B has a center of low density and appears tubular, while subfiber A, which bears the arms, has a homogeneously dense interior and therefore appears solid. Slight differences in density of the two subfibers have been observed in cilia and flagella as well as in sperm tails.

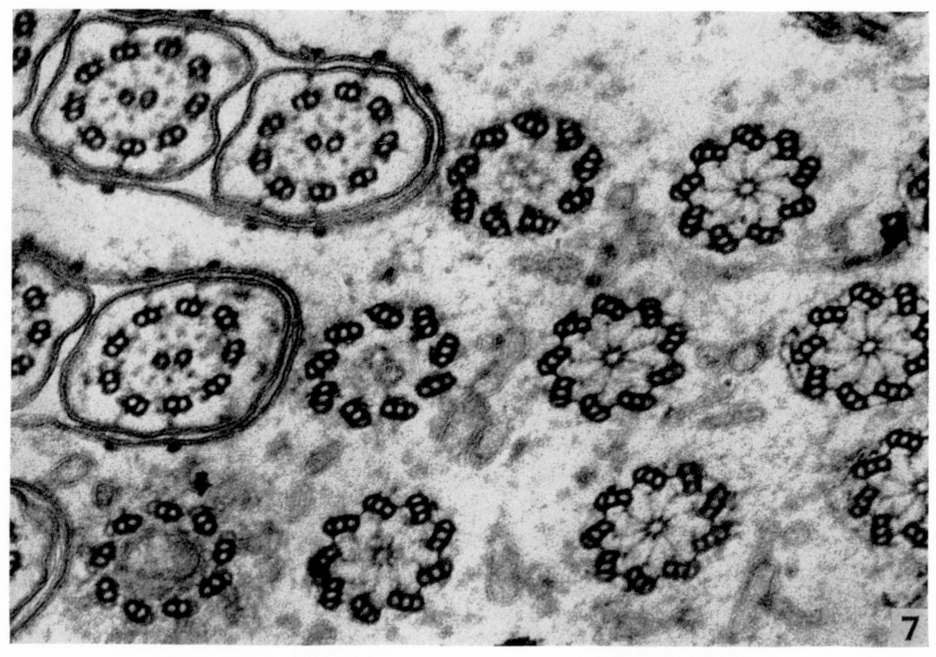
7

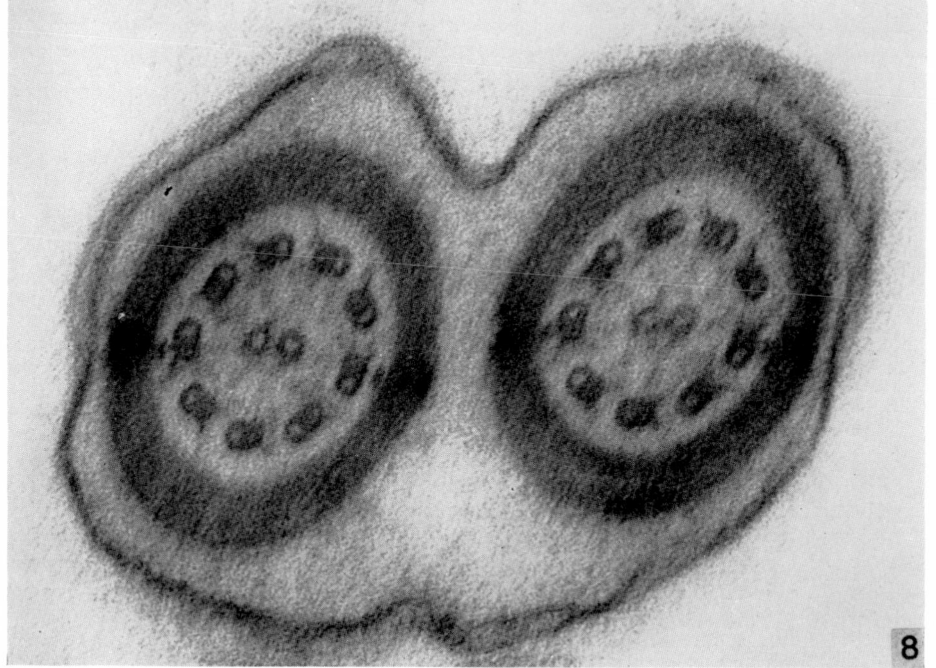
8

corresponding structures in thin sections. The *central sheath* of Gibbons and Grimstone is evidently the same as Manton and Clarke's *core sheath*, and the *arms* which recur every 150 A. along the length of the outer fibers apparently correspond to the periodic projections ("battlements") along one side of the dissociated fibers which led them to postulate a spiral-tube lining. These studies make it quite clear that, in addition to the eleven internal fibers, there are other structures of very small dimensions in the ciliary matrix, which our present methods are not able to reveal with sufficient clarity to warrant an interpretation of their functional significance.

The manner in which the fibers terminate at the tip is obviously important to an understanding of their function in ciliary motion. There has been no unanimity on this point. Manton reported that in dissociated plant cilia the outer fibers seemed to end at slightly different levels. The central pair ended before the outer fibers in some plant species, but in others they continued for some distance beyond, forming the core of a slender "whip lash tip." Cilia of the latter sort have not been observed in animal material. Rhodin and Dalhamn (1956) were of the opinion that, in the epithelial cilia of vertebrates, the central fibers end abruptly whereas the outer fibers converge at the tip and coalesce, but this view has not been supported by later work. Roth (1956, 1957), for example, found no evidence of fusion of fibers in protozoan cilia but found instead that the peripheral fibers end at slightly different levels and the central pair terminate a short distance beyond. The most detailed description of this region is to be found in the elegant paper by Gibbons and Grimstone (1960) on flagellar structure in certain flagellates. These authors find that, as the outer fibers converge toward the tip, the minute arms on subfiber A first disappear and then the even spacing and symmetrical arrangement of both the outer and the central fibers is gradually lost. The double fibers become single as a result of the earlier termination of one of their subfibers. The change from a double to a single fiber and the abrupt ending of the latter take place at different levels on different fibers. Thus, transverse sections through the terminal half-micron of the flagellum show variable numbers of mixed doublets and singlets. At this level the morphological distinction between central and outer fibers is lost, and ultimately all fibers seem to end independently and without establishing contact with the membrane covering the tip of the flagellum. The longitudinal fibers in the terminal piece of the mammalian sperm tail seem to end in much the same manner (Anberg, 1957; Schultz-Larsen, 1958).

Although the mode of termination of the longitudinal fibers at the tip of cilia and flagella is probably similar in different taxonomic groups within the animal kingdom, the relations of the fibers to the basal body is subject to considerable variation. In mollusks and amphibians, the outer fibers appear to end proximally in a distinct transverse *basal plate,* which is separated from the upper end of the basal body by a narrow clear zone (Figs. 10c, d, 11A). It is not possible to say whether this apparent interspace is occupied in life by a substance of very low density, or whether a denser material that bonds the basal plate of the cilium to its basal body has been extracted in specimen preparation. The appearance of this region in mammals and protozoans is quite different from that in mollusks and amphibians. In these, there is usually no clearly defined separation of the cilium and basal body (Fig. 10, a, b, e, and f). Their junction is marked only by an ill-defined band of greater density in the matrix, and this is traversed by the outer fibers of the cilium, which are clearly continuous with similar fibrous components in the wall of the basal body (Fig. 12).

In species wherein the cilia have a distinct basal plate, the central pair of fibers, as well as the outer nine, seem to terminate in or near it, but in those species where the outer fibers are continuous with the wall of the basal body, there is no uniformity in the manner of termination of the central pair. In longitudinal sections of mammalian cilia the central fibers become indistinct and appear to end a short distance above the basal body. Transverse sections of the cilium at this level show a central cavity, which is continuous with that of the basal body. Rhodin and Dalhamn (1956) report that one or both of the central fibers may bifurcate just before their termination so that in cross sections a group of three or four circular profiles may be found instead of the usual pair. However, this does not seem to be a regular occurrence. In ciliates of the families Isotrichidae and Ophryoscolecidae (Noirot-Timothée, 1958), in *Spirostomum* (Randall, 1956, 1957), and in the flagellate *Opalina* (Noirot-Timothée, 1959) the central fibers end proximally in a spheroidal osmiophilic body about 80 mμ in diameter situated at the upper limit of the basal body. This has been called the *axial granule.* In *Euplotes* (Roth, 1957) and *Tetrahymena* (Rudzinska, 1958) the central fibers pass through the dense transverse zone that marks the lower limit of the cilium and form a loop deep in the central cavity of the basal body (Fig. 10, e). In the flagellates, *Trichonympha* and *Pseudotrichonympha* (Gibbons and Grimstone, 1960), the central fibers end abruptly about 250 A. distal to the ill-defined basal plate, but one of them is attached to a dense, eccentrically placed crescentic body that seems to merge at its lower end with the substance of the plate (Fig. 9).

The differences between the central and the outer fibers of a cilium evidently go beyond the simple matter of singleness versus doubleness. A chemical difference is suggested by the observation that the central fibers are more sensitive to treatment with distilled water and are more readily digested by pepsin (Grigg and Hodge, 1949). In the central fibers, faint oblique cross striations with a repeating period of 130 A. can be seen in longitudinal sections (Gibbons and Grimstone, 1960), but in the outer fibers, signs of periodic substructure are rarely detectable. The suggestion that this striation is due to the presence of coiled fibrils that form a helix within the wall of each central fiber (Gibbons and Grimstone, 1960) requires further study.

The earlier workers who studied dissociated cilia speculated that the central fibers might twist around one another or that the outer fibers might have a helical course around the central pair. These suggestions have not been substantiated by more recent studies employing thin sections. In transverse sections, the outer fibers maintain exactly the same position relative to the central pair at all levels except very near the tip, where the symmetry of their arrangement is finally lost. Furthermore, in sections parallel or oblique to the surface of an epithelial cell, the orientation of the two central fibers is always essentially the same in all the cilia (Fig. 4). This would not be the case if either the central pair, or the axial filament complex as a whole, had a helical course. There seems no longer to be any reason to doubt that all the fibers run straight throughout the greater part of the length of a cilium or flagellum.

The orientation of the component fibers appears to have a constant relation to the direction of beat. On the laterofrontal cells of the gills of lamellibranchs there are only two rows of cilia on a cell and the direction of beat is along the rows. In electron micrographs of this epithelium, Fawcett and Porter (1954) found that lines drawn through the central fibers of the cilia were parallel to each other and approximately at right angles to the direction of beat (Fig. 4). The same relationship between the orientation of the central pair of fibers and the direction of beat has been found to hold for the swimming plates of ctenophores (Afzelius, 1961). Similar constancy of orientation of the center fibers is found in *Trichonympha* (Gibbons and Grimstone, 1960) and in other flagellates and ciliates which display considerable variation in the direction of beat of their motile processes.

A cilium or a flagellum is enclosed by a membrane which is continuous at its base with the plasma membrane of the cell. It is about 90 A. in thickness and, like other cell membranes that have been examined in high resolution micrographs, it has a trilaminar appearance

(Figs. 7 and 17) consisting of two opaque osmiophilic layers ~30 A. thick separated by a less-dense intermediate layer of approximately the same thickness. The membrane is normally smooth, but some unevenness or wrinkling is often observed near the base as though the membrane were slightly redundant in this segment where the greatest amount of bending occurs. In some protozoans a few short villous projections are found in this region (Roth, 1956). Their significance is obscure.

B. *The Basal Bodies*

The basal bodies of cilia[2] appear with the light microscope as spherical granules or short rods arranged in rows immediately beneath the cell surface. The electron microscope has provided decisive morphological evidence supporting the traditional view that basal bodies and centrioles are homologous structures.[3] In electron micrographs centrioles are hollow, cylindrical bodies 300–500 mμ in length and 120–150 mμ in diameter open at one or both ends. They have a dense wall that surrounds a central cavity containing protoplasm of relatively low density to electrons. Embedded within the wall of the centriole are straight tubules or groups of tubules. These usually appear double in cross section like the outer fibers of a cilium, but in exceptionally well-preserved specimens each may be seen to be made up of three tubular elements arranged in a row that is inclined at an angle of about 40° to a line drawn tangential to the circumference of the centriole (de Harven and Bernhard, 1956;

[2] The intracellular structure found at the base of every cilium or flagellum has been given many different names. In ciliated epithelia, it is usually called the *basal granule,* or *basal corpuscle.* In ciliate protozoans it is the *kinetosome,* and in flagellates, the *blepharoplast.* In the sperm tails of metazoa it is commonly called the *proximal centriole.* Since the origin and fine structure of this component of the ciliary apparatus is basically the same in all cell types and groups of organisms, there seems to be no reason for perpetuating this multiplicity of names. We have chosen, therefore, to use the simple descriptive term *basal body.*

[3] The cytologists Henneguy (1897) and Lenhossek (1898) independently studying epithelia that contained both ciliated and nonciliated cells, were able to demonstrate a pair of typical centrioles in the apical cytoplasm of the nonciliated members but did not find a corresponding "diplosome" in the ciliated cells. Moreover, they saw no mitoses in the ciliated cells. They concluded from these observations that the basal bodies of the cilia were essentially a group of centrioles derived by reduplication of the original diplosome and that the ciliated cells were unable to divide because their centrioles had been transformed into basal bodies. This generalization came to be known as the *Henneguy-Lenhossek theory.* Later investigations have shown that ciliated cells do retain a pair of unmodified centrioles and that they are capable of mitotic division. The Henneguy and Lenhossek concept of basal bodies and centrioles as homologous structures, however, has gained general acceptance and continues to be the prevailing view today.

Bernhard and de Harven, 1960). Rounded dense bodies about 70 mμ in diameter are often associated with the outer surface of centrioles and connected to them by linear densities (Bessis and Breton-Gorius, 1957). These pericentriolar *satellites* and their connecting *bridges* are not found regularly and may be transitory structures associated with certain phases of centriolar activity.

The basal bodies or kinetosomes in ciliates such as *Paramecium* are

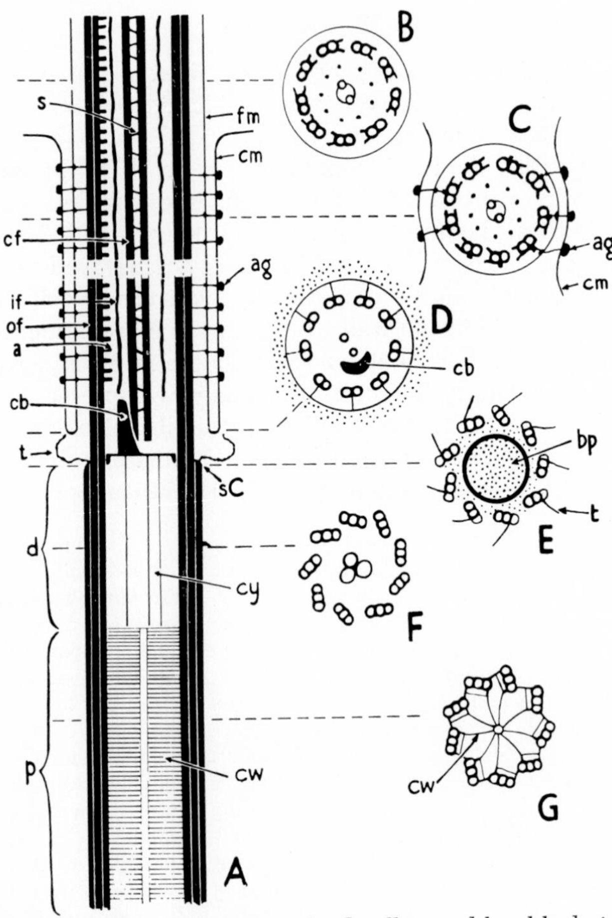

FIG. 9. Diagrammatic reconstruction of a flagellum and basal body in the anterior body region of *Pseudotrichonympha*. A. Features seen in median and tangential longitudinal sections. B–G. Transverse sections at the levels indicated. Key: *a*, arms; *ag*, anchor granule; *bp*, basal plate; *cf*, central fiber; *cb*, crescentic body; *cm*, cell membrane; *cw*, cartwheel structure; *cy*, cylinders; *d*, distal region of basal body; *fm*, flagellar membrane; *of*, outer fiber; *p*, proximal region of basal body; *s*, central sheath; *sC*, distal end of subfiber C; *sf*, secondary fiber; *t*, transitional fiber. From Gibbons and Grimstone (1960).

strikingly similar to centrioles. Each is cylindrical in form with its central cavity closed above by the *basal plate* of the cilium and open below to the cytoplasm (Fig. 12). The nine outer fibers of the cilium are continuous with nine uniformly spaced hollow fibers or tubules in the wall of the basal body (Sedar and Porter, 1955). These are described as being double in cross sections, like the fibers of the cilium, or triple. As the resolution and contrast of electron micrographs has improved, reports of

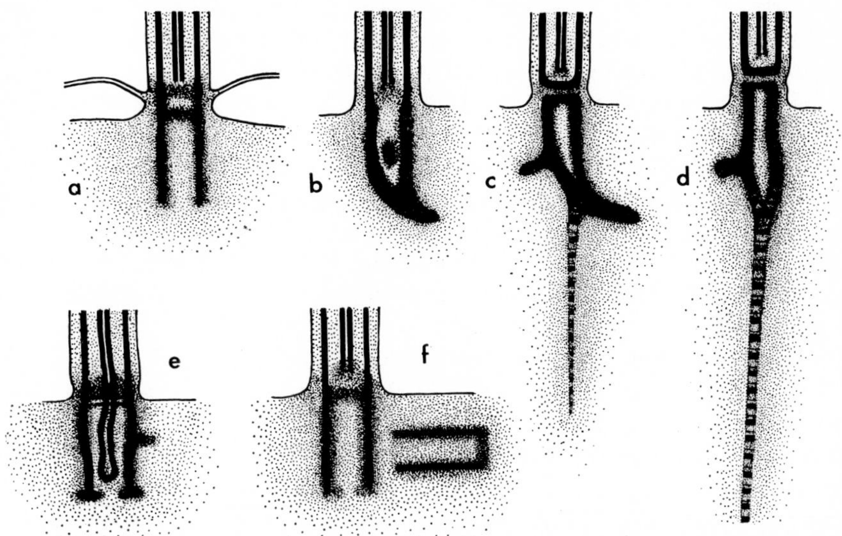

FIG. 10. Diagrammatic representation of some variations in the form of the basal apparatus of cilia and flagella. a. Simple cylindrical basal body of *Paramecium* (see also Fig. 12). The peripheral fibrils of the cilium are continuous with the wall of the basal body and the central cavity of the latter is open below to the cytoplasm. b. Closed, curved basal body of a mammalian cilium. c and d. Asymmetrical, closed basal bodies with fibrous rootlets as seen in the frog *Rana,* and the mussel *Elliptio* (see also Fig. 11). In these species the peripheral fibrils of the cilium do not appear to be continuous with the wall of the basal body. e. Basal apparatus of *Euplotes,* in which the central pair of fibrils continues into the cavity of the kinetosome. f. Base of a flagellum in the gastroderm of *Hydra,* where the second centriole of the original diplosome persists and retains its orientation perpendicular to the centriole which constitutes the basal body of the flagellum.

triple tubular fibers in the wall of basal bodies have become more frequent, and it is likely that this structure will prove to be general. The three elements of each triple fiber are in line (Figs. 7, and 9, F). In the upper part of the basal body the lines through the centers of the nine triplets lie approximately on the circumference of a circle, but lower down all the triplets are rotated 30–50 degrees and all in the same direction, thus forming a "pin-wheel" pattern (Noirot-Timothée, 1959; Gibbons and

Grimstone, 1960). It is the two inner elements, subfibers A and B, in each triplet that are continuous with the two subunits of the corresponding double fiber of the cilium. The outermost element, subfiber C, is an additional structure not represented in the cilium. In the flagellate *Trichonympha* (Figs. 7 and 9, G), delicate linear densities about 45 A. in thickness can be seen in transverse sections joining subfiber A of each triplet to subfiber C of the adjacent one (Gibbons and Grimstone, 1960). Similar fibrous or membranous connections between the triple fibers will no doubt be found in the basal bodies of other animal species when micrographs of comparable resolution are obtained.

In Metazoa the basal bodies have a similar basic structure; however, they depart in varying degree from the cylindrical form of the unmodified centriole. They are usually closed at their lower end and are assymmetrical around their vertical axis owing to the presence of various curvatures, knobs, and fibrous appendages, which are often characteristic of the particular species. In fresh-water mussels, for example, the basal bodies curve backward in the plane of ciliary beat. From the upper part of each, a short knob projects backward toward the next basal body in the same row (Fig. 10, d) and two fibrous rootlets arise from the lower pole and diverge at an acute angle in a plane perpendicular to the direction of beat (Fig. 11, A and B). In frogs, the cylindrical upper portion of each basal body is oriented perpendicular to the cell surface, but its tapering lower end curves back sharply so that its slender tip is directed parallel to the surface (Fig. 10, c). A short rounded protuberance projects forward and downward from the convex anterior surface of the basal body, and a single rootlet arises just below this knob and runs straight downward into the apical cytoplasm of the cell. In the mammalian ciliary apparatus, the basal body is usually devoid of projecting knobs and fibrous appendages, but its lower end is curved backward and often considerably flattened (Fig. 10, b). Transverse sections through the upper end of the basal body reveal nine discrete groups of parallel fibers in its wall, but sections through the recurving lower end reveal a complex cribriform appearance of the wall such as might result from transection of many highly contorted tubular structures. The nine discrete groups of hollow fibers identifiable in the upper end of the basal body thus appear to intermingle farther down, their component fibers losing their parallelism and pursuing a highly irregular course within the wall (Fawcett and Porter, 1954).

There is considerable variation in the contents of the lumen of the basal bodies in different organisms. In most ciliates the central cavity is occupied by a homogeneous material of very low density but contains no formed structures. In some, however, small dense bodies, termed *intrablepharoplastic granules* have been described (Noirot-Timothée, 1958). In others, the central pair of fibers of the cilium terminate in a dense sphere, the *axial granule,* which is situated at the level of the basal plate or in the upper part of the lumen of the basal body (Noirot-Timothée, 1958). In *Euplotes,* the central pair of fibers extends beyond the base of the cilium and form a loop in the cavity of the basal body (Roth, 1956). In *Stentor* (Randall and Fitton-Jackson, 1958) and in *Pseudotrichonympha* (Gibbons and Grimstone, 1960) basal bodies may vary in their structure in different regions of the same organism. In both of these flagellates, basal bodies in certain areas are unusually long and have the greater part of their lumen filled with large numbers of closely packed small granules (200–300 A.). No suggestion has been offered as to the functional significance of these granules. In the ciliated epithelia of mammals, the lumen of the basal bodies usually appears empty, but single spherical or reniform granules (Fig. 10, b) are occasionally observed (Rhodin and Dalhamn, 1956). These are indefinite in outline, inconstant in size and position and evidently are not essential components of the ciliary apparatus.

While cilia have retained the same internal structure over long periods of evolution, their basal bodies have come to vary considerably in form from species to species. The functional significance of the differences observed is poorly understood. It is well known that many Protozoa are able to change the plane of ciliary beat or reverse the direction of their effective stroke. It may be that this versatility of movement requires relatively simple cylindrical basal bodies that are symmetrical around their long axis. On the other hand, in the ciliated epithelia of Metazoa, each cilium beats in a single plane and the direction of the effective stroke is usually irreversible. This restricted capacity for movement in more than one plane may be related to the fact that the basal bodies in ciliated epithelia are directionally polarized and bilaterally symmetrical with respect to the plane of beat.

C. The Rootlets of Epithelial Cilia

As early as the middle of the nineteenth century cytologists observed in ciliated epithelia faint vertical striations running from the basal bodies downward into the cytoplasm (Friedreich, 1858; Eberth, 1866). Opinion differed as to the nature of these linear markings. A few investigators thought they were pathways through which nutritive materials reached the cilia (Engelmann, 1880; Parker, 1929). Some interpreted them as specialized nerve endings that entered the cell from below and terminated in close association with the basal bodies (Apathy, 1897). Others considered them to be contractile fibers that produced movements of the cilia by exerting traction on their basal bodies (Benda, 1899). The view that finally gained the widest acceptance was that they were anchoring fibers or *rootlets* that provided mechanical support for the cilia (Saguchi, 1917; Renyi, 1924). In mollusks, where the rootlets are particularly well developed, they often converge as they pass downward in the cell and form an acidophilic cone with its base at the cell surface and its apex adjacent to the upper pole of the nucleus. In some species, the apex of the "cone of rootlets" passes to one side of the nucleus and extends nearly to the cell base. Results of experimental studies involving micromanipulation of this region of the cell led some investigators to attribute to the rootlets a neuromotor function related to the coordination of ciliary beat (Grave and Schmitt, 1925; Worley, 1941).

In electron micrographs the rootlets are found to be cross-striated fibers 60–100 mμ in diameter with a major period of 550–700 A. (Fawcett and Porter, 1954). From two to four narrower lines are discernible between the prominent bands that delimit the major repeating period. The number of rootlets associated with each basal body appears to be quite constant for any given epithelium but varies from species to species. For example, on the lophophores of the bryozoan *Pectinatella,* three rootlets originate from each basal body, one running straight downward and the other two diverging at an angle of about 45° in a plane parallel to the cell surface. On the typhlosole of the fresh-water mussel *Elliptio,* two very long rootlets arise from the lower pole of each basal body and run downward toward the cell base, diverging at an angle of 10–15° in a plane at right angles to the direction of beat (Fig. 11, A, B). In the intestinal epithelium of the annelid worm *Lumbricus,* the basal bodies have a single large tapering rootlet (Fawcett, 1958). The rootlets are not so well developed in vertebrates. In the pharyngeal epithelium of the frog a single short fiber runs straight downward from each basal body. In the same animal, the very long cilia in the neck segment of the nephron have cross-striated rootlet fibers of varying length and diameter irregularly oriented in planes more-or-less parallel to the cell surface. In the

---

FIG. 11. Examples of ciliary rootlets from the epithelium covering the typhlosole of the fresh-water mussel *Elliptio.* The bases of three cilia are included at the top of A. The longitudinal fibrils of the cilium end in a distinct basal plate (*Bp*), separated by a narrow clear space from the upper surface of the basal body (*Bb*). The cross-striated rootlets (*Rt*) arise from lower pole of the hollow basal body and extend downward among the mitochondria (*M*) in the apical cytoplasm. Alternating broad and narrow dense bands are discernible in the rootlets of B. The distance between successive broad bands is about 660 A.

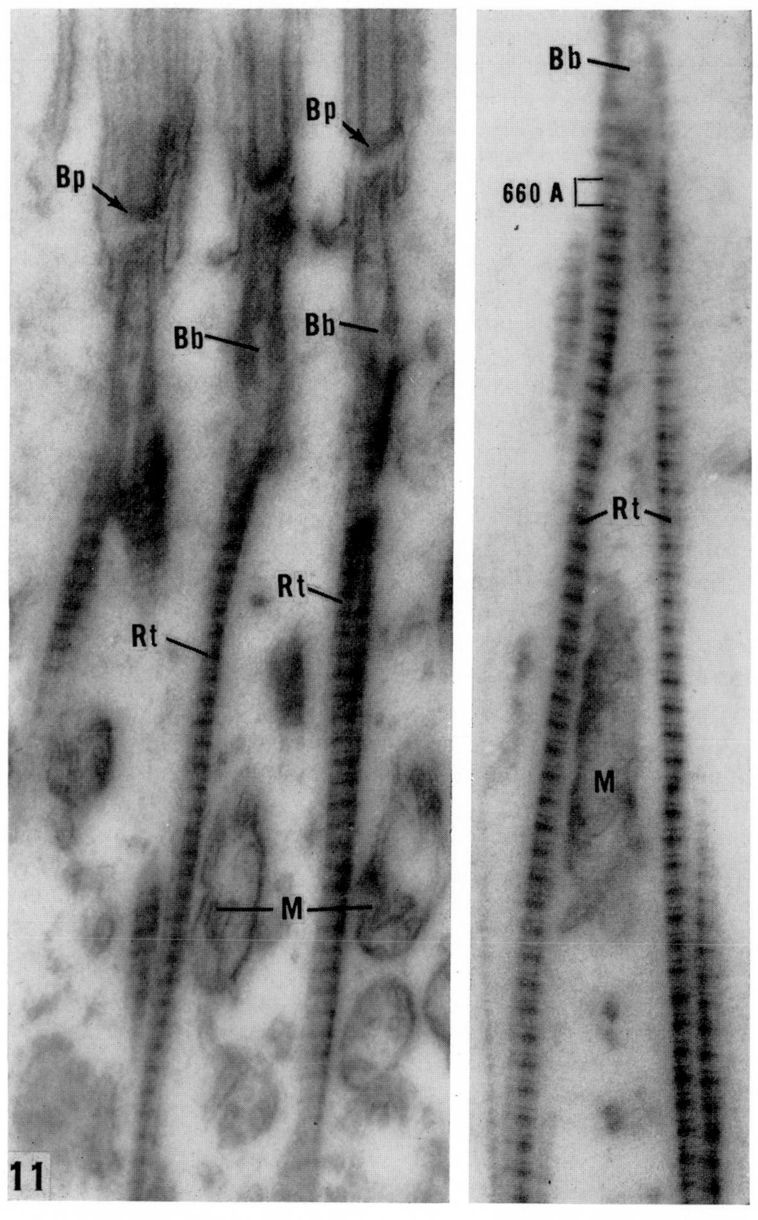

A B

several examples cited there is no apparent correlation between the orientation of the rootlets and the character of the ciliary movements.

In mammals, fine filaments may be found in the cytoplasmic matrix surrounding the basal bodies and occasionally there is a cross-banded appearance in the tip of the basal body, but cross-striated rootlets comparable to those of lower animals are rarely observed. It appears evident, therefore, that this component of the ciliary apparatus is not essential for motility or for the maintenance of coordinated metachronal rhythm. The observation that rootlets are best developed in invertebrates and in those epithelia of vertebrates that have particularly long cilia is consistent with the interpretation that they function principally as anchoring or stabilizing structures permitting the basal bodies to maintain a relatively fixed position in spite of the vigorous movements of the cilia. The rootlet fibers have not been isolated and examined by X-ray diffraction or subjected to chemical analysis, but their cross-banded structure suggests that they may belong to the group of collagen-like fibrous proteins. No fibers of this class have been shown to have irritable properties, and it does not seem likely that the ciliary rootlets are pathways for the conduction of nervelike impulses. In those early microdissection experiments which seemed to support this view (Worley, 1941), the incoordinate ciliary beat which followed severance of the rootlets could be explained on the basis of mechanical instability of the basal bodies after interruption of their anchoring fibers, or it might be attributed to changes in the physical properties of the surrounding cytoplasmic matrix induced by the micromanipulative procedure.

D. The Cortical Fibers of Protozoa

Three or more distinct types of fibers occur in the cortical cytoplasm of protozoa in association with the basal bodies of cilia and flagella. In ciliates examined with the light microscope, cilia are arranged in rows and basal bodies or kinetosomes of the same row appear to be connected by fibers. These are called *kinetodesmata*. Each row of kinetosomes and their associated kinetodesmal fibers constitute a structural unit commonly referred to as a *kinety*. In addition to the locomotor function of these units they have been found to have an important role in morphogenesis, controlling and integrating the differentiation of nearly all the cortical structure of the organism (Lwoff, 1950; Weisz, 1954; Tartar, 1956). This function is discussed in more detail by D. L. Nanney in Chapter 3, Volume IV. We are interested here in the morphological organization of the kinety and possible homologies of its various components with those of the metazoan ciliary apparatus.

Electron micrographs of *Paramecium* show that a single, tapered fiber arises from each ciliary basal body and curves laterally from its point of origin to run forward parallel to the surface of the organism (Metz *et al.*, 1953; Sedar and Porter, 1955). Contrary to the impression gained from light microscopy, this fiber does not attach to the next kinetosome in the row but passes to one side of it (Fig. 13). The fibers arising from successive kinetosomes overlap to form a fiber bundle that corresponds to the *kinetodesma* visible with the light microscope. The number of component fibers in the bundle is usually six. Since one fiber is added at each successive kinetosome along the row, it is evident that one also terminates and the length of each component fiber is therefore equal to about six interkinetosomal intervals (Sedar and Porter, 1955). A similar arrangement is found in other ciliates, but the length and number of the fibers vary. In *Tetrahymena* the kinetodesmata consist of overlapping pairs of fibers each of which is about one and a half kinetosomal intervals long (Metz and Westfall, 1954). The same appears to be true of *Colpidium* (Wohlfarth-Botterman and Pfefferkon, 1953). In all these ciliates, the kinetodesmal fibers are cross-striated and have a repeating period of about 350–400 A. Even though the parallel fibers from successive kinetosomes are not directly connected, some investigators still consider it likely that the kinetodesmata play some role in the coordination of ciliary activity. Their cross-banded structure, on the other hand, suggests that these fibers correspond to the striated rootlets of metazoan cilia and are probably supporting structures rather than conducting or contractile elements.

A second fibrillar system in *Paramecium* to which a neuromotor function has been attributed, is a polygonal network of nonstriated fibers called the *infraciliary lattice system* (Sedar and Porter, 1955). The meshes of this system surround the kinetosomes but do not seem to have any direct structural connection with them. It is difficult to imagine how this network could participate in the coordination of locomotion without establishing more intimate relations with the kinetosomes, but such a function cannot be excluded on the basis of the morphological findings.

A third type of fiber indirectly attached to ciliary basal bodies has been described in micrographs of ciliates of the family Ophryoscolecidae (Noirot-Timothée, 1958). In stained preparations of these organisms examined with the light microscope, argyrophilic bodies called *infraciliary rods* are visible in the adoral ciliated zone just beneath the basal bodies. Arising from the deep surface of these rods and running posteriorly in the ectoplasm are conspicuous strands called the *retrociliary fibers.* In electron micrographs the rods are homogeneous bars of appreciable density attached to the lower ends of the kinetosomes but apparently not an integral part of them. The retrociliary fibers consist of bundles of parallel fibrils which, in transverse section, have a tubular appearance and are arranged in several closely packed straight rows. Hollow fibrils of similar appearance have been found immediately beneath the pellicle of trypanosomes (Steinert, 1960), opalinids (Noirot-Timothée, 1959), and various other protozoans (Pyne, 1958, 1959). They have also been found radiating from the outermost kinetosomes of the cirri in *Euplotes* (Roth, 1957). They make up the ectoplasmic myoneme of *Spirostomum* (Randall, 1956; Fauré-Fremiet and associates, 1956). It has been suggested that tubular fibers of this kind are stiffening or possibly contractile elements, but actually nothing is known about the chemical nature, physical properties, or functional significance of these or either of the other types of cortical fibers in Protozoa. The cross-striated fibrils forming the kinetodesmata appear to be homologous with the rootlets of the metazoan ciliary apparatus. The two types of nonstriated fibers seem to have no counterpart in the ciliated cells of Metazoa.

III. Modified Cilia and Flagella

A. *Motile Cell Processes*

Although the basic structural plan of cilia and flagella has remained remarkably stable in the course of evolution, in a number of instances they have been altered in various ways to meet special functional requirements. In all modified cilia that have retained their motility a basal body is present and the axial filament complex remains essentially unchanged. The modifications consist of addition of fibrous components

---

Fig. 12. A section through a portion of the cortex of *Paramecium multimicronucleatum*. The pellicle (*Pl*) consists of two closely apposed membranes, which are continuous with the limiting membrane of the cilium near its base. The cilium projects from the center of a depression in the surface of the organism and is continuous, at its base, with a simple, cylindrical basal body (*Ks*), which is open to the cytoplasm at its lower end. Transverse sections of kinetodesmata (*Kd*) are found in the cortical ridges that bound the depression from which the cilia emerge. From Sedar and Porter (1955).

Fig. 13. Low-power electron micrograph of *Paramecium* cut parallel to the surface of the organism. The section passes through a rectilinear system of ridges that bound depressions in the cell surface. The bases of the cilia (*Cl*) are seen in the center of these depressions. The kinetodesmal fibers (*Kd*) are seen running in an anteroposterior direction in the ridges between the rows of cilia. The cross striations of the fibers are not visible at this magnification. (Courtesy Dr. K. R. Porter and Dr. R. Dippell.)

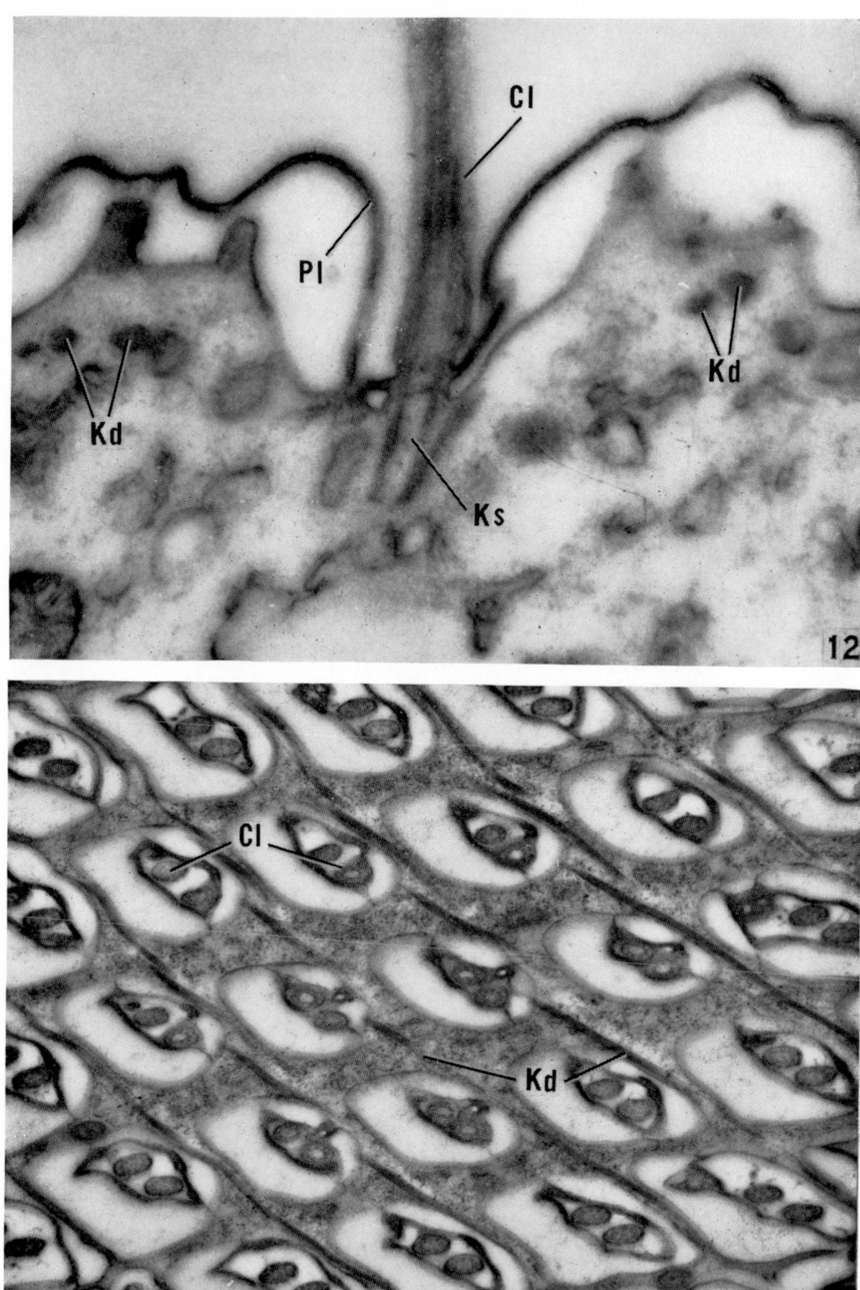

Cl
Pl
Kd
Kd
Ks
12
Cl
Kd
13

around the periphery of the eleven longtitudinal fibrils, or the development of specializations of the surface of the flagellum that alter the character of its movements, or increase its efficiency as an organ of propulsion.

1. Hispid Flagella

Hairlike lateral processes, or *mastigonemes,* were recognized with the light microscope on the flagella of some flagellates, on the spermatozoids of brown and yellow-green algae, and on the zoospores of certain aquatic fungi (Fischer, 1894; Mainx, 1928; Vlk, 1938).

The first of these "feathery," or "hairy" flagella to be studied with the electron microscope were those of *Euglena* and related flagellates (Brown, 1945; Houwink, 1951; Pitelka and Schooley, 1955). In micrographs of dried intact flagella, the mastigonemes appeared as a single row of fine hairs or filaments of uniform length projecting diagonally from the shaft of the flagellum. Manton and co-workers subsequently studied the corresponding structures on the spermatozoids of brown algae and on the zoospores of aquatic fungi. These biflagellate cells have an anterior flagellum which is hairy and a trailing flagellum which is smooth surfaced. On the spermatozoid of *Fucus* the hairs are arranged in two rows along opposite sides and they emerge nearly perpendicular to the surface of the flagellum. They generally occur in tufts of two or three, but near the tip they come off individually. On the zoospores of *Saprolegnia,* the straight lateral hairs of the anterior flagellum arise singly at close intervals along its entire length (Fig. 16). They are of the same thickness for most of their length, but near the tip they undergo an abrupt change in diameter and terminate in an exceedingly thin filament (Manton and Clarke, 1952). The same applies to the mastigonemes of flagellates (Pitelka and Schooley, 1955). It is not yet settled whether these appendages are simply specializations of the surface membrane or whether they arise within the shaft of the flagellum. Manton favors the latter interpretation.

Fig. 14. Electron micrograph of the tip of a front flagellum from a spermatozoid of the brown alga *Himanthalia.* The greater part of the length of the flagellum bears two rows of lateral hairs (*Lh*), but the tip end is bare. At the junction of these two segments is a large hook or spine (*Sp*) which may play a role in attachment to the egg. From Manton, Clarke and Greenwood (1953).

Fig. 15. A higher magnification of the hairy appendages on the flagellum of *Himanthalia.* The hairs are, in fact, much longer than shown here but they break off near the flagellum unless hardened in formalin. From Manton *et al.* (1953).

Fig. 16. Electron micrograph of the hairy flagellum of a first stage zoospore of *Saprolegnia.* Two lateral rows of hairs or mastigonemes extend the entire length of the flagellum. Magnification: × 8000. From Manton *et al.* (1952).

The functional significance of these relatively stiff lateral hairs is uncertain, but it seems likely that they increase the mechanical effectiveness of the flagellum by augmenting its surface (Mainx, 1928). Consistent

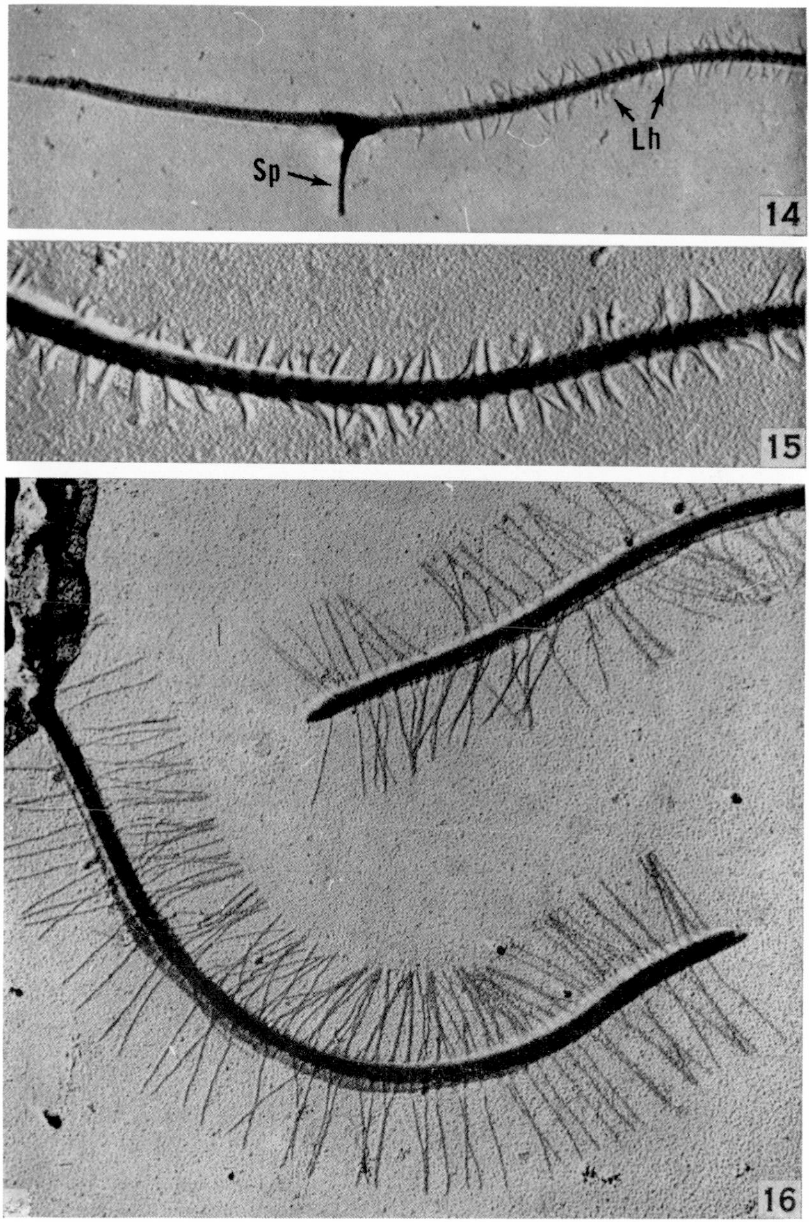

with this view, is the observation that among biflagellate protista, where only the anterior flagellum is hairy, it is this one which is active in swimming while the smooth-surfaced, trailing flagellum participates relatively little (Pitelka and Schooley, 1955).

Appendages of a rather different kind are found on flagella of the spermatozoids of certain brown algae. In *Himanthalia*, short lateral hairs are present on the proximal three-fourths of the front flagellum but absent on its distal one-fourth, and at the point where the hairs stop there is a large curved spine (Fig. 14). This curious armament appears to be attached at its base to one of the peripheral fibrils of the axial filament complex. It seems to be composed of a particularly resistant material, for it remains intact after the complete disintegration of other components (Manton and co-workers, 1953). In another brown alga, *Dictyota*, the distal half of the front flagellum carries about twelve short, conical spines uniformly spaced on a line midway between the two rows of lateral hairs. These spines are fixed at their base to one of the internal longitudinal fibrils, and they maintain their attachment even after extensive disintegration of the flagellum. These peculiar specializations are evidently unrelated to the locomotor functions of the flagellum. It is speculated that they may take part in the attachment of the spermatozoid to the ovum (Manton *et al.*, 1953).

2. *Sperm Tails*

The spermatozoa of many invertebrates and of some vertebrates are relatively simple in their structure, comprising a head, a single ring of mitochondria, and a tail that has the same internal structure as any other flagellum (Afzelius, 1955; Fujimura *et al.*, 1956). Mammalian sperm tails, however, are considerably more complex. The usual nine-plus-two fibrils have been retained as the core of the tail but this *axial filament complex* is surrounded, in its proximal portion, by an additional outer row of nine coarse longitudinal fibers that are not found in other flagella (Fig. 18). The basal body has been transformed into a conical *connecting piece* joining the tail to the head. The middle piece is enclosed in a long mitochondrial sheath, and distal to this, the tail is ensheathed for the greater part of its length in a wrapping of circumferentially oriented fibers forming the *fibrous sheath* of the principal piece. Only in the short end piece is the structure of the tail nearly the same as that of an unmodified flagellum. The finer structure and relations of these special components of the mammalian sperm tail must now be considered.

The modifications of the centriolar derivatives at the base of the tail are so extensive that their homologies can best be established by tracing

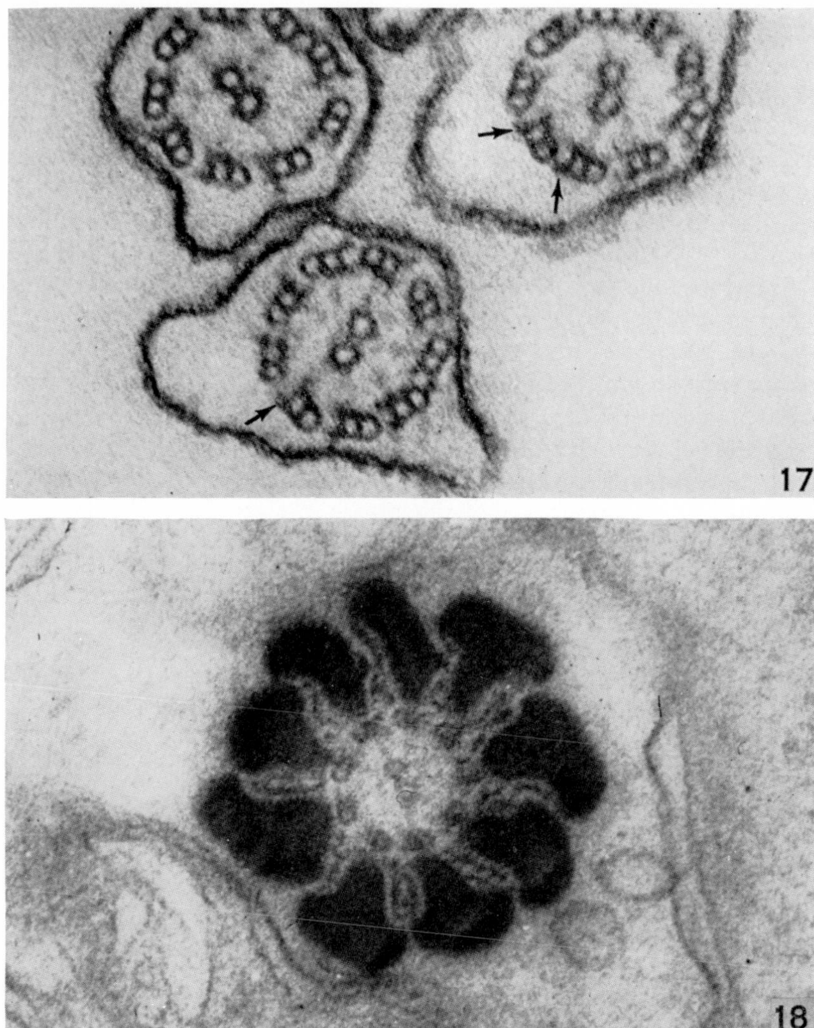

FIG. 17. Cross sections through the tails of three spermatozoa of the sea urchin *Psammechinus*. The double-barreled structure of the outer fibers is shown with unusual clarity, and one or two short, armlike projections can be seen extending from each fiber toward the next (at arrows). There is a faint suggestion of a radial pattern of organization in the matrix between the central and outer fibers. Magnification: × 170,000. From Afzelius (1959).

FIG. 18. Cross section through the proximal portion of the mid-piece of a late guinea pig spermatid showing the axial filament complex surrounded by an outer row of large dense fibers. The larger size of fibers 1, 5, and 6, which is conspicuous in the distal half of the mid-piece and in the principal piece (Fig. 19), is not evident here. At this level the outer coarse fibers are all nearly the same size.

their early development. The formation of the flagellum takes place early in spermiogenesis from one member of a diplosome situated in the peripheral cytoplasm of the spermatid. The two centrioles of the diplosome are cylindrical in form and are placed perpendicular to one another. The axial filament complex grows out from the end of that member which is oriented perpendicular to the cell surface. This corresponds to the *distal centriole* of classic cytology and constitutes the basal body of the flagellum. At this stage of development it resembles, in all respects, the basal body of any other flagellum. The other, or *proximal centriole,* seems to play no role in the induction of the axial filament complex. Later, the two centrioles and the base of the flagellum move deep into the cytoplasm of the spermatid and take up their definitive position at the caudal pole of the spermatid nucleus. The proximal centriole is then oriented transversely and is interposed between the basal body of the flagellum and the nuclear membrane. This centriole retains its original form throughout spermiogenesis, but the one that constitutes the basal body of the flagellum becomes so modified in the later stages of spermatid maturation that its centriolar nature is no longer evident. In the neck of the mature spermatozoan, one finds in its place a larger funnel-shaped structure with its upper rim closely applied to a specialized area of the nuclear membrane (Fawcett, 1958). The proximal centriole is entirely surrounded by the expanded upper end of this thick-walled *connecting piece*. The wall of the latter has a distinct cross-banded or segmented structure. The mode of formation of this component of the sperm tail is poorly understood. The earliest recognizable stage in its development is the emergence of a cross-banded appearance in dense material that accumulates around the periphery of the distal centriole. Whether the connecting piece arises by expansion and modification of the wall of the centriole itself or whether

---

FIG. 19. Cross sections through the principal piece of three guinea pig sperm tails. A prominent outer row of coarse, longitudinal fibers is present. In the mid-piece there are nine of these dense fibers, of which numbers 1, 5, and 6 are larger than the others. At the level illustrated here, numbers 3 and 8 have already terminated and their place is taken by the wedge-shaped inner edge of the two longitudinal columns of the fibrous sheath. The sperm tail at the upper right of the figure shows quite clearly the density difference between subfibers A and B of the nine double fibers of the axial filament complex.

FIG. 20. The mid-piece of a sperm tail of the bat *Myotis,* in cross section. In this species the pairs of crescentic mitochondria of the mid-piece always meet along the dorsal and ventral aspects of the tail approximately in the plane of the central pair of filaments. Instead of one large fiber on one side of this plane and two on the other, as in most mammalian species, there are four large fibers, two on each side.

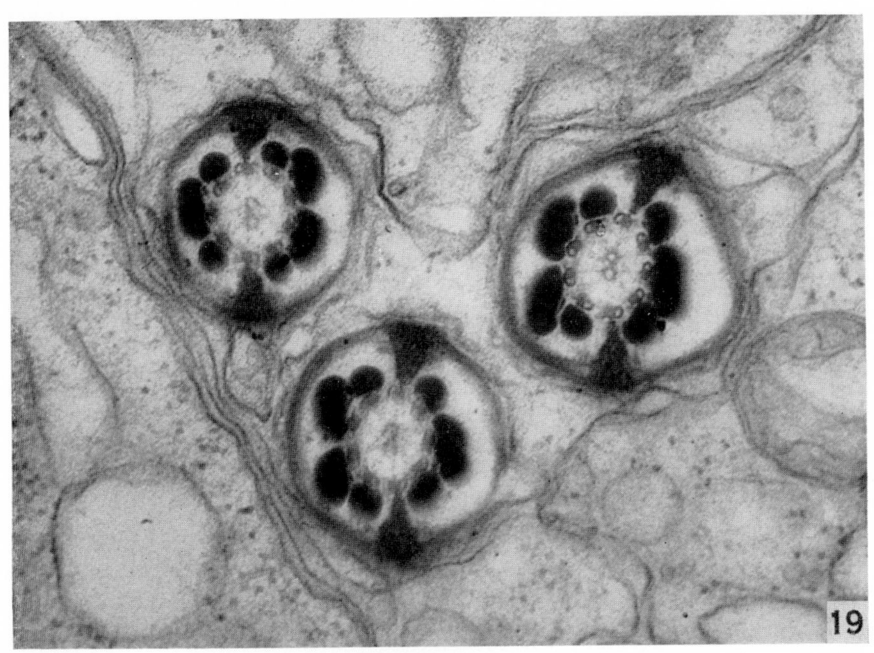
19

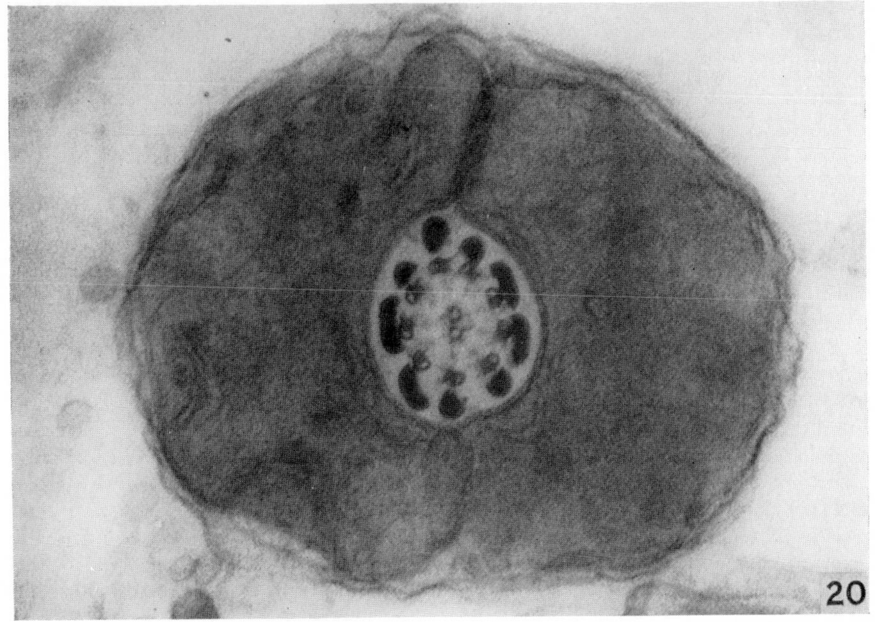
20

it is a new structure laid down around the centriole, is not known. In either case, the distal centriole, which is easily identified in early spermatid stages, can no longer be recognized as a separate entity after this new component of the kinetic apparatus has developed. When viewed in cross section, the wall of the connecting piece is composed of nine distinct columns separated by narrow interspaces. Thus the pattern of nine elements in the wall of the distal centriole is represented in the structure of the connecting piece which seems to develop from it as well as in the nine longitudinal fibrils that grow out from its end.

The nine columns making up the wall of the connecting piece continue caudally as coarse dense fibers that form a cylindrical bundle outside the row of nine double fibrils in the core complex (Figs. 18, 21). The segmentation or cross-banding observed in longitudinal sections is confined to the neck region. In the middle piece, outer fibers are uniformly dense and homogeneous. Their thickness is greatest in the upper part of the middle piece and progressively diminishes in the principal piece. The fibers terminate in this segment at different levels, depending upon the species. In cross sections at the level of the middle piece their shape is obovate or cuneate with their broad ends outward and the narrow ends inward (Fig. 18). Thus they present a radial petallike pattern around the periphery of the axial filament complex. The nine outer fibers are of approximately the same size in the human spermatozoön, but in many other mammalian species three of the nine are distinctly larger and also differ from the other six in their shape (Bradfield, 1955; Fawcett, 1959). One of these odd fibers is located on one side of the circular array of fibrils

---

FIG. 21. Diagram of the changing pattern of component fibers at different points along the rat sperm tail. A. Section through the mid-piece showing the mitochondrial sheath and the outer row of nine coarse fibers with numbers 1, 5, and 6 larger than the others. B. The upper end of the principal piece. The axial filament complex is surrounded by the fibrous sheath thickened on opposite sides where the bifid ends of the dense ribs of the sheath insert into two longitudinal columns that are roughly triangular in section. At points farther along the tail (C–F) the outer fibers terminate in a predictable sequence. First, 3 and 8 end and their place is taken by the inward projecting edges of the two longitudinal columns of the fibrous sheath (C). The disappearance of 3 and 8 is followed by 4 and 7 (D), then 2 and 9 (E), and finally near the end of the principal piece, 1, 5, and 6 (F). The terminal piece of the sperm tail (G) shows neither coarse outer fibers nor fibrous sheath.

The cross sections illustrated here are oriented so that they appear bilaterally symmetrical. Recent observations indicate, however, that the sperm tail in this species is, in fact, bilaterally asymmetrical, and the proper orientation should probably place the two longitudinal columns of the fibrous sheath along the dorsal and ventral aspects of the tail, and not along the sides as shown here.

and the other two are adjacent to one another on the opposite side (Fig. 21). Their relation to each other is such that, if the single large fiber be designated number 1 and the remaining fibers be numbered clockwise, the two other prominent fibers are numbers 5 and 6. It has been suggested that the assymetry introduced into the fiber pattern of the tail by the presence of two larger fibers on one side and one opposite, may have the effect of concentrating contractile power on one diameter and may thus be of considerable importance in the generation of two-dimensional waves by the tail. The outer row of fibers around the axial filament com-

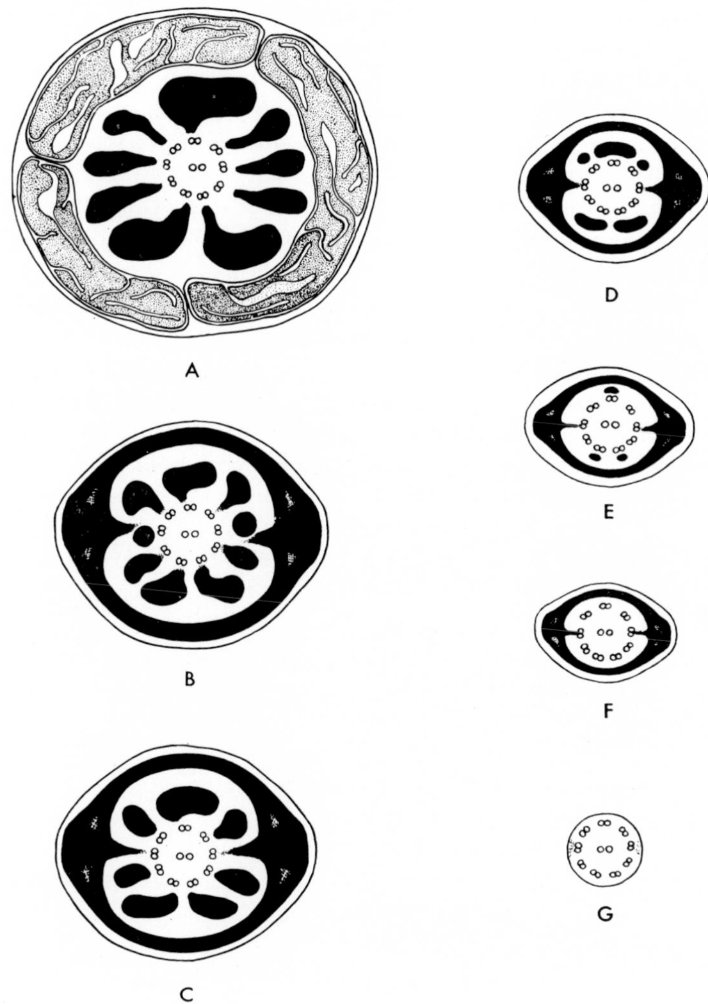

plex has been found in all mammalian species that have been studied with the electron microscope. However, it is not peculiar to mammals, as was formerly thought to be the case (Bradfield, 1955; Cleland and Rothschild, 1959). An outer row of thick fibers has also been reported in two avian species, *Passer montanus* (Yasuzumi *et al.*, 1956) and *Passer domesticus* (Sotelo and Trujillo-Cenóz, 1958), and in a gastropod, *Helix pomatia* (Grassé and co-workers, 1956).

In the middle piece, the double array of longitudinal fibers is surrounded by a relatively long mitochondrial sheath. It was long a matter of dispute among light microscopists whether the mitochondria form a continuous or a discontinuous strand. The majority of the early students of spermiogenesis reported that spheroidal mitochondria gathered in the postnuclear region of the spermatid and there began to elongate. It was generally believed that the resulting rods and filaments later coalesced end to end to form a single, long mitochondrial strand wound in a continuous helix around the axial filament (Stieve, 1930; Williams, 1950). The same impression was gained from the early electron micrographs of shadowed whole mounts of sperm (Schnall, 1952). Indeed, the mitochondrial sheath was described by several workers as a continuous double spiral (Randall and Friedlaender, 1950; Challice, 1954; Bradfield, 1955). Later investigations, using thin-sectioning techniques, have been unable to substantiate these earlier reports. In those mammalian species that have been studied by currently accepted methods, the sheath has not been found to be continuous but is made up of individual, elongated mitochondria wrapped around the longitudinal fibers in a helical manner. The sites of end-to-end contact between individual mitochondria can often be seen in subtangential sections of the middle piece. The number of mitochondrial turns varies from about 12 in the human to 40 or more in the guinea pig. In the bat where the individual mitochondria can be counted with considerable accuracy, they number about 115. The reason for the great species variation in the length of the middle piece is not apparent.

The principal piece of the mammalian sperm tail is enclosed by a sheath made up of circumferentially oriented dense fibers. This was formerly called the *spiral sheath* or *cortical helix* because electron micrographs of dried, intact specimens suggested that it consisted of one or more fibers wound in a tight helix around the axial fibrils (Randall and Friedlaender, 1950; Bretschneider, 1950). The term *fibrous sheath* is now preferred because more recent electron microscope studies have clearly demonstrated that the circumferential fibers frequently branch and anastomose and that the successive turns are joined together on opposite sides of the tail by two slender longitudinal columns which run the full length of the principal piece (Figs. 19, 21, 23, and 24). Thus, instead of a continuous spiral wrapping, the fibers form a series of semicircular ribs, which appear to merge with two longitudinal columns that are of similar density. The encircling fibers are not of uniform diameter, but are thicker near their fixed ends. In longitudinal sections of the tail, which show the ribs in cross section, they present a variety of shapes (Fig. 23) but are generally thicker in their radial dimension than they are in width. In some species the fibrous sheath is rather thick at its upper end, but farther back it gradually becomes somewhat thinner and ends abruptly at the point that marks the junction of the principal piece and end piece.

In some species the fibrous sheath is thickened where the ribs join the longitudinal columns. These two opposite thickenings of the sheath give the sperm tails an elliptical profile in the cross sections, with the long axis of the ellipse always on the diameter that passes through the centers of the two central fibrils of the axial filament complex (Figs. 19, 21). The longitudinal components of the fibrous sheath of human and monkey sperm are less conspicuous than those of other species and are often overlooked. Anberg (1957) and Schultz-Larsen (1958) both failed to find the two longitudinal columns, and the latter suggested that the thickenings of the fibrous sheath reported by other workers, were sectioning artifacts resulting from compression of the tail in a direction parallel to the knife edge. This explanation cannot be accepted, however, for if one examines cross sections, their long axes, through opposite thickenings of the sheath, are often found to be at right angles to one another in neighboring sperm tails in the same micrograph—a finding that could not be explained on the basis of sectioning artifact, since deformations due to knife compression would, of necessity, be consistent in direction throughout the same section (Telkka *et al.*, 1961). The two longitudinal columns thus appear to be real components of the fibrous sheath present in most, if not all, mammalian species.

The function of the fibrous sheath is, at present, a subject of speculation. Bradfield (1955) considers it to be a resistant fibrous wrapping merely serving to prevent the axial filament complex from disintegrating during the vigorous movements of the tail. Challice (1953), on the other hand, suggests that it probably has elastic properties and may act as a spring, offering resistance to the waves of contraction that pass along the axial fibrils, thus providing the necessary couple for the transverse bending movements of the tail.

The final 10–12 μ of the tail closely resembles a simple flagellum consisting of the axial filament complex enclosed only by the cell membrane.

Examination of cross sections using improved methods of specimen preparation has revealed certain additional details of fine structure in the axial filament complex that were overlooked in earlier studies of mammalian spermatozoa. The arms described by Afzelius projecting from one side of the double fibers in flagella of sea urchin sperm have now been observed in sperm tails of several mammalian species (Telkka *et al.*, 1961). It has also been found that one member of each doublet (subfiber A) has a dense interior and appears *solid* in section whereas the other member (subfiber B) has a dense wall but a content of low density and thus appears hollow or *tubular* (Fig. 8). It is always subfiber A of the doublet which is provided with arms. A slight difference in the density of the two subunits of the double fibers was also noted by Gibbons and Grimstone in flagella of *Pseudotrichonympha.* This difference is more marked in the mid-piece and principal piece than it is in the end piece.

With all its well developed middle piece, its additional row of coarse fibers, and its long fibrous sheath, the mammalian spermatozoon executes undulant movements that do not seem very different from those of spermatozoa with less complex tail structure. The significance of its additional structural elements is not clear at present, but possibly, when the ultrastructure of sperm tails has been studied in a greater variety of mammals, it may be possible to define the essential features common to all and to correlate the structural peculiarities of certain forms with particular internal mechanical problems, or with special modes of progression. In this way more insight may be gained into the function of the various tail components. The promise which this comparative approach

---

FIGS. 22–24. Electron micrographs of the mid-piece and principal piece of the bat sperm tail. (Courtesy Dr. S. Ito and J. Raziano.)

FIG. 22. A longitudinal section of the distal portion of the mid-piece illustrating the intimate relation of the mitochondrial sheath (*Ms*) to the axial filament complex. The matrix of the mitochondria is unusually dense and obscures their internal membrane structure.

FIG. 23. A longitudinal section of the principal piece showing the axial filament complex (*Af*) flanked by the irregularly shaped cross sections of the circumferential ribs of the fibrous sheath (*Fs*).

FIG. 24. A surface view of the principal piece, showing the successive turns of the fibrous sheath connected by one of the two longitudinal columns (*Lb*) that run along opposite surfaces of the tail. The circumferential fibers or ribs are observed to branch and join (arrows) in a manner incompatible with the earlier interpretation of the sheath as a continuous spiral wrapping.

holds is exemplified by the work of Cleland and Rothschild (1959) on the bandicoot spermatozoon. Their description of the atypical structure of the sperm tail in this marsupial has produced suggestive evidence bearing upon the functional significance of the outer row of longitudinal

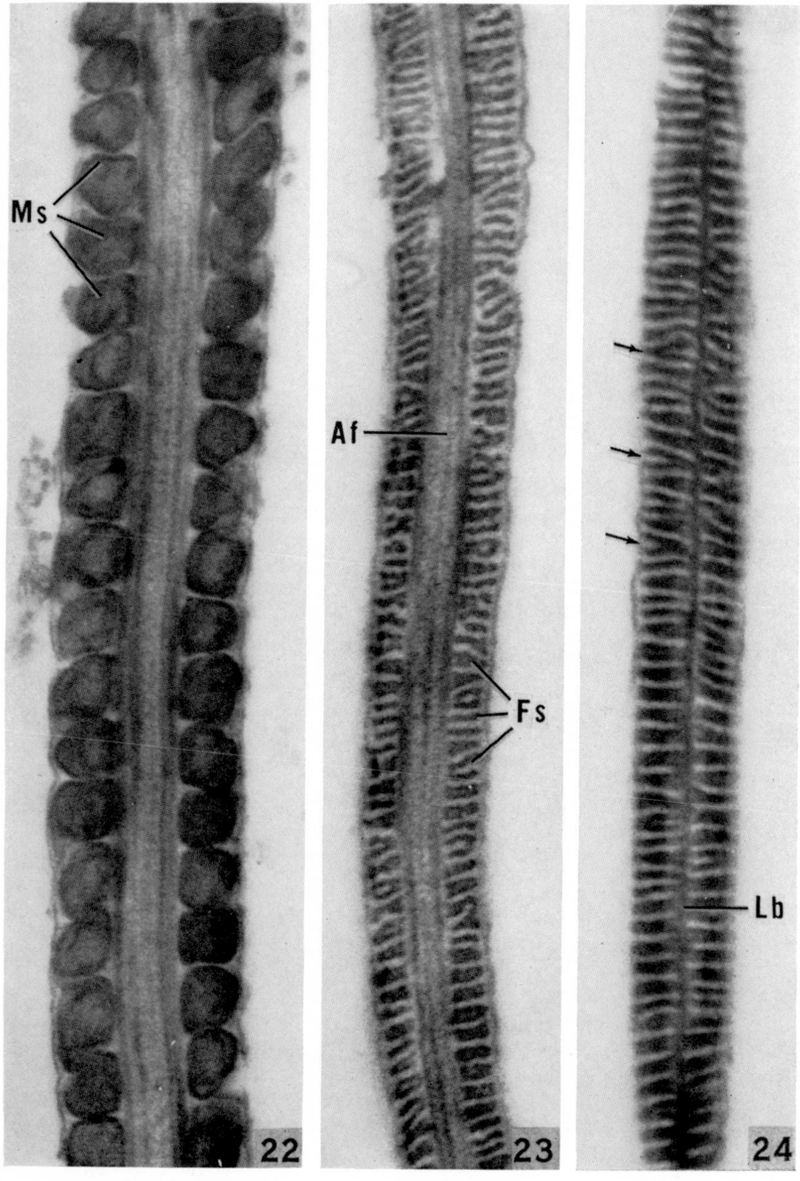

fibers. The organization of the sperm tail in this species differs from that of other mammals in several respects, but the most noteworthy of these is the fact that the outer longitudinal fibers are situated at a considerable distance from the axial filament complex and are connected with the corresponding members of the inner row by thin laminae composed of fine filaments (Fig. 25). Furthermore, the fibers along one diameter are at a

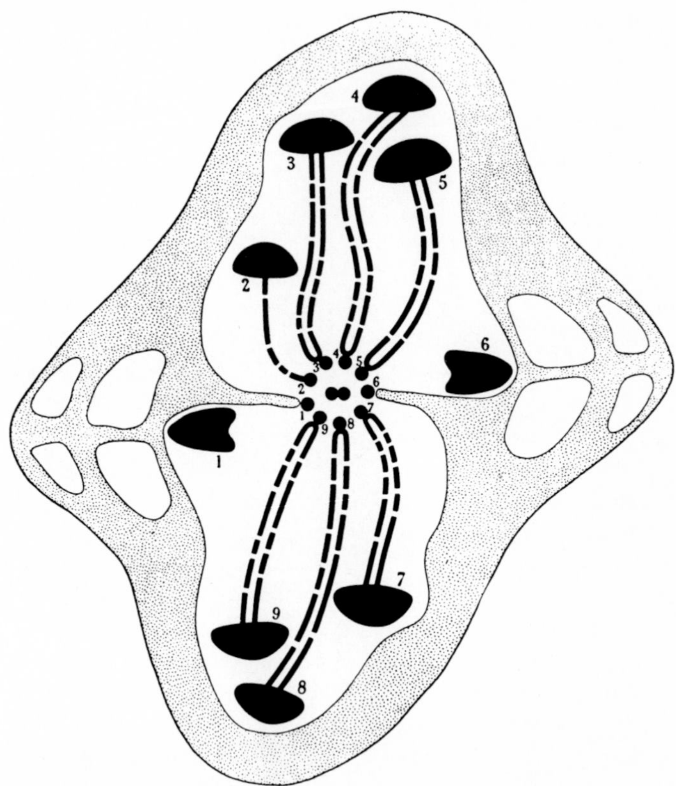

FIG. 25. Diagrammatic cross section of the bandicoot sperm tail. The outer dense fibers, unlike those of other mammalian sperm tails, are a considerable distance away from the axial filament complex and are connected to the corresponding fibers in the latter by thin laminae represented here by interrupted lines. The coarse fibers are farthest away on the diameter perpendicular to the axis of bending. Assuming that these fibers are contractile, this arrangement may make them more effective in overcoming the resistance to bending in the relatively thick tail. From Cleland and Rothschild (1959). (The numbering of the fibers here is not consistent with that shown elsewhere. Number 8 in Cleland and Rothschild's system of numbering evidently corresponds to number 1 in the system introduced by Bradfield and adopted in the text of this chapter.)

greater distance from the axis of the tail than those along the diameter perpendicular to it. Assuming that these fibers are contractile, this arrangement would provide greater mechanical advantage for the generation of bending movements of the tail in the plane of the longer diameter. Cleland and Rothschild (1959) therefore postulate that the peripheral row of fibers in the middle piece of mammalian spermatozoa is an adaptation to overcome the considerable resistance to bending introduced by the long mitochondrial sheath. In the case of the bandicoot sperm, which has an exceptionally thick middle piece, it is evidently necessary for the peripheral fibers to be situated farther from the axial fibrils in order to retain their efficiency as tensile elements.

3. *Undulating Membranes*

The spermatozoa of the various species of amphibia differ greatly in their structure and mode of locomotion. The most interesting are those in which the tail possesses an *undulating membrane*. In the toad *Bufo arenarum*, the sperm tails, under the light microscope, appear to consist of a flagellum with a thin ruffled fin along one margin. Observed in the living state, the tail as a whole executes relatively slow flexuous movements of long wavelength while rapid undulations of shorter wavelength are passing along its membrane. The relative importance of the flagellum and the undulating membrane in locomotion is not well understood. Some investigators are of the opinion that the undulating membrane *passively* increases the effectiveness of the tail by increasing its surface. Other workers, observing that its undulations are more rapid than are the waves of the tail as a whole, insist that the membrane is endowed with contractile properties of its own. The latter view derives support from phase-contrast cinemicrographic studies which demonstrate that the undulating membrane may remain active after the flagellum itself has become quiescent (Burgos and Fawcett, 1955).

Light microscopists recognized in these sperm tails two longitudinal filamentous components: an axial filament forming the core of the flagellum, and a slender refractile strand in the free edge of the undulating membrane, called the *membrane filament*. The fine structure of these components has now been established by electron microscopic studies (Burgos and Fawcett, 1955). The "axial filament" of the light microscopists is composed of the usual complex of two central and nine peripheral fibrils, and these are embedded in a homogeneous matrix of low density (Fig. 26, C and D; and 27). The undulating membrane projects from the surface of the flagellum in the plane of the central pair of fibrils and consists of a ribbonlike sheet of electron dense material enclosed in

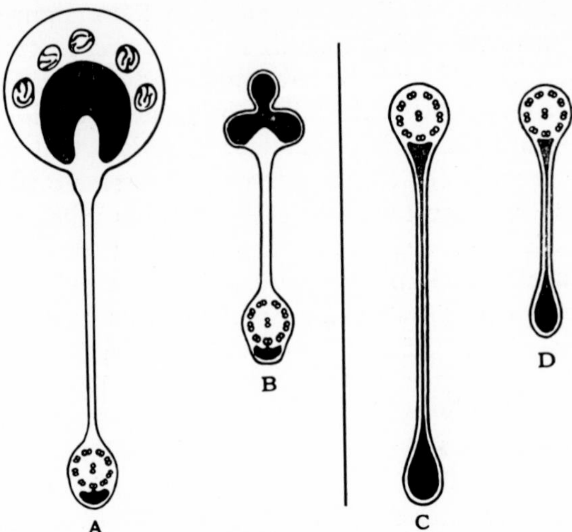

Fig. 26. Schematic cross sections through the sperm tails of a salamander (A and B) and of a toad (C and D). A and C are representative of the disposition of components in the major portion of the tail in the two species. B and D are a few microns from the tip of the tail. The uppermost portion of each drawing is in the axis of the sperm head, and the undulating membrane extends down toward the bottom of the figure. In the salamander (A and B), a large dense rod, shaped like a horseshoe or a trefoil, is in the axis of the head, and the flagellum is in the free edge of the undulating membrane. In the toad, the relations are reversed. The flagellum continues back in the longitudinal axis of the head, and between the leaves of the undulating membrane there is a layer of dense material with a conspicuous thickening along the free edge.

Fig. 27. Transverse section of spermatid mid-pieces and tails of the toad *Bufo arenarum*. Above, are the bases of the tails, surrounded by a layer of cytoplasm that contains numerous mitochondria (*M*). Below, are several tails showing a typical axial filament complex (*Af*) and a thin undulating membrane (*Um*) in the plane of the central pair of filaments. A thickening in the free edge of the undulating membrane contains the so-called marginal filament (*Mf*). From Burgos and Fawcett (1956).

Fig. 28. Transverse sections through the distal third of several sperm tails of the salamander *Triturus viridescens*. The trefoil-shaped objects in the lower half of the figure are cross sections of a dense, supporting rod (*Sf*), which continues back into the tail in the axis of the sperm head (see diagram Fig. 26). The broad undulating membrane (*Um*) is attached to the margins of a groove in this odd-shaped rod and carries, in its free margin, the fiber complex of a typical flagellum (*Af*). In this species, in contrast to the toad, the flagellum is not in the axis of the tail, but in the edge of the undulating membrane.

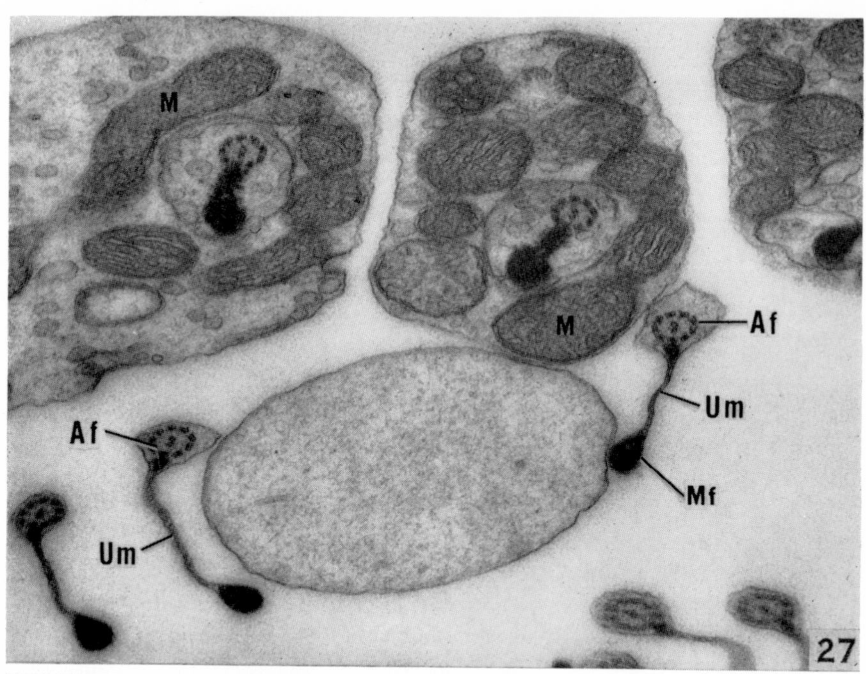
M
M
Af
Af
Um
Um
Mf
27

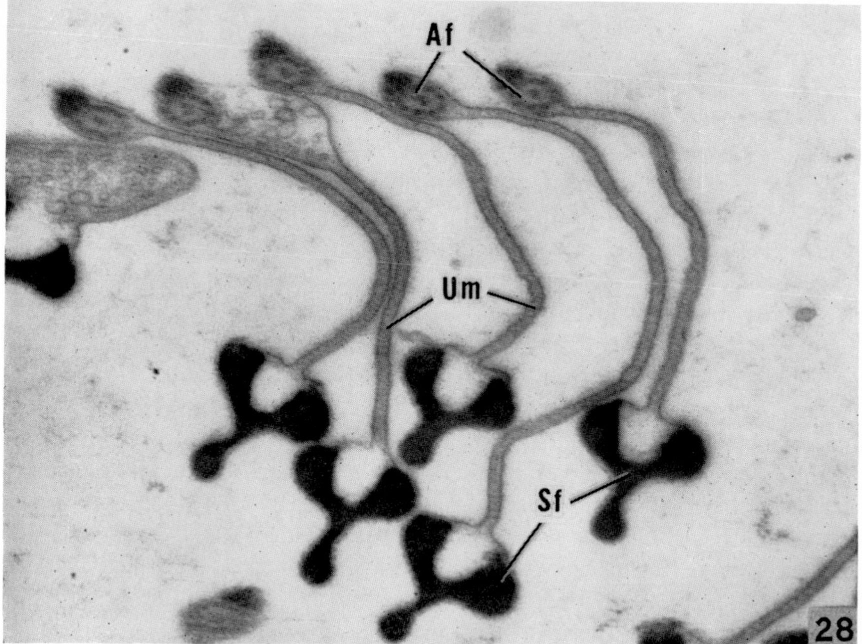
Af
Um
Sf
28

a thin fold of the flagellar membrane. The dense layer within the membrane is thickened along both its margins. The thickening of the free edge (Fig. 26, C) corresponds to the "membrane filament" of the light microscopists. The relatively inconspicuous thickening along the fixed margin is in close apposition to one of the peripheral fibrils of the axial filament complex and is not resolved as a separate structure with the light microscope. At the base of the sperm tail the undulating membrane narrows abruptly and converges upon a centriole which is set at an angle to the basal body of the flagellum proper. In the mammalian sperm the distal centriole becomes the basal body of the flagellum while the unmodified proximal centriole appears to have no kinetic function. In the tail of the toad sperm, where the undulating membrane as well as the flagellum, seems to be contractile, it seems likely that the two centrioles function as more-or-less independent kinetic centers for the two major components of the tail movements.

The spermatozoa of salamanders also have a prominent undulating membrane, but the character of its movements is quite unlike that of the toad. When living sperm of *Triturus viridescens* are observed, the tail as a whole exhibits no active flexion and the forward progress of the cell appears to depend entirely upon waves that sweep in rapid succession along the undulating membrane. Its movements resemble the graceful rippling of the ventral fin of the knifefish (*Gymnotus*), which can propel the animal forward without flexuous movements of the body. Early investigators of salamander sperm were in accord in believing that only one of the two tail filaments discernible with the light microscope was contractile. The other was thought to serve as a stiffening or supporting element. One was therefore called the *motile filament* (*Bewegungsfaden*), and the other was referred to as the *supporting filament* (*Stützfaden*).

In electron micrographs of these sperm, the supporting filament is found to continue backward in the axis of the sperm head as a thick, dense rod that gradually diminishes in diameter and terminates a short distance from the tip of the tail. At its anterior end, the rod is reniform or U-shaped in cross section owing to the presence of a deep longitudinal groove along its ventral surface (Fig. 26, A). The broad undulating membrane attaches at the margins of the groove. On the side opposite the membrane, the rod is covered for three-fourths of its length by a sheath of cytoplasm containing numerous mitochondria. In the terminal one-fourth of the tail, the cytoplasmic sheath is lacking and the cross-sectional shape of the rod is that of a trefoil (Figs. 28, 26, B). The undulating membrane consists of a thin amorphous layer of protoplasm of low density between two leaves of a narrow fold of the tail membrane. In its free edge is a motile flagellum with the usual internal structure. An additional dense fiber or slender rod, which escaped detection with the light microscope, runs longitudinally along the outermost of the nine peripheral fibrils of the flagellum (Figs. 28, 26, A, B).

Although the sperm tails of these two groups of amphibia are alike in having undulating membranes, the arrangement of their components and the mechanism of their locomotion is very different. In the toad, a typical motile flagellum forms the axial component of the sperm tail. In the salamander, on the other hand, the axial structure of the tail is a nonmotile rod and the flagellum runs in the free edge of the undulating membrane. In this instance, the membrane appears to be passive, its undulations being produced by the movements of the flagellum in its edge. The origin, composition, and properties of the marginal filament of toad sperm and of the supporting rod of the salamander sperm tail are in need of further study. Their density and fine structural appearance in electron micrographs suggest that there may be homologies between these structures and the outer longitudinal fibers or the fibrous sheath of mammalian sperm tails.

B. *Nonmotile Modified Cilia*

The specialized cilia and flagella considered in the foregoing section all had developed additional internal structural elements or external devices for increasing their effective surface. The internal complex of longitudinal fibrils remained unmodified in all these motile cell processes. The organelles to be considered in this section originate as cilia or flagella but, in their subsequent differentiation, the axial fiber complex is extensively modified or lost. The disturbance of this basic pattern of internal organization is accompanied, in all instances, by a loss of motility.

1. *The Crown-Cell Processes of the Saccus Vasculosus*

The epithelium of the saccus vasculosus of the fish brain includes cells that have short, club-shaped processes projecting from their free surface into the third ventricle. These so-called *crown cells* (*Krönchenzellen*) were first described by Dammerman (1910). He believed that they had a sensory function, because he was able to see in their cytoplasm in silver-impregnated preparations, fine vertical striations, which he interpreted as neurofibrillae. More recently, Bargmann (1954) has demonstrated in these cells, masses of colloidlike material which stains with the periodic acid-Schiff reaction. Fine granules or droplets with similar staining properties are found in the tuft of blunt processes on the cell surface and also free in the lumen of the third ventricle. It was concluded from these observations that the crown cells probably have a secretory function. Under the electron microscope the processes forming the apical tuft are found to be modified cilia (Bargmann and Knoop, 1955). At the base of each is a typical basal body with two short, cross-striated rootlets. Fibrils can be traced upward a short distance into the process, but there they seem to end blindly (Fig. 31). The expanded tip of each of these modified cilia is occupied by a mass of small vesicles, some of which appear empty while others have a content of appreciable density (Fig. 31). These vesicles are regarded by Bargmann as additional evidence of a secretory function. Porter (1957) has made similar observations on these processes in the same organ of another species, but does not share Bargmann's interpretation as to their secretory function. Until more precise information is available on the physiology of the saccus vasculosus, the processes on the crown cells can only be cited as nonmotile modified cilia of unknown function.

2. *Sensory Processes of Photoreceptor Organs*

The rod cells of the vertebrate retina consist of a cylindrical outer segment connected by a slender stalk to a thicker inner segment. Several early investigators of the histogenesis of the retina noted that the first indication of the development of the outer segment was the protrusion of a slender filament from one of a pair of centrioles situated at the apex of the formative cell (Lebouca, 1909; Seefelder, 1910). Mann (1928) interpreted this filament as a cilium or short flagellum, and concluded that the outer segments of the visual cells arose by enlargement and modification of cilia. This theory seems not to have gained favor among contemporary students of retinal cytology. Forty years later, however, it was revived by Willmer (1955), soon after the publication of the first descriptions of the ultrastructure of the retinal rod (Sjöstrand, 1953a, b) and of the internal organization of cilia (Fawcett and Porter, 1954). It was Wilmer who focused the attention of electron microscopists upon the resemblances between the fine structure of a cilium and the stalk connecting the inner and outer segments of the retinal rod. Subsequent electron microscopic studies on the histogenesis of the retina have fully substantiated the concept that the photoreceptor components of rods and cones are modified cilia (De Robertis, 1956; Carasso, 1958; Tokuyasu and Yamada, 1959; Lasansky and De Robertis, 1960). The initial stage in the differentiation of their outer segment is the development of an abortive cilium which contains the usual nine peripheral fibers but appears to lack the central pair. In later stages of its differentiation, the fibrils persist at its base, but the distal portion of the cilium becomes filled with

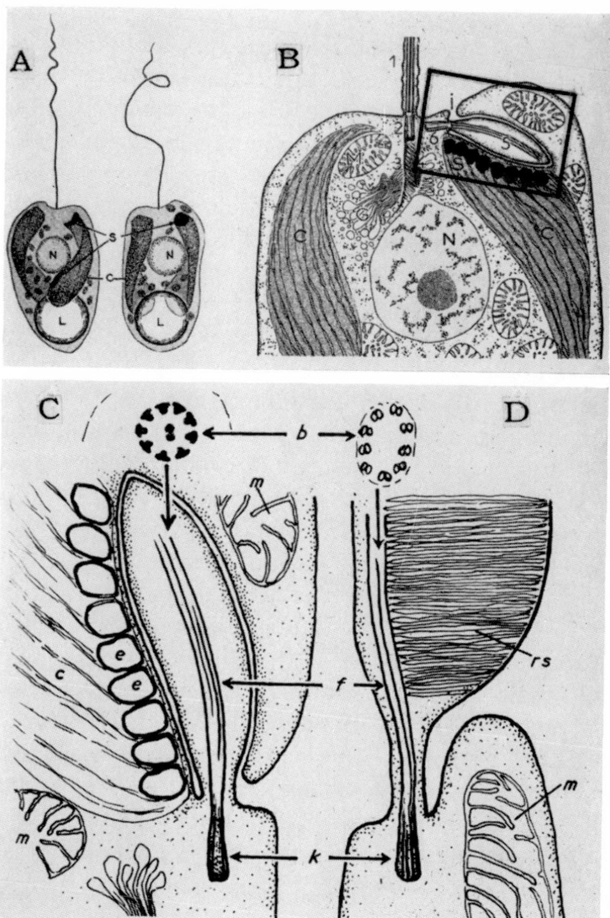

Fig. 29. Composite diagram comparing the stigma of the phytoflagellate, *Chromulina psammobia* with the rod cell of the vertebrate retina. A. Appearance of *Chromulina* with the light microscope. B. Enlarged view of the anterior end of the organism showing the base of the external flagellum and the adjacent internal flagellum lodged in a depression in the membrane of the chromoplast. The area included in the rectangle is presented in C in juxtaposition to a drawing (D) of the base of the outer segment and connecting segment of the vertebrate retinal rod. The external segment of the vertebrate visual cell and the eye spot of the phytoflagellate are comparable in that both involve the coaptation of a ciliary organelle and a vesicular system containing a carotenoid pigment. *c*, Chromoplast of *Chromulina; e,* eye spot chambers; *f,* fibrillar bundle; *k,* kinetosomes; *m,* mitochondria; *rs,* flattened rod sacs of the retinal cell. Modified from Fauré-Fremiet and Rouiller (1957) and Fauré-Fremiet (1958).

small vesicles. These subsequently coalesce into disklike membranous units, oriented transverse to the axis of the cilium and stacked one above the other in parallel array (Sjöstrand, 1953a, b; Porter, 1957). The unmodified basal portion of the cilium forms the slender stalk between the outer and inner segments (Fig. 30). Since this is the only connection between the two, one must assume that the cilium or its membrane is involved in conduction of the state of excitation from the outer photoreceptor element to the inner segment. This evidence of irritable properties in a modified cilium is of interest in relation to "hairlike" processes on other sensory cells such as those of the olfactory epithelium, taste buds, lateral-line organs, etc. The finer structure and functional mechanisms of these are yet to be elucidated.

The bivalve mollusk *Pecten* has a large number of blue-green eyes, each a millimeter or so in diameter, located along the free margins of the mantle. The ultrastructure of the sensory cells of the retina in these has been studied by Miller (1958). They possess complex organelles consisting of concentrically arranged double lamellae. Each of the latter takes origin from a short ciliary shaft that arises from a typical basal body. Thus, unlike the outer segment of the vertebrate retinal rod, which is derived from a single cilium, the lamellar appendage of the sensory cell of *Pecten* appears to arise from multiple cilia. In view of their origin from cilia and their lamellar structure, it is assumed that these structures are photoreceptor elements comparable to the retinal rods of vertebrates.

The participation of modified cilia in sensory receptor mechanisms is not confined to the Metazoa. A number of phytoflagellates possess a sensory organelle called the *stigma* or *eye spot* which contains an orange-red carotenoid pigment and is believed to be a photoreceptor (Rothert,

---

FIG. 30. Electron micrograph of a portion of a rod cell of the retina, including part of its outer segment (*Os*), its slender connecting stalk (*Cs*), and the upper end of the inner segment (*Is*). The connecting segment is clearly a modified cilium, with a typical basal body (*Bb*) and vestiges of nine longitudinal fibrils (*Fb*). In the outer segment, the fibrils are replaced by a system of parallel membranes (*Lm*), which constitute the photosensitive component of the receptor cell. The cilium, which gives rise to the outer segment, originates from one member of a diplosome in the apical cytoplasm of the primitive epithelial cell. The other centriole (*Ce*) persists near the basal body of the fully differentiated outer segment. Electron micrograph contributed by Dr. K. R. Porter.

FIG. 31. Two of the clavate processes on a crown cell of the saccus vasculosus of a teleost fish. That these curious surface specializations originate from cilia is evident from the basal body (*Bb*) at their inner end, and the presence of rudimentary fibrils (*Fb*) in their interior. The bulk of the ciliary matrix is occupied by membrane-bounded vesicles of varying size. Electron micrograph contributed by Dr. K. R. Porter.

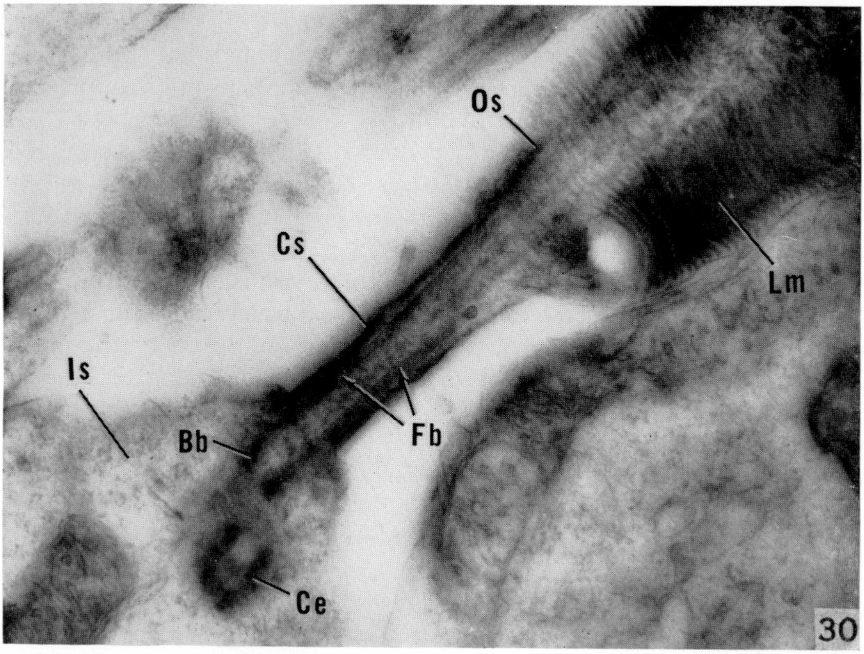

Os
Cs
Lm
Is
Fb
Bb
Ce
30

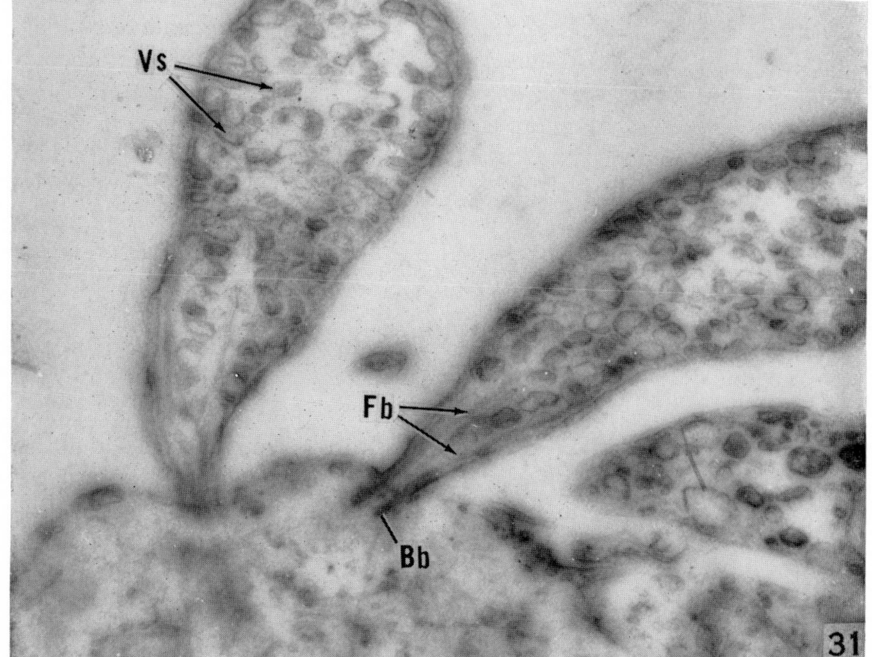
Vs
Fb
Bb
31

1914; Mast, 1927, 1938). In *Euglena,* the bases of the flagella are in close relationship to the eye spot, and it has been suggested that this may be the basis for the phototaxic movements of the organism (Wolken and Palade, 1953; Wolken, 1956). In the biflagellate spermatozoid of the brown alga *Fucus,* the recurrent flagellum is flattened near its base and closely adheres to the cell membrane overlying the eye spot (Manton and Clarke, 1955). A somewhat more complex association of flagellum and stigma is found in the marine chrysomonad *Chromulina psammobia.* Electron micrographs of this organism reveal a very short internal flagellum situated at the base of the external flagellum and oriented perpendicular to it (Fig. 29, B). This modified flagellum has a typical basal body and the usual complement of internal fibrils including the central pair, but instead of projecting from the cell surface it occupies an invagination of the cell membrane, and is closely applied to the surface of the chromoplast in the region of the stigma (Fauré-Fremiet and Rouiller, 1957).

In the development of the retinal cells of Metazoa, cilia give rise to lamellar systems of membrane-bounded compartments containing a carotenoid visual pigment. In the light-sensitive organs of the phytoflagellates, on the other hand, the pigment-containing component arises from a chromoplast, and a ciliary organelle secondarily becomes associated with it. The parallelism between these two photosensitive systems in organisms as far apart in the evolutionary scale as *Euglena* and man is indeed striking (Fig. 29, C, D) and has prompted Wolken (1956) to state that the eyespot-flagellum complex can be looked upon as a "primitive eye" or the most "elementary nervous system." Lacking chromoplasts, the Protozoa apparently have not developed the equivalent of the eyespot of the Protophyta. Fauré-Fremiet (1958) finds it tempting to interpret these findings as evidence supporting the hypothesis of Baker (1948) and of Hardy (1953) that the Metazoa may have originated from the Protophyta rather than from the Protozoa.

3. *Cnidocils Associated with Nematocysts*

In many of the Cnidaria, each cnidoblast is provided with a hairlike sensory process called the *cnidocil* which projects from the surface of the cell near the operculum of its nematocyst. This stiff, nonmotile process together with a complex of rodlike structures surrounding its base constitute a trigger device controlling the opening of the operculum and the discharge of the long slender tube which either penetrates and poisons the animal's prey or helps to immobilize it by entanglement. The development of the several components of this apparatus has not been studied in detail, but electron micrographs have revealed that the shaft of the cnidocil itself consists of a dense core surrounded by a mantle of less-dense protoplasm and the whole is enclosed in an extension of the plasma membrane of the cnidoblast (Bouillon *et al.*, 1958). At its base is a typical cylindrical basal body. In cross sections, however, the familiar pattern of eleven internal fibrils is lacking and is replaced by a dense core with an amorphous center and nine uniformly spaced ridges around its periphery, which appear to represent slender tubular or rodlike elements embedded in the amorphous core substance (Chapman and Tilney, 1959). The general appearance of the cnidocil, its origin from a centriole-like structure, and the occurrence of nine units of some kind in its core, strongly suggest that it is a modified flagellum. It is not known whether this receptor is sensitive to chemical or mechanical stimuli.

4. *Stalk Fibers of Colonial Protozoa*

The fine structure of the colonial ciliates *Opercularia, Campanella,* and *Zoothamnium* has been studied by Rouiller and associates (1956). The cilia are usually confined to a single spiral tract associated with the oral groove. During the brief, free-swimming ciliospore stage of the life cycle, however, an aboral group of cilia is also present. At the time when these migratory forms settle down and become attached to a solid substrate, the aboral cilia are lost, but the basal bodies persist and participate in the formation of the stalk of the sessile stage of the organism. In *Opercularia* and *Campanella,* the stalk is noncontractile and consists of a bundle of parallel fibrous elements embedded in an amorphous matrix and enclosed in a membranous sheath. The contractile stalk of *Zoothamnium* is similar but, in addition to the fibrous supporting elements, it has in its interior a slender prolongation of cytoplasm which contains the myoneme. This latter structure is responsible for the prompt retraction of the cells when the colony is disturbed. In all three of these colonial organisms, stiffening or supporting elements in the interior of the stalk grow out from the group of persisting aboral basal bodies and thus appear to be highly modified cilia. In electron micrographs the fine structure of the stalk fibers is quite different in the three species. In *Campanella,* thin-walled tubes are formed which are devoid of internal longitudinal fibrils, but are bounded by a membrane rather like that of a cilium, and this is reinforced on its inner aspect by an interlacing fabric of exceedingly fine filaments. In *Opercularia,* each of the supporting elements visible with the light microscope is found by electron microscopy to consist of nine fibrils having the same arrangement as those of a typical cilium. In this case, however, the fibrils are single and are homogeneous in cross section. In longitudinal section they show distinct cross striations. The central pair of fibrils is absent. In *Zoothamnium*, separate fibrils are not found; instead these seem to have coalesced to form a thick-walled tube that exhibits cross striations with a regular period of about 480 A. (Fig. 32). The central pair of fibrils are present but show no periodic structure (Rouiller *et al.*, 1956).

It is interesting in relation to the varied morphogenetic potentialities of centriolar organelles that in two of these colonial ciliates, the basal bodies associated with the contractile fibers of cilia in the free-swimming stage of the life cycle give rise, in the sessile phase of the same organism, to cross-banded fibrous proteins possessing no contractile properties.

5. *Taillike Appendages of Isopod Sperm*

A striking example of the development of cross-striated structure in a modified flagellum has been reported by Blanchard *et al.* (1958) in the sperm tails of the isopod *Cyathura*. The spermatozoa of this crustacean have a slender fusiform head joined at an acute angle by a long threadlike appendage which bears a close resemblance to the flagellum of vertebrate spermatozoa but which is nonmotile. In describing this taillike structure, Retzius (1909) stated that it had a thick sheath staining heavily with aniline dyes, and a lighter "axial filament." He noted, however, that the axial strand did not fray into finer filaments, as was often observed in similar preparations of motile sperm. Electron micrographs of the isopod sperm tail show that it is a cylindrical shaft, 0.3 μ in diameter, consisting of a thick wall (1200 A.) enclosing a slender central canal about 450 A. wide. In longitudinal section, the heavy wall is cross-striated with a major repeating period of ~660 A. There are in addition at least three

---

FIG. 32. An electron micrograph of a longitudinal section of the stalk of a colonial ciliate, *Zoothamnium*, showing the thick-walled tubular supporting elements that seem to develop in place of the peripheral fibrils of cilia. The walls of these noncontractile, tubular structures have a cross-banded appearance with a regular repeating period of 440–470 A. The central pair of fibrils persists in the interior but shows no periodic structure. Magnification: × 44,000. Electron micrograph contributed by Dr. Charles Rouiller.

FIG. 33. Longitudinal section of several of the nonmotile tails of the sperm of an isopod, *Cyathura*. These modified flagella have a discontinuous central cavity, visible in the two tails at the left of the figure. They seem to be thick-walled tubular structures formed by coalescence and modification of the peripheral fibrils of a flagellum. Accompanying this coalescence of fibrous components there is apparently a profound alteration at the molecular level, resulting in a cross-striated fibrous protein without contractile properties. Electron micrograph contributed by Drs. R. Blanchard, R. Lewin, and D. Philpott.

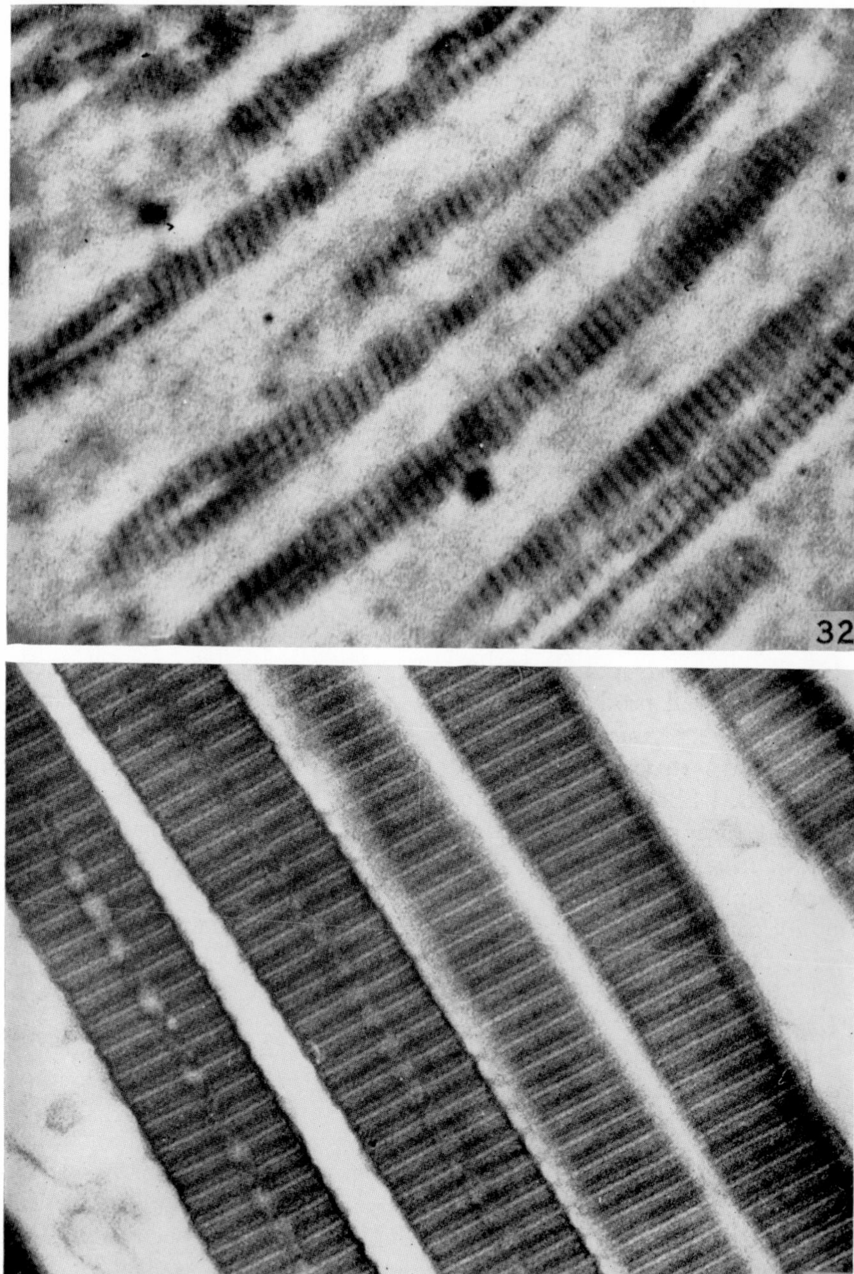
32
33

thin, intraperiod bands (Fig. 33). The central canal appears to be interrupted by transverse septa recurring at regular intervals along its length. It is probably this central canal that Retzius interpreted as a "light-staining axial filament." It is likely that the thick, cross-striated wall is a tubular outgrowth from a centriole, or basal body, situated at the base of the sperm head. It bears a superficial resemblance to the cross-striated, tubular structures found in the stalk of *Zoothamnium* and is probably homologous with the peripheral fibrils of the axial filament complex in motile sperm flagella.

C. *Summary*

The foregoing section has presented examples of motile processes that retain the universal pattern of internal fibrils, but in which the surface has been modified by the development of bristlelike lateral appendages, undulating membranes, and the like. Some sperm flagella have acquired a long mitochondrial wrapping, an additional row of longitudinal fibers and a complex sheath enclosing the basic core of eleven fibrils. All these added structural components and specializations of the surface, it is assumed, increase the efficiency of these flagella as organs of locomotion. Where cilia or flagella are adapted to some function for which motility is not a requirement, then one may find the internal fiber complex replaced by such diverse structures as stiff rods with a sensory function, vesicles containing a cell product, or lamellar systems of membranes containing photosensitive pigments. Where stiffness or tensile strength are needed in supportive structures, the fibrils of the cilium may be replaced by cross-striated fibers or by a hollow cylinder with a thick cross-striated wall. The examples of nonmotile modified cilia and flagella which have been described here emphasize the great versatility of the basal bodies as morphogenetic centers and serve indirectly to reinforce the belief that motility of vibratile cell processes depends upon the nine-plus-two complex of internal fibrils.

IV. Functional Aspects of the Ciliary Apparatus

A. *Characteristics of Ciliary Motion*

1. *Types of Ciliary Movement*

Owing to the small size of cilia and the rapidity of their motion, it is difficult to determine by direct observation whether each cilium is beating in a plane, describing an ellipse, or making rotary movements. The observational problem is somewhat simpler in the case of flagella, but here too it is difficult to establish whether the waves are flat or three dimensional. Therefore the early efforts at classification of ciliary motion are to be accepted with some reservations and, as yet, modern cinematographic and stroboscopic techniques have been applied to only a few types of ciliated cells.

Valentin (1842) described four types of motion: a pendular movement (*motus vacillans*); a hooklike flexion (*motus uncinatus*); an undulant motion (*motus undulatus*); and a funnellike rotary movement (*motus infundibularis*). Gray (1928) could recognize only three distinct kinds of movement, *pendular, uncinate,* and *undulant,* and believed that all other patterns could be explained as combinations of two or more of these three basic types.

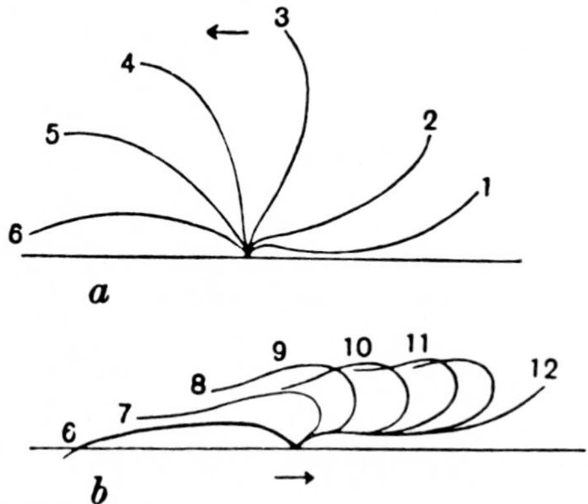

FIG. 34. Diagram showing in (a) the form of the cilium during the effective stroke, and in (b) its form in successive stages of the recovery stroke. From Gray (1928).

In pure pendular motion, the cilium is rigid except at the base and shows little change in shape in the different phases of the beat. In uncinate movement the cilium, as a whole, does not swing through an arc but exhibits a cycle of bending which begins at the tip and moves toward the base, causing the cilium to assume a hook-like shape. During the return stroke, the process is reversed and the cilium straightens from base to tip. These two types of ciliary motion are seldom seen in pure form. In the ciliated epithelia of most metazoans, the motion commonly observed appears to combine certain features of both pendular and uncinate movement. There is a rapid *effective stroke* during which the cilium is stiff, and a slower *recovery stroke* during which it bends and then progressively stiffens from base to tip (Fig. 34). This common type of ciliary motion has been studied in detail by stroboscopic and high-speed cinematographic techniques. For further information the reader is referred to the papers by Gray (1930), Lucas (1932), and Dalhamn (1956).

Undulant motion is seen mainly in flagella that propel single cells through a fluid medium. Waves arise at the base of the flagellum and proceed to the tip. In spermatozoa the flagellum is at the posterior end and pushes the cell forward. In Protozoa it is more common for the flagella to project from the anterior end. Some organisms show great versatility in the employment of a single anterior flagellum. They can move forward by executing rapid pendular movements between the forward position and one at right angles to the direction of progression (Fig. 35). They can swim backward by means of undulatory movement with the flagellum directed forward, or may move to the side by means

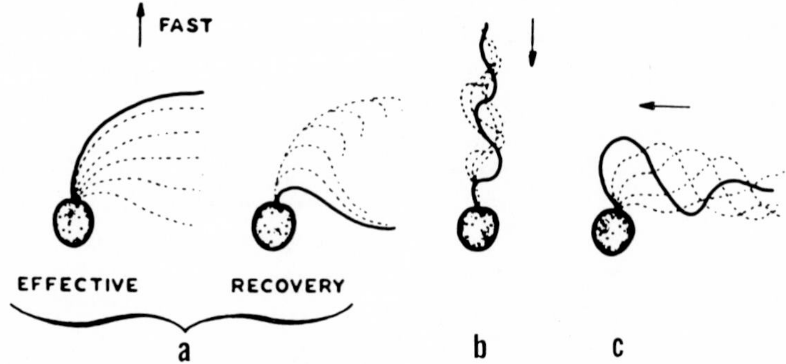

FIG. 35. Three different modes of progression used by the flagellate *Monas*. (a) Rapid pendular movements propelling the organism in the direction of the arrow, (b) swimming backward by means of undulant movements with the flagellum anterior, (c) swimming to the left using an undulant motion, with the tail tip directed to the right. From Prosser *et al.* (1950); redrawn from Krijgsman (1925).

of undulations of the flagellum with the latter at a right angle to the anteroposterior axis of the body. By directing the tip of the flagellum backward its waves will propel the organism forward (Krijgsman, 1925). Contrary to earlier impressions, undulatory activity of a flagellum directed anteriorly cannot pull an organism through the medium. The waves always travel from base to tip and exert pressure tending to drive the surrounding medium in the direction of the waves. Thus they can only exert a pushing action. For further discussion of the movements of protozoan flagella the papers of Lowndes (1943) and Brown (1945) may be consulted.

2. *Movements of Sperm Tails*

The undulant movements of spermatozoa have often been likened to the swimming of fish because both make progress through a fluid medium by causing waves of lateral displacement to travel along their length from head to tail end. It has recently been pointed out, however, that for microscopic organisms the hydrodynamics of self propulsion through liquid are rather different from those that apply to larger animals (Taylor, 1951). Fish take advantage of the *inertia* of the surrounding water, the viscosity of the medium being negligible for an animal of this size (Gray, 1955). For Protozoa and spermatozoa the magnitude of the stresses due to *viscosity* of the surrounding liquid may be several thousand times greater than those due to inertia and therefore different principles apply (Taylor, 1951). For a discussion of the general theory of self-propelling undulatory systems and mathematical analysis of the movements of sperm tails and their hydrodynamic effects, the reader may consult the scholarly papers of Taylor (1951, 1952), Hancock (1953), and Gray (1955).

Early investigators of sperm locomotion were of the opinion that the tail movements were spiral. Rothschild (1953) has presented a novel kind of evidence for this view by demonstrating that if the movements of active bull spermatozoa are photographed with dark-ground illumination, using long exposures, a regular but intermittent bright path is recorded on the photographic film. This was interpreted as meaning that the flattened sperm head rotates during swimming and produces a flash of scattered light whenever the larger surface is normal to the direction of vision. Rothschild inferred from the rotation of the head that the waves produced by the tail were three dimensional. On the other hand, Gray (1955), using ciné-dark-ground microscopy to analyze the swimming movements of sea urchin spermatozoa, concluded that they propagate two-dimensional waves along their tails. This interpretation was based upon observation and photographic recording of the shape of the light area behind the head in the dark-field image of the active spermatozoön. If photographed with a very short exposure the undulations of the tail are "stopped" and the tail appears as a wavy luminous line. If longer exposures are employed the motion is not "stopped," but an elliptical luminous area is recorded which represents the area within which the tail movements are occurring (Fig. 37). In direct visual observation of the sperm, the eye of the observer cannot follow the rapid oscillations of the tail, but the shape of the luminous "optical envelope" of the tail can be seen. In an actively swimming sperm the visual image of the tail envelope alternates rhythmically between an elliptical outline and a relatively straight bright line (Fig. 37). No such change in shape of the optical envelope would be expected if the waves were three dimensional. According to Gray, these observations can only be explained on the basis that the oscillations of the tail of the sea urchin sperm are largely restricted to a single morphological plane while the cell as a whole rolls about its long axis.

Application of similar methods of analysis to the movements of bull spermatozoa (Gray, 1958) yielded results somewhat different from those on sea urchins sperm. Gray attributes the "flashing" of the head of the bull sperm in dark field illumination to "rocking" or "rolling" around its median longitudinal axis, as did Rothschild. He notes however that when the flattened head is seen edge on, the "optical envelope" of the tail is narrow, indicating that the flagellum must execute transverse movements in a plane coinciding, in the main, with the plane of the flattened head. Points near the distal end of the tail, however, seem to depart slightly

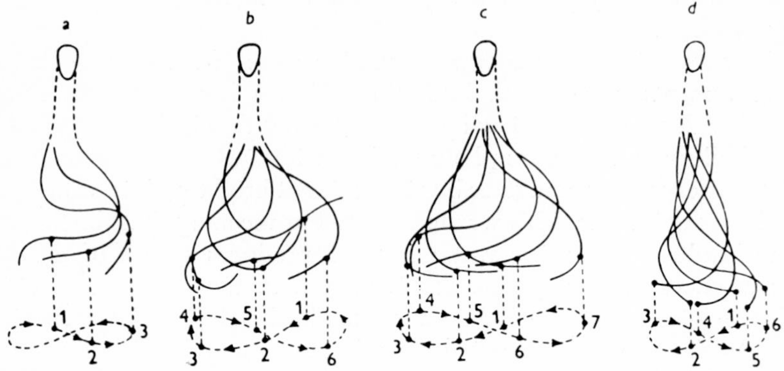

FIG. 36. Diagram illustrating the figure-of-eight paths, relative to the head, followed by points lying towards the distal end of the tail of a bull spermatozoön. In *a–c* the point is approximately 10 μ from the tip of the tail. In *d* it is at the tip of the tail. The numbers indicate successive stages of the bending cycle. From Gray (1958).

from this plane at certain phases of the contractile cycle in such a way as to describe a flat figure-of-eight relative to the head (Figs. 36, 38). The tendency of the sperm to rotate about its longitudinal axis thus appears to be related to the fact that all elements of the tail are not executing their transverse movements in exactly the same plane during the whole of their contractile cycle.

The common impression that sperm locomotion depends upon three-dimensional helical gyrations of the tail thus appears to be erroneous. The fact that rotation of the head is observed with predominantly two-dimensional undulations or "sculling" movements of the tail suggests that the same may apply to the rotary progression of flagellate protozoa, and leaves us in some doubt as to whether "propeller-like" spiral wave motion occurs in cilia or flagella anywhere in nature.

3. *Frequency of Beat*

Before the introduction of stroboscopic methods for determining frequency, it was expedient to judge the direction of movement and the degree of ciliary activity by observing the rate of transport of small particles across the surface of the epithelium. Sharpey (1836) used carbon particles for this purpose, as did many later investigators (Ritter, 1859; Hill, 1928). The results, however, were highly variable, depending upon

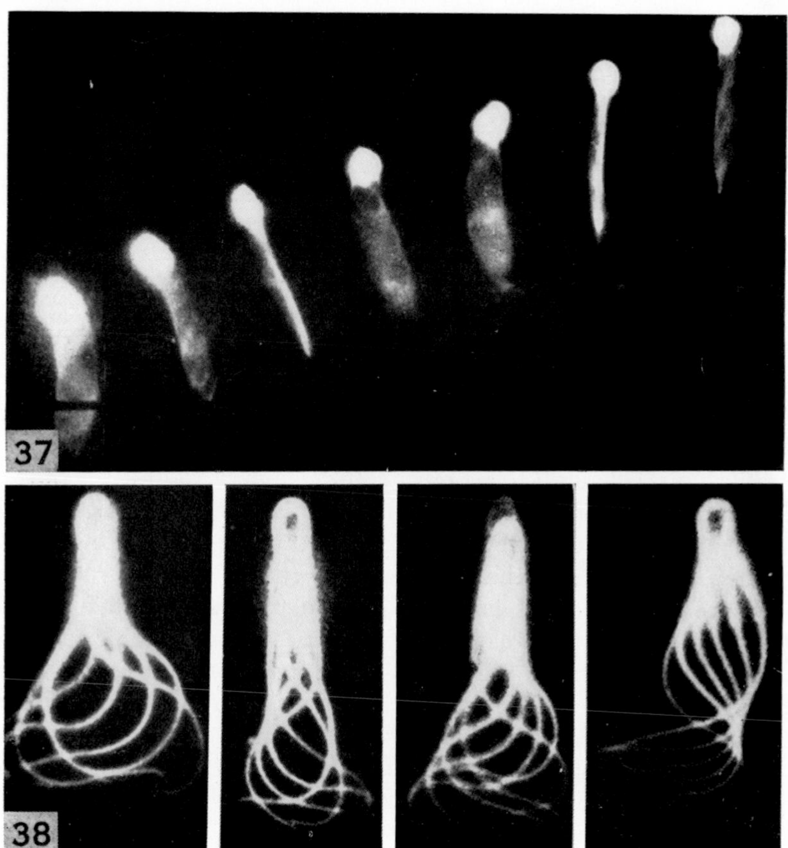

FIG. 37. Seven successive photographs from a cinematographic record (12 frames per second) of a sea urchin spermatozoön rolling about its longitudinal axis two times per second. Note that the "optical envelope" changes alternately from an elliptical area to a narrow bright line—a finding which is interpreted as meaning that the undulations are planar. From Gray (1955).

FIG. 38. Dark-ground photographs of bull sperm taken on stationary film. Interval between flashes 1/80 second. For interpretation of the path of points in the distal portion of the tail during the cycle, see Fig. 36. From Gray (1958).

the nature of the particles and their tendency to clump. Rates of transportation observed in living animals were usually greater than those from excised strips of epithelium (Lucas and Douglas, 1934). The lack of agreement is attributable to the fact that the amount of mucus and its physical consistency may be very different under these two experimental conditions and this factor can profoundly alter the rate of transport of indicator particles, without any change in activity of the underlying cilia. Measurements of the rate of mucus transport by these methods are therefore of questionable value as indicators of rate of ciliary beat. The following figures are illustrative of the variation in the estimates of rate of mucus transport from *in vivo* observations of mammalian ciliated epithelia using particulate indicators: 2.7–5 mm./min. (Gordonoff and Manderli, 1936); 4.6 mm./min. (Hach, 1925); 10 mm./min. (Hilding, 1932); 15–20 mm./min. (Henderson and Taylor, 1910); 35 mm./min. (Florey *et al.*, 1932).

The stroboscope was first used in the study of cilia by Martius (1884), who determined by this means that the frequency of beat in the frog pharynx was 600–950 per minute. Gray (1930), Lucas (1932), and Jennison and Bunker (1934), using cinematographic methods on the gill epithelium of lamellibranchs, have reported frequencies from 500 to 1200 per minute. Similar studies on mammalian ciliated epithelia have been relatively few. The most carefully controlled of these are the observations of Dalhamn (1956) on the epithelium of the rat trachea. The living anesthetized animals were maintained in a chamber that permitted control of temperature and humidity. The rate of mucus flow was determined by registering the time required for shed cells, or other naturally occurring particles in the mucus, to traverse a certain distance. The rate of ciliary beat was calculated from cinematographic recordings of the oscillations in the reflection from a light directed vertically onto the mucosa through an Ultrapak microscope. The rate of mucus transport in normal rat trachea was found to be ~13 mm./min. and the frequency of beat was ~1300 per minute. Under the experimental conditions, these values varied relatively little from animal to animal.

Little is known concerning the factors that normally influence the rate of ciliary beat. This hiatus in our knowledge has been due, in large measure, to the lack of accurate methods of measuring frequency in higher mammals. Using Dalhamn's equipment, Borell *et al.* (1957) have investigated the lining of the rabbit oviduct at different times in the estrous cycle and found a rate of about 1500 per minute in estrus. After copulation, ciliary activity gradually rose. The increase amounted to about 20% on the second and third day and persisted through the early stages of gestation. This appears to be the first suggestion of a hormonal effect on ciliary activity in the reproductive tract.

Several investigators have reported evidence of nervous control of ciliary motion in invertebrates. The cilia on the swimming plates of ctenophores (Parker, 1905a), on the nudibranch veliger (Carter, 1926), on the gills of lamellibranchs (Lucas, 1931a, b), and on the lophophores of bryozoans have been observed to cease all movement suddenly and, after a quiescent period, resume vigorous activity. Such stoppages have been attributed to an "inhibitory" nervous impulse originating outside the cells (Carter, 1926). Nerve fibers to the ciliated regions of these organisms have not been demonstrated, however, and the nervous origin of the inhibitory impulse remains in doubt.

In vertebrates, epithelial cilia appear to have no spontaneous periods of quiescence but beat continuously for the life span of the cell. The ability of excised strips of epithelium to maintain their beat at near-normal rates suggests that epithelial cilia are independent of the nervous system. Nevertheless, some workers continue to believe that nerves can regulate the rate of beat to some extent. MacDonald *et al.* (1928) reported that transport of particles over the frog pharynx was accelerated 100% by stimulation of the sympathetic nerves. On the other hand, Lommel (1908) and Lucas and Douglas (1934) were unable to detect any significant alteration in ciliary activity of the trachea after electrical stimulation of the vagus nerve or after nerve section. The apparent increase in ciliary activity reported by MacDonald and associates is probably explained by nervous stimulation of secretion resulting in alterations in the character of the mucous layer which permitted more rapid transport without increased frequency of beat. At the present time there is little indication of a direct effect of nerves upon ciliary action in vertebrates.

4. *Direction of Beat*

Among metazoans, instances of ciliary reversal are rare. Valentin (1842) and Engelmann (1868) reported spontaneous reversal in the gills of the mussel. Parker (1929; Parker and Marks, 1928) was able to induce experimentally a reversal of ciliary beat on the lips of the sea anemone *Metridium*. The cilia normally beat outward from the mouth, but by bathing them with extracts of crab meat or mussel flesh they could be made to change the direction of their effective stroke and beat inward. Glycogen, peptone, and a variety of other substances were also effective. Since the ciliary action in the resting animal is directed outward, it is suggested that normal feeding probably requires reversal. A similar condition has been described by Carlgren (1905) for certain other actinians.

To date, the only example of this phenomenon in a vertebrate is a reversal of ciliary action on the surface of early amphibian embryos, reported by Twitty (1928). Local reversal occurred in response to nonspecific mechanical stimuli, or following transplantation of ectoderm. After closure of the neural tube, physiological polarity of the cells was determined and rotation of a transplant was no longer followed by ciliary reversal.

Flagellates often exhibit great versatility of movement, but the prevailing direction of beat in ciliates is usually rather stable. It is well known, however, that various inorganic ions can cause reversal of beat. Merton (1923, 1932, 1935) compared a variety of anions and cations with regard to their capacity to induce reversal and found that the monovalent cations have this capacity whereas bivalent cations do not. Tartar (1957) reports similar results in experiments on *Stentor*. Lithium chloride at 1% concentration produced backward swimming, and ammonium acetate, at the same concentration, caused the most prolonged and continuous reversal of ciliary beat of any of the compounds tested. It would be of interest to study the fine structure of protozoa during experimental reversal of ciliary beat to see whether there are any detectable changes in orientation or symmetry of the various components of the ciliary apparatus.

B. *Mechanism of Ciliary Movement*

1. *Early Speculations*

Without knowledge of the internal structure of cilia, the early cytologists lacked the morphological basis for a reasonable explanation of ciliary motion and the early theories were therefore highly speculative. Schäfer (1891) visualized the cilium as a slightly curved hollow process limited by a semielastic membrane and suggested that a rhythmic flow of hyaloplasm into and out of the interior of the cilium might produce an alternate extension and flexion. Other investigators of that period attributed ciliary motion to active shortening of rootlets or some other contractile element in the ectoplasm and considered the cilia themselves to be passive processes possessing a certain degree of inherent rigidity (Henneguy, 1897; Peter, 1899). Heidenhain (1911) assumed that the cilium consisted of an elastic axial filament surrounded by a protoplasmic sheath which had a contractile component on the side facing in the direction of the effective stroke. The forward motion of the cilium was believed to be due to contraction of the sheath on this side, and its return to the resting position was attributed to the elastic recoil of the axial filament. A structural or physicochemical difference between the two sides of the cilium was also postulated by Gray (1930), who speculated that the beat might be due to alternate changes in the distribution of water within the cilium. It was reasoned that if a change were to take place on one side of the shaft making it capable of absorbing more water than the other, the cilium would bend into an arc, convex on the side containing the excess water—a process comparable to the curling of a strip of paper moistened on one side. It was thought that the changing affinity for water might be related to changes in ionization of the proteins.

The structure of the cilium that has been revealed by electron microscopy is clearly not compatible with any of these proposed mechanisms of ciliary motion, and they are now of historical interest only. It must be admitted, however, that our more detailed knowledge of the organization of the ciliary apparatus has not yet produced an entirely satisfactory alternative to these earlier theories.

2. *Current Hypotheses*

Bradfield (1955) has offered an ingenious hypothesis which seems to be consistent with many of the newer observations on the fine structure of the ciliary apparatus, but one which certainly cannot be regarded as a definitive explanation of the mechanism of ciliary motion. In developing this hypothesis it was assumed that (1) the stiffness of a cilium is dependent upon an appreciable turgor in its interfibrillar matrix; (2) the nine peripheral fibrils are capable of contracting locally and of propagating waves of local contraction along their length from base to tip; (3) the central pair of fibrils are not contractile but are specialized for rapid conduction; and (4) the impulse initiating the beat arises rhythmically at a point in the basal body beneath one of the outer fibrils and spreads from there to involve the central pair and the eight other contractile fibrils. Referring to the diagram in Fig. 39, the sequence of events in the beat of a cilium is explained in the following manner. At the beginning of the effective stroke an impulse arising under fibril 1 sets off a wave of contraction that is propagated up that fibril. At the same time, the impulse spreads radially, firing off fibrils 2 and 9. When the impulse reaches the base of the conducting central pair (10 and 11), it sweeps rapidly upward stimulating all points along fibrils 1, 2, 9, 3, and 8 almost instantaneously. The shortening of this group of fibrils simultaneously throughout most of their length, causes the cilium to bend forward as a stiff rod. In the recovery phase, this group of fibrils relaxes while the continued spread of the impulse around the basal corpuscle initiates local contractions in fibrils 4 and 7, 5 and 6. Contractions spreading along these fibrils at their normal slower rate of propagation tend to pull the

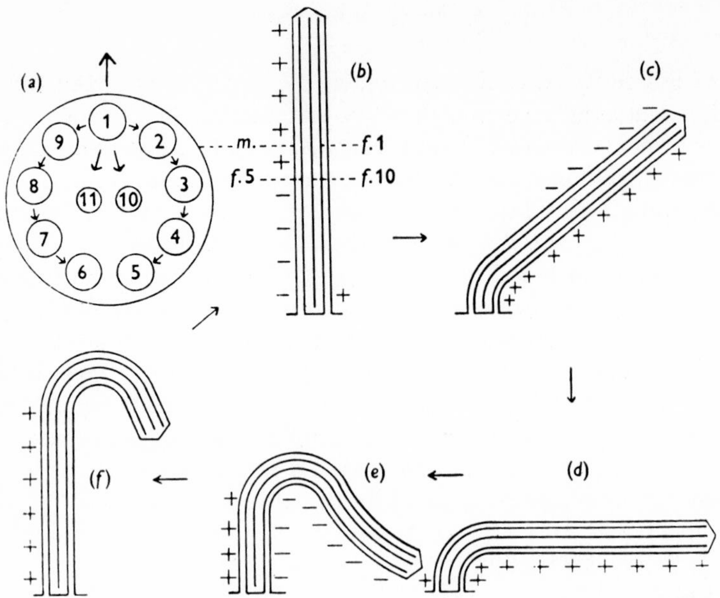

FIG. 39. Hypothetical mechanism for ciliary movement proposed by Bradfield (1955). (a) Transverse section of a cilium. The impulse causing contraction of the fibers is assumed to originate under fiber 1 and to spread both round the ring of 9 fibers and to the central pair, as indicated by arrows inside the cilium. The external arrow indicates the direction of beat. (b) Side view of cilium beating in the plane of the paper, showing membrane (*m*), fiber 1 (*f. 1*), fiber 5 (*f. 5*), and fiber 10 (*f. 10*). The cilium is at the beginning of the effective stroke. An impulse arising under fiber 1 has just produced a propagated contraction (indicated by plus sign) at the base of this fiber. (c) The impulse is spreading round the basal apparatus (not indicated) and has also spread to the central fibers, which are specialized for conduction and which rapidly convey it up the cilium causing fibers 1, 2, 9, 3, and 8 to contract (plus signs) almost simultaneously throughout their length (only fiber 1 is shown). As a result, the cilium is bent forward as a stiff rod and is not deformed by the resistance of the medium so as to be convex forward. (d) The end of the effective stroke. The impulse spreading around the basal ring is reaching fibers 4, 7, 5, and 6 and initiating contraction at their base (plus sign). (e) The contractions in 4, 7, 5, and 6 spread slowly up these fibers at their natural rate of propagation and are not spread by the rapidly conducting central fibers because the latter are now in their refractory state. Fibers 1, 2, 9, 3, and 8 are relaxing (minus sign) and hence also refractory. Owing to the contraction being propagated more slowly than in the effective stroke, the cilium returns to the vertical through a series of curves and not as a stiff rod. (f) Contraction continuing to spread up fibers 4, 7, 5, and 6; completion of recovery stroke. Relaxation complete (no sign) in fibers 1, 2, 9, 3, and 8, which are therefore ready to commence the effective stroke in (b). In (b) and (c) fibers 4, 7, 5, and 6, which have brought about the recovery stroke, are relaxing (except perhaps in their distal regions) and are hence refractory and not affected by the impulse conducted by the central fibers. The cilium is arbitrarily shown as vertical at the end of the recovery stroke and beginning of the effective stroke. In fact, however, it is likely, on the mechanism postulated here, that the cilium may be pulled over to the left of vertical during recovery.

cilium back to its original vertical position and probably beyond. A new impulse would then arise under number one, and the cycle would be repeated. With the fibrils on opposite sides thus contracting alternately, the cilium would execute two-dimensional pendular movements.

Electron micrographs do not bear out the assymetry in the structure of the cilium postulated by Heidenhain and by Gray. A line drawn through the central pair of fibrils may divide the outer fibrils unevenly, leaving 5 on one side and 4 on the other, but it is doubtful whether this would give one side a significant power advantage. The mechanism proposed by Bradfield does not require a visible structural difference on the two sides of the cilium to account for the difference between the effective stroke and the recovery stroke. The stiffening of the cilium and its more rapid movement during the effective stroke is attributed to the nearly simultaneous contraction of the entire length of the fibrils on the leading edge, due to unusually rapid impulse conduction along the central pair. The slower propagation of a wave of local contraction in the other fibrils is sufficient to account for the recovery stroke.

The development of a satisfactory theory to explain flagellar motion is hampered by lack of general agreement concerning the nature of the movements. Although recent evidence favors a predominantly two-dimensional wave motion (Gray, 1955), speculations in the past concerning the mechanism of action of flagella and sperm tails have nearly all been based upon the belief that their undulations are helical. Astbury *et al.* (1955) saw two possibilities for transmitting a helical line of contraction along a flagellum. One would be for the fibrils to have a spiral course and for them to shorten and lengthen simultaneously throughout their length (Fig. 40, a). The other would be for the fibrils to be straight but have waves of local contraction propagated along their length from base to tip, out of phase, and in sequence (Fig. 40, b). The first of these alternatives is excluded by electron microscopic observations, which show quite clearly that the outer fibrils are not spiral. The second offers a more attractive possibility.

Developing a scheme first suggested by Brown (1945), and later elaborated by Hodge (1949), Astbury *et al.* (1955) proposed a mechanism of flagellar motion in which a spirally intertwined central pair of fibrils would be specialized for conduction and would form a "kind of rotary impulse transmittor" or "multiple contact commutator system" in the center of the flagellum, distributing the contractile impulse along its length in proper sequence to produce helical undulations. He regarded the interfibrillar space seen in electron micrographs as an artifact of specimen preparation and believed that, in life, the fibrils are in contact.

It was assumed that the central fibrils twist around each other and, as a result of their helical course, they would contact all the outer fibrils in succession. An impulse conducted by them might therefore cause a spiral line of local contraction to move along the straight peripheral fibrils. Standing in the way of acceptance of this attractive theory is the morphological evidence that the central pair run straight and parallel.

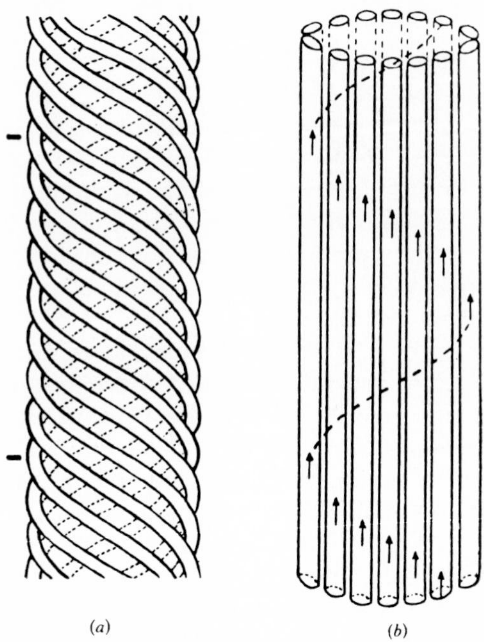

FIG. 40. Two modes of transmitting a helical line of contraction along a flagellum: (a) helical subfibrils; (b) straight subfibrils. For simplicity the diagrams represent the initial state and not the actual form produced by passage of the contractile wave. From Astbury *et al.* (1955).

Bradfield's interpretation of flagellar motion does not assign to the central pair any special role in distribution of the impulse but assumes that the timing or triggering mechanism resides in the basal body. Three-dimensional undulations would result from his model if the impulse arising under fibril 1 spread in only one direction around the ring, firing the fibrils one after another.

3. *Biochemical Aspects of Motility*

A full explanation of the mechanism of flagellar motion will require additional information on the chemical nature of the contractile elements and on the reactions specifically concerned with the conversion of chemical energy into mechanical work. Tibbs (1957, 1958) has studied the properties of algal flagella and fish sperm tails purified by precipitation or sedimentation. The sedimented preparations seemed to be more free of contaminating cytoplasmic material. The results of these analyses are presented in Table I. The samples consisted mainly of a noncollagenous protein similar to myosin. Little or no carbohydrate was present, but an appreciable amount of lipid was found. The cilia of *Tetrahymena* have been isolated by Child (1959). The isolated material when examined with the electron microscope was found to consist mainly of the axial bundles of fibrils. These were not soluble in water, salt solutions, dilute acid, or urea. Their substance was largely protein but appeared also to contain small amounts of adenine and uracil nucleotides.

All recent discussions of possible mechanisms of ciliary motion have emphasized analogies with contraction of muscle. Cilia, flagella, and sperm tails have all been shown to have adenosinetriphosphatase activity (Mann, 1945; Tibbs, 1958; Child, 1959), and numerous observations indicate that adenosine triphosphate (ATP) is as important in ciliary and flagellar motion as it is in muscular contraction. Dramatic demonstrations of its role have come from studies on glycerin-extracted models of flagella (Hoffman-Berling, 1956), cilia (Alexandrow and Arronet, 1956), and sperm tails of various species (Bishop, 1958). These nonliving preparations have had most of their soluble inorganic and organic components extracted, but their contractile and structural proteins remain. When ATP is added to glycerin-extracted flagella or sperm tails, they begin a *rhythmic* activity which persists, in the case of sperm tails, for minutes or even hours and may equal or exceed the normal frequency of beat (Bishop, 1958). This repetitive activity is in striking contrast to the behavior of similar preparations of muscle which contract but once upon addition of ATP (Szent-Györgyi, 1949). With all their vigorous bending movements, ATP-treated sperm models do not show wave propagation along the flagellum and fail to make any significant forward progress. Bishop has interpreted this as evidence that wave formation and wave propagation constitute two distinct components of flagellar motion depending upon different biochemical mechanisms.

The capacity for coordination of ciliary beat is preserved for weeks if the glycerol-extracted cell models are kept in glycerol, but is abolished by digitonin or saponin, agents that have no deleterious effect on similarly extracted models of flagella. This finding suggests that an additional structural element exists in ciliated cells which is responsible for coordination. An essential lipid or lipoprotein component may account for its sensitivity to surface-active agents.

TABLE I
RESULTS OF ANALYSIS OF ALGAL FLAGELLA AND FISH SPERM TAILS[a]

	Precipitated *Polytoma* preparations	Precipitated sperm (various) preparations	Sedimented *Polytoma* preparations	Sedimented sperm (perch) preparations
Ash content (%)	0–3	—	3–4	0–3
Nitrogen (%)	14.8–15.2	—	10.1	7.6–9.8
Nitrogen of lipid-free material (%)	14.8–15.2	—	12.7	14.5
Lipid	Absent	Absent	About 20%	Some variation around 40%
X-Ray	Consistent with protein. No information about form	Consistent with protein. No information about form	Consistent with protein. No information about form	—
Infrared absorption	C = O absorption maximum about 1650 cm^{-1}. Little indication of β	—	C = O absorption maximum about 1650 cm^{-1}. Little indication of β	C = O absorption maximum about 1650 cm^{-1}. Little indication of β
Paper chromatography	Common mixture of amino acids. Proline present, hydroxyproline absent	—	Common mixture of amino acids. Proline present, hydroxyproline absent	—
Tyrosine (gm. acid/100 gm. protein)	3.35–3.92	3.73–4.26	3.44–3.85	3.38–3.85

TABLE I (*continued*)

	Precipitated *Polytoma* preparations	Precipitated sperm (various) preparations	Sedimented *Polytoma* preparations	Sedimented sperm (perch) preparations
Tryptophan (gm. acid/100 gm. protein)	2.22–2.61	2.63–2.68	2.35–2.84	2.63–3.06
Cystine (gm. acid/100 gm. protein)	0.81–0.87	1.12	0.40–0.41	—
Hexosamine (%)	0.2–0.5	—	—	Not detectable i.e. 0.2 or less
Hexose (%)	0.6–6.2	—	6.6–8.4	1.3–1.8
Adenosine triphosphatase activity	Positive	—	Positive	Positive

[a] From Tibbs (1958).

In describing a set of small secondary fibers between the nine outer and two inner primary fibers, Gibbons and Grimstone (1960) drew attention to certain superficial similarities between flagella and muscle. For example, both appear to have a parallel array of large and small fibers and in both the larger fibers bear small lateral projections and show evidence of helical substructure. Although it is tempting to look for a two-fiber sliding system in flagella corresponding to that now widely accepted for striated muscle (Huxley and Hanson, 1955, 1957), it would be premature to press these points of similarity too far. The differences are equally impressive, and as yet the reality of the secondary filaments in cilia and flagella is not established beyond question and no component similar to the actin of muscle has been found by chemical analysis.

C. *Coordination of Ciliary Beat*

Cilia may have an *isochronal rhythm* in which all beat together, or, more commonly, they may exhibit a *metachronal rhythm,* in which the successive cilia in each row start their beat in sequence so that each is slightly more advanced in its cycle than the preceding one. This sequential activation of the cilia results in the formation of waves that

FIG. 41. Diagram illustrating the form of a metachronic wave moving across a ciliated epithelium from left to right. (After Verworn, from Borradaile *et al.* 1958. "The Invertebrata." Cambridge Univ. Press.)

sweep over the surface of the epithelium (Fig. 41). In surface view under the microscope, these resemble the waves that move before the wind across a field of grain, except that the ciliary waves are perfectly regular in their recurrence and pass over the epithelium at a uniform rate. Biologists have long been interested in the mechanism of coordination responsible for metachronism. Historically there have been two opposing views. The *neuroid theory,* first proposed by Engelmann (1868, 1880), postulates a nervelike impulse passing from cell to cell and activating the cilia in regular sequence. The mechanical theory, advanced by Verworn (1890), holds that the movement of one cilium mechanically stimulates the next to action.

1. *The Theory of Neuroid Transmission*

The neuroid theory derives support from several independent experimental approaches. If the metachronal rhythm is the result of propagation of an excitatory stimulus through the epithelium, it is reasonable to expect that the coordinating mechanism would be able to function independently of the contractile activity of the cilia. Experimental evidence that such is indeed the case was provided as early as 1880 by Kraft, who showed that a stimulus causing increased activity in a ciliated epithelium was transmitted across a region where the cilia were completely inactive. Parker (1905a), a strong proponent of the neuroid theory, later demonstrated in the swimming plates of ctenophores that if local cessation of ciliary beat was produced by low temperature or mechanical intervention there was no appreciable disturbance of the metachronal rhythm beyond the quiescent region.

In seeking a structural basis for the conduction of a coordinating impulse, attention focused early upon the intracellular fibrous appendages of the cilia. In ciliate Protozoa, the fibrillar "silver-line" system was long believed to have this function. Taylor (1920) showed that coordination was disturbed in the oral membranelle of *Euplotes* if the subjacent fibers were severed. There were corresponding observations for Metazoa. For example, Grave and Schmitt (1925) drew attention to a system of fibers in the gill epithelium of mollusks that seemed to join the basal bodies together in rows. In addition to these fibrous interconnections, the ciliary rootlets were believed to anastomose, forming a complex of impulse-conducting fibers that was continuous throughout the epithelium. Worley (1941) observed incoordinate ciliary movement after severing the fibrous rootlets in the gut epithelium of a mollusk and interpreted this finding as evidence that the rootlets are involved in coordination of beat. This argument is not convincing, however, for if the rootlets merely serve as anchoring structures, the disturbance of the normal mechanical support of the basal bodies might well result in disordered movements without involving the coordinating mechanism directly.

From electron microscopic observations it now appears unlikely that a fiber system is responsible for conduction of a coordinating impulse. Micrographs of Protozoa reveal that the kinetodesmal fibrils do not actually connect one kinetosome or basal body with another, nor is there any direct connection between the kinetosomes and the infraciliary lattice system (Sedar and Porter, 1955). Investigations of the fine structure of ciliated epithelia in mollusks have also failed to confirm the existence of the fibrous interciliary connections reported by Grave and Schmitt. Although blunt processes extend from each basal body toward the next, they do not make contact. The cross-striated rootlets of adjacent cilia run in close proximity but do not anastomose. In no invertebrate epithelium so far studied do these structures have the relations that would be expected if they were to constitute uninterrupted pathways for intra- or intercellular transmission of coordinating impulses. Furthermore, metachronal rhythm occurs in ciliated epithelia of mammals in which neither ciliary rootlets nor connections between basal bodies have been demonstrated. If there is a neuroid transmission mechanism in these epithelia, then the basis for the horizontal polarity of the cells and for their capacity to conduct an impulse does not reside in the visible fibrous components of the basal apparatus but must depend upon irritable properties of the cell membrane or of the cytoplasmic matrix.

Certain biochemical and pharmacological similarities between ciliated cells and nerve cells have been cited as indirectly supporting a neuroid mechanism of coordination. Acetylcholine and acetylcholinesterase occur in both. A relation between the presence of these substances and ciliary motion is suggested by the fact that acetylcholine is present in the motile protozoan *Trypanosoma rhodesiense,* but not in the nonmotile *Plasmodium gallinaceum* (Bülbring *et al.,* 1949). Moreover, it has been reported that there is a sharp rise in acetylcholinesterase in the developing sea urchin at the time when ciliated tufts appear on its surface. Experimental suppression of ciliation prevents the expected increase in acetylcholinesterase activity (Augustinsson and Gustafson, 1949). Lucas (1932) postulated a myoneural type of junction between the conduction system and the contractile elements of the cilia. It is interesting, in this connection, that low concentrations of acetylcholine are reported to increase ciliary activity in isolated strips of ciliated epithelium, whereas atropine or *d*-tubocurarine depress or arrest it. Since the presence of nerve cells in such preparations can be ruled out, it is suggested that acetylcholine is produced in the epithelial cells and that it participates in the control of ciliary movement (Kordik and associates, 1952). That essentially the same biochemical mechanism is operative in ciliary motion over a wide range of animal forms, is indicated by the observation that motility of *Tetrahymena,* and the ciliary activity of frog esophagus and clam gill are all inhibited by approximately the same concentrations of eserine and diisopropyl fluorophosphate (Seaman and Houlihan, 1951).

2. *The Mechanical Theory*

Three main considerations led Gray (1930) to reject the theory of neuroid transmission. First, the rate of transmission is slower than that of any known nervous impulse. Secondly, the metachronal waves cannot be traced to any center from which the excitatory stimulus can be shown to arise. Thirdly, there is the observation that the activity of the cilia and the physiological polarity of the cells is not seriously disturbed when cells are isolated from one another. Instead they continue to beat and to display a metachronal wave that travels in the same direction as it did in the intact epithelium. Gray found it difficult to explain these findings on the basis of the passage of a nervelike impulse and therefore favored a mechanical triggering mechanism whereby one cilium would excite the next when it reached a certain stage in its cycle. The work previously cited implicating acetylcholine in ciliary activity suggests a synapselike mechanism at the level of the basal bodies. The slow rate of transmission observed might be explained if a neuroid impulse had to pass in succession through a field of synapses. The continued activity of isolated cells would seem to pose difficulties for the mechanical theory as great as those that stand in the way of acceptance of the neuroid theory.

Cells exhibiting metachronal rhythm usually have their cilia closely spaced in straight rows, and several observations suggest that a linear arrangement of cilia and a certain degree of proximity of their basal bodies may be necessary for coordinated activity. For example in the endoderm of *Hydra,* where flagella occur in two or three widely separated areas of the cell surface, they beat independently (McConnell, 1931). In tissue cultures where the stereotaxic response of epithelial cells causes them to become greatly flattened, cilia continue to beat in unison as long as their basal bodies are close together in rows. When the base bodies are separated either spontaneously or with microneedles, the beating of the cilia slows and becomes disorganized (Duryee and Doherty, 1954). Such observations have been cited as evidence supporting the mechanical theory since they suggest that if the interciliary distance becomes too great the triggering effect of one cilium upon another may be inoperative. It might also be argued, however, that separation of the basal bodies interrupted paths of neuroid transmission.

3. *The Pacemaker and Conduction Hypothesis*

From quantitative studies on the relation between ciliary activity and the coordination process in the peristomial cilia of *Stentor,* Sleigh (1956, 1957) has arrived at a theory of ciliary coordination which combines certain features of the mechanical and the neuroid theories, but differs basically from both. Using a stroboscopic method to establish the frequency of beat, and measuring the length of the metachronal wave by means of photographs, he calculated the rate of propagation of the waves under different conditions. In a series of carefully conducted experiments, he found that the frequency of beat and the wave velocity have different temperature characteristics. The presence of magnesium chloride in low concentrations caused an increase in frequency but no significant change in wave velocity. Other experiments demonstrated that an increase in viscosity of the medium caused a decrease in frequency, but no appreciable change in wave velocity. These results seemed to him to indicate that the mechanical process involved in ciliary beat functions independently of the coordination process and the conduction of metachronal waves is independent of mechanical factors. The findings were consistent with a continuous neuroid conduction stimulating the cilia to contract as the impulses pass. In later studies, however, results were encountered which could not be fully explained on this basis (Sleigh, 1957). After transection of the peristomial row of cilia, differences in frequency were noted on the two sides of the cut. The frequency on the distal side was sometimes greater and sometimes less than on the other side, which seemed to indicate that conduction can start at any place along the row. Moreover, in the intact organism, variations were found in wave velocity in different parts of the row, and these could be correlated with differences in the size and the spacing of the cilia. There were clear indications that the wave velocity depended, in some way, upon the number of cilia involved in transmission over a given distance. These, and other findings, suggested a mechanism in which the cilia themselves take part in the transmission. A two-step process was therefore proposed, an *intra*ciliary excitation process and an *inter*ciliary conduction process. The intraciliary process was believed to consist of the building up of an "excitatory state." Discharge of this excitatory state would then set off both the contraction of the cilium and the next conduction phase. It was inferred from the results of the micrurgical experiments that the frequency of the group was determined by a "pacemaker," presumably a cilium at the beginning of the row. The theory of metachronal coordination proposed by Sleigh is represented diagrammatically in Fig. 42. In the row of three cilia depicted there, the frequency would be determined by the rate of contraction and rate of excitation of the pacemaker cilium. The wave velocity would depend upon the rate of *inter*ciliary conduction and the time required for the intraciliary excitation process in each of the other cilia. Sleigh points out, that if the excitation process were long relative to the rate of transmission, the actual conduction process could well be of a speed comparable to that of nerve or cardiac muscle. Interestingly enough, he has shown that wave velocity is increased by digitoxin, a drug that lowers the threshold of excitability of heart muscle, and it is easy to speculate that it may similarly affect the intraciliary excitation process.

D. Concluding Comment

When we relied entirely upon the light microscope for information about the structure of motile cell processes it was confidently supposed that the mechanism of ciliary beat would become clear if only a little

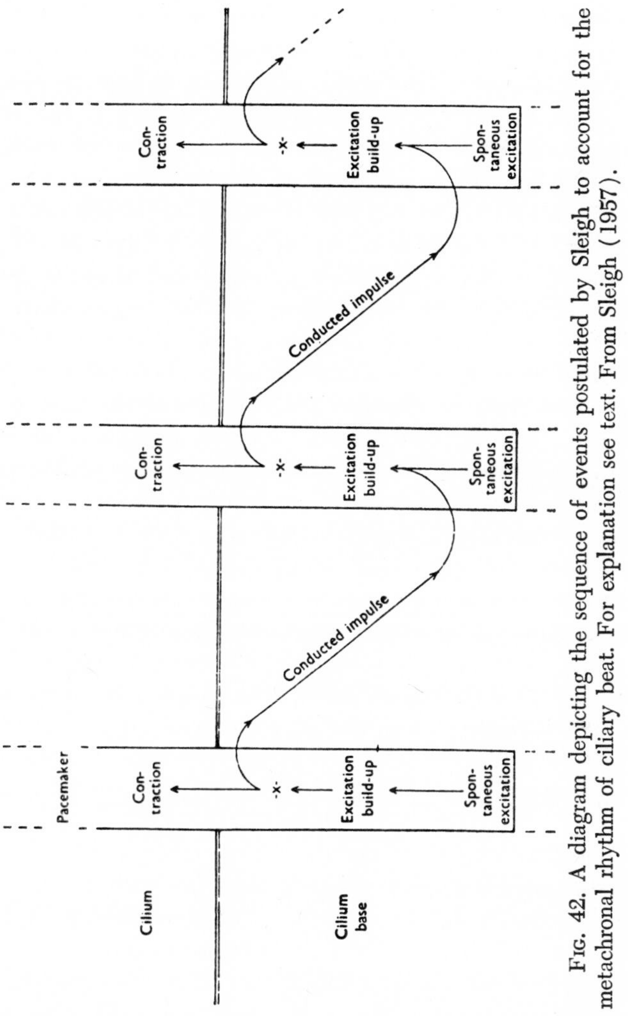

FIG. 42. A diagram depicting the sequence of events postulated by Sleigh to account for the metachronal rhythm of ciliary beat. For explanation see text. From Sleigh (1957).

more magnification and a little better resolution could be obtained. The electron microscope has now taken us far beyond the barrier to higher resolution imposed by the wavelength of light and has revealed in cilia and flagella an entirely unexpected complexity of internal structure. But, the mechanism of their vibratile motion still eludes us. The evolutionary and functional significance of the universal nine-plus-two pattern of internal fibers found throughout the plant and animal kingdoms is still an intriguing mystery. Improved methods of preservation, embedding, and staining for electron microscopy will, no doubt, continue to bring out additional details of the structure of the fibrils and will disclose new components in the ciliary matrix. We would again be deluding ourselves, however, to expect from morphological methods alone a full answer to the physiological problem of ciliary and flagellar motion. The microscopist can only define the components of the ciliary apparatus and describe their spatial relations. The ultimate explanation of their mechanism of action lies in the chemistry of those components. For the present, the descriptive science of morphology with its modern instruments for research in ultrastructure has gained a commanding lead over the more analytical and quantitative disciplines of physiology and biochemistry. An understanding of the exact chemical nature and functional significance of the minute structures already revealed may have to await instrumental and technological advances in biophysical and microchemical analysis that will have an impact in these fields comparable to that which the electron microscope has had upon morphology.

References

Afzelius, B. (1955). *Z. Zellforsch. u. mikroskop. Anat.* **42**, 134.

Afzelius, B. (1959). *J. Biophys. Biochem. Cytol.* **5**, 269.

Afzelius, B. (1961). *J. Biophys. Biochem. Cytol.* **9** (in press).

Alexandrow, W. J., and Arronet, N. Y. (1956). *Doklady Akad. Nauk. S.S.S.R.* **110**, 457.

Anberg, A. (1957). *Acta Obstet. Gynecol. Scand.* **36**, *Suppl.* **2**, 1.

Apathy, S. (1897). *Mitt. zool. Stat. Neapel* **12**, 495.

Astbury, W. T., Beighton, E., and Weibull, C. (1955). *Symposia Soc. Exptl. Biol. No.* **9**, 282.

Augustinsson, K. B., and Gustafson, T. (1949). *J. Cellular Comp. Physiol.* **34**, 311.

Baker, J. R. (1948). *Nature* **161**, 548 and 587.

Ballowitz, E. (1888). *Arch. mikroskop. Anat. u. Entwicklungsmech.* **32**, 401.

Ballowitz, E. (1890). *Arch. mikroskop. Anat. u. Entwicklungsmech.* **36**, 253.

Bargmann, W. (1954). *Z. Zellforsch. u. mikroskop. Anat.* **40**, 49.

Bargmann, W., and Knoop, A. (1955). *Z. Zellforsch. u. mikroskop. Anat.* **43**, 184.

Benda, C. (1899). *Arch. Anat. u. Physiol. Anat. Abt.* 376.

Bernhard, W., and de Harven, E. (1960). *Proc. 4th Intern. Conf. on Electron Microscopy,* **B-2**, 217.

Bessis, M., and Breton-Gorius, J. (1957). *Rev. hématol.* **12**, 43.
Bishop, D. W. (1958). *Nature* **182**, 1638.
Blanchard, R., Lewin, R., and Philpott, D. (1958). Personal communication.
Borell, U., Nilsson, O., and Westman, A. (1957). *Acta Obstet. Gynecol. Scand.* **36**, 22.
Bouillon, J., Castiaux, P., and Vandermeersche, G. (1958). *Bull. microscop. appl.* **8**, 61.
Bradfield, J. R. G. (1955). *Symposia Soc. Exptl. Biol. No.* **9**, 306.
Bretschneider, L. H. (1950). *Koninkl. Akad. Wetenschap. Amsterdam* **53**, 531.
Brown, H. P. (1945). *Ohio J. Sci.* **45**, 247.
Bülbring, E., Lourie, E. M., and Pardoe, U. (1949). *Brit. J. Pharmacol.* **4**, 290.
Bülbring, E., Burn, J. H., and Shelley, H. J. (1953). *Proc. Roy. Soc.* **B141**, 445.
Burgos, M. H., and Fawcett, D. W. (1955). *J. Biophys. Biochem. Cytol.* **1**, 287.
Burgos, M. H., and Fawcett, D. W. (1956). *J. Biophys. Biochem. Cytol.* **2**, 223.
Carasso, N. (1958). *Comp. rend. acad. sci.* **247**, 527.
Carlgren, O. (1905). *Biol. Zentr.* **25**, 308.
Carter, G. S. (1926). *Brit. J. Exptl. Biol.* **4**, 1.
Challice, C. E. (1953). *J. Roy. Microscop. Soc.* **73**, 115.
Challice, C. E. (1954). *J. Roy. Microscop. Soc.* **74**, 27.
Chapman, G. B., and Tilney, L. G. (1959). *J. Biophys. Biochem. Cytol.* **5**, 69.
Child, F. M. (1959). *Exptl. Cell Research* **18**, 258.
Cleland, K. W., and Rothschild, L. (1959). *Proc. Roy. Soc.* **B150**, 24.
Colwin, A. L., Colwin, L. H., Philpott, D. E. (1957). *J. Biophys. Biochem. Cytol.* **3**, 489.
Dalhamn, T. (1956). *Acta Physiol. Scand.* **36**, *Suppl.* **123**, 1.
Dammerman, K. W. (1910). *Z. wiss. Zool.* **96**, 654.
Dellinger, O. P. (1909). *J. Morphol.* **20**, 171.
de Harven, E., and Bernhard, W. (1956). *Z. Zellforsch. u. mikroskop. Anat.* **45**, 378.
De Robertis, E. (1956). *J. Biophys. Biochem. Cytol.* **4**, 227.
Duryee, W. R., and Doherty, J. K. (1954). *Ann. N.Y. Acad. Sci.* **58**, 1210-1230.
Eberth, C. J. (1866). *Arch. pathol. Anat. u. Physiol. Virchow's* **35**, 447.
Engelmann, T. W. (1868). *Jena. Z. Naturw.* **4**, 321.
Engelmann, T. W. (1880). *Arch. ges. Physiol. Pflüger's* **23**, 505.
Engstrom, H. (1951). *Acta Oto Laryngol.* **39**, 364.
Erhard, H. (1910). *Arch. Zellforsch.* **4**, 309.
Fauré-Fremiet, E. (1958). *Quart. J. Microscop. Sci.* **99**, 123.
Fauré-Fremiet, E., and Rouiller, C. (1957). *Compt. rend. acad. sci.* **244**, 2655.
Fauré-Fremiet, E., and Rouiller, C. (1958). *Bull. microscop. appl.* **8**, 117.
Fauré-Fremiet, E., Rouiller, C., and Gauchery, M. (1956). *Arch. Anat. Microscop.* **45**, 139.
Fawcett, D. W. (1954). *Laryngoscope* **64**, 557.
Fawcett, D. W. (1958). *In* "Frontiers of Cytology" (S. Palay, ed.), p. 19. Yale Univ. Press, New Haven, Connecticut.
Fawcett, D. W. (1959). *Intern. Rev. Cytol.* **8**, 195.
Fawcett, D. W., and Porter, K. R. (1954). *J. Morphol.* **94**, 221.
Fischer, A. (1894). *Jahrb. wiss. Botan.* **26**, 187.
Florey, H., Carleton, H. M., and Wells, A. Q. (1932). *Brit. J. Exptl. Pathol.* **13**, 269.
Friedreich, N. (1858). *Arch. pathol. Anat. u. Physiol. Virchow's* **15**, 535.
Fujimura, W., Harutsugu, M., Nishiki, T., and Ito, K. (1956). *Nara Igaki Zassi* **7**, 122.

Gibbons, I. R., and Grimstone, A. V. (1960). *J. Biophys. Biochem. Cytol.* **7**, 697.
Gordonoff, T., and Manderli, H. (1936). *Z. ges. exptl. Med.* **98**, 265.
Grassé, P. (1956). *Arch. biol. (Liège)* **67**, 595.
Grassé, P., Carasso, N., and Favard, P. (1956). *Ann. sci. nat. Zool. et biol. animale* [II] **18**, 339.
Grave, C., and Schmitt, F. O. (1925). *J. Morphol. and Physiol.* **40**, 479.
Gray, J. (1928). "Ciliary Movement." Macmillan, New York.
Gray, J. (1930). *Proc. Roy. Soc.* **B107**, 313.
Gray, J. (1955). *J. Exptl. Biol.* **32**, 775.
Gray, J. (1958). *J. Exptl. Biol.* **35**, 96.
Grigg, G. W., and Hodge, A. J. (1949). *Australian J. Sci. Research Ser. B* **2**, 271.
Hach, I. W. (1925). *Z. ges. exptl. Med.* **46**, 558.
Hancock, G. J. (1953). *Proc. Roy. Soc.* **A217**, 96.
Hardy, A. C. (1953). *Quart. J. Microscop. Sci.* **94**, 441.
Harvey, E. N., and Anderson, T. (1943). *Biol. Bull.* **85**, 151.
Heidenhain, M. (1911). "Plasma und Zelle." Gustav Fischer, Jena.
Henderson, V. E., and Taylor, A. H. (1910). *J. Pharmacol. Exptl. Therap.* **2**, 153.
Henneguy, L. F. (1897). *Arch. Anat. Microscop.* **1**, 481.
Hilding, A. C. (1932). *Am. J. Physiol.* **100**, 664.
Hill, L. (1928). *Lancet* **215**, 802.
Hodge, A. J. (1949). *Australian J. Sci. Research Ser.* **B2**, 368.
Hoffmann-Berling, H. (1955). *Biochim. et Biophys. Acta* **16**, 146.
Hoffmann-Berling, H. (1956). *In* "Cell, Organism and Milieu" (D. Rudnick, ed.), p. 45. Ronald Press, New York.
Houwink, A. L. (1951). *Koninkl. Ned. Akad. Wetenschap. Proc. Ser.* **C54**, 132.
Huxley, H., and Hanson, J. (1955). *Symposia Soc. Exptl. Biol. No.* **9**, 228.
Huxley, H., and Hanson, J. (1957). *Biochim. et Biophys. Acta* **23**, 229.
Jackson, S. F., and Randall, J. T. (1958). *Proc. Roy. Soc.* **B148**, 290.
Jennison, M. W., and Bunker, J. W. M. (1934). *J. Cellular Comp. Physiol.* **5**, 189.
Jensen, O. S. (1887). *Arch. mikroskop. Anat. u. Entwicklungsmech.* **54**, 329.
Klein, B. M. (1932). *Ergeb. Biol.* **8**, 75.
Kordik, P., Bülbring, E., and Burn, J. H. (1952). *Brit. J. Pharmacol.* **7**, 67.
Korschikov, A. (1923). *Arch. Soc. russe Protistol.* **2**, 195.
Kraft, H. (1880). *Arch. ges. Physiol. Pflüger's* **47**, 196.
Krijgsman, B. J. (1925). *Arch. Protistenk.* **52**, 478.
Lasansky, A., and De Robertis, E. (1960). *J. Biophys. Biochem. Cytol.* **7**, 679.
Lebouca, G. (1909). *Arch. Anat. Microscop.* **10**, 555.
Lenhossek, M. (1898). *Verhandl. deut. anat. Ges., Jena* **12**, 106.
Lommel, F. (1908). *Deut. Arch. klin. Med.* **94**, 365.
Lowman, F. G. (1953). *Exptl. Cell Research* **5**, 335.
Lowndes, A. G. (1941). *Proc. Zool. Soc. London Ser. A* **111**, 111.
Lowndes, A. G. (1943). *Proc. Zool. Soc. London Ser. A* **113**, 99.
Lucas, A. M. (1931a). *J. Morphol. and Physiol.* **51**, 147.
Lucas, A. M. (1931b). *J. Morphol. and Physiol.* **51**, 195.
Lucas, A. M. (1932). *J. Morphol. and Physiol.* **53**, 243.
Lucas, A. M., and Douglas, L. C. (1934). *A.M.A. Arch. Otolaryngol.* **21**, 285.
Lwoff, A. (1950). "Problems in Morphogenesis of Ciliates." Wiley, New York.
McConnell, C. H. (1931). *J. Morphol.* **52**, 249.
MacDonald, J. R., Leisure, C. E., and Lenneman, E. E. (1928). *Trans. Am. Acad. Ophthalmol. Otolaryngol.* **33**, 318.

Mainx, F. (1928). *Arch. Protistenk.* **60**, 305.

Mann, I. (1928). "The Development of the Human Eye." Cambridge Univ. Press, London.

Mann, T. (1945). *Biochem. J.* **39**, 451.

Mann, T. (1954). "The Biochemistry of Semen." Methuen, London.

Manton, I. (1952). *Symposia Soc. Exptl. Biol.* **6**, 306.

Manton, I., and Clarke, B. (1950). *Nature* **166**, 973.

Manton, I., and Clarke, B. (1951a). *J. Exptl. Botany* **2**, 125.

Manton, I., and Clarke, B. (1951b). *J. Exptl. Botany* **2**, 242.

Manton, I., and Clarke, B. (1952). *J. Exptl. Botany* **3**, 265.

Manton, I., and Clarke, B. (1956). *J. Exptl. Botany* **7**, 416.

Manton, I., Clarke, B., and Greenwood, A. D. (1951). *J. Exptl. Botany* **2**, 321.

Manton, I., Clarke, B., Greenwood, A. D., and Flint, E. A. (1952). *J. Exptl. Botany* **3**, 204.

Manton, I., Clarke, B., and Greenwood, A. D. (1953). *J. Exptl. Botany* **4**, 319.

Manton, I., Clarke, B., and Greenwood, A. D. (1955). *J. Exptl. Botany* **6**, 126.

Martius (1884). Cited by Lucas, A. M. (1931).

Mast, S. O. (1927). *Arch. Protistenk.* **60**, 197.

Mast, S. O. (1938). *Biol. Revs. Cambridge Phil. Soc.* **13**, 186.

Merton, H. (1923). *Arch. ges. Physiol. Pflüger's* **198**, 1.

Merton, H. (1932). *Arch. Protistenk.* **77**, 491.

Merton, H. (1935). *Biol. Zentr.* **55**, 268.

Metz, C. B., and Westfall, J. A. (1954). *Biol. Bull.* **107**, 106.

Metz, C. R., Pitelka, D. R., and Westfall, J. A. (1953). *Biol. Bull.* **104**, 408.

Miduno, T. (1934). *J. Fac. Sci. Univ. Tokyo* **4**, 367.

Miller, W. H. (1958). *J. Biophys. Biochem. Cytol.* **4**, 227.

Mühldorf, A. (1930). *Beih. botan. Zentr.* **47**, 169.

Newman, S. B., Borysko, E., and Swerdlow, M. (1950). *J. Appl. Phys.* **21**, 67.

Noirot-Timothée, C. (1958). *Compt. rend. acad. sci.* **246**, 2293.

Noirot-Timothée, C. (1959). *Ann. Sci. nat. Zool. et biol. animale* [12] **1**, 266.

Palade, G. (1952). *J. Exptl. Med.* **95**, 285.

Parker, G. H. (1905a). *J. Exptl. Zool.* **2**, 407.

Parker, G. H. (1905b). *Am. J. Physiol.* **13**, 1.

Parker, G. H. (1905c). *Am. J. Physiol.* **14**, 1.

Parker, G. H. (1929). *Quart. Rev. Biol.* **4**, 155.

Parker, G. H., and Marks, A. P. (1928). *J. Exptl. Zool.* **52**, 1.

Peter, K. (1899). *Anat. Anz.* **15**, 271.

Pitelka, D., and Schooley, C. N. (1955). *Univ. of Calif. (Berkeley) Pubs. Zool.* **61**, 79.

Porter, K. R. (1957). *Harvey Lectures Ser.* **51**, 175.

Porter, K. R., and Blum, J. (1953). *Anat. Record* **117**, 685.

Porter, K. R., and Dippell, R. (1957). Personal communication.

Prosser, C. L., Bishop, D. W., Brown, F. A., Jahn, T. L., and Wulff, V. J. (1950). "Comparative Animal Physiology," p. 645. Saunders, Philadelphia.

Purkinje, J. F., and Valentin, G. (1834). *Arch. Anat. u. Physiol.* **1**, 391.

Pyne, C. K. (1958). *Exptl. Cell Research* **14**, 388.

Pyne, C. K. (1959). *Compt. rend. acad. sci.* **248**, 1410.

Randall, J. T. (1956). *Nature* **178**, 9.

Randall, J. T. (1957). *Symposia Soc. Exptl. Biol.* **10**, 185.

Randall, J. T., and Jackson, S. F. (1958). *J. Biophys. Biochem. Cytol.* **4**, 807.

Randall, J. T., and Friedlaender, M. G. (1950). *Exptl. Cell Research* **1**, 1.
Renyi, G. (1924). *Z. Anat. Entwicklungsgeschichte* **73**, 338.
Retzius, G. (1909). *Biol. Untersuch.* [N.F.] **14**, 1.
Rhodin, J., and Dalhamn, T. (1956). *Z. Zellforsch.* **44**, 345.
Ritter, S. (1859). *Deut. Klin.* **11**, 27.
Roth, L. E. (1956). *J. Biophys. Biochem. Cytol.* **2** *Suppl.* 235.
Roth, L. E. (1957). *J. Biophys. Biochem. Cytol.* **3**, 985.
Rothert, W. (1914). *Ber. deut. botan. Ges.* **32**, 91.
Rothschild, L. (1953). *Mammalian Germ Cells. Ciba Foundation Symposium,* p. 122.
Rouiller, C., and Fauré-Fremiet, E. (1957). *Exptl. Cell Research.*
Rouiller, C., Fauré-Fremiet, E., and Gauchery, M. (1956). *Exptl. Cell Research* **11**, 527.
Rudzinska, M. (1958). Personal communication.
Saguchi, S. (1917). *J. Morphol. and Physiol.* **29**, 217.
Schäfer, E. A. (1891). *Proc. Roy. Soc.* **49**, 193.
Sharpey, W. (1836). "Cilia," Todd's Cyclopaedia of Anatomy and Physiology. Volume 1, p. 606.
Schmitt, F. O., Hall, C., and Jakus, M. (1943). *Biol. Symposia* **10**, 261.
Schnall, M. (1952). *Fertility and Sterility* **3**, 62.
Schultz-Larsen, J. (1958). *Acta Pathol. Microbiol. Scand. Suppl.* 128.
Seaman, G. R., and Houlihan, R. K. (1951). *J. Cellular Comp. Physiol.* **37**, 309.
Sedar, A. W., and Porter, K. R. (1955). *J. Biophys. Biochem. Cytol.* **1**, 583.
Seefelder, R. (1910). *Arch. Ophthalmol. Graefe's* **73**, 419.
Sjöstrand, F. S. (1953a). *J. Cellular Comp. Physiol.* **42**, 15.
Sjöstrand, F. S. (1953b). *J. Cellular Comp. Physiol.* **42**, 45.
Slautterback, D. L., and Fawcett, D. W. (1959). Unpublished.
Sleigh, M. A. (1956). *J. Exptl. Biol.* **33**, 15.
Sleigh, M. A. (1957). *J. Exptl. Biol.* **34**, 106.
Sotelo, J. R., and Trujillo-Cenóz, O. (1958). *Z. Zellforsch.* **48**, 565.
Steinert, E. (1960). Personal communication.
Stieve, H. (1930). *In* "Handbuch der mikroskopischen Anatomie des Menschen" (W. von Möllendorff, ed.), Volume 7, Pt. 2, p. 103. Springer, Berlin.
Szent-Györgyi, A. (1949). *Biol. Bull.* **96**, 140.
Tartar, V. (1956). *In* "Cellular Mechanisms in Differentiation and Growth" (D. Rudnick, ed.). Princeton Univ. Press, Princeton, New Jersey.
Tartar, V. (1957). *Exptl. Cell Research* **13**, 317.
Taylor, C. V. (1920). *Univ. Calif. (Berkeley) Publs. Zool.* **19**, 403.
Taylor, G. (1951). *Proc. Roy. Soc.* **A209**, 447.
Taylor, G. (1952). *Proc. Roy. Soc.* **A211**, 225.
Telkka, A., Fawcett, D. W., and Christensen, A. K. (1961). In press.
Tibbs, J. (1957). *Biochim. et Biophys. Acta* **23**, 275.
Tibbs, J. (1958). *Biochim. et Biophys. Acta* **28**, 636.
Tokuyasu, K., and Yamada, E. (1959). *J. Biophys. Biochem. Cytol.* **6**, 225.
Twitty, V. C. (1928). *J. Exptl. Zool.* **50**, 319.
Valentin, G. (1842). *In* "Wagner's Handwörterbuch der Physiologie," Vol. 1, p. 484.
Verworn, M. (1890). *Arch. ges. Physiol. Pflüger's* **48**, 149.
Vlk, W. (1938). *Arch. Protistenk.* **90**, 448.
Weisz, P. (1954). *Quart. Rev. Biol.* **29**, 207.
Williams, W. W. (1950). *Fertility and Sterility* **1**, 199.
Willmer, E. N. (1955). *Ann. Rev. Physiol.* **17**, 339.

Wohlfarth-Bottermann, K. E., and Pfefferkon, G. (1953). *Protoplasma* **42**, 227.
Wolken, J. J. (1956). *J. Protozool.* **3**, 211.
Wolken, J. J., and Palade, G. (1953). *Ann. N.Y. Acad. Sci.* **56**, 873.
Worley, L. G. (1941). *J. Cellular Comp. Physiol.* **18**, 187.
Wyman, J. (1925). *J. Gen. Physiol.* **7**, 545.
Yasuzumi, G., Fujimura, W., Tanaka, A., Ishida, H., and Masuda, T. (1956). *Okajimas Folia Anat. Japon.* **29**, 133.

CHAPTER 5

Mitochondria (Chondriosomes)

By ALEX B. NOVIKOFF

I. Historical Survey

As in many branches of biology, the growth in knowledge of cytoplasmic particles was slowed not through lack of men with keen powers of observation and imaginative intelligence, but by the limited state of available techniques. Not until apochromatic lenses were available, in the last quarter of the nineteenth century, were cytoplasmic granules the object of intensive study [see Newcomer (1940) for a review of the first observations on mitochondria]. In his classic review, Cowdry (1918) lists the many names given to mitochondria during the first outburst of interest in their study when, because of inadequate fixation, they could hardly be distinguished clearly from other granules. Flemming's "fila" (1882) included mitochondria and other particles; their description rested upon the appearance of cells fixed in fluids containing acetic acid, which destroys many mitochondria.

Altmann (1890) improved the fixative and thus could see mitochondria more clearly, but even his technique left them insufficiently delineated from other granules. In this period of growing interest in mitochondria, Altmann developed his concept of "bioblast" (from the Greek words for "life" and "germ"). The view that mitochondria were the ultimate in living things, "elementary organisms" like bacteria, met with a negative response and made it more difficult for the cytoplasmic particles to compete with the nucleus and heredity for the interest of investigators. The fixatives used by cytologists to bring out nuclear detail usually destroyed the mitochondria.

It was almost ten years before Benda (1897-98), lowering the acetic acid content of Flemming's fixative and introducing crystal violet staining, revived interest in mitochondria. He introduced the term mitochondria (singular, *mitochondrion*, from the Greek for "thread" and "grain") to describe the "fädenkornern" ("thread granules") which he saw in cells during spermatogenesis. He demonstrated the presence of mitochondria in both egg and sperm, and in the cleaving egg. He showed that they passed from cell to cell during mitosis. He concluded (1902) that mitochondria were permanent cell organs which played an important part in heredity and which differentiated during histogenesis into many specialized structures such as myofibrils and the basal bodies of cilia. Among the strongest supporters of Benda's views was Meves. Coining new terms for varying mitochondrial shapes, Meves (1918) described their transformation into fibrils of muscle, nerve, connective-tissue cells, glandular epithelia, and ciliated cells; and into secretory granules, pigment granules, and yolk spheres. Although overenthusiastic, the claims of Benda and Meves, particularly that they were part of the material basis of heredity, brought attention to the mitochondria in an era when chromatin still held the fascination of investigators.

There ensued a period in which cytologists like Regaud, Duesberg, Hoven, Fauré-Fremiet, and Guilliermond brought support to some of the ideas of Benda and Meves, but other equally excellent observers, like Retzius, Vejdovsky, Cowdry, and Mottier, were skeptical of their basic concepts. For the most part reliance was placed on the same cytological techniques, with two important exceptions: the introduction by Michaelis (1900) of Janus green as a supravital stain for mitochondria, and the observation of mitochondria in living cells grown in tissue culture (Champy, 1912; Lewis and Lewis, 1914-15). These techniques finally dispelled the notion that mitochondria were fixation artifacts, and helped lead to the conclusion that mitochondria were present in practically all living cells, plant as well as animal.

Concerning the biochemical nature and basic physiological role of mitochondria little could be learned by the techniques then available. True, as early as 1908 Regaud (1908) had drawn attention to their phosphatide and protein nature, and Kingsbury in 1912 had suggested that they were "a structural expression of the reducing substances concerned in cellular respiration." Cowdry's (1926) concept of "the surface film theory of the function of mitochondria," while it described an important aspect of mitochondria structure, told little of its biochemical function, and the view of Marston, supported by Horning (see Horning, 1933), that mitochondria were the sites of enzymatic syntheses was little more than inference from indirect morphologic evidence, no better established than the now-discarded view that mitochondria are bacteria living symbiotically with the cells of higher organisms.

The situation was essentially the same in 1934 when Sharp wrote in the third edition of his "Introduction to Cytology," "Owing to the difficulties attending the observation of such minute objects and the determination of their relation to other cytoplasmic constituents, opinion regarding the origin, behavior, and biological significance of chondriosomes is still in a very unsettled state. . . . Our knowledge of chondriosomes is far too incomplete to warrant categorical assertions concerning their function. The literature is not only complicated by conflicting statements regarding their observed behavior, but it is further encumbered with a variety of hypotheses, some of which rest on very narrow foundations."

In that same year the paper of Bensley and Hoerr (1934) appeared in which they described the isolation of mitochondria from the cells of guinea pig liver by the process of differential centrifugation of ground tissues. Shortly thereafter, Claude reported the separation, by higher centrifugal forces, of submicroscopic granules (for which he adopted the old term "microsomes") after first sedimenting the "large granules" (Claude, 1943). Identification of the "large granules" of Claude with mitochondria was achieved by Hogeboom and co-workers in 1948, on the basis of their stainability with Janus green B and their elongate forms when isolated from hypertonic sucrose homogenates.

The next decade saw the growth of an enormous literature devoted to the biochemistry of isolated mitochondria. Outstanding was the demonstration that the mitochondria were the chief, if not the exclusive, sites of oxidative phosphorylation, the process by which energy of foodstuffs is made available to cell metabolism and cell function. No longer could it be said that "our knowledge of chondriosomes is far too incomplete to warrant categorical assertions concerning their function."

During this period, too, the technique of electron microscopy of tissue sections developed so dramatically that mitochondria are no longer the "minute objects" Sharp described them, and we may confidently expect that soon the "origin, behavior, and biological significance" will be largely settled rather than "unsettled." Improved methods of tissue fixation, embedding, and sectioning [for reviews see Claude (1954), Sjöstrand (1956a, b), and Selby (1959)] led to the description of intramitochondrial structure. The first electron microscopists to refer to the internal mitochondrial fine structure are listed by Palade (1952), but it was mainly through the work of Palade (1952, 1953), Sjöstrand (1953), and Sjöstrand and Rhodin (1953) that the main features of mitochondrial fine structure became quickly and firmly established.

Current research is aimed at relating the multienzyme systems described by biochemists to the intramitochondrial membranes seen in electron micrographs of thin sections. Electron microscopists are also approaching the problem of mitochondrial "origin." Ten years from today we may still repeat Sharp's statement of 1934, "This places the problem of ultimate origin in the realm of the invisible," but the invisible of 1970 will be considerably closer to the level of molecules and atoms than was that of 1934. In 1970 we may be able to relate the dynamic state of mitochondria, as seen in living cells, to the biochemical events "in the realm of the invisible" within and outside the mitochondria. Already biochemists are focusing on this problem. They are examining the mitochondrion as a biological machine in which the transduction of chemical energy into utilizable form occurs at the level of macromolecules (Lehninger, 1959b; Green and Hatefi, 1961). Encouraging, too, is the development of microspectrophotometric methods; they have already yielded data on large mitochondrial masses in specialized cells—intact (and living?) (Perry *et al.*, 1959; Chance and Thorell, 1959) and on small areas of cytoplasm in more typical cells (Thorell and Chance, 1959). Also of great importance is the increasing attention given by electron microscopists to the effects of different fixatives and other preparative steps, and their meaning for revealing the organization of macromolecules in cells.

When reviewing the field of mitochondria, E. B. Wilson (1928) wrote of Benda, Meves, and their contemporaries: "With these authors arose a new terminology, which, as has so often happened before, contributed to the impression that the chondriosomes represented a newly discovered cell-component. But, as most of the leading investigators in this field have clearly recognized, what was new was not the thing itself or even its theoretical treatment but only an impulse to its further investigation."

We may say the same of contemporary cytologists studying electron micrographs and isolated fractions. Yet with "the impulse to its further investigation" come new data leading to the rejection or modification of old ideas and the birth of new concepts. Establishing as fact the past conjecture of a brilliant mind is itself an act of novelty.

Our task in this chapter is to attempt a summary of the new investigations in the perspective of the old. Over forty years ago, when faced with a similar task, Cowdry (1918) wrote, "In a field so large it has been necessary to choose and select, to elaborate some points and to leave others almost untouched, so that many important contributions have been passed without mention." Such omissions are even more inevitable today!

Some readers may find these general reviews of value: for morphology, Cowdry (1918, 1924), Wilson (1928), Sharp (1934), Guilliermond (1941), Newcomer (1940), Steffen (1955), Dangeard (1958), Rouiller (1960); for biochemistry, Schneider (1959), Ernster and Lindberg (1958), Hogeboom *et al.* (1957), Schneider (1956), Hackett (1959), Millerd (1956), Hackett (1955), Goddard and Stafford (1954); and the correlation of structure and biochemistry, Green and Hatefi (1961), Lehninger (1960), Palade (1956a) and "Symposium: The Structure and Biochemistry of Mitochondria" (1953).

II. Mitochondria in Living Cells: Their Plasticity and Polymorphism

Early cytologists deduced the plasticity of mitochondria from their highly variable size and form in fixed preparations. A variety of granules, rods, and filaments were beautifully illustrated in the third edition of E. B. Wilson's magnificent "The Cell in Heredity and Development," (1928) as well as highly specialized forms such as "chondriospheres" and "nebenkern." He wrote: "More recent studies have shown that they consist of a specific material, showing definite cytologic and microchemical characters but morphologically highly plastic, so that it may appear under many forms, which are probably to be regarded as only different phases of the same material."

Reconstruction of *process* from static preparations is unsatisfactory at best, and Wilson recognized the importance of observations on living cells. Mitochondria can readily be seen, particularly in thin cells like those of animal tissues grown in culture or of plant tissues like leaf epidermis. The early observations on mitochondria in tissue culture are cited by Newcomer (1940). Among the most extensive studies were those of Lewis and Lewis (1914-15).

The Lewises described the remarkable changes in position and shape of mitochondria within these animal cells. Later investigators, armed with phase contrast micrography (see Frederic, 1958), produced cinemaphotomicrographs which are much more vivid and perhaps more generally acceptable than the simple line drawings of the Lewises, but the major aspects of movement and plasticity were already described in 1914 by the Lewises. They concluded: "The mitochondria are extremely variable bodies, which are continually moving and changing shape in the cytoplasm. There are no definite types of mitochondria, as any one type may change into another. They appear to arise in the cytoplasm and to be used up by cellular activity. They are, in all probability, bodies connected with the metabolic activity of the cell." They described the fusion of two or more granules to form a single mitochondrion, the division of a mitochondrion into two or more, and the increase and decrease of mitochondrial size without fusion or division. The shape may change fifteen to twenty times in 10 minutes; it can be changed by heat, hypertonic and hypotonic media; "the mitochondria are extremely plastic bodies and often react more rapidly than any other cell structure" to heat, carbon dioxide, acids, fat solvents, potassium permanganate, and osmotic change.

The laboratories in which mitochondria of living animal cells were studied most actively by phase contrast microscopy included those of Chèvremont (1956); Frederic (1958); Gey (1956); Gey *et al.* (1956); Zollinger (1950); Pomerat *et al.* (1954); Rose (1957); Lettré (1954); Biesele (1955); and Tobioka and Biesele (1956); for other references see Murray and Kopech (1953). Among the interesting observations recorded were the separation of lateral branches from mitochondria (Frederic, 1958), tendency of mitochondria to stick to one another (Gey *et al.*, 1956), formation of loops by end to end fusion (Gey *et al.*, 1956; Frederic, 1958; Tobioka and Biesele, 1956), fusion of rows of granules to form filaments (Frederic, 1958; Gey *et al.*, 1956), side-to-side fusion of mitochondria (Tobioka and Biesele, 1956), and the apparently passive division of mitochondria trapped by pinocytosis vacuoles advancing through the cell (Gey *et al.*, 1956) or by cell membranes forming furrows during mitoses (Frederic, 1958). Of particular interest are the observations by Frederic and Chèvremont (1952) of intimate contact between mitochondria and nucleus (Fig. 1) (see page 395), and of Frederic (1958) that despite manifest changes in individual mitochondria the total mitochondrial volume remains constant.

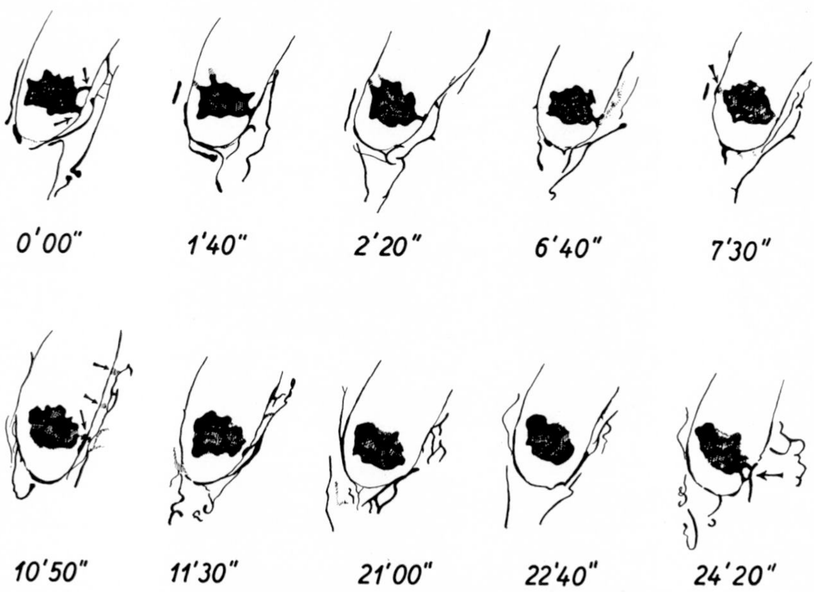

FIG. 1. Movement of mitochondria to nucleus, from cinemaphotography of living chick fibroblast in tissue culture. Mitochondria may be seen to make intimate contact with the nuclear membrane. Arrows indicate apparent movement of nucleolar material through the nuclear membrane. From Frederic (1958).

The effects of a wide variety of chemical and physical agents on mitochondrial behavior in cultured cells have been studied by Frederic (1958) and others (Lettré, 1954; Tobioka and Biesele, 1956) [cf. observations of Buvat (1948), who described reversible mitochondrial changes upon addition of water to plant cells]. Some materials, e.g., detergents (Frederic, 1958), have a similar effect *in vivo* as on mitochondria isolated from homogenates. On the other hand, it is difficult at the moment to correlate the effects of other agents with the known action of these agents on the structure or metabolism of isolated mitochondria. Thus dinitrophenol (Frederic, 1958) and coenzyme A (Biesele, 1955; Tobioka and Biesele, 1956) produce similar effects: a speeding of mitochondrial movement and an exaggeration of their end-to-end fusion so that the cell comes to possess a few very long filaments. Frederic (1958) finds support from his extensive investigations for two important views: (1) that mitochondrial substance is in dynamic equilibrium with submicroscopic cytoplasmic substance ("la substance fondamentale") (Fig. 2), and (2) that mitochondria possess two types of movement, an active

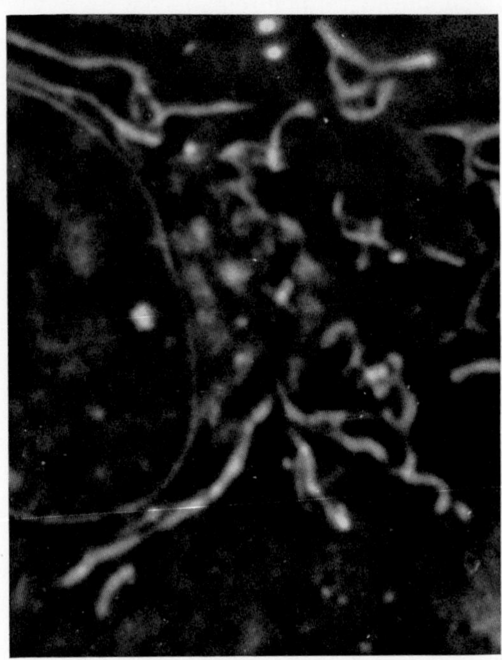

FIG. 2. Living chick fibroblast in tissue culture, photographed by Anoptral phase contrast microscopy. To the right of the nuclear membrane is the Golgi zone, in which "la substance fondamentale" and refractile droplets are present. Magnification: × 3000. From Frederic (1958).

one in which they change their form, and a passive one by which they are moved about in the cell.

In a later section (page 358) we shall again discuss the fate of mitochondria during division of cells in culture. Here we will note that in some instances the plasticity of mitochondria is dramatically displayed. They round out and become much smaller and, when division is complete, they revert to their original size and shape. Whether any disappear entirely, to have new ones reappear after division, cannot be established by light microscopy because of its limited resolving power.

The study of living cells cannot be expected to answer such questions as: what are the mechanisms by which mitochondrial number is kept relatively constant although they are not seen to divide in mitosis?; is mitochondrial movement passive or active? (see Tobioka and Biesele, 1956; Frederic, 1958); what is the functional meaning of mitochondrial movement, or their apparent contacts with the nucleus?; is there a submicroscopic mitochondrial "material" or "substance fondamentale" with which the mitochondria are in equilibrium? (Frederic, 1958).

Some caution concerning observations on tissue culture cells may not be out of order. Wilson (1928) long ago suggested that such cells may be living under stress. They may be demonstrating capacities never exhibited in the intact organism where they are influenced by neural and hormonal factors and probably directly by adjacent cells. Thus, some cells undergo "abnormal" chromosomal changes in tissue culture (Hsu, 1959). Many cells show extensive surface bubbling at late anaphase and telophase, which Barer and Joseph (1956) have shown can be prevented, in insect spermatocytes, when the cells are constrained by a mass of neighboring cells and spermatid fibrils. Because of their sensitivity, mitochondria might well be exhibiting "abnormal" behavior when fragmenting or fusing. It will be of great interest to learn how mitochondria behave in cells (human and opossum) grown under more optimal conditions in which the chromosomes remain stable for hundreds of generations (Tjio and Puck, 1958). Naturally, such concern about the effect of stress of mitochondrial behavior does not apply to the same extent to observations of mitochondrial movements in living Protozoa (Fauré-Fremiet, 1910), ova, or certain plant tissues. Guilliermond (1941) has described changing mitochondrial shapes, thickenings, and branchings in motion pictures of cells from a variety of fungi and higher plants. Sorokin (1941) described mitochondrial division in plant cells.

Not all cells need be alike in the extent of mitochondrial plasticity. In mammalian cardiac muscle or kidney tubule cells, in snail spermatozoa, and in other cells the mitochondria are probably firmly anchored, but in others they appear to be free to move about. Palade (1956a) has suggested that when mitochondria move through a cell they may have better access to substrates used in oxidative phosphorylation and other biochemical processes, and when they are more firmly situated they probably have important relationships to cell membranes or intracellular areas of specialized activity.

III. Identification of Mitochondria in Microscopic Preparations

A. *Light Microscopy: Staining Methods*

When well preserved in fixation, the long rods ("chondrioconts") are readily identified as mitochondria. But in some cells, smaller more spherical mitochondria may be difficult to distinguish from other cytoplasmic particles such as lipid spheres and yolk droplets. The diagnostic value of the traditional mitochondrial stains is limited because the underlying chemistry is largely unknown.

The most useful fixatives for mitochondria are chromium- or osmium-containing fluids with little or no acid. Altmann, Benda, Bensley, Regaud, Champy, and others have developed such fixatives [see Cowdry (1924), Jones (1950), and Gatenby and Beams (1950) for details of their preparation; and Baker (1958) for a discussion of some principles of fixation and dyeing]. Palade (1952) found that when osmium tetroxide was buffered at neutral pH, mitochondrial fine morphology was better preserved (see also Ornstein and Pollister, 1952), and it may be expected that buffered fixatives will now be employed for light microscopy as well (but see Baker, 1958). Traditional mitochondrial stains have included iron hematoxylin, crystal violet, and acid fuchsin [see Cowdry (1924), Gatenby and Beams (1950) for preparation and use; and Baker (1958) for an analytical discussion; for staining mitochondria in frozen-dried sections, see Chang (1956)].

The acid-hematin test, as employed by Baker (1946), stains mitochondria presumably because of their high phosphatide content. If conducted as Baker outlines, this test is apparently quite specific for phosphatides (Deane, 1958), but whether it will stain all mitochondria has not yet been established; it may lack sufficient sensitivity, particularly if the phosphatide is present as a mixture with other lipids (Deane, 1958).

Better understood are the chemical bases of three widely used reactions, which, in theory, are capable of staining mitochondria of *unfixed* tissues, cells, or isolated fractions. The staining reactions reflect the presence of oxidative enzymes in the mitochondria: presumably cytochromes a and a_3 (cytochrome oxidase) in the "G-Nadi" reaction; enzymes of the electron transport chain, more or less intimately associated with dehydrogenases, in the tetrazolium reaction; and transport chain enzymes, including the cytochrome system, in Janus green B staining, the most widely employed and until recently the most useful of the three reactions.

Our present knowledge of enzyme localization suggests that a stain dependent upon cytochrome a and a_3 would appear the best suited for the exclusive demonstration of mitochondria. Yet the use of the "G-Nadi" reaction has until very recently been largely abandoned, despite evidence that it probably demonstrates cytochrome oxidase (for references, see Deane *et al.,* 1960; Nachlas *et al.,* 1958b; and Burstone, 1959). The original procedure involves the oxidation of dimethylphenylenediamine in the presence of α-naphthol, to produce "indophenol blue." Unfortunately, indophenol blue is both unstable and lipid soluble so that the method is unsuitable for intracellular localization. However, improvements have recently been achieved by substituting other amines for the dimethylphenylenediamine and by using other coupling agents, more complex than the α-naphthol, which are less toxic to the enzyme and

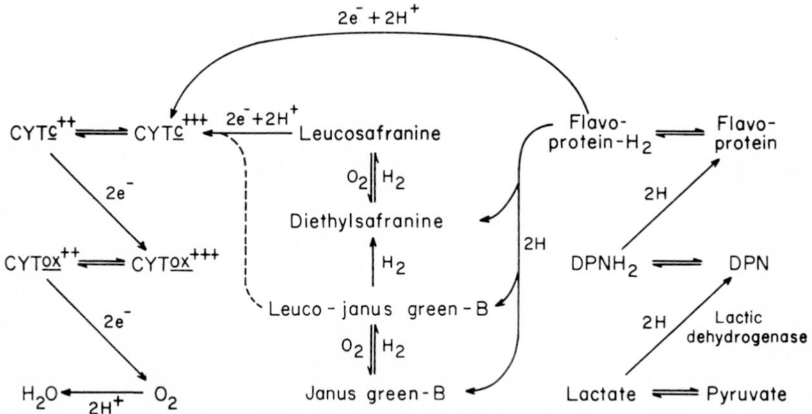

FIG. 3. Interaction of Janus green B with respiratory enzymes. From Lazarow and Cooperstein (1953).

produce more "substantive" dyes (i.e., dyes that adhere to the proteins at their sites of formation) (Burstone, 1959). It may yet be possible to use this reaction routinely to stain mitochondria.

Although Janus green B staining is sometimes too faint and fickle [see Newcomer (1940) for special difficulties with plant cells], most investigators report no difficulty in supravital staining of mitochondria with this dye, as suggested by Michaelis (1900) and Bensley and Bensley (1938). The chemical basis of its use has been studied intensively by Bensley's student, Lazarow. Lazarow and Cooperstein (1953) conclude that Janus green links with flavoprotein enzymes as shown in Fig. 3. These flavoprotein enzymes are present in other cytoplasmic structures as well as mitochondria (isolated microsomes and supernatant fluid reduce the dye to the leuco form). But only in the mitochondria,

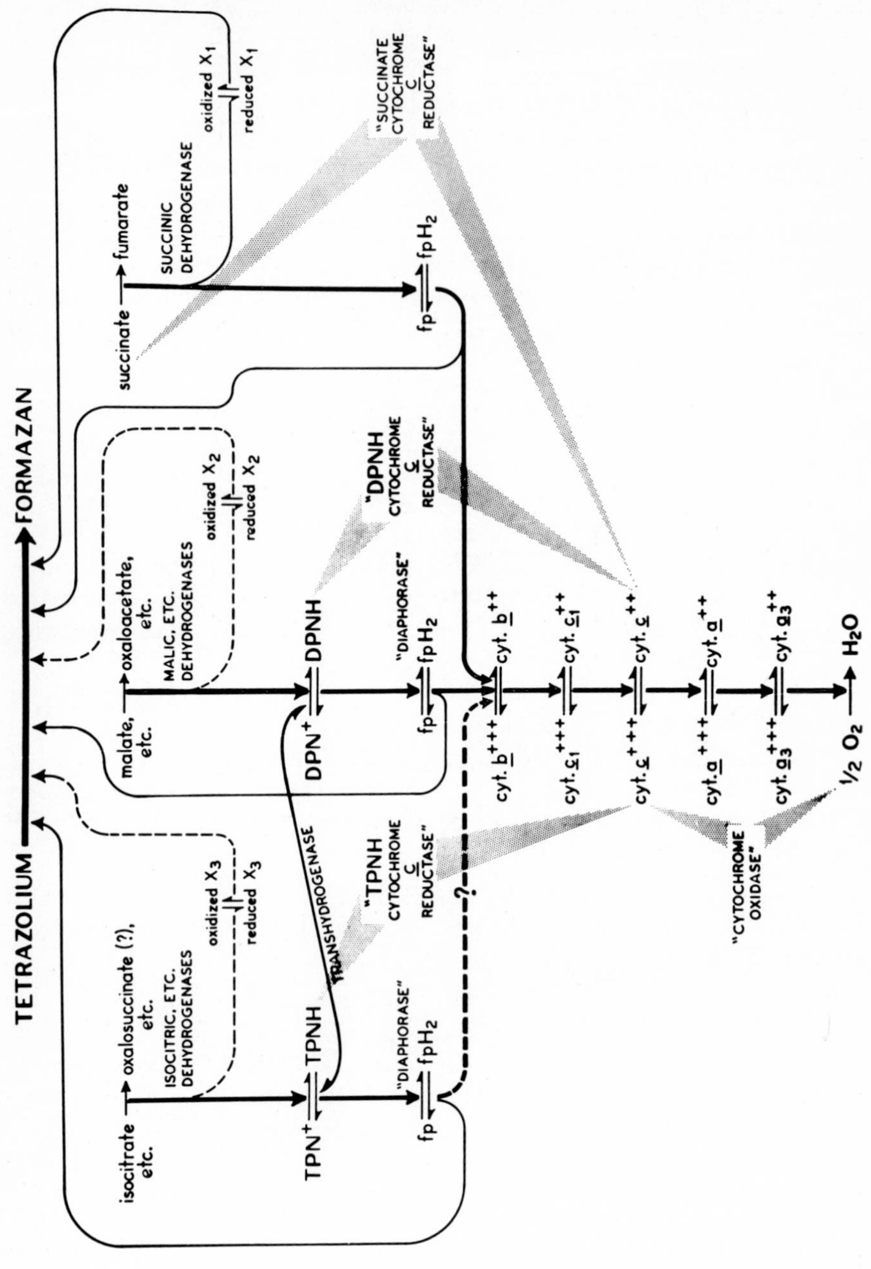
TETRAZOLIUM
FORMAZAN
isocitrate etc.
oxalosuccinate (?), etc.
ISOCITRIC, ETC. DEHYDROGENASES
oxidized X_3
reduced X_3
TPN^+
TPNH
TRANSHYDROGENASE
"DIAPHORASE"
fp
fpH_2
"TPNH CYTOCHROME c REDUCTASE"
?
malate, etc.
oxaloacetate, etc.
MALIC, ETC. DEHYDROGENASES
oxidized X_2
reduced X_2
DPN^+
DPNH
"DPNH CYTOCHROME c REDUCTASE"
succinate
fumarate
SUCCINIC DEHYDROGENASE
oxidized X_1
reduced X_1
"SUCCINATE CYTOCHROME c REDUCTASE"
cyt. b+++
cyt. b++
cyt. c1+++
cyt. c1++
cyt. c+++
cyt. c++
cyt. a+++
cyt. a++
cyt. a3+++
cyt. a3++
1/2 O_2
H_2O
"CYTOCHROME OXIDASE"

with their high level of cytochromes c_1, c, a, and a_3 (see page 339), is the dye reoxidized to the blue form.

Brenner (1953) reported that the more basic the phenosafranine dye the lower the concentration necessary to stain lymphocyte mitochondria supravitally. He considered that this, like his earlier observations (Brenner, 1949) on the effects of cyanide on Janus green-stained cells, controverts the views of Lazarow and Cooperstein. Frederic (1958) also has some relevant observations on the decolorization of the blue form of the dye in mitochondria.

Tetrazolium staining is the most versatile of the reactions that are used to stain mitochondria. In Fig. 4 we indicate the likely points of interaction of tetrazolium salts (such as "nitro-BT") and the electron transport system of mitochondria. The biochemical studies of Singer and co-workers (Singer and Kearney, 1954, 1957; Singer *et al.*, 1956) with succinic dehydrogenase and the histochemical work of Farber and colleagues (Farber *et al.*, 1956a, b; Farber and Bueding, 1956; Farber and Louviere, 1956; Sternberg *et al.*, 1956) with coenzyme-linked dehydrogenases demonstrate that tetrazolium salts accept electrons not from the dehydrogenases but from subsequent steps in the transport chain. Recent work (Wattenberg and Leong, 1960; Wattenberg and Gronvall, 1960) dealing with the influence on electron transport of added coenzyme Q (Crane *et al.*, 1957, 1959) and with the differential effects of inhibitors on the ability of different tetrazolium salts to accept electrons underlines the fact that the link of tetrazolium salts to the chain may be several steps removed from the dehydrogenases. For some tetrazolium salts the linkage points seem further removed from the dehydrogenases than the one shown in the diagram (Kamin *et al.*, 1957; Smith and Lester, 1960; Nachlas *et al.*, 1960). The primary succinic dehydrogenase does not, under normal circumstances, separate from the section of the electron transport chain that links to nitro-BT or other tetrazolium salts. However, in the oxidations involving DPN- or TPN-dependent dehydrogenases a ready separation of substrate-oxidizing and tetrazolium-reducing activities occur, thus complicating problems of localization (Novikoff, 1955, 1959c, in press).

FIG. 4. Probable steps in the mitochondrial electron transport system (Novikoff, 1959c). Acceptors shown are (a) the cytochrome system, (b) tetrazolium salts, and (c) phenazine methosulfate. Abbreviations: *cyt.*, cytochrome; *fp*, flavoprotein; DPN, diphosphopyridine nucleotide; TPN, triphosphopyridine nucleotide; X_1, phenazine methosulfate; X_2 and X_3, postulated dyes that may act in analogous fashion to phenazine methosulfate, which accepts electrons directly from dehydrogenase (Singer and Kearney, 1954; Singer *et al.*, 1956).

The most useful of the commercially available tetrazolium salts is "nitro-BT" [2,2′-di-*p*-nitrophenyl-5,5′-diphenyl-3,3′-(3,3′-dimethoxy-4,4′-biphenylene)ditetrazolium dichloride]. Tsou *et al.* (1956) and Nachlas *et al.* (1957), who developed the salt, showed that it did not suffer from major weaknesses of previous tetrazoliums: marked lipid solubility of their reduced forms (formazans), tendency of their formazans to crystallize in the tissue, and their limited ability to compete successfully with oxygen for electrons. If the mitochondria are well preserved and if the substrate used is not oxidized appreciably by nonmitochondrial components (succinate, glutamate, etc.), then it may be expected that the formazan will stain the mitochondria. On the other hand, if the substrate is DPNH or TPNH, or one that leads to the formation of the reduced pyridine nucleotides in the medium (e.g., lactate in the presence of DPN or isocitrate in the presence of TPN), stained mitochondria may or may not be seen, depending upon the tissue. In rat kidney, where there is little nonmitochondrial reductase, both DPNH and TPNH will produce stained mitochondria (Fig. 5). In rat heart, DPNH does so, but not TPNH (Fig. 6). Finally, in tissues like rat liver and pancreas the abundant formazan produced by nonmitochondrial (ergastoplasmic) reductase obscures most mitochondria.

There is another reason for stressing that not all sites of formazan formation with "nitro-BT" are mitochondria. Not only does the formazan form crystals upon the surface of lipid droplets of the adrenal cortex, as Nachlas *et al.* (1957) showed, but we have found that it does the same on lipofuscin granules of human liver, on intracellular lipid droplets in rat ascites hepatoma, proximal tubules of kidneys in nephrotic rats, and a variety of other cells, and on the large fat vacuoles of fat cells. In these cases the formazan has the same blue color as in mitochondria rather than the pink color it shows when dissolved in lipid structures (e.g., in the Golgi apparatus of hepatoma cells and lipid droplets sometimes produced in frozen sections of liver). In the past, attention of histochemists has been focused on the solubility of formazans within lipid as a source of

---

FIGS. 5 and 6. Frozen sections of formol-calcium-fixed tissues stained with nitro-BT (Novikoff, 1959a, c; 1960b).

FIG. 5. Rat kidney, with TPNH as substrate. Note, in the cells of the convoluted tubules (*c*) the elongate mitochondria mostly at the base of the cell, and in the macula densa cells (*m*) the strongly stained mitochondria. The stain in the glomerulus (*g*) is mostly in the epithelial cells. Magnification: × 1000.

FIG. 6. Rat heart, with DPNH as substrate. Note linearly arranged mitochondria. The irregular "intercalated disks" are unstained. Magnification: × 1180.

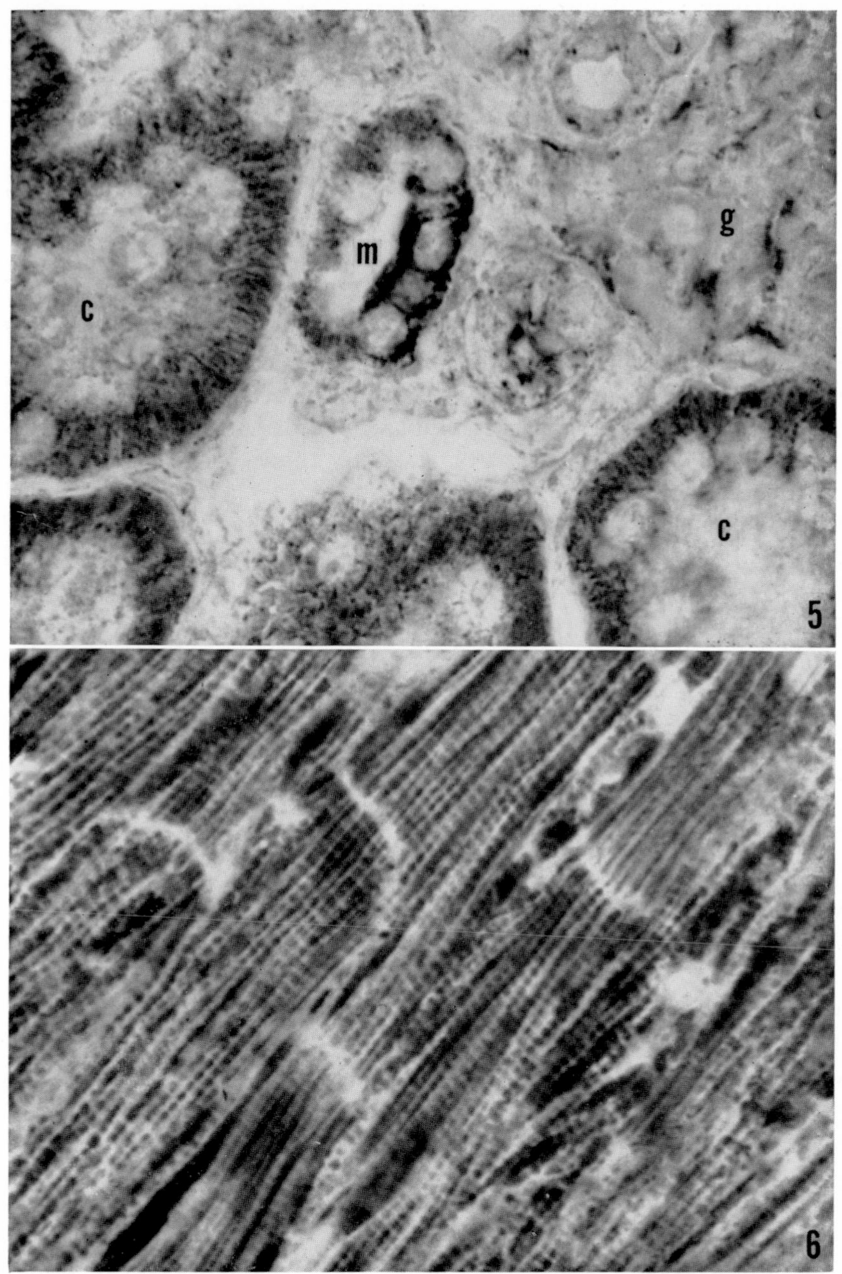
c
m
g
c
5
6

artifact; now with improved tetrazoliums the deposition of formazans upon interfaces between lipid and cytoplasm needs also to be recognized (Novikoff *et al.*, 1961). On the level of electron microscopy such attraction to interfaces may prove as restrictive as the solubility within lipid was with earlier tetrazolium salts on the level of light microscopy.

Pearse (1957) has proposed that the tetrazolium reaction be conducted with the monotetrazolium salt (MTT) [3-(4,5-dimethylthiazolyl-2)-2,5-diphenyl tetrazolium], in the presence of metallic ions which chelate with the formazan as it forms. By incubating sections successively with different metals and substrates, Pearse *et al.* (1958) have obtained results which they claim differentiate mitochondria with different oxidative enzymes within the same cell. Pearse (1958) believes the newer tetrazolium techniques, with unfixed cryostat-cut sections, to be so excellent that he asserts: "Nobody should further consider for an instant the employment of older staining techniques for these organelles." However, we have demonstrated that, in contrast to nitro-BT, the use of MTT is not reliable for the intracellular localization of oxidative enzymes (Novikoff *et al.*, 1961). Even with nitro-BT such localization is not possible unless cell structures are better preserved than they are in the usual cryostat-cut sections of unfixed tissues. Fortunately, it has been found (Novikoff and Masek, 1958; Novikoff and Arase, 1958) that visualization of both DPNH- and TPNH-tetrazolium reductases is still possible after cold formol-calcium fixation. With such fixation mitochondrial morphology is well enough preserved to permit use of these methods as mitochondrial stains in those cells where, as already indicated, nonmitochondrial staining does not obscure the mitochondrial stain (Figs. 5, 6). Were succinate or glutamate dehydrogenases to survive such fixation sufficiently, exclusive mitochondrial staining would be possible. Thus far we have found phenazine methosulfate (Fig. 4) to be helpful with only a few tissues like rat adrenal and human liver (Novikoff and Essner, 1960).

B. *Electron Microscopy: Fine Structure*

So similar is the fine structure of mitochondria in all plant and animal cells, when adequately studied by electron microscopy, that it seems best to define a mitochondrion as a cytoplasmic particle which, when properly fixed in osmium-containing fluids, adequately sectioned and examined in the electron microscope, shows a smooth outer membrane (see page 316) and an infolded inner membrane. It is generally assumed that all such bodies contain the oxidative phosphorylation enzymes and others we will later enumerate. However, even if some be deficient in a few or many of these enzymes we should still consider them mitochondria, as, for example, in the yeast "petites" where they lack succinic dehydrogenase, cytochrome oxidase, and other respiratory enzymes of normal yeast mitochondria (Slonimski, 1953; Ephrussi, 1953) yet show folded inner membranes and smooth external membranes (Yotsuyanagi, 1959). Conversely, we should not do so for "microbody" (Rhodin, 1954), "dense body" (Rouiller and Bernhard, 1956; Novikoff, 1957a), or other cytoplasmic structure, irrespective of chemical composition or developmental relationship to mitochondria, if it lacks a folded inner membrane.

At the level presently attainable in electron microscopy, the outer membrane appears essentially similar in all mitochondria but the inner ones show extraordinary variability in structural detail. The inner folds most frequently appear in sections as shelflike extensions of varying width ("internal ridges" or "cristae mitochondriales," Palade, 1952), tubules, or concentrically arranged layers. The functional meaning of these variations is not clear but one may appropriately apply to this near-molecular level inside the mitochondria the words, already quoted, used by E. B. Wilson (1928) to describe the organelles themselves: "They consist of a specific material, showing definite cytologic and microchemical characters but morphologically highly plastic, so that it may appear under many forms, which are probably to be regarded as only different phases of the same material."

What is the nature of this "material" and what are the "different phases"? Only rudimentary answers can now be given. Before we can decipher the composition and organization of macromolecules we will need to know the effects of osmium fixation and other preparative steps on the appearances of "membranes," "matrix," and "granules." Differences in these appearances are to be expected both from the manner in which macromolecules are "set" and the degree of "staining" by the fixative (Sjöstrand, 1956a, b; Gersh, 1959). Sjöstrand early placed emphasis on the need to compare fixation by osmium and other chemicals with freeze-drying (see Sjöstrand, 1956a, b). Sjöstrand and Baker (1958) found that the mitochondrial image obtained after freeze-drying was the reverse of that seen after osmium fixation. In this "negative image" there is a less opaque zone on the surface and the cristae are less opaque. After staining with phosphotungstic acid, the cristae give a triple-layered appearance. [There is no evidence of ribosomes in the cytoplasm of frozen-dried material (see Finck, 1958). This finding has been confirmed and extended by Hanzon *et al.* (1959). If the tissue (pancreas) is homogenized before freeze-drying, the ribosomes *are* visible. On the other hand, after osmium fixation the ribosomes are visible in tissue as well as in homogenate. They conclude "that the nucleoprotein is probably not organized as 150 Å particles in the living cell. The particles seem to be formed rapidly by a change in the internal milieu of the cell as e.g. by homogenization or osmium fixation."[1]] "Negative" mitochondrial images are also obtained by Stäubli (1960), after formalin fixation and dehydration in a water-soluble epoxy resin, and by Rebhun (personal communication), who froze *Spisula* eggs in Freon at —150°C., dehydrated in acetone-OsO_4 at —80°C. (Feder and Sidman, 1958), embedded in Araldite 502, and stained sections in potassium permanganate. (In these "freeze-substituted" eggs the ribosomes are readily seen.) On the other hand, the mitochondria in the material of Hanzon *et al.* (1959) shows the typical "positive" images of membranes and cristae. The same obtains after potassium permanganate fixation followed by the customary dehydration and embedding (Luft, 1956). For other observations on the mitochondrial matrix, see page 328.

Although a long road still lies ahead, improvements in resolution coupled with progress in analysis of model systems and development of staining methods now make it possible to express less pessimism than Sjöstrand was in position to do in 1956, when he wrote, "It is desirable to be able to differentiate between protein and lipid structures, and a reliable and specific electron staining of different chemical bonds would satisfy our most daring expectations. The chances for success in this respect are, however, rather minute. It is not only a question of working out a staining reaction without changing the localization of the *individual molecules.* Our experience from light microscopy, where we demand only a staining of cell *regions,* provoke great pessimism" (Sjöstrand, 1956b).

An underlying assumption in current work is that the electron-opaque lines of the mitochondria are basically similar to those of the cell membrane. Yet almost all we can say is that both are "unit membranes" in the sense of Robertson (1959): each may be resolved into two opaque lines separated by a light central zone, and the dimensions of these are comparable in cell membrane and mitochondrial membranes. Although other views have been proposed (Sjöstrand, 1956a, b; 1959), general opinion is that each line represents one or two bimolecular layers of oriented lipid molecules with each surface covered by a layer of protein. This is based on X-ray diffraction and electron microscopic analyses of model systems and of myelin sheaths, which may be considered compounded cell membranes (Finean, 1954; Fernández-Morán and Finean, 1957; Engström and Finean, 1958; Robertson, 1957, 1958, 1959, 1960; Stoeckenius, 1959; Stoeckenius *et al.*, 1960). This protein-lipid arrangement is the same as that suggested by Danielli, Schmidt, and Schmitt and Bear (see Danielli,

[1] Bullivant (1960) offers another explanation consistent with the presence of the particles *in vivo* [see also Seno and Yoshizawa (1960) and Beams *et al.* (1960)].

1952) and is consistent with the high lipid and protein content of mitochondria. However, within this basic structure there will surely prove to be great molecular variation with important influence on permeability, specificity of receptor sites, and other properties of the membrane. Thus there may be extensive chemical differences between the outer and inner mitochondrial membranes (see discussion in André, 1959a). We believe that at the moment it may be profitable to conceive of the "inner membrane" as the limit of the mitochondrial body and the "outer membrane" as the true external membrane (see discussion following Palade, 1959; also Rouiller, 1960).

Despite our present ignorance regarding their detailed molecular structure, it seems better not to speak generally of "mitochondrial membranes" without distinction between outer smooth membrane and inner folded membrane (Ziegler *et al.*, 1958; Lehninger *et al.*, 1958). André (1959a) emphasizes the relative stability of the outer membrane while the inner membranes are undergoing profound changes during spermatogenesis in the butterfly. It seems to us that in many electron micrographs, particularly in altered mitochondria of isolated fractions that we have studied (Novikoff, 1957a), the outer mitochondrial membrane appears quite different from the crista-forming inner membrane. The outer membrane, but not the inner one, may disappear when a mitochondrion is intimately associated with a lipid droplet (Palade, 1959). The results of Malamed and Recknagel (1959) may tentatively be interpreted to mean that the outer membrane is permeable to sucrose and the inner crista-forming membrane is not. Among the possibilities which Tedeschi (1959) considers to explain the osmotic phenomena that he observes is an outer highly permeable membrane and an inner folded membrane with restricted permeability (also see Lehninger, 1960c). It is likely that functional differences in the mitochondria are reflected in the dramatic structural differences seen in the inner membranes of the adrenal mitochondria (De Robertis and Sabatini, 1958; Zelander, 1959) or the kinetonucleus-associated mitochondria in *Trypanosoma mega* (Steinert, 1960) where typical tubules or cristae are seen in some cells and concentrically arranged membranes in others (Figs. 31, 32). It is possible, as Belt and Pease (1956) have suggested, that the tubular nature of the inner membranes may be peculiar to certain functions—in the cases studied by them, functions common to steroid-secreting cells. The appearance of numerous "tubular" inner membranes during differentiation of plant meristem cells may also be associated with changed functions (Lance, 1958; Buvat, 1958; Caporali, 1959). Napolitano and Fawcett (1958) describe elongated mitochondria in the brown fat of newborn mice and rats in which the usual closely spaced cristae are totally absent at the ends (also see Pappas and Brandt, 1959). They consider that this is not due either to artifact or to the geometry of the membranes in relation to the plane of section, and they suggest "that these are incompletely differentiated regions of mitochondria that were in the process of elongating," portions that "have not yet developed their internal structure." It is not surprising that in the remarkable transformations of mitochondria that take place during spermatogenesis among invertebrates striking changes are found in the inner membranes (see page 363). Having described those that occur in the scorpion, André (1959a) writes, ". . . ces mécanismes ne peuvent s'opérer que grâce à une grande mobilité des molécules constituant les membranes. Les profondes modifications subies par les crêtes nous conduisent à la même conclusion. Les crêtes sont capables de bourgeonner, de s'allonger, de se raccourcir, de se rejoindre (là encore avec disparition au point de contact), avec la plus grande facilité. . . . Les liaisons entre les molécules des dites «double membranes» sont lâches et sans cesse revisables. Les images statiques du matériel fixé demandent toujours à être interprétées en tenant compte de ces constantes migrations moléculaires."

Generally, the mitochondrial cristae run transversely, more or less completely, across the mitochondrion. However, in some cells [*Helix* spermatids (Grassé *et al.*, 1956), *Otala* spermatids (Powers *et al.*, 1956), transplantable rat hepatoma (Novikoff, 1959c), neurons (Palay and Palade, 1955) (Fig. 13), and the U-shaped mitochondria around the I bands of certain striated muscles (Palade, 1956a)], the cristae may be oriented parallel to the long axis of the mitochondrion. In the spermatocytes of *Viviparus*, transverse, longitudinal, and oblique orientations are found in the same mitochondrion (Kaye, 1958). The same is true of mitochondria in skeletal muscle (Fig. 10) and in the distal convolutions of the renal tubule (Rhodin, 1958) (Fig. 14). In some cells [human leucocytes (Low, 1956), rat tracheal epithelium (Rhodin and Dalhamn, 1956), striated muscle (Palade, 1956a), Novikoff hepatoma (Fig. 7), scorpion spermatocytes (André, 1959a)], the cristae may branch and interconnect. In other cells (Fig. 9), the membranes appear to be long

---

FIG. 7. Ascites form of Novikoff hepatoma. Nucleus of cell is at lower left, cell membrane with microvilli at upper right. Interconnecting cristae may be seen in several mitochondria. Magnification: × 17,500. Courtesy of Dr. Edward Essner.

FIG. 8. Brown fat of newborn mouse. Note the closely packed cristae in many mitochondria. Magnification: × 17,500. From Napolitano and Fawcett (1958); courtesy of Dr. Don Fawcett.

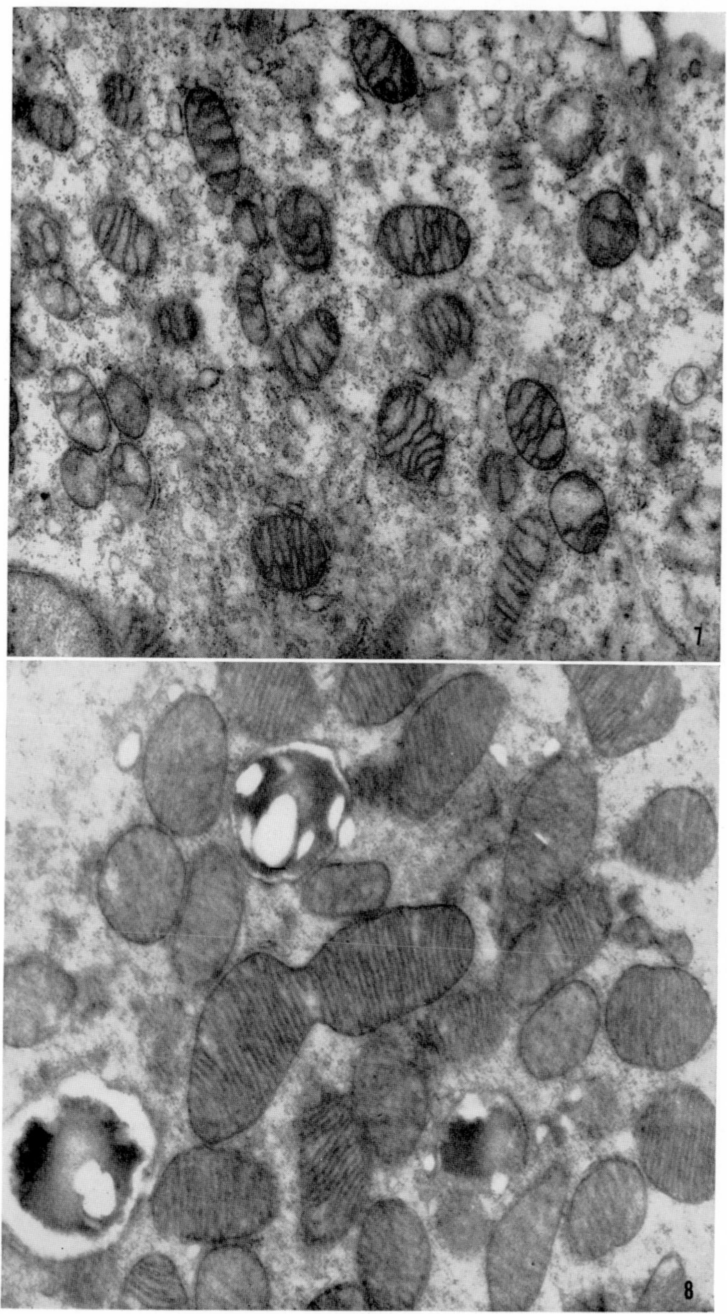
7
8

tortuous tubules or villi [protozoa (Rudzinska and Porter, 1953; Sedar and Porter, 1955; Powers *et al.*, 1956; Dalton and Felix, 1957; Rouiller and co-workers, 1957; Brandt and Pappas, 1959; Noirot-Timothée, 1959), plant cells (Buvat, 1958; Caporali, 1959; Lance, 1958), giant axon of squid (Geren and Schmitt, 1956), Malpighian tubules of insects (Beams *et al.*, 1955; Bradfield, 1956), mammalian adrenal (Palade, 1953; Belt and Pease, 1956; De Robertis and Sabatini, 1958), and other steroid-secreting cells (Belt and Pease, 1956)]; in the interstitial cells of the opossum testis, the villi appear to be very short (Fawcett, 1959). Some mitochondria in cardiac muscle show concentrically arranged mitochondria (Moore and Ruska, 1957), as do the spermatid mitochondria in snails (Grassé *et al.*, 1956) and in several mammals, particularly the opossum (Fawcett, 1958, 1959) (Fig. 12). During spermatogenesis in the rat (Palade, 1953), cat (Burgos and Fawcett, 1955), and butterfly (André, 1959b), still another type of crista is present, a broad shelf that is so flat that the mitochondria appear "empty" or "hollow" (Fig. 29).

That the inner membranes of plant mitochondria (Fig. 11) will also show interesting variations related to function is indicated by recent studies (Lance, 1958; Buvat, 1959; Caporali, 1959; Whaley *et al.*, 1960).[2]

Within the variability in detail of inner membranes there is an impressive over-all similarity. As Palade (1956a) has written: "It appears, therefore, that we are dealing with a common pattern of mitochondrial structure which was presumably developed at an early stage in evolution and subsequently transmitted without considerable modifications from protozoa to mammals and from algae to flowering plants." In 1927, Du Noüy and Cowdry wrote: "The mitochondrial substance is evidently arranged in such a way as to give a surface film of maximum extent, with the minimum amount of material." In 1956, Palade wrote, "the primary function of the infoldings is the increase in the internal surface";

[2] Other electron micrographs of plant mitochondria will be found in the publications of Lund *et al.* (1958), Sitte (1958), Hodge *et al.* (1957), Sager and Palade (1957) (Fig. 11), Turian and Kellenberger (1956), Farrant *et al.* (1956), Mühlethaler (1955), Wolken and Palade (1953). For additional recent references see Lance (1958) and Hackett (1959).

FIG. 9. *Paramecium multimicronucleatum.* The tortuous tubular nature of the cristae is evident. Magnification: × 24,000. Courtesy of Dr. Albert W. Sedar and Keith R. Porter.

FIG. 10. Sartorius muscle of the rat. This demonstrates the "syncytial reticulum" formed by the branching mitochondria (in contact with the I band). Note the high concentration of cristae. Magnification: × 32,000. Courtesy of Dr. George E. Palade.

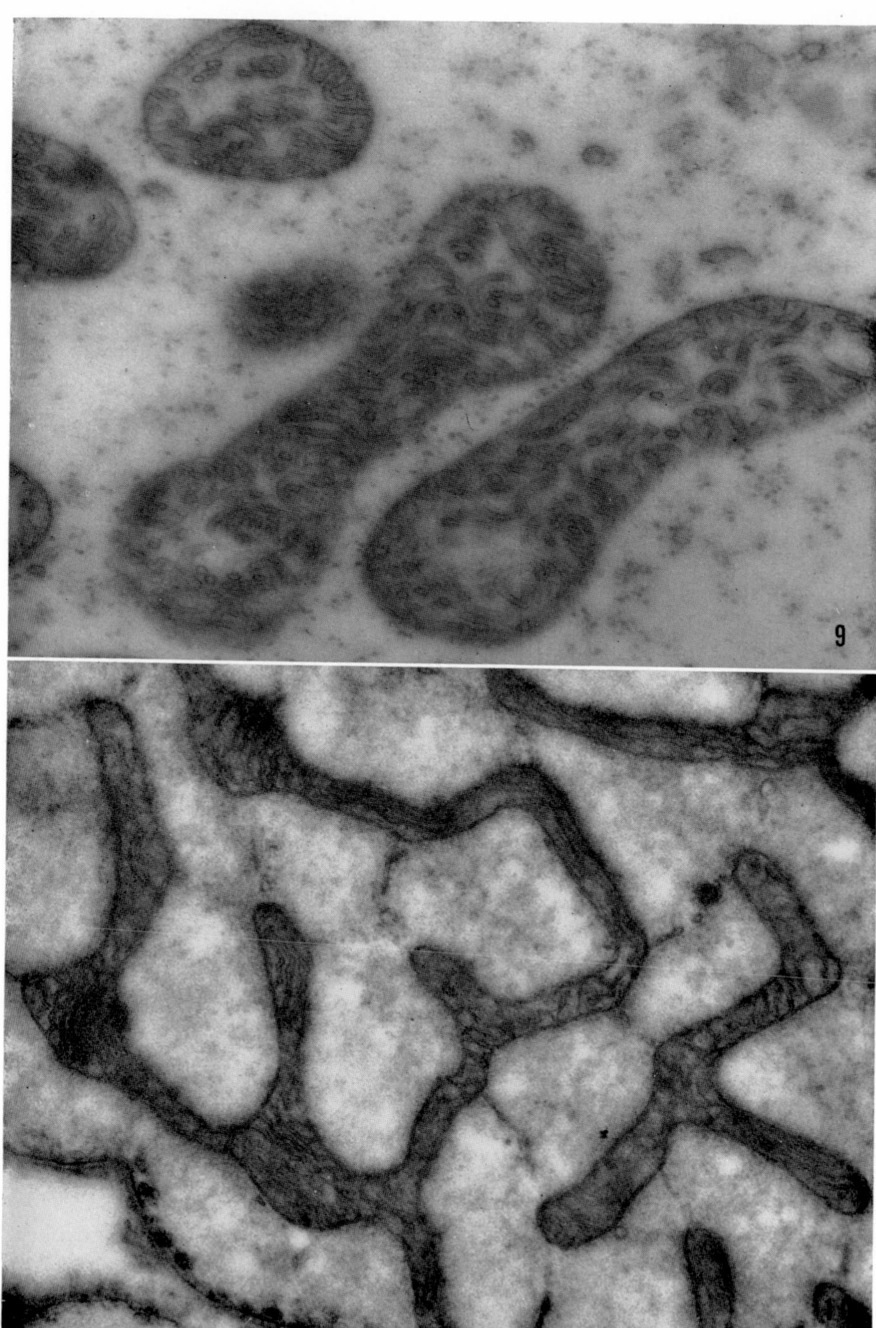
9

he pointed to the highest concentration of cristae in insect flight muscle (Fig. 33) and mammalian cardiac muscle, two tissues with high rates of oxidative metabolism (Palade, 1956a). Other examples of cells with high concentrations of cristae are *Viviparus* spermatozoa (Fig. 23, B) (Kaye, 1958), brown fat (Fig. 8) (Napolitano and Fawcett, 1958), mammalian skeletal muscle (Fig. 10) (Palade, 1956a), and cells in the distal convolutions and ascending limbs of Henle in the mammal kidney (Fig. 14) (Rhodin, 1958). The complex arrangement of villi in the mitochondria of the giant ameba *Pelomyxa* similarly provides a high ratio of membranes to matrix (Pappas and Brandt, 1959).

Two aspects of crista structure, matters of sharp disagreement a few years ago, are now being resolved. These concern the precise relation of the internal membranes to the external surface of the mitochondrion, and the existence of individual compartments within the mitochondrion. There seems little reason any longer to question the continuity of the inner membranes and the external surface of the mitochondrion. Such continuity is seen in high-resolution pictures of many tissues and cells.[3] The *number* of such continuities in a single mitochondrion, and the interpretation to be placed upon them, are still unsettled; for the most complete recent expression of the Sjöstrand school, see Andersson-Cedergren (1959). It is difficult to know to what degree the "central free channel" described in his earlier work by Palade (1953) was due to a rounding of the mitochondria during processing [this seems to be true of the liver mitochondria shown in Figs. 1 and 2 of Palade (1953)]. Similarly, we do not know the extent to which the septa-like arrangement of cris-

[3] Among these tissues and cells are liver (Palade, 1956a), leucocytes (Freeman, 1956; Low, 1956), kidney tubules (Palade, 1953; Rhodin, 1954; Gansler and Rouiller, 1956; Dalton and Felix, 1957), skeletal and cardiac muscle (Palade, 1956a; Porter and Palade, 1957; Moore and Ruska, 1957; Andersson-Cedergren, 1959), intestinal epithelium (Palade, 1956a), pancreas (Sjöstrand, 1956a, b), retinal rods (Sjöstrand, 1956b), thyroid (Ekholm and Sjöstrand, 1957), adrenal cortex (Zelander, 1959), arteriole endothelium (Rouiller, 1960), transplantable liver tumor (Howatson and Ham, 1955), sea urchin egg (Afzelius, 1957a), lobster axon (Geren and Schmitt, 1956), snail spermatids (Grassé *et al.*, 1956; Kaye, 1958), *Paramecium* (Sedar and Porter, 1955), isolated mitochondria of liver (Novikoff, 1957a; Dalton and Felix, 1957), *Chironomus* salivary glands (Bradfield, 1956), root cells of maize (Whaley *et al.*, 1959), apical meristem cells of *Chrysanthemum segetum* (Lance, 1958), and the fungus *Allomyces macrogynus* (Blondel and Turian, 1960).

FIG. 11. *Chlamydomonas reinhardi.* Mitochondria possess typical crista structure. Magnification: × 60,000. Courtesy of Dr. Ruth Sager and Dr. George E. Palade.

FIG. 12. Spermatid of the opossum. Cristae of mitochondria are arranged concentrically. Magnification: × 60,000. Courtesy of Dr. Don Fawcett.

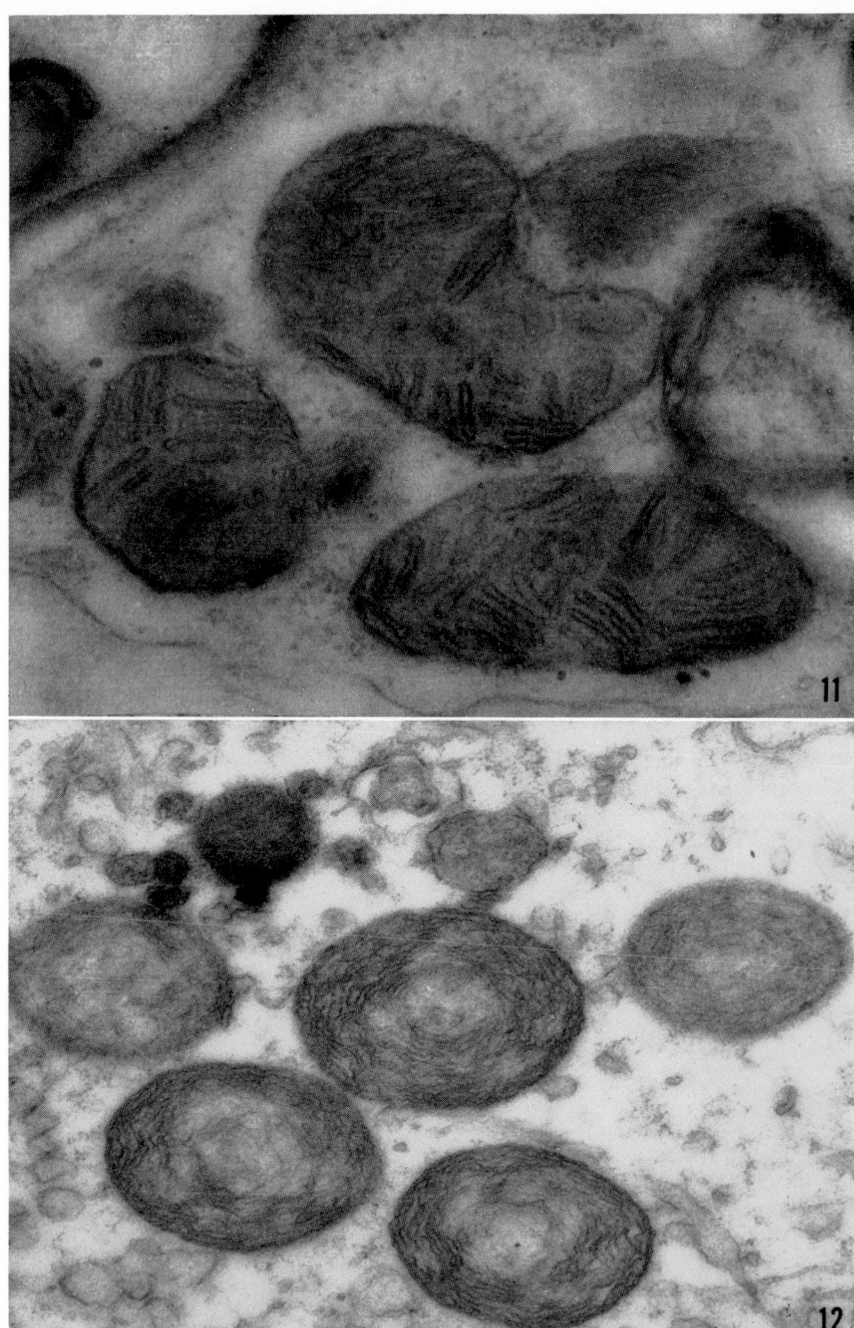
11
12

tae apparent in published micrographs is due to the plane of sectioning, as suggested by Palade (1953). It appears to us unlikely that the "free channel" is a large central structure, for then one would not expect to see as many completely traversing cristae as one does in electron micrographs.[4]

Neither observations of apparently complete septa nor those of apparently blind-ending cristae in one section traversing the mitochondrion in the next serial section and joining the mitochondrial surface at the other side in the third (e.g., Gelber, 1957) rule out the existence of a continuous "channel" or "chamber." They do, however, suggest that the chamber is likely to be a narrow, tortuous one. This is the picture which emerges from the three-dimensional reconstructions, based on serial sections, of Andersson-Cedergren (1959). It may be that studies on the extent of the "osmotically dead space" (Tedeschi and Harris, 1955; Werkheiser and Bartley, 1957; Amoore and Bartley, 1958; Malamed and Recknagel, 1959) may also permit an approximation of the chamber's volume. From electron micrographs it seems evident that cells vary in the extent and complexity of baffling of the chamber by cristae. The baffling may conceivably undergo change within a given cell under

[4] Examples of this may be seen in kidney tubule cells (Palade, 1953, 1956a; Rhodin, 1954; Dalton and Felix, 1957; Ruska *et al.*, 1957), retinal rods (Sjöstrand, 1956b), submaxillary gland (Pease, 1956), thyroid (Ekholm and Sjöstrand, 1957), parathyroid (Lever, 1957), pancreas (Palade, 1956b; Sjöstrand, 1956b), leucocytes (Low, 1956; Freeman, 1956), liver (Schulz *et al.*, 1957; Gansler and Rouiller, 1956), neurons (Palay, 1956), jejunum epithelium (Zetterquist, 1956), gall bladder epithelium (Yamada, 1955), skeletal and cardiac muscle (Moore and Ruska, 1957; Porter and Palade, 1957; Palade, 1953, 1956a; Andersson-Cedergren, 1959), alveolar epithelial cells of the lung (Karrer, 1956), pigeon breast muscle (Howatson, 1956), insect muscle (Chapman, 1954), *Helix* ovotestis, male portion (Powers *et al.*, 1956), brown fat (Napolitano and Fawcett, 1958), transplantable liver tumor (Howatson and Ham, 1955), *Viviparus* and *Murgantia* spermatids (Kaye, 1958), *Helix* spermatids (Grassé *et al.*, 1956), lobster axon (Geren and Schmitt, 1954), salivary glands of *Chironomus* larvae (Bradfield, 1956), sea urchin egg (Afzelius, 1957a), and *Allomyces* gametes (Turian and Kellenberger, 1956).

FIG. 13. Section through two boutons terminaux abutting upon a dendrite in the abducens nucleus of the rat. Note: (1) Closely packed mitochondria with high concentration of cristae; (2) numerous synaptic vesicles; and (3) within the dendrite, the longitudinally oriented "canaliculi" of endoplasmic reticulum. Magnification: × 13,000. Courtesy of Dr. Sanford Palay.

FIG. 14. Cell from distal convoluted tubule of mouse. Note: (1) the proximity of mitochondria and cell membranes (see Rhodin, 1958), and (2) the variable directions of the cristae within the mitochondria. Magnification: × 118,000. Courtesy of Dr. Johannes Rhodin.

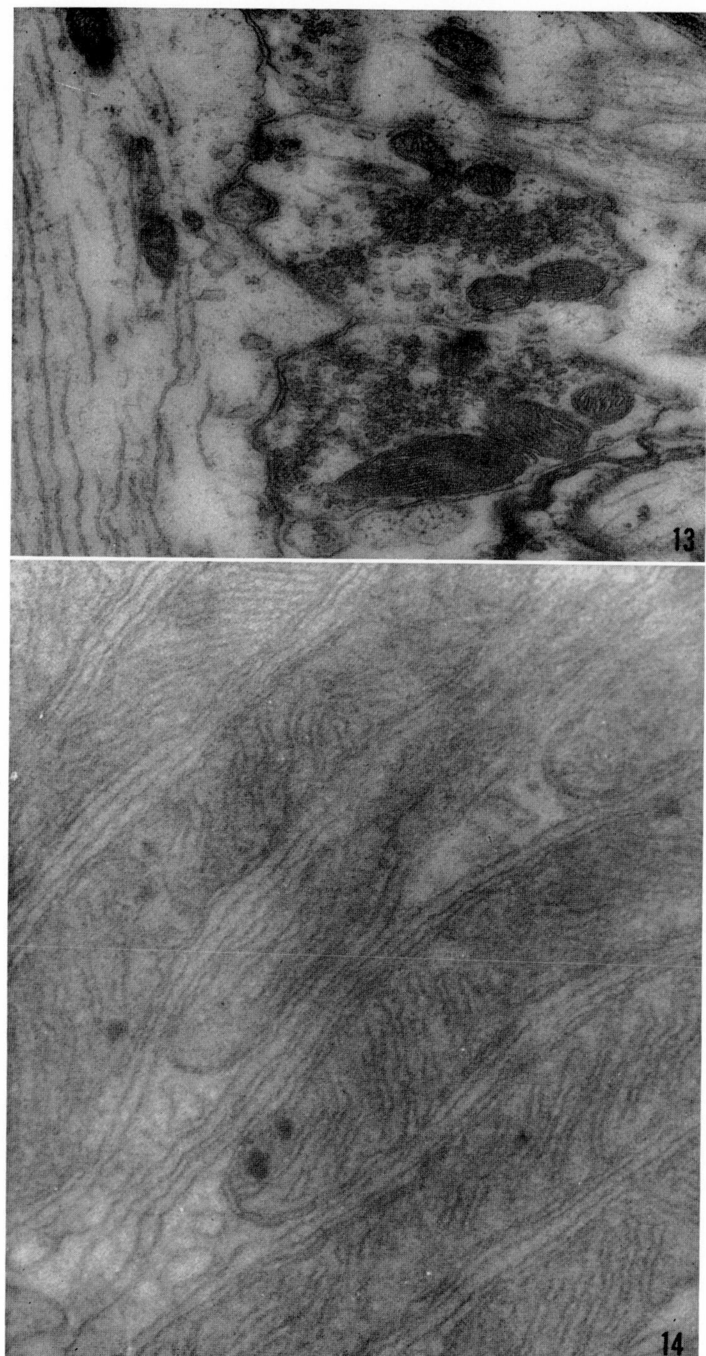
13
14

certain states of activity, as observed by Sager and Palade (1957) in *Chlamydomonas* grown under different nutritional conditions. Observations made on mitochondria which round out following ligation of the blood vessels to the rat kidney suggest the presence of both complete septa and incomplete cristae in the same mitochondrion (Fig. 3, Novikoff *et al.*, 1961).

The resolution of these unsettled questions will contribute to understanding the functional implications of mitochondrial fine structure. Molecular movement, and perhaps conduction, would be facilitated by a continuous inner chamber (even if it were not fluid filled; see page 328), and the material between the mitochondrial surface and external membrane (see below) is brought closer to this chamber if the cristae are continuous with the mitochondrial surface. Both these functional implications were early suggested by Palade (1953).

The existence of an outer mitochondrial membrane has long been postulated to account for certain properties of isolated mitochondria: their apparent osmotic behavior (Claude, 1946a, b; Hogeboom *et al.*, 1948); their high content of "soluble" enzymes like adenylate kinase (Kielley and Kielley, 1951), small molecules like citrate (Schneider, 1953), and "soluble" proteins that were lost when the mitochondria were disrupted (Hogeboom and Schneider, 1950); and apparent increase in permeability to DPNH upon brief exposure to water (Lehninger, 1951). Whittam and Davies (1954) considered that the turnover values for sodium and potassium ions in cells indicated the presence of semipermeable mitochondrial membranes in addition to the plasma membrane. This was similarly suggested by investigators (Bartley and Davies, 1954; Bartley *et al.*, 1954; MacFarlane and Spencer, 1953; Stanbury and Mudge, 1953) who found the concentration of certain ions to be higher in the mitochondria than in the medium. Werkheiser and Bartley (1957) demonstrated that neither nucleotides nor polyglucose penetrate into mitochondria whereas sucrose, potassium, sodium, and chloride do. Tedeschi and Harris (1955) have shown that isolated rat liver mitochondria follow osmotic laws (Boyle-van't Hoff) fairly exactly. From their permeability studies of dissolved nonelectrolytes they conclude that "there is a true semipermeable barrier which is characterized by a high degree of permeability to lipide-soluble substances. In accord with the interpretation of similar data for the erythrocyte and other cells [see, for example, Davson and Danielli (1952)], these findings may be taken to mean that the barrier is essentially lipide-like." Recknagel and Malamed (1958) agree that osmotic laws govern the swelling of mitochondria. They conclude that the presence of phosphate ions or carbon tetrachloride in the isotonic sucrose increases the mitochondrial membrane permeability and results in the movement of water into the mitochondria.

Tedeschi and Harris (1955) considered the osmotic dead space of rat liver mitochondria, about 50% of its volume, to be solid and thus impermeable to sucrose. However, Werkheiser and Bartley (1957) and Amoore and Bartley (1958) demonstrated that the mitochondria are partially permeable to sucrose. Malamed and Recknagel (1959) confirmed the partial permeability to sucrose. In addition, they resolved the apparent contradiction with the basic conclusions of Tedeschi and Harris by showing that the sucrose-inaccessible space behaved as a true osmometer while the sucrose-accessible space was the osmotic dead space. Correlation of such results with mitochondrial structure will require more precise and complete electron microscopic studies than are yet available.

Suggestive morphological evidence for an outer mitochondrial membrane came from phase contrast microscope studies of Zollinger (1948) and electron micrographs of isolated mitochondria in dried smears (in lymphosarcoma cells, Claude and Fullam, 1945; liver, Dalton *et al.*, 1949; and kidney, Mühlethaler *et al.*, 1950). However, definitive evidence for the membrane's existence came only from electron micrographs of thin sections, beginning with those of Palade (1952) and Sjöstrand (1953) and now exceedingly numerous.

There are few investigators who still doubt the existence of the outer mitochondrial membrane. Harman (1956) has presented his views in a recent review. To us the review lacks critical evaluation of the technical quality of the electron micrographs [see Sjöstrand's criticism in the same volume (1956b)] and shows too great a bias in choice of literature cited and in quotations selected from that which is cited. We feel that Harman's basic contentions must be considered as not proved.

Having beautifully described the changes in mitochondrial forms induced by the addition of detergents to living cells in culture, Frederic (1954) compares the outer "coat" thus produced with the external membrane seen in electron micrographs. He considers that both are "coats" which, "although it has, perhaps, osmotic properties, is an artifact and represents only one of the two phases resulting from denaturation of the mitochondrion . . . one of these phases is formed on the surface of the other, looking like a coat." We find it hard to conceive of such massive rearrangement of mitochondrial materials without simultaneous change in mitochondrial shape, as indeed occurs in the mitochondria of Frederic's detergent-treated cells. No such alteration occurs when cells (Porter *et al.*, 1945; Ornstein and Pollister, 1952) or isolated mitochondria (Novikoff, 1957a and unpublished observations) are properly fixed in osmium tetroxide (see also Frederic, 1958), dehydrated in alcohols, and embedded in methacrylate. The contention that the mitochondrial membrane is an artifact of osmium fixation is reminiscent of the early claim that mitochondria themselves were such artifacts. As Lewis and Lewis (1914-15) described it, "The mitochondria are so well fixed by osmic vapor or by a fixing solution which contains osmic acid that it has been suggested that the mitochondria may be artifacts due to osmic fixation."

Palade (1953) suggested that the "mitochondrial matrix" within the central "channel," like the electron-lucid area between mitochondrial body and external membrane, is fluid-filled in life. However, in some published figures of osmium-fixed mitochondria *in situ* (e.g., Palade, 1956a, Fig. 3; Schulz *et al.*, 1957, Fig. 1) and in our micrographs of isolated mitochondria (see also Rouiller, 1960), these areas generally have a considerably greater electron opacity than the electron-lucid areas of the cristae, and we have seen (page 315) that negative images are produced after freeze-drying and other methods of killing tissue. After uranium-methacrylate embedding of osmium-fixed tissue (Ward, 1958), the mitochondria in hepatoma cells (Essner, unpublished) and in *Viviparus* spermatids (Kaye, 1958) show a clearly granular "matrix." Furthermore, small spheres of extremely high electron opacity are frequently encountered within this "matrix." Generally called intramitochondrial granules, they have attracted little attention, and nothing is known of their biochemical nature (but see Weiss, 1955). These granules have been observed to increase in number in kidney tubule cells after injection of mice with albumin (Rhodin, 1954) and in the pancreas cells of fasted guinea pigs where the mitochondria make intimate contact with lipid droplets (Palade, 1959). Crystalline inclusions have also been found within the mitochondrial "matrix" (see page 385). All this suggests that the "matrix" is fairly rigid or viscous.

Finally, it should be commented that movement in and out of mitochondria must involve factors other than membrane permeability. Amoore and Bartley (1958) have suggested that manganese uptake by isolated mitochondria involves chelation of the metal ion. Gamble (1957) has found that, like intact mitochondria, submitochondrial fragments presumably devoid of outer (or intact?) membranes concentrate potassium ions from solution, thus suggesting "active transport." Lehninger (1959a, b; 1960) stresses the importance of the respiratory enzyme assemblies (Section V, A) in these processes.

IV. Morphological Form: Diversity and Uniformity

A. *Multicellular Animals and Plants*

In Section II we suggested that the extent of mitochondrial movement, plasticity, and polymorphism varied from cell to cell. In plant cells, where protoplasmic streaming is conspicuous, mitochondria are swept along in the current, as early described by N. H. Cowdry (1917) and Guilliermond (1941), but it is unlikely that this occurs to a similar degree in the cells of higher animals. And it seems unlikely that such mitochondrial movement occurs at all in those cells where the mitochondria always have an intimate relation with the plasma membrane [distal and proximal tubules of the renal tubule (Figs. 14, 18); synaptic junctions of axons (Fig. 13)] or other cell structures [skeletal and cardiac muscle (Fig. 33); spermatids]. Palade (1956a) believes that the more permanent arrangement of mitochondria is related to the constant demand for energy in these specialized areas of the cell. This view is shared by Geren and Schmitt (1956), who found mitochondria concentrated immediately below the surface of the Schwann cell; by Fawcett (1959), who discussed a similar concentration beneath the cell membrane of early rat spermatids (Palade, 1952) during the formation of the acrosome; and by Kaye (1958) who studied the fine structure of the mitochondria surrounding the axial filament of *Viviparus* sperm.

The polarized arrangement of mitochondria was discussed by Cowdry (1918) and interpreted in relation to the distribution of secretory granules. Pollister (1941) later reinterpreted the significance of this polarity. He noted that in cells which had no stored material or visible fibrillar differentiations the mitochondria were also polarized. In both epithelial and leucocyte type of cells, "the threadlike mitochondria are parallel to the course of diffusion in the cell—on a line between capillary and lumen in the former—and radial to the approximate center of the main mass of cytoplasm, which is organized into an aster, in the latter." Pollister interpreted this as a reflection of the parallel orientation of "long protein molecules—the structure-proteins" in the hyaloplasm. This may be related to orientation at a higher level of organization, like that of endoplasmic reticulum or myofibrils. A striking instance of relationship to endoplasmic reticulum is that described by Copeland and Dalton (1959) in the cells of the pseudobranch "gland" of *Fundulus*. There the mitochondria, polarized in the direction of the blood vessels, are surrounded by similarly polarized tubules of endoplasmic reticulum (Fig. 15). In the flight muscle of the dragonfly, the large oriented mitochondria, with closely applied endoplasmic reticulum, are wedged between myofibrils

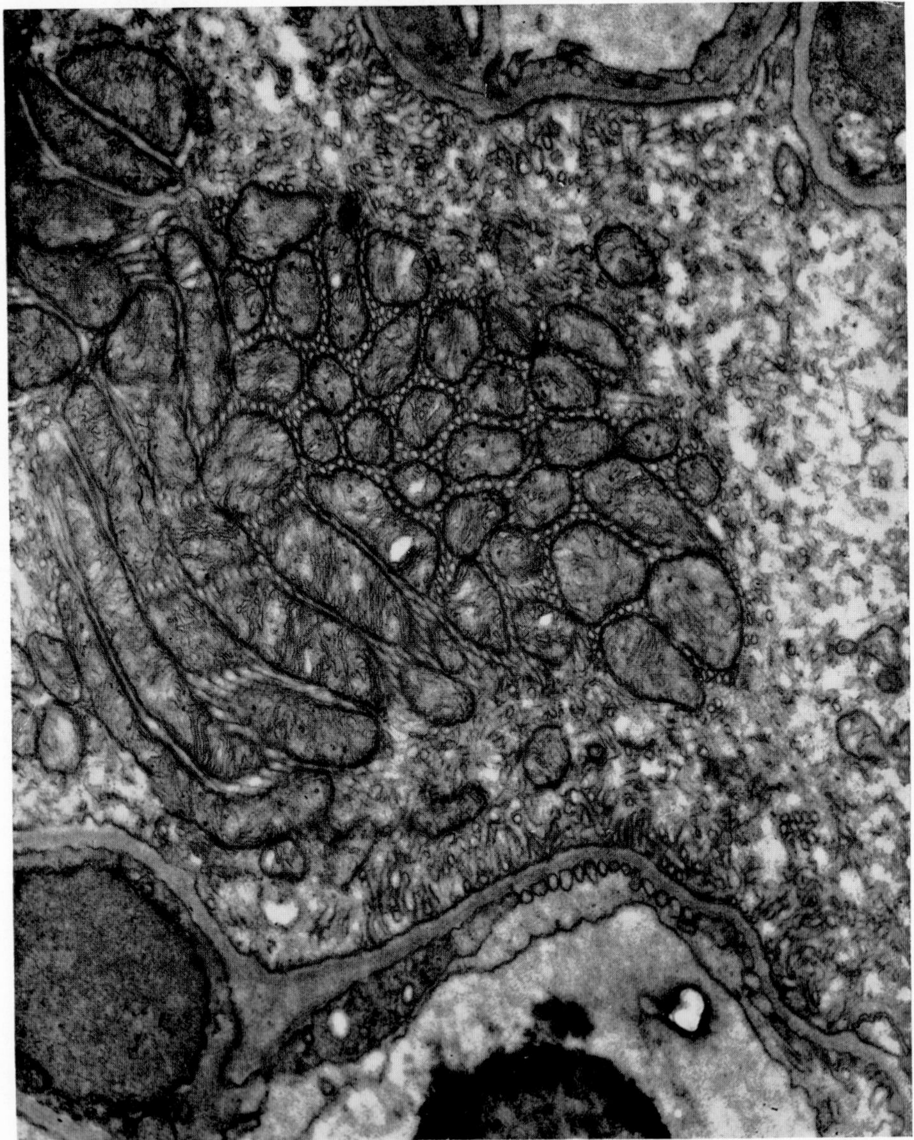

Fig. 15. Portion of cell from pseudobranch gland of *Fundulus heteroclitis*. Note concentration and orientation of the mitochondria. Surrounding each mitochondrion are the similarly oriented arrays of endoplasmic reticulum. Magnification: × 19,000. Courtesy of Dr. D. E. Copeland and Dr. A. J. Dalton.

(Smith, 1961) (Fig. 33). On the other hand, it may be that in many cells mitochondrial polarity is unaccompanied by other polarized structures visible at the electron microscope level. For example, in a transplantable hepatoma cell the "hyaloplasm" shows no oriented arrays of endoplasmic reticulum, but only small ribonucleoprotein granules scattered through the cell, yet the mitochondria are polarized in relation to the blood supply (Novikoff, 1957b). When the tumor cells grow freely

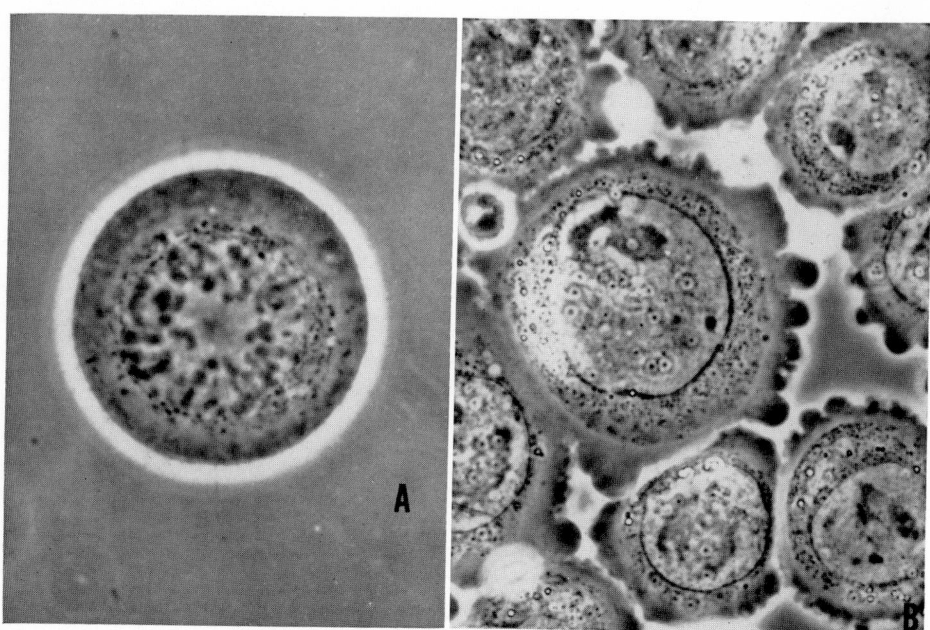

FIG. 16. Phase contrast photographs of ascites form of Novikoff hepatoma. (Courtesy of Dr. Monroe Birns.) A. Dividing cell mounted in serum, unflattened. Most mitochondria are localized in a zone surrounding the nucleus, the membrane of which has disappeared as the chromosomes (dark, ribbonlike structures) formed. Magnification: × 1400. B. Cells mounted in saline, considerably flattened. The perinuclear arrangement of mitochondria is more evident in the flattened cells. Flattening, however, produces irregular blebs at the cell surface. Magnification: × 1400.

in the abdominal cavity, as an ascites tumor, the mitochondria no longer show this orientation. Most are now concentrated in a zone around the nucleus, lying roughly parallel to the nuclear membrane (Fig. 16) (Birns *et al.*, 1959; Novikoff, in press) (cf. Ludford, 1952; Cowdry, 1924).

Barer and Joseph (1956) consider the possible relation of mitochondria to the energetics of mitosis, and Andressen (1956) to that of vacuole contraction in amebae (cf. Danielli, 1958, Plate 2).

The early cytologists were impressed with how frequently the shape and size of mitochondria were characteristic of a given cell type. Cowdry (1918) wrote: "The several tissues of the higher plants and animals possess mitochondria of characteristic form; that is to say, in some of them filaments predominate, in others granules, and so on, but in similar tissues of different animals they are much the same. . . . This general constancy of mitochondria, where the function is similar, is, I think, of considerable importance, because it must surely indicate that their morphology is a fundamental property ingrained in the very organization of the cell in phylogeny and that it is not always a passing trivial affair which varies from moment to moment." Sorokin (1941) stressed the constancy of size and shape of mitochondria in any given tissue.

According to Cowdry, filaments are the most common form of mitochondria. They vary in size, attaining a length of 10–12 μ in pancreas exocrine cells (Cowdry, 1918) and 20–40 μ in the oöcytes of *Rana pipiens* (Ward, personal communication). In homogenates made in polyvinylpyrrolidone-sucrose, rat liver mitochondria attain a length of 5.1 μ (average length = 3.3 μ) whereas in the transplantable hepatoma mitochondria are not above 1.6 μ (average 1.0 μ) (Novikoff, 1957b). The mitochondria of *Aeshna* sp. flight muscle are slablike structures up to 8 μ in length and about 0.6 μ in width (Smith, 1961) (Fig. 33). Cowdry emphasizes that within one cell type the width of mitochondria shows surprising uniformity despite wide variation in length. This suggested to him that mitochondria add substance at their extremities. (In tissue culture cells, as we have seen (page 304), thickening in width and side-to-side fusions may occur occasionally.) Spherical mitochondria generally vary in diameter from about 0.2 to 1 μ or more.

Figure 17 is taken from Cowdry (1924) to show representative mito-

FIG. 17. Mitochondria in plant and animal cells (Cowdry, 1924). 1. Filament of *Spirogyra maxima*, after Guilliermond, containing typical rodlike and filamentous mitochondria. 2. A diatom, after Guilliermond, containing similar mitochondria. 3. A fungus, *Pustularia vesiculosa*, after Guilliermond. 4. A spermatophyte, *Narcissus poeticus*, after Guilliermond. 5. A myxomycete, *Arcyria denudata*, after N. H. Cowdry. 6. A protozoon, *Glaucoma piriformis*, after Fauré-Fremiet. 7. A coelenterate, *Aurelia aurita*, ovarian egg, after Tsukaguchi. Note perinuclear accumulation of mitochondria. 8. An arachnid, *Amblyomma americana*, Malpighian tubule. 9. An insect, *Cimex lectularius*, intestinal epithelium. 10. An amphibian, *Rana esculenta*, pharyngeal epithelial cells after Saguchi, showing distal condensation. 11. A selachian, *Scyllium canicula*, cell from choroid plexus after Grynfeltt and Euziere, illustrating perinuclear clumping of mitochondria. 12. Kidney cells of a white mouse with mitochondria in proximal cytoplasm. 13. Small cell of locus coeruleus and large cell of mesencephalic nucleus of the fifth nerve of a white mouse to indicate difference in amount of mitochondria. 14. Human spinal ganglion cell.

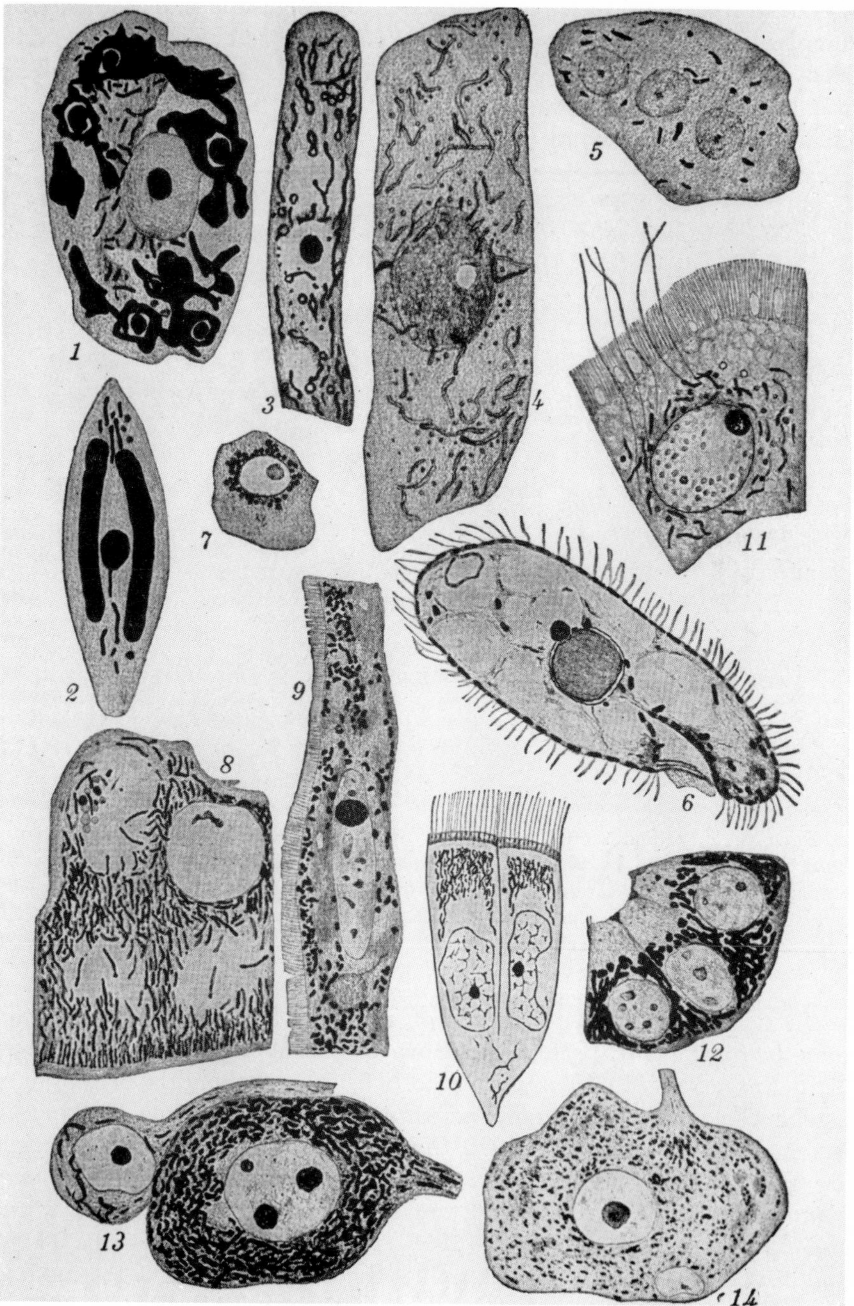
1
2
3
4
5
6
7
8
9
10
11
12
13
14

chondrial forms and distributions in stained preparations. These may be compared with phase contrast photographs of unfixed cells (Fig. 16) and fixed cells (Fig. 18). For other drawings of cells stained for mitochondria, the reader may consult Cowdry (1918, 1924), Wilson (1928), Sharp (1934), and Guilliermond (1941).

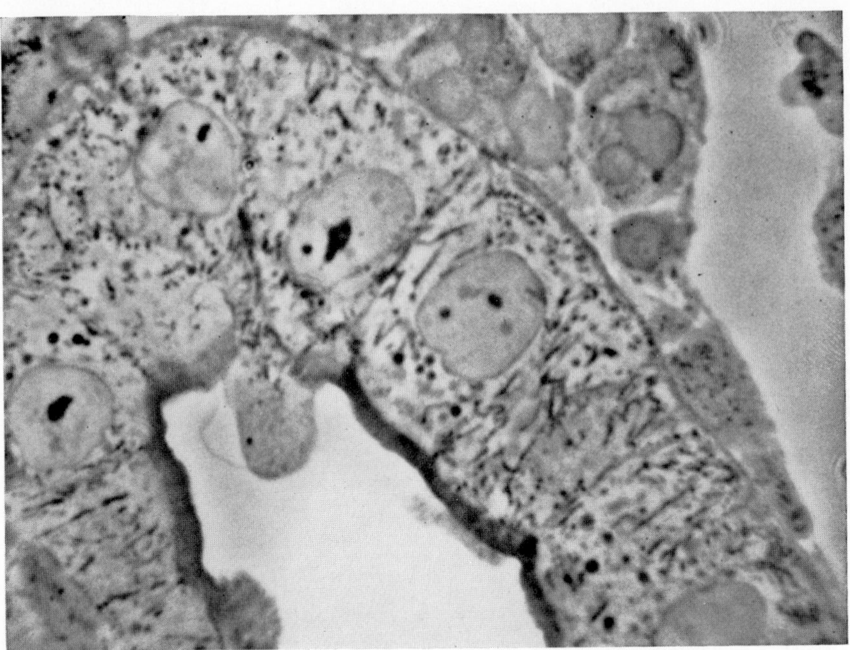

FIG. 18. Unstained section through proximal tubule of the kidney of the frog *Rana pipiens*, viewed by phase contrast microscopy. Tissue fixed briefly in neutral osmium tetroxide and overnight in neutral formalin; embedded in butyl methacrylate; and sectioned at approximately 0.5 μ. The mitochondria appear gray. Their filamentous form and polarized arrangement are evident. Courtesy of Dr. Leonard Ornstein.

In a given cell, the number of mitochondria appears to remain roughly constant. In protozoa and ova the number may be exceedingly high. Andressen (1956) found up to 500,000 mitochondria in the giant ameba *Chaos chaos*, the number varying with the volume of the cell. Afzelius (1957b) has estimated, from electron micrographs, that there are about 150,000 mitochondria in the egg of the sea urchin *Psammechinus miliaris*. In two other sea urchins, *Strongylocentrotus purpuratus* and *Lytechinus pictus*, Shaver (1956) reports the number to be 14,000 and 33,000, respectively, as determined by a modification of the method of Shelton *et al.* (1953). Shelton *et al.* (1953) and Allard *et al.* (1952)

independently developed essentially the same procedure. Nuclei and mitochondria are counted in tissue homogenates examined with phase contrast microscopy, and the number of mitochondria per nucleus is used as a measure of mitochondria per cell. Naturally, this is an approximation since it does not take into account factors like binucleate cells and variations in mitochondrial number among different cell types. The reported results for rat liver show a wide spread: about 500 (Striebich *et al.*, 1953), 825 (Lowe *et al.*, 1955), and 2500 and 1400 (Allard *et al.*, 1952, 1957). In rapidly growing liver, the number is about the same as in sham-operated controls, about 1000 per cell (Allard *et al.*, 1957). In liver tumors the number is reduced, to about 800 in primary tumors and about 175 in a transplantable tumor (Allard *et al.*, 1952).

When we consider the transformations of mitochondria (Section VI) we will see instances of remarkable constancy of mitochondrial number during spermatogenesis in some species.

B. *Unicellular Organisms: The Problem of Mitochondria in Bacteria and Blue-Green Algae*

Long before the advent of electron microscopy, typical mitochondria had been described in unicellular organs: in protozoa, slime molds, and all algae except the blue-green or Cyanophyceae (see Guilliermond, 1941; Sharp, 1934; Wilson, 1928). As was to be expected, these mitochondria possess the typical fine structure: an inner membrane folded into cristae and a smooth outer membrane. For studies of protozoa see Rudzinska and Porter (1953), Sedar and Porter (1955), Powers *et al.* (1956), Dalton and Felix (1957), Brandt and Pappas (1959), Rouiller *et al.* (1957); and of algae, Sager and Palade (1957), Wolken and Palade (1953).

In yeast cells, the situation was not much different [see Perner (1958) and Agar and Douglas (1957) for literature references], although precise delineation of mitochondria from other granules presented some difficulty. Yotsuyanagi (1955) distinguishes true mitochondria from "refractile granules" on this basis: the former are stained in living cells by Janus green and in fixed cells by Altmann's technique (see also Ephrussi *et al.*, 1956). The "refractile granules" are unstained in both procedures, despite their staining with the indophenol oxidase and tetrazolium methods. Their positive reaction with the latter methods is misleading: the indophenol blue and formazan diffuse, probably from the mitochondria where they are formed, and dissolve in the "refractile granules." Yeast mitochondria possess the typical fine structure (Agar and Douglas, 1957; Hagedorn, 1957; Yotsuyanagi, 1959); this permits their ready delineation from the "refractile granules" (Yotsuyanagi, 1959).

The existence of mitochondria in bacteria has been the subject of much discussion in recent years. Only a few years ago it could be asserted that bacterial cytoplasm was characterized by "the absence of all kinds of intra-cytoplasmic membranes" (Bradfield, 1956) since even the best of the thin sections revealed only small granules, about 100 A. in diameter (e.g., Maaløe and Birch-Anderson, 1956). By early 1959, the situation had changed dramatically. Cytoplasmic organelles, approximately the size of bodies seen in light micrographs and electron micrographs of entire cells (Mudd *et al.*, 1956; Georgi *et al.*, 1955; Niklowitz, 1958), had been reported by many electron microscopists. These organelles, 1000–1500 A. in diameter, were sometimes brought out more clearly by pretreating the sections, e.g., with lanthanum nitrate (Caro *et al.*, 1958) or uranyl acetate (Niklowitz, 1958), or by fixation in potassium permanganate (Tokuyasu and Yamada, 1959). The fine structure of these organelles was not adequately established. In some electron micrographs only spheres were seen within them (Chapman *et al.*, 1959); in others it was difficult to tell whether there were spheres or truly membranous structures (Shinohara *et al.*, 1957, 1958). Their membranous nature was most clearly indicated in the micrographs of Tokuyasu and Yamada (1959), Niklowitz (1958), and Ryter and Kellenberger (1958). However, the study of Glauert and Hopwood (1960) is dramatic demonstration that the preparative procedures used in the earlier studies profoundly alter the details of these membranes. Minimizing fixation artifacts (by the procedure of Kellenberger *et al.*, 1958) and sectioning artifacts by a newly developed microtome (Huxley, 1957), these authors studied the fine structure of the membranes in *Streptomyces coelicolor*, where they attain a high level of complexity. The membranes are "single"; i.e., with the "unit" membrane structure of Robertson (1959), and not double, i.e., with a smooth outer "unit" membrane and a folded inner "unit" membrane, as in mitochondria. Well-developed membranous systems are present in the mycobacteria (Briegert *et al.*, 1959; Glauert and Hopwood, in press) and, generally in somewhat simpler form, in the eubacteria (Glauert and Hopwood, 1960, in press; Murray, 1960; Glauert *et al.*, 1961; van Iterson, in press). There is another common feature in these membrane systems, their continuity with the plasma membrane.

Are these structures mitochondria? On morphological grounds, the single rather than double nature of the membranes would dictate a negative answer (see also Fauré-Fremiet and Rouiller, 1958). Functionally, firm evidence is lacking. Marr (1960) discusses the difficulties in interpretation of results with bacteria from the two techniques capable of yielding functional information—specific *in situ* staining procedures and biochemical analyses of isolated subcellular particles. He stresses the desirability of applying these techniques and electron microscopy to the same bacteria.

Most valuable for linking the membranous particulates with the sites of oxidative enzymes would be work like that of Georgi *et al.* (1955) with isolated fractions containing particles about 1000 A. in diameter. Thin sections of the sedimented fractions could determine whether these particles were identical with the membranous structures seen *in situ.* There would then remain the formidable problem of obtaining this particulate fraction sufficiently free of cell membrane fragments to associate unequivocally the high levels of oxidative enzymes with the particles (for the evidence indicating the presence of oxidative enzymes in the cell membrane, see Weibull *et al.*, 1959; Weibull, 1953b, 1956; Mitchell, 1959; Mitchell and Moyle, 1956a, b; Storck and Wachsman, 1957; Mudd *et al.*, 1960; also see the review of Marr, 1960).

Valuable information may also come from staining methods for oxidative enzymes, but their application to bacteria illustrates dramatically their limitations on the intracellular level. Mudd and collaborators (Mudd, 1953a, b, 1954, 1956; Mudd *et al.*, 1956) have vigorously supported the view that bacteria possess mitochondria, even through the earlier period when prevailing opinion was unequivocally opposed to this view. They have demonstrated structures of appropriate size range by staining living bacteria with tetrazolium, Nadi, or Janus green B reagents. However, the reliability of each of these techniques for intracellular localization has been questioned: tetrazolium by Weibull (1953a) and by Cota-Robles and co-workers (1958); Nadi by Grula and Hartsell (1954); and Janus green B by Glauert and Brieger (1955) and by Bradfield (1956). Most recently, Mudd *et al.* (in press) have used the new tetrazolium salt, nitro-BT, to reveal "the reducing sites" (mitochondria) and also the cell membrane. Although this tetrazolium salt is in general much superior to those used earlier, misleading results are not excluded since, as indicated on page 312, its formazan may readily deposit upon lipid droplets or lipid-containing granules. We have no experience with "TTC" (5,5′,3,3′-tetraphenyl 2,2′-di-*p*-stilbenditetrazolium chloride) used by Drews (1955), but our results with nitro-BT and other tetrazolium salts (Novikoff *et al.*, 1961; Novikoff, 1959c) suggest the desirability of critical studies before concluding that artifactual deposition of formazan on nonenzymatic sites does not occur.

If one reasons, with Glauert and Hopwood (1960), that the plasma membrane and cytoplasmic membranous structures of *Streptomyces coelicolor* form a unified system, and if one accepts the evidence that the plasma membranes of bacteria contain oxidative enzymes, the suggestion of these authors regarding the cytoplasmic structures seems reasonable: "Thus, in the absence of typical mitochondria, this system would perform the functions of the mitochondria of the cells of higher organisms." On the other hand, Glauert and Hopwood point to the resemblances between the cytoplasmic membranes and the endoplasmic reticulum, and conclude: "It may be suggested that the intracytoplasmic membrane system of *Streptomyces coelicolor* performs the functions of one or other of the two chief membranous systems of the cells of higher organisms, the endoplasmic reticulum and the mitochondria, or it may conceivably combine the functions of both." It seems likely that less equivocal statements regarding the states of these membranes will soon be warranted from accumulating observations. It is likely that more attention will be given to the presently unorthodox view that in higher cells mitochondria may arise from infoldings of the cell membrane (see page 401). Could it be that mitochondria have developed from the cell membrane in the course of evolution as Robertson (1959) and Geren and Schmitt (1954) have suggested that they may in morphogenesis? This possibility has similarly been raised for the membranous systems found in the malarial parasite, *Plasmodium berghei,* by Rudzinska and Trager (1959). In these cells, unlike those of *P. lophurae* (Rudzinska and Trager, 1957), there are no mitochondria. Thus the authors assume that the concentric membranous structures perform mitochondrial functions, and they write, "Should further study confirm this hypothesis, a significant step in the evolution of a cell organelle might have been found."

The newer techniques of biochemical cytology are only beginning to be applied to the blue-green algae (see Niklowitz and Drews, 1957). To our knowledge, there is as yet no unequivocal evidence for the existence of mitochondria in the cells of these algae. They are said to be absent also from red algae and photosynthetic bacteria [for references see Mühlethaler and Frey-Wyssling (1959), who interpret this absence of mitochondria to suggest their secondary origin, following the origin of plastids].

V. Biochemical Features: Complexity and Integration

The extensive body of data on the biochemistry of mitochondria has been amassed almost exclusively with fractions isolated by the technique of differential centrifugation. The mitochondrial localization of oxidative enzymes, particularly succinic dehydrogenase, has been confirmed by independent techniques—by the staining techniques already discussed (Section III, A); in the case of amebae, by the subdivision of the living cell after sedimenting most of the mitochondria to one end (Holter, 1956); and, most recently, by direct absorbancy measurements on the nebenkern of intact grasshopper spermatids (Perry *et al.*, 1959; Chance and Thorell, 1959).

Although the mitochondria of rodent liver have been the most widely studied, those isolated from many other tissues of vertebrates, invertebrates, and plants have also been analyzed. The description which follows is based upon results from animal tissues; the biochemical roles of plant mitochondria have been reviewed by Millerd (1956) and Hackett (1959); for special problems of isolated particles of plants, see Hackett (1955, 1959) and Millerd (1956), and for bacteria and yeasts see Vendrely (1955), Marshak (1955) and Marr (1960). We have already discussed the question whether the isolated bacterial particles capable of oxidative phosphorylation should be considered mitochondria.

A. *Enzymes*

The supreme fact to emerge from the biochemical investigations is that mitochondria are the chief, if not the exclusive, sites of oxidative phosphorylation. Cytochromes a and a_3 (cytochrome oxidase), and a number, perhaps not all (Schneider, 1956; Schneider and Hogeboom, 1956; Schneider, 1959) of the Krebs cycle enzymes appear to be localized in these structures exclusively. The mitochondria are probably the chief sites of the enzymes linking phosphorylation to oxidation (Siekevitz, 1952), of cytochrome c (Beinert, 1951; Schneider and Hogeboom, 1950), and the enzymes catalyzing the oxidation of fatty acids (Schneider, 1948; Kennedy and Lehninger, 1949; see Ernster and Lindberg, 1958; Green and Wakil, 1960), amino acids (Claude, 1944; Paigen, 1954), and choline (Kensler and Langemann, 1951; Williams, 1952). It is generally considered that mitochondria possess the full complement of enzymes, cofactors, and accessory substances required for these reactions (Green and Järnefelt, 1959; Lehninger, 1959b). However, Schneider (1959) stresses that demonstration is still lacking for the mitochondrial localization of some of the key enzymes. We have discussed elsewhere the wisdom of leaving open the possibility that mitochondria are neither the only "furnaces" of the cell nor complete ones (Novikoff, 1959c).

From the evidence that mitochondria retain their capacity for electron transport and oxidative phosphorylation when broken into small fragments (Section V, C), it can be deduced that the enzymes involved in these processes are arranged in repetitive arrays (Green, 1958a, 1959a; Green and Lester, 1959; Green and Hatefi, 1961; Lehninger *et al.*, 1958; Lehninger, 1960a, b). It is of considerable satisfaction to the cell biologist that those biochemists who are engaged most actively in unravelling the sequences of these reactions are the ones who most stress the inseparability of structure and function in the mitochondrion. Rigid proof is still lacking that the cristae contain "polymeric" arrangements of multienzyme assemblies. Yet there is sufficient reason to consider special properties of the enzymes to derive from their spatial organization in these membranes, where they function in a nonaqueous medium as "solid-state" arrays, with perhaps as many as 5,000 or 20,000 such assemblies in a single liver or heart mitochondrion (Lehninger, 1959b, 1960a; Green and Hatefi, 1961). Other enzymes may also be present in the membranes, and there is some supporting evidence for the reasonable view that other, more-readily solubilized enzymes are localized in the inner "matrix" of the mitochondrion (see Fig. 8 of Lehninger, 1960b, for a list of such enzymes).

By refined spectrophotometric methods, Chance and Williams (Chance and Williams, 1956; Chance, 1956) have been able to follow the oxidative sequence, step by step, in isolated mitochondrial suspensions. Their elegant evidence suggests that the flow of electrons from substrate to oxygen occurs along the chain shown in reaction sequence (1).

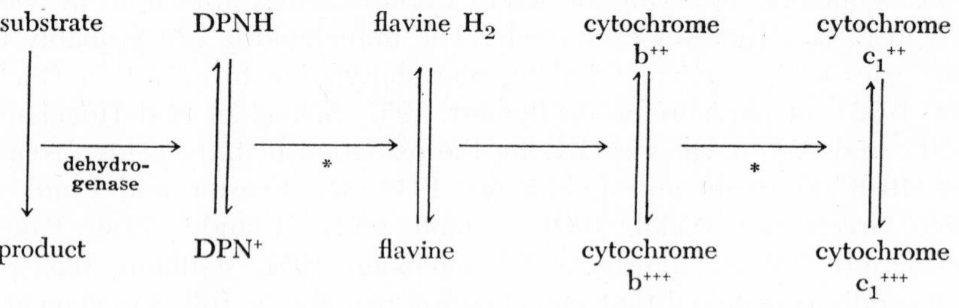

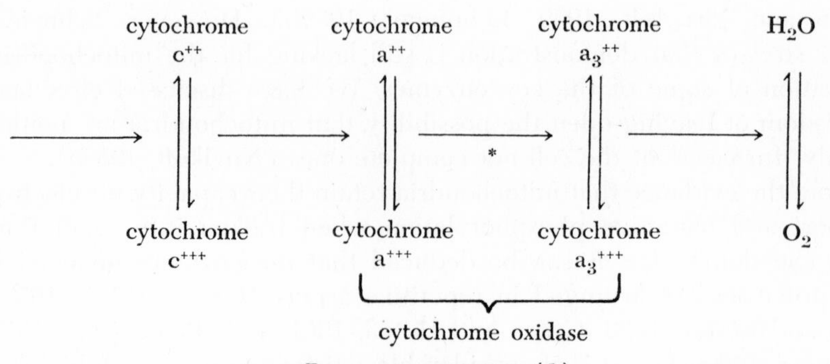

REACTION SEQUENCE (1)

They were also able to follow sequence (1) in suspensions of *living* yeast and ascites tumor cells. Chance (1956) writes: "It is, of course, of considerable reassurance to biochemists that the isolated material does not involve a serious artifact." We have already referred to the work of Chance and colleagues on the nebenkern of intact cells (Perry *et al.*, 1959; Chance and Thorell, 1959).

Chance and his colleagues (see Chance and Hess, 1959) were able to show too that in living cells, as in isolated mitochondria, respiration is controlled by the level of ADP, the acceptor of the high-energy phosphate bonds produced by oxidation:

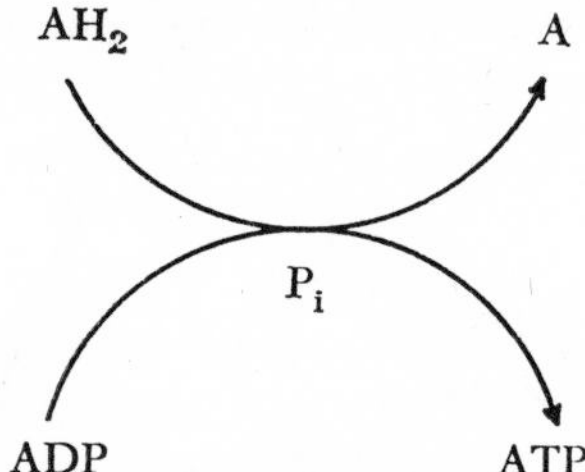

As Chance and co-workers indicate, their work confirms and extends the earlier concepts of Johnson, Lynen, Potter, and Lardy. These indicated that either ADP or inorganic phosphate controls the respiration rate of mitochondria [see Lardy (1956) for review].

Finally, the work of Chance and collaborators has not only helped define where in the respiratory chain oxidation and phosphorylation are linked (shown by asterisks in the respiratory chain diagram), but also where specific respiratory inhibitors act (see Chance, 1956, Figs. 3 and 10). Figure 19 is a diagram of Lehninger *et al.* (1958) showing the likely mechanism of oxidative phosphorylation (see also Lehninger, 1959b); more recent findings on the "M-factor" and "C-factor" have led Lehninger (1960a, b) to add details to each of the three paths to ATP formation.

Although the individual steps of the electron transport system are generally established, future studies are likely to bring some modifications. Green and co-workers (Green, 1959a, b; Green and Lester, 1959) consider that cytochrome b is not on the main pathway (Fig. 21). From the work of the Madison and the Liverpool groups it is probable that coenzyme Q (see Section V, B) is an integral part of the transport chain, but its precise location is still being discussed by these workers (Green, 1959a, b, Green and Lester, 1959; Readfarn and Pumphrey, 1960; Doeg *et al.*, 1960). An area of considerable uncertainty is the pathway of TPNH oxidation; does it occur in mitochondria?; if so, is it coupled to phosphorylation as is the pathway of DPNH oxidation? (see Schneider, 1959).

ATP generated in the mitochondria probably plays an important role in maintaining the morphologic integrity of the mitochondria. It is utilized in a great many chemical reactions, some outside and others within the mitochondria. Among the latter, at least in liver and kidney, are reactions involved in the apparent secretion and absorption of water and certain electrolytes by mitochondria. Bartley and Davies (1954) and Bartley *et al.* (1954) consider the mitochondria to be the cell's "basic units of secretory and absorptive activities" (Bartley and Davies, 1954),

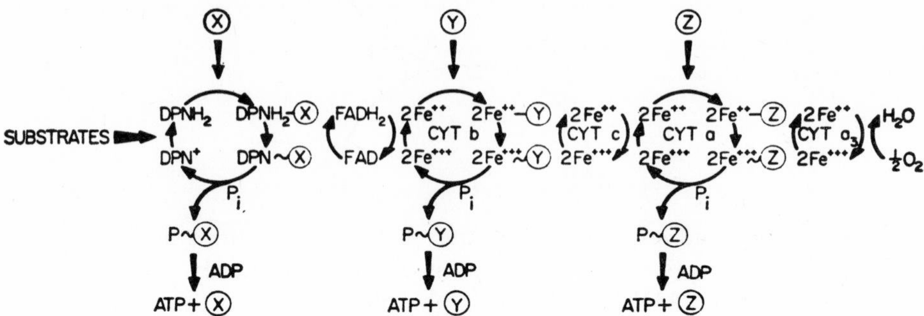

FIG. 19. Schematic representation of phosphorylating electron transport. At the three phosphorylating sites in the respiratory chain the liberated energy is conserved in "high-energy" intermediates of carriers labeled, *X*, *Y*, and *Z*. These intermediates then donate their "high-energy" phosphate groups to ADP to form ATP. From Lehninger *et al.* (1958).

and Lehninger (1956) ranks this function of mitochondria as second only to oxidative phosphorylation. The work of Gamble (1957) and Lehninger (1959a, b; 1960) suggests a functional link between the two processes which will be considered when mitochondrial swelling and contraction are discussed (Section V, D). On page 326 we have considered observations on mitochondrial permeability and membrane structure. It is of interest that in several segments of the mammalian renal tubule the cell membranes lie close to the mitochondria (Rhodin, 1958). By staining methods, we have demonstrated that these membranes possess a high level of apparent ATPase activity (Spater *et al.*, 1958). R. N. Robertson (1957) has discussed the probable role of mitochondria in ion accumulation by plant cells.

Other ATP-utilizing reactions that occur within the mitochondria are the syntheses of phosphatides (Kennedy, 1957; McMurray *et al.*, 1957), protein (Bates *et al.*, 1958; McLean *et al.*, 1958; Bates *et al.*, 1959), hippuric acid (Leuthardt and Nielson, 1951), *p*-aminohippuric acid (Kielley and Schneider, 1950), and citrulline (Siekevitz and Potter, 1953); the phosphorylation of nucleoside diphosphates (Herbert and Potter, 1956); carboxylations (Lardy and Adler, 1956); and the catabolism of protein (Penn, 1959a, b). The review of Schneider (1959) should be consulted for a critical discussion of the role of mitochondria in most of these reactions, as well as the conflicting evidence concerning the mitochondrial localization of enzymes participating in fatty acid syntheses (also see Green and Wakil, 1960).

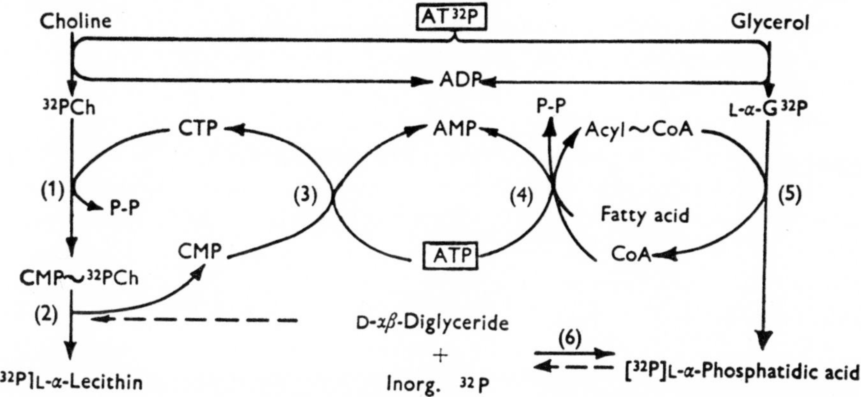

FIG. 20. Scheme representing P^{32}-labeling of lecithin in brain preparations. Abbreviations: ATP, ADP, AMP, adenosine tri-, di-, and monophosphate; CMP, cytidine monophosphate; CMP-P^{32}Ch, cytidine diphosphate choline; P^{32}-Ch, phosphorylcholine; CoA, coenzyme A; L-α-GP^{32}, L-α-glycerophosphate. From McMurray *et al.* (1957).

Mitochondrial synthesis of phosphatide has been studied largely in the laboratories of Kennedy, with liver, and of Rossiter, with brain. Figure 20 shows the reaction sequences as drawn by McMurray *et al.* (1957). It should be pointed out, however, that the mitochondria are not established as the sole sites of phosphatide syntheses; it is still possible that the microsomes, for example, are also capable of such synthesis. Many synthetic reactions involving mitochondrial-generated ATP occur in other subcellular fractions (see Lindberg and Ernster, 1954).

Simpson and colleagues suggest that "all highly organized structures of the cell play a role in protein synthesis" (McLean *et al.*, 1958). They have been able to demonstrate an ATP-dependent amino acid incorporation into cytochrome c by isolated liver mitochondria (Bates *et al.*,

1958) and into the proteins of isolated mitochondria from liver, pancreas, kidney, and muscle (McLean *et al.*, 1958). McLean *et al.* found no inhibition of incorporation by liver mitochondria in the presence of pancreatic ribonuclease. Indeed, at high levels, ribonuclease strongly stimulated incorporation. More recently, Bates *et al.* (1959), with a more stable mitochondrial preparation from calf heart, have been able to obtain a 20% *net increase* in cytochrome c content during a 12-hour incubation. Sonically disrupted mitochondria yielded a nonsedimentable preparation which could incorporate valine-*1-C*^{14} into total protein and into cytochrome c. In contrast to intact mitochondria, the submitochondrial preparation was sensitive to ribonuclease treatment (Kalf *et al.*, 1959).

Penn (1959, 1960) has found that mitochondrial preparations from rat liver and brain will catabolize exogenous serum albumin. This protein degradation process is an energy-requiring system and is markedly stimulated by the addition of ATP and coenzyme A. Microsome fractions are inactive in this system. Indeed their addition to the mitochondrial system markedly inhibits the formation of acid-soluble products from the albumin. The rate at which the mitochondrial reaction occurs appears to be sufficient to account for most of the albumin turnover of the intact animal. Penn (1960) suggests the likelihood that in the catabolic process activated intermediates are formed which can be reutilized in the formation of new protein.

Mitochondria are probably important loci of amino acid formation since they have high levels of glutamic dehydrogenase (Hogeboom and Schneider, 1953) and transaminases (Hird and Rowsell, 1950) (see Schneider, 1959). The evidence suggests that they are the exclusive sites of adenylate kinase (Kielley and Kielley, 1951; Novikoff *et al.*, 1952), transhydrogenase (Humphrey, 1957; Reynafarje and Potter, 1957), and rhodanese (Ludewig and Chanutin, 1950), and they have high levels of glutaminase I (Müller and Leuthardt, 1950; Shepherd and Kalnitsky, 1951). They appear to be involved in three steps of porphyrin and heme synthesis (Sano *et al.*, 1958).

Hesselbach and DuBuy (1953) and DuBuy and Hesselbach (1956, 1958) are of the opinion that mitochondria possess the full complement of *glycolytic* enzymes as well as those of the Krebs cycle. These, they believe, are more readily lost in preparing mitochondria from tissues like liver than from brain. They suggest that the washing procedure commonly employed to remove microsomes from liver mitochondria actually removes breakdown products of these mitochondria. However, neither morphology (electron microscopy) nor biochemistry (enzyme markers) is used to eliminate the more widely accepted view that the material removed by washing consists, not of mitochondrial derivatives but of supernatant fluid and of ergastoplasmic derivatives rich in RNA, glucose-6-phosphatase, and other enzymes, i.e., microsomes. In our laboratory the microsomes of liver and of ascites hepatoma have been found to possess high levels of lactic dehydrogenase activity (Novikoff *et al.*, in preparation; Novikoff, 1960d). The "instability" which the authors ascribe to liver mitochondria is yet to be demonstrated. Moreover, Aldridge (1957) has recently discussed the conflicting chemical data on brain fractions and concludes that brain mitochondria cannot be said to possess glycolytic enzymes. Hochstein (1957) has reported that the mitochondrial fraction from a mouse melanoma possessed glycolytic enzymes which required boiled supernatant fluid for activity, and that, unlike the glycolytic enzymes of the supernatant fluid, these were sensitive to insulin and anti-insulin hormones. However, the conclusion is weakened because the mitochondrial fraction was not washed.

Schneider (1959) reviews other oxidations that occur in mitochondria, such as those of mesotartrate, itaconate (methylene succinate), xylitol and other polyols, sarcosine, cholesterol and other steroids, kynurenine, thyroxine, and triiodothyronine. He concludes his review with the statement: "The diversity and complexity of mitochondrial functions, revealed by the experiments of the past decade, cannot help but be a source of amazement, even to those who have been closely associated with this field since its inception."

B. *Other Constituents*

In rat liver, mitochondria account for about 30% of the total nitrogen and about 35% of the total protein of the tissue (Schneider and Hogeboom, 1951); in the blowfly, Levenbook and Williams (1955) estimate that the sarcosomes make up about one-third of the muscle mass on a dry weight basis and about 40% on a wet weight basis. Mitochondria possess considerable lipid, about 20—30% of their dry weight. In rat liver, about half (Harel *et al.*, 1957) or more (Spiro and McKibbin, 1956) of this is phosphatide. The phosphatides have been studied with newer chromatographic techniques by Marinetti *et al.* (1957); 70% of the total consists of lecithin and phosphatidylethanolamine. In pig heart muscle, Marinetti *et al.* (personal communication, 1960) find about 30% of the mitochondria to be lipid; of this 26.2% is phosphatide and 0.71% free and esterified cholesterol. In this tissue, too, most of the phosphatide is lecithin and phosphatidylethanolamine (Marinetti *et al.*, 1958; cf. Edwards and Ball, 1954).

The high phosphatide content of mitochondria accounts for the positive reaction they give in the Baker acid hematin stain (Baker, 1946). It probably accounts for the difficulty experienced by earlier cytologists in preserving the mitochondria well with the acid fixatives then in vogue. The phosphatide may play a vital role in enzymatic processes, either as structural framework or electron acceptor (Green, 1958a, 1959a, b; Petrushka *et al.*, 1957; Nygaard, 1953).

Green and his colleagues at the Enzyme Institute in Wisconsin have stressed the importance of mitochondrial lipid beyond that of a structural framework (1958a, 1959a, b; Green and Järnefelt, 1959). They have isolated a lipoprotein named coenzyme Q, which they believe to be part of the electron transport chain, functioning between flavoprotein and cytochrome c_1. This is based upon (1) the rapidity of the coenzyme's reduction by reduced flavoprotein and its oxidation by oxidized cytochrome c_1, and (2) the analytical composition of various fragments of the electron transport particles obtained from heart mitochondria. Green and co-workers have isolated different lipoproteins which they consider to be integral parts of units containing flavoprotein, cytochromes, and possibly other components of the electron transport chain. The lipoproteins are considered to be essential for the native activities of these components (Fig. 21). Green concludes that electrons thus pass along the highly organized transport chain in much more efficient fashion than would be the case if reactions were dependent upon thermal collisions as in free solutions. He also suggests that "the lipoprotein may provide a built-in nonaqueous segment of the electron transfer chain for the stabilization of the unstable intermediates which are formed in the process of oxidative phosphorylation" (Green, 1959a); similarly, Lehninger (1960a) speaks of the "association of the enzymes of electron transport and phosphorylation in 'solid-state' lipoprotein arrays in the mitochondrial membranes" and of the possibility that "intermediate reactions . . . may take place where the chemical potential of water molecules is low."

Crawford and Morrison (1959) have found that the inactivation of the DPNH-cytochrome c reductase activity of pig heart particles resulting from isooctane extraction could be restored by a number of lipids such as phytol and vitamin K_1. From the fact that such lipids could not be shown to play a role in a *soluble* succinate-cytochrome c reductase preparation the authors suggest that "lipides function only as structural factors in particulate preparations."

Bouman and Slater (1957) have shown that the Keilin-Hartree

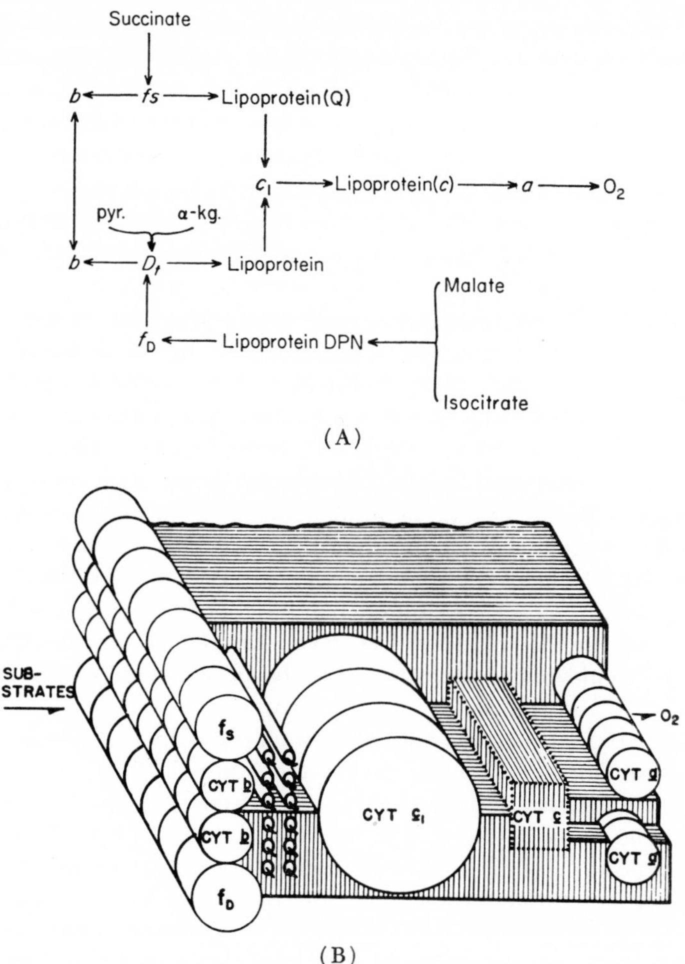

FIG. 21. A. The mitochondrial electron transfer chain, as formulated by Green (1959a). The arrows indicate the direction of electron flow as well as structural links. Abbreviations: DPN, diphosphopyridine nucleotide; f_D, DPNH dehydrogenase (flavoprotein); f_s, succinic dehydrogenase (flavoprotein); D_t, a thioctamide-requiring dehydrogenase; *b*, c_1, *c*, *a*, cytochromes b, c_1, c, a, respectively; Q, coenzyme Q; *pyr.*, pyruvate; α-kg., α-ketoglutarate. The lipoproteins shown are those that contain DPN, coenzyme Q, and lipid cytochrome c. B. Representation of electron transport chain showing possible distribution of lipid (shaded area) and electron carriers by Green and Lester (1959). Abbreviations as in A. The possible polymeric nature of the structure is indicated in the third dimension.

heart-muscle preparation contains about twice the amount of α-tocopherol (or compounds containing it) as it does cytochrome c. All the vitamin E is in the sarcosomes. The authors support the suggestion that vitamin E is one of the links between oxidation and phosphorylation. This suggestion had been offered by Nason *et al.* (1957), who believe that the vitamin is a cofactor in the cytochrome c reductase portion of the DPNH oxidase and succinate oxidase systems. However, it appears that its effects in the *in vitro* situations stem from its antioxidant nature rather than from participation in electron transport (Pollard and Bieri, 1960).

Among the soluble materials found in isolated mitochondria are coenzymes of oxidative enzymes, vitamins, and citrate (Schneider, 1959; Lindberg and Ernster, 1954; Schneider, 1953). The importance of mitochondrial magnesium for the structure and respiration of the organelles is underlined by the studies of Baltscheffsky (1957).

Analyses by Werkheiser and Bartley (1957), using an improved technique, show the dry weight of mitochondria to be about 34% of their wet weight, considerably higher than the usual 20% reported.

Mitochondria lack DNA and possess little, if any, RNA. Schneider and Hogeboom (1951) considered that 15% or less of the total RNA of rat liver was in the mitochondria. Even this figure, and surely higher ones to be found in the literature, may be due in good measure to contamination of the fraction by RNA-rich microsomes. We have reported (Novikoff, 1957a) experiments in which mitochondria isolated from PVP-sucrose homogenates of liver were studied. Their RNA content and esterase activity decreased in parallel fashion during five successive washes. The final mitochondrial residue showed 3–6% of the homogenate esterase activity and RNA content. Since we consider esterase to be a marker for microsomes of this tissue, we believe that most, if not all, of the RNA of the fraction resided in microsomes which still contaminated the fraction. Feigelson and Feigelson (1959) injected tryptophan into rats and followed the turnover of P^{32} in isolated mitochondrial and microsomal fractions of the liver prior to, during, and after the period of tryptophan peroxidase induction. At all times, the specific activities of the two fractions were identical. This implies that the RNA of the mitochondria has an identical function as microsomal RNA or, as seems more likely to us, that the RNA of the mitochondrial fraction is due to microsomal contamination.

Pollister (1957) has called attention to the failure of mitochondria to stain with basic dyes, under conditions where these dyes are bound by nucleic acid, and also to the ultraviolet transparency of mitochondrial nebenkern in spermatids, where they may be as much as 20 μ in diameter (see also Moriber, 1956). Both observations indicate that RNA, if present, is not in high concentration; Moriber (1956) calculates it to be less than 0.1%. A similar conclusion is reached by Rudkin and Schultz (1956), who found that sarcosomes of *Drosophila* muscle show negligible absorption at 2570 A.

Birbeck and Reid (1956), using a raffinose-dextran-Versene-heparin medium, obtained subfractions of mitochondria with 8–9% of the RNA in the original homogenate. However, much, if not all, of this RNA could be due to microsomal contamination since roughly the same proportion of the original glucose-6-phosphatase activity was recovered in the fractions and electron micrographs showed the presence of "ribonucleoprotein granules." Harel *et al.* (1957), from a comparison of the lipid and RNA contents of various fractions obtained from frozen mitochondria, have concluded that although variable and of small amount, the RNA is intrinsic to mitochondria, and not the result of contamination. In unpublished experiments, Kuff and Dalton (personal communication) find about 5% of the total RNA of liver homogenates in mitochondrial fractions packed at high speeds and showing no contaminating granules in electron micrographs. [For the older RNA chemical data and staining observations, see Lindberg and Ernster (1954).]

It should be emphasized that although the concentration of RNA in mitochondria may be low its significance may be great. In the viruses of equine encephalitis, fowl plague, Newcastle disease, and influenza, the RNA content is less than 5% (in influenza it is 1% or less) (Schäfer, 1959). An infecting particle of equine encephalitis or influenza has but one molecule of RNA (Schäfer, 1959). Similarly, the low levels of RNA (1.0 mg./100 mg. protein) in the mitochondrial preparations of McLean *et al.* (1958) may be highly important in protein synthesis by these organelles. The difficulty of reconciling this possibility with the apparent insensitivity of the mitochondria to ribonuclease has now been removed. Kalf *et al.* (1959) find that the mitochondria isolated from calf heart acquire such sensitivity when disrupted by sonic oscillation. Rendi (1959) reports the separation of a fraction from deoxycholate-treated rat liver mitochondria, which he calls "intramitochondrial ribonucleoprotein particles." It has an RNA content of about 15%, and its ability to incorporate C^{14}-leucine into protein is impaired by ribonuclease treatment. As in the studies by Simpson's group, intact mitochondria show no such ribonuclease sensitivity. To us these observations of Kalf *et al.* and of Rendi constitute compelling evidence that at least some of the RNA in the isolated fractions is intrinsic to mitochondria and is not due to contaminating microsomes. It will be of great interest to learn what electron microscopy of these preparations will reveal. It will be recalled that mitochondria *in situ* frequently show a few internal electron-opaque granules (page 328).

C. *Structural-Functional Interrelations*

In recent years biochemists have stressed the integrated nature of the multienzyme systems of mitochondria. In Lehninger's (1956) words, mitochondria "possess a common biochemical denominator: an organized chain of carrier enzymes which transfer electrons from substrate to oxygen and the still mysterious auxiliary enzymes which couple phosphorylation of ADP to the exergonic electron transport system" (see also Green, 1954, 1958a). Attempts have been made to correlate altered biochemical function and changed mitochondrial morphology, both *in vivo* and *in vitro*. While on the one hand efforts were made to isolate mitochondria in the best-preserved condition, on the other a variety of means was tried to disrupt the mitochondria—in order to unravel the individual steps of biochemical sequences.

When liver mitochondria are isolated from hypertonic solutions, their mitochondria are rod-shaped, as they are in the intact tissues. However, their capacity for oxidative phosphorylation is limited, and much of their ATPase activity is manifest rather than latent. On the other hand, those isolated in isotonic sucrose have rounded into small spheres, yet their phosphorylation is not uncoupled from oxidation and their ATPase is latent. This led Harman and co-workers to suggest that these biochemical properties, attributed by most workers in the field to a "native" morphology, were actually products of "structural deformity" (Kaltenbach and Harman, 1955). However, we found that in PVP-sucrose the elongate nature of mitochondria is beautifully preserved, yet they are as effective in oxidative phosphorylation as the spherical isotonic mitochondria and their ATPase is latent (Novikoff, 1956, 1957a).

Mitochondria exposed briefly to distilled water are still capable of linking phosphorylation to the oxidation of DPNH (Lehninger, 1951). Indeed, this treatment increases the phosphorylation level, perhaps because the changed mitochondrial permeability permits the reduced nucleotide to gain access to the enzymes of the respiratory chain inside the mitochondrion (see Lardy, 1956). At first sight, this might appear contradictory to the observations of Schneider *et al.* (1948), recently confirmed by Siekevitz and Watson (1956), that endogenous cytochrome c is no longer functional in the respiratory chain when mitochondria are prepared in distilled water (Schneider *et al.*, 1948) or suspended in it after preparation in sucrose (Siekevitz and Watson, 1956). Yet it may be that the duration of stay in water or the nature of pretreatment affects differently the extent to which the sites of the respiratory chain (mitochondrial cristae?) remain intact.

Digitonin has been used by Cooper and Lehninger (1956a, b) and Devlin and Lehninger (1956) to break mitochondria into particles estimated to be 1/2000th the size of the original mitochondria. These particles are lipid rich and contain all the constituents of the respiratory chain. However, they differ from mitochondria in several respects—the particle fraction is incapable of oxidizing any of the Krebs cycle intermediates except succinate and it cannot catalyze reactions of the fatty acid cycle, indicating a loss of dehydrogenases and other enzymes. It can remain in distilled water for hours without loss of activity, suggesting to Lehninger that the particles are not surrounded by semipermeable membranes as the mitochondria are. However, when these particles are supplemented with β-hydroxybutyrate as substrate and ADP as acceptor, P:O ratios approaching the 3.0 of intact mitochondria are obtained (up to 2.8). Unlike intact mitochondria, this does not require addition of magnesium ions; apparently magnesium is present in the preparation.

Digitonin-produced mitochondrial fragments have been studied with the electron microscope by Siekevitz and Watson (1957). They consisted mostly of aggregates of small vesicles, 200–500 A. in diameter. Their succinoxidase activity suggests to the authors that the membranes have arisen from the original mitochondrial membranes, but morphological evidence concerning their origin is lacking. It is of interest that the fragments contain bound nucleotides, approximately in the same proportion as in intact mitochondria.

High-frequency sonic vibrations of short duration have been used to break mitochondria into particles retaining the capacity for oxidative phosphorylation (Kielley and Bronk, 1958; McMurray *et al.*, 1958). Unlike Cooper and Lehninger's digitonin preparation, these particles can efficiently couple phosphorylation with added DPNH and succinate, as well as β-hydroxybutyrate. They also differ in not possessing DPN^+ or magnesium ions (Kielley and Bronk, 1958) although the latter are required for coupled phosphorylation. They can catalyze P_i-ATP and ADP-ATP exchanges (Bronk and Kielley, 1958); unlike intact mitochondria, these particles require neither ADP nor P_i nor magnesium ions for oxidation (Kielley and Bronk, 1958).

The most complete correlated biochemical and electron microscopic study of disrupted mitochondria is that of Watson and Siekevitz (1956) and Siekevitz and Watson (1956). They found that by deoxycholate treatment a fraction could be obtained of much-altered "mitochondrial membranes." This possessed considerable succinoxidase activity. Perhaps its inability to couple phosphorylation to oxidation is due to the drastic alteration of the "membranes" (cristae?). Siekevitz *et al.* (1958) have shown this membrane preparation to contain a Mg^{++}-activated ATPase which "is as much a part of the membrane as is the succinoxidase activity." Harel *et al.* (1957) have obtained a succinoxidase-rich fraction from ruptured frozen mitochondria.

Green and colleagues have used beef heart mitochondria, which they consider to have a more stable phosphorylating system than tissues like liver, to obtain a variety of interesting respiratory preparations. Recently, these have been studied with the electron microscope (Green *et al.*, 1957; Ziegler *et al.*, 1958). According to their description, the isotonic sucrose homogenate (macroblendor) has three types of mitochondria and derivatives. "Spherical mitochondrial shells or vesicular forms of varying diameter with few or no cristae," when sedimented together with a small number of what they call "sarcosomes" (but others call "lipid droplets") and myofibrils, yield a fraction, "ETP," capable of electron transport but not of oxidative phosphorylation. "Mitochondria-free masses of dense cristae," yield a fraction, "PETP," which gives high P:O ratios with all Krebs cycle substrates tested. "Intact mitochondria with dense cristae" constitute the third constituent of the homogenate; when sedimented, they yield a "residue" fraction. This has a partial or complete requirement for added cytochrome c, whereas PETP does not require added cytochrome c. The most novel of these important findings is the suggestion that the ETP particles are derived from the *external* mitochondrial membrane, which thus shares with the cristae the possession of electron transport enzymes.

The presence in the external membrane of such enzymes might be of considerable significance to mitochondrial activities such as their movement through the cytoplasm and the translocation of molecules to and from the surrounding milieu. However, the electron micrographs (Ziegler *et al.,* 1958) can at the moment be considered as only suggestive. Taken at their face value, they show extensive morphological alteration of the membranes as they form "vesicles," either as a result of chemical treatment or of homogenization, and it may well be that this is accompanied by significant enzyme redistribution within the mitochondria. Because the micrographs do not demonstrate their formation clearly, the origin of the vesicles from cristae or other mitochondrial material is not excluded. It is regrettable that the publication provides no basis for assessing the sampling adequacy for both the "ETP" (external membrane) and "PETP" (cristae-containing) fractions. It may be relevant to note that the observations of Palade (1959) suggest that the outer membrane is not indispensable. Apparently it may disappear when mitochondria make intimate contact with fat droplets.

Ball and Barrnett (1957) have published electron micrographs of a purified succinate and DPNH oxidase system from beef heart. From a slightly modified Keilin and Hartree (1947) preparation, the authors sediment an "original enzyme pellet." This consists of vesicles of various sizes, generally bounded by double membranes; but, unfortunately, their precise mode of origin from mitochondria was not investigated. Treatment of this preparation with 0.5% deoxycholate solution removes a large part of its lipid content. By centrifugation a pellet is now obtained which retains a fair proportion of the cytochrome b, c_1, a, and a_3 in the original preparation but is lacking most of its succinic dehydrogenase activity, all its cytochrome c reductase activity, and its cytochrome c. Morphologically, the preparation now consists of closely packed membranes which the authors assume to have arisen by disruption of the vesicles in the original preparation and a thinning of their walls.

These observations on submitochondrial preparations are not inconsistent with the view that the respiratory chain and associated oxidative phosphorylation enzymes are localized in the cristae. However, this view still rests more on inferences from biochemical work than from direct morphological evidence. It is well to point to the absence of electron micrographs in the work of Lehninger and co-workers, the uncertainty regarding the origin of the small vesicles in Siekevitz and Watson's digitonin preparation and of the "membranes" in Ball and Barrnett's "original enzyme preparation," the inability of the membrane preparations (deoxycholate) of Watson and Siekevitz to phosphorylate, and the doubts left unsettled by the paper of Ziegler *et al.* concerning the relation of ETP to the external membrane or other parts of the native mitochondrion. In his review, Green (1959a) suggests that submitochondrial particles capable of phosphorylation invariably show a "double-membrane or double-layer" structure, and that to carry out the complete citric acid cycle the particles must retain both external membrane and cristae in which there is bound pyridine nucleotide. It may be relevant to recall that Yotsuyanagi (1959) found no detectable difference in fine structure between normal yeast mitochondria and mitochondria of the "petites" in which succinic dehydrogenase, cytochromes a and b, cytochrome oxidase, and other enzymes are lacking.

For cell physiology and cell pathology the greatest interest, in the area of structural-functional interrelations, centers on the relation between the extent of coupling of phosphorylation to oxidation and mitochondrial form and permeability. We have reviewed the earlier literature on "uncoupling" agents elsewhere (Novikoff, 1959c); here we will briefly summarize some newer studies. For earlier reviews of uncoupling phenomena see Chance and Williams, 1956; Lardy, 1956; Hunter, 1951; and Potter *et al.,* 1951; for reviews of the morphological and chemical effects of aging and the effect of ATP and other conditions see Ernster and Lindberg (1958), Ernster (1959), and Schneider (1959).

In the conception that is emerging the catalysts coupling oxidation and phosphorylation are seen as also controlling swelling and contraction of the mitochondria. These enzymes may thus be intimately involved in changes of mitochondrial form that occur *in vivo* (Lehninger, 1959a, b; 1960a, b; Packer and Golder, 1959; Packer, 1960) and perhaps in active transport and water movement in cells (Lehninger, 1960b). As Ernster (1956) had done earlier, Lehninger compares the contractile properties of mitochondria with those of muscle. He adopts the term "mechano-enzyme," used for actomyosin, for enzymes presumed to couple oxidation with phosphorylation and also to change their physical state in the presence of ATP, thus leading to permeability changes of the membrane and shapes of the mitochondria.

Lehninger stresses the specificity of ATP for mitochondrial contraction and ascribes the effects described for ADP (Packer, 1960) to ATP formation via adenylate kinase. On the other hand, Packer considers that under conditions where oxidation and phosphorylation are "tightly coupled," only ADP causes reversal of substrate-induced swelling; ATP needs first to be converted to ADP. The resolution of this question, like many others, should come more readily now. Proteins are currently being isolated from mitochondria; when purified, they have effects on specific reactions. Some are capable of catalyzing reversible "partial reactions" of oxidative phosphorylation, such as the interchange of P^{32}-labeled ADP with unlabeled ATP in the terminal steps of ATP formation, shown in Fig. 19 (Wadkins, 1959; Chiga and Plaut, 1959). Others can restore phosphorylating ability to submitochondrial fragments devoid of such ability (Pullman *et al.,* 1958; Lehninger, 1960a, b). One, the "C-factor," can function in the swelling-contraction cycle; it may be identical with the "M-factor" linking the ADP-ATP exchange enzyme and the respiratory chain enzyme (Lehninger and Gotterer, 1959; Lehninger, 1960a, b). Polis and Shmukler (1957) isolated a hemoprotein from liver mitochondria. When added to fresh liver mitochondria this "mitochrome" inhibited oxidative phosphorylation (i.e., decreased the P:O ratio) and activated ATP dephosphorylation. However, it has been shown that the active component is not the protein but a mixture of fatty acids bound to it (Hülsmann *et al.*, 1958, Hülsmann *et al.*, 1960). Wojtczak and Wojtczak (1960) have demonstrated that the ability of serum albumin to enhance oxidative phosphorylation by isolated mitochondria of insect tissues (see, e.g., Sacktor *et al.*, 1958) is due to the removal by the albumin of fatty acids released from mitochondria during their isolation. Such binding of uncoupling fatty acids may also be involved in the protective effect of serum albumin on ATPase latency in isolated tumor mitochondria (Blecher, 1960). Apparently, tumor mitochondria, like insect mitochondria, do not survive the homogenization, isolation, and incubation procedures to the same degree as mammalian tissues like heart and liver (see Section VI, E; also Novikoff, 1960d). Free fatty acids present in the microsomal fraction of liver had been shown by Pressman and Lardy (1956) to uncouple oxidative phosphorylation. Another nonmitochondrial substance to have uncoupling effect is bilirubin (Zetterström and Ernster, 1956).

Among the physiological substances which uncouple oxidative phosphorylation and induce mitochondrial swelling none is more potent or more interesting than thyroxine (for reviews of the literature see Lehninger, 1960c; Ernster and Lindberg, 1958; Lehninger, 1956; Lardy, 1956; also see Bronk, 1958, and Park *et al.*, 1958). Aebi and Abelin (1953) and Tapley have shown that liver mitochondria isolated from hyperthyroid animals swell more readily than do normal mitochondria, even in the presence of ATP. Tapley (1956) showed that the mitochondria of *hypo*thyroid animals were more resistant than those of normal animals. Schulz *et al.* (1957) have shown profound changes in fine structure of the swollen liver mitochondria in hyperthyroid animals. Maley and Lardy (1955) found that P:O ratios were considerably lower in mitochondria isolated from hyperthyroid rats than in those of normal animals. Tapley and Cooper (1956) were able to correlate the susceptibility of isolated mitochondria to thyroxine-induced swelling with the extent of I^{131}-thyroxine uptake *in vivo*. Lehninger (1960c) believes that thyroxine-induced mitochondrial swelling occurs before uncoupling of phosphorylation, and suggests that the change from "tightly coupled" to "loosely coupled" respiration that occurs in the early stages of swelling may be mediated through the activation or release of "R-factor" (Remmert and Lehninger, 1959). Ernster *et al.* (1959) found coupling to be "loose" in mitochondria isolated from sartorius muscle of two patients with thyrotoxicosis and one with hypermetabolism without demonstrable thyroid disorder.

Although triiodothyronine enhances ATPase activity of liver mitochondria (Maley and Johnson, 1957) it inhibits the activity of a solubilized heart mitochondrial ATPase (Penefsky *et al.*, 1960). Deoxycorticosterone has similar effects, enhancing the ATPase activity of mitochondria from lymphosarcoma or liver and inhibiting the activity of fractions derived from disrupted mitochondria (Blecher and White, 1960a, b).

Ernster and Lindberg (1958) review the reported effects of other hormones, pharmacological agents, and other substances on the biochemical reactions of isolated mitochondria. They note the particular difficulties in relating these to the specific changes in the animal, and they caution against a literal transfer of such effects to the physiology of the animal.

D. *Heterogeneity of Mitochondria*

Differential centrifugation as developed in Claude's laboratory has generally been used to separate four fractions: nuclear, mitochondrial, microsomal, and soluble phase or supernatant fluid. However, variations in centrifugation schedules were used increasingly to shift the point of separation of mitochondria and microsomes (de Duve and Berthet, 1954), and also to subfractionate the fractions (Novikoff *et al.*, 1953). This led to the concept that both mitochondria and microsomes were biochemically heterogeneous classes. However, the data available did not establish the concept (Novikoff, 1957a). De Duve and collaborators, in a series of important papers, emphasized an alternative interpretation: the granules of a given group are chemically homogeneous, and the apparent heterogeneity results from distribution of a special class of granules with intermediate sedimentation properties between those of mitochondria and microsomes (de Duve and Berthet, 1954; de Duve, 1959; de Duve *et al.*, 1955). De Duve and co-workers proposed the name "lysosome" for this new particle because it contained high concentrations of a number of hydrolytic enzymes. Lysosomes and related particles are discussed in Chapter 6.

The possibility remains, however, that the mitochondria of liver are not alike biochemically. Their morphology in the centrolobular cells differs from that in the peripheral cells of the lobule (Figs. 24, 25). There is some suggestive evidence from staining results with tetrazolium (nitro-BT) and substrates considered to be oxidized by the mitochondria: succinate, β-hydroxybutyrate, and glutamate: the peripheral cells stain more deeply with succinate, but the centrolobular cells do so with β-hydroxybutyrate and glutamate. However, staining is not strong in any cells with β-hydroxybutyrate or glutamate, and there are important limitations to the staining observations (discussed in Novikoff, 1959d). More definitive information regarding cell heterogeneity in the hepatic lobule has been published by Shank *et al.* (1959). Two of the nine enzymes that they studied are generally considered as mitochondrial: malic and glutamic dehydrogenases. Malic dehydrogenase activity is more concentrated in the peripheral cells; and glutamic dehydrogenase, in the centrolobular cells. However, the differences are small; and the possibility exists that not all these enzyme activities are mitochondrion bound. The most decisive data presently available are those of Beaufay *et al.* (1959). The distributions of four mitochondrial enzyme activities—glutamic dehydrogenase, malic dehydrogenase, alkaline deoxyribonuclease, and cytochrome oxidase—were studied in subfractions obtained by density-gradient centrifugation. In all cases the distributions of activities were the same, leading the authors to conclude: "In view of the numerous factors, size, density, osmotic behaviour, uptake of deuterium oxide, which intervene in determining the distributions observed, such results provide strong additional support to the hypothesis that mitochondria are essentially homogeneous in enzymic content." In discussing these results, de Duve (1959) writes: "It would indeed be very astonishing if all the mitochondria from all the liver cells had exactly the same ratio of, let us say, glutamic dehydrogenase to cytochrome oxidase. In fact, such a possibility seems almost incompatible, at least for some enzymes, with the histochemical demonstration of zonal differences within hepatic lobules, as with the existence of several types of cells within the liver. The homogeneity is therefore a statistical one, and the crux of the matter is to appreciate at what stage our techniques become sensitive enough to show up the heterogeneity which undoubtedly exists."

Tetrazolium (nitro-BT) staining results are suggestive of mitochondrial differences among cells of other organs as well. Nachlas *et al.* (1958a) report that in the kidney of the mouse all tubule cells stain with malate as substrate but only the cells of the proximal convolutions and thick limbs of Henle do so with β-hydroxybutyrate. A similar situation is reported for different areas of gastric mucosa in the rat. Himmelhoch and Karnovsky (1961) find that in the kidney of *Necturus* the cells of the proximal convolutions possess mitochondria differing in staining reactions as well as fine structure from those of the distal convolutions. They are smaller and are not oriented at the base of the cells as are the mitochondria of distal convolution cells. They have fewer and less regularly arranged cristae. Cells of the proximal convolutions are unstained for succinic dehydrogenase and cytochrome oxidase activities, and those of the distal convolutions are unstained for the hexose monophosphate enzyme activities studied.

Cells of the developing embryo may possess different numbers of mitochondria (Section VI, C), and evidence begins to accumulate for changes in mitochondrial form and fine structure. Karasaki (1959) has followed the mitochondria during development of the urodele *Triturus*. In the gastrula they are spherical or ovoid and have inconspicuous cristae. In passing from presumptive neural cells to neural plate to spinal cord cells, the mitochondria become more elongate, the cristae more complex, and the matrix more electron opaque in osmium-fixed tissue (see also Eakin and Lehmann, 1957). Bellairs (1959) finds similar changes in the development of neuroblasts in the chick embryo. Such fine structure changes may contribute to the increased activity of cytochrome oxidase and other mitochondrial enzymes known to occur during development. The possibility that during embryogenesis some enzymes move from the "soluble phase" to mitochondria has recently been re-examined by Solomon (1959), who studied glutamic dehydrogenase activity in chick embryos.

Biochemical differences in mitochondria are known among mutant forms. The "petites" of yeast lack cytochrome oxidase and other respiratory enzymes despite apparently normal fine structure (page 315). Differences in cytochromes and other constituents are found among *Neurospora* mutants (page 376).

There is no acceptable evidence of which we are aware demonstrating biochemical differences among mitochondria of the same cell. As already indicated (page 314), Pearse *et al.* (1958) claim that by using different metal chelators with the tetrazolium, MTT, mitochondria containing succinic dehydrogenase only and those with DPNH-diaphorase only can be demonstrated in the same cell. However, it has been found that MTT is unreliable for intracellular localization (Novikoff *et al.*, 1961).

VI. Mitochondrial Changes

A. *In Dividing Cells*

That mitochondria divided during mitosis, and that one half of each passed into the daughter cells, was believed by many early cytologists (Benda, Duesberg, Meves). Meves' suggestion that mitochondria played a part in inheritance was based on his belief that at fertilization the sperm brought a complement of mitochondria into the egg, and that each of the cells of the embryo then received descendants (by division) of fused sperm and egg mitochondria. Wilson (1928) said of this view: "Not the slightest proof has been produced of a fusion between the paternal and maternal chondriosomes." He also pointed to instances where all the sperm mitochondria passed into one of the many blastomeres of the developing ovum (see also Lillie and Just, 1924).

Lewis and Lewis (1914-15) could find no evidence of mitochondrial division in dividing cells in culture. As cells prepared to divide, the mitochondria decreased in size and became more evenly distributed in the cell. When the cell divided, the mitochondria happening to be on either side of the cleavage plane were carried into the respective daughter cells. They then increased in number, either during mitosis or as the daughter cells grew in size. However, at no time were any signs of mitochondrial division seen.

Frederic (1958) has briefly summarized his critical studies of the mitochondria during division of living fibroblasts. He describes three phases in their evolution. The *first phase,* which lasts until completion of metaphase, is divided into two periods. Period 1 precedes the morphological signs in the nucleus of ensuing mitosis; the total volume of mitochondrial material ("chondriome") begins to decrease but the mitochondria remain active. Period 2 coincides with signs of mitosis in the nucleus. The continuing decrease in "chondriome" volume is now accompanied by reduced mitochondrial movement, pronounced thinning, fragmentation into small spheres, loss of optical density and even complete disappearance of some mitochondria: a progressive "melting" into the cytoplasm. Since these changes precede nuclear division, Frederic suggests that the chondriome may participate in the initiation of division. In the *second phase,* when the cell divides in two, the modified mitochondria are separated, passively, into the daughter cells; never have either transverse or longitudinal divisions of the mitochondria been observed. In the *third phase,* beginning when the chromosomes reach the poles of the mitotic spindle, the modified mitochondria, now skeleton-like ("squelette"), are reconstituted by the addition of "éléments constitutifs dans le cytoplasme" sometimes accompanied by fusion of thickened mitochondria to produce filaments. Frederic stresses that the reconstitution process starts at the end of the second phase and proceeds rapidly, suggesting that the mitochondrial material is in a form invisible with phase contrast, having been synthesized in advance of morphological reconstitution of the chondriome. The time required for reconstitution of the mitochondria continues considerably beyond that for nuclear reconstitution. This leads Frederic to suggest that "interphase" might better be considered in terms of mitochondria than the nucleus.

Even from sectioned material there is little doubt that in certain special instances, particularly in spermatogenesis, a precise pattern of mitochondrial distribution into the daughter cells is followed. Yet, as Wilson (1928) pointed out, these can be misleading and, in any case, generalizations to all cells would be hazardous. Wilson wrote: "The most careful studies seem to show that wide differences exist between different species in respect to the precision and orderliness of the distribution. In this regard numerous gradations exist, beginning with a condition in which the chondriosomes show no orientation to the centers or the spindle-poles and seem to be segregated into two groups passively, without themselves undergoing division during mitosis." In this group would fall the mitotic divisions of cells studied in tissue culture.

Wilson described the mitochondrial distributions during spermatogenesis in *Opisthacanthus*. In about three-quarters of the cells, the mitochondrial number is reduced in the two meiotic divisions from 24 to 12 to 6; in the remaining quarter of the cells 7 or 5 mitochondria remain rather than 6. For recent phase contrast studies of insect spermatogenesis see Barer and Joseph (1956).

In the snail *Viviparus*, recently studied with phase contrast microscopy by Kaye (1958), there are about 22 mitochondria. At prometaphase of the first meiotic division there is apparently end-to-end fusion, in threes, so that 8 or 9 long rods are produced. Each daughter cell obtains 4 or 5 mitochondria (Fig. 22, A). At the second meiotic division, these are divided transversely and approximately equally, so that the daughter cells still have 4 or 5.

In the hemipteran *Gerris*, Pollister (1930) described the arrangement of mitochondria into a well-defined compact ring, but without fusion. However, such fusion does occur in three closely related species of scorpions, *Centrurus* (Wilson, 1916), *Centruroides* (Wilson and Pollister, 1937), and *Buthus* (Nath and Gill, 1950). Wilson (1928) described events in *Centrurus* as follows: "All the chondriosome-material aggregates into a single ring-shaped body, which is placed tangentially to the spindle in first spermatocyte and is cut across transversely by the division accurately into two half-rings. Each half-ring now breaks apart to form two parallel rods which in the second mitosis are again cut across transversely into two shorter rods. The original ring thus is divided into eight equal parts, of which each resulting cell (spermatid) receives two, a process comparable in precision with the division of a heterotypic chromosome-ring though very different in detail."

Wilson concluded: "It is certain from the foregoing that in some cases the chondriosomes are actually divided in the course of mitosis; but on the whole the present evidence points to the conclusion that the division is a passive and mechanical result of the cell-constriction. It must, however, be borne in mind that in many of these cases the larger chondriosomes seen during the actual divisions arise by the growth and aggregation of much smaller bodies; and we should keep clearly in view the possibility that the latter may be capable of division." Sharp (1934) describes instances in spore formation among plants in which the mitochondria show a marked orientation in relation to the mitotic figure; there is, however, no indication of mitochondrial division.

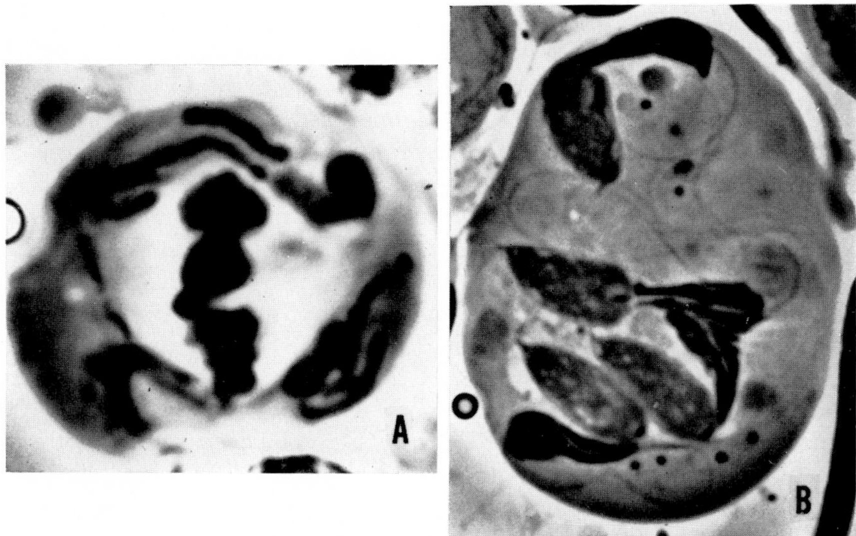

FIG. 22. Phase contrast photomicrographs of living cells in the testis of *Viviparus contectoides*. A. A spermatocyte in metaphase of first meiotic division. The long mitochondria are lined up along the light spindle area. The chromosomes are in the form of tetrads on the equatorial region of the spindle. Magnification: × 2200. B. A cell in mid-spermatid stage. The nebenkerns have elongated and have begun to twist around the axial filament. Magnification: × 2200. From Kaye (1958); courtesy of Dr. Jerome Kaye.

Modern light-microscopic studies, and even electron-microscopic investigations, have yet to provide firm conclusions regarding mitochondrial division (not to be conceived as synonymous with mitochondrial "fragmentation" in tissue culture cells) during mitosis or at other times. Payne (1952) noted that in fowl adrenal cortex the mitochondria frequently appeared as pairs and sometimes showed constrictions. This suggested to Payne that division, rather than fusion, was occurring. Later, Payne (1957) described a rounding and constriction of mitochondria in dividing and nondividing thyroid cells. Miller (1953) studied the mitochondria of the rat adrenal gland (outer third of the fascicularis) after ACTH or epinephrine administration. There is a transient increase in the number of elongated mitochondria, followed by a marked increase in the number of spherical ones. Miller concludes from these counts and from the constrictions present in some elongated mitochondria that the latter segment and produce the spherical mitochondria. Leon and Cook (1956) studied the mitochondria of the "fluffy layer" isolated from 0.25 *M* or 0.6 *M* sucrose homogenates of liver. As they were observed under the coverslip, with phase contrast microscopy, divisions were seen to occur. We have never observed division of mitochondria in 0.25 *M* or 0.88 *M* sucrose, including those in the "fluffy layer." Even if Leon and Cook's observations were confirmed, the question would remain whether this has any counterpart in the living cell. Isolated mitochondria, devoid of metabolites and much else, could surely behave differently in this respect from those in the living cell.

Chapman (1954) considered that the shape of some mitochondria in electron micrographs of insect flight-muscle cells suggested transverse division. Fawcett (1955) reported somewhat more suggestive indications of mitochondrial division in electron micrographs of frog liver, although he cautioned that "the possibility that the double mitochondrial structures encountered in electron micrographs represent mitochondria in process of fusion certainly cannot be excluded." Essner (unpublished) has observed mitochondrial pairs like those of Fawcett, occasionally in transplantable hepatoma cells and frequently in human liver cells. Weber (1958) has described elongate mitochondria, presumably in the process of division, in electron micrographs of cleaving *Tubifex* eggs. Until a more complete study is reported we will be unable to judge whether this is indeed mitochondrial division and whether it is temporally related to cell division.

B. In Spermiogenesis

In the transformation of spermatids into spermatozoa (spermiogenesis) striking mitochondrial transformations may occur. According to Franzen (1956a, b), in sperm shed directly into water with the ova, the mitochondria generally fuse into four or five spheres which do not always stain with Janus green. In sperm discharged in a more viscous medium, these spheres go through a variety of further changes. Frequently, the four spheres fuse into two, each of which transforms into an elongate ribbon which moves with the cytoplasm along the axial filament. The elongate, usually helical mitochondria inside the midpiece may attain lengths of 50–70 μ, as in the Bryozoan *Triticella* and the archiannelid *Protodrilus*. The mitochondria are usually wrapped around the axial filament, but occasionally, as in the cephalopods *Sepietta* and *Loligo,* the mitochondrial mass is separate from the axial filament, at another side of the sperm head. In some species the mitochondria fuse to form a nebenkern, as described by early cytologists (see Wilson; André, 1959a, b). Fused mitochondria may attain the length of 200 μ or more, as in the snail *Helix* (Grassé *et al.,* 1956).

Impressive electron microscopic studies of mitochondria during sperm formation have been reported by Afzelius (1955), Fawcett (1959), Grassé *et al.* (1956), Kaye (1958), and André (1959a-c). Afzelius studied spermiogenesis in *Strongylocentrotus* and other sea urchins. Here the mitochondria have the cristae and external membrane characteristic of usual mitochondria. In the rat, Palade (1952) and Fawcett (1959) described transient congregations of large numbers of mitochondria just beneath the cell membrane, when the acrosome is being formed, followed by their movement to the base of the flagellum where they elongate, wrap around the axial bundle of fibrils, and form the mitochondrial sheath. The spermatid mitochondria no longer have the typical cristae; instead, the cristae are flattened against the inner aspect of the mitochondrial membrane and have an irregularly concentric arrangement. This concentric arrangement is best developed in the opossum, among the mammals studied (Fawcett, 1958, 1959) (Fig. 12).

In the studies of Grassé *et al.* (1956) and Kaye (1958) on mitochondrial nebenkern we encounter a remarkable series of changes at or near the macromolecular level. The reader is urged to read the original articles for the exciting detail uncovered with the electron microscope in a structure which to light microscopists appeared "homogeneous."

Grassé *et al.* (1956) studied these changes in the snail *Helix pomatia.* The initial mitochondria possess longitudinal cristae, generally 4 or 5 in number. These cristae separate from the external membrane, arrange themselves concentrically within the mitochondria and fuse end to end. This produces the concentric structure described earlier by Beams and Tahmisian (1954) in low-resolution photographs. When the axial filament forms, the mitochondria with concentric membranes (those still with typical cristae not participating) move toward it and align themselves along and around it. Those in contact with the filament elongate and fuse end to end to form the mitochondrial sheath, in which the transformed cristae run parallel to the length of the axial filament. Cavities then appear among the membranes and fuse to form larger spaces. The axial filament and its surrounding membranes move to an eccentric position in the cell. Now the mitochondrial macromolecules, which to this point appear in electron micrographs as typical membranes, assume a more crystalline array—in a sheath which in the mature sperm is more than 200 μ long! This remarkable pattern has been studied more intensively by André (1959c) in another snail, *Testacella;* it has a complex paracrystalline arrangement rather than that of a classic crystal. André calls attention to the continued Janus green stainability of the macromolecules: "Il parais dès lors vraisemblable que l'équipement enzymatique mitochondrial soit encore présent dans le réseau paracristallin."

When the residual cytoplasm is cast off as the sperm matures, it carries with it those mitochondria that have retained their original cristae and those with concentric membranes. The latter generally have undergone degeneration, forming an electron-dense material in the center or giving rise to irregular swellings of the membranes.

Kaye's studies were done with the freshwater snail *Viviparus contectoides* (cf. Yasuzumi and Tanaka, 1958). When spermiogenesis begins, the four or five mitochondrial rods round out and surround the axial filament. These "nebenkern spheres" now elongate and twist around the axial filament to form the mitochondrial sheath (Fig. 22, B). The final sheath consists of helical strands enclosing the major part of the axial filament. Up to metaphase of the second meiotic division, the mitochondria have typical transverse cristae without special order. At this time they become grouped in blocks of three to five membrane pairs (Fig. 23, A). As the mitochondria shorten to form the nebenkern spheres, the blocks may contain up to eight or nine membrane pairs. They now become oriented in parallel fashion, with all plates perpendicular to the long axis of the nebenkern rod. The cristae are now packed very tightly; as Kaye says, "leaving little room in which to fit more membranes." They are also about one-third the height of those in the early spermatid. In the final elongated helix, the membranes are very regularly spaced along its length, the distance between adjacent cristae being less than 100 A.; each membrane runs uninterruptedly from the inner surface of the nebenkern sheath to the outer surface (Fig. 23, B). Kaye proposes a possible mechanism by which more concentrated cristae are produced without loss of membrane continuity. This and alternative mechanisms are discussed by André (1959a), who describes the impressive fusions and membrane changes that occur in spermatogenesis of the scorpion *Euscorpius.* These changes, occurring as the total mitochondrial volume increases appreciably, are summarized by André as follows: "Pendant la croissance du spermatocyte, les mitochondries goniales, d'ultrastructure classique, se soudent, et édifient de gros chondriosomes cupuliformes à crêtes anastomosées et nids d'abeilles, pour lesquelles l'auteur a proposé le nom de *culichondries.* Par la suite, ces dernières augmentent de taille, s'arrondissent, et deviennent les *chondriosphères* des spermatides, dont les crêtes sont anastomosées dans les trois plans de l'espace, selon au réseau serré ne laissant que peu de place pour la matrice." We have already (page 318) referred to André's conclusions regarding the dynamic nature of the inner membranes of mitochondria.

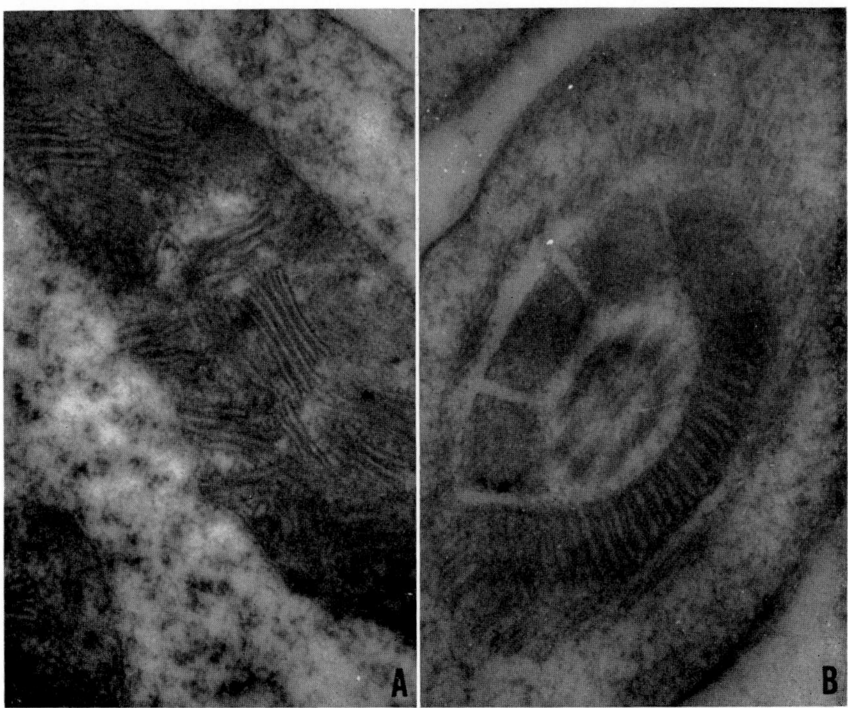

FIG. 23. Spermiogenesis in *Viviparus contectoides.* A. A portion of a spermatid. Most of the cristae are arranged in groups of three to five, in close proximity and parallel to each other. Magnification: × 63,000. B. An oblique section of a mature nebenkern in the spermatozoon. The nebenkern spiral around the axial filament. The latter is cut obliquely and does not show clearly. Note the platelike arrangement of the cristae in one nebenkern strand. Three other strands are cut almost perpendicularly so that the crista membranes are not visible. Magnification: × 85,000. Courtesy of Dr. Jerome Kaye.

Nebenkern formation has been studied most recently by André (1959b) in the butterfly *Pieris brassicae,* where another remarkable series of mitochondrial transformations occurs. These include an increase in mitochondrial size and number in the spermatocyte, a "clarification" of the matrix to produce "hollow" mitochondria, their end-to-end fusion, their apparently centriole-oriented movements during meiosis, and their distribution in specific relationships to the spindle as they form the large nebenkern which changes its shape in characteristic fashion. The nebenkern is "a single, huge chondriosoma, which is permeated with cytoplasm: it is a sort of three-dimensional puzzle" (Fig. 29).

Yasuzumi *et al.* (1958) show the appearance of an electron-opaque body within the nebenkern of spermatids of *Drosophila* and *Gelastorrhinus;* they describe it as "indistinguishable from a lipid dropet."

With regard to the need for mitochondria in cell metabolism, perhaps the most interesting instances of spermiogenesis are those where complete elimination of mitochondria have been reported. Franzen describes this in *Phoronis pallida,* where the mitochondrial mass loses its Janus green stainability, does not participate in midpiece formation, and is finally shed. It may occur in the mollusk *Aplysia depilans,* although Franzen did not see it actually shed. S. Hughes-Schrader (1946), using Janus green and a variety of fixatives and stains, followed the mitochondria in spermatogenesis of the iceryine coccids. They are plentiful in the spermatocytes and spermatids. They "remain unchanged in the cytoplasm and are discarded *in toto* with the rest of the cell body." The sperm consists of chromosomes inside the tail, with only a delicate sheath of cytoplasm. Moses (1958) studied spermiogenesis in the crayfish *Cambarus clarkii* by electron microscopy. The mature aflagellate spermatozoon has no mitochondrial structures.

Nath (1956) claims this occurs among many groups of animals, including vertebrates: rhabdocoeles, acoeles, polyclads, triclads, trematodes, spiders and snakes (*Natrix piscator*). In the silver fish, *Lepisma,* a typical nebenkern forms, but it disappears instead of forming the midpiece filament. In the dragonfly *Sympetrum hypomelas,* Nath (1956) states "the mitochondria are conspicuous by their absence throughout spermatogenesis" ["from the earliest spermatagonia to the ripe sperm" (Nath and Rishi, 1953)]. At least among the motile forms of these spermatozoa, electron microscopic studies may reveal mitochondria or membranous "equivalents" (see discussion in Section IV, B). The need for critical evaluation of the methods and interpretations of Nath and his colleagues may be suggested by recalling the story of the arthropod *Peripatus.* Montgomery (1912) reported that in this species all mitochondria are lost with the cytoplasm stripped from the late spermatid. However, Gatenby (1925), using better techniques, found that "finer," "less resistant" mitochondria were retained in the midpiece. There is always a possibility that investigators, relying on one fixation and staining technique and in a few cases phase contrast microscopy, may miss some mitochondria, particularly if they are "less resistant" than usually. On the other hand, a recent study with the electron microscope by Gatenby and Dalton (1959) again illustrates the difficulty of following small cytoplasmic particles by light microscopy. These authors find no support for the view of Chatton and Tuzon (1941) that in *Lumbricus* one group of mitochondria is eliminated but another is not. It will be of great interest to see electron microscopic studies of thin sections of dragonfly spermatocytes, spider spermatids, and those of *Lepisma, Phoronis,* and other species in which mitochondria are said by Nath to be lacking.

C. *In Embryonic Development*

The precise role of mitochondria in embryonic development remains to be established. Apparently, mitochondria of ova and embryos possess the same fine structure as in adult cells (in sea urchin, Afzelius, 1957a, b; Gross *et al.,* 1960; in *Tubifex,* Weber, 1958; Lehman, 1958; in ascidians, Berg and Humphreys, 1960), but changes may occur during development (page 358). There is little reason to doubt that most possess essentially the same oxidative, phosphorylating, and other activities as found in mitochondria of mature tissues (Brachet, 1957, 1960). The unanswered question is whether they serve exclusively as suppliers of energy and metabolites, or have additional, more specific directive roles in development.

Meves (1912) described the localization of mitochondria in bundles in the unfertilized sea urchin egg; after fertilization they disaggregate and disperse through the cytoplasm. Afzelius (1957b) showed these bundles in the unfertilized egg to contain up to sixty mitochondria; he considered that their aggregation accounted for the egg's low respiratory rate.

It is clear that in the mosaic egg of *Phallusia* some blastomeres normally possess few mitochondria and relatively low levels of oxidative enzyme activity. Reverberi (1957) stained the unfertilized egg with Janus green and followed the movement of the green-stained mitochondria after fertilization. They became concentrated first in the vegetal pole, then the yellow crescent area, the posterior vegetal blastomeres, the prospective mesoderm cells, and, finally, the muscle cells in the tadpole tail. Direct biochemical analyses performed on eggs of *Ciona,* another ascidian, by Berg (1956, 1957), showed that both anterior and posterior blastomeres possessed oxidative enzymes. Presumably the failure of anterior cells to stain with Janus green is due not to the absence of mitochondria, but to their reduced number. This quantitative difference has been vividly demonstrated by Berg and Humphreys (1960) in an electron microscopic study of the four-cell stage in *Ciona* and another ascidian, *Styela*. A gradient in mitochondrial number exists, from a high concentration at the vegetal pole to very few at the animal pole, in the posterior biastomeres of both species and in the anterior blastomere in *Ciona*. The paucity of mitochondria in the anterior chorda-neuroplasm is impressive. No qualitative differences are noted in mitochondria of anterior and posterior blastomeres.

Does the normal development of the posterior cells require the presence and activity of this many mitochondria? Reverberi (1957) believes it does because treatment of ascidian eggs with azide or other oxidative enzyme inhibitors led to abnormal tail development. Ries (1939) had found that centrifuging ascidian eggs redistributed the granules, presumably mitochondria, that gave cytochemical reactions for oxidases; the muscles differentiated in the regions where they were concentrated. La Spina (1958) found that centrifuging ascidian eggs gave some embryos with abnormal tail development. He attributed this to irregular distribution of mitochondria to the cells that develop into muscle. Berg and Humphreys (1960) suggest that the difference between these experiments and earlier ones of Conklin, in which centrifugation did not lead to abnormal muscle development, is due to the extent of diminution of mitochondrial number in the posterior cells. The yellow crescent area may be provided "with more mitochondria than necessary for normal development and low centrifugal forces may not displace a sufficient number of mitochondria to affect development."

Electron microscopy has been combined with biochemical analyses of the different blastomeres of the annelid *Tubifex*, by Weber (1958) (see also Lehmann, 1958). The Nadi-negative 4D cell and endoderm-forming macromeres show about 40% of the level of cytochrome oxidase activity (expressed on a nitrogen basis) as the Nadi-positive somatoblasts 2d, 4d and micromeres; corrections for probable nitrogen content of the yolk would raise the level to 70%. They contain ample mitochondria, which tend to be located more deeply in the cytoplasm and show a higher degree of polymorphism than in the Nadi-positive cells.

The work of Clement and Lehmann (1956) on the egg of the snail *Ilyanassa* is of interest because in this case the area of highest morphogenetic significance has relatively few mitochondria. On the other hand, in another mollusk, *Dentalium*, Reverberi (1958) finds that an accumulation of mitochondria in the polar lobe does occur (see Reverberi's paper for observations on other eggs).

Changes in the number or kind of mitochondria are suggested by some biochemical observations. Nakano and Monroy (1958) have described an increased uptake of S^{35}-methionine into the mitochondria-containing fraction in the course of sea urchin development. They consider this to correlate with Gustafson's curves on the relative number of mitochondria during development (Gustafson, 1954). Boell and Weber (1955) found that from the early larva to the late developmental stages of *Xenopus* there was more than a threefold increase in the cytochrome oxidase activity, expressed on a nitrogen basis.

It has long been known that there is an early period of development which is cyanide insensitive. In the frog, it lasts until the late blastula stage. Presumably mitochondrial oxidative metabolism is less essential at this period than later (see review by O'Connor, 1957).

The observations thus suggest that different blastomeres require varying number of functioning mitochondria for their normal development. Some early investigators (Beckwith, 1914; E. B. Harvey, 1946, 1953) concluded from the results of centrifugation experiments that a blastomere or part of an egg could be deprived of *all* mitochondria without serious impairment of development, including *de novo* formation of new mitochondria. However, there is strong reason to believe (page 398) that not all mitochondria had been removed from these portions of the eggs as these authors believed.

Differences in mitochondrial content of blastomeres have been given a directive meaning for development by Gustafson (1954). He described a gradient in mitochondrial number in the sea urchin egg, with the animal pole possessing the highest number and the vegetal pole the lowest. This gradient persists until actively differentiating areas appear. The latter contain many mitochondria; these are said to produce inhibitors of mitochondrial development in adjacent areas. Mitochondria are also considered to produce fibrous proteins that are responsible for shape transformations like cell stretching and for the appearance of the apical tuft.

However, this hypothesis cannot yet be accepted. Gustafson's procedure of differentiating and enumerating mitochondria was reported by Shaver (1955-1957) to be unreliable. With the standard mitochondrial stains on fixed preparations, Shaver could find no evidence of a gradient in the egg [but see the demonstration by Hörstadius (1952) of a "reduction gradient" in Janus green-stained eggs]. Shaver points out that biochemical evidence currently favors the submicroscopic ribonucleoprotein particles (microsomes) rather than the mitochondria as the major site of protein synthesis. Shaver also found differential distributions of mitochondria, but he preferred to correlate these with levels of general metabolic activity rather than specific directive actions.

The reviews of Brachet (1957) and Pasteels (1958) may be consulted for references to other studies on mitochondria and other cell particulates in developing ova. These, and especially the recent "The Biochemistry of Development" by Brachet (1960) will fill the many gaps in this necessarily brief section.

In reflecting upon the meaning of the changes in mitochondrial fine structure that appear to occur during development (page 358) the study of Hay (1958, 1959) may be of relevance. During the dedifferentiation of cartilage and muscle cells that follows amputation of the salamander limb, marked changes occur in Golgi apparatus, endoplasmic reticulum, ribosomes, and nucleoli but the mitochondria remain essentially unchanged. The only changes noted are a slight alteration in crista regularity and some diminution in matrix opacity.

D. *In Different Physiological and Pathological States*

1. *Alterations in Number, Size, Shape, and Biochemical Properties*

Cowdry (1918, 1924) early stressed the great sensitivity of the mitochondria to altered cell activity, normal and abnormal; his papers should be read by all embarking on a study of mitochondrial changes. He remarked that even holding tissue between forceps may cause mitochondria to break into granules. He recognized three modes of reaction —"qualitative, quantitative, and topographical, which may occur singly or in combination." The sensitivity of plant mitochondria to external agents was similarly emphasized by Guilliermond (1941), who described their progressive vesiculation with cell necrobiosis (see also Dangeard, 1958). Many authors have pointed to mitochondrial swelling and fragmentation as an early sign of adverse conditions in tissue culture (see Ludford, 1952; Manuelidis, 1958), and we have already described the work of Frederic (1958) and others who followed such mitochondrial changes induced by the addition of various substances to the culture medium.

In the 1920's, Noël (1923) studied the mitochondria in the mouse liver lobule at various intervals after feeding, and after ingestion of diets rich in fat, carbohydrate, or protein. The reader will find Noël's paper of great value, both for his observations and his review of the early literature on the cytology of the hepatic cell. Some of his conclusions regarding a "cycle évolutif sécrétoire du chondriome" may be largely attributed to limitations of light microscopy, but his association of different mitochondrial forms with the position of a given cell in the hepatic lobule reflects the thoroughness of his investigations. Figure 24 is Noël's representation of his findings. Kater (1933) described the mitochondrial patterns in the hepatic lobules of eight species of mammals and two of birds; impressive species differences were noted in both form and sensitivity to changes in blood sugar level.

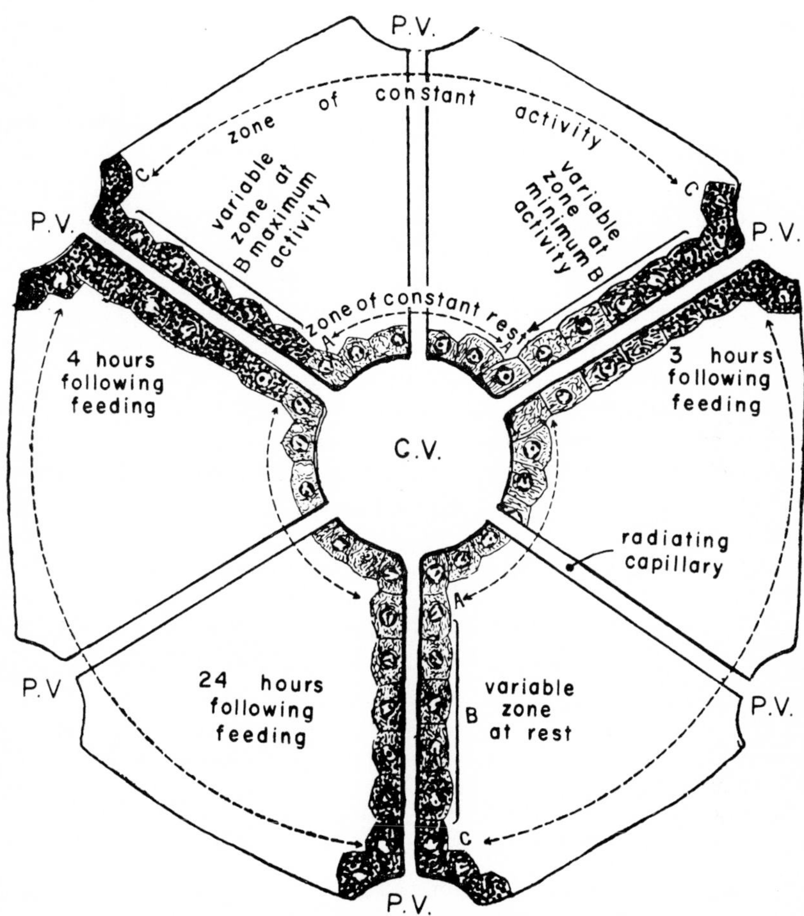

FIG. 24. Diagrammatic representation of a hepatic lobule of the mouse (Noël, 1923). The different shapes of the mitochondria are represented in the cells of the "permanently functional," "variable," and "permanently resting" areas of the lobule; P.V., portal vein; C.V., central vein.

Deane (1944) also studied the heterogeneity of mitochondrial form in the hepatic lobule of the mouse: large spherical mitochondria in the peripheral zone; thin filaments, fewer in number, in the central cells; and intermediate forms between the two areas. She found no regular change in the mitochondria in relation to the diurnal cycle as she did with the Golgi apparatus. She concluded that there are no indications that either the mitochondria or Golgi substance play direct roles in the secretion of bile acids or in the storage of glycogen or fat: "All related zonation phenomena are more probably dependent upon the quality of the blood bathing the cells."

J. Walter Wilson (1953) commented upon the work of Noël and of Deane: "Since the portal blood, carrying products of digestion, arrives at the peripheral zone first, Noël concluded that this zone is a zone of permanent activity and the central zone a zone of permanent repose, the width of the active zone depending on the amount of food materials arriving in the liver at the time. In mice that have been subjected to partial inanition or to prolonged starvation, we have shown that the spherical mitochondria of the peripheral zone tend to be replaced by rods, or fine filaments. The filaments may become beaded or other pleomorphic forms may appear. In general the large spheres stain more vigorously in Regaud's hematoxylin than the filamentous forms. It would seem reasonable to conclude, therefore, that liver mitochondria may be quite different in different parts of the same liver, and in different livers under different physiological conditions, and this should be taken into

---

Fig. 25. Diagrammatic representation of a hepatic lobule of the rat (Novikoff, 1959d).

In the upper left quadrant, a photograph shows a portion of the hepatic lobule stained to demonstrate apparent ATPase activity in the bile canaliculi (from Novikoff *et al.*, 1958).

The quantitative differences between centrolobular cells and peripheral cells are indicated by plus marks (+) and by schematic representation of the cytological structures. In the lower right corner, the substrates are arranged, from left to right, in decreasing order of staining produced with tetrazolium (nitro BT). Kupffer cells are indicated where they contain high levels of the enzyme activities or large numbers of the cytological structures drawn.

The diffuse cytoplasmic staining obtained with unfixed frozen sections when 5′-nucleotidase activity is visualized is more intense (++) in the centrolobular area than peripherally (+). It has not been included in the diagram since we have not established whether this represents cytoplasmic (microsomal?) activity or diffusion (enzyme?) from the sinusoidal aspects of the cell.

Abbreviations: ATP-ase, adenosinetriphosphatase; A5′P-ase, 5′-nucleotidase; Alk. P-ase, alkaline phosphatase; Am. Pept., aminopeptidase; Acid P-ase, acid phosphatase; G-6-P-ase, glucose-6-phosphatase; cent., centrolobular cells; periph., peripheral cells; DPNH, reduced diphosphopyridine nucleotide; TPNH, reduced triphosphopyridine nucleotide; DPN-linked oxidations: La, lactate; βOHB, β-hydroxybutyrate; Gl, glutamate; Ma, malate; αGP, α-glycerophosphate; and TPN-linked oxidations: G-6-P, glucose-6-phosphate; Is, isocitrate; and Su, succinate; Cyt Ox, cytochrome oxidase.

consideration as more refined methods are developed for isolating mitochondria from liver homogenates and for studying their biochemical properties."

The same conclusion was reached by MacCardle and Congdon (1955), who, however, believed that most mitochondria of the peripheral cells were short, thick rods rather than spheres. They also noted that the midzonal cells were the first to respond to lethal doses of total body irradiation.

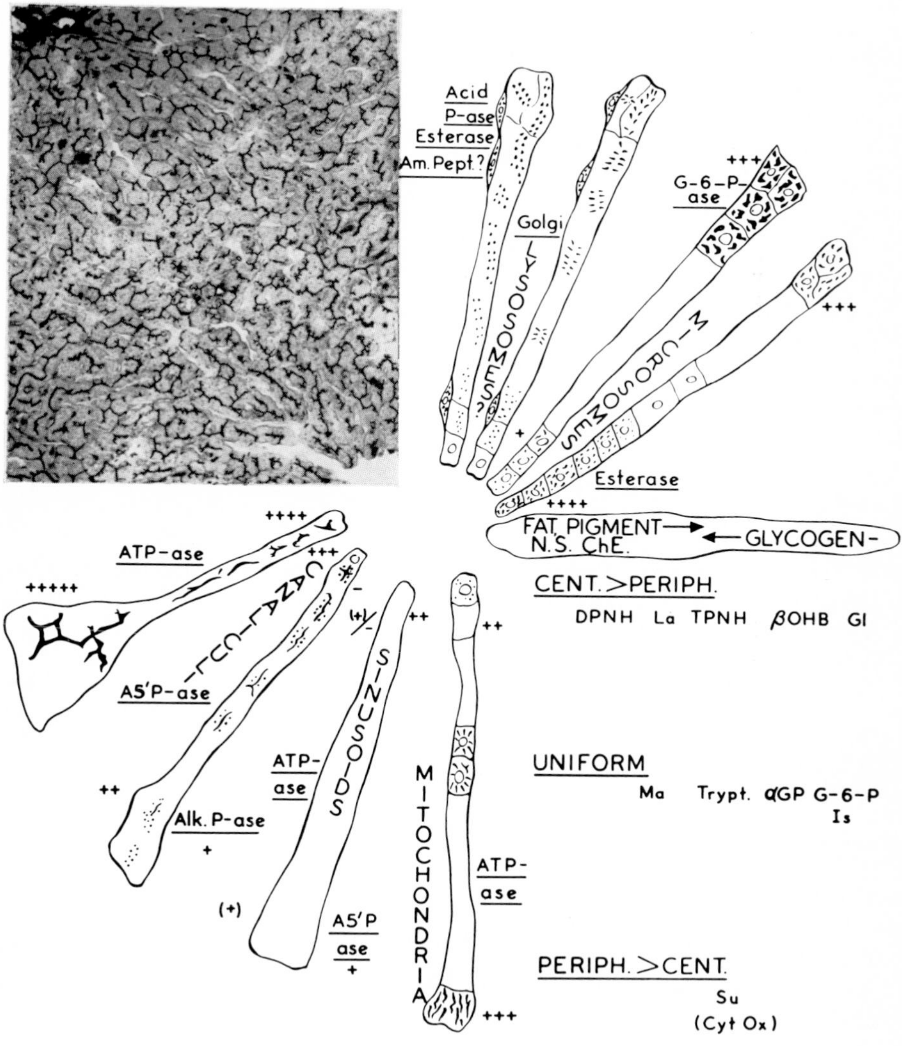

We have elsewhere (1959b, d) summarized the cytological and cytochemical heterogeneity of the cells within the hepatic lobule of the rat as shown in Fig. 25 (also see Schumacher, 1957). The numerous long mitochondria present in the peripheral cells and the high succinic dehydrogenase activity (as shown by the tetrazolium staining methods) suggest a high level of oxidative respiration (Krebs cycle) in these cells whereas in the central cells the scanty rounded mitochondria and low succinic dehydrogenase activity are suggestive of low oxidative respiration. However, glycolysis and other oxidative pathways may occur at a relatively high rate in these central vein cells since the staining resulting from the activities of lactic, β-hydroxybutyric, and glutamic dehydrogenases, and of DPNH- and TPNH-tetrazolium reductases is somewhat higher than in other cells of the lobule.

Dalton (1934) studied the effect of fasting upon the hepatic mitochondria of young chicks. He reported that in fed animals the mitochondria were longer and fewer in number than in fasted animals. He also observed a polarized arrangement of the mitochondria within the cell similar to that in the urodele *Amphiuma* (Pollister, 1932) and the rat (Novikoff, 1959a, b, d).

Gansler and Rouiller (1956) used the electron microscope to study mitochondrial changes in the liver and kidney of rats deprived of food for 3–5 days, and of fasted animals refed a protein-rich diet. Starvation leads to a swelling and rounding of the mitochondria, as Noël (1923) and others had described, a decrease in electron opacity of the matrix, and a loss of all but the peripheral portions of the cristae. After refeeding, most mitochondria become normal, the process being essentially complete by 24 hours. Some mitochondria degenerate completely in the process. Particularly in the kidney, there is an increased number of "microbodies." Similar changes in liver mitochondria were described by Rouiller and Bernhard (1956) following carbon tetrachloride administration or partial hepatectomy. Electron microscopic studies of the livers of fasted and refed animals were also reported by Bernhard *et al.* (1952) and Fawcett (1955). According to Bassi (1960), the first consequence of carbon tetrachloride poisoning is an increased uptake of water, leading to the swelling of the endoplasmic reticulum; mitochondrial swelling is a subsequent event.

Rojas *et al.* (1934) described mitochondrial changes in the liver of the fish *Cnesterodon decemmaculatus* (Fig. 26). At the time the fishes eat, the mitochondria are filaments. Nine hours later they have shortened and become "club-shaped and racket-type forms." At 24 hours they are short rods, mostly near the nucleus. At 48 hours they are long filaments once more.

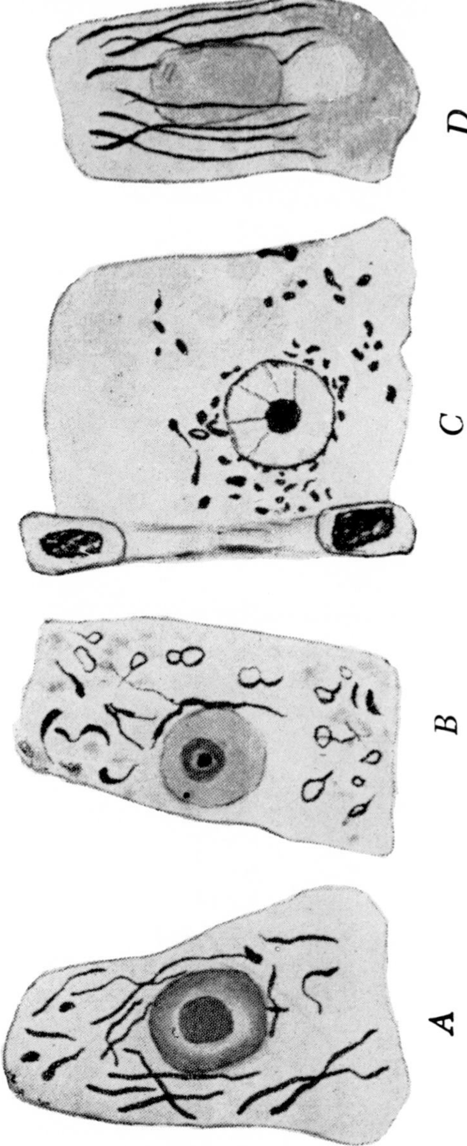

Fig. 26. Hepatic cells of the fish *Cnesterodon decemmaculatus*. A, at time of eating; B, 9 hours later; C, 24 hours after eating; and D, 48 hours after eating. From De Robertis *et al.* (1954).

Scharrer (1945) observed that in those neuropils (i.e., areas of interwoven nerve fibers and synaptic endings) with poor vascularity the cells showed fewer mitochondria. In those with rich vascularity the cells were rich in mitochondria.

Yotsuyanagi (1955) has reported striking changes in the form of mitochondria in yeast, induced by varying the conditions of growth.

Marked mitochondrial hypertrophy has been described by Hovasse (1948) in the green *Euglena gracilis* grown in the dark, as well as a colorless mutant. Striking mitochondrial changes were also noted when the culture medium was altered. Hovasse concluded that the "chondriome" is not a fixed entity, but may change with the physiological state of the organism and the conditions of the medium. A similar situation in another alga, *Chlamydomonas,* was studied by Sager (1959), using the electron microscope. In the pale green mutant, in which the chlorophyll content is only 5% of that in the green wild-type and the plastid lamellae or disks are markedly reduced in number, mitochondria are apparently hypertrophied. Sager suggests that "if much of the high-energy phosphate in *Chlamydomonas* is normally of photosynthetic origin, then this hypertrophy may be an expression of an ATP lack." We have already cited other examples of alterations in fine structure of the inner membranes of animal and plant mitochondria that undoubtedly reflect, in a manner yet to be elucidated, altered function of the organelles (pages 317-320).

Differences in cytochromes and other constituents associated with mitochondria have been found to result from both genic and extragenic factors in yeast and *Neurospora.* These are summarized by Wagner and Mitchell (1955); see also Lindegren *et al.* (1958). Caspari and co-workers claim to have demonstrated constitutional differences among liver mitochondria of different inbred strains of mice (see Caspari and Blomstrand, 1956). This is based on: (1) consistent but small differences in ratios of total phosphorus to total nitrogen in isolated mitochondrial fractions, and (2) differences in color pattern of the sedimented fractions. However, no attempt was made to assess the degree to which both differences might be due to variation in the extent of microsomal contamination of the fractions. Succinoxidase was the only enzyme activity studied and it showed no significant difference between the strains tested. Clearly, more definitive evidence is required.

Harkness (1957) has recently reviewed the extensive literature on so-called regeneration of the liver which follows partial hepatectomy in the rat. He notes that in general the levels of mitochondrial enzymes increase more slowly, relative to total liver protein, than do ribonucleic acid or the enzymes localized in the microsomes or supernatant fluid.

We have already described (pages 355-356) some biochemical studies, particularly the altered effectiveness of oxidative phosphorylation, of isolated mitochondria in hormonally altered animals. Schulz *et al.* (1957) have used electron microscopy to study the swollen liver mitochondria of thyroxine-treated rats. In many of these mitochondria the "matrix" appears much more electron-lucid. It seems that as the mitochondria swell the cristae tear in the middle, remaining intact at the periphery, and material is lost from the interior. Measurements by Schulz *et al.* indicate that there is appreciable swelling of the electron-lucid material between outer and inner membranes at the periphery and in the parts of the cristae that remain.

Lowe and Williams (1953), Lowe *et al.* (1955), and Lowe and Lehninger (1955) have reported cytological and biochemical studies of the liver of rats injected intramuscularly with very large doses of cortisone (25 mg. cortisone acetate daily for 5 days in 100–200-gm. rats) and then starved for 24 hours. The mitochondria become greatly swollen, clear spheres; in homogenate counts their number is less than 300 per nucleus as compared to over 800 in the controls. It is surprising that despite these striking cytological changes, both succinoxidase active and oxidative phosphorylation (with succinate or glutamate as substrate) are unaltered. (The authors report a complete disappearance from the mitochondria of ethanol-precipitable ribonucleic acid ("presumably highly polymerized forms of RNA"), but for reasons already discussed (pages 348-350) it would be preferable to speak of its disappearance from the mitochondrial *fraction* rather than from mitochondria (see also Petermann and Hamilton, 1958).

On the other hand, diminution of phosphorylating ability has been reported for liver mitochondria isolated from animals which have received total body irradiation (1100r.) (van Bekkum, 1955), or have been exposed to cold (Panagos *et al.*, 1958) or which have a thiamine deficiency (Frei and Ryser, 1956) or fatty livers (Dianzani, 1954). Weinbach and Garbus (1956) have reported that in old rats (2–3 years) the levels of both oxidation and accompanying phosphorylations are lower than in young animals, in isolated liver (but not brain) mitochondria. Weiss and Lansing (1953) have reported that in old mice (1 year) many mitochondria of anterior pituitary cells are swollen, their cristae are reduced to "small, inward-directed, stumps," and they appear to have lost much of their matrix.

Okada and Peachey (1957) have shown electron micrographs of rat liver mitochondria exposed to X-ray (10 million r). The mitochondria are swollen, the cristae appear torn and there is apparently a loss of "matrix" material. The authors consider that this may explain the increased deoxyribonuclease activity (DNAase II) observed in liver mitochondria isolated from irradiated animals. However, this study is complicated both by the long exposures (up to 3–4 hours) of the isolated mitochondria to room temperature and by the indications from the work of de Duve *et al.* (1955) that the enzyme studied is lysosomal rather than mitochondrial. Ryser *et al.* (1954) have reported that after either total body irradiation (1000 r) or local (liver field) irradiation, the isolated liver mitochondria had unaltered succinoxidase activity and their ability to oxidize pyruvate was diminished by about 10%. MacCardle and Congdon (1955) found that after lethal doses of total body irradiation (900 and 1200 r.) administered to mice, the mitochondria of the liver were altered first in the midzonal cells of the hepatic lobule. Their long, thick rods become globular and fragment, and they are packed more closely against the blood-vascular (sinusoid) surface of the cell [see Grynfeltt and Lafont (1921) for similar effects of sulfonal poisoning in rabbit liver].

Luft and Hechter (1957) have reported interesting preliminary studies with cow adrenal glands. When examined some 20 minutes after the animal is bled, or after storage in iced saline 60–90 minutes after death, the mitochondria are rounded and show many internal vesicles. However, this damage is reversible, as seen when the glands are perfused for 1–2 hours with liver-filtered, warmed, oxygenated, citrated beef blood. Now, many of the mitochondria are elongated and they display the typical cristae.

Other observations on the variations in mitochondrial form are reviewed by Dempsey (1956) and Rouiller (1960).

Maturation of erythrocytes, even in species where they remain nucleated, is accompanied by a decreased number of mitochondria (deRobertis *et al.*, 1954). These nucleated erythrocytes do not have an appreciably longer life span than nonnucleated ones. It may be of interest that Defendi and Pearson (1955) have reported staining data, and Rubinstein and Dendstedt (1953) chemical data to indicate the presence of oxidative enzymes in the nuclei of bird erythrocytes (but see Stern and Timonen, 1954).

The effects of a number of toxic agents on mitochondrial biochemistry have been studied by Judah and his colleagues (see Cameron, 1956). Carbon tetrachloride is considered by Christie and Judah (1954) to alter mitochondrial integrity so that their capacity to retain their stores of pyridine nucleotides is diminished, thus leading to reduced effectiveness of the Krebs cycle. This occurs within 10–15 hours after administration of the poison whereas the mechanisms for oxidative phosphorylation and phosphatide synthesis persist for a long time (however, see Recknagel *et al.*, 1958). Gallagher *et al.* (1956) report that in rats made copper-deficient the mitochondria become physically altered so that they lose co-enzymes and manganese ions more rapidly than do normal mitochondria. Their ability to synthesize phosphatide is markedly diminished and their cytochrome oxidase activity is moderately lowered. MacFarlane and Datta (1954) attribute the action of Welch α-toxin to breakdown of mitochondrial phosphatide and modification of activities of various enzymes [for effects of phospholipases on isolated mitochondria, see Nygaard (1953), Petrushka *et al.* (1957)]. Gallagher *et al.* (1956) find that the calcium content of necrotizing liver of rats given a single injection of thioacetamide is some six times that of normal liver. The mitochondrial fractions isolated from such livers have about twenty-five times the calcium content than those of normal liver. The authors attribute the impaired ability of these mitochondria to oxidize Krebs cycle intermediates to this high calcium concentration. They think the effect of the thioacetamide operates through alteration of the cell membrane permeability so that calcium (also sodium) enters the liver cells while magnesium and potassium leave them. After 24 hours, there is histological evidence of tissue repair, the calcium concentration returns to normal and so does the respiratory rate of tissue slices and isolated mitochondria.

We have already mentioned a number of situations in which the mitochondria respond by a rounding and swelling. Manuelidis (1958) lists many others. Such mitochondrial swelling is apparently the basis of what pathologists usually call "cloudy swelling." Although other changes are sometimes included under this term (see Manuelidis, 1958; Cameron, 1952), we would support Altmann's (1955) proposal that the term be restricted to those cellular changes associated with enspherulation and swelling of the mitochondria. Electron microscopic examination of such mitochondrial alteration has been reported by Bernhard *et al.* (1952), Fawcett (1955), and Gansler and Rouiller (1956) who, as described earlier, studied these changes in rats deprived of food (also see Rouiller, 1960). The studies of Opie, Robinson, and others relating mitochondrial swelling to the movement of water into tissues are summarized by Manuelidis (1958); see Zollinger (1950) and also the brief comments of Roberts (1953). Fonnesu and Severi (1956) reported inhibition of phosphorylation associated with the oxidation of succinate or α-ketoglutarate in livers with cloudy swelling induced by injection of *Salmonella typhimurium.*

2. *Deposition of Materials; Transformation into Other Cell Granules*

Classical cytologists often faced insurmountable difficulty in establishing mitochondria as the sites of specific deposits or as precursors of newly formed granules. Because so little was known about cell metabolism, a decrease in mitochondrial number while other particles, like lipid droplets, were accumulating could too quickly be assumed to signify a causal relation between the two. In living cells unstained particles were often poorly visualized and many particles were too small to be resolved by light microscopy. With fixed cells there were the hazards of translating static photographs into dynamic process. With the advent of phase contrast microscopy most mitochondria became observable in living cells, but often the decisive "precursor" granules remained below the limit of its resolving power. The resolving capacity of the electron microscope could not be fully utilized owing to difficulties in preserving fine structure through the many steps from fixation to photography. Even when individual molecules like ferritin and hemocyanin had already been described in sections, the structure of membranes and granules often remained elusive, as we have seen in the discussion of bacteria (page 336). There were also the problems of properly interpreting images produced by different planes of sectioning (Napolitano and Fawcett, 1958) and of recognizing lysosomes and other particles in the size range of mitochondria (see Chapter 6).

We will briefly review the electron microscopic observations on reports, old and new, of mitochondrial transformation into plastids of plant cells, lipid droplets in animal cells, yolk platelets of ova, various pigments in animal cells, specific granules of blood cells, and, under pathological or experimental conditions, into "protein droplets" in the mammalian kidney tubules and keratohyaline granules in A-avitaminosis, as well as the depositions of materials like ferritin. Careful electron microscopy will be required for evaluating the claimed roles of mitochondria in secretion and enzyme production, or in the histogenesis of myofibrils and neurofibrils. In 1918, Cowdry published a list of eighty substances in the formation of which mitochondria were said to be concerned. For examples of mitochondrial roles and transformations claimed from light microscopy see Cowdry (1918), Wilson (1928), Newcomer (1940, 1951), Guilliermond (1941), and Bourne (1951); also Noël (1923) and du Buy *et al.* (1949). For examples of such claims stemming from electron microscopy see Rouiller (1960).

Wilson (1928) wrote: "There are few of the cytoplasmic formed bodies in the cytoplasm which have not been supposed to be products of the chondriosomes; but few of these conclusions have not been contradicted by other observers." Yet he felt that, despite some contrary opinion, the evidence from the work beginning in 1910 with Lewitsky and Pensa and continuing with that of Maximow, Meves, Nassonov, and others, especially Guilliermond, was substantial ground for accepting the derivation of plastids from chondriosomes. When Newcomer reviewed the literature in 1940, he wrote: "The problem at present stands essentially where Lewitsky left it in 1910, when he adequately demonstrated by photographs of fixed and fresh material, the transformation of some of the chondriome into plastids." In his review ten years later (Newcomer, 1951), he reiterated this viewpoint, "that plastids in plants can and do originate from mitochondria was perhaps best demonstrated by Lewitsky in 1911." Sharp (1934) took the position that mitochondria were not the primordia of plastids. Until recently, electron microscopy has brought mostly evidence for Sharp's view. This evidence suggests that plastids develop from self-perpetuating bodies which, although hardly distinguishable from mitochondria by light microscopy, possess a three-dimensional crystalline array of granules (100–250 A.) rather than the cristae typical of mitochondria. As these "proplastids" mature into plastids, the granules develop into the characteristic plastid lamellae or disks (see Granick, 1955; Sjöstrand, 1956b; Sager and Palade, 1957; Wolken, 1959). These are more linear and more precisely arranged than the cristae of mitochondria; also, in *Chlamydomonas* the plastid envelope has been reported to be continuous at times with the endoplasmic reticulum (Sager and Palade, 1957). Thus it seemed that Cowdry was correct when in 1918 he stressed that neither close topographical association of mitochondria and developing plastids nor the diminution in mitochondrial number as plastids increase in number establishes the "chemical transformation of mitochondria into plastids." However, it now appears that the question will not be satisfactorily settled until improved methods of section preparation are applied to differentiating plant meristems. Mercer (1956) was correct in asserting that this is a question "which may be answered by the electron microscope, but which certainly will never be by the light microscope"; yet technical limitations have delayed the answer (see Whaley *et al.,* 1960).

The recent work from Buvat's laboratory shows the fine structure of the earliest proplastids to be essentially indistinguishable from mitochondria, including an outer membrane, cristae, and a few small electron-opaque granules in the matrix (Lance, 1958; Buvat, 1958; Caporali, 1959). Buvat himself emphasizes the differences that appear to be present: the proplastids are somewhat larger, their cristae are shorter, and their electron-opaque granules appear to be absent from *Elodea* meristem mitochondria: "L'existence constante de différences, au moins quantitatives, est de plus favorable à leur indépendance génétique." Whaley *et al.* (1960) also stress the differences between the cristae of mitochondria and inner membranes of the proplastids in meristem of maize; the proplastid membranes are more uniform and stain more darkly with potassium permanganate. Yet, Caporali, Buvat's student, considers that the inability to distinguish some particles as mitochondria or proplastids "même suggère une parenté d'origine entre les deux lignées de chondriosomes." To Mühlethaler and Frey-Wyssling (1959) the similarity between early proplastids and mitochondria "seems to indicate that there exists some phylogenetic relationship between the two cell organelles."

It may be possible to distinguish more definitively between early proplastids and mitochondria on biochemical grounds. If so, the definition of mitochondria given on pages 314-315 would need qualification to permit this biochemical differentiation. (For a discussion of both resemblances and differences in biochemistry as well as fine structure between plastids and mitochondria, see Chapter 7 by Granick.)

In contrast to the hesitation displayed by electron microscopists in accepting the uncertain granules as "transitions" between mitochondria and proplastids, they are often ready, in animal cells, to accept tenuous resemblances as the basis for claiming depositions in mitochondria or transformations of mitochondria into other particles. Our own attitude toward these claims is more conservative and critical than that recently expressed by Rouiller (1960).

The transformation of mitochondria into *fat droplets* during cell degeneration was claimed by numerous cytologists (see Cowdry, 1918; Kater, 1933; Ludford, 1952, for references). Among more recent reports are those of Ludford (1952) in tissue culture cells and of de Robertis *et al.* (1954) in hepatic cells of hypophysectomized toads.

In a beautiful study of brown fat cells in newborn mice and rats, Napolitano and Fawcett (1958) demonstrate that uncritical analysis of electron micrographs has led several investigators recently to the erroneous conclusion that, in the cells of brown fat and of adrenal cortex, mitochondria transform into lipid. The studies of Napolitano and Fawcett, as those of Palade and Schidlowsky (1958) and Palade (1959) on liver and pancreas cells of fasted or cortisone-treated rats, and of de Robertis and Sabatini (1958) and Zelander (1959) on adrenal cortex, provide no evidence of such transformation. There is, however, intimate contact between mitochondria and lipid droplets (Fig. 27); the physiological or developmental signficance of this relation remains to be established. Palade (1959) has speculated that in the pancreas of the fasted guinea pig, mobilized fat is probably being metabolized by the fatty acid oxidases of the mitochondria, but the direct biochemical evidence

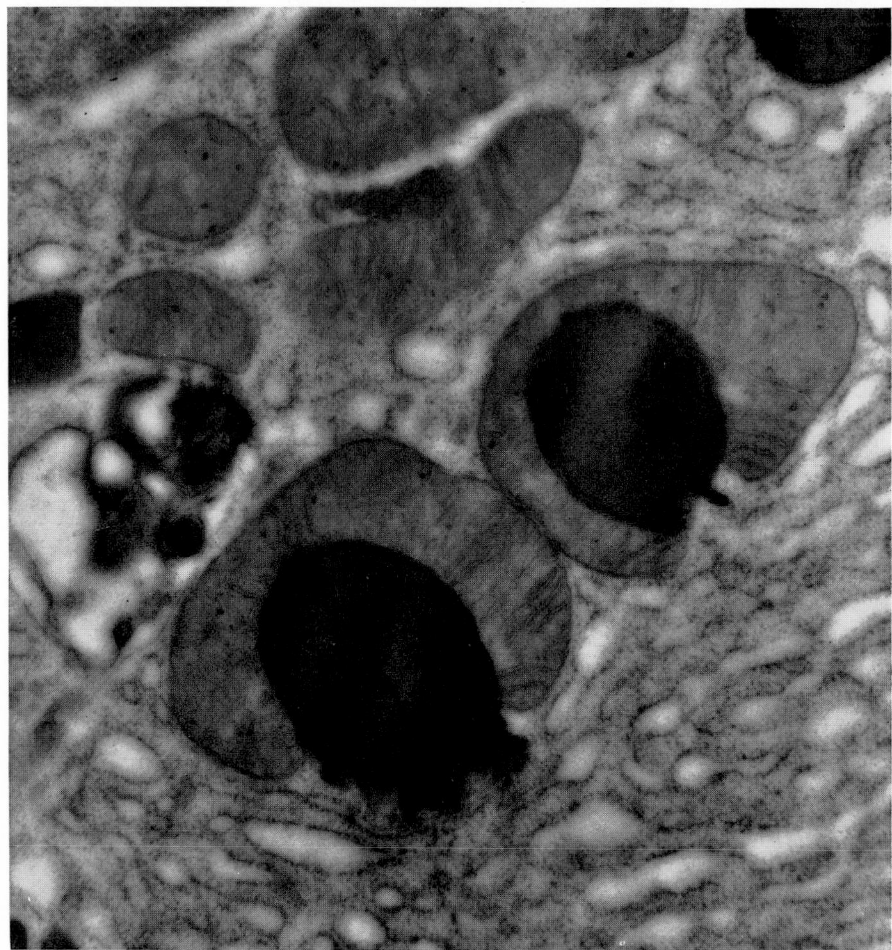

FIG. 27. Intimate relation of mitochondria to lipid droplets in guinea pig pancreas. Animal starved for 48 hours. Magnification: × 26,000. Courtesy of Dr. George E. Palade.

is still lacking. Two mitochondrial changes which may occur in this situation have already been described: an increase in number of intramitochondrial granules and the disappearance of the outer membrane in regions of contact. Another is a marked disturbance in the cristae pattern

(Palade, 1959). A relation between mitochondria and lipid metabolism of the sea urchin egg is suggested by Pasteels (1958), who observed (Pasteels *et al.*, 1958) that each lipid droplet of the *Paracentrotus* egg is surrounded by a sheath of mitochondria. Berg and Humphreys (1960) find a similar arrangement of mitochondria on the yellow lipid droplets in the eggs of *Styela.*

Gansler and Rouiller (1956) express the view that mitochondria produce the fat droplets of fatty liver, but the evidence (Rouiller, unpublished) is not presented (see Rouiller, 1960).

Rinehart (1955), in a preliminary report of electron microscopic study of thin sections of leucocytes and platelets, wrote: "The series of transition forms between mitochondria and *'specific' granules,* which we have described and illustrated, would appear to afford convincing evidence that the granules are differentiated derivatives of mitochondria." Similar evidence has been reported by Bernhard *et al.* (1955) for the granules of eosinophiles and by Bernhard and Leplus (1955) for blood platelet granules. However, no such transition forms are seen in the higher resolution electron micrographs of Low and Freeman (1958). Bessis (1956) cites no evidence in favor of the mitochondrial origin of specific granules of leucocytes. The work of Cohn and Hirsch (1960) indicates a lysosomal rather than mitochondrial nature for the granules of rabbit polymorphonuclear leucocytes (see Chapter 6).

Dalton and Felix (1953) investigated cells of the mouse melanoma S-91 with the electron microscope and found no evidence for the view of Woods *et al.* (1949) and du Buy *et al.* (1949) that *melanin granules* are formed by direct transformation of mitochondria (see also Wellings and Siegel, 1959).

Sheldon and Zetterquist (1956) have published electron micrographs of cells from the "intermediate strata" of the keratinizing epithelium of vitamin A-deficient rats. They show electron-opaque areas within the mitochondria which are interpreted as being early keratohyaline bodies.

Hess (1955) interpreted electron micrographs to mean that *lipofuscin granules* of neurons were derived from mitochondria. From the resemblance of these granules (see also Bondareff, 1957) to those in human liver (Essner and Novikoff, 1960a) and from other evidence, we consider their origin from lysosomes more likely than that from mitochondria (see Chapter 6).

As discussed in Chapter 6, cytochemical and electron microscopic studies suggest a connection to lysosomes for the so-called "*protein droplets*" in the proximal convolution cells of rat kidney (Novikoff, 1959e; 1960c), although numerous electron microscopic reports had concluded that they arose from mitochondria (Rhodin, 1954; Gansler and Rouiller, 1956; Miller and Sitte, 1955; Farquhar *et al.*, 1957; Wallace, 1960; Rouiller, 1960). It seems that the mitochondrial changes that occur are secondary to the overloading of cells with protein; they are not morphologically involved in the formation of the droplets.[5]

The deposition of ferritin and another more "powdery" form of iron described as "*micelles ferrugineuses*" in the mitochondria of erythroblasts has been demonstrated by Bessis and Breton-Gorius (1959) (see also Bessis, 1959, 1961) in patients with hypochromic-hypersideremic anemias. These authors believe that in these anemias the disturbance in normal iron utilization leads to the accumulation of the "powdery" iron material to the point that the mitochondria swell and burst. It will be of interest to know how frequent, in these and other conditions studied by the authors, such ferritin-containing mitochondria are, relative to the other cytoplasmic ferritin "clusters." These "clusters," in some photographs, appear to be membrane bound, and they may, like the "siderosomes" described by Richter (1957, 1959), be lysosomes. Of relevance to this question would be any evidence that the small vesicles (which the authors prefer to call "ropheocythosis" rather than the commonly used term, "micropinocytosis") that take ferritin into the cell also transport them to the "clusters" (see Chapter 6 for a discussion of micropinocytosis vacuoles and lysosomes).

Ward (1959) has shown small yolk platelets of hexagonal form to appear and grow in size within the mitochondria of the oöcytes of the frog *Rana pipiens* (Fig. 28) [see Lanzavecchia and Le Coultre (1958), who interpret micrographs of dense material within cristae as evidence for the origin of mitochondria from yolk platelets]. They have the same crystal-like nature as the yolk platelets of mature eggs which Karasaki (1959) studied by electron diffraction. In the snail *Planorbis corneus*, the crystalline array of molecules, probably ferritin, in the yolk platelets have been beautifully demonstrated by Favard and Carasso (1958). They suggest, on the basis of presumed transitional forms, that the crystalline arrays are deposited within mitochondria to convert them into the simpler of two types of yolk platelets found in these eggs.

Crystalline electron-opaque arrays of unknown nature and function have been found within mitochondria of brown fat by Napolitano and Fawcett (1958). Fibrillar material has been described in mitochondria of the giant ameba, *Pelomyxa* (Pappas and Brandt, 1959). Nilsson (1958) has found that the mitochondria in the uterine epithelium of the mouse

[5] For a brilliant electron microscopic study reported while this volume was in press see Miller (1960).

show interesting changes during estrus. As the cells elongate, the mitochondria become longer. Some acquire membrane-limited dense bodies, 0.15 μ in diameter, that displace the cristae.

Electron-opaque material, more lamellar than crystalline, has been found in the mitochondria of a 52-year-old male with chronic, familial, nonhemolytic acholuric jaundice (Fig. 30). Jézéquel (1959) has reported similar findings in two cases, one with a bile duct carcinoma and another with prolonged viral hepatitis. In all three cases, the mitochondria are enormous and bizarre; they have an unusually large number of cristae and intramitochondrial granules.

Occasional large mitochondria with numerous cristae and granules have been described in cells of a rat liver tumor (de Man, 1960). Oberling and Bernhard (see Chapter 7, Volume V) cite a number of instances in rodent and human tumors of deposition of materials within mitochondria, sometimes obscuring most cristae. Setälä *et al.* (1960) describe the appearance of electron-opaque spherical bodies within the mitochondria in cells of mouse epidermis painted with carcinogens. The authors consider these bodies to resemble the "dégénérescence mitochondriale" described in a spontaneous mouse sarcoma (Rivière *et al.*, 1959). The significance of such mitochondrial deposits is not known. It remains to establish which are purely degenerative, which have other functional meanings, and which may disappear with continued life of the cell.

In erythroblastosis of fowl some virus-containing bodies in erythroblasts are considered to be mitochondria (Benedetti and Bernhard, 1958; Bonar *et al.*, 1959). In avian myeloblastosis, evidence for the mitochondrial nature of virus-containing bodies in myeloblasts (Bonar *et al.*, 1959) seems less suggestive. Beard's group believes that the initial virus infection in both myeloblastosis and erythroblastosis occurs in the mitochondria (Bonar *et al.*, 1959).

FIG. 28. Portion of mitochondrion in an oöcyte of *Rana pipiens*. The yolk crystal, with its ordered arrangement of electron-opaque lines, is clearly within the mitochondrion. Typical cristae may also be seen in the mitochondrion. Magnification: × 105,000. Courtesy of Dr. Robert T. Ward.

FIG. 29. Portion of elongated spermatid in the butterfly *Pieris brassicae*. At this stage the nebenkern has developed into two elongated cylindrical structures alongside the flagellum (*F*). The mitochondria which constitute these cylinders have a "hollow" appearance with the cristae projecting for only a short distance into the matrix. Within the matrix is a series of interconnected vesicles ("droplets") of hyaloplasm; these are apparent in the cylinders sectioned longitudinally (extreme left) and transversely (extreme right). Magnification: × 17,000. Courtesy of Dr. Jean André.

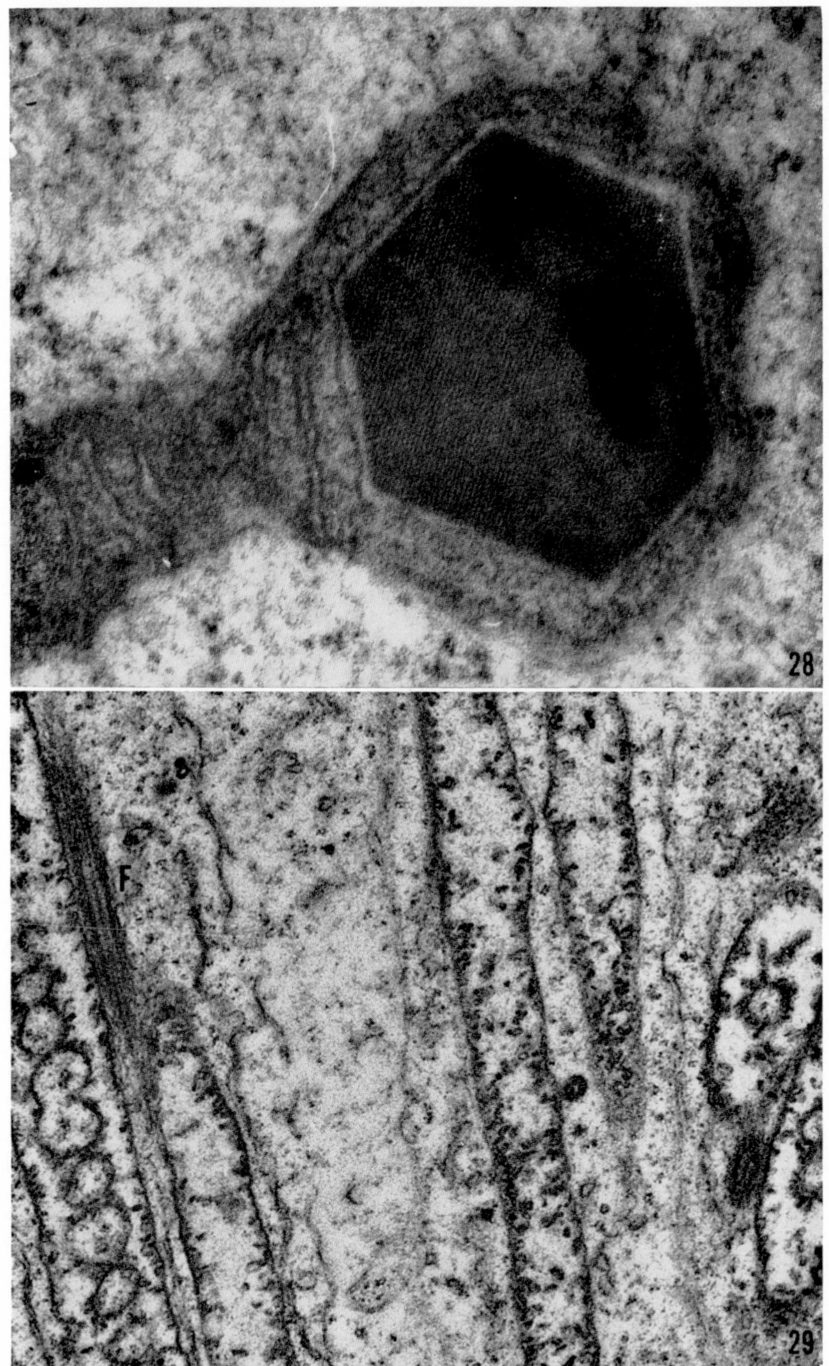
28
F
29

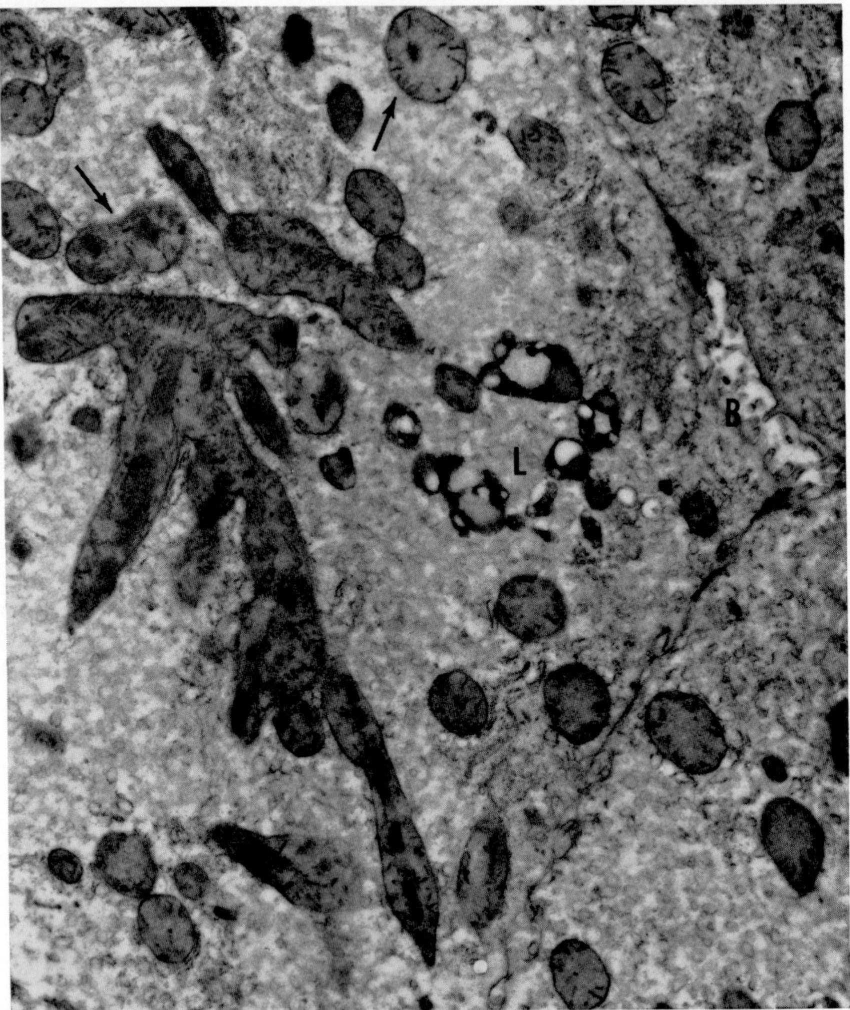

FIG. 30. Section through liver of 52-year-old man with chronic, familial, nonhemolytic acholuric jaundice. (Courtesy of Dr. Edward Essner.)

The most prominent feature of the cell to the left of the bile canaliculus (*B*) is the very large mitochondrion of bizarre shape with numerous "intramitochondrial granules," numerous cristae, and an inner material of high electron opacity. This material is also evident in the mitochondria indicated by arrows. Other features of interest are: (1) the scarcity of granular reticulum (ergastoplasm) in the cell with the bizarre mitochondria, and (2) the lipofuscin granules (*L*) near the bile canaliculus (see Essner and Novikoff, 1960a; Novikoff and Essner, 1960). Magnification: × 18,000.

E. Carcinogenesis and Cancer Cells

Warburg, the first to isolate subcellular fractions containing mitochondria ("grana"), has long maintained that all carcinogens act by permanently altering their structure and respiratory function: "there is only one common cause into which all other causes of cancer merge, the irreversible injuring of respiration. There is today no other explanation for the origin of cancer cells" (Warburg, 1956). However, as Weinhouse (1955, 1956) has stressed, all the Krebs cycle enzymes are present in cancer cells and isotope experiments show them to be operative in the same fashion as in normal cells. Potter (1958; 1960) also agrees that the altered metabolism of cancer cells (manifested by a high aerobic glycolysis, as demonstrated by Warburg, and a high rate of DNA synthesis) does not result from a simple impairment of respiratory enzymes; see also Woods (1956).

Chance and Hess (1959) stress the importance of structural relationships of enzyme systems within the cell. With the availability of ascites tumor cells, decisive measurements could be made of changes in concentrations of possible "control substances" at the site of their action within the cell. "These results clearly indicate the high capabilities of the respiratory chain of the ascites tumor cell for rapid electron transfer and efficient phosphorylation." Respiration is not "impaired." Enzyme systems are not "unbalanced." Rather it appears that the predominance of glycolytic over respiratory activity in the ascites tumor is due to the limitations imposed by the intracellular concentration of ADP.

Even the matter of mitochondrial volume in cancer cells is not simple. In some cancers, like human toxic adenomas and myxosarcomas, the mitochondria may be more numerous than in the normal cell prototype (Cowdry, 1955). Ludford (1952) was of the opinion that mitochondria in malignant cells were generally more numerous; however, they were also smaller. But even in the numerous cancers in which there appear to be both fewer and smaller mitochondria in the cells [e.g., in rat liver tumors (Allard *et al.*, 1957; Howatson and Ham, 1955; Novikoff, 1957b); and others (Bernhard, 1958)] this may be related to the general diminution in cytoplasmic volume.

The fine structure of cancer cell mitochondria is essentially like those of normal cells, with a cristae-containing inner membrane and smooth outer membrane. However, many variations are seen, so that, Bernhard (1958) wrote, "one is struck by the extraordinary variations in their number, size, form, and density, and by the frequent lesions they present." This is dealt with more extensively in Volume V, Chapter 7 by Oberling and Bernhard. The difficulty in isolating tumor mitochondria with high levels of oxidative phosphorylation or low levels of ATPase activity, and the protective effect of serum albumin added to the incubation medium (Blecher, 1960) probably reflects the damage incurred by these organelles during homogenization, isolation, and incubation (see page 355). It is not clear whether this is due to a relative "fragility" of these mitochondria as compared to those of some mammalian tissues or to the drastic methods required for disrupting tumor cells (Novikoff, 1960d). The same problem applies to the responses to swelling agents reported for tumor mitochondria (Mutolo and Abrignani, 1957; Emmelot *et al.*, 1956).

Changes in the number and biochemical properties of rat liver mitochondria have been studied during the feeding of carcinogenic dyes. Striebich *et al.* (1953) counted the number of mitochondria in homogenates of the tissue. Feeding the carcinogen, 3′-methyl-4-dimethylaminoazobenzene, for 4 weeks produced a marked decrease in number, whereas with the noncarcinogenic dye, 2-methyl-4-dimethylaminoazobenzene, there was an increase. These results are consistent with biochemical findings on isolated mitochondrial fractions. In addition, biochemical assays (Price *et al.*, 1950; Schneider *et al.*, 1953) show that after feeding of the carcinogen the loss in nitrogen content of the fraction is paralleled by a loss in riboflavine content and decreases in activity of succinoxidase and other oxidative enzymes. Chang *et al.* (1958) have reported a decrease in mitochondrial number in the peripheral cells of the hepatic lobule in rats fed 3′-methyl-4-dimethylaminoazobenzene for 1 and 2 weeks; this is not accompanied by a loss in succinic dehydrogenase activity to the point that it produces a significant decrease in tetrazolium staining. It is in these peripheral cells of the lobule that Weiler (1956) has reported a marked decrease in antigen found in the mitochondrial-microsomal fractions isolated from normal rat liver. It may be relevant that Hogeboom and Schneider (1951) found one of four normal protein components missing from mitochondria isolated from the transplantable mouse hepatoma 98/15. Porter and Bruni (1959) have described changes in fine structure of rat liver cells following the feeding of 3′-methyl-4-dimethylaminoazobenzene. Arcos *et al.* (1960) report a decreased response to swelling agents by mitochondria (and microsomes) isolated from such liver; the decrease occurs at the critical time for tumor determination. Of much interest is the finding of Price *et al.* (1950) that feeding 4′-fluoro-4-dimethylaminoazobenzene, a potent carcinogen, decreased the protein and riboflavine levels of the mitochondrial fraction only slightly.

Following application of the carcinogen, methylcholanthrene, to the skin of mice, there is a marked rise in cytochrome oxidase activity (Carruthers, 1950). Histological studies may show a parallel increase in mitochondria-richer cells. The contrast between the increased level of this mitochondrial enzyme during mouse skin carcinogenesis and the decreased level during rat liver carcinogenesis is probably an indication that rapidly growing tumors "tend chemically to resemble each other more than they do normal tissues or than normal tissues resemble each other" (Greenstein, 1954).

The view of Woods and du Buy (1951) that mitochondrial mutation is the "continuing cause" of neoplasia has not had support from others.

VII. General Considerations

A. *Interrelations with Other Cell Structures*

Among the most interesting intracellular relationships reported for mitochondria is that with the *nucleus.* Many cells show a perinuclear concentration of mitochondria (page 331). In fibroblast cells in culture, Frederic and Chèvremont (1952) (see also Frederic, 1958) described the rhythmic movement of the mitochondria to the nucleus and their intimate contact with the nuclear membrane, at a time when nucleolar material appeared to be migrating into the cytoplasm (Fig. 1). In electron micrographs of early oöcytes of *Rana clamitans,* Ornstein (1956) described "processes" extending from the nucleus through "pore-like rings." These "processes converge on and seem to 'fuse' with" aggregates of mitochondria and electron-dense material on the other side of the nuclear membrane. "The mitochondria are closely applied to the denser material which seems to be continuous with the nuclear processes—*as if* we have caught the nucleus and mitochondria in the process of cooperatively producing the denser masses." Brandt and Pappas (1959) have studied the nuclear-mitochondrial associations that "most commonly appear at and shortly after nuclear division" of the giant ameba *Pelomyxa.* Electron microscopy shows the associations to be intimate; indeed the authors consider that the outer membranes of mitochondria and nuclei are continuous. Such continuity has been demonstrated unequivocally in trypanosomes (Steinert, 1960; Ris, private communication; see also Clark and Wallace, 1960). The kinetoplast (or kinetonucleus) may be considered as a combined nucleus and mitochondrion, not only from the electron microscopic evidence (Figs. 31, 32), but from cytochemical as well. Thymidine incorporation and other evidence shows the kinetoplast to contain DNA, synthesized synchronously with that in the conventional nucleus; Janus green and DPNH-tetrazolium reductase staining show it to possess mitochondrial enzyme activity (Clark and Wallace, 1960; Steinert, 1960). Obviously great interest attaches to these observations. Is this a means of transferring genetic information from nucleus to mitochondria? Does this association have morphogenetic or evolutionary meaning? Does it have implications for cells generally?

Of relevance to the question of mitochondrial-nuclear relations are the interesting observations of Chèvremont *et al.* (1959a,b). In chick embryo fibroblasts cultured in the presence of acid deoxyribonuclease there is a marked inhibition of mitotic activity and profound changes occur in the mitochondria. DNA (Feulgen-positive) is synthesized in the cytoplasm (tritiated-thymidine incorporation) and accumulates there (see Brachet, 1960, for an account of the so-called reserve DNA in the cytoplasm of eggs). According to Chèvremont *et al.*, the accumulation occurs in the altered mitochondria. The authors see the evidence as support for their view that under normal circumstances DNA, or some precursor, is produced in the cytoplasm, and the extension of the view to involve the mitochondria in the passage of this material to the nucleus. They consider that the acid deoxyribonuclease treatment alters DNA synthesis and mitochondria so that the process becomes evident in the accumulation of DNA in mitochondria rather than nuclei.

In Section IV, A several instances were described of mitochondria showing a definite relationship to the *cell membrane*. These should be mentioned here. In the distal and proximal tubules of the mammalian renal tubule, the long mitochondria are arranged within compartments of the cytoplasm, each one close to the cell membrane (see Rhodin, 1958) (Fig. 14); we have shown (Spater *et al.*, 1958; Kaplan and Novikoff, 1959) that the cell membrane delimiting these compartments has a high level of ATPase activity as demonstrated by staining methods (see Novikoff, 1960b, c).

Palay (1956) described a marked concentration of mitochondria in the presynaptic expansion of the axon, the end foot or bouton terminal (Fig. 13). Clusters of small vesicles (200–650 A. in diameter) are also present there. Palay wrote: "Although the significance of these structures in the physiology of the synapse is still unknown, two suggestions are made: that the mitochondria, by means of the relation between their

FIG. 31. Section through the crithidia form of *Trypanosoma mega*. Note: (1) the concentrically arranged inner membranes in the mitochondrion (*M*) associated with the kinetonucleus (*KN*), and (2) the outer membrane enclosing both mitochondrion and kinetonucleus. Also seen are the nucleus (*N*), Golgi apparatus (*G*), and flagellum (*F*). Magnification: × 37,000. Courtesy of Dr. Maurice Steinert.

FIG. 32. Section through the crithidia form of *Trypanosoma mega*. Note: (1) the typical crista structure of the inner membranes of the mitochondrion (*M*), and (2) the outer membrane enclosing both mitochondrion and kinetonucleus (*KN*); it is seen most distinctly at the junction of mitochondrial and kinetonuclear substance (arrow). Nucleus is seen at *N*. Magnification: × 55,000. Courtesy of Dr. Maurice Steinert.

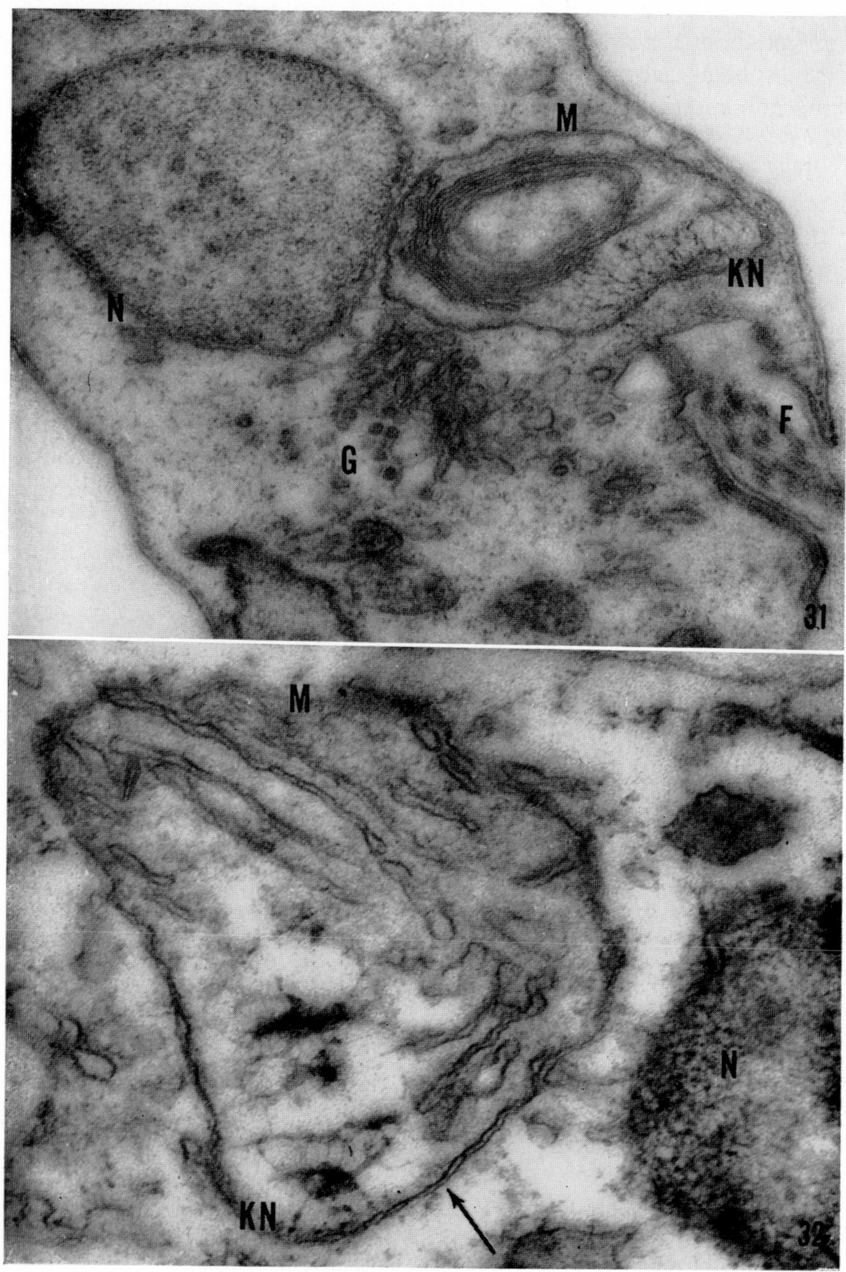
M
N
KN
F
G
31
M
N
KN
32

enzymatic activity and ion transport, participate in the electrical phenomena about the synapse; and that the small synaptic vesicles provide the morphological representation of the prejunctional, subcellular units of neurohumoral discharge at the synapse demanded by physiological evidence."

The proximity of mitochondria and cell membrane has also been described in the Schwann cell (Geren and Schmitt, 1956) and in spermatids of the rat (Palade, 1952; Watson, 1952; Fawcett, 1959) and of the guinea pig (Fawcett, 1959). This is considered to reflect energy demands, in the first case as the myelin sheath is formed (Geren and Schmitt, 1954) and in the second as the acrosome is being synthesized (Fawcett, 1959).

One of the most striking associations of mitochondria and other cell structures is to be seen in mammalian skeletal muscle. The mitochondria are arranged in "rings or braces around the I bands of the myofibrils" [Palade, 1956a; also see Andersson-Cedergren (1959) for a study of these mitochondria]. Their branching reticulum-like arrangement (Fig. 10) provides a wide area of contact between mitochondria and the region where the myofibrils undergo profound alteration during muscular contraction (Palade, personal communication).

In the Malpighian tubule cells of the grasshopper, Beams *et al.* (1955) described the concentration of mitochondria within the protoplasmic processes constituting the brush border. There the mitochondria are in close approximation with the cell membrane. It is not clear whether this is related to dynamic interchange between the lumen and the cell or, as the authors suggest, whether it is the terminal stage in the movement of the mitochondria into the protoplasmic processes, which then pinch off into the lumen.

A relationship of mitochondria and *endoplasmic reticulum* was described by Bernhard and Rouiller (1956). When rats are fasted, the endoplasmic reticulum disappears except around the mitochondria; upon refeeding it reappears. Bernhard and Rouiller propose that the mitochondria provide energy for its regrowth. "Mitochondria seem to play an important part in the elaboration of hepatic ergastoplasm." On the other hand, Hay (1958) finds a close relation between the two organelles when the amount of endoplasmic reticulum is decreasing during dedifferentiation of cartilage following limb amputation in *Amblystoma* larvae. Where the membranes of endoplasmic reticulum and mitochondria are in contact, there is sometimes an accumulation of electron-opaque material. The author suggests that this indicates a "transfer of energy or metabolic products."

The striking association between mitochondria and tubular endoplasmic reticulum in the pseudobranch "gland" of *Fundulus heteroclitus* has already been described (page 329; Fig. 15). Each mitochondrion is surrounded by twenty to thirty tubules running parallel to its length. This intimate relation is considered by Copeland and Dalton (1959) to reflect the high rate of carbonic anhydrase synthesis by these cells. We have also referred to the intimate association of endoplasmic reticulum to the sarcosomes of the flight muscle cells in the dragonfly. It is seen from Fig. 33 that sarcosomes and myofibrils are separated by two layers of reticulum: circumscribed convoluted portions lying within indentations of the sarcosomes, and perforated curtainlike portions running the length of the myofibrils (Smith, 1961).

Special indentations of *hyaloplasm* into the mitochondrial structure (nebenkern), appearing as a series of linked vesicles (Fig. 29), is found in spermatids of the butterfly *Pieris* (André, 1959b).

The intimate relation of the mitochondria to the *centriole,* and its derivative in sperm cell, the *axial filament,* was described by the early cytologists. As we have already seen, the development of these mitochondria has been studied in detail, particularly by Grassé *et al.* (1956), Kaye (1958), and André (1959a, b). André (1959b) discusses the polarizing action of these structures. Kaye writes of the close relation of the motile filament and the mitochondria with their densely packed cristae: "It is apparent that the mature nebenkern, with the cristae packed together so closely, has the structural basis for higher oxidative rates per unit volume than the mitochondria of previous stages. Thus the nebenkern sheath utilizes the space it occupies as efficiently as possible while adequately supplying the energy needs of the cell."

The intimate relationship between mitochondria and developing lipid droplets of some cells has already been described (pages 382-384). The meaning of this relation is not yet clarified. There does not appear to be any particular relationship between mitochondria and the Golgi apparatus as has sometimes been claimed (see Bourne, 1952). The function of the mitochondria frequently found in yolk "droplets" and "nuclei" (e.g., Bellairs, 1958; Favard and Carasso, 1958; André and Rouiller, 1957) remains to be elucidated.

It seems reasonable to conclude that mitochondria are frequently localized in those areas of the cell with high rates of dynamic or molecular activity.

B. *Origin of Mitochondria*

Since there is no definitive demonstration that in cells generally all mitochondria arise from pre-existing mitochondria, there has been much conjecture that mitochondria may arise "*de novo.*" Even when not recognized explicitly, this meant in essence that mitochondria arose from cellular material below the resolution limits of light microscopy. Obviously this is an area where the electron microscope can bring considerable new information since it can detect particles down to macromolecular dimension and can reveal mitochondria which, because of special staining or optical properties, may go unnoticed in light microscopy.

Undoubtedly the most widely quoted evidence for the "*de novo*" origin of mitochondria are the classic experiments of Beckwith (1914) and E. B. Harvey (1946, 1953). (For a discussion of the earlier suggestions of "*de novo*" origin, see Newcomer, 1940.)

Beckwith centrifuged the eggs of *Hydractinia echinata* to produce an oil-cap at one end, a clear protoplasmic mass below it, and a "mingled mass of yolk and mitochondria" at the other end. Cleavage sometimes separated the yolk and mitochondria into one blastomere and the oil into the other blastomere. Although many eggs died, considerable numbers developed into half larvae. When these "ciliated planulae" were fixed in Meves' fluid and stained with Benda's method, the larvae containing the yolk showed apparently unchanged mitochondria, "while the other contains none." "It is apparent, then, that up to this point of development the mitochondria are not essential for development"; ". . . they can hardly be vital constituents of the protoplasm since they may be centrifuged out of the protoplasm like any metaplastic body, such as yolk." "They arise *de novo* throughout the egg" . . . "there is no indication of their multiplication by division. . . . Their origin and behavior . . . indicate that they may be either precociously differentiated portions of the protoplasm (Vedjovsky '07) or metaplastic bodies."

Harvey centrifuged the eggs of *Arbacia* and obtained two half-eggs; upon further centrifugation these were broken into quarter eggs. When fertilized, the "clear quarter, which apparently lacked mitochondria," developed into a small pluteus larva. "During the first cleavages, there are no mitochondria present, as can easily be told by the lack of stain with methyl green. The blastulae, even when 24 hours old, also contain no mitochondria, but the plutei *do* have mitochondria since they now

---

FIG. 33. A representation of the fine structure of flight muscle in the dragonfly, *Aeshna* sp., by Dr. David S. Smith (labeling added). A portion of the nucleus (*N*) is at upper right. Slablike sarcosomes (*S*) alternate with myofibrils (*F*), the details of which are not shown. Note that the sarcosomes terminate at the Z bands (*Z*) and that they have indentations (arrows) in which branches of the convoluted portion of the endoplasmic reticulum (*ER*) are situated. Also lying between the sarcosomes and the myofibrils is a portion of the reticulum that takes the form of a perforated curtainlike cisterna (*P*).

stain with methyl green or Janus green. The mitochondria have been replaced between the blastula and pluteus stage."

However, when thin sections of centrifuged *Arbacia* eggs were studied with the electron microscope by Lansing *et al.* (1952) and Lansing (1953),

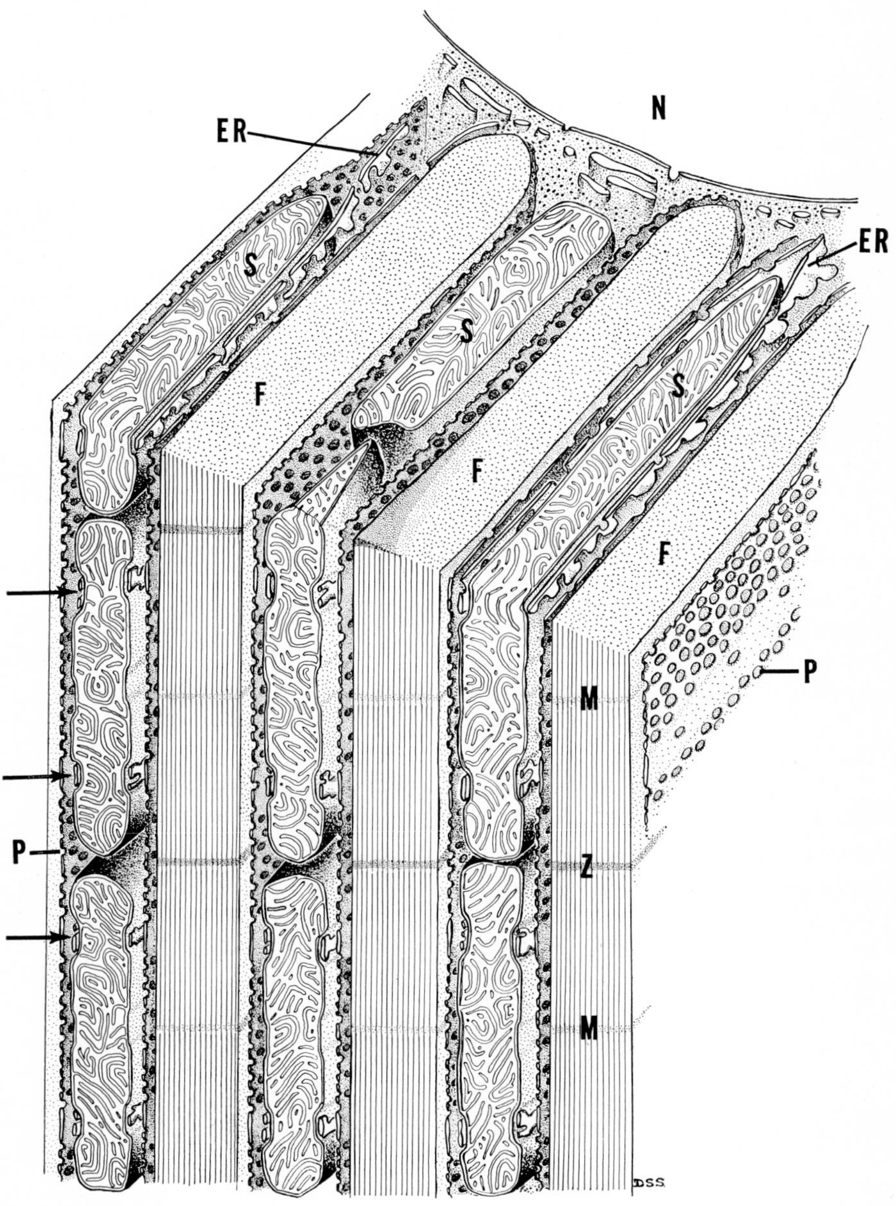

two layers of mitochondria were found. One, that seen by Harvey, was at the centrifugal end. The other was in the "centripetal lipid layer" where Harvey had thought no mitochondria to be present. The observation that among the centrifuged lipid droplets mitochondria are always to be found has been confirmed by Parpart (unpublished) and Gross *et al.* (1960). The same results have been obtained with eggs of other sea urchins by Afzelius (unpublished) and Pasteels *et al.* (1958). Berg and Humphreys (1960) similarly found that centrifugation of *Styela* eggs did not dislodge the mitochondria intimately associated with the yellow lipid droplets. Although these investigators have all studied whole eggs centrifuged at relatively low force, it is unlikely that the situation would be different at the higher centrifugal force used by Harvey to produce quarter eggs.

One may only guess why Harvey was unable to see these mitochondria among the oil droplets. Their width, as judged by electron micrographs, is above the limit of resolution of light microscopy, although in the unfixed living cell they may be narrower [cf. the observations of Dalton *et al.* (1949) that in spontaneous mouse hepatomas and in melanoma S-91 some mitochondria, although quite long, are less than 0.2 μ in diameter and thus invisible by light microscopy]. The work of Alfert (1952) demonstrates that factors such as combination with protein would affect methyl staining; in any case, methyl green would not be expected to stain mitochondria (see Novikoff, 1955). Janus green staining is sometimes capricious, and it is even possible that the mitochondria in the lipid zone lack enzyme sufficient to keep the dye in oxidized state [compare the Nadi-negative macromeres of *Tubifex* (Weber, 1958) and the enzyme-deficient mitochondria of yeast "petites" (Ephrussi and Slonimski, 1955)].

Gustafson and colleagues (Gustafson and Lenicque, 1952; Gustafson, 1954) have supported the "*de novo*" origin of mitochondria because of their observations (Gustafson and Lenicque, 1952, 1955) that the number of mitochondria increases markedly at the "mesenchyme blastula" stage, the same stage at which Hultin (1953) found an increased isotope uptake into mitochondrial protein. They considered that the new mitochondria developed from smaller stainable nonmitochondrial particles. However, Shaver (1957), reinvestigating this problem, could find no evidence for aggregation of such particles. "Mesenchyme blastulae" should be valuable objects for electron microscopy; they might help resolve the problem of mitochondrial origin and mitochondrial division.

Several attempts have already been made to investigate the problem of mitochondrial origin by electron microscopy. Hartmann (1954) reinvestigated his earlier light microscopic observations that following section of their axons, motor nerve cells formed new mitochondria. "The evidence available in this study," Hartmann (1954) writes, "suggests that the mitochondria arise by accretion of submicroscopic cytoplasmic particles at or near the nucleo-cytoplasmic interface." "The granules of the thickened nuclear membrane" . . . are . . . "of the same size and electron density as the constituent granules of the mitochondria." However, the low resolution of the micrographs precludes acceptance of this far-reaching conclusion. Low-resolution micrographs have also been used by Hoffman and Grigg (1958) to suggest a mitochondrial origin from the nuclear membrane. An essentially similar view has been proposed by Brandt and Pappas (1959) from studies with the giant ameba *Pelomyxa.* The micrographs are excellent for mitochondrial structure, but continuity of mitochondrial membranes and those of the nuclei are not unequivocal, and, naturally, direction of movement of membranes and organelles cannot be determined from static pictures. This limitation also restricts interpretation regarding origin in the case of trypanosomes where the continuity of mitochondria and kinetonucleus has been established beyond doubt (see page 391). Steinert (1960) suggests that trypanosomes may be favorable cells in which to study mitochondrial morphogenesis.

Superficial resemblance in electron micrographs has been the basis of two other suggestions of mitochondrial origin. Lever (1956a, b) found that some mitochondria in ACTH-stimulated adrenal glands of rats and hamsters showed peripheral "polylaminated" membranes resembling those of the Golgi membranes, and suggested that the Golgi membranes developed into mitochondria. However, considerable damage is apparent in the sections. Also, the possibility has not been eliminated that the "polylaminated" membranes represent longitudinal sections of the tubular cristae of these mitochondria. Ehret and Powers (1955) suggested a nucleolar origin of mitochondria because nucleoli and mitochondria had a similar appearance in thin sections of *Paramecium* after removal of the methacrylate. However, methacrylate removal under these circumstances apparently causes such extensive damage that significance cannot be attached to this similarity in appearance.

Electron microscopic studies of high quality by Gansler and Rouiller (1956) and Rouiller and Bernhard (1956) have led these authors to suggest another mode of mitochondrial origin. They studied regenerating hepatic cells of the rat after various experimental procedures. After partial hepatectomy, carbon tetrachloride poisoning, refeeding of starved animals, or several injections of egg albumin, there is a marked increase in the number of cytoplasmic particles which they call "microbodies" although they differ somewhat from the kidney "microbodies" of Rhodin (1954) (see also Howatson, 1956). The liver particles are 0.1–0.5 μ in diameter, are delimited by a single membrane, and contain a dense, finely granulated substance. Their cores, usually homogeneous, sometimes show double membranes resembling cristae. These are considered by the authors as transitional forms between "microbodies" and mitochondria, indicating that mitochondria arise from "microbodies."

Engfeldt *et al.* (1958) have studied the cells of the proximal convoluted tubule in rats injected with large doses of parathyroid hormone. They describe an accumulation of "microbodies" and a decrease in number of mitochondria, and conclude that "the normal development of microbodies into mitochondria" has been disturbed.

These studies do not, in our opinion, establish this role for microbodies in these situations or in more physiological circumstances; nor do those of Weissenfels (1958) in normal embryonic development. Essner and Novikoff (1960b) have demonstrated that after infusion of bilirubin into rats large numbers of microbodies appear in hepatic cells. They have the same fine structure as the hepatic microbodies studied by Gansler and Rouiller. From their possession of single outer membranes and acid phosphatase activity they appear to be lysosomes. We have seen no evidence in any tissue of the transformation of lysosomes into mitochondria. Biochemical data combined with electron microscopy of isolated fractions suggest to de Duve and colleagues (private communication) that in untreated rat liver the microbodies represent a distinct class of particle containing catalase and related enzymes but not acid phosphatase (see Chapter 6). These differences in results between untreated and bilirubin-infused rats require clarification, but no evidence of mitochondrial structure or enzymes has been found in these bodies.

Some time ago a popular view was that mitochondria arose from microsomes (see, e.g., de Robertis *et al.*, 1954). This was originally proposed by the Brachet school (Chantrenne, 1947; Jeener, 1948; Brachet, 1950, 1952) but is no longer held by them (Brachet, 1957). When the Belgian workers proposed this developmental role, microsomal morphology, even in liver, was totally unknown. Microsomes were simply granules too small to be sedimented at certain centrifugal forces or to be visible with the light microscope. Knowledge of microsomal chemistry was extremely limited—they were considered to be rich in RNA and deficient in enzyme activity. Thanks to rapid development of cell fractionation techniques and electron microscopy, we now possess sufficient information concerning liver microsomes (those that led Chantrenne to his suggestion) to evaluate the claim. In this tissue, the transformation of microsomes into mitochondria would involve not a mere addition of substance, as Chantrenne thought, but the simultaneous loss of virtually all its known constituents (high RNA content, high levels of glucose-6-phosphatase, esterase, DPNH-cytochrome c reductase, and marked abilities to synthesize protein and cholesterol). Structurally, it would involve the complete dissolution of the ergastoplasm and its transformation into the complex mitochondrial structure. There is no suggestion available that such transformation occurs.

Finally, one should mention the view of Gey (1956) that in tissue culture cells some mitochondria may arise from pinocytosis vacuoles. Although to our knowledge this observation has not been reported by other investigators, Holter (1959) does not dismiss the possibility. He is "struck by the morphological similarity" between mitochondria and large complex pinocytosis vacuoles described by Brandt (1958). Also, there is the related unorthodox speculation of Robertson (1959) who suggests that mitochondria may arise from the cell membrane and may still retain their connections to it as a kind of tail. Earlier, Geren and Schmitt (1954) had published electron micrographs suggesting the participation of the plasma membrane of the Schwann cell in the formation of mitochondria in squid and lobster axons.

C. *The "Organelle" Status of Mitochondria*

As recognized early by Dalton *et al.* (1949), electron microscopy should enable us better to evaluate views such as expressed by Bensley (1953) (see also Cowdry, 1918; Tennent *et al.*, 1931; Hartmann, 1954) that "mitochondrial substance is expendable" and "to erect the mitochondria to the dignity of an organelle or cell organ . . . is a blunder." Bensley (1947) elsewhere suggested that "mitochondria were simply temporary aggregates of complex composition, consisting of a number of major components and many trace substances including enzymes and vitamins" and "that phospholipins are the trigger substances in this process as in many of the similar coacervations studied *in vitro* by DeJong and his associates." These views of Bensley derived from the fact that when guinea pigs were starved, the mitochondria "practically disappeared from the acinous cells. The loss was not complete but the fragments that remained were profoundly altered." But, as Palade (1953) has pointed out, "Dr. Bensley observed in the light microscope that mitochondria 'disappear' after feeding, but such observations may simply describe variations in refractive index or variations in dimensions below and above the limit of resolution of the light microscope. The problem should be reexamined with the electron microscope." Palade then goes on to commit himself to a stability of mitochondria he considers requisite for status as a cell organelle. "Dr. Bensley's experiments evidently indicate that mitochondria lose or gain materials under certain conditions. But this does not preclude their being recognized as cell organs, because such changes are true for any organ on any scale. What is stable in a given organ is only a certain pattern of organization in which the various compounds fit while continuously turning over. This apparently applies to mitochondria and thus justifies the view that they are actually cell organs."

Electron microscopy should also help evaluate other conditions in which the disappearance of mitochondria has been reported. Frederic (1958) considers that mitochondria may disappear and reappear during cell division of fibroblasts in culture, or after the addition of substances like trihydroxy-*N*-methylindole. Gustafson (1954) suggests that the mitochondria in the vegetal portion of the early sea urchin blastula are "consumed for morphogenetic purposes," i.e., the production of structural proteins presumably required for the invagination of gastrulation. Wasserman (1954) has observed that in the rootlets of germinating *Pisum sativum* "the mitochondria are not preserved during the period of dehydration and are reformed with participation of nuclear substances in the beginning of germination" [see Hertwig (1929) for the 1920 drawing of Wasserman; compare Figs. 189 and 190]. Dangeard (1958) reviews instances of what he believes is mitochondrial "neoformation" from small, normally invisible granules "du type des 'microsomes' ou *cytogranula*" remaining in cells of plants treated with dilute acid or other agents [but see Buvat (1948) who concluded from such studies that the mitochondria do not disappear]. We have already considered, in Section VI, B, instances where spermatozoa are said to be without mitochondria, but this may be true only in nonmotile, essentially terminal cells. Mitochondria disappear in other terminal cells, such as erythrocytes and squamous epithelium.

As we view it, there is little reason to question the "organelle" status of mitochondria even should they disappear and reappear within a cell. Do not the nucleoli and nuclear membrane disappear at each mitotic division? Mitochondria need not always retain their organization pattern or biochemical composition. Thus, the mitochondria in the "petite" mutants of yeast look normal, with present electron microscopy, yet they lack cytochrome oxidase and other respiratory enzymes (Slonimski, 1953; Ephrussi, 1953; Ephrussi *et al.*, 1956; Yotsuyanagi, 1955, 1959); and we have seen the profound yet reversible alterations of which mitochondria are capable in altered physiological states or in pathological conditions.

Possibly related to this discussion is the situation in bacteria and the malarial parasite, *P. berghei,* where typical mitochondria are not present; their functional equivalents may be represented by membranous structures continuous with the plasma membrane (Section IV, B).

D. A General "Chondriome" Hypothesis

The biochemical data that we summarized earlier have removed much of the uncertainty from the view described by Wilson (1928), ". . . the chondriosome is regarded as a *localized center of specific chemical transformation,* a view urged especially by Regaud ('09, '11), who compared the chondriosomes in this respect to the plastids of plant-cells, which undoubtedly are such centers of action. Physiologically, in Regaud's view, the chondriosomes are 'eclectosomes' which have a specific selective action upon the surrounding cytoplasm and are centers of specific chemical elaboration and accumulation."

It remains for future research to establish the role of mitochondria in the cell's elaboration of products such as specific cell granules and lipid droplets. But with the supreme importance of mitochondria in the energetics of cell metabolism established, the question really is how direct or indirect is their role in a given cell activity. The high resolving power of the electron microscope, by establishing topographical relationships, may help attain the question's answer, if coupled with biochemical and cytochemical analysis. We may anticipate the resolution, before long, of important details of mitochondrial biochemistry. What are the precise mechanisms of oxidative phosphorylation? What is the extent and nature of protein synthesis in mitochondria? If there is RNA in mitochondria, where is it localized, and is its role in protein synthesis similar to that of ergastoplasmic RNA? How do the permeability properties of outer membrane and inner crista-forming membrane differ and what is the macromolecular basis for these differences? What is the nature and role of the mitochondrial "matrix"? What physiological roles involve the "contractile process" in mitochondria (Lehninger, 1960)?

Future biochemists will face a problem even more formidable than relating their data to essentially static electron micrographs. This will be the description in chemical terms of the constant dynamism of the cell, so vividly expressed in phase contrast cinemaphotography of cells in culture. [For the beginnings of such descriptions see Lehninger (1959a, b; 1960) and Packer (1960).] This will be inextricably related to the old questions of mitochondrial origin, mitochondrial disappearance and appearance. Also awaiting further analysis is the significance, morphogenetic and evolutionary, of membrane continuities such as those between cell membrane and possible mitochondrial equivalents in bacteria, malarial parasites and blue-green algae, and that between the kinetonucleus and mitochondria in trypanosomes.

It is worth pausing to ask whether the term "chondriome" is simply a convenient collective name for all mitochondria within a cell. Or, is there, as Frederic (1958) believes, a pool of mitochondrial substance in the cell, constantly changing in form but remaining essentially unchanged quantitatively, a "chondriome" in dynamic equilibrium with the rest of the cell? With highly developed electron microscopic and biochemical techniques at hand, we may hope for a firm answer to the question, firmer than was possible for Altmann, Benda, Meves, Guilliermond, and other early cytologists who, impressed by the ubiquity of mitochondria, considered them as persistent autonomous cell components.

There are still great difficulties in evaluating the role of mitochondria in cell heredity. For discussions of this problem, as it regards cases of extrachromosomal inheritance, the reader might consult Sonneborn (1951), Ephrussi (1953), Wagner and Mitchell (1955), and Rhoades (1955). "There are, it seems to me," writes Sonneborn (1951), "reasons for suspecting that the molecular organization of the cytoplasm may be a hereditary property of the cytoplasm, comparable to the hereditary arrangement of the genes in the chromosomes. The rapid and efficient operation of enzyme systems with many enzymes participating in a regular sequence, seems to require a precision of localization on enzyme-bearing particles such as mitochondria; and this arrangement is most easily conceived as a consequence of the surface pattern on which the enzymes are adsorbed. As the mitochondria are probably self-duplicating, the pattern too may be perpetuated."

Acknowledgments

The work done in our laboratories has been greatly aided by generous and sustained support from the American Cancer Society, the United States Public Health Service, the National Science Foundation, and the Damon Runyon Memorial Fund.

It is a pleasure to acknowledge the assistance of my colleagues: Mr. L. J. Walker for the photography; Mrs. Joan Drucker and Mr. Woo-Yung Shin for preparation of the sections shown in Figs. 5–8; and Miss Sheila Schrift for typing the several versions of the manuscript.

Our sincere thanks go also to the scientists who generously provided us with original photographs, and to the many others who sent us reprints of their work and, we hope, will continue to do so.

References

Aebi, H., and Abelin, I. (1953). *Biochem. Z.* **324**, 364.
Afzelius, B. A. (1955). *Z. Zellforsch. u. mikroskop. Anat.* **42**, 134.
Afzelius, B. A. (1957a). *Z. Zellforsch. u. mikroskop. Anat.* **45**, 660.

Afzelius, B. A. (1957b). "Electron Microscopy on Sea Urchin Eggs," 20 pp. Almqvist and Wiksells, Uppsala.
Agar, H. D., and Douglas, H. C. (1957). *J. Bacteriol.* **73**, 365.
Aldridge, W. N. (1957). *Biochem. J.* **67**, 423.
Alfert, A. (1952). *Biol. Bull.* **103**, 145.
Allard, C., de Lamirande, G., and Cantero, A. (1957). *Cancer Research* **17**, 862.
Allard, C., Mathieu, R., de Lamirande, G., and Cantero, A. (1952). *Cancer Research* **12**, 407.
Altmann, H. W. (1955). *In* "Handbuch der allgemeinen Pathologie" (Büchner, F., ed.), p. 419. Springer, Berlin.
Altmann, R. (1890). "Die Elementarorganismen und ihre Beziehungen zu den Zellen," p. 145. Veit Co., Leipzig.
Amoore, J. E., and Bartley, W. (1958). *Biochem. J.* **69**, 223.
Andersson-Cedergren, E. (1959). *J. Ultrastruct. Research Suppl.* **1**, pp. 191.
André, J. (1959a). *J. Ultrastruct. Research* **2**, 288.
André, J. (1959b). *Ann. sci. nat. zool. et biol. animale* [12] **1**, 283.
André, J. (1959c). *Compt. rend. acad. sci.* **249**, 1264.
André, J., and Rouiller, C. (1957). *J. Biophys. Biochem. Cytol.* **3**, 977.
Andressen, N. (1956). *Compt. rend. trav. lab. Carlsberg, Ser. Chim.* **29**, 435.
Arcos, J. C., Griffith, G. W., and Cunningham, R. W. (1960). *J. Biophys. Biochem. Cytol.* **7**, 49.
Baker, J. R. (1946). *Quart. J. Microscop. Sci.* **87**, 441.
Baker, J. R. (1958). "Principles of Biological Microtechnique." Methuen, London.
Ball, E. G., and Barrnett, R. J. (1957). *J. Biophys. Biochem. Cytol.* **3**, 1023.
Baltscheffsky, H. (1957). *Biochim. et Biophys. Acta* **25**, 382.
Barer, R., and Joseph S. (1956). *Symposia Soc. Exptl. Biol.* **10**, 160.
Bartley, W., and Davies, R. E. (1954). *Biochem. J.* **57**, 37.
Bartley, W., Davies, R. E., and Krebs, H. A. (1954). *Proc. Roy. Soc.* **B142**, 187.
Bassi, M. (1960). *Exptl. Cell Research* **20**, 313.
Bates, H. M., Craddock, V. M., and Simpson, M. V. (1958). *J. Am. Chem. Soc.* **80**, 1000.
Bates, H. M., Kalf, G. F., and Simpson, M. V. (1959). *Federation Proc.* **18**, 187.
Beams, H. W., and Tahmisian, T. N. (1954). *Exptl. Cell Research* **6**, 87.
Beams, H. W., Tahmisian, T. N., Levine, R. L. (1955). *J. Biophys. Biochem. Cytol.* **1**, 197.
Beams, H. W., Tahmisian, T. H., Anderson, E., and Devine, R. (1960). *J. Biophys. Biochem. Cytol.* **8**, 793.
Beaufay, H., Bendall, D. S., Baudhuin, P., Wattiaux, R., and de Duve, C. (1959). *Biochem. J.* **73**, 628.
Beckwith, C. J. (1914). *J. Morphol.* **25**, 189.
Beinert, H. (1951). *J. Biol. Chem.* **190**, 287.
Bellairs, R. (1958). *J. Embryol. Exptl. Morphol.* **6**, 149.
Belt, W. D., and Pease, D. C. (1956). *J. Biophys. Biochem. Cytol.* **2**, Suppl., 369.
Benda, C. (1897/8). Ueber die Spermatogenese der Vertebraten und höherer Evertebraten. *Verhandl. physiol. Ges. Berlin,* pp. 14-17 (Aug. 11, 1898).
Benda, C. (1902). Die Mitochondria. *Ergeb. Anat. u. Entwicklungsgeschichte* **12**, 743.
Benedetti, E. L., and Bernhard, W. (1958). *J. Ultrastruct. Research* **1**, 309.
Bensley, R. R. (1947). *Anat. Record* **98**, 609.
Bensley, R. R. (1953). *J. Histochem. and Cytochem.* **1**, 179.
Bensley, R. R., and Bensley, S. H. (1938). "Handbook of Histological and Cytological Technique." Univ. of Chicago Press, Chicago, Illinois.

Bensley, R. R., and Hoerr, N. L. (1934). *Anat. Record* **60**, 449.
Berg, W. E. (1956). *Biol. Bull.* **110**, 1.
Berg, W. E. (1957). *Biol. Bull.* **113**, 365.
Berg, W. E., and Humphreys, W. J. (1960). *Develop. Biol.* **2**, 42.
Bernhard, W. (1958). *Cancer Research* **18**, 491.
Bernhard, W., and Leplus, R. (1955). *Schweiz. med. Wochschr.* **85**, 897.
Bernhard, W., and Rouiller, C. (1956). *J. Biophys. Biochem. Cytol.* **2** Suppl., 73.
Bernhard, W., Haguenau, F., Gautier, A., and Oberling, C. (1952). *Zeit. f. Zellforsch. u. mikroskop. Anat.* **37**, 281.
Bernhard, W., Haguenau, F., and Leplus, R. (1955). *Rev. hématol.* **10**, 267.
Bessis, M. (1956). "Cytology of the Blood and Blood-Forming Organs," p. 629 (translated by E. Ponder). Grune and Stratton, New York.
Bessis, M. (1959). *In* "The Kinetics of Cellular Proliferation" (F. Stohlman, Jr., ed.) pp. 22-29. Grune and Stratton, New York.
Bessis, M. (1961). *In* "The Cell" (J. Brachet and A. E. Mirsky, eds.), Vol. V, Chapter 3. Academic Press, New York (in press).
Bessis, M., and Breton-Gorius, J. (1959). *J. Biophys. Biochem. Cytol.* **6**, 231.
Biesele, J. J. (1955). *J. Biophys. Biochem. Cytol.* **1**, 119.
Birbeck, M. S. C., and Reid, E. (1956). *J. Biophys. Biochem. Cytol.* **2**, 609.
Birns, M., Essner, E., and Novikoff, A. B. (1959). *Proc. Am. Assoc. Cancer Research* **3**, 7.
Blecher, M. (1960). *Federation Proc.* **19**, 53.
Blecher, M., and White, A. (1960a). *J. Biol. Chem.* **235**, 12.
Blecher, M., and White, A. (1960b). *Biochem. Biophys. Research Communs.* **3**, 471.
Blondel, B., and Turian, G. (1960). *J. Biophys. Biochem. Cytol.* **7**, 197.
Boell, E. J., and Weber, R. (1955). *Exptl. Cell Research* **9**, 559.
Bonar, R. A., Parsons, D. F., Beaudreau, G. S., Becker, C., and Beard, J. W. (1959). *J. Natl. Cancer Inst.* **23**, 199.
Bondareff, W. (1957). *J. Gerontol.* **12**, 364.
Bouman, J., and Slater, E. C. (1957). *Biochim. et Biophys. Acta* **26**, 624.
Bourne, G. H. (1952). *In* "Cytology and Cell Physiology"(G. H. Bourne, ed.), 2nd ed., Chapter 6. Oxford Univ. Press, London and New York.
Brachet, J. (1950). "Chemical Embryology," p. 533. Interscience, New York.
Brachet, J. (1952). "Le rôle des acides nucléiques dans la vie de la cellule et de l'embryon." Masson, Paris.
Brachet, J. (1957). "Biochemical Cytology," p. 516. Academic Press, New York.
Brachet, J. (1960). "The Biochemistry of Development," 320 pp. Pergamon Press, New York.
Brachet, J., Decrolez-Briers, M., and Hoyez, J. (1958). *Bull. soc. chim. biol.* **40**, 2039.
Bradfield, J. R. G. (1956). *Symposium Soc. Gen. Microbiol.* **6**, 296.
Brandt, P. W. (1958). *Exptl. Cell Research* **15**, 300.
Brandt, P. W., and Pappas, G. D. (1959). *J. Biophys. Biochem. Cytol.* **6**, 91.
Brenner, S. (1949). *S. African J. Med. Sci.* **14**, 13.
Brenner, S. (1953). *Biochim. et Biophys. Acta* **11**, 480.
Briegert, E. M., Glauert, A. M., and Allen, J. M. (1959). *Exptl. Cell Research* **18**, 418.
Bronk, J. R. (1958). *Biochim. et Biophys. Acta* **27**, 667.
Bronk, J. R., and Kielley, W. W. (1958). *Biophys. et Biochim. Acta* **29**, 369.
Bullivant, S. (1960). *J. Biophys. Biochem. Cytol.* **8**, 639.
Burgos, M. H., and Fawcett, D. W. (1955). *J. Biophys. Biochem. Cytol.* **1**, 287.
Burstone, M. S. (1959). *J. Histochem. and Cytochem.* **7**, 112.
Buvat, R. (1948). *Rev. de Cytol. et de Cytophysiol. Végét.* **10**, 5.

Buvat, R. (1958). *Ann. Sci. Nat. Botan. et biol. végétale* [11] **19**, 121.
Cameron, G. R. (1952). "Pathology of the Cell," p. 840. Oliver and Boyd, London.
Cameron, G. R. (1956). "New Pathways in Cellular Pathology," p. 90. Arnold, London.
Caporali, L. (1959). *Ann. Sci. Nat. Botan. et biol. végétale* [11] **20**, 215.
Caro, L. G., van Tubergen, R. P., and Forro, F., Jr. (1958). *J. Biophys. Biochem. Cytol.* **4**, 491.
Carruthers, C. (1950). *Cancer* **10**, 255.
Caspari, E., and Blomstrand, I. (1956). *Cold Spring Harbor Symposia Quant. Biol.* **21**, 291.
Champy, C. (1912). *Compt. rend. soc. biol.* **72**, 987.
Chance, B. (1956). *In* "Enzymes, Units of Biological Structure and Function" (O. H. Gaebler, ed.), pp. 447-463. Academic Press, New York.
Chance, B., and Hess, B. (1959). *Science* **129**, 700.
Chance, B., and Thorell, B. (1959). *Nature* **184**, 931.
Chance, B., and Williams, G. R. (1956). *Advances in Enzymol.* **17**, 65.
Chang, J. P. (1956). *Exptl. Cell Res.* **11**, 643.
Chang, J. P., Spain, J. D., and Griffin, A. C. (1958). *Cancer Research* **18**, 670.
Chantrenne, H. (1947). *Biochim. et Biophys. Acta* **1**, 437.
Chapman, G. B. (1954). *J. Morphol.* **95**, 237.
Chapman, G. B., Hanks, J. H., and Wallace, J. H. (1959). *J. Bacteriol.* **77**, 205.
Chatton, E., and Tuzon, O. (1941). *Compt. rend. acad. sci.* **213**, 373.
Chèvremont, M. (1956). "Notions de Cytologie et Histologie," p. 994. Editions Desoer, Liège.
Chèvremont, M., Chèvremont-Comhaire, S., and Baeckeland, E. (1959a). *Arch. biol. (Liège)* **70**, 811.
Chèvremont, M., Chèvremont-Comhaire, S., and Baeckeland, E. (1959b). *Arch. biol. (Liège)* **70**, 833.
Chiga, M., and Plaut, G. W. E. (1959). *J. Biol. Chem.* **234**, 3059.
Christie, G. S., and Judah, J. D. (1954). *Proc. Roy. Soc.* **B142**, 241.
Clark, T. B., and Wallace, F. G. (1960). *J. Protozool.* **7**, 115.
Claude, A. (1943). *Science* **97**, 451.
Claude, A. (1944). *Am. Assoc. Advance. Sci. Research Conference on Cancer* p. 223.
Claude, A. (1946a). *J. Exptl. Med.* **84**, 51.
Claude, A. (1946b). *J. Exptl. Med.* **84**, 61.
Claude, A. (1954). *Proc. Roy. Soc.* **B142**, 177.
Claude, A., and Fullam, E. F. (1945). *J. Exptl. Med.* **81**, 51.
Cleland, K. W. (1952). *Nature* **170**, 497.
Clement, A. C., and Lehmann, F. E. (1956). *Naturwissenschaften* **43**, 478.
Cohn, Z. A., and Hirsch, J. G. (1960). *J. Exptl. Med.* **112**, 983.
Cooper, C., and Lehninger, A. L. (1956a). *J. Biol. Chem.* **219**, 489.
Cooper, C., and Lehninger, A. L. (1956b). *J. Biol. Chem.* **219**, 519.
Copeland, D. E., and Dalton, A. J. (1959). *J. Biophys. Biochem. Cytol.* **5**, 393.
Cota-Robles, E. H., Marr, A. G., and Nilson, E. H. (1958). *J. Bacteriol.* **75**, 243.
Cowdry, E. V. (1918). "The Mitochondrial Constituents of Protoplasm," pp. 39-160. Carnegie Institution of Washington, Washington, D.C.
Cowdry, E. V. (1924). *In* "General Cytology" (E. V. Cowdry, ed.), Section VI. Univ. of Chicago Press, Chicago, Illinois.
Cowdry, E. V. (1926). *Am. Naturalist* **60**, 157.
Cowdry, E. V. (1955). "Cancer Cells," p. 677. Saunders, Philadelphia, Pennsylvania.
Cowdry, N. H. (1917). *Biol. Bull.* **33**, 196.

Crane, F. L., Hatefi, Y., Lester, R. L., and Widmer, C. (1957). *Biochim. et Biophys. Acta* **25**, 220.

Crane, F. L., Lester, R. L., Widmer, C., and Hatefi, Y. (1959). *Biochim. et Biophys. Acta* **32**, 73.

Crawford, R. B., and Morrison, M. (1959). *Federation Proc.* **18**, 209.

Dalton, A. J. (1934). *Anat. Record* **58**, 321.

Dalton, A. J., and Felix, M. D. (1953). *In* "Pigment Cell Growth" (M. Gordon, ed.), pp. 267-276. Academic Press, New York.

Dalton, A. J., and Felix, M. D. (1957). *Symposia Soc. Exptl. Biol. No.* **10**, 148.

Dalton, A. J., Kahler, H., Kelly, M. G., Lloyd, B. J., and Striebich, M. J. (1949). *J. Natl. Cancer Inst.* **9**, 439.

Dangeard, P. (1958). "Le Chondriome de la Cellule Végétale: Morphologie du Chondriome" (L. V. Heilbrunn and F. Weber, eds.) Band III, A1. Protoplasmatologia Handbuch der Protoplasma forschung. Springer, Wien.

Danielli, J. F. (1952). *In* "Cytology and Cell Physiology" (G. H. Bourne, ed.), 2nd ed., Chapter 4. Oxford Univ. Press, London and New York.

Danielli, J. F. (1958). *In* "Surface Phenomena in Chemistry and Biology" (J. F. Danielli, K. G. A. Pankhurst, and A. C. Riddiford, eds.), p. 246. Pergamon Press, New York.

Davson, H., and Danielli, J. F. (1952). "The Permeability of Natural Membranes," p. 365. Cambridge Univ. Press, London and New York.

Deane, H. W. (1944). *Anat. Record* **88**, 39.

Deane, H. W. (1958). *In* "Frontiers of Cytology" (S. L. Palay, ed.), p. 227. Yale Univ. Press, New Haven, Connecticut.

Deane, H. W., Barrnett, R. J., and Seligman, A. M. (1960). *In* "Handbuch der Histochimie" (W. Graumann and K. Neumann, eds.), Vol. VII. Enzymes. Part 1. Histochemical Methods for the Demonstration of Enzymatic Activity, p. 202. Gustaf Fischer, Stuttgart.

de Duve, C. (1959). *In* "Subcellular Particles" (T. Hayashi, ed.), p. 128. Ronald Press, New York.

de Duve, C., and Berthet, J. (1954). *Intern. Rev. Cytol.* **3**, 225.

de Duve, C., Pressman, B. C., Gianetto, R., Wattiaux, R., and Appelmans, F. (1955). *Biochem. J.* **60**, 604.

Defendi, V., and Pearson, B. (1955). *Experientia* **11**, 355.

de Lamirande, G., and Allard, C. (1957). *Can. Cancer Conf.* **2**, 83.

de Man, J. C. H. (1960). *J. Natl. Cancer Inst.* **24**, 795.

Dempsey, E. W. (1956). *J. Biophys. Biochem. Cytol.* **2** Suppl., 305.

de Robertis, E., and Sabatini, D. (1958). *J. Biophys. Biochem. Cytol.* **4**, 667.

de Robertis, E., Nowinski, W. W., and Saez, F. A. (1954). "General Cytology," 2nd ed., p. 456. Saunders, Philadelphia, Pennsylvania.

Devlin, T. M., and Lehninger, A. L. (1956). *J. Biol. Chem.* **219**, 507.

Dianzani, M. U. (1954). *Biochim. et Biophys. Acta* **14**, 514.

Doeg, K. A., Krueger, S., and Ziegler, D. M. (1960). *Biochim. et Biophys. Acta* **41**, 491.

Drews, G. (1955). *Arch. Mikrobiol.* **23**, 1.

du Buy, H. G., and Hesselbach, M. L. (1956). *J. Histochem. and Cytochem.* **4**, 363.

du Buy, H. G., and Hesselbach, M. L. (1958). *J. Natl. Cancer Inst.* **20**, 403.

du Buy, H. G., Woods, M. W., Burk, D., and Lackey, M. D. (1949). *J. Natl. Cancer Inst.* **9**, 325.

du Noüy, P. L., and Cowdry, E. V. (1927). *Anat. Record* **34**, 313.

Eakin, R. M., and Lehmann, F. E. (1957). *Wilhelm Roux Arch. Entwicklungsmech. Organ.* **150**, 177.

Edwards, S. W., and Ball, E. G. (1954). *J. Biol. Chem.* **209**, 619.

Ehret, C. F., and Powers, E. L. (1955). *Exptl. Cell Research* **9**, 241.

Ekholm, R., and Sjöstrand, F. S. (1957). *In* "Electron Microscopy" (F. S. Sjöstrand and J. Rhodin, eds.), pp. 171-173. Academic Press, New York.

Emmelot, P., Bos, C. J., and Brombacher, P. J. (1956). *Brit. J. Cancer* **10**, 188.

Engfeldt, B., Gardell, S., Hellström, J., Ivemark, B., Rhodin, J., and Strandh, J. (1958). *Acta Endocrinol.* **29**, 15.

Engström, A., and Finean, J. B. (1958). "Biological Ultrastructure." Academic Press, New York.

Ephrussi, B. (1953). "Nucleo-Cytoplasmic Relations in Microorganisms," p. 127. Oxford Univ. Press, London and New York.

Ephrussi, B., and Slonimski, P. P. (1955). *Nature* **176**, 1207.

Ephrussi, B., Slonimski, P. P., Yotsuyanagi, Y., and Tavlitski, J. (1956). *Compt. rend. lab. Carlsberg, Sér. physiol.* **26**, 87.

Ernster, L. (1956). "The enzyme organization of mitochondria and its role in the regulation of metabolic activities in animal tissues." Almquist & Wiksells, Uppsala.

Ernster, L. (1958a). *Federation Proc.* **17**, 216.

Ernster, L. (1958b). *Proc. Swedish Biochem. Soc.*, March 8, 1958, p. 6.

Ernster, L. (1959). *Biochem. Soc. Symposia.* **16**, 54.

Ernster, L., and Lindberg, O. (1958). *Ann. Rev. Physiol.* **20**, 13.

Ernster, L., and Navazio, F. (1956). *Exptl. Cell Research* **11**, 483.

Ernster, L., and Navazio, F. (1957). *Biochim. et Biophys. Acta* **26**, 408.

Ernster, L., Ikkos, D., and Luft, R. (1959). *Nature* **184**, 1851.

Essner, E., and Novikoff, A. B. (1960a). *J. Ultrastruct. Research* **3**, 374.

Essner, E., and Novikoff, A. B. (1960b). *J. Histochem. Cytochem.* **8**, 318.

Farber, E., and Bueding, E. (1956). *J. Histochem. and Cytochem.* **4**, 357.

Farber, E., and Louviere, C. D. (1956). *J. Histochem. and Cytochem.* **4**, 347.

Farber, E., Sternberg, W. H., and Dunlap, C. E. (1956a). *J. Histochem. and Cytochem.* **4**, 254.

Farber, E., Sternberg, W. H., and Dunlap, C. E. (1956b). *J. Histochem. and Cytochem.* **4**, 284.

Farquhar, M. G., Vernier, R. L., and Good, R. A. (1957). *Schweiz. med. Wochschr.* **87**, 501.

Farrant, J. L., Potter, C., Robertson, R. N., and Wilkins, M. J. (1956). *Australian J. Botany* **4**, 117.

Fauré-Fremiet, E. (1910). *Arch. anat. microscop.* **11**, 457.

Fauré-Fremiet, E., and Rouiller, C. (1958). *Exptl. Cell Research* **14**, 29.

Favard, P., and Carasso, N. (1958). *Arch. anat. microscop. et morphol. exptl.* **47**, 211.

Fawcett, D. W. (1955). *J. Natl. Cancer Inst.* **15**, 1475.

Fawcett, D. W. (1958). *Intern. Rev. Cytol.* **7**, 195.

Fawcett, D. W. (1959). *In* "Developmental Cytology" (D. Rudnick, ed.), Chapter 8. Ronald Press, New York.

Feder, N., and Sidman, R. L. (1958). *J. Biophys. Biochem. Cytol.* **4**, 593.

Feigelson, P., and Feigelson, M. (1959). *Biochim. et Biophys. Acta* **32**, 430.

Fernández-Morán, H., and Finean, J. B. (1957). *J. Biophys. Biochem. Cytol.* **3**, 725.

Finck, H. (1958). *J. Biophys. Biochem. Cytol.* **4**, 291.

Finean, J. B. (1954). *Exptl. Cell Research* **6**, 283.

Flemming, W. (1882). "Zellsubstanz, Kern und Zelltheilung." Leipzig.

Fonnesu, A., and Severi, C. (1956). *J. Biophys. Biochem. Cytol.* **2**, 293.

Franzen, Å. (1956a). "Investigations into Spermiogenesis and Sperm Morphology among Invertebrates," p. 13. Almquist and Wiksells, Uppsala.
Franzen, Å. (1956b). *Zool. Bidr. Uppsala* **31**, 356.
Frederic, J. (1954). *Ann. N.Y. Acad. Sci.* **58**, 1246.
Frederic, J. (1958). *Arch. biol. (Liège)* **69**, 167.
Frederic, J., and Chèvremont, M. (1952). *Arch. biol. (Liège)* **63**, 109.
Freeman, J. A. (1956). *J. Biophys. Biochem. Cytol.* **2**, Suppl. 353.
Frei, J., and Ryser, H. (1956). *Experientia* **12**, 105.
Gallagher, C. H., Gupta, N. H., Judah, J. D., and Reese, K. R. (1956). *J. Pathol. Bacteriol.* **72**, 193.
Gamble, J. L., Jr. (1957). *J. Biol. Chem.* **228**, 955.
Gansler, H., and Rouiller, C. (1956). *Schweiz. Z. Pathol. u. Bakteriol.* **19**, 217.
Gatenby, J. B. (1925). *Quart. J. Microscop. Sci.* **69**, 629.
Gatenby, J. B., and Beams, H. W. (eds.). (1950). "The Microtomist's Vade-Mecum (Bolles Lee)," 11th ed., p. 753. Churchill, London.
Gatenby, J. B., and Dalton, A. J. (1959). *J. Biophys. Biochem. Cytol.* **6**, 45.
Gelber, D. (1957). *J. Biophys. Biochem. Cytol.* **3**, 311.
Georgi, C. E., Militzer, W. E., and Decker, T. S. (1955). *J. Bacteriol.* **70**, 716.
Geren, B. B., and Schmitt, F. O. (1954). *Proc. Natl. Acad. Sci. U.S.* **40**, 863.
Geren, B. B., and Schmitt, F. O. (1956). *In* "Symposium on The Fine Structure of Cells," p. 251. Interscience, New York.
Gersh, I. (1959). *In* "The Cell" (J. Brachet and A. E. Mirsky, eds.), Vol. I, pp. 22-66. Academic Press, New York.
Gey, G. (1956). *Harvey Lectures Ser.* **50**, 154.
Gey, G. O., Shapras, P., Bang, F. B., and Gey, M. K. (1956). *In* "Symposium on the Fine Structure of Cells," p. 38. Interscience, New York.
Glauert, A. M., and Brieger, E. M. (1955). *J. Gen. Microbiol.* **13**, 310.
Glauert, A. M., Brieger, E. M., and Allen, J. M. (1961). *Exptl. Cell Research* **22**, 73-85.
Glauert, A. M., and Hopwood, D. A. (1960). *J. Biophys. Biochem. Cytol.* **7**, 479.
Glauert, A. M., and Hopwood, D. A. (1961). Proceed. European Regional Conf. Electron Microscopy, Delft, 1960. *Ned. Ver. Electronenmicroscopie* (in press).
Goddard, D. R., and Stafford, H. A. (1954). *Ann. Rev. Plant Physiol.* **5**, 115.
Granick, S. (1955). *In* "Handbuch der Pflanzenphysiologie" (W. Ruhland, ed.), Vol. 1, p. 507. Springer, Berlin.
Grassé, P-P., Carasso, N., and Favard, P. (1956). *Ann. sci. nat. Zool. et biol. animale* [11] **18**, 339.
Green, D. E. (1954). *In* "Chemical Pathways of Metabolism" (D. M. Greenberg, ed.), Vol. 1, pp. 27-65. Academic Press, New York.
Green, D. E. (1958a). *Harvey Lectures Ser.* **58**, 177.
Green, D. E. (1958b). *Sci. American* **199**, 56.
Green, D. E. (1959a). *Advances in Enzymol.* **21**, 73.
Green, D. E. (1959b). *In* "Subcellular Particles" (T. Hayashi, ed.), p. 84. Ronald Press, New York.
Green, D. E., and Hatefi, Y. (1961). *Science* **133**, 13.
Green, D. E., and Järnefelt, J. (1959). *Perspectives in Biol. Med.* **2**, 163.
Green, D. E., and Lester, R. L. (1959). *Federation Proc.* **18**, 987.
Green, D. E., Lester, R. L., and Ziegler, D. M. (1957). *Biochim. et Biophys. Acta* **23**, 516.
Green, D. E., and Wakil, S. J. (1960). *In* "Lipide Metabolism" (K. Bloch, ed.), pp. 1-40. Wiley, New York.

Greenstein, J. P. (1954). "Biochemistry of Cancer," 2nd ed., p. 653. Academic Press, New York.

Gross, P. R., Philpott, D. E., and Nass, S. (1960). *J. Biophys. Biochem. Cytol.* **7**, 135.

Grula, E. A., and Hartsell, S. E. (1954). *J. Bacteriol.* **68**, 498.

Grynfeltt, E., and Lafont, R. (1921). *Compt. rend. soc. biol.* **85**, 292, 406.

Guilliermond, A. (1941). "The Cytoplasm of the Plant Cell," translated by L. R. Atkinson, p. 247. Chronica Botanica, Waltham, Massachusetts.

Gustafson, T. (1954). *Intern. Rev. Cytol.* **3**, 277.

Gustafson, T., and Lenicque, P. (1952). *Exptl. Cell Research* **3**, 251.

Gustafson, T., and Lenicque, P. (1955). *Exptl. Cell Research* **8**, 114.

Hackett, D. P. (1955). *Intern. Rev. Cytol.* **4**, 143.

Hackett, D. P. (1959). *Ann. Rev. Plant Physiol.* **10**, 113.

Hagedorn, H. (1957). *Naturwissenchaften* **44**, 641.

Hanzon, V., Hermodsson, L. H., and Toschi, G. (1959). *J. Ultrastruct. Research* **3**, 216.

Harel, L., Jacob, A., and Moulé, Y. (1957). *Bull. soc. chim. biol.* **39**, 819.

Harkness, R. D. (1957). *Brit. Med. Bull.* **13**, 87.

Harman, J. W. (1956). *Intern. Rev. Cytol.* **5**, 88.

Hartmann, J. F. (1954). *Anat. Record* **118**, 19.

Harvey, E. B. (1946). *J. Exptl. Zool.* **102**, 253.

Harvey, E. B. (1953). *J. Histochem. and Cytochem.* **1**, 265.

Hay, E. D. (1958). *J. Biophys. Biochem. Cytol.* **4**, 583.

Hay, E. D. (1959). *Develop. Biol.* **1**, 555.

Herbert, E., and Potter, V. R. (1956). *J. Biol. Chem.* **222**, 453.

Hertwig, G. (1929). Allgemeine mikroskopische Anatomie der lebenden Masse. *In* "Handbuch der mikroskopischen Anatomie" (W. v. Möllendorff, ed.) Vol. I, Part 1, pp. 1-420. Springer, Berlin.

Hess, A. (1955). *Anat. Record* **123**, 399.

Hesselbach, M. L., and Du Buy, H. G. (1953). *Proc. Soc. Exptl. Biol. Med.* **83**, 62.

Himmelhoch, S. R., and Karnovsky, M. J. (1961). *J. Biophys. Biochem. Cytol.* (in press).

Hird, F. J. R., and Rowsell, E. V. (1950). *Nature* **166**, 517.

Hochstein, P. (1957). *Science* **125**, 496.

Hodge, A. J., Martin, E. M., and Morton, R. K. (1957). *J. Biophys. Biochem. Cytol.* **3**, 61.

Hoffman, H., and Grigg, G. W. (1958). *Exptl. Cell Research* **15**, 118.

Hogeboom, G. H., and Schneider, W. C. (1950). *Nature* **166**, 302.

Hogeboom, G. H., and Schneider, W. C. (1951). *Science* **113**, 355.

Hogeboom, G. H., and Schneider, W. C. (1953). *J. Biol. Chem.* **204**, 233.

Hogeboom, G. H., Kuff, E. L., and Schneider, W. C. (1957). *Intern. Rev. Cytol.* **6**, 425.

Hogeboom, G. H., Schneider, W. C., and Palade, G. E. (1948). *J. Biol. Chem.* **172**, 619.

Holter, H. (1956). *In* "Symposium on the Fine Structure of Cells," p. 71. Interscience Publishers, New York.

Holter, H. (1959). *Ann. N.Y. Acad. Sci.* **78**, 524.

Horning, E. S. (1933). *Ergeb. Enzymforsch.* **2**, 336.

Hörstadius, S. (1952). *J. Exptl. Zool.* **120**, 421.

Hovasse, R. (1948). *New Phytologist* **47**, 68.

Howatson, A. F. (1956). *J. Biophys. Biochem. Cytol.* **2**, Suppl., 363.

Howatson, A. F., and Ham, A. W. (1955). *Cancer Research* **15**, 62.
Hsu, T. C. (1959). *In* "Developmental Cytology" (D. Rudnick, ed.), p. 47. Ronald Press, New York.
Hughes-Schrader, S. (1946). *J. Morphol.* **78**, 43.
Hülsmann, W. C., and Slater, E. C. (1957). *Nature* **180**, 372.
Hülsmann, W. C., Elliot, W. B., and Rudney, H. (1958). *Biochim. et Biophys. Acta* **27**, 663.
Hülsmann, W. C., Elliot, W. B., and Slater, E. C. (1960). *Biochim. et Biophys. Acta* **39**, 267.
Hultin, T. (1953). "Studies on the Structural and Metabolic Background of Fertilization and Development." Emil Kihströms, Stockholm.
Humphrey, G. F. (1957). *Biochem. J.* **65**, 546.
Hunter, F. E. (1951). *In* "Phosphorus Metabolism" (W. D. McElroy and B. Glass, eds.), Vol. 1, p. 297. Johns Hopkins Press, Baltimore, Maryland.
Huxley, A. F. (1957). *J. Physiol. (London)* **137**, 73P.
Jeener, R. (1948). *Biochim. Biophys. Acta* **2**, 633.
Jézéquel, A. M. (1959). *J. Ultrastruct. Research* **3**, 210.
Jones, R. M. (ed.). (1950). "McClung's Handbook of Microscopical Technique," 3rd ed., Revised, p. 790. Hoeber, New York.
Kalf, G. F., Bates, H. M., and Simpson, M. V. (1959). *J. Histochem. and Cytochem.* **7**, 245.
Kaltenbach, J. C., and Harman, J. W. (1955). *Exptl. Cell Research* **8**, 435.
Kamin, H., Gibbs, R. H., and Merritt, A. D. (1957). *Federation Proc.* **16**, 202.
Kaplan, N. O., Swartz, M. N., Frech, M. E., and Ciotti, M. M. (1956). *Proc. Natl. Acad. Sci. U.S.* **42**, 481.
Kaplan, S. E., and Novikoff, A. B. (1959). *J. Histochem. and Cytochem.* **7**, 295.
Karasaki, S. (1959). *Embryologia* **4**, 247.
Karrer, H. E. (1956). *J. Biophys. Biochem. Cytol.* **2**, 241.
Kater, J. McA. (1933). *Z. Zellforsch. mikroskop. anat.* **17**, 217.
Kaye, J. (1958). *J. Morphol.* **102**, 347.
Keilin, D., and Hartree, E. F. (1947). *Biochem. J.* **41**, 500.
Kellenberger, E., Ryter, A., and Séchaud, J. (1958). *J. Biophys. Biochem. Cytol.* **4**, 671.
Kennedy, E. P. (1957). *Federation Proc.* **16**, 847.
Kennedy, E. P., and Lehninger, A. L. (1949). *J. Biol. Chem.* **179**, 957.
Kensler, C. J., and Langemann, H. (1951). *J. Biol. Chem.* **192**, 551.
Kielley, R. K., and Schneider, W. C. (1950). *J. Biol. Chem.* **185**, 869.
Kielley, W. W., and Bronk, J. R. (1958). *J. Biol. Chem.* **230**, 521.
Kielley, W. W., and Kielley, R. K. (1951). *J. Biol. Chem.* **191**, 485.
Kingsbury, B. F. (1912). *Anat. Record* **6**, 39.
Lance, A. (1958). *Ann. sci. nat. Botan. et biol. végétale* [11] **19**, 167.
Lansing, A. I. (1953). *J. Histochem. and Cytochem.* **1**, 265.
Lansing, A. I., Hillier, J., and Rosenthal, T. B. (1952). *Biol. Bull.* **103**, 294.
Lanzavecchia, G., and Le Coultre, A. (1958). *Arch. ital. anat. e Embriol.* **63**, 445.
Lardy, H. A. (1956). *Proc. Intern. Congr. Biochem. 3rd Congr. Brussels* p. 71.
Lardy, H. A., and Adler, J. (1956). *J. Biol. Chem.* **219**, 933.
Lardy, H. A., and Elvehjem, C. A. (1945). *Ann. Rev. Biochem.* **14**, 1.
La Spina, R. (1958). *Acta Embryol. Morphol. Exptl.* **2**, 66.
Lawn, A. M. (1960). *J. Biophys. Biochem. Cytol.* **7**, 197.
Lazarow, A., and Cooperstein, S. J. (1953). *J. Histochem. and Cytochem.* **1**, 234.

Lehmann, F. E. (1958). *In* "The Chemical Basis of Development" (W. D. McElroy and B. Glass, eds.), p. 73. Johns Hopkins Press, Baltimore, Maryland.

Lehninger, A. L. (1951). *J. Biol. Chem.* **190**, 345.

Lehninger, A. L. (1956). *In* "Enzymes, Units of Biological Structure and Function" (O. H. Gaebler, ed.), pp. 217-233. Academic Press, New York.

Lehninger, A. L. (1959a). *J. Biol. Chem.* **234**, 2187.

Lehninger, A. L. (1959b). *Rev. Modern Phys.* **31**, 136.

Lehninger, A. L. (1960a). *Federation Proc.* **19**, 952.

Lehninger, A. L. (1960b). *Pediatrics* **26**, 466.

Lehninger, A. L. (1960c). *Ann. N.Y. Acad. Sci.* **86**, 484.

Lehninger, A. L., and Gotterer, G. S. (1959). *J. Biol. Chem.* **235**, PC 8.

Lehninger, A. L., and Ray, B. L. (1957). *Biochim. Biophys. Acta* **26**, 643.

Lehninger, A. L., Wadkins, C. L., Cooper, C., Devlin, T. M., and Gamble, J. L., Jr. (1958). *Science* **128**, 450.

Leon, H. A., and Cook, S. F. (1956). *Science* **124**, 123.

Lettré, H. (1954). *In* "Symposium on the Fine Structure of Cells," p. 141. Interscience, New York.

Leuthardt, F., and Nielsen, H. (1951). *Helv. Chim. Acta* **34**, 1618.

Levenbook, L., and Williams, C. M. (1955). *J. Gen. Physiol.* **39**, 497.

Lever, J. D. (1956a). *Endocrinology* **58**, 163.

Lever, J. D. (1956b). *J. Biophys. Biochem. Cytol.* **2**, Suppl., 313.

Lever, J. D. (1957). *J. Anat.* **91**, 73.

Lewis, M. R., and Lewis, W. H. (1914-15). *Am. J. Anat.* **17**, 339.

Lillie, F. R., and Just, E. E. (1924). *In* "General Cytology" (E. V. Cowdry, ed.), Section VIII. Univ. of Chicago Press, Chicago, Illinois.

Lindberg, O., and Ernster, L. (1954). *In* "Protoplasmatologia" (L. V. Heilbrunn and F. Weber, eds.), Band III/A/4: Chemistry and Physiology of Mitochondria and Microsomes. Springer, Vienna.

Lindegren, C. C., Nagai, S., and Nagai, H. (1958). *Nature* **182**, 446.

Löw, H., Siekevitz, P., Ernster, L., and Lindberg, O. (1958). *Biochim. et Biophys. Acta* **29**, 392.

Low, F. N. (1956). *J. Biophys. Biochem. Cytol.* **2**, Suppl. 337.

Low, F. N., and Freeman, J. A. (1958). "Electron Microscopic Atlas of Normal and Leukemic Blood." McGraw-Hill, New York.

Lowe, C. U., and Lehninger, A. L. (1955). *J. Biophys. Biochem. Cytol.* **1**, 89.

Lowe, C. U., and Williams, W. L. (1953). *Proc. Soc. Exptl. Biol. Med.* **84**, 70.

Lowe, C. U., MacKinney, D., and Sarkaria, D. (1955). *J. Biophys. Biochem. Cytol.* **1**, 237.

Ludewig, S., and Chanutin, A. (1950). *Arch. Biochem.* **29**, 441.

Ludford, R. J. (1952). *In* "Cytology and Cell Physiology" (G. H. Bourne, ed.), 2nd ed., Chapter 9. Oxford Univ. Press, London and New York.

Luft, J. H. (1956). *J. Biophys. Biochem. Cytol.* **2**, 799.

Luft, J. H., and Hechter, O. (1957). *J. Biophys. Biochem. Cytol.* **3**, 615.

Lund, H., Vatter, A. E., and Hanson, J. B. (1958). *J. Biophys. Biochem. Cytol.* **4**, 87.

Maaløe, O., and Birch-Anderson, A. (1956). *Symposium Soc. Gen. Microbiol.* **6**, 261.

MacCardle, R. C., and Congdon, C. C. (1955). *Am. J. Pathol.* **31**, 725.

MacFarlane, M. G., and Datta, N. (1954). *Brit. J. Exptl. Pathol.* **35**, 191.

MacFarlane, M. G., and Spencer, A. G. (1953). *Biochem. J.* **54**, 569.

McLean, J. R., Cohn, G. L., Brandt, I. K., and Simpson, M. V. (1958). *J. Biol. Chem.* **233**, 657.

McMurray, W. C., Maley, G. F., and Lardy, H. (1958). *J. Biol. Chem.* **230**, 219.
McMurray, W. C., Strickland, K. P., Berry, J. F., and Rossiter, R. J. (1957). *Biochem. J.* **66**, 634.
Malamed, S., and Recknagel, R. O. (1959). *J. Biol. Chem.* **234**, 3027.
Maley, G. F., and Johnson, D. (1957). *Biochim. Biophys. Acta* **26**, 522.
Maley, G. F., and Lardy, H. A. (1955). *J. Biol. Chem.* **215**, 377.
Manuelidis, E. E. (1958). *In* "Frontiers of Cytology" (S. L. Palay, ed.), pp. 417-446. Yale Univ. Press, New Haven, Connecticut.
Marinetti, G. V., Erbland, J., Albrecht, M., and Stotz, E. (1957). *Biochim. et Biophys. Acta* **26**, 130.
Marinetti, G. V., Erbland, J., and Kochen, J. (1958). *Federation Proc.* **17**, 269.
Marr, A. G. (1960). *In* "The Bacteria" (I. C. Gunsalas and R. Y. Stanier, eds.), Vol. I, pp. 35-96. Academic Press, New York.
Marshak, A. (1955). *Intern. Rev. Cytol.* **4**, 103.
Mercer, F. V. (1956). *Proc. Linnean Soc. N.S. Wales* **81**, 4.
Meves, F. (1912). *Arch. mikroskop. Anat. u. Entwicklungsmech.* **80**, 81.
Meves, F. (1918). *Arch. mikroskop. Anat. u. Entwicklungsmech.* **92**, 41.
Michaelis, L. (1900). *Arch. Mikroskop. Anat. u. Entwicklungsmech.* **55**, 558.
Miller, R. A. (1953). *Am. J. Anat.* **92**, 329.
Miller, F. (1960). *J. Biophys. Biochem. Cytol.* **8**, 689.
Miller, F., and Sitte, H. (1955). *Verhandl. deut. Ges. Pathol.* **39**, 183.
Millerd, A. (1956). *In* "Handbuch der Pflanzenphysiologie." (W. Ruhland, ed.), Vol. 2, p. 573. Springer, Berlin.
Mitchell, P. (1959). *Symposium Biochem. Soc.* **16**, 73.
Mitchell, P., and Moyle, J. (1956a). *Discussions Faraday Soc.* **21**, 258.
Mitchell, P., and Moyle, J. (1956b). *Symposium Soc. Gen. Microbiol.* **6**, 150.
Montgomery, T. H. (1912). *Biol. Bull.* **22**, 309.
Moore, D. H., and Ruska, H. (1957). *J. Biophys. Biochem. Cytol.* **3**, 261.
Moriber, L. (1956). *J. Morphol.* **99**, 271.
Moses, M. (1958). *Anat. Record* **130**, 343.
Mudd, S. (1953a). *J. Histochem. and Cytochem.* **1**, 248.
Mudd, S. (1953b). *Symposium Soc. Gen. Microbiol.* **6**,
Mudd, S. (1954). *Ann. Rev. Microbiol.* **8**, 1.
Mudd, S. (1956). *Bacteriol. Revs.* **20**, 268.
Mudd, S., Takeya, K., and Henderson, H. J. (1956). *J. Bacteriol.* **72**, 767.
Mudd, S., Kamata, T., Payne, J. I., Sall, T., and Takagi, A. (1960). *Nature* (In press).
Mühlethaler, K. (1955). *Protoplasma* **45**, 264.
Mühlethaler, K., and Frey-Wyssling, A. (1959). *J. Biophys. Biochem. Cytol.* **6**, 507.
Mühlethaler, K., Müller, A. F., and Zollinger, H. U. (1950). *Experientia* **6**, 16.
Müller, A. F., and Leuthardt, F. (1950). *Helv. Chim. Acta* **33**, 268.
Murray, M. R., and Kopech, G. (1953). "A Bibliography of the Research in Tissue Culture 1884 to 1950," Vol. I. Academic Press, New York.
Murray, R. G. E. (1960). *In* "The Bacteria" (I. C. Gunsalas and R. Y. Stanier, eds.), Vol. I, pp. 443-468. Academic Press, New York.
Myers, D. K., and Slater, E. C. (1957). *Biochem. J.* **67**, 558.
Mutolo, V., and Abrignani, F. (1957). *Brit. J. Cancer* **11**, 590.
Nachlas, M. M., Tsou, K.-C., de Souza, E., Cheng, C.-S., and Seligman, A. M. (1957). *J. Histochem. and Cytochem.* **5**, 420.

Nachlas, M. M., Walker, D. C., and Seligman, A. M. (1958a). *J. Biophys. Biochem. Cytol.* **4**, 29.
Nachlas, M. M., Crawford, D. T., Goldstein, T. P., and Seligman, A. M. (1958b). *J. Histochem. and Cytochem.* **6**, 445.
Nachlas, M. M., Margulies, S. I., and Seligman, A. M. (1960). *J. Biol. Chem.* **235**, 2739.
Nakano, E., and Monroy, A. (1958). *Exptl. Cell Research* **14**, 236.
Napolitano, L., and Fawcett, D. (1958). *J. Biophys. Biochem. Cytol.* **4**, 685.
Nason, A., Donaldson, K. O., and Lehman, I. R. (1957). *Trans. N.Y. Acad. Sci.* [2] **20**, 27.
Nath, V. (1956). *Intern. Rev. Cytol.* **5**, 395.
Nath, V., and Gill, G. K. (1950). *Research Bull. East Panjab Univ.* **1**, 1.
Nath, V., and Rishi, R. (1953). *Research Bull. East Panjab Univ.* **31**, 67.
Newcomer, E. H. (1940). *Botan. Rev.* **6**, 85.
Newcomer, E. H. (1951). *Botan. Rev.* **17**, 53.
Niklowitz, W. (1958). *Zentr. Bakteriol. Parasitenk.* **173**, 12.
Niklowitz, W., and Drews, G. (1957). *Arch. Mikrobiol.* **27**, 150.
Nilsson, O. (1958). *J. Ultrastruct. Research* **1**, 375.
Noël, R. (1923). *Arch. anat. microscop.* **19**, 1.
Noirot-Timothée, C. (1959). *Ann. sci. nat. Zool. et biol. animale* [12] **1**, 265.
Novikoff, A. B. (1955). *In* "Analytical Cytology" (R. C. Mellors, ed.), Chapter 2. McGraw-Hill, New York.
Novikoff, A. B. (1956). *Proc. Intern. Congr. Biochem. 3rd Congr. Brussels,* p. 315.
Novikoff, A. B. (1957a). *Symposia Soc. Exptl. Biol. No.* **10**, 92.
Novikoff, A. B. (1957b). *Cancer Research* **17**, 1010.
Novikoff, A. B. (1959a). *In* "Subcellular Particles" (T. Hayashi, ed.), p. 1. Ronald Press, New York.
Novikoff, A. B. (1959b). *Bull. N.Y. Acad. Med.* [2] **35**, 67.
Novikoff, A. B. (1959c). *In* "Analytical Cytology" (R. C. Mellors, ed.), 2nd ed., p. 69. McGraw-Hill, New York.
Novikoff, A. B. (1959d). *J. Histochem. and Cytochem.* **7**, 240.
Novikoff, A. B. (1959e). *Biol. Bull.* **117**, 385.
Novikoff, A. B. (1959f). *J. Histochem. and Cytochem.* **7**, 301.
Novikoff, A. B. (1960a). *Acta Union contra le Cancer* **16**, 966.
Novikoff, A. B. (1960b). *In* "Developing Cell Systems and Their Control" (D. Rudnick, ed.), p. 167. Ronald Press, New York.
Novikoff, A. B. (1960c). *In* "Biology of Pyelonephritis" (E. Quinn and E. Kass, eds.), p. 113. Little, Brown, Boston, Massachusetts.
Novikoff, A. B. (1960d). *In* "Cell Physiology of Neoplasia" The University of Texas Press, p. 219. Austin, Texas.
Novikoff, A. B. (*in press*). *Proc. 1st Intl. Congr. Histochem. Cytochem., Paris, 1960.*
Novikoff, A. B., and Arase, M. M. (1958). *J. Histochem. and Cytochem.* **6**, 397.
Novikoff, A. B., and Essner, E. (1960). *Am. J. Med.* **29**, 102.
Novikoff, A. B., and Masek, B. (1958). *J. Histochem. and Cytochem.* **6**, 217.
Novikoff, A. B., Hecht, L., Podber, E., and Ryan, J. (1952). *J. Biol. Chem.* **194**, 153.
Novikoff, A. B., Podber, E., Ryan, J., and Noe, E. (1953). *J. Histochem. and Cytochem.* **1**, 27.
Novikoff, A. B., Beaufay, H., and de Duve, C. (1956). *J. Biophys. Biochem. Cytol.* **2**, Suppl., 179.

Novikoff, A. B., Hausman, D. H., and Podber, E. (1958). *J. Histochem. and Cytochem.* **6**, 61.
Novikoff, A. B., Shin, W.-Y., and Drucker, J. (1961). *J. Biophys. Biochem. Cytol.* (in press).
Novikoff, A. B., Jedeikin, L., Arase, M., and Pritzker, M. (in preparation).
Nygaard, A. P. (1953). *J. Biol. Chem.* **204**, 655.
Nygaard, A. P. (1954). *Exptl. Cell Research* **6**, 453.
O'Connor, R. J. (1957). *Intern. Rev. Cytol.* **6**, 343.
Okada, S., and Peachey, L. D. (1957). *J. Biophys. Biochem. Cytol.* **3**, 239.
Ornstein, L. (1956). *J. Biophys. Biochem. Cytol.* **2**, 351.
Ornstein, L., and Pollister, A. W. (1952). *Trans. N.Y. Acad. Sci.* [2] **14**, 194.
Packer, L. (1960). *J. Biol. Chem.* **235**, 242.
Packer, L., and Golder, R. H. (1959). *Biochim. et Biophys. Acta* **32**, 281.
Paigen, K. (1954). *J. Biol. Chem.* **206**, 945.
Palade, G. E. (1952). *Anat. Record* **114**, 427.
Palade, G. E. (1953). *J. Histochem. Cytochem.* **1**, 188.
Palade, G. E. (1956a). *In* "Enzymes: Units of Biological Structure and Function" (O. H. Gaebler, ed.), Chapter 9. Academic Press, New York.
Palade, G. E. (1956b). *J. Biophys. Biochem. Cytol.* **2**, Suppl. 85.
Palade, G. E. (1959). *In* "Subcellular Particles" (T. Hagashi, ed.), p. 64. Ronald Press, New York.
Palade, G. E., and Schidlowsky, G. (1958). *Anat. Record* **130**, 352.
Palay, S. L. (1956). *J. Biophys. Biochem. Cytol.* **2**, Suppl., 193.
Palay, S. L., and Palade, G. E. (1955). *J. Biophys. Biochem. Cytol.* **1**, 69.
Panagos, S., Beyer, R. E., and Masoro, E. J. (1958). *Biochim. Biophys. Acta* **29**, 204.
Pappas, G. D., and Brandt, P. W. (1959). *J. Biophys. Biochem. Cytol.* **6**, 85.
Park, J. H., Meriwether, B. P., and Park, C. R. (1958). *Biochim. Biophys. Acta* **28**, 662.
Pasteels, J. J. (1958). *In* "The Chemical Basis of Development" (W. D. McElroy, and B. Glass, eds.), p. 381. Johns Hopkins Press, Baltimore, Maryland.
Pasteels, J. J., Castiaux, P., and Vandermeerssche, G. (1958). *J. Biophys. Biochem. Cytol.* **4**, 575.
Payne, F. (1952). *J. Morphol.* **91**, 555.
Payne, F. (1957). *J. Morphol.* **101**, 89.
Pearse, A. G. E. (1957). *J. Histochem. and Cytochem.* **5**, 515.
Pearse, A. G. E. (1958). *J. Clin. Pathol.* **11**, 520.
Pearse, A. G. E., Scarpelli, D. G., and Brown, S. (1958). *Biochem. J.* **68**, 18.
Pease, D. C. (1956). *J. Biophys. Biochem. Cytol.* **2**, Suppl., 203.
Penefsky, H. S., Pullman, M. E., Datta, A., and Racker, E. (1960). *J. Biol. Chem.* **235**, 11.
Penn, N. W. (1959). *Federation Proc.* **18**, 301.
Penn, N. W. (1960). *Biochim. Biophys. Acta* **37**, 55.
Perner, E. S. (1958). *In* "Protoplasmatologia" (L. V. Heilbrunn and F. Weber, eds.), Band III, A2 "Die Sphärosomen der Pflanzenzelle." Springer, Wien.
Perry, R. P., Thorell, B., Åkerman, L., and Chance, B. (1959). *Nature* **184**, 929.
Perry, S. V. (1960). *In* "Comparative Biochemistry. A Comprehensive Treatise" (M. Florkin and H. S. Mason, eds.), pp. 245-340. Academic Press, New York.
Petermann, M. L., and Hamilton, M. G. (1958). *J. Biophys. Biochem. Cytol.* **4**, 771.
Petrushka, E., Quastel, J. H., and Scholefield, P. G. (1957). *Can. Cancer Conf.* **2**, 106.

Plaut, G. W. E., and Sung, S.-C. (1954). *J. Biol. Chem.* **207**, 305.
Polis, B. D., and Shmukler, H. W. (1957). *J. Biol. Chem.* **227**, 419.
Pollard, C. J., and Bieri, J. G. (1960). *J. Biol. Chem.* **235**, 1178.
Pollister, A. W. (1930). *J. Morphol. and Physiol.* **49**, 455.
Pollister, A. W. (1932). *Am. J. Anat.* **50**, 179.
Pollister, A. W. (1941). *Physiol. Zool.* **14**, 268.
Pollister, A. W. (1957). *Ann. N.Y. Acad. Sci.* **69**, 580.
Pomerat, C. M., Lefeber, G., and Smith, Mc D. (1954). *Ann. N.Y. Acad. Sci.* **58**, 1311.
Porter, K. R., and Bruni, C. (1959). *Cancer Research* **19**, 997.
Porter, K. R., and Palade, G. E. (1957). *J. Biophys. Biochem. Cytol.* **3**, 269.
Porter, K. R., Claude, A., and Fullam, E. (1945). *J. Exptl. Med.* **81**, 233.
Potter, V. R. (1958). *Federation Proc.* **17**, 691.
Potter, V. R. (1960). *Acta Unio Intern. contra Cancrum* **16**, 27.
Potter, V. R., Recknagel, R. O., and Hurlbert, R. B. (1951). *Federation Proc.* **10**, 646.
Powers, E. L., Ehret, C. F., and Roth, L. E. (1955). *Biol. Bull.* **108**, 182.
Powers, E. L., Ehret, C. F., Roth, L. E., and Minick, O. T. (1956). *J. Biophys. Biochem. Cytol.* **2**, Suppl., 341.
Pressman, B. C., and Lardy, H. A. (1956). *Biochim. Biophys. Acta* **21**, 458.
Price, J. M., Miller, E. C., Miller, J. A., and Weber, G. M. (1950). *Cancer Research* **10**, 18.
Pullman, M. E., Penefsky, H., and Racker, E. (1959). *Arch. Biochem. Biophys.* **76**, 227.
Raaflaub, J. (1953). *Helv. Physiol. et Pharmacol. Acta* **11**, 142.
Readfarn, E. R., and Pumphrey, A. M. (1960). *Biochem. J.* **76**, 64.
Rebhun, L. I. (1956). *J. Biophys. Biochem. Cytol.* **2**, 93.
Recknagel, R. O., and Malamed, S. (1958). *J. Biol. Chem.* **232**, 705.
Recknagel, R. O., Stadler, J., and Litteria, M. (1958). *Federation Proc.* **17**, 129.
Regaud, C. (1908). *Compt. rend. soc. biol.* **65**, 718.
Remmert, L. F., and Lehninger, A. L. (1959). *Proc. Natl. Acad. Sci. U.S.* **45**, 1.
Rendi, R. (1959). *Exptl. Cell Research* **17**, 585.
Revel, J. P., Ito, S., and Fawcett, D. (1958). *J. Biophys. Biochem. Cytol.* **4**, 495.
Reverberi, G. (1957). *In* "The Beginnings of Embryonic Development" (A. Tyler, R. C. von Borstel, and C. B. Metz, eds.), pp. 319-340. Am. Assoc. Advance. Sci., Washington, D.C.
Reverberi, G. (1958). *Acta Embryol. Morphol. exptl.* **2**, 79.
Reynafarje, B., and Potter, V. R. (1957). *Cancer Research* **17**, 1112.
Rhoades, M. M. (1955). *In* "Handbuch der Pflanzenphysiologie" (W. Ruhland, ed.), Vol. I, p. 19. Springer, Berlin.
Rhodin, J. (1954). "Correlation of Ultrastructural Organization and Function in Normal and Experimentally Changed Proximal Convoluted Tubule Cells of the Mouse Kidney," 76 pp. Aktiebolaget Godvil, Stockholm.
Rhodin, J. (1958). *Intern. Rev. Cytol.* **7**, 485.
Rhodin, J., and Dalhamn, T. (1956). *Z. Zellforsch. u. mikroskop. Anat.* **44**, 345.
Richter, G. W. (1957). *J. Exptl. Med.* **106**, 203.
Richter, G. W. (1959). *J. Exptl. Med.* **109**, 197.
Ries, (1939). *Arch. exptl. Zellforsch. Gewebezücht.* **23**, 95.
Rinehart, J. F. (1955). *Am. J. Clin. Pathol.* **25**, 605.

Rivière, M. R., Chouronlinkov, I., and Guérin, M. (1959). *Bull. Assoc. Cancer* **46**, 736.
Roberts, H. S., Jr. (1953). *J. Histochem. and Cytochem.* **1**, 268.
Robertson, J. D. (1957). *J. Biophys. Biochem. Cytol.* **3**, 1043.
Robertson, J. D. (1958). *J. Biophys. Biochem. Cytol.* **4**, 349.
Robertson, J. D. (1959). *Symposium Biochem. Soc.* **16**, 3.
Robertson, J. D. (1960). *Anat. Record* **136**, 346.
Robertson, R. N. (1957). *Endeavour* **16**, 193.
Rose, G. G. (1957). *J. Biophys. Biochem. Cytol.* **3**, 697.
Rouiller, C. (1960). *Intern. Rev. Cytol.* **9**, 227.
Rouiller, C., and Bernhard, W. (1956). *J. Biophys. Biochem. Cytol.* **2**, Suppl., 355.
Rouiller, C., Fauré-Fremiet, E., and Gauchery, M. (1957). *In* "Electron Microscopy" (F. S. Sjöstrand and J. Rhodin, eds.), pp. 216-218. Academic Press, New York.
Rubinstein, D., and Denstedt, O. F. (1953). *J. Biol. Chem.* **204**, 623.
Rudkin, G. T., and Schultz, J. (1956). *Cold Spring Harbor Symp. Quant. Biol.* **21**, 303.
Rudzinska, M. A., and Porter, K. R. (1953). *Anat. Record* **115**, 363.
Rudzinska, M. A., and Trager, W. (1957). *J. Protozool.* **4**, 190.
Rudzinska, M. A., and Trager, W. (1959). *J. Biophys. Biochem. Cytol.* **6**, 103.
Ruska, H., Moore, D. H., and Weinstock, J. (1957). *J. Biophys. Biochem. Cytol.* **3**, 249.
Ryser, H., Aebi, H., and Zuppinger, A. (1954). *Experientia* **10**, 304.
Ryter, A., and Kellenberger, E. (1958). *Z. Naturforsch.* **13b**, 597.
Sacktor, B., O'Neill, J. J., and Cochran, D. G. (1958). *J. Biol. Chem.* **233**, 1233.
Sager, R. (1959). *Brookhaven Symposia in Biol. No.* **11**.
Sager, R., and Palade, G. E. (1957). *J. Biophys. Biochem. Cytol.* **3**, 463.
Sano, S., Inoue, S., Tanabe, Y., Sumiya, C., and Koike, S. (1958). *Science* **129**, 275.
Schäfer, W. (1959). *In* "The Viruses. Biochemical, Biological, and Biophysical Properties" (F. M. Burnet and W. M. Stanley, eds.). Vol. 1, Chapter 8. Academic Press, New York.
Schneider, W. C. (1948). *J. Biol. Chem.* **176**, 259.
Schneider, W. C. (1953). *J. Histochem. and Cytochem.* **1**, 212.
Schneider, W. C. (1956). *In Proc. Intern. Congr. Biochem. 3rd Congr. Brussels* p. 305.
Schneider, W. C. (1959). *Advances in Enzymol.* **21**, 1.
Schneider, W. C., and Hogeboom, G. H. (1950). *J. Biol. Chem.* **183**, 123.
Schneider, W. C., and Hogeboom, G. H. (1951). *Cancer Research* **11**, 1.
Schneider, W. C., and Hogeboom, G. H. (1956). *Ann. Rev. Biochem.* **25**, 201.
Schneider, W. C., Claude, A., and Hogeboom, G. H. (1948). *J. Biol. Chem.* **172**, 451.
Schneider, W. C., Hogeboom, G. H., Shelton, E., and Striebich, M. J. (1953). *Cancer Research* **13**, 285.
Scharrer, E. (1945). *J. Comp. Neurol.* **83**, 237.
Schulz, H., Löw, H., Ernster, L., and Sjöstrand, F. S. (1957). *In* "Electron Microscopy: Proceedings of the Stockholm Conference September 1956" (F. S. Sjöstrand and J. Rhodin, eds.), pp. 134-137. Academic Press, New York.
Schumacher, H.-H. (1957). *Science* **125**, 501.
Sedar, A., and Porter, K. R. (1955). *J. Biophys. Biochem. Cytol.* **1**, 583.
Selby, C. C. (1959). *In* "Analytical Cytology" (R. C. Mellors, ed.), 2nd ed., p. 273. McGraw-Hill, New York.

Seno, S., and Yoshizawa, K. (1960). *J. Biophys. Biochem. Cytol.* **8**, 617.

Setälä, K., Merenmies, L., Niskanen, E. E., Nyholm, M., and Stzernvall, L. (1960). *J. Natl. Cancer Inst.* **25**, 1155.

Shank, R. E., Morrison, G., Cheng, C. H., Karl, I., and Schwartz, R. (1959). *J. Histochem. and Cytochem.* **7**, 237.

Sharp, L. W. (1934). "Introduction to Cytology," 3rd Ed. McGraw-Hill, New York.

Shaver, J. R. (1955). *Experentia* **11**, 351.

Shaver, J. R. (1956). *Exptl. Cell Research* **11**, 548.

Shaver, J. R. (1957). *In* "The Beginnings of Embryonic Development" (A. Tyler, R. C. von Borstel, and C. B. Metz, eds.), pp. 263-290. Am. Assoc. Advance. Sci., Washington, D.C.

Sheldon, H., and Zetterquist, H. (1956). *Exptl. Cell Research* **10**, 225.

Shelton, E., Schneider, W. C., and Striebich, M. J. (1953). *Exptl. Cell Research* **4**, 32.

Shepherd, J. A., and Kalnitsky, G. (1951). *J. Biol. Chem.* **192**, 1.

Shinohara, C., Fukushi, K., and Suzuki, J. (1957). *J. Bacteriol.* **74**, 413.

Shinohara, C., Fukushi, K., Suzuki, J., and Sato, K. (1958). *J. Electronmicroscopy (Chiba)* **6**, 47.

Siekevitz, P. (1952). *J. Biol. Chem.* **195**, 549.

Siekevitz, P., and Potter, V. R. (1953). *J. Biol. Chem.* **201**, 1.

Siekevitz, P., and Watson, M. (1956). *J. Biophys. Biochem. Cytol.* **2**, 653.

Siekevitz, P., and Watson, M. L. (1957). *Biochim. et Biophys. Acta* **25**, 274.

Siekevitz, P., Löw, H., Ernster, L., and Lindberg, O. (1958). *Biochim. et Biophys. Acta* **29**, 378.

Singer, T. P., and Kearney, E. B. (1954). *Biochim. et Biophys. Acta* **15**, 151.

Singer, T. P., and Kearney, E. B. (1957). *Methods of Biochem. Anal.* **4**, 307.

Singer, T. P., Kearney, E. B., and Massey, V. (1956). *In* "Enzymes: Units of Biological Structure and Function" (O. H. Gaebler, ed.), Chapter 20. Academic Press, New York.

Sitte, P. (1958). *Protoplasma* **49**, 447.

Sjöstrand, F. S. (1953). *Nature* **171**, 30.

Sjöstrand, F. S. (1956a). *Intern. Rev. Cytol.* **5**, 455.

Sjöstrand, F. S. (1956b). *In* "Physical Techniques in Biological Research" (G. Oster and A. W. Pollister, eds.), Vol. 3, Chapter 6. Academic Press, New York.

Sjöstrand, F. S. (1959). *J. Ultrastruct. Research* **3**, 210.

Sjöstrand, F. S., and Baker, R. F. (1958). *J. Ultrastruct. Research* **1**, 239.

Sjöstrand, F. S., and Rhodin, J. (1953). *Exptl. Cell. Research* **4**, 426.

Slonimski, P. P. (1953). "Formation des enzymes respiratoires chez le levure." Masson, Paris.

Smith, A. L., and Lester, R. L. (1960). *Federation Proc.* **19**, 34.

Smith, C. A., and Dempsey, E. W. (1957). *Am. J. Anat.* **100**, 337.

Smith, D. S. (1961). *J. Biophys. Biochem. Cytol.* **2**, *Suppl.* (In press).

Sonneborn, T. M. (1951). *In* "Genetics in the 20th Century" (L. C. Dunn, ed.), p. 291. Macmillan, New York.

Solomon, J. B. (1959). *Develop. Biol.* **1**, 182.

Sorokin, H. (1941). *Am. J. Botany* **28**, 476.

Spater, H. W., Novikoff, A. B., and Masek, B. (1958). *J. Biophys. Biochem. Cytol.* **4**, 765.

Spiro, M. J., and McKibbin, J. M. (1956). *J. Biol. Chem.* **219**, 643.

Stanbury, S. W., and Mudge, G. H. (1953). *Proc. Soc. Exptl. Biol.* **82**, 675.
Stäubli, W. (1960). *Compt. rend. acad. sci.* **250**, 1137.
Steffen, K. (1955). *In* "Handbuch der Pflanzenphysiologie" (W. Ruhland, ed.), Vol. 1, p. 574. Springer, Berlin.
Steinert, M. (1960). *J. Biophys. Biochem. Cytol.* **8**, 542.
Stern, H., and Timonen, S. (1954). *J. Gen. Physiol.* **38**, 41.
Sternberg, W. H., Farber, E., and Dunlap, C. E. (1956). *J. Histochem. and Cytochem.* **4**, 266.
Stoeckenius, W. (1959). *J. Biophys. Biochem. Cytol.* **5**, 491.
Stoeckenius, W., Schulman, J. H., and Prince, L. M. (1960). *Kolloid-Z.* **169**, 170.
Storck, R., and Wachsman, J. T. (1957). *J. Bacteriol.* **73**, 784.
Striebich, M. J., Shelton, E., and Schneider, W. C. (1953). *Cancer Research* **13**, 279.
Symposium: The Structure and Biochemistry of Mitochondria. (1953). *J. Histochem. and Cytochem.* **1**, 179.
Tapley, D. F. (1956). *J. Biol. Chem.* **222**, 325.
Tapley, D. F., and Cooper, C. (1956). *J. Biol. Chem.* **222**, 341.
Tapley, D. F., Cooper, C., and Lehninger, A. L. (1955). *Biochim. Biophys. Acta* **18**, 597.
Tedeschi, H. (1959). *J. Biophys. Biochem. Cytol.* **6**, 241.
Tedeschi, H., and Harris, D. L. (1955). *Arch. Biochem. Biophys.* **58**, 52.
Tennent, D. H., Gardiner, M. S., and Smith, D. E. (1931). *Carnegie Inst. Wash. Publ. No.* **413**, 1.
Thorell, B., and Chance, B. (1959). *Nature* **184**, 934.
Tjio, J. H., and Puck, T. T. (1958). *J. Exptl. Med.* **108**, 259.
Tobioka, M., and Biesele, J. J. (1956). *J. Biophys. Biochem. Cytol.* **2**, Suppl., 319.
Tokuyasu, K., and Yamada, E. (1959). *J. Biophys. Biochem. Cytol.* **5**, 123.
Tsou, K. C., Cheng, C. S., Nachlas, M. M., and Seligman, A. M. (1956). *J. Am. Chem. Soc.* **78**, 6139.
Turian, G., and Kellenberger, E. (1956). *Exptl. Cell. Research* **11**, 417.
Van Bekkum, D. W. (1955). *Biophys. Biochim. Acta* **16**, 437.
van Iterson, W. (1961). Proceed. European Regional Conf. Electron Microscopy, Delft, 1960. *Ned. Ver. voor Electronenmicroscopie* (in press).
Vendrely, R. (1955). *Intern. Rev. Cytol.* **4**, 115.
Wadkins, C. L. (1959). *Federation Proc.* **18**, 346.
Wagner, R. P., and Mitchell, H. K. (1955). "Genetics and Metabolism," pp. 444. Wiley, New York.
Wallace, B. J. (1960). *J. Histochem. Cytochem.* **8**, 105.
Warburg, O. (1956). *Science* **123**, 309.
Ward, R. T. (1958). *J. Histochem. and Cytochem.* **6**, 398.
Ward, R. T. (1959). *J. Appl. Phys.* **30**, 2040.
Wasserman, F. (1954). *Ann. N.Y. Acad. Sci.* **58**, 1256.
Watson, M. L. (1952). Univ. of Rochester Atomic Energy Project, unclassified report—U. R. 185.
Watson, M. L., and Siekevitz, P. (1956). *J. Biophys. Biochem. Cytol.* **2**, 639.
Wattenberg, L. W., and Gronvall, J. A. (1960). *Proc. Soc. Exptl. Biol. Med.* **104**, 394.
Wattenberg, L. W., and Leong, J. L. (1960). *J. Histochem. Cytochem.* **8**, 296.
Weber, R. (1958). *Wilhelm Roux Arch. Entwicklungsmech. Organ.* **150**, 542.
Weibull, C. (1953a). *J. Bacteriol.* **66**, 137.
Weibull, C. (1953b). *J. Bacteriol.* **66**, 696.

Weibull, C. (1956). *Symposium Soc. Gen. Microbiol.* **6**, 111.

Weibull, C., Beckman, H., and Bergström, L. (1959). *J. Gen. Microbiol.* **20**, 519.

Weibull, C., and Thorsson, K. G. (1957). *In* "Electron Microscopy. Proceedings of the Stockholm Conference September 1956" (F. S. Sjöstrand and J. Rhodin, eds.), p. 266. Academic Press, New York.

Weiler, E. (1956). *Z. Naturforsch.* **11b**, 31.

Weinbach, E. C., and Garbus, J. (1956). *Nature* **178**, 1225.

Weinhouse, S. (1955). *Advances in Cancer Research* **3**, 269.

Weinhouse, S. (1956). *Science* **124**, 267.

Weiss, J. M. (1955). *J. Exptl. Med.* **102**, 783.

Weiss, J., and Lansing, A. I. (1953). *Proc. Soc. Exptl. Biol. Med.* **82**, 460.

Weissenfels, N. (1958). *Z. Naturforsch.* **136**, 182.

Wellings, S. R., and Siegel, B. V. (1959). *J. Ultrastruct. Research* **3**, 147.

Werkheiser, W. C., and Bartley, W. (1957). *Biochem. J.* **66**, 79.

Whaley, W. G., Kephart, J. E., and Mollenhauer, H. H. (1959). *Am. J. Botany* **46**, 743.

Whaley, W. G., Mollenhauer, H. H., and Leech, J. H. (1960). *Am. J. Botany* **47**, 401.

Whittam, R., and Davies, R. E. (1954). *Biochem. J.* **56**, 445.

Williams, J. N., Jr. (1952). *J. Biol. Chem.* **194**, 139.

Wilson, E. B. (1916). *Proc. Natl. Acad. Sci. U.S.* **2**, 321.

Wilson, E. B. (1928). "The Cell in Development and Heredity," 3rd ed. with corrections, p. 1232. Macmillan, New York.

Wilson, E. B., and Pollister, A. W. (1937). *J. Morphol.* **60**, 407.

Wilson, J. W. (1953). *J. Histochem. and Cytochem.* **1**, 267.

Wojtczak, L., and Wojtczak, A. B. (1960). *Biochim. et Biophys. Acta* **39**, 277.

Wolken, J. J. (1959). *Ann. Rev. Plant Physiol.* **10**, 71.

Wolken, J. J., and Palade, G. E. (1953). *Ann. N.Y. Acad. Sci.* **56**, 873.

Woods, M. (1956). *J. Natl. Cancer Inst.* **17**, 615.

Woods, M. W., and du Buy, H. G. (1951). *J. Natl. Cancer Inst.* **11**, 1105.

Woods, M. W., du Buy, H. G., Burk, D., and Hesselbach, M. L. (1949). *J. Natl. Cancer Inst.* **9**, 311.

Yamada, E. (1955). *J. Biophys. Biochem. Cytol.* **1**, 445.

Yasuzumi, G., Fujimura, W., and Ishida, G. (1958). *Exptl. Cell Research* **14**, 268.

Yasuzumi, G., and Tanaka, H. (1958). *J. Biophys. Biochem. Cytol.* **4**, 621.

Yotsuyanagi, Y. (1955). *Nature* **176**, 1208.

Yotsuyanagi, Y. (1959). *Compt. rendus acad. sci.* **248**, 274.

Zelander, T. (1959). *J. Ultrastruct. Research, Suppl.* **2**, pp. 111.

Zetterquist, H. (1956). The Ultrastructural Organization of the Columnar Epithelial Cells of Mouse Intestine. Thesis, Karolinska Institute, Stockholm.

Zetterström, R., and Ernster, L. (1956). *Nature* **178**, 1335.

Ziegler, D. M., Linnane, A. W., Green, D. E., Dass, C. M. S., and Ris, H. (1958). *Biochim. Biophys. Acta* **28**, 524.

Zollinger, H. U. (1948). *Am. J. Pathol.* **24**, 569.

Zollinger, H. U. (1950). *Rev. hematol.* **5**, 696.

CHAPTER 6

Lysosomes and Related Particles

By ALEX B. NOVIKOFF

I. Introduction

In writing this chapter we have the feeling that a delay of a year or two would permit rapidly accumulating observations to validate or invalidate the *ideas* that we will express. Yet a modern book on the cell must include the body of *facts* already gathered on these newly discovered cytoplasmic particles. And even speculations that prove unfounded will have served their purpose if they help focus attention on these interesting particles.

This Chapter is unique in another sense. The field it covers has a short history, the inception of which can be stated unequivocally. It began less than ten years ago in the laboratories of Christian de Duve in Louvain. De Duve and his colleagues, until the present, have obtained the basic quantitative data upon which the field rests.

II. Lysosomes: A Biochemical Concept

The development of the lysosome concept illustrates both the limitations and the power of the differential centrifugation technique. In the hands of some investigators acid phosphatase and uricase activities of rat liver were concentrated in the mitochondrial fraction (Palade, 1951; Berthet and de Duve, 1951; Schein *et al.*, 1951; Schneider and Hogeboom, 1952), but in the hands of others they were localized in the microsome fraction (Tsuboi, 1952; Novikoff *et al.*, 1953). An arbitrary technical matter was responsible for the difference. If the centrifugal force was somewhat higher, these enzyme activities were recovered in the mitochondrial fraction; if the force was a bit lower, the activities were in the microsome fraction. Aided greatly by the finding that the particles containing acid phosphatase are impermeable to their substrate (Berthet and de Duve, 1951; Berthet *et al.*, 1951), the Louvain group was led to interpret their data and those from other laboratories in terms of multiplicity of cell particles rather than heterogeneity of mitochondria and of microsomes (see Novikoff, 1957). Instead of the single classic mitochondrial fraction they obtained a heavy M and a lighter L fraction (de Duve *et al.*, 1953; Appelmans *et al.*, 1955). The small L fraction, with only about 4% of the total homogenate nitrogen but the highest concentration of acid phosphatase, was assumed to include a concentration of new particles rich in phosphatase but without cytochrome oxidase. When the L fraction was found (de Duve *et al.*, 1955) also to contain high concentrations of cathepsin, acid ribonuclease, acid deoxyribonuclease, and β-glucuronidase (Fig. 1), the name lysosomes was given to the particles assumed to be concentrated in it. "For practical purposes, it is proposed to refer to these granules as lysosomes, thus calling attention to their richness in hydrolytic enzymes."

Figure 2 summarizes the essential properties of these postulated particles. Assuming a spherical shape, in 0.25 *M* sucrose the particles are calculated, from their sedimentation characteristics, to have a mean diameter of about 0.4 μ and an average density of about 1.15. They possess no oxidative enzymes but are characterized by the presence of easily soluble hydrolases with acid pH optima. As the diagram shows, a variety of experimental procedures make enzymes and substrates accessible to each other. From the actions of lecithinase and proteolytic enzymes it is deduced that there is a lipoprotein membrane around the particle. After alteration of the membrane by any of the treatments indicated, all enzyme activities are released simultaneously, in fully active form.

The number of acid hydrolases recovered in high concentration in the L fraction has grown considerably since 1955. It now stands at ten. In addition to the original acid phosphatase, cathepsin, acid deoxyribonuclease, acid ribonuclease, and β-glucuronidase, there are arylsulfatases A and B (Viala and Gianetto, 1955; Roy, 1958); cathepsins A and B, and probably C (Finkelstaedt, 1957); phosphoprotein phosphatase (Paigen and Griffiths, 1959); β-galactosidase; and β-N-acetylglucosaminidase; and α-mannosidase (Sellinger *et al.*, 1960). De Duve (1961) considers it "reasonable, unless direct evidence to the contrary is obtained, to consider the lysosomes as forming a single group, all of which contain all the acid hydrolases recognized as lysosomal, but in variable proportion."

Recent work from de Duve's laboratory (de Duve *et al.*, 1960) largely removes the uncertainty that had attached to uricase activity (see, e.g., de Duve, 1959a) and, even more important, points to the existence of another biochemically distinct type of granule in rat liver. When the mitochondrial fraction is centrifuged in a sucrose gradient in D_2O three distinct peaks of activity are obtained. These are, in increasing density, the mitochondria, lysosomes, and uricase-containing particles. When the mitochondrial fraction is centrifuged in a glycogen gradient in 0.5 *M* sucrose the uricase particles show an unusually low density; the sediment contains lysosomal enzymes but very little uricase activity. In the uricase-containing particles, or in closely related particles sedimenting similarly but not identically, are two other enzymes, catalase and D-amino acid oxidase. The authors point to the great interest, with respect to cellular metabolism and mechanisms of cell protection against radiation damage, attaching to the fact that uricase and D-amino acid oxidase activities result in the formation of H_2O_2 and that catalase activity destroys it.

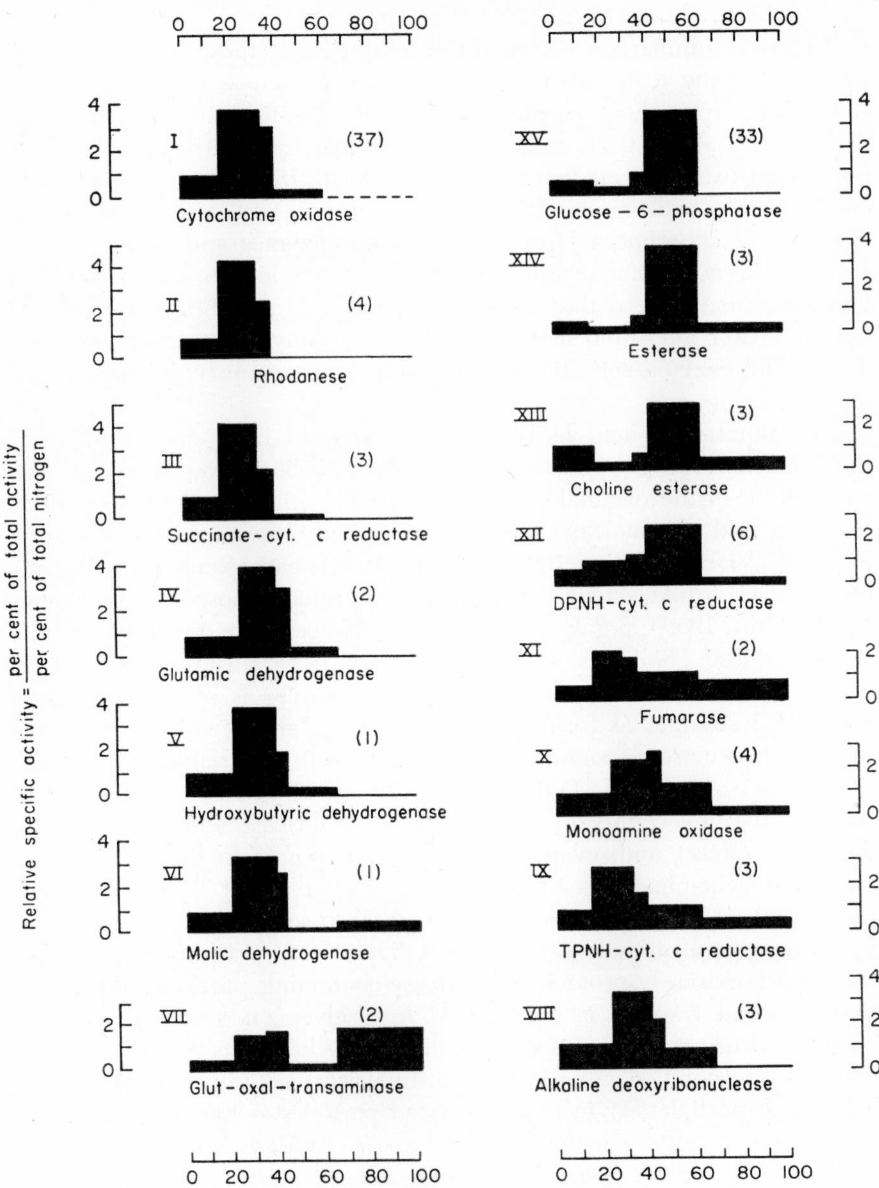

FIG. 1. Distribution of enzyme activities among subcellular fractions isolated from liver. Ordinate: mean relative specific activity of fractions. Abscissa: fractions are represented by their relative nitrogen content, from left to right: nuclear fraction, heavy mitochondrial fraction, light mitochondrial fraction, microsomal fraction, and final "supernatant fluid."

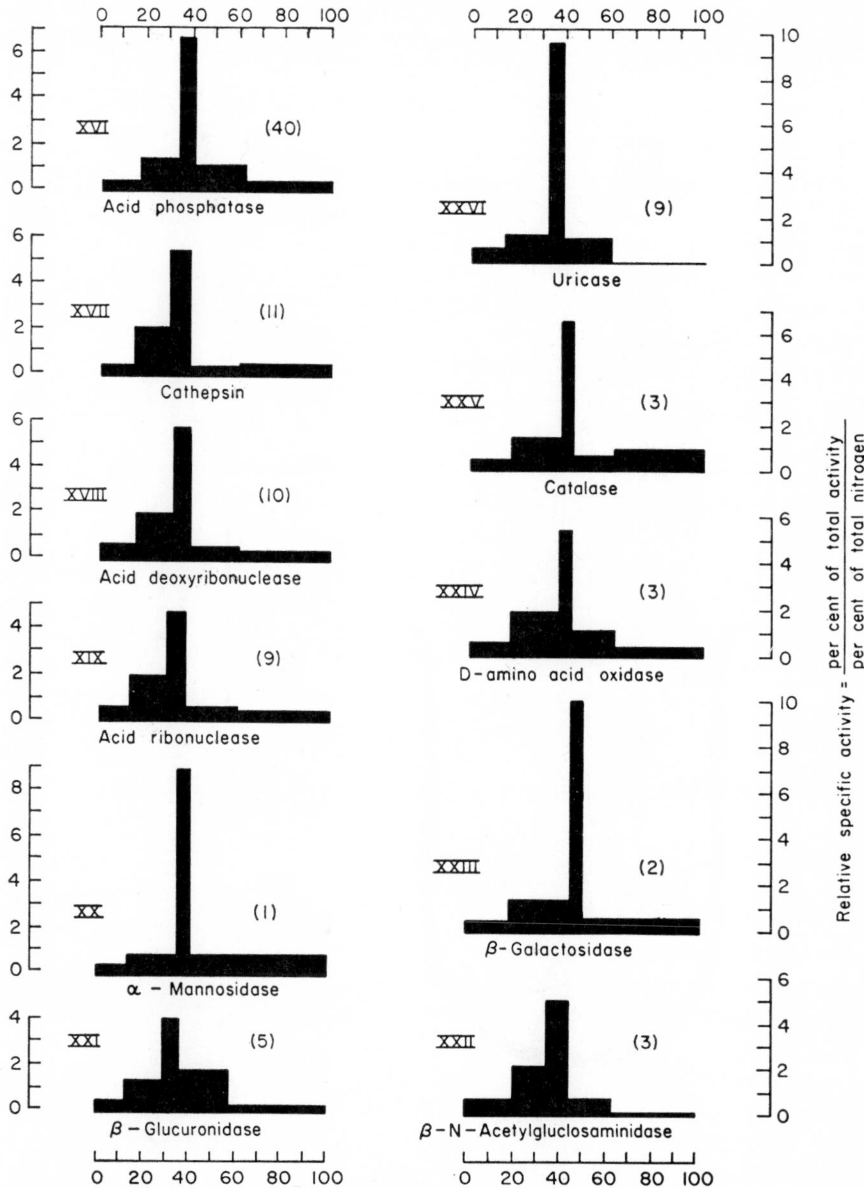

Mitochondrial patterns are shown by enzymes I–V and, to a lesser extent, VI; microsomal patterns by enzymes XIII–XV; and lysosomal patterns by enzymes XVI–XXIII, with XXI showing high activity also in the microsome fraction. Similar distributions are shown by enzymes XXIV–XXVI. Combination of patterns are shown by VII–XII. From de Duve *et al.* (1960a); courtesy of Dr. C. de Duve.

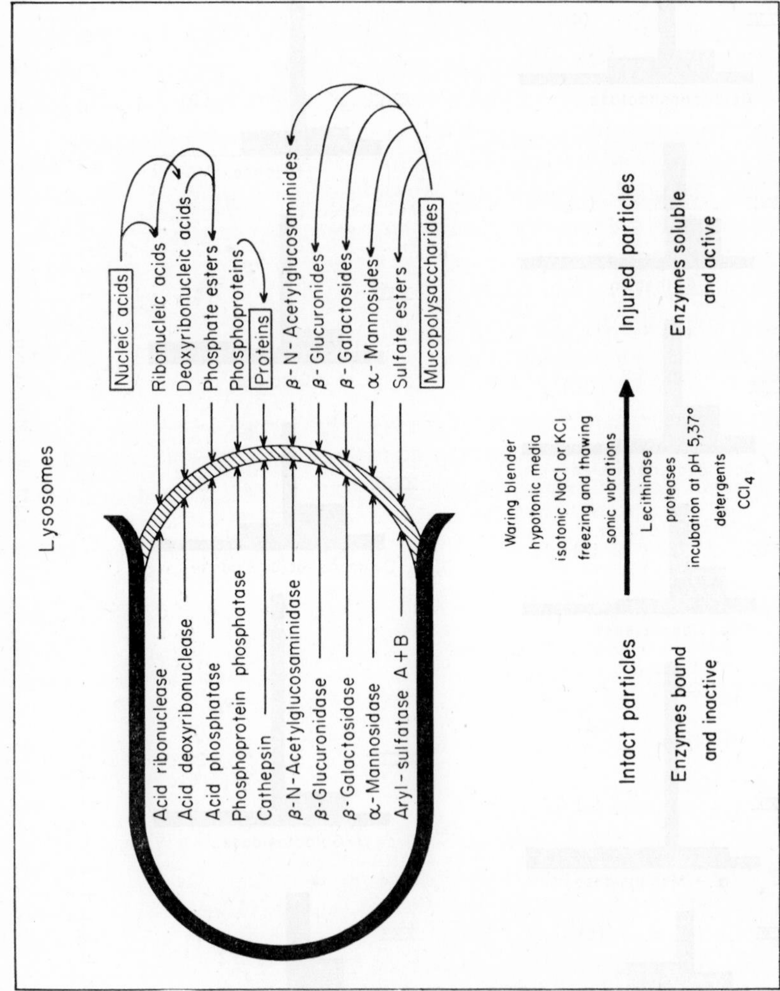

FIG. 2. Schematic representation of the lysosome concept; courtesy of Dr. C. de Duve.

There is also evidence suggesting the presence of other cytoplasmic particles closely related to but not identical with lysosomes. These are particles in which intravenously injected peroxidase is concentrated (de Duve, 1961; Jacques *et al.*, 1961). These peroxidase-containing particles, for which the name "phagosome" (Straus, 1959) is adopted, when isolated shortly after injection of the protein behave very much like lysosomes. They remain with lysosomes and apart from other particles in the several centrifugation procedures employed. Procedures that cause release of hydrolases from lysosomes also free the peroxidase from their particles. However, as the time interval between injection of peroxidase and isolation of the particles is lengthened, they become less like lysosomes in their centrifugation behavior while retaining their similar manner of releasing their contents after various treatments, including a significant sex difference in their sensitivity to incubation at pH 5 and 37°C. This release, presumably due to an autolytic type of disruption, occurs much more rapidly in females than in males. Furthermore, even newly formed phagosomes show a slightly sharper density distribution when centrifuged in a sucrose gradient in water than lysosomes do and they have a higher average density when centrifuged in a sucrose-D_2O gradient. As de Duve (1961) indicates, this might be interpreted as nonidentity of the two particles or, in an alternate fashion, that peroxidase is in fact present in lysosomes but only in those that form after injection of the protein and that the relationship of phagosomes and lysosomes is further complicated by factors such as active uptake by the Kupffer cells (Straus, 1959; Wachstein *et al.*, 1959; Novikoff *et al.*, 1960a). We will return to this matter later, when discussing morphological studies on lysosomes and phagosomes.

There is strong biochemical evidence for the existence of lysosomes in kidney and suggestive evidence in several other tissues. Straus (1954) used differential centrifugation to separate a fraction rich in so-called droplets from rat kidney homogenates. This droplet fraction showed a high level of acid phosphatase activity and considerable catheptic activity but low levels of cytochrome oxidase and succinic dehydrogenase activities. Later, Straus (1956) subdivided the droplet fraction into three and studied them for acid phosphatase and the other hydrolases shown to be present in hepatic lysosomes by de Duve *et al.* (1955). High concentrations of acid phosphatase, cathepsin, acid deoxyribonuclease, acid ribonuclease and β-glucuronidase appear to characterize kidney droplets as they do hepatic lysosomes. Furthermore, they appear to be activated by changes in the droplet membrane resulting from exposure to hypotonic media (Straus, 1957a). Straus also showed that the droplet fraction acquired high levels of proteins injected into rats, both egg white (Straus and Oliver, 1955; Straus, 1957a) and peroxidase (Straus, 1957b, 1958).

De Duve (1959a) summarizes the data from his laboratory that suggest the presence in homogenates of brain, spleen, and thyroid of lysosome-like particles that contain acid hydrolases and require structural alteration for accessibility of enzymes and substrates. Greenbaum *et al.* (1960) present similar data for rat mammary gland. Van Lancker and Morrill (1958) and Van Lancker and Holtzer (1959a) obtained a fraction from mouse pancreas homogenate in which acid phosphatase and deoxyribonuclease activities are concentrated to a greater extent than cytochrome oxidase activity or ribonucleic acid. The work of Brachet *et al.* (1958) shows that in the Müllerian duct of the chick embryo acid phosphatase, cathepsin, and acid ribonuclease activities occur in sedimentable form. Acid phosphatase and cathepsin activities are also granule bound in the giant ameba *Chaos chaos* (Holter, 1954, 1956; Holter and Lowy, 1959). Unlike the case of the mitochondrial succinic dehydrogenase, homogenization releases the two hydrolases in unsedimentable form (Holter, 1956); the same is apparently true of acid phosphatase in *Amoeba proteus* (Quertier and Brachet, 1959).

III. The Cytological Identification of Lysosomes

A. *Hepatic Lysosomes and Pericanalicular Dense Bodies*

In 1956, lysosome-rich fractions isolated from rat liver were examined in the electron microscope and found to contain numerous distinctive particles seldom seen in fractions poor in acid phosphatase activity (Novikoff, Beaufay, and de Duve, 1956). In the osmium-fixed sections the particles were characterized by "single" outer membranes ("unit membranes" of Robertson, 1959), the presence of many electron-opaque grains resembling ferritin molecules, and, occasionally, internal cavities. Attention was directed to their resemblance to the "dense bodies" seen along the bile canaliculi *in situ*. However, the identification of lysosomes with pericanalicular dense bodies could only be provisional since the fractions also contained many mitochondria. Some direct support for the presence of ferritin in the lysosomes came from studies of Beaufay *et al.* (1959a) indicating that lysosome-rich fractions contain more easily detachable iron than the other particulate fractions.

The firmer identification of the liver lysosomes was pursued along two lines: through greater purification of the lysosome fractions in de Duve's laboratory, and by demonstration of acid phosphatase activity *in situ* in our laboratory and in those of Holt (1959), Barka (1960a), Barka *et al.* (1960), and de Man *et al.* (1960).

As we have already indicated, by centrifugation procedures of high resolving power, particles with uricase, D-amino acid oxidase, and catalase activities were separated from lysosomes (de Duve and co-workers, 1960). The results of an electron microscopic examination of fractions obtained from sucrose-D_2O gradients (p. 425) are consistent with the view that the ferritin-containing particles are the lysosomes, and they suggest that the particles containing uricase and related activities may be "microbodies." Microbodies, first described *in situ* by Gansler and Rouiller (1956) and Rouiller and Bernhard (1956), resemble lysosomes in size and in possession of an outer "unit" membrane. However, there are no ferritin-like grains, and the particles are best identified by the presence of an inner lamellated (crystalline?) body (see Figs. 25 and 26, Novikoff and Essner, 1960).

The oldest staining method for demonstrating acid phosphatase activity in tissue sections, that of Gomori (1952), gives reliable information when applied to frozen sections of tissue fixed in cold formol-calcium (Baker, 1946) rather than to paraffin sections of acetone-fixed tissue (Novikoff, 1959a; 1960a, b, d). The confidence in this technique resulted from observations on many tissues examined in the light microscope and several in the electron microscope.[1] They may be summarized as follows: (1) With this procedure there are no obvious diffusion artifacts that are apparent, for example, with the early azo dye methods for acid phosphatase activity (see Gomori, 1956).[2] (2) The changes in the size, number, and distribution of characteristically shaped cytoplasmic granules in stained preparations of experimentally or pathologically altered tissues are highly reproducible. These changes will be referred to in appropriate places later in this chapter. (3) There is good correlation of electron microscopic observations in thin sections with light microscope observations of acid phosphatase preparations in the granules of rat kidney, normal

---

[1] Another reason for such confidence is the similarity in localizations obtained by other techniques, those employing modified indoxyl phosphate (Holt, 1954) and modified naphthyl phosphates (Burstone, 1960; Barka, 1960b).

[2] It is not known that calcium ions in the fixative are important in the preservation of lysosomes (see Discussion after Holt, 1956; Holt, 1959; but see Deane, 1958; Novikoff and Goldfischer, 1961b). It is difficult to account for the failure of Gomori (1956) to note the superiority of his incubation medium for intracellular localization. He wrote: "While the histochemical methods for acid phosphatase are of definite usefulness, their relative merits are hard to evaluate. Their merits cannot be interpreted with the same degree of confidence as those of methods for some other enzymes (dehydrogenases, alkaline phosphatase, and esterase). On the basis of data now available, it is impossible to disentangle some of the diffusion artifacts from genuine differences in enzyme localization."

and hydronephrotic; parenchyma cells of rat liver and human liver; Kupffer cells of rat liver; and rat erythrophagocytes. In all but the last instance, where it was not tried, direct visualization of the product of acid phosphatase activity in the electron microscope has confirmed the light microscope observations.

A similar confidence in the Gomori method applied to frozen sections of formol-calcium fixed rat liver and kidney has been expressed by Holt (1959). He found no indication of cytoplasmic adsorption artifacts with inactive sections exposed to medium in which lead phosphate was produced, or diffusion of either enzyme or reaction product from active to inactivated sections. The detergent Triton X-100 and the enzyme lecithinase produced effects on the staining of tissue sections that paralleled those the Louvain group had shown these reagents to have on the biochemical properties of isolated liver lysosomes. Adopting the view that these results can be explained on the basis of the phospholipid nature of the lysosomal membrane, Holt shows that fixatives which lead to the loss of phospholipid also result in loss of acid phosphatase activity from the pericanalicular granules in liver and the droplets in kidney. Holt cautions against assuming that this staining method is reliable when used with other tissues in the absence of biochemical assay and study of detergent effects. With all 25 cell types studied we find Triton X-100 to affect acid phosphatase-rich granules as in liver and kidney.

The most direct support for the link between lysosomes and pericanalicular dense bodies comes from studies in which small pieces of liver, fixed overnight in cold formol-calcium, were incubated in the Gomori medium containing 15–25% sucrose and then treated in osmium tetroxide; thin sections were then prepared for electron microscopy in the usual fashion (Essner and Novikoff, 1960b, 1961; Novikoff and Essner, 1960). The material included normal rat liver, liver of bilirubin-infused rats, and human liver containing lipofuscin granules which, from earlier studies (Essner and Novikoff, 1960a; Ehrlich *et al.*, 1960), appear to be altered lysosomes. In all three types of liver the identification of pericanalicular dense bodies and acid phosphatase-rich lysosomes can be suggested on the basis of size and distribution of the granules showing accumulated reaction product (lead phosphate). In the case of the

---

FIG. 3. Section of human liver incubated for acid phosphatase. Electron opaque material is lead phosphate, resulting from acid phosphatase activity when incubated as described in the text. In favorable areas (arrows) it may be seen that the lead phosphate is restricted to the periphery of the lipofuscin granules. This is the region that retains the fine structure of the original dense bodies (see Essner and Novikoff, 1960a). Magnification: × 25,000. From Essner and Novikoff (1961).

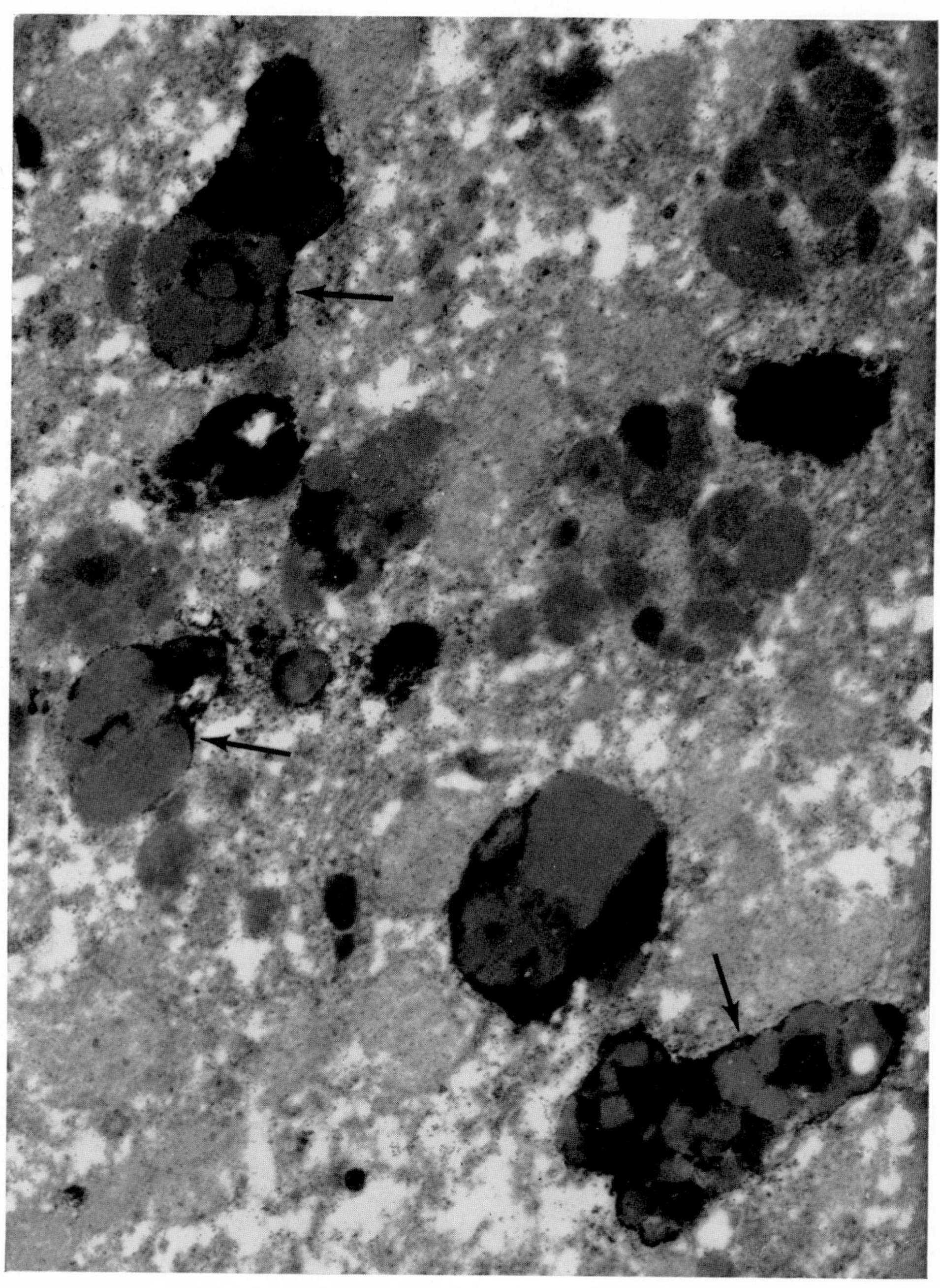

altered dense bodies or lipofuscin granules the identification has strong additional support in identity of fine structure (Fig. 3). In the bilirubin-infused rats, there is additional support in the great increase in number of pericanalicular acid phosphatase-rich bodies. Because they possess an outer "unit" membrane and high levels of acid phosphatase activity, we consider these bodies to be lysosomes. Most, if not all, of the bodies that appear after bilirubin infusion are microbodies (Novikoff and Essner, 1960; Essner and Novikoff, 1961). Clarification is required of the apparent discrepancy between this conclusion that these microbodies are lysosomes and the suggestion that, in normal liver, the microbodies contain uricase and related enzymes rather than acid phosphatase and other lysosomal enzymes (p. 431) (de Duve, 1960b). Until de Duve's group has isolated the microbodies from bilirubin-injected rats, we may accept the evidence that most, if not all, of the cytoplasmic bodies stained in acid phosphatase preparations of frozen sections of liver, are lysosomes.

From electron microscopic studies of isolated subcellular fractions of liver and sections of kidney and renal adenocarcinoma, Claude (1960a, b) concludes that the microbodies of these tissues are "cytosomes-lysosomes." He stresses the likelihood that they serve as a waste disposal system for the cell.

B. "Droplets" in Kidney Tubules

On one hand, the lysosomes of rat kidney cells can readily be distinguished morphologically from mitochondria; on the other hand, the heterogeneity problem is undoubtedly great since the cell architecture varies in different regions of the tubule. The droplets in the cells of the proximal convolutions are large spheres readily distinguished by light microscopy from the elongated mitochondria of these cells. It is likely that many of the droplets, probably most of the larger ones, seen by Straus (1954, 1956) and Davies (1954) in isolated fractions were derived from these cells not only in kidneys from protein-injected rats, but also in normal animals. Many of the small droplets in these fractions probably originated from other parts of the tubules. Aside from those in the glomerular epithelium, which, as we shall see, have recently gained

---

FIG. 4. Portion of cell in proximal convolution of the unoperated kidney of a rat in which the blood vessels to the contralateral kidney had been ligated for 2 hours. Note the canalicular structures (*c*) at the base of the microvilli (*mv*) constituting the brush border. Micropinocytosis vacuoles (*p*) appear to come from the canalicular structures and fuse with the larger luminal vacuoles (*v*) (compare Figs. 5 and 17). In the basal part of the cell are the mitochondria (*m*) and lysosomes (*L*). Arrow indicates the outer membrane of the lysosome. Magnification: × 21,000.

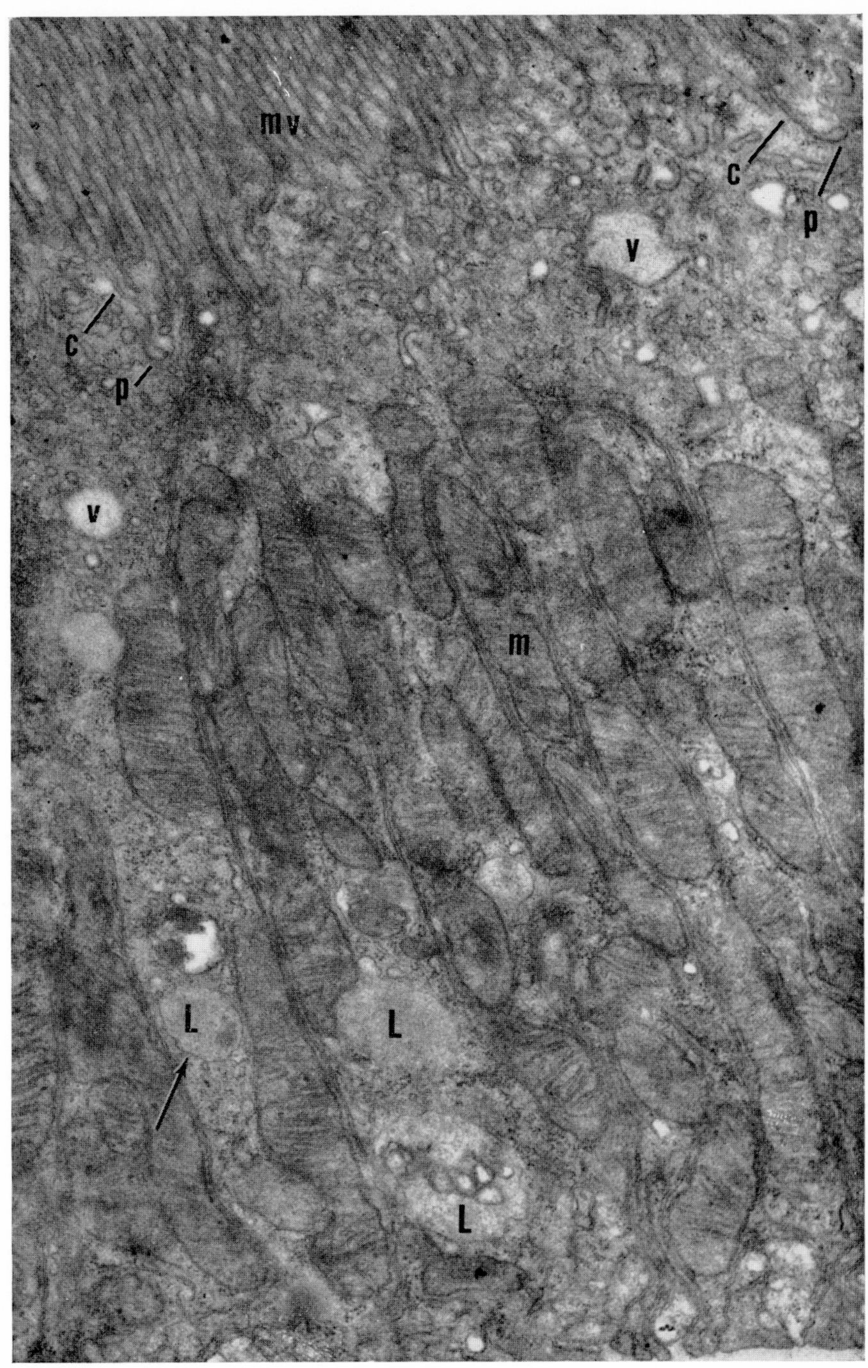
mv
c
p
v
c
p
v
m
L
L
L

attention, little is generally said about the droplets in the rest of the nephron, where they may have characteristic sizes and intracellular distributions. For example, in the cells of the distal convolutions and thick limbs of Henle the droplets are small and are situated between nucleus and lumen, in contrast to droplets in the cells of the proximal convolutions, which are large and are situated in the basal part of the cell (Figs. 4 and 17).

Phase contrast microscopy was used independently by Holt (1959) and by us (unpublished) on acid phosphatase preparations of rat kidney to establish that all the droplets in the basal parts of the cell in the proximal convolutions show phosphatase activity. Electron microscopists appear to have overlooked the droplets, possibly because they are difficult to preserve, because their acid phosphatase activity was either unknown or unappreciated, or because the mouse has been used for some of the most intensive studies. From the fluorescence studies of Sjöstrand (1944) it appears that the droplets of mouse and rat may be differently distributed.[2a] As Fig. 4 (arrow) shows, the rat droplets have single outer membranes and are filled with material of moderate electron opacity. Particularly after experimental alteration of the kidney, they may contain material of greater opacity, including cytoplasmic organelles (see Novikoff, 1959b). The product of acid phosphatase activity has been demon-

[2a] See the recent paper of Miller (1960) describing "vacuolated bodies" in the mouse kidney.

FIG. 5. Frozen section of formol-calcium-fixed rat kidney stained by the benzidine method of Mitsui and Ikeda (1951) (× 400). Incubation: 20 minutes, 25°C. Five minutes after intravenous injection of horseradish peroxidase. The injected peroxidase is seen in luminal vesicles within the cells of the proximal convolutions (cf. Fig. 17).

FIG. 6. Frozen section of formol-calcium-fixed rat kidney processed as in Fig. 5 (× 400). Incubation: 10 minutes, 25°C. Two hours after intravenous injection of horseradish peroxidase. The injected enzyme is now in the lysosomes of the cells in the proximal convolutions (cf. Fig. 17). Erythrocytes are dark by virtue of the reaction of hemoglobin with the benzidine reagent.

FIG. 7. Frozen section of formol-calcium-fixed rat kidney stained for acid phosphatase activity (× 400). Incubation: 15 minutes, 37°C. Kidney 4 hours after ligation of renal artery and renal vein. Note, especially at arrows, small lysosomes. These are much smaller than the normal lysosomes of the cells in the proximal convolutions.

FIG. 8. Frozen section of formol-calcium-fixed rat kidney stained for acid phosphatase activity (× 400). Incubation: 15 minutes, 37°C. Unligated kidney from the rat used for Fig. 7. Note the larger, more darkly stained lysosomes. The increase in their size over normal is presumably due to the increased protein load to this kidney.

strated directly in electron micrographs, but by a procedure that produces considerable cell damage (Novikoff, 1959b).

C. *Other Tissues*

As we have already indicated, many tissues of the rat, mouse, and man have been examined by the Gomori acid phosphatase method in

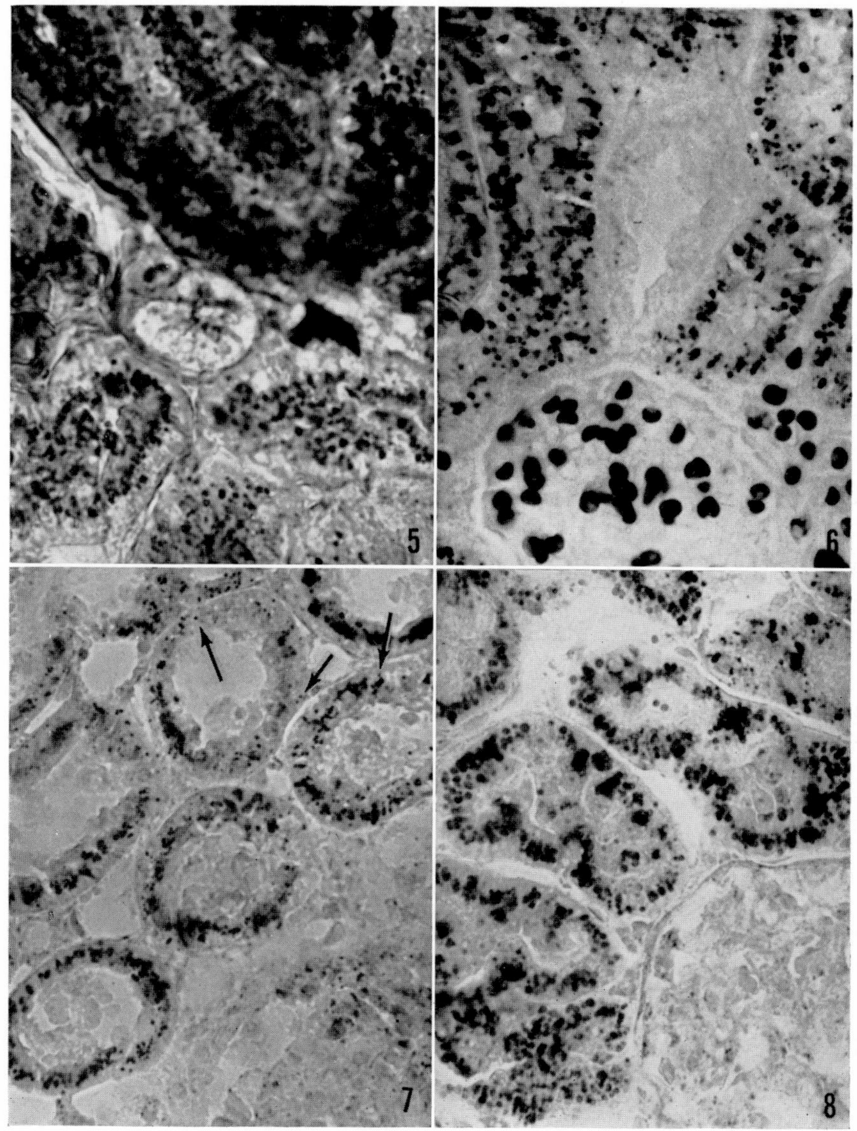

our laboratory. So have some tissues of the frog tadpole and chick embryo and of *Amoeba proteus* and *Chaos chaos*. Some of the results are shown in Figs. 5–16 and 25–27; others are published elsewhere (Novikoff, 1960a, c; Becker *et al.*, 1960). All tissues studied show cytoplasmic granules with acid phosphatase activity, excepting rat skeletal and cardiac muscle (which have not been analyzed by varying fixation time or other steps in the procedure). Barka (1960a, b) has used an azo dye method to study acid phosphatase-rich granules in liver and reticulo-endothelial cells. Our colleagues Rosenbaum and Rolon (1960) have studied them in the food-digesting phagocytes of planarians. The work of Avers and King (1960) suggested that in plants, too, acid phosphatase is localized in granules. Frozen sections of formol-calcium-fixed onion root tips and smears of formol-calcium-fixed molds (unidentified) show abundant acid phosphatase-rich granules (Novikoff and Goldfischer, unpublished observations). Acid phosphatase activity is present in bacteria, apparently in the cell membrane (Mitchell and Moyle, 1956), as are oxidative enzymes (see Chapter 5).

In mammals, we have found very large acid phosphatase granules in phagocytes (Figs. 13, 14), cerebral capillary endothelium (Fig. 15), choroid epithelium (Fig. 9), and urinary bladder epithelium (Fig. 11). These granules sometimes show oriented distributions, perhaps reflecting their movement through the cell (see below). They are linearly arranged in the basal half of cells in the proximal convolutions of the rat kidney (see Section IV, A); they are most concentrated in the ends of the choroid epithelium cells exposed to the ventricles; and they are most numerous

---

Fig. 9. Choroid plexus in a frozen section of formol-calcium-fixed rat brain stained for acid phosphatase activity (× 600). Incubation: 45 minutes, 37°C. Note that the epithelial cells contain numerous large lysosomes concentrated at the ends of the cells away from the blood capillary (*C*).

Fig. 10. A portion of an atretic follicle in a frozen section of formol-calcium-fixed rat ovary stained for acid phosphatase activity (× 310). Incubation: 20 minutes, 37°C. Note the enlarged lysosomes ("cytolysomes") in the degenerating epithelial cells (Novikoff, 1959a, 1960a).

Fig. 11. A portion of the urinary bladder in a frozen section of formol-calcium-fixed rat bladder stained for acid phosphatase activity (× 300). Incubation: 45 minutes, 37°C. Note the numerous lysosomes, some quite large, in the epithelial cells of this normal bladder.

Fig. 12. A keratinized portion of a frozen section of formol-calcium-fixed urinary bladder of rat exposed to methylcholanthrene for 6 months (× 300). Incubation: 45 minutes, 37°C. Note the intense diffuse acid phosphatase activity in the keratinized layers (see p. 440). Courtesy of Dr. A. A. Angrist.

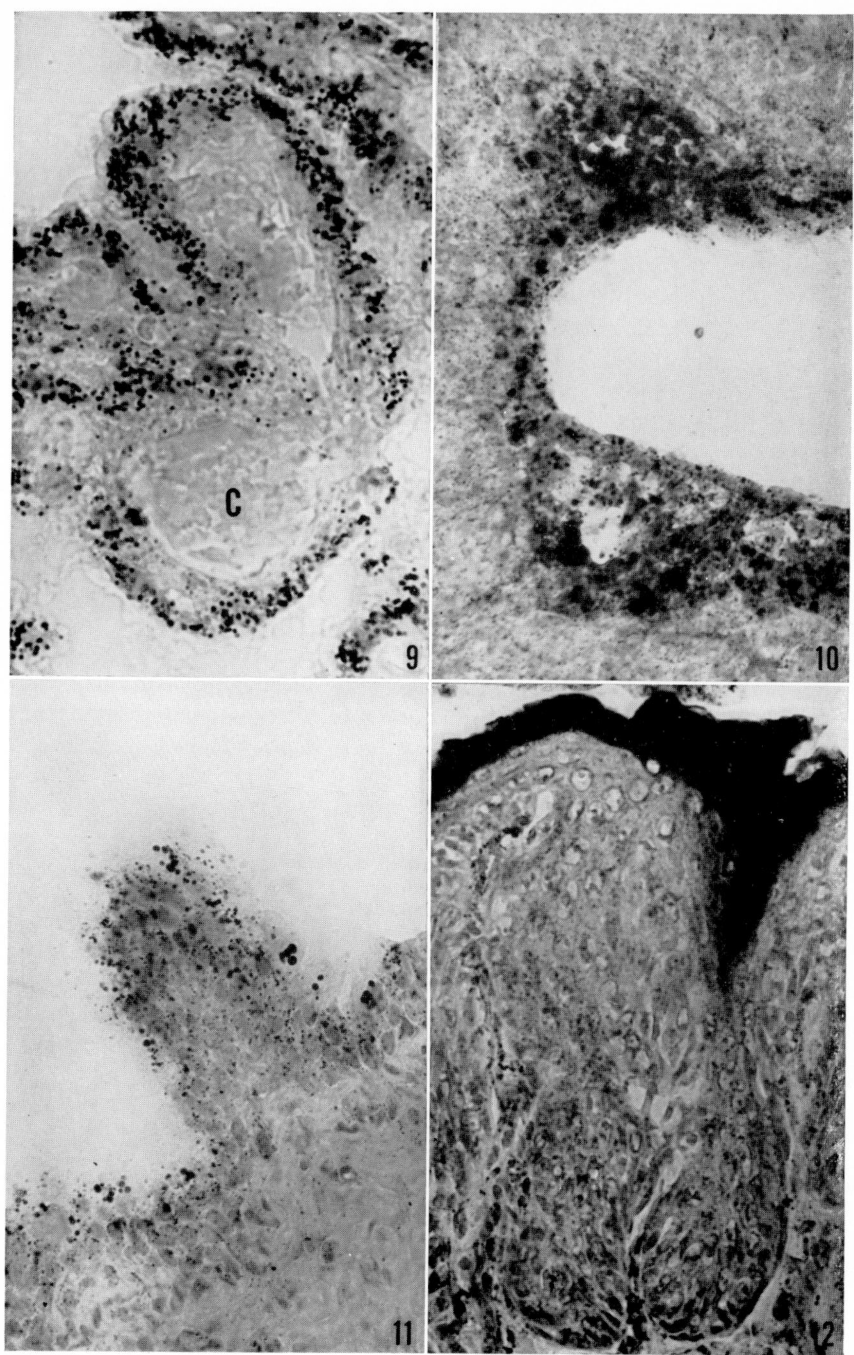
C
9
10
11
12

toward the nucleus (Golgi zone?) in the endothelial cells of the cerebral capillaries (Fig. 15).

We have observed unusually high levels of acid phosphatase activity in endocrine glands [beta cells of the pancreas (Fig. 16), thyroid gland cells, and adrenaline-secreting cells of the adrenal medulla] and neurosecretory cells in the brain (compare Eränkö, 1951; Sloper, 1955; Kobayashi *et al.*, 1960), but the extent to which these are associated with granules needs further study. Earlier studies in which intense staining of endocrine cells has been observed include those on the pituitary (Abolins and Abolins, 1949), thyroid (Rutenburg and Seligman, 1955), and pancreatic islets (Lazarus, 1959). Chemical analyses of homogenates by Charvat *et al.* (1959) and Schreiber and Kmentová (1959) revealed a marked rise in acid phosphatase activity following methylthiouracil treatment; histochemical observations consistent with this finding and localizing much of the activity to the Golgi zone were made by Lojda *et al.* (1960) (see also Schreiber *et al.*, 1960).

It may be that the high levels of acid phosphatase activity in the beta cells of the pancreatic islets (as suggested by Eränkö, 1951), in the neurosecretory cells (as suggested by Pearse, 1960), and in the keratinizing epithelial cells may be related to the synthesis of S—S proteins (insulin, neurosecretory material, keratin). We are currently studying these cells, to determine how much of this activity is associated with cytoplasmic granules. It is of interest that the neurosecretory cells also have high levels of esterase activity (see Section VII, D).

As a tentative hypothesis we consider that the granules staining for

---

FIG. 13. Frozen section of formol-calcium-fixed rat spleen stained for acid phosphatase activity (× 125). Incubation: 20 minutes, 37°C. Phagocytic cells are readily distinguished by their numerous, large lysosomes. There are many in the red pulp (above) and relatively few in the white pulp or Malpighian corpuscle (below). Rectangle indicated by lines is enlarged in Fig. 14.

FIG. 14. Portion of Fig. 13, enlarged (× 250). Note the large lysosomes in the cytoplasm of the phagocytes.

FIG. 15. Frozen section of formol-calcium-fixed rat brain stained for acid phosphatase activity (× 1000). Incubation: 30 minutes, 37°C. Below are three neurons in the cerebral cortex; note lysosomes in the cell processes as well as cell bodies. In the center, a portion of a capillary is seen, with large lysosomes concentrated toward the nuclei of the cell bodies.

FIG. 16. Frozen section of formol-calcium-fixed rat pancreas stained for acid phosphatase activity (× 350). Incubation: 45 minutes, 37°C. Note the lysosomes in the exocrine cells and the intense acid phosphatase activity in the island of Langerhans (beta cells).

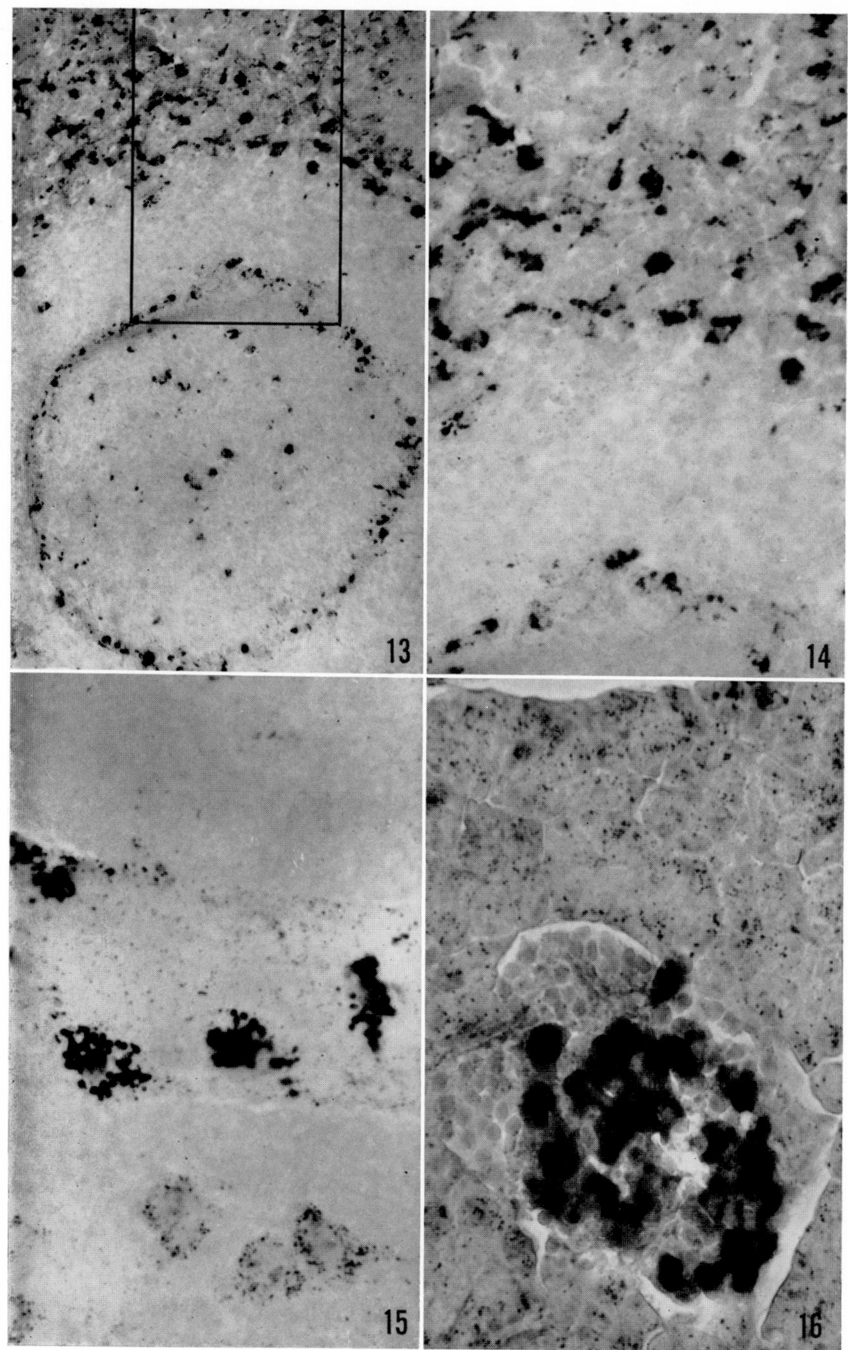
13
14
15
16

acid phosphatase activity by the methods indicated are likely to be lysosomes. The presumption is strengthened where electron microscopy shows the particular granule to have a single ("unit") external membrane.

D. Lysosome, A Biochemical Concept, and Lysosome, A Morphological Description

The heading of this section emphasizes the present uncertainty regarding the cytological identification of lysosomes, an uncertainty that is hardly surprising considering (1), the brief time since their discovery and (2), their apparent heterogeneity. Centrifugation methods of sufficient resolution for separation of cytoplasmic particles like lysosomes but not identical with them have become available only very recently, and they have been used only by de Duve and his colleagues. Coupled with electron microscope examination of the isolated fractions, these studies need to be extended to kidney, where heterogeneity is likely to be even greater, and to other tissues.

Quantitative data on highly purified fractions are obviously more firm than qualitative observations based on a staining procedure in which enzyme activity is drastically reduced by fixative. With future developments of freeze substitution or other methods it may no longer be necessary to sacrifice full enzyme activity in the interest of good intracellular preservation. It may soon be possible to demonstrate the intracellular localization of other lysosomal and related enzyme activities (e.g., cathepsin, β-glucuronidase, ribonuclease), but at the moment reliable methods are not available (Novikoff, 1959c; Novikoff *et al.*, 1960b). At present we can only presume that other lysosomal enzymes are present in many or all of the phosphatase-containing particles seen in the sections.

Knowledge of lysosomal fine structure may be expected to grow rapidly. Our present description is a negative one in the sense that it is more precise in what it excludes, i.e., mitochondria, Golgi apparatus, endoplasmic reticulum, ribosomes, lipid droplets. On the positive side, the presence of a single "unit" membrane is its main characteristic. Lysosomes have a fairly narrow size range, but there are exceptions such as the large droplets in kidney cells and phagocytes. Some lysosomes, as in liver parenchyma and in reticuloendothelial cells, contain ferritin-like grains; those of the erythrophagocyte are clear except for the ingested material; in the cells of the proximal convolution they contain a material of variable, but generally moderate, electron opacity.

It is encouraging that thus far whenever there has been evidence from both procedures, this fine structure and acid phosphatase activity occurred together. Both techniques have been used in our laboratory with the large lysosomes of erythrophagocytes (Essner, 1960), Kupffer cells (Essner and Novikoff, 1960a; 1961; Novikoff and Essner, 1960), and cells of the proximal convolutions (Novikoff, 1959b, 1960b). Localizations seen by us in stained preparations correspond to structures possessing the characteristic fine structure as reported by others: choroid plexus (Case, 1959; Tennyson, 1961); tracheal epithelium (Rhodin and Dalhamn, 1956); spleen macrophages (Palade, 1956); lung macrophages (Karrer, 1958, 1960; Schulz, 1958); uterine macrophages (Lindner, 1958; Nilsson, 1958); and urinary bladder epithelium (Walker, 1960) (see Fig. 11). Exceptions are to be expected, but thus far none has been encountered.

E. Nonenzymatic Components

For the morphologist the two most useful of the nonenzymatic components of lysosomes, at least in liver and kidney, are phospholipid and periodic acid-Schiff (PAS)-positive material. Experiments in de Duve's laboratory (see page 425) suggest a high phospholipid content of the external membrane of hepatic lysosomes. The acid hematin staining procedure, specific for phospholipid (Deane, 1958), gives a positive reaction throughout the granules, both in kidney (Oliver *et al.*, 1954; see also Fig. 9 in Holt, 1959, and Fig. 15, Novikoff, 1960b) and liver. The presence of high levels of PAS-positive material resistant to salivary digestion in some, if not all, droplets in the cells of the proximal convolutions was described by Huber (1953) and Davies (1954) in kidneys of embryos as well as adults. It was considered to represent a glycoprotein, filtered through the glomerulus, undergoing absorption in the proximal convolution. Liver lysosomes, in both parenchymatous cells and Kupffer cells, also contain a saliva-resistant PAS-positive material (Novikoff and Essner, 1960; Novikoff *et al.*, 1960a). This PAS-positive material can be demonstrated best in paraffin sections of formol-calcium-fixed tissue.

Sjöstrand (1944) has investigated sections of frozen-dried kidney with the fluorescent microscope. The fluorescent "vacuoles" described in the rat kidney are identical in size, shape, and distribution with the PAS-positive droplets rich in phospholipid and acid phosphatase activity. The fluorescent granules described in liver (see Popper and Schaffner, 1957) may also be lysosomes.

Although iron appears to be present in hepatic lysosomes, the usual staining procedures give insufficient color for its visualization, except where it is in excessive amount (Essner and Novikoff, 1960; Novikoff and Essner, 1960a). The staining method also demonstrates iron in the lysosomes of Kupffer cells and spleen macrophages (unpublished).

In macrophages, liver parenchyma, and perhaps other cells, the lysosomes are sites at which insoluble materials like hemosiderin and lipofuscin (residues of metabolism?) are deposited.

F. Relation to Golgi Apparatus

Thanks to the definition of their fine structure (see Chapter 8 by Dalton) and to methods of sufficient resolution for the localization of acid phosphatase activity [and esterase activity (see below)] the relations of the Golgi membranes and vacuoles to the sites of phosphatase (and esterase) activity and of PAS-positive material will probably be clarified before long.

Owing to methodological inadequacies little more can be concluded from earlier studies than that in many cells a correspondence appears to exist between the Golgi zone and areas with high levels of acid phosphatase activity (Deane and Dempsey, 1945; Deane, 1947; Montagna, 1952; Weiss and Fawcett, 1953). Recent results confirm this correspondence in some cells but not in others. Thus, in parenchymal cells of liver both Golgi apparatus and acid phosphatase-rich lysosomes are multiple pericanalicular structures; three experimental procedures that change the distribution of the Golgi apparatus change the lysosome distribution correspondingly (Novikoff, 1959d). Lojda (1960) found acid phosphatase-rich bodies to be concentrated in the Golgi zones of basophiles and other cells of the rat pituitary. On the other hand, most lysosomes in the cells of the proximal convolutions in the rat kidney do *not* lie in the Golgi zone (see Fig. 17). Even in these cells, however, a functional relation may exist between Golgi elements and lysosomes (page 450). It is of interest that both phospholipid and acid phosphatase activity were concentrated in the Golgi fraction isolated from homogenates of epididymis (Kuff and Dalton, 1959). In macrophages and other cells (Fig. 15) the acid phosphatase granules accumulate in the Golgi area. As we shall see later, there is evidence that suggests the appearance of lipofuscin granules (altered lysosomes) in the Golgi zone; in neurons of the aged the Golgi material apparently fragments and disappears as the lipofuscin accumulates.

Because reliable methods have been available for a considerable time, it is known that in many cells PAS-positive granules are found in the Golgi zone (Gersh, 1949; Leblond, 1950; Arzac and Flores, 1952; Aterman, 1952a; Weiss and Fawcett, 1953). These have been considered to contain mucopolysaccharide or mucoprotein, but the possibility should be considered that oxidized cephalin (Mildvan and Strehler, 1960) or other lipid contributes to the stain. We have already seen that phase contrast microscopy shows that there are no "droplets" in the cells of the proximal convolutions that are not rich in acid phosphatase activity; thus PAS-positive droplets must contain also the phosphatase. This has been demonstrated directly in the kidneys of rats receiving intraperitoneal injection of egg white (Novikoff, 1960b). In liver, the identity of PAS-positive granules and lysosomes has been demonstrated with the electron microscope in the Kupffer cells but not yet in the parenchymatous cells (Novikoff and Essner, 1960).

One of the most direct bits of evidence of a relationship between lysosome-like granules and the Golgi apparatus is the work of Benedetti and Leplus (1958). In the "para-erythroblasts" of chickens infected with erythroblastosis virus, vacuoles or granules like hepatic lysosomes in appearance are found in or adjacent to the highly developed Golgi zone. There is some indication, from Fig. 8 of their publication, of continuity between Golgi membrane and membrane of the granule—i.e., the granule appears like an enlarged Golgi vacuole. This indication needs to be established by repeated observation; and acid phosphatase studies are needed before we may consider these granules to be lysosomes. Confirmation of these suggestions would help to establish the Golgi apparatus as a site for segregation of lysosomal hydrolases and insoluble materials like the ferritin-like material observed within the granules of these "para-erythroblasts."

Similar suggestions of a direct relationship of lysosome-like granules and the Golgi apparatus may be seen in: (1) Fig. 23 of Dalton and Felix (1957) of a mouse neuron (this will be considered in Section V where lipofuscin granules are discussed); (2) Fig. 17 of Lindner (1958), in which one of the iron-containing "cytosomes" in the connective tissue cells of the rat uterus appears like an enlarged Golgi vesicle; (3) Fig. 2 of Kuff and Dalton (1959) of an epithelial cell in rat epididymis (see below); (4) Fig. 166 of Palay (1958) of an acinar cell of mouse pancreas; and (5) Fig. 167 of Palay (1958) of a cell in the rat adenohypophysis.

We may ask questions such as these: Are the pericanalicular dense bodies enlarged Golgi vesicles? Do the kidney vacuoles (see below) acquire high levels of acid phosphatase activity in the Golgi apparatus? If, as the evidence to be summarized later suggests, pinocytosis vacuoles are lysosomes, how are these related to the Golgi apparatus? The answers will not be forthcoming until electron microscopy is combined with the identification of PAS-positive structures and sites of acid phosphatase activity [and other lysosomal enzymes, including esterase activity (see below)] in normal and experimentally altered circumstances. Particularly interesting for such studies would be cells of epididymis and others in which the Golgi apparatus is very large (Fig. 25). Golgi material isolated from epididymis, consisting mostly of membranous arrays, was found by Kuff and Dalton (1959) to have high levels of acid phosphatase activity. However, equally high activity was present in a fraction with few such arrays. The discrepancy may find its explanation in our observations on stained sections. Most acid phosphatase activity is concentrated in lysosome-like granules (Fig. 27) rather than in the lamellar structures seen in both osmium or silver preparations (Fig. 25) and in unfixed cells when examined by phase microscopy (Dalton and Felix, 1957).[3] The membranous arrays of electron microscopy apparently correspond to the lamellar structures of light microscopy. In the electron micrographs of epididymis *in situ* electron-dense granules with single outer membranes are seen adjacent to the membrane arrays [e.g., Fig. 2 of Kuff and Dalton (1959)]. It is tempting to suggest that these correspond to the acid-phosphatase granules. However, the identification of their lysosome nature will require electron microscopic study of tissue first incubated for acid phosphatase activity. Their relation to the secretory granules also requires clarification. This is also true of the lysosome-like bodies in other cells. It is interesting that, in discussing "neutral red or lipoidal bodies," Lacy and Challice (1957) point to their close topographical relation with the Golgi apparatus and suggest that they may also be functionally related, probably helping "in the formation of many kinds of secretion products" (see also Fig. 27 in Fawcett, 1959). A report by Ogawa *et al.* (1960) is of considerable interest: in cultured neural cells the cytoplasmic granules that stain supravitally with neutral red give positive staining reactions for acid phosphatase, alkaline phosphatase, and lipase activities, suggesting the possibility that they are lysosomes.

IV. Pinocytosis, Phagocytosis, and Lysosomes

In 1931 Warren H. Lewis wrote, "Pinocytosis may be a much more universal process than we at present suspect." Thirty years later, we

[3] Studies with Sidney Goldfischer in our laboratory indicate that whereas acid phosphatase activity is concentrated in the granules of the Golgi *zone*, the lamellar structures of the Golgi *apparatus* show a phosphatase hydrolyzing adenosine diphosphate (Fig. 26). It is of interest that adenosine triphosphate (ATP) and adenosine monophosphate (AMP) are not split under similar conditions (at pH 7.1 in a lead-containing medium and at pH 9.4 in a calcium-containing medium). From our staining observations it seems likely that the phosphatase(s?) reported in the isolated Golgi fractions by Schneider and Kuff (1954) have arisen from tissues at the base of the epithelial cells (basement membrane, lamina propria, smooth muscle, etc.), and possibly from the luminal surface of some epithelial cells. These readily split glycerophosphate, AMP, and ATP at alkaline pH, whereas the Golgi apparatus does not. For a summary of more recent studies see Novikoff and Goldfischer (1961a, b) and Novikoff *et al.* (1961).

know from electron microscopy that many cells may form vacuoles resembling pinocytosis vacuoles, such as those formed by amebae (Edwards, 1925; Mast and Doyle, 1934), but with dimensions below the resolving power of the light microscope. The present trend is to minimize apparent differences in underlying mechanisms (but cf. Gosselin, 1956, and Holtzer and Holtzer, 1960) or structural appearances—whether pseudopodia form or not, whether the pseudopodia are individual or "ruffled" as in Lewis's tissue culture cells—and in dimension—whether above or below the limit of resolution of the light microscope—and to speak of the process as "pinocytosis" or "micropinocytosis" (Yamada, 1955; Odor, 1956). As Holter (1961) expresses it: "Regardless of these differences in mechanisms and dimensions, the morphologically essential feature is the same: a certain area of the surface membrane of the cell encloses a droplet of the surrounding medium, separates from the surface and migrates into the cell." By visualizing *molecules* of ferritin *in solution,* the electron microscope has removed the sharp distinction that had been made between pinocytosis, where ingested material is invisible, and phagocytosis, where ingested material is visible (Holter, 1959). Also, the acid phosphatase observations we will discuss suggest that the enzymatic properties of phagocytosis and pinocytosis vacuoles are basically similar, confirming the notion expressed by Lewis and others that both are "mechanisms of cell nutrition" (Lewis, 1931). In the interest of brevity we shall use the term *cytosis.* Although this term is noncommittal regarding dimensions, underlying mechanisms of binding to cell membrane and vacuole formation, and nature of ingested material, it separates these processes under consideration from the continuous intake into cells by diffusion or "active transport" across the cell membrane. Vacuole formation is thus viewed as a discontinuous process by which material is taken up by the cell in gulps or quanta.

We do not wish to imply, by the examination of cells that follows, that *all* cells are capable of pinocytosis. Perhaps they are, at a microscopic level. But even continuities, in electron micrographs, between plasma membrane and membranes of apparent vacuoles cannot establish movement, let alone *direction* of movement. For this, markers are required. On the light microscope level, with the fluorescence microscope Holtzer and Holtzer (1960) were able to find pinocytosis vacuoles in only a minority of the cells of rabbit, mouse, and chick tested with fluorescein-labeled rabbit globulin, rabbit antimyosin, and bovine albumin.

We also recognize that from the few cases yet studied it is too early to generalize that all cytotic vacuoles of the large variety are lysosomes, i.e., that they possess acid phosphatase activity and a "unit" outer membrane. This has been demonstrated in a rat erythrophagocyte (Essner, 1960) and in the Kupffer cells (Novikoff and Essner, 1960), and there is suggestive evidence for it in phagocytes of lung, spleen, uterus, and other tissues. Acid phosphatase activity has been studied in only one cell with large pinocytosis vacuoles, *Chaos chaos;* here the vacuoles also have acid phosphatase activity (Novikoff, 1960a; Birns, 1960).

It is not possible to do more than speculate regarding the lysosomal nature of "microcytotic" vacuoles. The situation will remain thus until acid phosphatase reaction product can be demonstrated in electron micrographs by methods which preserve all fine structure and are capable of revealing low levels of enzyme activity such as might be expected in the small vacuoles. As we shall see, from the work of Cohn and Hirsch (1960a, b) it is possible to think of the specific granules of rabbit polymorphonuclear leucocytes as lysosomes that, following ingestion of particulate matter by the cells, empty their contents into the phagocytic vacuoles. In HeLa cells, Rose (1957) considers the "microkinetospheres" as lysosomes that transport acid hydrolases to newly formed pinocytosis vacuoles.

A. *Cells in the Proximal Convolutions of the Rat*

Figure 17 illustrates our present concept of the cell in the proximal convolutions of the rat kidney (Novikoff, 1960b). It is based on electron microscopy of cells in which the apical portions are well preserved and on tracer studies with injected egg white and peroxidase (Novikoff, 1959a, 1960b; Novikoff *et al.,* 1960a) and colloidal particles (Burgos, 1960). These materials enter the cells in microcytotic vacuoles at the ends of "canaliculi" extending into the cell between adjacent microvilli. Burgos suggests that the mucopolysaccharide coat of the cell membrane contains binding sites for material entering the cell, as suggested in general form by Bennett (1956a). A coat on the cell membrane of amebae has been shown by histochemistry (Bairati and Lehmann, 1953) and electron microscopy (Pappas, 1959; Brandt and Pappas, 1960; Nachmias and Marshall, 1960, 1961; Roth, 1960). Evidence of "receptor sites" has come from fluorescence studies (see Holter, 1959) [see Weidel (1958) for a discussion of receptor sites in bacteria]. Holter (1959) and Gosselin (1956) discuss the evidence of a reversible adsorption phase preceding the irreversible ingestion phase. The early work of Höber (1940) is of interest since it led him to the conclusion that, in kidney and liver, dyes with particular molecular configurations were able to "anchor with greater or smaller forces, in the phase boundary between the cell and its aqueous surrounding and, subsequently, to release or to change physiological activity." For comments on dif-

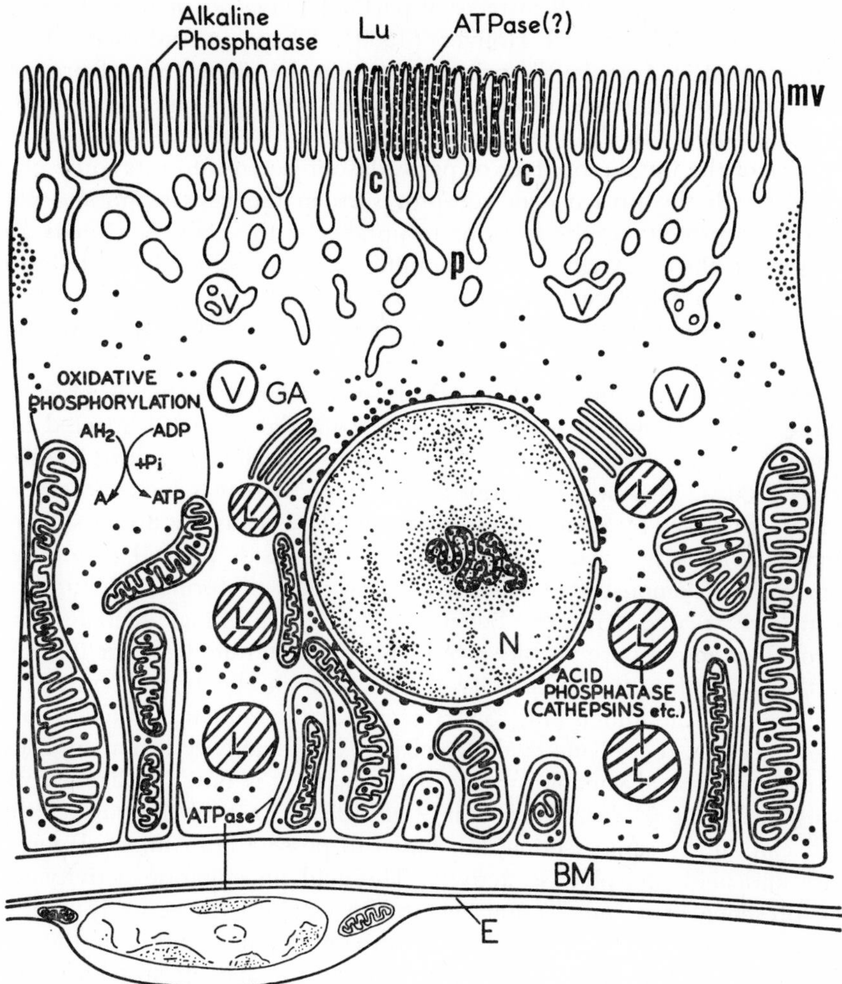

FIG. 17. Diagrammatic representation of a cell in the proximal convolution of the rat kidney (Novikoff, 1960b). Micropinocytosis vacuoles (*p*) are shown forming at the ends of the canalicular structures (c) extending into the cell between adjacent microvilli (*mv*) of the brush border. These are shown fusing into larger vacuoles (*V*); it is probably these vacuoles that are visible in peroxidase preparations shortly after injection of peroxidase (Fig. 5). Later, the injected peroxidase is in the lysosomes (*L*) (Fig. 6). The relations to the Golgi apparatus (*GA*) are uncertain. Other abbreviations: A, oxidized substrate; ADP, adenosine diphosphate; AH_2, reduced substrate; ATP, adenosine triphosphate; ATPase, adenosinetriphosphatase; *BM*, basement membrane; *E*, endothelial cell of blood capillary; *Lu*, lumen; and Pi, orthophosphate.

ferences in binding to cell surfaces of particulate matter as against soluble protein, see Holtzer and Holtzer (1960). Chapman-Andresen (1960) suggests that in *Amoeba* the primary stage in the pinocytosis of proteins, the binding of the protein to the cell surface, "depends on a precipitation of mucoproteins in the mucus coat." Jacques *et al.* (1961) consider the kinetics of the uptake by liver of peroxidase injected into rats to be compatible with the presence of receptor sites in the cell membrane with which the protein combines prior to uptake by the cells. Nachmias and Marshall (1961) describe the uncoupling, *in Amoeba,* of the first step in pinocytosis, a selective binding of positively charged substances to the mucopolysaccharide carrier substance, from the next step, the formation of vacuoles.

Microcytotic vacuoles apparently fuse into larger vacuoles (Fig. 4). These can be shown to contain not only colloidal particles injected into the rat (Burgos, 1960), but also intravenously injected proteins (Novikoff *et al.,* 1960a). The most useful of the proteins tried was horseradish peroxidase, first used by Straus (1957b, 1958, 1959). Because, as Straus showed, enzymatic activity is retained by the peroxidase inside the cell, its localization can be demonstrated when frozen sections of formol-calcium-fixed tissue are treated with the benzidine reagent. Luminal vacuoles with peroxidase activity are visible at 5 minutes and 30 minutes after injection of the peroxidase, when activity is also seen in the brush border and sometimes in the lumen (Fig. 5). However, if the animal is killed 2 hours after injection, there is no evidence of peroxidase activity in lumen, brush border, or luminal vacuoles. Instead the activity is now localized in the lysosomes (Fig. 6) (cf. Wachstein *et al.,* 1959). Each of the basal "droplets" can, in the same section, be shown to possess both injected peroxidase and acid phosphatase activity. The acid phosphatase activity and the presence of a "unit" outer membrane establish them as lysosomes. We must presume that a polarized streaming takes place in the cell so that the microcytotic vacuoles are brought together in larger vacuoles that move toward the base of the cell. Apparently in the region of the Golgi apparatus, the vacuoles acquire sufficient acid phosphatase activity to be demonstrable by the staining procedure; whether any activity is present in microcytotic vacuoles or larger luminal vacuoles is not known since the method may not be adequate for its demonstration (see page 448). From the work of Straus (1954, 1956) it is safe to assume that other hydrolytic enzymes are present in the lysosomes. Presumably the ingested protein is digested, but we can only surmise that if the protein load is low the lysosomes become smaller and disappear whereas they accumulate under conditions of excess protein load (see Oliver and MacDowell, 1958).

We should note that in the view of "protein droplet" formation here presented, the mitochondria play no direct role. Elsewhere (Novikoff, 1960b; see also Chapter 5) we have briefly discussed another view, once widely held, that much of the substance of "protein droplets" derive from mitochondria; see also Straus (1958) and Farquhar and Palade (1960), who give references to the older literature on "athrocytosis," a term used by Gérard and Cordier (1934) to describe the absorption, concentration, and retention of colloidal particles below 0.1 μ by the cells of the proximal convolution—cells that they showed were also capable of taking up larger particles like melanin and cinnabar by "athrophagocytosis."[3a]

The distributions of alkaline phosphatase and ATPase shown in Fig. 17 may here call attention to the fact, sometimes overlooked in the current emphasis on pinocytosis, that transport *across* cell membranes remains vital in the entry of molecules into cells. Indeed, little is known concerning the permeability properties of the pinocytosis vacuoles (see Nachmias and Marshall, 1961). Furthermore, the quantitative aspects of pinocytosis have yet to be established (Palade, 1960).

It should be observed that cells in other parts of the renal tubules also take up injected protein and in these cells an increase in the number of acid phosphatase-rich lysosomes is evident following injection. Worthy of note is that the lysosomes retain a characteristic size and position in each cell type. Thus, in the straight portion of the proximal convolution (in the outer stripe of the medulla) the lysosomes, although enlarged after protein administration, remain very much smaller than those of the cells in the convoluted portion; and in the distal convolutions and thick limbs of Henle's loops the lysosomes remain in the luminal rather than basal portions of the cells. It is hardly surprising that both protein uptake and lysosome variation are inherent in the nature of the cell type.

B. *Epithelial Cells in the Rat Glomerulus*

Farquhar and Palade (1960) have brilliantly studied the uptake of ferritin molecules in glomerular epithelium cells of normal and nephrotic rats. These molecules were traced at various time intervals in the basement membrane, micropinocytotic invaginations and vacuoles, and large vacuoles and dense bodies. "The findings suggest that ferritin molecules —and presumably other proteins which penetrate the basement membrane—are picked up by the epithelium in pinocytotic vacuoles and transported *via* the small vacuoles to larger vacuoles which are subse-

[3a] Conclusions in general harmony with those expressed here are reached by Miller (1960) from an impressive study of hemoglobin uptake in the mouse kidney. This study underlines the desirability of investigating the relations to lysosomes of the morphologically different cytoplasmic bodies.

quently transformed into dense bodies by progressive condensation. The content of the dense bodies may then undergo partial digestion and be extruded into the urinary spaces where it disperses." The latter observation is qualified by the authors: "We cannot decide whether the discharge is a normal event in the final disposal of incorporated material, or a sign that the amount of material with which the cell is confronted exceeds its capacity for disposal, or, finally, an indication that the ability of the cell to deal with its ingesta is impaired." The authors consider that, as in phagocytosis, the movement of the micropinocytosis vacuoles is toward the central Golgi region.

In Fig. 19 we have portrayed the dense bodies as lysosomes because (1) we have concluded, from parallel frozen sections, that the PAS-positive bodies of glomerular epithelium (in aminonucleoside-nephrotic rats) have acid phosphatase activity and (2) these bodies have "single" outer membranes. The diagram shows no large vacuoles, simply because the difference between them and dense bodies is not striking in the studies of Farquhar and Palade and we have seen microcytotic vacuoles in contact with dense bodies, as if fusing with them. The resolution of this matter requires further study before we can say that relations of microcytotic vacuoles, large vacuoles, and lysosomes are the same in these cells as in those of the proximal convolutions.

C. *Trypanosoma mega*

Figure 20 is based on studies of ferritin uptake in the trypanosome *T. mega* (Steinert and Novikoff, 1960). Binding sites in the plasma membrane are apparently restricted to a localized area at the mouth of the

---

FIG. 18. Diagrammatic representation of a rat erythrophagocyte. Linear arrays of micropinocytosis vacuoles are seen leading from the cell surface to the phagocytic vacuole or lysosome (*L*). The diagonal lines indicate that acid phosphatase activity has been demonstrated in the vacuole (Essner, 1960). Also shown are a few mitochondria (*M*) and, at upper right, two erythrocyte fragments being engulfed into the cell by a pseudopod.

FIG. 19. Diagrammatic representation of a glomerulus epithelial cell in a rat kidney. Linear arrays of micropinocytosis vacuoles are shown carrying ferritin (Farquhar and Palade, 1960) to the lysosomes (*L*). Diagonal lines indicate that acid phosphatase activity has been demonstrated in the lysosome. Also shown are mitochondria (*M*) and, to the left, the basement membrane.

FIG. 20. Diagrammatic representation of *Trypanosoma mega*. Near the base of the flagellum (*F*) micropinocytosis vacuoles are shown engulfing ferritin molecules and moving down the cytostome and toward the posterior end of the cell, as indicated by arrows (Steinert and Novikoff, 1960). Large vacuoles or inclusion bodies (*L*?) are shown with ferritin molecules brought to it by the micropinocytosis vacuoles. Also shown are mitochondria (*M*); nucleus (*N*); Golgi apparatus, at the bottom of the nucleus; and combined kinetonucleus (*KN*)-mitochondrion (*M*) (Steinert, 1960).

cystostome, for ferritin molecules do not attach elsewhere on the cell surface. An electron-lucid material apparently coats the plasma membrane since the ferritin molecules are always 100 A. or more from the outer electron-opaque layer of the membrane (cf. discussion on page 448 of the cells in the proximal convolutions and amebae).

The intracellular distributions of the ferritin molecules suggest that the cell membrane carrying the bound ferritin invaginates to form microcytotic vacuoles, which move in a course indicated in Fig. 20 to the posterior part of the cell where they flow into inclusion bodies. At the earliest "feeding" time (10 minutes) the inclusion bodies already contain a great many ferritin molecules. These bodies, like the plasma membrane and

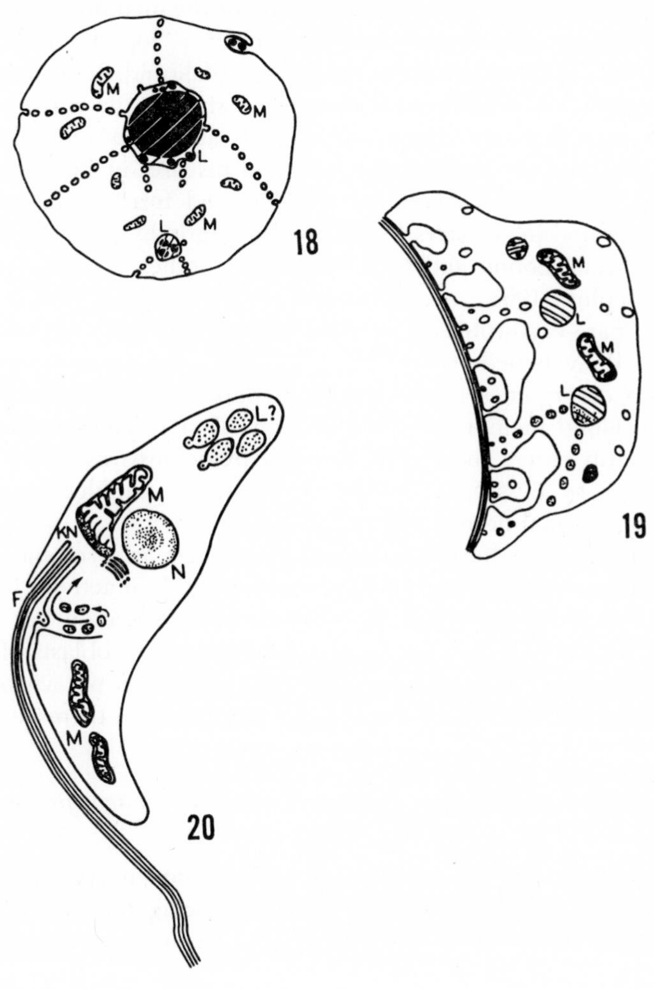

microcytotic vacuoles, are delimited by a "unit" membrane. In the absence of acid phosphatase studies, the suggestion that they may be lysosomes is only tentative.

D. Ferritin Uptake by Erythroblasts

Bessis and Breton-Gorius (1959; see also Bessis, 1959 and Chapter 3 in Volume V) have described in patients with hypochromic-hypersideremic anemias, "erythroblastic islets" in which ferritin molecules may be seen as if passing from reticular cells to erythroblasts. The plasma membrane indents to form a vacuole, without formation of pseudopodia [the photographs give some suggestions of a plasma membrane "coat" (see page 448) and an increased electron opacity of the membrane as it indents (see page 455)]. The authors suggest the term "ropheocytosis" for the ferritin uptake without apparent "drinking" of liquid. Like most others, we will employ the term "micropinocytosis" since in its current usage it is neutral regarding underlying mechanisms and because in this cell as in other instances of "cytosis" the mechanisms are, in fact, not known. Bessis and Breton-Gorius find that pinocytosed ferritin accumulates in mitochondria and in "clusters." In some photographs these clusters seem to be membrane bound and to resemble the bodies described by Benedetti and Leplus (1958).

In the "para-erythroblasts" of chickens infected with erythroblastosis virus studied by Benedetti and Leplus (1958), the ferritin accumulates in vacuoles or granules that resemble hepatic lysosomes and may separate from Golgi membranes (see Section III, F). The authors suggest that, if ferritin enters these cells through pinocytosis as in the erythroblasts studied by Bessis and Breton-Gorius, these granules represent "en réalité, le stade final de pinocytose."

To establish these ferritin-containing granules as lysosomes requires demonstration of phosphatase or other acid hydrolase activity in them.

It is of much interest that identical vacuole formation has been described by Bessis (see Volume V, Chapter 3) in erythroblasts of patients with "ferritin-poor" anemias. In these cells ferritin is not visible. Thus, the possibility is raised that ferritin molecules may not themselves be the inducers of micropinocytosis.

E. Erythrophagocytes, Polymorphonuclear Leucocytes, and Other Phagocytes

Essner (1960) has described a submicroscopic surface fragmentation of erythrocytes in the ascitic fluid of rats bearing the Novikoff ascites hepatoma and has followed the fragments when ingested by phagocytes.

The fragments accumulate selectively at the surface of the phagocytes; they are never seen on tumor cells or other cells in the ascitic fluid. Other observations are relevant to our present discussion: (1) Erythrocyte fragments appear to be separated from the electron-opaque cell membrane by an electron-lucid material. This may be compared to the separation of ferritin molecules from the electron-opaque layer of the cell membrane in *T. mega* (Section IV, C). (2) In the region where the fragments attach, the electron opacity of the cell membrane appears to increase. This, like the appearance in *Amoeba proteus* of high levels of acid phosphatase activity in the cell membrane areas to which food particles attach before engulfment (Novikoff, 1960a), may prove of interest with respect to the question, currently unanswered, of the *origin* of acid phosphatase and other enzyme activities in newly formed lysosomes.

In these cells, cytotic vacuoles containing erythrocyte fragments form in two ways. The usual process appears to be an active "engulfing" one in which a pseudopod rises over the fragment and meets the plasma membrane on the other side, where vacuole membrane and plasma membrane fuse. The other process is rare: the cell membrane is indented until, presumably, the edges of the "pits" ("caveolae intracellulares" of Yamada, 1955) are brought together and fuse. Whether external pressure or internal "suction" is involved in the second process is, of course, not known, but its appearance is similar to that called rhopheocytosis by Bessis and Breton-Gorius (1959). As a consequence of both processes, the fragment-containing vacuoles are thus delimited by portions of the plasma membrane. In addition, Essner has shown that the vacuoles visible by light microscopy have high levels of acid phosphatase activity. Thus, by our definition they are lysosomes. No attempt was made to apply electron microscopy to cells incubated in the Gomori medium in order to determine if the small cytotic vacuoles, about 80 mμ in diameter, also have acid phosphatase activity. All that can be said of the small vacuoles is that frequently they are seen in linear arrays (Fig. 18), presumably moving from cell membrane to the lysosome in which the erythrocyte fragments are undergoing digestion. Essner considers the possibility that the small vacuoles bring acid hydrolases to the large vacuole, in the way Rose (1957) suggests that the "microkinetospheres" do to the large pinocytosis vacuoles in HeLa cells. Neither electron microscopy adequate for study of micropinocytotic vacuoles nor staining methods for acid phosphatase have yet been applied to these HeLa cells; much can be learned from these studies against the background of Rose's observations of the living cell.

As we have indicated, it has been presumed that the small vacuoles move from cell membrane to lysosome, but the possibility exists that the small vacuoles *form* at the surface of the large ones, as on the digestive vacuoles of malarial parasites (Rudzinska and Trager, 1957, 1959) and of *Pelomyxa* (Roth, 1960) and apparently on the pinocytosis vacuoles of *Amoeba* (Nachmias and Marshall, 1961). A firm choice is not possible without studies with tracers like ferritin. In *Pelomyxa* the small vacuoles do not persist all the way to the cell membrane, but apparently they may in *Plasmodium berghei* (Rudzinska and Trager, 1959). In the erythrophagocyte nothing is seen that resembles the cloud of small vacuoles that form on the food vacuole of *Pelomyxa,* and the linear arrangement of vacuoles is different from the distribution of small vacuoles in *Plasmodium.* Indeed, its arrangement of small and large vacuoles is quite similar to that seen in the cells of the proximal convolution and glomerular epithelium and in *T. mega,* in which tracer studies show the direction of movement to be from cell membrane to large vacuole.

Cohn and Hirsch (1960a, b) and Hirsch and Cohn (1960) have reported important experiments on the polymorphonuclear leucocytes in peritoneal exudates of rabbits. These cells possess few mitochondria and large numbers of specific cytoplasmic granules. A particulate fraction can therefore be obtained from sucrose lysates in which fewer than 10% of the particles are mitochondria and the remainder are the specific granules. The granules, when examined with the electron microscope, show a single outer membrane; and they possess acid phosphatase and other enzyme activities characteristic of lysosomes. The lysosome nature of the granules is also indicated by the effects of acidity, freezing and thawing, and saponin. These treatments release the granule-bound enzymes in soluble form; freezing and thawing or saponin treatment also greatly increases the levels of activities of these enzymes. In addition to the usual lysosomal enzymes, acid phosphatase, ribonuclease, deoxyribonuclease, β-glucuronidase, and cathepsin, the granules contain 5′-nucleotidase (see Novikoff, 1959c), lysozyme, and alkaline phosphatase. They also contain the bactericidal agent, phagocytin, which is solubilized together with the hydrolytic enzymes by the different treatments tested. Ribonucleic acid and deoxyribonucleic acid are present in the granule fraction in trace amounts, too low to be considered part of the granules rather than of contaminating structures.

Because they are readily resolved in the light microscope, the relation of these lysosomal granules to the larger cytotic vacuoles are better studied in these cells than in other phagocytes. Hirsch and Cohn (1960) find that within 30 minutes following phagocytosis of microorganisms or yeast cell walls by the rabbit leucocytes (also by human neutrophiles) degranulation occurs; electron microscopic studies are in progress to determine whether the granule contents are released directly into the vacuoles. The extent of degranulation varies with the number of ingested organisms. Similarly, chemical redistributions can be shown to occur when larger numbers of organisms are ingested. Within 15 minutes of such ingestion a decreased proportion of acid phosphatase and β-glucuronidase activities are recovered in the granule fraction and an increased proportion in the supernatant fluid. This redistribution reaches its maximum at 30 minutes after ingestion. In the case of cathepsin, maximum redistribution occurs earlier, at 15 minutes; it is not known if this reflects an enzymatic heterogeneity of granules coupled with differential breakdown following phagocytosis or a difference in the extent to which the enzymes are structure-bound. It should be noted that no alteration in level of enzyme activities occurs following phagocytosis. Because the evidence needed to distinguish between the alternatives is still unavailable, Cohn and Hirsch (1960b) entertain two possibilities whereby the released hydrolytic enzymes and phagocytin interact with the ingested bacteria: "either in the cytoplasm of the cell or within the phagocytic vacuole." It is presumed that an increased level of lactic acid following phagocytosis initiates such intracellular redistributions *in vivo.*

Before considering the Kupffer cell, attention should be called to other phagocytes, listed on page 443, in which vacuoles or bodies delimited by "single" membranes have been described by electron microscopists and in which bodies of corresponding size and distribution have been found by us in acid phosphatase preparations. Staining studies would be interesting in other phagocytes, such as those described by Sampaio (1956), Felix and Dalton (1956), and Odor (1956).

It would probably be rewarding to apply staining reactions and electron microscopy to invertebrates in which the situation resembles that in *Echinus:* "The use of amoebocytes in the lumen and epithelium of the gut for carrying out absorption is correlated with a poorly developed hemal system which lacks a pumping mechanism" (Stott, 1954). Rosenbaum and Rolon (1960) have shown that, after feeding, previously starved planarians acquire high levels of acid phosphatase, cathepsin, and β-glucuronidase in the food-digesting phagocytes. Müller *et al.* (1960) find the digestive vacuoles of *Tetrahymena* to have high levels of esterase activity (see Section VII, D). When digestion is complete the esterase activity falls.

Food vacuoles of protozoa have been studied with the electron microscope as already indicated. In the ameba, *Pelomyxa,* they have been described by Pappas (1959), Mercer (1959), and Roth (1960). The invagination of the plasma membrane has been described by Brandt and Pappas (1960), Nachmias and Marshall (1960), and Roth (1960). In the excellent study of *Plasmodium berghei* (Rudzinska and Trager, 1959), two interesting observations were made. In this cell the plasma membrane is double as seen under the electron microscope. When it invaginates to form a digestive vacuole, the vacuole, too, has a double membrane. Secondly, the electron-opaque digestion products (hematin?) appear not in the large food vacuole, but in smaller vacuoles which apparently have separated from it. Acid phosphatase preparations might be useful in testing the suggestion of the authors that in this cell digestion does not occur in the large food vacuoles but that it does so in *Plasmodium lophurae* (Rudzinska and Trager, 1957). Levels of acid phosphatase activity are high in the food vacuoles of *Amoeba proteus,* and there is a suggestion that the enzyme activity appears in the cell membrane as it invaginates (Novikoff, 1960a).

F. Kupffer Cells

The situation in the Kupffer cells is basically like that in the erythrophagocyte: the large vacuoles with ingested material have "single" outer membranes and high levels of acid phosphatase activity (Novikoff and Essner, 1960; Essner and Novikoff, 1961); microcytotic vacuoles occur in linear arrays between plasma membrane and large vacuoles or lysosomes (Fig. 21). The studies of Hampton (1958) show that colloidal mercuric sulfide or Thorotrast (thorium dioxide) particles adhere to the cell surface (there may be an electron-lucid area of several hundred angstroms separating the two; this is interpreted as fibrin, but it may be the plasma membrane "coat" of which we have spoken), which then invaginates to form vacuoles. These apparently move into the cell and fuse into larger vacuoles. The latter may then condense to form more electron-opaque vacuoles. [For references to earlier studies, with light and electron microscopy, see Hampton (1958).] We have found (Novikoff *et al.*, 1960a) that injected enzyme (peroxidase, acid phosphatase, or alkaline phosphatase) is readily demonstrable in the large vacuoles by light microscopy (see also Wachstein *et al.*, 1959); these are the lysosomes, rich in PAS-positive material as well as acid phosphatase activity. In electron micrographs, linearly arranged microcytotic vacuoles are always seen between cell membrane and lysosomes. They resemble those found by Hampton to transport colloidal material, and it seems reasonable to assume that they constitute the means by which injected proteins (enzymes) are transported to the lysosomes, where they are concentrated (and possibly digested; see page 450). Parenthetically, it should be noted that within 5 minutes after intravenous injection all sinusoidal lining cells have taken up enough injected enzyme to be stained; in less than 20 minutes they have begun to enlarge, the large lysosomes increasing in number. These observations with enzymatic proteins parallel earlier ones made with proteins labeled by linkage to colored dyes (Sabin, 1939;

FIG. 21. Diagrammatic representation of a Kupffer cell and parenchymatous cell of rat liver. In the Kupffer cell (above), micropinocytosis vacuoles are seen leading into the lysosomes (*L*). The diagonal lines indicate that acid phosphatase activity has been demonstrated in the lysosomes. Two mitochondria are also shown. In the parenchymatous cell, micropinocytosis vacuoles are shown forming between microvilli at the upper surface of the cell (i.e., in the space of Disse). They are shown in linear arrays directed toward the bile canaliculi (*BC*); some are shown fusing with lysosomes (*L*). Some mitochondria (*M*) are also shown.

Smetana, 1947; Kruse and McMaster, 1949) or to fluorescein (Schiller *et al.*, 1953; Holtzer and Holtzer, 1960). Kruse and McMaster showed that the dye azoglobulin retained its antigenicity for at least 2 days after uptake. The antigen concentration in the blood fell rapidly so that, by the twelfth day, only a trace remained "but these traces continue to be present for many more days. Is it possible that the traces represent a slow output of stored antigen from the R-E cells, as the latter slowly lose their color? If so, the prolonged formation of antibody may be consequent on sustained stimulation in this way." The blue color remains for as long as three-and-one-half months, but the antigenicity of the material was not tested. Coons (1956) reviews other reports of persistence of antigenicity in the Kupffer cells.

Barka (1960) and Barka *et al.* (1960) have shown that the lysosomes of Kupffer cells and other phagocytes will store vital dyes. Aterman (1952a) showed that after partial hepatectomy there was a marked increase in number and size of PAS-positive spheres in the Kupffer cells. These closely resemble lysosomes; their increase may reflect the increased protein load to the liver lobes remaining after surgery. After azo-dye feeding and many other experimental procedures, there is a similar increase in Kupffer cell lysosomes (Novikoff and Essner, 1960; Barka, 1960). De Man *et al.* (1960) find a marked increase in number of lysosomes of parenchymatous cells in mouse liver following intraperitoneal injection of dextran. In both parenchymatous cells and Kupffer cells, particularly the latter, the electron dense material of the lysosomes appears to form part of the walls of the large vacuoles containing the dextran. Previous studies (van Duijn *et al.*, 1959) show a twofold increase in acid phosphatase activity of liver homogenates (fully activated) following such dextran injection. Golberg *et al.* (1960) find marked increase in levels of acid phosphatase and other lysosomal enzymes of mouse liver following injection of excessive doses of iron. Judging from changes in enzymes associated with other cell organelles (increased alkaline phosphatase; increased ATPase; decreased glucose-6-phosphatase and esterase; decreased catalase) the situation is probably complicated and morphological studies would be desirable for assessing the alterations in both parenchymal and Kupffer cells.

G. *Parenchymatous Cells of Liver*

It has long been known that parenchymal cells of liver can take up dyes and particulate matter from the blood and transport them into the bile. The paper of Weatherford (1932) may be consulted for a fine example of light microscope studies and for the early literature. Weatherford wrote: "Particulate matter is stored in the liver cells and eliminated into the bile in the same manner as intravitam dyes, but elimination is much slower." The process has been followed by Hampton (1958) with the electron microscope. At 30 seconds after intravenous injection, particles of colloidal mercuric sulfide or Thorotrast are present in submicroscopic tubular structures or vacuoles at the cell surface. At 1 hour, the particle-containing vacuoles are dispersed in the cytoplasm. They then increase in size and contain more particles. Aggregates of particles may now be found in the pericanalicular dense bodies. Particles are excreted into the bile as early as 30 minutes after injection. The increasing size of aggregates suggests to Hampton "that the hepatic cell cannot excrete the particles so rapidly as it can phagocytose them." He considers the dense bodies as sites of temporary storage "until such time as the particles can be excreted via the bile canals." Bennett (1956b) suggests that in normal liver the pericanalicular dense bodies "may represent segregated phagocytosed or pinocytosed material. . . . The enzymes associated with the granules may have a role in breaking down some of the contents of the bodies." Palade (1956) writes of similar bodies in splenic macrophages: "It is assumed that these . . . represent the terminal appearance of phagocytic vacuoles . . . ." De Duve (1961) points to the presence in rat bile of significant amounts of all lysosomal hydrolases, as support for the idea that the particles may finally be secreted into the bile. The cytochemical and electron microscopic studies (de Man *et al.*, 1960) of dextran uptake by mouse liver have already been summarized (p. 460).

The evidence regarding protein uptake is more equivocal. Holtzer and Holtzer (1960) find no indication of uptake of fluorescein-labeled rabbit globulin, rabbit antimyosin, or bovine albumin by uninjured parenchymatous cells of mouse and chick liver slices (cf. views of Coons, 1956). The authors refrain from generalizing their results to all proteins. They also recognize that negative results with this and other cells may reflect unfavorable *in vitro* conditions, where shift of pH, absence of hormones, or low oxygen tension may inhibit pinocytosis. If the acid phosphatase-rich lysosomes of rat liver reflect the extent of pinocytosis, it may be relevant that they are larger and more numerous in the peripheral cells of the hepatic lobule, where there is reason to suggest that oxidative processes *via* the Krebs cycle may operate at a higher level than in the more centrolobular cells (Novikoff and Essner, 1960). In Kupffer cells and other macrophages, pinocytosis may be less sensitive than it is in parenchymatous cells to factors like lowered oxygen tension.

When horseradish peroxidase, wheat germ acid phosphatase, or calf alkaline phosphatase is injected intravenously into rats, the specific enzyme can be demonstrated in the pericanalicular lysosomes by appropriate staining reactions, but the intensity of stain is much less than in the Kupffer cell lysosomes (Novikoff *et al.*, 1960a). Very soon after injection, the enzymes appear in the bile. Such positive evidence is more compelling than the failure of earlier investigators to note protein-containing bodies within parenchymal cells after injection of proteins linked to dyes (Sabin, 1939; Smetana, 1947; Kruse and McMaster, 1949) or to fluorescein (Schiller *et al.*, 1953). The lysosomes would not be expected to survive the fixation and paraffin embedding procedures employed by these investigators. Insufficient description is given for evaluation of the fresh and frozen sections employed by Kruse and McMaster; it is of interest that in their work, in contrast to ours, neither bile nor urine gave evidence of injected protein. Schiller *et al.* emphasized the need for caution in interpretation because of "possible effects of minor modifications of histological procedure." The cytological deficiencies even of their freeze-dried preparations is shown by the absence of droplets in the cells of the proximal convolutions.

At the moment we can say only that injected enzyme is present in pericanalicular lysosomes; the *amount* of enzyme that enters the parenchymal cells and the *manner of entry* are not yet known. Jacques *et al.* (1961) recover a large part of injected peroxidase (up to 75% when low doses are injected) in homogenates of liver, but the extent to which the Kupffer cells contribute is not established. Both Jacques *et al.* and we find injected enzyme in the bile.

After injection of enzyme, the number of lysosomes, judged by acid phosphatase activity and PAS-staining and by electron microscopy, appears to increase (Novikoff *et al.*, 1960a), but there is no increase in the phosphatase level in homogenates (Jacques *et al.*, 1961). This apparent increase in lysosome number occurs not only near the bile canaliculi, but elsewhere in the cell, even close to the sinusoidal surfaces. The electron microscope reveals a marked increase in the number of vacuoles, measuring 0.1 μ to 0.2 μ or more in diameter. These may be seen in linear arrays in sinusoid-canalicular directions; they often contain a material of moderate electron opacity. Such vacuole arrays are present in normal rat liver, but in smaller number and without the inner material of moderate opacity. In both normal and injected rats suggestions may be found of membrane continuity of forming vacuoles and plasma membrane. The possibility that the vacuoles are emptying materials into the blood plasma cannot be excluded without markers but there is no evidence that products of hepatic secretion are packaged into membrane-bound vacuoles as, for example, in the pancreas. The presence of material within the vacuoles of injected rats is consistent with the transport of injected material.

Some microcytotic vacuoles may appear to be fusing with lysosomes, as shown in Fig. 21. We do not have evidence to suggest the fate of the majority. Perhaps they transport material (injected protein?) directly into the canaliculi. Perhaps they lead to the Golgi membranes or vesicles: some of the latter come to contain a material of moderate electron opacity.

The complexity of the situation in the parenchymal cells is especially unfortunate since liver is the tissue for which de Duve and colleagues have obtained quantitative data. As we saw on page 429, the quantitative data on fractions isolated from the livers of rats shortly after injection of peroxidase may be interpreted in two ways; of these, the one which more readily fits the morphological observations is that the peroxidase accumulates in those lysosomes that appear after the injection whereas the lysosomes already present acquire little or no injected protein. Why the newly formed protein-containing granules ("phagosomes") are more like lysosomes than the "older" ones is not clear. It should be noted, however, that until the isolated fractions are studied by electron microscopy, it will not be known whether the isolated "phagosomes" correspond to the microcytotic vacuoles. Perhaps electron microscopy of the fractions may also help to determine the extent to which Kupffer cell lysosomes, which undergo such rapid development in these animals, contribute to the biochemical picture.

Before leaving the liver parenchymal cells, we wish to comment briefly on two observations. The first concerns a feature already referred to, the quantitative variation within the hepatic lobules in number and size of pericanalicular lysosomes as demonstrated in acid phosphatase preparations. They are larger and more numerous in the peripheral cells. This may reflect a higher rate of pinocytosis by which proteins and other materials normally enter the cells from the nutrient-laden blood in the periphery. The second observation is the presence, in some rats receiving enzyme injections, of occasional cells with extremely large PAS-positive spheres in the cytoplasm that approach the size of the nucleus. The spheres may represent accumulations of enzyme-containing plasma. Similar bodies have been described in the livers of rats after partial hepatectomy (Price and Laird, 1950; Aterman, 1952a, b; Novikoff and Noe, 1955). Doniach and Weinbren (1952) showed them to contain protein and not to appear in the regenerating liver of starved rats or of rats fed only dextrose. They were thus regarded as protein storage bodies.

H. Multivesicular Bodies

The studies by Dalcq, Pasteels, and Mulnard of the metachromatic granules of eggs have recently been reviewed (Dalcq, 1960; Pasteels, 1958). Two types of granules, α and β, have been described, the latter being larger, lighter, less numerous, and less metachromatic than the former. Both apparently have acid phosphatase activity. When eggs are centrifuged the metachromatic granules separate from mitochondria and other formed elements; the zone containing the granules shows high acid phosphatase activity. In the stained preparations, the reaction product (lead phosphate) is not clearly attached to granules, but this may result from the type of fixation or other procedural detail (see Discussion after Pasteels, 1958). The phosphatase is clearly sedimentable both in living eggs, as the Belgian workers show, and in homogenates (Rebhun, 1959).

Mulnard *et al.* (1959) suggest that the β-granules are related to the Golgi complexes of the eggs. This suggestion is based on similarities in sedimentation when eggs are centrifuged, in vital staining properties, in their possession of PAS-positive material and acid phosphatase activity, and in structural appearance. Electron microscopic examination of centrifuged eggs shows that the zone containing the β-granules also contains Golgi elements that may be impregnated by the usual silver and osmium techniques (Pasteels *et al.*, 1958; Pasteels and Mulnard, 1960). Rebhun (1961) also finds many small vesicles resembling Golgi vesicles in the areas where the metachromatic granules (β-granules) of *Spisula* eggs are concentrated, e.g., in the asters during early cleavages. Also present are larger bodies, about ¼ to ½ μ in diameter, containing a great many of the smaller (20–50 mμ) vesicles within them (Fig. 23). These bodies re-

---

FIG. 22. A portion of thin limb of Henle in rat kidney (Novikoff, 1960b). Vacuoles that appear to be micropinocytosis vacuoles (*p*) are numerous. The arrow indicates an area which may be interpreted to show micropinocytosis vacuoles moving into the larger multivesicular body. Magnification: × 22,000.

FIG. 23. A portion of an egg of *Spisula*. Note the similarity to Fig. 22 of: (1) the multivesicular bodies (arrows), and (2) the relation of small vacuoles and multivesicular bodies. Astral rays are seen at *a*. Magnification: × 25,000. Courtesy of Dr. L. I. Rebhun.

FIG. 24. Pinocytosis vacuoles within *Chaos chaos*, seen with the fluorescence microscope. The vacuole membranes fluoresce by virtue of adsorbed fluorescein-globulin (Brandt, 1958). Note the resemblance of these vacuoles to the multivesicular bodies shown in Figs. 22 and 23. Magnification: × 1200. Courtesy of Dr. P. W. Brandt.

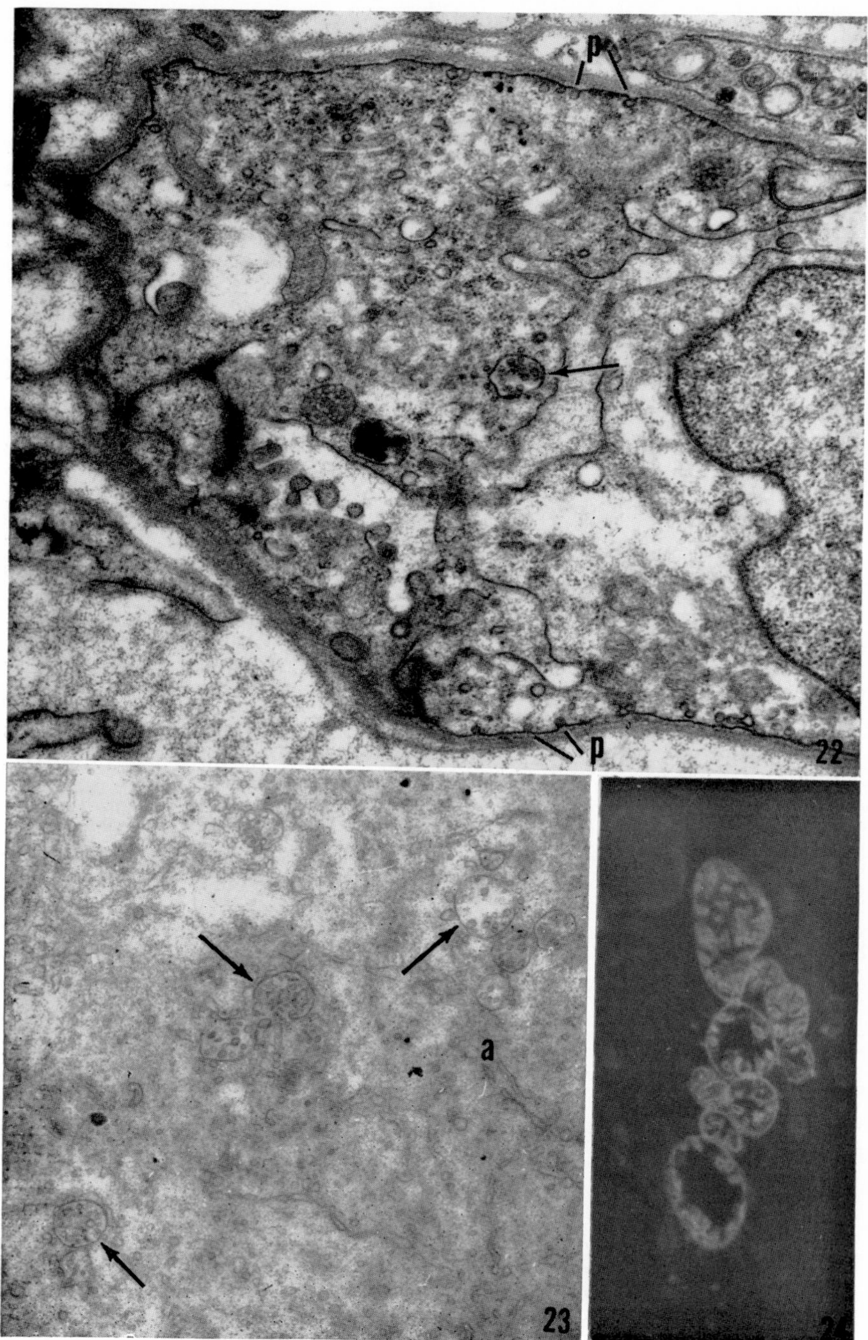
p
p
22
a
23
24

semble the *multivesicular bodies* described in mammalian eggs by Sotelo and Porter (1959) and present in many different cell types.[4]

Several observations had earlier inclined us to the view that the small vacuoles in the multivesicular bodies might be micropinocytotic vacuoles held within a membrane by an electron-lucid material such as mucopolysaccharide. (1) As already indicated, in the cells of the proximal convolution micropinocytotic vacuoles appear to merge into larger apical vacuoles. Not uncommonly, however, some of the smaller vacuoles appear to be enclosed within the larger vacuole (Fig. 17). (2) Cells in the glomerular epithelium show multivesicular bodies comparable in size to the PAS-positive acid phosphatase-rich bodies in these cells. (3) In the cells of the thin limb of Henle in the rat and in cells in or near the synovial membrane of man there are a great many micropinocytosis vacuoles. These cells also have numerous multivesicular bodies enclosing vacuoles of the size and shape of the micropinocytosis vacuoles. Images suggestive of smaller vesicles merging into larger ones may be found (Fig. 22), as in the glomerular epithelium and in human leucocytes (Low and Freeman, 1958). (4) Brandt (1958) has described the formation in *Chaos chaos* of unusually large pinocytosis vacuoles resembling multivesicular bodies in appearance (Fig. 24). (5) Quertier and Brachet (1959) find that the metachromatic granules of *Amoeba proteus* form twice as rapidly when pinocytosis is induced by bovine albumin.

In Rebhun's electron micrographs of *Spisula* eggs, as in other cells enumerated above, the outer membrane of the multivesicular body frequently seems incomplete, as if the linearly arranged small vesicles were being separated off from the rest of the cytoplasm or were moving into the larger vacuole. We have referred several times to the risk in inferring direction of movement from electron micrographs in the absence of markers. Furthermore, Sotelo and Porter (1959) have proposed a movement opposite to the one we suggest. Yet we consider it likely that the small vacuoles come together to form the collection, the multivesicular body, not only for the reasons enumerated in other tissues but because of observations with eggs. The most important is that living eggs con-

[4] In our laboratory we have seen multivesicular bodies in cells of the glomerular epithelium and thin limbs of Henle in the rat kidney, the synovial membrane of man, and rat hepatomas. Sotelo and Porter (1959) list literature reports of their presence in neurons, basophiles, regenerating nerve fibers, intestinal epithelial cells, ciliary epithelium, sperm cells of several invertebrates and vertebrates, and in the alga *Chlamydomonas*. They have been reported also by Benedetti and Bernhard (1958) in cultured embryonic fibroblasts, by Bernhard *et al.* (1955) in mouse mammary tumor, by Granboulan (1960) in mouse lymphocytes, by Low and Freeman (1958) in human leucocytes, and by Nilsson (1959) in mouse uterine epithelium.

centrate dyes such as neutral red and toluidine blue in the aggregates. Secondly, Pasteels and Mulnard (1957) report the reappearance of metachromatic granules in blastomeres freed of granules by centrifugation, and Mulnard *et al.* (1959) believe that "when an egg is stained in the usual procedure, the pre-existent β-granules are not demonstrated, and only those formed after the staining can be observed subsequently." Finally, Rebhun (1961) finds evidence only for aggregation of small vesicles into the multivesicular bodies, and not vice versa.[4a]

The possibility of uptake of macromolecules by "phagocytosis at the submicroscopic level" in eggs, particularly during the period of rapid growth, has been discussed by Schechtman (1956); see note by Telfer and Koch (1958) for a report of blood protein movement into the oöcyte of *Cecropia* and its collection into spheres at the periphery of the egg. The primary oöcytes in the guinea pig ovary (Anderson and Beams, 1960) reveal numerous vacuoles near the cell membrane, some in linear arrays, appearing like micropinocytosis vacuoles. The authors suggest that pinocytosis "may account for the many vesicular elements seen in the peripheral cytoplasm of the primary oocyte as well as inclusion bodies which may be related to yolk material." In *Spisula,* too, micropinocytosis-like vacuoles may be seen near the cell membrane (Rebhun, personal communication). However, in the absence of direct evidence that these vacuoles move into the multivesicular bodies, another speculative alternative might be considered. If Dalton and Felix (1957) are correct in suggesting that the Golgi apparatus functions "as a hydrostatic mechanism for removing or segregating water and possibly other fluids" from the cell interior, and if water enters the cell by a mechanism which does not involve pinocytosis, is it possible that formation of membranes of the α-granule (and hydrolytic enzymes?) is related to this function of the Golgi material?

Rebhun (1959, 1961) calls attention to the work of (1) Marsland (1958) and Marsland *et al.* (1960) indicating that external pressure induces cleavage of sea urchin eggs by causing these metachromatic vacuoles to break down; (2) Kojima (1959a, b) who showed that only blastomeres with vitally stained granules cleaved; and (3) Holt (1957) who observed that granules rich in acid phosphatase (and esterase) activity showed the astral localization in dividing cells of rat liver after partial hepatectomy. Brachet (1960) also adopts the view that the β-granules may play a role in cell division.

The publications of Dalcq (1960), Brachet (1960), and Rebhun

[4a] Important evidence is provided by the finding of Farquhar *et al.* (1961) that the multivesicular bodies of glomerular epithelium cells acquire ferritin molecules injected into the rat.

(1961) may be consulted for thoughts concerning possible morphogenetic roles of the metachromatic granules (multivesicular bodies), particularly in the mosaic eggs where they are concentrated (along with the mitochondria) in the posterior blastomeres. In assessing these, it might be recalled that Berg (1957) found in *Ciona* more acid phosphatase activity in the anterior than in the posterior blastomeres. On the other hand, yolk and nonparticulate phosphatase may complicate chemical analyses; and it is possible that, when better preserved, the "large vesicles" concentrated in the posterior blastomeres (Berg and Humphreys, 1960) may prove to be multivesicular bodies. It is of interest that in mammals the metachromatic granules are concentrated in the trophoblast (see Dalcq, 1960), a finding perhaps related to a role of lysosomal enzymes in the digestion of maternal substances.

I. General Comments

It is apparent from the preceding sections that much remains to be clarified both with regard to pinocytosis and phagocytosis and to the roles of lysosomes in these processes. What is the nature of the binding sites in the cell membrane? Are the forces involved different for particulates and for soluble materials? Is PAS-positive material present generally on cell membranes? What are the precise mechanisms of vacuole formation? How is new membrane synthesized? Do acid phosphatase and other acid hydrolases appear or become more highly concentrated in the invaginating membrane? Do all pinocytosis and phagocytosis vacuoles satisfy our definition of lysosomes, as do those we have studied? Will the linear arrangement of micropinocytosis vacuoles prove to be a general feature of cells? Are these vacuoles the means by which both ingested materials and acid hydrolases are transported to the larger vacuoles? Is this true also of the parenchymatous cells of liver—the only organ for which quantitative biochemical data are available—and other cells in which Holtzer and Holtzer (1960) saw no signs of pinocytosis with fluorescence microscopy? How shall we interpret the observation that in the Novikoff ascites hepatoma all cells have a full complement of acid phosphatase particles (Birns *et al.*, 1959; Novikoff, 1960d), yet, as in the experience of Holtzer and Holtzer (1960) with fluorescein-labeled protein uptake by two other ascites tumors, only a small percentage ($<10\%$) give evidence of uptake of enzymatic protein (peroxidase) (unpublished)? Even in cells with a high rate of apparent pinocytosis, what is its quantitative importance in the cell's economy? Does it vary for different chemical constituents? Are acid hydrolases synthesized in cytotic vacuoles, large or small? If so, does this synthesis require ribonucleic acid? Is ribonucleic acid intrinsic to the vacuoles? What are the fates of the materials engulfed within the vacuoles? By what mechanisms do they reach the cytoplasm interior?

Despite such unanswered questions, the observations we have summarized amply bear out the suggestion of de Duve *et al.* (1955), more fully stated by de Duve (1959a, b), that lysosomes may function as organelles of intracellular digestion. Indeed, at the moment, this seems to be the most likely role of lysosomes in tissues generally (as we saw in Section III, C, almost all cells have granules with acid phosphatase activity).[5] So great is the temptation to find a common function for a cell organelle that we find ourselves disregarding the sound advice implicit in the paper of Holtzer and Holtzer (1960), who note that many cells appear not to have the ability to pinocytose, and minimizing the observation, already noted, that on the light microscope level acid phosphatase granules (lysosomes) are more abundant in ascites cells than are vacuoles with ingested protein. We note that the suggestive evidence relating lysosomes to the Golgi apparatus (see Section III, F; also Fig. 15) is consistent with the appearance of accumulated pinocytosis vacuoles in the Golgi zone (Lewis, 1931; Rose, 1957; Holtzer and Holtzer, 1960). We tend to view lipofuscin and other insoluble materials (Section V) as indigestible substances, possibly end products of metabolism, that accumulate in the lysosomes. We see the contradiction between the presence of digestive enzymes in lysosomes and the retention of enzymatic activity, at least for a time, in engulfed proteins and of antigenic activity, presumably for a long period (Section IV, F), but we seek recourse to the possibility of "protection" (Sabin, 1939) and to absence of quantitative information: *how much* of the protein retains its "native" properties?

We come now to consider the relation of pinocytosed materials to other cell organelles. In none of the cells studied is there evidence of direct morphological involvement of mitochondria in either protein uptake or formation of "protein droplets" (see page 451; also Chapter 5); nor does this appear to be the case in the uptake of lipid droplets (Palay and Karlin, 1959, and Palay, 1960).

The situation regarding the endoplasmic reticulum is less clear. It is sometimes stated that pinocytosis vacuoles empty into the endoplasmic reticulum (see Haguenau, 1961, and discussion following it). Yet the only

[5] Since this was written the report by Tennyson (*Anat. Record* **136**, 290, 1960) on choroid plexus cells has come to our attention. As we have seen (page 438), these cells have large acid phosphatase-rich cytoplasmic bodies. Tennyson finds that the cells ingest thorium dioxide from the cerebrospinal fluid and segregate it into cytoplasmic vacuoles.

evidence of which we are aware that suggests this are the reports of lipid transport in intestinal mucosa cells by Palay and Karlin (1959) and Palay (1960). There is little question that lipid from the intestinal lumen gains access to the endoplasmic reticulum, through which it may move out of the cell or accumulate at the nuclear membrane. Although Palay and Karlin's interpretation, that the lipid reaches the reticulum as droplets engulfed in pinocytosis vacuoles, seems reasonable, the authors suggest that a more complete and critical study is required before this is considered as established. It should be indicated that, with rare exceptions such as in lung macrophages (Karrer, 1960), it is presently not possible to distinguish the membrane structure of "channels," i.e., long indentations of the cell membrane, from that of the endoplasmic reticulum. Moreover, if it should prove that lipid does indeed enter by pinocytosis vacuoles and that the vacuoles empty the lipid directly into the endoplasmic reticulum, one may ask if the situation is unique to the intestinal mucosa or to the lipid nature of the material moving through the cell. Although there may be functional connections between pinocytosis vacuoles and endoplasmic reticulum in cells generally and with substances other than lipids, evidence is lacking. With injected proteins (enzymes and ferritin) and with electron-opaque colloidal material, no accumulation is seen at the nuclear membrane. On the other hand, as we have seen, there is clear evidence of accumulation of these materials in lysosomes. If there is indeed a difference between proteins and lipids in their modes of entry and transit in the cell, can this be related to the apparent presence in the endoplasmic reticulum of esterase (see Section VII, D), but not of acid nucleases and cathepsin? Perhaps lipid may be digested in the endoplasmic reticulum (and in lysosomes too?) whereas proteins and polysaccharides are broken down in the lysosomes.

V. Lysosomes, the Storage of Insoluble Materials, and Aging

Essner and Novikoff (1960a) suggested that, in the parenchymatous cells of human liver, lipofuscin is deposited in the pericanalicular lysosomes, converting them to golden brown pigment granules. They also suggested that under pathological conditions there may be deposited in these lysosomes the unknown pigment of chronic idiopathic jaundice (Dubin-Johnson), iron, or biliary pigment. It is possible that pigments, in these and other cells, are insoluble residues of materials brought to the lysosomes by pinocytosis (page 460). Although all are of interest to the cell pathologist, one, lipofuscin, holds special interest for the gerontologist.

Of the several cytological changes described as correlated with aging the most widely accepted is the accumulation of lipofuscin granules (see de Robertis *et al.*, 1960; Bondareff, 1959; Bourne, 1957; Wilcox, 1957). That lipofuscin accumulates with age in the human myocardium has been shown quantitatively by Strehler *et al.* (1959), who consider that, "because of its absence in the very young, its presence without exception in the aged hearts studied, its lack of correlation with specific cardiac diseases or heart failure, and its large displacement of myocardial volume, the accumulation of lipofuscin in the human myocardium seems to meet the criteria set forth for a basic biological aging process." Mildvan and Strehler (1960; 1961) have isolated a pigment granule fraction by repeated centrifugations of 0.25 *M* sucrose homogenates of human myocardium. The granules contain a brightly fluorescent phospholipid with chromatographic properties of autoxidized cephalin. The fraction has little or none of the mitochondrial and microsomal enzyme activities and pigments studied. It has high levels of acid phosphatase and cathepsin activities; it also shows esterase[6] activity and trace amounts of acid ribonuclease and acid deoxyribonuclease activities. This confirms and extends the earlier reports of Heidenrich and Siebert (1955) on the enzymatic properties of a lipofuscin granule fraction isolated from human heart. It provides a consistent story with the electron microscopy and cytochemical studies from our laboratory on liver (Novikoff, 1959a; Essner and Novikoff, 1960a, b; Novikoff and Essner, 1960), the earlier histochemical studies of Gedick and Bontke (1956) showing the presence of acid phosphatase and esterase activity in the lipofuscin granules of a variety of tissues (heart, liver, ganglion cells, etc.), and the available electron micrographs of pigment granules in neurons (Hess, 1955; Bondareff, 1957). Lipofuscin granules may then be conceived as being altered lysosomes in which oxidized cephalin and other materials have accumulated.[7]

Various observations have been recorded that suggest a relation between the Golgi apparatus and lipofuscin granules. Gatenby and Moussa (1949, 1951) found the granules to be closely associated with the Golgi "canals" in sympathetic neurons of rat, mouse, rabbit, and man. They believed that the granules arose from the canals (they considered the granules to contain a form of neurosecretion). Gatenby (1951) noted

[6] The question of esterase is discussed on page 478. Esterase activity was also found in lipofuscin granules by Gomori (1955) and Gössner (1955).

[7] It may be more than coincidence that we have found numerous large lysosomes (PAS-positive granules with acid phosphatase activity) in cells of the choroid plexus (Fig. 9), and Wilcox (1959) finds it one of the best areas of the nervous system in which to demonstrate lipofuscin granules.

that the granules stained vitally with neutral red; this may be support for a relation to pinocytosis and/or Golgi apparatus (page 467).[8] Sulkin and Kuntz (1952) noted that as lipofuscin accumulated in the autonomic ganglia of aging men and dogs the Golgi apparatus fragmented and largely disappeared. Dalton and Felix (1957) show an electron micrograph of a neuron, in the mouse cerebral cortex, with granules adjacent to Golgi membranes and vacuoles. The granules are described as "presumably lipoidal pigment granules." It is not difficult to imagine them as altered Golgi vesicles, perhaps lysosomes in transition to lipofuscin-like granules. Bondareff (1957) published electron micrographs to demonstrate a close proximity of lipofuscin granules and Golgi apparatus in spinal ganglia cells, but the definition is insufficient to relate the two definitively or to support the view that the Golgi apparatus gives rise to the granules.

These observations, like those considered on page 445, suggest that more than a spatial relation exists between Golgi apparatus and lysosomes (Figs. 25-27) or their altered form, lipofuscin granules. Evidence should be sought to test the suggestion that lysosomes separate from the Golgi membranes (i.e., are enlarged Golgi vesicles) and somehow come to acquire molecules like oxidized cephalin (Mildvan and Strehler, 1960). If Golgi structures could be separated from cells with abundant lipofuscin, as Kuff and Dalton (1959) have done for rat epididymis, informative assays would be possible for such materials as the fluorescent oxidation products of cephalin.

VI. Lysosomes and Cell Death

The term "suicide-bags" (de Duve, 1959a) emphasizes two aspects of lysosomes: (1) By segregation into granules with impermeable outer membranes the hydrolytic enzymes are kept from degrading cell constituents in uncontrolled fashion. (2) Lysosomes may be expected to undergo change when autolysis and cell death occur (see also de Duve, 1959b).

Glücksmann (1951) has emphasized the widespread occurrence and likely ontogenetic importance of cell death in normal embryogenesis. This includes the well-known instances of involution, such as occur in the amphibian tadpole tail, chick Müllerian duct, and mammalian mesonephros, but also numerous less studied instances of circumscribed

[8] Behnsen (*Z. Zellforsch. u. mikroskop. Anat.* **4**, 515, 1926) is quoted by Cowdry [*In* "Cowdry's Problems of Aging" (A. E. Lansing, ed.). Williams and Wilkins, Baltimore, Maryland, 1952] as noting that nerve cells easily stain with trypan blue in young animals but that this property is lost in adults.

regions of necrosis (see also Saunders, 1960). It is of considerable interest that involutions in the adult (atretic ovarian follicles, thymus, etc.) as well as in the embryo are under hormonal control. The possibility is thus raised that lysosomes are target organelles for hormones and similar diffusible materials (de Duve, 1959a; Brachet *et al.*, 1958). It should be noted, however, that neither the biochemical nor morphological studies thus far establish a primary role for lysosome alteration in the *causation*

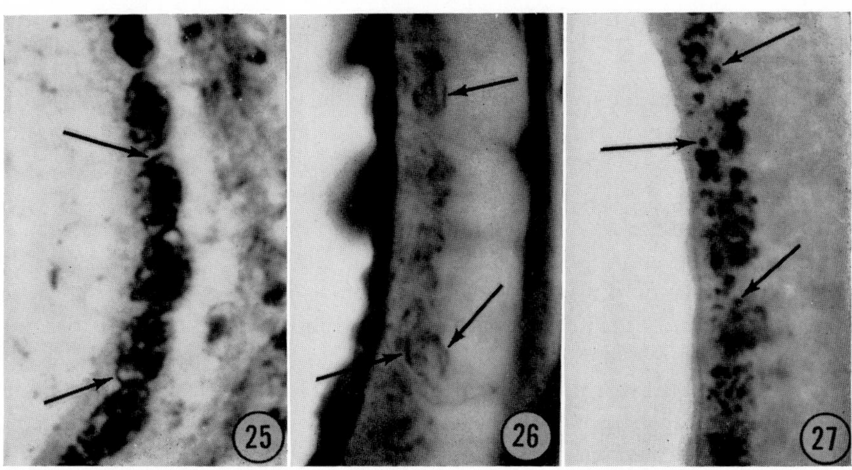

FIG. 25. Paraffin section of rat epididymis (head), Aoyama silver preparation (× 1500). Arrows indicate regions in which the canalicular nature of the Golgi apparatus is evident.

FIG. 26. Frozen section of formol-calcium-fixed rat epididymis (head), stained for adenosine diphosphate-dephosphorylation activity (× 1500). Incubation: 30 minutes, 37°C, in the lead-containing medium of Wachstein and Meisel (1957) at pH 7.1. At the arrows the localization of reaction product in canalicular structures is evident.

FIG. 27. Another section from same block as Fig. 26, stained for acid phosphatase activity (× 1500). Incubation: 15 minutes, 37°C. At the arrows the localization of reaction product in granules of variable size is evident. (Cf. Novikoff and Goldfischer, 1961b.)

of cell death (see de Duve, 1961, for preliminary results of Wattiaux suggesting that lysosome rupture precedes rather than results from cell death). However, dramatic changes in lysosomes have been encountered in all the instances of cell death—physiological (embryo and adult) and pathological—studied either biochemically or morphologically.

Brachet *et al.* (1958) have demonstrated that when the Müllerian ducts of the chick embryo regress (on both sides in the male and on the right side in the female) there is a marked rise in lysosomal enzyme activities (acid phosphatase, cathepsin, and acid ribonuclease) and also in the proportions of these activities that are unsedimentable. Wolff (1953) had earlier shown by cytochemical tests that regression of the Mullerian duct in organ culture could be induced by testosterone. Such regressing ducts show increased proteolytic and nucleolytic activities in stained preparations. The localization of acid phosphatase activity in cytoplasmic granules in the Müllerian duct is readily demonstrated by staining methods (Novikoff, 1960a). Weber (1957) has shown that when the tail of the amphibian *Xenopus* is regressing there is a marked rise in specific activity of cathepsin. The high level of acid phosphatase activity in the regressing *Rana* tadpole tail has been shown by the staining procedure (Novikoff, 1960a). By the staining procedure also, large lysosomes have been observed in the dorsal roots of the spinal ganglia of the chick embryo (Novikoff, 1960a); Saunders (personal communication) suggests the possibility that this area of the spinal ganglion may be one that undergoes cell degeneration.

Enlarged lysosomes (cytolysomes) have been observed in dying cells under a variety of physiological and pathological circumstances: atretic ovarian follicles (Fig. 10) (Novikoff, 1959d, 1960a), keratinizing squamous epithelia (Fig. 12) (Section VII, B), rat liver after ligation of the bile duct (Novikoff and Essner, 1960), transplantable tumors (Novikoff, 1960d), rat kidney after ligation of the ureter (Novikoff, 1959b), and neurons in rat brain in anoxia (Becker and Barron, 1960). Four hours after ligation of the vasculature to the kidney, the lysosomes in the cells of the proximal convolutions are smaller; perhaps if studies were done earlier, enlargement of lysosomes might be seen in these cells too. Whether or not enlargement into "cytolysomes" occurs, it is presumed that acid hydrolases and their substrates in cell substance become more readily accessible to each other under these conditions. This may be reflected in an increase in the proportion of acid hydrolases recovered in the "soluble fraction" isolated from homogenates.

An increase of lysosomal enzyme content of the "soluble fraction" has been shown to follow ligation of the blood vessels to rat liver (Beaufay and de Duve, 1957; de Duve and Beaufay, 1959) or rat kidney (Novikoff, 1960a), removal of liver fragments from the body of the mouse (Van Lancker and Holtzer, 1959b), perfusion of dog liver with nonaerated blood (Mason *et al.*, 1959), and treatment of rats by a variety of hepatotoxic substances (Deckers-Passau *et al.*, 1957; Beaufay *et al.*, 1959c; Martini and Dianzani, 1959). After ligation of the ureter, interesting changes occur in the lysosomes of cells in the proximal convolutions, but the proportion of acid phosphatase activity recovered in the "soluble fraction" is unchanged (Novikoff, 1959b, 1960a). The staining procedure should not be expected to demonstrate "soluble" enzyme except under unusual circumstances in which the movement of enzyme from tissue into fixative is curtailed. However, the appearances suggest a "solubilization" of lysosomal acid phosphatase in the keratinized layers of dead epithelial cells (Fig. 12; see also Novikoff, 1960a), in regressing tadpole tail (Novikoff, 1960a), and in the kidney after ligation of the vasculature. In the latter case, staining studies of a temporal sequence suggest a solubilization of enzyme and a release, in the dying cells, of phosphate ions which react with calcium ions (of the plasma?) to form the calcium phosphate deposits known to pathologists (cf. Becker and Barron, 1961).

According to de Duve (1959a), the observations on pathologically induced alterations in the sedimentability of lysosomal hydrolases suggest that partial or total anoxia makes the lysosomal membrane more permeable, in effect "rupturing" the lysosomes.

More detailed study seems indicated for the reported increases in lysosomal enzyme activity of spleen, thymus, and liver following total body irradiation (see Errera, 1959, p. 713 for summary). It would also be desirable to study the effects of radiation on the new class of particles described in rat liver (de Duve *et al.*, 1960) since all three enzymes thus far localized in it are related to hydrogen peroxide (see Section II).

We will conclude this section by emphasizing the advantage, for cytological studies of cell death, of the simple but sensitive staining method for acid phosphatase activity. Striking changes in lysosomes are evident at a time when routine preparations show little or no change in either nuclei or cytoplasm.

VII. Unsettled Problems; Future Perspectives

The tentative and speculative nature of the ideas expressed regarding lysosomes and related particles reflect the recency of the discovery of the particles and the restriction, until recently, of intensive study of their biochemistry to de Duve's laboratory, of their staining properties to two or three laboratories, and of their fine structure to our laboratory. More widespread study is, however, assured, and with it will come data with which to answer many questions.

A. *Micropinocytosis and Larger Cytotic Vacuoles*

Will other cells show the pattern of small vesicles, assumed to be micropinocytosis vacuoles, in linear arrays leading to lysosomes? Do any of these arrays arise, not from the cell membrane, but from Golgi membranes? Is this a mechanism by which hydrolytic enzymes, as well as proteins and other ingested materials, reach the lysosomes or digestive vacuoles? The situation is likely to be more complex, as indicated by recent unpublished observations with the Novikoff ascites hepatoma cells. All these cells show abundant acid phosphatase-rich bodies or lysosomes (Birns *et al.*, 1959; Novikoff, 1960d), yet, with the light microscope, only a few cells show vacuoles containing either horseradish peroxidase or fluorescein-labeled rabbit globulin added to the suspension medium (cf. Holtzer and Holtzer, 1960).

Will the current trend continue toward minimizing differences that may exist between pinocytosis and phagocytosis with respect to mechanisms of adsorption to the cell membrane, vacuole formation, and other aspects (page 447)? If so, there would be value in the generalized term, cytosis, that we have used to embrace both pinocytosis and phagocytosis. Cytosis is the uptake of materials (solid, liquid; large, small) from the environment in a discontinuous process, by "gulps" or "quanta." The future will tell whether cytosis occurs in all cells.

B. *Cell Death; Hormone and Vitamin Action*

Does enlargement of lysosomes into "cytolysomes" and "solubilization" of acid hydrolases occur in all cells as they begin to die? To what extent is such enlargement reversible? Do lysosome changes participate causally in events leading to cell death or are they manifestations of such events? How are observations of increased cell permeability with cell injury (e.g., Holtzer and Holtzer, 1960; Lewis and McCoy, 1922) to be fitted into this picture?

Will future work provide evidence for the suggestion that lysosomes may be target organelles for hormones and other diffusible materials, perhaps by alteration of their membrane structure (Willmer, 1960)? Is this related to the observation (Novikoff, 1960a) of unusually high levels of acid phosphatase activity in endocrine glands (Section III, C)?

The addition of vitamin A to cultures of the cartilaginous limb-bone rudiments of chick embryos, causes loss of metachromasia from the cartilage (Fell and Mellanby, 1952), and increases the proteolytic action of the explants on the culture medium to twice that of the controls without added vitamin A (Dingle *et al.*, 1961). Similar effects on the matrix can be produced by adding papain protease to the culture medium (Fell and Thomas, 1960). These results suggested that vitamin A acts on the matrix by increasing the proteolytic activity of the chondrocytes. That normal embryonic cartilage *ex vivo* contains a protease capable of producing the vitamin A changes was shown by Lucy *et al.* (1961); they placed normal cartilage from 10-day-old chick embryos, in distilled water at 4°C. for 30 minutes to disrupt the cells and their organelles, and then incubated the rudiments in buffers of different pH for 2 hours at 37°C.

The question may be raised whether vacuoles, particularly those that turn over rapidly, can rightly be called "-somes." Does the word "soma" signify more highly structured persistent bodies? Even though, etymologically, soma may have this meaning, in cell physiology there seems to us to be little fundamental difference between a rapidly changing vacuole and a longer-lasting "condensed vacuole" or a long-lasting lipofuscin granule. Furthermore, other cell organelles such as mitochondria may turn over more rapidly than is sometimes thought [recently, Fletcher and Sanadi (unpublished) have suggested that the half-life for rat liver mitochondria is only 10½ days].

The more rapid turnover rate of lysosomes in many cells, perhaps decisive to their short-term metabolic functions, may contribute to the difficulty of elucidating the biochemical properties of these particles. Together with the small size of the particles, their rapidly changing character may account both for the fact that they were largely overlooked in the earlier cytology and for the difficulty encountered in purifying them by centrifugation.

Acknowledgments

The work done in our laboratories has been greatly aided by generous and sustained support from the American Cancer Society, the United States Public Health Service, the National Science Foundation, and the Damon Runyon Memorial Fund.

It is a pleasure to acknowledge the assistance of my colleagues: Mr. L. J. Walker in the photography; Mrs. Joan Drucker and Mr. Woo-Yung Shin in the preparation of the sections shown in Figs. 5–16; Miss Barbro Runling in the electron microscopy; and Drs. Edward Essner and Maurice Steinert for their studies of the erythrophagocytes and trypanosomes, respectively.

My sincere thanks go to Dr. Christian de Duve for the opportunity to work in his laboratory and for stimulating collaboration that has continued since 1955.

References

Abolins, L., and Abolins, A. (1949). *Nature* **164**, 455.
Anderson, E., and Beams, H. W. (1960). *J. Ultrastruct. Research* **3**, 432.
Appelmans, F., Wattiaux, R., and de Duve, C. (1955). *Biochem. J.* **59**, 438.
Arzac, J. P., and Flores, L. G. (1952). *Stain Technol.* **27**, 9.
Aterman, K. (1952a). *A.M.A. Arch. Pathol.* **53**, 197.
Aterman, K. (1952b). *A.M.A. Arch. Pathol.* **53**, 209.
Avers, C. J., and King, E. E. (1960). *Am. J. Botany* **47**, 220.
Bairati, A., and Lehmann, F. E. (1953). *Exptl. Cell Research* **5**, 220.
Baker, J. R. (1946). *Quart. J. mic. Sci.* **87**, 441.
Barka, T. (1960a). *J. Histochem. and Cytochem.* **8**, 320.
Barka, T. (1960b). *Nature* **187**, 248.
Barka, T., Schaffner, F., and Popper, H. (1960). *Federation Proc.* **19**, 187.
Barrows, C. H. (1960). *Abstr. Intern. Congr. Gerontol. 5th Congr.* p. 30. San Francisco, California.

Beaufay, H., and de Duve, C. (1957). *Arch. intern. physiol. et biochem.* **65**, 156.

Beaufay, H., and de Duve, C. (1959). *Biochem. J.* **73**, 604.

Beaufay, H., Bendall, D. S., Baudhuin, P., and de Duve, C. (1959a). *Biochem. J.* **73**, 623.

Beaufay, H., Bendall, D. S., Baudhuin, P., Wattiaux, R., and de Duve, C. (1959b). *Biochem. J.* **73**, 628.

Beaufay, H., Van Campenhout, E., and de Duve, C. (1959c). *Biochem. J.* **73**, 617.

Becker, N. H., and Barron, K. D. (1960). *Am. J. Pathol.* **38**, 161.

Becker, N. H., Goldfischer, S., Shin, W.-Y., and Novikoff, A. B. (1960). *J. Biophys. Biochem. Cytol.* **8**, 649.

Benedetti, E. L., and Bernhard, W. (1958). *J. Ultrastruct. Research* **1**, 309.

Benedetti, L. E., and Leplus, R. (1958). *Rev. hématol.* **13**, 199.

Bennett, H. S. (1956a). *J. Biophysic. Biochem. Cytol.* **2** Suppl., 99.

Bennett, H. S. (1956b). *J. Biophysic. Biochem. Cytol.* **2** Suppl., 185.

Berg, W. E. (1957). *Biol. Bull.* **113**, 365.

Berg, W. E., and Humphreys, W. J. (1960). *Develop. Biol.* **2**, 42.

Bernhard, W., Bauer, A., Guérin, M., and Oberling, Ch. (1955). *Bull. du Cancer* **62**, 163.

Berthet, J., and de Duve, C. (1951). *Biochem. J.* **50**, 174.

Berthet, J., Berthet, L., Appelmans, F., and de Duve, C. (1951). *Biochem. J.* **50**, 182.

Bessis, M. (1959). *In* "The Kinetics of Cellular Proliferation" (F. Stohlman, Jr., ed.), pp. 22-29. Grune and Stratton, New York.

Bessis, M. (1960). *Rev. hématol.* **15**, 233.

Bessis, M., and Breton-Gorius, J. (1959). *Rev. hématol.* **14**, 165.

Birns, M. (1960). *Exptl. Cell Research* **20**, 202.

Birns, M., Essner, E., and Novikoff, A. B. (1959). *Proc. Am. Assoc. Cancer Research* **3**, 7.

Bondareff, W. (1957). *J. Gerontol.* **12**, 364.

Bondareff, W. (1959). *In* "Handbook of Aging and the Individual" (J. E. Birren, ed.), pp. 136-172. Univ. of Chicago Press, Chicago, Illinois.

Bourne, G. H. (1957). *In* "Modern Trends in Geriatrics" (W. Hobson, ed.), pp. 22-49. Hoeber, New York.

Brachet, J. (1960). "The Biochemistry of Development," 320 pp. Pergamon Press, New York.

Brachet, J., Decroly-Briers, M., and Hoyez, J. (1958). *Bull. soc. chim. biol.* **40**, 2039.

Brandt, P. W. (1958). *Exptl. Cell Research* **15**, 300.

Brandt, P. W., and Pappas, G. D. (1960). *Anat. Record* **136**, 16.

Burgos, M. (1960). *Anat. Record* **137**, 171.

Burstone, M. S. (1960). *J. Histochem. and Cytochem.* **8**, 341.

Case, N. M. (1959). *J. Biophys. Biochem. Cytol.* **6**, 527.

Chapman-Andresen, C. (1960). *Congr. intern. biol. cellulaire, 10th Congr., Paris* p. 100.

Charvát, J., Schreiber, V., and Kmentovà, V. (1959). *Rev. Czech. Med.* **5**, 1.

Claude, A. (1960a). *Arch. Intern. Physiol. Biochem.* **68**, 672.

Claude, A. (1960b). "Berliner Symposium über Fragen der Carcinogenese," pp. 241-242. Akademische-Verlag, Berlin.

Cohn, Z. A., and Hirsch, J. G. (1960a). *J. Exptl. Med.* **112**, 983.

Cohn, Z. A., and Hirsch, J. G. (1960b). *J. Exptl. Med.* **112**, 1015.

Coons, A. H. (1956). *Intern. Rev. Cytol.* **5**, 1.

Dalcq, M. (1960). *In* "Fundamental Aspects of Normal and Malignant Growth" (W. W. Nowinski, ed.), pp. 305-494. Elsevier, New York.
Dalton, A. J., and Felix, M. D. (1957). *Symposium Soc. Exp. Biol.* **10**, 148.
Davies, J. (1954). *Am. J. Anat.* **94**, 45.
Deane, H. W. (1947). *Am. J. Anat.* **80**, 321.
Deane, H. W. (1958). *In* "Frontiers of Cytology" (S. L. Palay, ed.), pp. 227-263, Yale Univ. Press, New Haven, Connecticut.
Deane, H. W., and Dempsey, E. W. (1945). *Anat. Record* **93**, 401.
Deckers-Passau, L., Maisin, J., and de Duve, C. (1957). *Acta Unio Intern. contra Cancrum* **13**, 822.
de Duve, C. (1959a). *In* "Subcellular Particles" (T. Hayashi, ed.), pp. 128-159. Ronald Press, New York.
de Duve, C. (1959b). *Exptl. Cell Research Suppl.* **7**, 169.
de Duve, C. (1960a). *Bull. Soc. Chim. biol.* **42**, 11.
de Duve, C. (1960b). *Nature,* **187**, 836.
de Duve, C. (1961). *In* "Biological Aspects of Cancer Chemotherapy" (R. J. C. Harris, ed.). Academic Press, New York (in press).
de Duve, C., and Beaufay, H. (1959). *Biochem. J.* **73**, 610.
de Duve, C., and Berthet, J. (1954). *Intern. Rev. Cytol.* **3**, 225.
de Duve, C., Gianetto, R., Appelmans, F., and Wattiaux, R. (1953). *Nature* **172**, 1143.
de Duve, C., Pressman, B. C., Gianetto, R., Wattiaux, R., and Appelmans, F. (1955). *Biochem. J.* **60**, 604.
de Duve, C., Berthet, J., and Beaufay, H. (1959). *Progr. Biophys.* **9**, 325.
de Duve, C., Beaufay, H., Jacques, P., Rahman-Li, Y., Sellinger, O. Z., Wattiaux, R., and de Coninck, S. (1960). *Biochim. Biophys. Acta* **40**, 186.
de Man, J. C. H., Daems, W. Th., Willighagen, R. G. J., and van Rijssel, Th. G. (1960). *J. Ultrastruct. Research* **4**, 43.
De Robertis, E. D. P., Nowinski, W. W., and Saez, F. A. (1960). "General Cytology," 3rd ed. p. 555. Saunders, Philadelphia, Pennsylvania.
Deuchar, E. M. (1960). *Develop. Biol.* **2**, 129.
Dingle, J. T. (1961). *Biochem. J.* (in press).
Dingle, J. T., Lucy, J. A., and Fell, H. B. (1961). *Biochem. J.* (in press).
Doniach, I., and Weinbren, K. (1952). *Brit. J. Exptl. Pathol.* **33**, 499.
Edwards, G. (1925). *Biol. Bull.* **48**, 236.
Ehrlich, J., Novikoff, A. B., Platt, R., and Essner, E. (1960). *Bull. N.Y. Acad. Med.* **36**, 488.
Eränkö, O. (1951). *Acta Physiol. Scand.* **24**, 1.
Errera, M. (1959). *In* "The Cell" (J. Brachet and A. E. Mirsky, eds.), p. 695. Academic Press, New York.
Essner, E. (1960). *J. Biophys. Biochem. Cytol.* **7**, 329.
Essner, E., and Novikoff, A. B. (1960a). *J. Ultrastruct. Research* **3**, 374.
Essner, E., and Novikoff, A. B. (1960b). *J. Histochem. Cytochem.* **8**, 318.
Essner, E., and Novikoff, A. B. (1961). *J. Biophys. Biochem. Cytol.* (in press).
Farquhar, M. G., and Palade, G. E. (1960). *J. Biophys. Biochem. Cytol.* **7**, 297.
Farquhar, M. G., Wissig, S. L., and Palade, G. E. (1961). *J. Exptl. Med.* **113**, 47.
Fawcett, D. W. (1959). *In* "Developmental Cytology" (D. Rudnick, ed.), pp. 161-189. Ronald Press, New York.
Felix, M. D., and Dalton, A. J. (1956). *J. Biophys. Biochem. Cytol. Suppl.* **2**, 109.
Fell, H. B., and Mellanby, E. (1952). *J. Physiol.* **116**, 320.

Fell, H. B., and Thomas, L. (1960). *J. Exptl. Med.* **111**, 719.
Finkelstaedt, J. T. (1957). *Proc. Soc. Exptl. Biol. Med.* **95**, 302.
Gansler, H., and Rouiller, C. (1956). *Schweiz. Z. Pathol. u. Bakteriol.* **19**, 217.
Gatenby, J. B. (1951). *Nature* **167**, 185.
Gatenby, J. B., and Moussa, T. A. A. (1949). *J. Roy. Microscop. Soc.* **69**, 185.
Gatenby, J. B., and Moussa, T. A. A. (1951). *J. Physiol. (London)* **114**, 252.
Gedick, P., and Bontke, E. (1956). *Z. Zellforsch. u. mikroskop. Anat.* **44**, 495.
Gérard, P., and Cordier, R. (1934). *Biol. Revs. Biol. Proc. Cambridge Phil. Soc.* **9**, 110.
Gersh, I. (1949). *A.M.A. Arch. Pathol.* **47**, 99.
Glücksmann, A. (1951). *Physiol. Revs.* **26**, 59.
Gössner, W. (1955). *Verhandl. deut. Ges. Pathol.* **39**, 193.
Golberg, L., Martin, L. E., and Batchelor, A. (1960). *Biochem. J.* **77**, 252.
Gomori, G. (1952). "Microscopic Histochemistry: Principles and Practice." Univ. Chicago Press, Chicago, Illinois.
Gomori, G. (1955). *J. Histochem. and Cytochem.* **3**, 479.
Gomori, G. (1956). *J. Histochem. and Cytochem.* **4**, 453.
Gosselin, R. E. (1956). *J. Gen. Physiol.* **39**, 625.
Granboulan, N. (1960). *Rev. hématol.* **15**, 52.
Greenbaum, A. L., Slater, T. F., and Wang, D. Y. (1960). *Nature* **188**, 318.
Haguenau, F. (1961). *In* "Biological Aspects of Cancer Chemotherapy" (R. J. C. Harris, ed.). Academic Press, New York (in press).
Hampton, J. C. (1958). *Acta Anat.* **32**, 262.
Heidenreich, A., and Siebert, G. (1955). *Arch. pathol. Anat. u. Physiol. Virchow's* **327**, 112.
Hess, A. (1955). *Anat. Record* **123**, 399.
Hess, R., and Pearse, A. G. E. (1958). *Brit. J. Exptl. Pathol.* **39**, 292.
Hirsch, J. G., and Cohn, Z. A. (1960). *J. Exptl. Med.* **112**, 1005.
Höber, R. (1940). *Cold Spring Harbor Symposia Quant. Biol.* **8**, 40.
Holt, S. J. (1954). *Proc. Roy. Soc.* **B142**, 160.
Holt, S. J. (1956). *J. Histochem. and Cytochem.* **4**, 541.
Holt, S. J. (1959). *Exptl. Cell Research Suppl.* **7**, 1.
Holter, H. (1954). *Proc. Roy. Soc.* **B142**, 140.
Holter, H. (1956). *In* "Fine Structure of Cells," pp. 71-76. Noordhof, Groningen.
Holter, H. (1959). *Intern. Rev. Cytol.* **8**, 481.
Holter, H. (1961). *In* "Biological Aspects of Cancer Chemotherapy" (R. J. C. Harris, ed.). Academic Press, New York (in press).
Holter, H., and Lowy, B. A. (1959). *Compt. rend. trav. lab. Carlsberg* **31**, 105.
Holtzer, H., and Holtzer, S. (1960). *Compt. rend. trav. lab. Carlsberg* **31**, 373.
Huber, P. (1953). *Helv. Physiol. Pharmacol. Acta* **11**, C 41.
Jacques, P., Straus, W., and de Duve, C. (1961). *Biochem. J.* (in press).
Karrer, H. E. (1958). *J. Biophys. Biochem. Cytol.* **4**, 693.
Karrer, H. E. (1960). *J. Biophys. Biochem. Cytol.* **7**, 357.
Kobayashi, H., Wolfson, A., Wise, M. A., and Haubrich, D. R. (1960). *Anat. Record* **137**, 372.
Kojima, M. K. (1959a). *Embryologia* **4**, 191.
Kojima, M. K. (1959b). *Embryologia* **4**, 211.
Kruse, H., and McMaster, P. D. (1949). *J. Exptl. Med.* **90**, 425.
Kuff, E. L., and Dalton, A. J. (1959). *In* "Subcellular Particles" (T. Hayashi, ed.), pp. 114-127. Ronald Press, New York.

activities (Novikoff *et al.*, 1953; Underhay *et al.*, 1956). Unfortunately, esterase determinations on the "droplet fractions" of rat kidney appear not to have been done. Lipofuscin granule fractions isolated from human heart have esterase as well as acid phosphatase and catheptic activities (Heidenreich and Siebert, 1955; Mildvan and Strehler, 1961).

Two approaches have been used to reconcile the apparent microsomal localization of esterase in biochemical studies of rat liver with the apparent lysosomal localization in stained preparations of this and other organs. Holt has observed that placing frozen-dried sections in cold sucrose solution results in the loss from the lysosomes of esterase activity but not of acid phosphatase activity (see de Duve, 1960b). He considers esterase to be an exclusively lysosomal enzyme (perhaps situated on its surface) which in the course of homogenization is eluted from the lysosomes and is adsorbed by the microsomes. On the other hand, Wachstein and Meisel (1960) and Shnitka and Seligman (1960) observe that exclusively lysosomal localization in stained preparations is obtained only in the presence of inhibitors like the organophosphorus compound E-600 or the ferro-ferricyanide system employed in Holt's staining procedure. They consider that there is an inhibitor-resistant esterase in the lysosomes and an inhibitor-sensitive esterase in the general cytoplasm.

We have also observed the inhibition by E-600 of generalized cytoplasmic esterase activity, as first noted in kidney and other cells by Hess and Pearse (1958). With Miss Greta Heus, we have studied the effects of E-600 on the esterase activities of isolated subcellular fractions, employing α-naphthyl acetate as substrate. When compared with the original level of activity in the homogenate, the light mitochondrial fraction, isolated by the method of de Duve *et al.* (1955), had a six- to eightfold enrichment of acid phosphatase activity and a three- to fourfold enrichment of E-600-resistant esterase activity. The microsomal esterase activity was 98–100% inhibited by the E-600 (1×10^{-6} *M*). However, even the light mitochondrial fraction was inhibited to the extent of 90–94%. This high inhibition may reflect the presence of microsomes in the fraction. It should be noted that the E-600-resistant activity makes a small contribution, less than 3%, to the total activity in the homogenate. It would be without important effect on the distribution patterns of total esterase activity among isolated fractions. On the other hand, its presence would be of considerable functional significance since it would mean that lipids, like other cell constituents, could be hydrolyzed by lysosomal enzymes (see page 469). Pearse (1960) discusses the possible relation of E-600-resistant esterase activity to cathepsin C.

E. Origin of Lysosomal Enzymes; Their "Solubilization"; Their Role in Embryonic Development; The Value of Staining Reactions

Increasing use may be anticipated of cell smears and tissue sections in which both the morphology of the lysosomes and sufficient enzyme activities are preserved. This is due to two factors. The staining methods are so simple that a person can learn to do them in a day or two. Secondly, despite their qualitative nature staining methods can make significant contributions.

It is apparent that staining methods, because they lack true quantitation, cannot *establish* the occurrence of induction of lysosomal enzymes like acid phosphatase. Yet, already they suggest either the appearance, or a marked increase in level, of acid phosphatase activity as phagocytosis vacuoles appear in amebae (Section IV, E), planaria (Rosenbaum and Rolon, 1960), macrophages (Montagna, 1952; Weiss and Fawcett, 1953), and other phagocytes (pages 443). Staining methods cannot establish changes in levels of total and "soluble" enzyme activity, yet they suggest such changes, e.g., after ligation of the renal vessels (page 475).

Expanding application of these staining methods to eggs and embryos may be expected. In addition to work on metachromatic granules (multivesicular bodies) that was discussed earlier, there is chemical evidence for the presence of granule-bound catheptic activity (Lundblad and Lundblad, 1954). Important changes in catheptic activity have been noted not only during regressions (page 473), but also when chick embryos are treated with the amino acid analog, ω-bromoallylglycine (Deuchar, 1960). Dalcq (1960) suggests that embryonic induction may involve the movement of ribonucleoprotein into reactor cells, where it is "attacked first by a proteinase, then by a cell ribonuclease, and the various fragments of the molecule introduced into chains of unknown but imaginable reactions."

F. Intracellular Self-Regulation; Turnover of Organelles

Among the likely homeostatic mechanisms operating on a subcellular level there may be those that control accessibility of cell constituents and the acid hydrolases segregated in lysosomes. These mechanisms may be varied and subtle. The digestive vacuoles of protozoa, and perhaps of phagocytes in higher organisms, turn over rapidly; the same is apparently true of pinocytosis vacuoles. Yet in the Kupffer cells some materials remain within these vacuoles for a long time, perhaps throughout the life of the cell. This seems to be true of lipofuscins that accumulate within lysosomes. The influence of genetic factors on the distribution of a lysosomal enzyme has been described by Paigen (1959).

Lacy, D., and Challice, C. E. (1957). *Symposia Soc. Exptl. Biol.* **10**, 62.
Lazarus, S. S. (1959). *Proc. Soc. Exptl. Biol. Med.* **102**, 303.
Leblond, C. P. (1950). *Am. J. Anat.* **86**, 1.
Lewis, W. H. (1931). *Bull. Johns Hopkins Hosp.* **49**, 17.
Lewis, W. H., and McCoy, C. C. (1922). *Bull. Johns Hopkins Hosp.* **33**, 284.
Lindner, E. (1958). *Erg. allg. Path. u. path. Anat.* **38**, 46.
Lojda, Z. (1960). *Morfologie* **8**, 148.
Lojda, Z., Schreiber, V., and Kmentovā, V. (1960). *Českoslov. fysiol.* **9**, 173.
Logothetopoulos, J., and Weinbren, K. (1955). *Brit. J. Exptl. Pathol.* **36**, 402.
Low, F. N., and Freeman, J. A. (1958). "Electron Microscopic Atlas of Normal and Leukemic Human Blood." pp. 347. McGraw-Hill, New York.
Lucy, J. A., Dingle, J. T., and Fell, H. B. (1961). *Biochem. J.* (in press).
Lundblad, G., and Lundblad, I. (1954). *Arkiv. Kemi* **6**, 387.
Marsland, D. (1958). *Anat. Record* **132**, 473.
Marsland, D., Zimmerman, A. M., and Auclair, W. (1960). *Exptl. Cell Research* **21**, 179.
Martini, E., and Dianzani, M. U. (1959). *Experientia* **14**, 285.
Mason, E. E., Lee, R. A., Smith, J., and Dierks, C. (1959). *Surgery* **45**, 765.
Mast, S. O., and Doyle, W. L. (1934). *Protoplasma* **20**, 555.
Mercer, E. H. (1959). *Proc. Roy. Soc.* **B150**, 216.
Mildvan, A. S., and Strehler, B. L. (1960). *Abstr. Intern. Congr. Gerontol., 5th Congr. San Francisco, California.*
Mildvan, A. S., and Strehler, B. L. (1961) (in press).
Miller, F. (1960). *J. Biophys. Biochem. Cytol.* **8**, 689.
Mitchell, P., and Moyle, J. (1956). *Discussions Faraday Soc.* **21**, 258.
Mitsui, T., and Ikeda, S. (1951). *Okajimas Folia Anat. Japon.* **23**, 331.
Montagna, W. (1952). *Ann. N.Y. Acad. Sci.* **55**, 629.
Müller, M., Tóth, J., and Törö, I. (1960). *Nature* **187**, 65.
Mulnard, J., Auclair, W., and Marsland, D. (1959). *J. Embryol. Exptl. Morphol.* **7**, 223.
Nachmias, V. T., and Marshall, J. M., Jr. (1960). *Anat. Record* **136**, 250.
Nachmias, V. T., and Marshall, J. M., Jr. (1961). *In* "Symposium on Biological Structure and Function, Stockholm." In press.
Nilsson, O. (1958). *J. Ultrastruct. Research* **2**, 73.
Nilsson, O. (1959). *J. Ultrastruct. Research* **2**, 331.
Novikoff, A. B. (1957). *Symposium Soc. Exptl. Biol.* **10**, 92.
Novikoff, A. B. (1959a). *Biol. Bull.* **117**, 385.
Novikoff, A. B. (1959b). *J. Biophys. Biochem. Cytol.* **6**, 136.
Novikoff, A. B. (1959c). *In* "Analytical Cytology" (R. C. Mellors, ed.), 2nd ed., pp. 69-168. McGraw-Hill, New York.
Novikoff, A. B. (1959d). *Bull. N.Y. Acad. Med.* **35**, 67.
Novikoff, A. B. (1959e). *J. Histochem. and Cytochem.* **7**, 240.
Novikoff, A. B. (1959f). *In* "Subcellular Particles" (T. Hayashi, ed.), pp. 1-22. Ronald Press, New York.
Novikoff, A. B. (1960a). *In* "Developing Cell Systems and Their Control" (D. Rudnick, ed.), pp. 167-203. Ronald Press, New York.
Novikoff, A. B. (1960b). *In* "Biology of Pyelonephritis" (E. Quinn and E. Kass, eds.), pp. 113-144. Little, Brown, Boston, Massachusetts.
Novikoff, A. B. (1960c). *Acta Unio. Intern. contra Cancrum* **16**, 966.

Novikoff, A. B. (1960d). *In* "Cell Physiology of Neoplasia," p. 219. The Univ. of Texas Press, Austin, Texas.
Novikoff, A. B., and Essner, E. (1960). *Am. J. Med.* **29**, 102.
Novikoff, A. B., and Goldfischer, S. (1961a). *5th Intern. Congr. Biochem., Moscow* (in press).
Novikoff, A. B., and Goldfischer, S. (1961b). *Proc. Natl. Acad. Sci. U.S.* (in press).
Novikoff, A. B., and Noe, E. F. (1955). *J. Morphol.* **96**, 189.
Novikoff, A. B., Podber, E., Ryan, J., and Noe, E. (1953). *J. Histochem. and Cytochem.* **1**, 27.
Novikoff, A. B., Beaufay, H., and de Duve, C. (1956). *J. Biophys. Biochem. Cytol.* **2**, Suppl., 179.
Novikoff, A. B., Runling, B., Drucker, J., and Kaplan, S. E. (1960a). *J. Histochem. and Cytochem.* **8**, 319.
Novikoff, A. B., Shin, W.-Y., and Drucker, J. (1960b). *J. Histochem. and Cytochem.* **8**, 37.
Novikoff, A. B., Goldfischer, S., Essner, E., and Iaciofano, P. (1961). *J. Histochem. and Cytochem.* (in press).
Odor, D. L. (1956). *J. Biophys. Biochem. Cytol.* **2** Suppl., 105.
Ogawa, K., Mizuno, N., Hashimoto, K., Fujii, S., and Okamoto, M. (1960). *Proc. Dept. Anat., Kyôto Univ., School Med.* **4**, 1.
Oliver, J., and MacDowell, M. (1958). *J. Exptl. Med.* **107**, 731.
Oliver, J., Moses, M. J., MacDowell, M. C., and Lee, Y. C. (1954). *J. Exptl. Med.* **99**, 605.
Paigen, K. (1959). *J. Histochem. and Cytochem.* **7**, 248.
Paigen, K., and Griffiths, S. K. (1959). *J. Biol. Chem.* **234**, 299.
Palade, G. E. (1951). *Arch. Biochem.* **30**, 144.
Palade, G. E. (1956). *J. Biophys. Biochem. Cytol.* **2** Suppl., 85.
Palade, G. E. (1960). *Anat. Record,* **136**, 254.
Palay, S. L. (1958). *In* "Frontiers of Cytology" (S. L. Palay, ed.), pp. 305-342. Yale Univ. Press, New Haven, Connecticut.
Palay, S. L. (1960). *J. Biophys. Biochem. Cytol.* **7**, 391.
Palay, S. L., and Karlin, L. J. (1959). *Biophys. Biochem. Cytol.* **5**, 373.
Pappas, G. D. (1959). *Ann. N.Y. Acad. Sci.* **78**, 448.
Pasteels, J. J. (1958). *In* "Chemical Basis of Development" (W. D. McElroy and B. Glass, eds.), pp. 383-403; 414-415. Johns Hopkins Press, Baltimore, Maryland.
Pasteels, J. J., and Mulnard, J. (1957). *Arch. biol. (Liège)* **68**, 115.
Pasteels, J. J., and Mulnard, J. (1960). *Compt. rend. acad. sci.* **250**, 190.
Pasteels, J. J., Castiaux, P., and Vandermeersche, G. (1958). *J. Biophys. Biochem. Cytol.* **4**, 575.
Pearse, A. G. E. (1960). "Histochemistry, Theoretical and Applied," 998 pp. Little, Brown, Boston, Massachusetts.
Popper, H., and Schaffner, F. (1957). "Liver: Structure and Function." McGraw-Hill, New York.
Price, J. M., and Laird, A. K. (1950). *Cancer Research* **10**, 650.
Quertier, J., and Brachet, J. (1959). *Arch. biol. (Liège)* **70**, 153.
Rebhun, L. I. (1959). *Biol. Bull.* **117**, 518.
Rebhun, L. I. (1961). *Ann. N.Y. Acad. Sci.* (in press).
Rhodin, J., and Dalhamn, T. (1956). *Z. Zellforsch. u. mikroskop. Anat.* **44**, 345.
Robertson, J. D. (1959). *Biochem. Soc. Symposia (Cambridge, Engl.)* **16**, 3.
Rose, G. G. (1957). *J. Biophys. Biochem. Cytol.* **3**, 697.

Rosenbaum, R., and Rolon, C. I. (1960). *Biol. Bull.* **118**, 315.

Roth, L. E. (1960). *J. Protozool.* **7**, 176.

Rouiller, C., and Bernhard, W. (1956). *J. Biophys. Biochem. Cytol.* **2**, Suppl. 355.

Roy, A. B. (1958). *Biochem. J.* **68**, 519.

Rudzinska, M. A., and Trager, W. (1957). *J. Protozool.* **4**, 190.

Rudzinska, M. A., and Trager, W. (1959). *J. Biophys. Biochem. Cytol.* **6**, 103.

Rutenburg, A. M., and Seligman, A. M. (1955). *J. Histochem. Cytochem.* **3**, 455.

Sabin, F. (1939). *J. Exptl. Med.* **70**, 67.

Sampaio, M. M. (1956). *Anat. Record* **124**, 501.

Saunders, J. W., Jr. (1960). *Congr. intern. biol. cellulaire, 10th Congr., Paris,* p. 139.

Schechtman, A. M. (1956). *Intern. Rev. Cytol.* **5**, 303.

Schein, A. H., Podber, E., and Novikoff, A. B. (1951). *J. Biol. Chem.* **190**, 331.

Schiller, A. A., Schayer, R. W., and Hess, E. L. (1953). *J. Gen. Physiol.* **36**, 489.

Schneider, W. C., and Hogeboom, G. H. (1952). *J. Biol. Chem.* **195**, 161.

Schneider, W. C., and Kuff, E. L. (1954). *Am. J. Anat.* **94**, 209.

Schreiber, V., and Kmentová, V. (1959). *Endokrinologie* **38**, 69.

Schreiber, V., Charvát, J., Kmentová, V., and Rybák, M. (1959). *Nature* **183**, 473.

Schreiber, V., Rybák, M., and Kmentová, V. (1960). *Physiol. Bohemosloven.* **9**, 303.

Schreiber, V., Charvát, J., Lojda, Z., Rybák, M., and Jirgl, V. (1960). *First Intern. Congr. Endocrinol., Copenhagen,* p. 91.

Schulz, H. (1958). *Beit. pathol. Anat. u. allgem. Pathol.* **119**, 71.

Sellinger, O. Z., Beaufay, H., Jacques, P., Doyen, A., and de Duve, C. (1960). *Biochem. J.* **74**, 450.

Shnitka, T. K., and Seligman, A. M. (1960). *J. Histochem. and Cytochem.* **8**, 344.

Sjöstrand, F. (1944). *Acta Anat. Suppl. No.* **1**.

Sloper, J. C. (1955). *J. Anat.* **89**, 301.

Smetana, H. (1947). *Am. J. Pathol.* **23**, 255.

Sotelo, J. R., and Porter, K. R. (1959). *J. Biophys. Biochem. Cytol.* **5**, 327.

Steinert, M. (1960). *J. Biophys. Biochem. Cytol.* **8**, 542.

Steinert, M., and Novikoff, A. B. (1960). *J. Biophys. Biochem. Cytol.* **8**, 563.

Stott, F. C. (1954). *Proc. Zool. Soc. (London)* **125**, 63.

Straus, W. (1954). *J. Biol. Chem.* **207**, 745.

Straus, W. (1956). *J. Biophys. Biochem. Cytol.* **2**, 513.

Straus, W. (1957a). *J. Biophys. Biochem. Cytol.* **3**, 933.

Straus, W. (1957b). *J. Biophys. Biochem. Cytol.* **3**, 1037.

Straus, W. (1958). *J. Biophys. Biochem. Cytol.* **4**, 541.

Straus, W. (1959). *J. Biophys. Biochem. Cytol.* **5**, 193.

Straus, W., and Oliver, J. (1955). *J. Exptl. Med.* **102**, 1.

Strehler, B. L., Mark, D. D., Mildvan, A. S., and Gee, M. V. (1959). *J. Gerontol.* **14**, 430.

Sulkin, N. M., and Kuntz, A. (1952). *J. Gerontol.* **7**, 533.

Tappel, A. L., Zalkin, H., Desai, I., and Caldwell, K. (1961). *Federation Proc.* (in press).

Telfer, W. H., and Koch, W. E. (1958). *Anat. Record* **132**, 513.

Tennyson, V. M. (1961). *J. Biophys. Biochem. Cytol.* (in press).

Tsuboi, K. K. (1952). *Biochim. Biophys. Acta* **8**, 173.

Underhay, E., Holt, S. J., Beaufay, H., and de Duve, C. (1956). *J. Biophys. Biochem. Cytol.* **2**, 635.

Van Duijn, P., Willighagen, R. G. J., and Meijer, A. E. F. H. (1959). *Biochem. Pharmacol.* **2**, 177.

Van Lancker, J. L., and Holtzer, R. L. (1959a). *Am. J. Pathol.* **35**, 563.
Van Lancker, J. L., and Holtzer, R. L. (1959b). *J. Biol. Chem.* **234**, 2359.
Van Lancker, J. L., and Morrill, G. A. (1958). *Federation Proc.* **17**, 463.
Viala, R., and Gianetto, R. (1955). *Can. J. Biochem. Physiol.* **33**, 839.
Wachstein, M., and Meisel, E. (1957). *Am. J. Clin. Pathol.* **27**, 13.
Wachstein, M., and Meisel, E. (1960). *J. Histochem. Cytochem.* **8**, 317.
Wachstein, M., Meisel, E., and Falcon, C. (1959). *J. Histochem. Cytochem.* **7**, 428.
Walker, B. E. (1960). *J. Ultrastruct. Research* **3**, 345.
Weatherford, H. L. (1932). *Z. Zellforsch. u. mikroskop. Anat.* **15**, 343.
Weber, R. (1957). *Experientia* **13**, 153.
Weidel, W. (1958). *Ann. Rev. Bacteriol.* **12**, 27.
Weiss, L. P., and Fawcett, D. W. (1953). *J. Histochem. and Cytochem.* **1**, 47.
Wilcox, H. H. (1959). *In* "The Process of Aging in the Nervous System" (J. E. Birren and W. F. Windle, eds.), pp. 16-23, C. C Thomas, Springfield, Illinois.
Willmer, E. N. (1960). *Congr. intern. biol. cellulaire, 10th Congr. Paris* p. 93.
Wolff, E. (1953). *Experientia* **9**, 121.
Yamada, E. (1955). *J. Biophys. Biochem. Cytol.* **1**, 551.

CHAPTER 7

The Chloroplasts: Inheritance, Structure, and Function

By S. GRANICK

I. Introduction

In this review the origin, differentiation, structure, and functions of the plastids will be discussed from both a biological and a chemical point of view.

The life history of the chloroplast of a higher plant may be divided into three phases. In the first phase the predominant enzymatic activities are those connected with multiplication (i.e., self-duplication); it is clear that some reproducing templates are present in proplastids, perhaps in connection with ribonucleic acid (RNA), but little is known of this phase. In the second phase the predominant activities are those concerned with differentiation to form the structures and enzymes required for photosynthesis. Knowledge of this phase is also slight although progress is being made in studies of chlorophyll and carotenoid biosynthesis. In the third phase the activities of the enzymes that predominate are those concerned with photosynthesis, and here considerable progress has been made. At the same time morphological studies with the aid of the electron microscope have revealed some aspects of the origin and development of the liproprotein lamellae concerned with photosynthesis. Still to be clarified on a molecular level are the structural arrangements of the pigments and enzyme compartmentation that make photosynthesis such a highly efficient process.

Although little is known about the processes of self-duplication and differentiation, there are many data available on the effects of internal and external environments on these processes. The internal environment results from the action of the inheritable factors of the genome, plasmon, and plastidome. The external, or culture medium, environment includes the products of neighboring cells, light, temperature, etc. A plastid with an apparent potentiality to form a functional chloroplast may be affected by one or a number of factors of the internal or external environment. For example, the plastid may remain tiny, undifferentiated, and colorless; or it may enlarge as a colorless body specializing in the storage of starch (amyloplast), or protein (aleurone plastid), or fat and steroid (elaioplast); or it may lose chlorophyll and protein and consist predominantly of carotenoids (chromoplast). Part of the answer to the problem of cellular differentiation toward specialization may lie in an understanding of how bodies like the plastids, mitochondria, sphaerosomes, etc., become specialized. A major task that lies before the physiologist is to analyse these various influences on plastids, perhaps through attempts to grow the plastid in culture.

The plastids are large cytoplasmic bodies the morphological features of which may be readily observed directly by examination of the intact cell, or the plastids may be readily isolated and their composition studied. In contrast to the smaller cytoplasmic granules such as the mitochondria, important physical and chemical differences in the plastids (even within the same cell) can be easily recognized, such as changes in size or shape, and the presence or absence of the chlorophylls, carotenoids, and starch. The studies on self-duplication of the plastids, the inheritable factors within them, and the changes that occur in them during cellular differentiation are not only of importance in themselves, but may well serve to illuminate the properties of cytoplasmic granules at or below the limits of visibility of the light microscope. The various kinds of cytoplasmic granules represent complex organizations of enzymes that perform specialized functions. One may ask similar questions about them as about the chloroplast. Are they self-duplicating entities? What type of organization for self-reproduction do they possess? In what way do internal and external factors influence the development and activities of these bodies?

Because the plastid is a self-duplicating body, the problems connected with its inheritance may well be similar to those connected with virus multiplication within the cytoplasm. For example, if plastids could infect cells or if hosts such as sucking insects could transmit plastids, the analogy between virus and plastid would be close. The analogy would be still closer if a plastid that had a chlorophyll defect and could outgrow normal plastids were transmitted to a normal cell, resulting in a decrease in the photosynthetic ability of this cell.

In this review no attempt has been made to cite all the literature, nor to present a historical development of the subject. Rather, references have been selected with a view toward illustrating specific points of the discussion. Several reviews of the older literature with extensive references are those by Weier (1938a), Küster (1935), and Schürhoff (1924). Some more recent reviews are those of Weier and Stocking (1952a), Rabinowitch (1945, 1951, 1956), Frey-Wyssling (1953), Thomas (1955), Mühlethaler (1955), Hodge *et al.* (1956), Leyon (1956), von Wettstein (1958, 1959), and Mühlethaler and Frey-Wyssling (1959). Chloroplast structure and function are discussed in a series of papers in the *Brookhaven Symposia of Biology* (**11**, 1958); and photosynthesis, in the Emerson Memorial Issue of *Plant Physiology* (**34**, 1959).

During the latter half of the nineteenth century, excellent studies were made on chloroplast structure and inheritance. It was then discovered that the chloroplasts of higher plants contain grana and that algal plastids arise from pre-existing plastids—facts that have been confirmed and more firmly established by recent techniques. Particularly noteworthy were the contributions of Meyer (1883), Schimper (1883), Schmitz (1884), and Sachs (1887).

II. Classification and General Morphology of Plastids

Plastids may be divided into two general groups; the chromoplasts, or chromatophores or colored plastids; and the leucoplasts, or colorless plastids. The chromoplasts may also be divided into two subgroups:

those that are photosynthetically active and those that are not. By way of introduction, a brief survey of the various kinds of plastids is presented without considering their detailed structures.

A. *Photosynthetically Active Chromoplasts*

With the exception of bacterial plastids, these all contain chlorophyll a and, depending on the phylum, may contain in addition some other chlorophyll (Smith and Benitez, 1955), a number of characteristic carotenoids, and phycobilins (Strain, 1949).

1. *Chloroplasts*

The most common plastid of this kind is the green plastid of the green algae and higher plants. It contains two chlorophylls: a and b, usually in the ratio of 2–3:1.

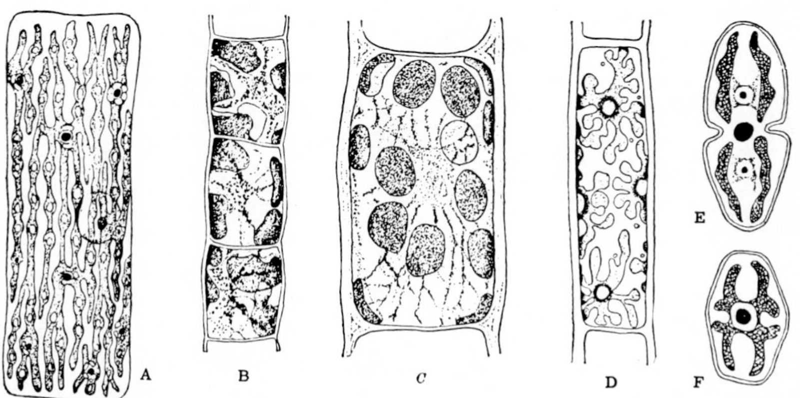

FIG. 1. Various types of algal chloroplasts. A. *Oedogonium* (green alga) showing pyrenoids (as dark dots) surrounded by starch sheaths, and starch grains (as white circles) lying within the stroma (i.e., protein matrix) of the single strap-shaped branching plastid. B. *Leptonema fasciculatum* (brown alga). C. *Pylaiella varia* (brown alga). D. *Rhodochorton floridulum* (red alga) with one pyrenoid per plastid. E and F. *Euastrum dubium* (green alga), front and side views with one pyrenoid per plastid. From L. W. Sharp, "An Introduction to Cytology," 3rd ed. McGraw-Hill, New York, 1934.

The chloroplasts, especially among the algae, are greatly diversified as to number per cell and form (Fig. 1). In some species of algae there may be one or two chloroplasts in a cell; in others, ten or more. The algal chloroplasts are usually placed parietally just beneath the cell wall, embedded in cytoplasm, as in the higher plants. The chloroplasts may be bell-shaped as in *Chlorella* or *Chlamydomonas,* or spiral-shaped as in *Spirogyra,* or form an irregular network, as in *Oedogonium;* or they may be suspended in the center of the cell as in the stellate-shaped chloroplast of *Zygnema*. The shape of the chloroplast depends in part on its method of growth. In some chloroplasts, pseudopodial-like extensions occur, as in *Zygnema* or *Rhodochorton,* to form a stellate-shaped chloroplast; or the pseudopods may extend to form a plastid network, as in *Oedogonium.* In other types of chloroplasts, holes may be formed by differential rates of growth of the expanding plastid. In still others, color-

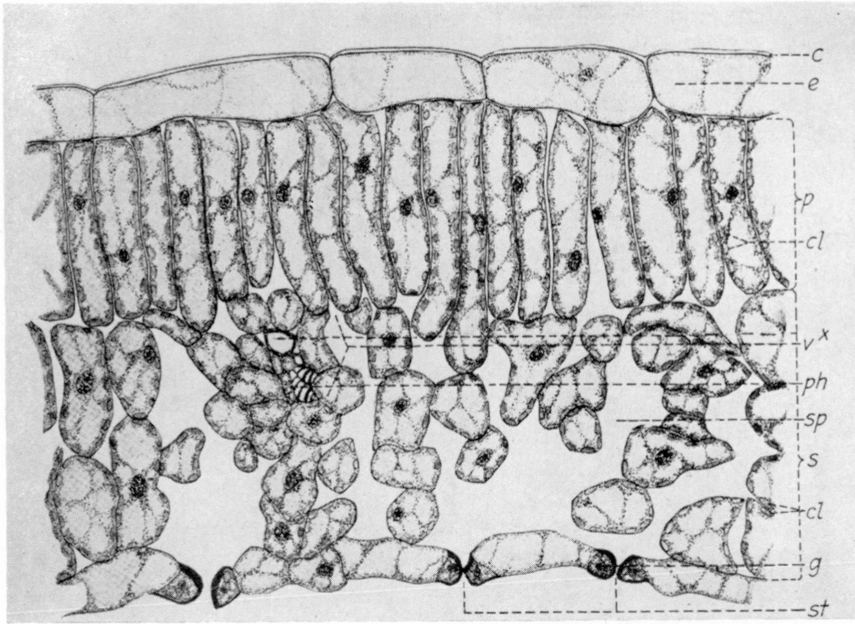

FIG. 2. Cross section of the leaf of *Lonicera tatarica* illustrating cuticle, *c;* epidermis, *e;* palisade mesophyll, *p,* with chloroplasts, *cl;* vein, *v,* with xylem, *x,* and phloem, *ph;* spongy mesophyll, *s,* with air spaces, *sp,* and chloroplasts, *cl;* lower epidermis with stomata, *st,* and guard cells, *g.* From J. B. Hill, L. D. Overholts, and H. W. Popp, "Botany, a Textbook for Colleges," 1st ed. McGraw-Hill, New York, 1936.

less regions of the plastid stroma may separate the green areas, as in *Chlorella* growing in the dark in the presence of glucose.

In higher plants the chloroplasts of the mesophyll cells of green leaves may be considered typical (Fig. 2). On a dry weight basis, the chloroplast composition is as follows: protein 40–50%, lipid 23–35%, chlorophylls a + b 5–10%, and carotenoids 1–2% (see Section VII). The chloroplast is shaped like a planoconvex lens and is approximately 5 μ in diameter and 2–3 μ thick. It is embedded in cytoplasm and lies appressed with its broad side parallel to the cell wall. Within the chloroplast, the flat lamellae of the grana are arranged so that they also face the adjacent cell wall (Fig. 8). In general, chloroplasts tend to lie in those regions of the cell which are adjacent to intercellular air spaces (Haberlandt, 1914), thus facilitating gaseous exchange (Fig. 2). The size and shape may vary, depending on inheritable and environmental conditions, from a large number of small chloroplasts per cell to a small number of large chloroplasts and from a more or less compact body to an ameboid-like body as in certain mutant chloroplasts of *Polypodium* (Knudson, 1940). In 215 species of plants, 75% of the plastids have a long diameter between 4 and 6 μ (Möbius, 1920). In *Tropaelum majus* the average dimensions of the chloroplast are 3.9 by 2.9 μ with an average volume of approximately 9.4 μ^3 (Meyer, 1917). In *Ricinus communis* the average palisade mesophyll cell contains about 36 chloroplasts and the spongy mesophyll cell about 20. Per square millimeter of leaf area, there are a total of 403,000 chloroplasts of which 72,000, or 18%, are in the spongy mesophyll cells (Haberlandt, 1914). In the palisade cells, the chloroplasts may often be so closely packed as to form an almost flat compact layer lining all the cell walls of the cell. The cells of many plants appear to have the ability to regulate the number of plastids per cell. If the chloroplasts are few, division of the chloroplasts may take place. Under adverse physiological conditions, do all the chloroplasts degenerate equally or do a few survive at the expense of the others? This question has yet to be answered. In certain strains of maize, Eyster (1929) noted that if cells have large plastids, there are fewer plastids per cell than if cells contain smaller plastids. In the immature leaf, the proplastids and young chloroplasts multiply by fission, but fission ceases at an early stage. In the tomato leaf the number of chloroplasts increases only by about 30% from the time the leaves are one-third their maximum size to the fully expanded leaves, but on the other hand, protein content of the chloroplasts increases over tenfold during this time (Granick, 1938b).

2. *Pheoplasts*

These are brown or yellow plastids which occur in the brown algae, the diatoms, and dinoflagellates. Here the brown carotenoids mask the chlorophyll color which consists of chlorophyll a and small amounts of c. According to Strain (1949, 1951) the yellow or brown color of the pheoplasts appears to result from the physical state or condition of the pigments or from their geometrical arrangement in the plastids rather than from a preponderance of the carotenoids. For example, exposure of the plants to heat causes the plastids to turn green. Extraction of heated plants or of fresh plants with alcohol yields green solutions in which the chlorophylls predominate. The unequivocal evidence that in the brown algae fucoxanthin can absorb light and transfer the energy to chlorophyll a (Dutton *et al.*, 1943) suggests that this yellow pigment must be situated spatially quite close to the chlorophyll molecules.

3. *Rhodoplasts*

These are red-colored plastids which often occur in the Rhodophyceae, especially the Florideae. The red color is due to the phycobilin protein phycoerythrin (Strain, 1949; Lemberg and Legge, 1949) and may be in sufficient abundance to mask the green color, which is due to chlorophyll a with traces of d. Phycocyanin, containing the blue phycobilin, may also occur to a smaller extent in some species of red algae. The color of plastids of such species may be the result of light absorption by chlorophyll, by phycoerythrin, and by phycocyanin, as they may occur in varying proportions under differing environmental conditions (Rabinowitch, 1945). From studies of photosynthesis, it is known that the light which is absorbed by the phycobilins is transmitted rather efficiently to chlorophyll a (Haxo and Blinks, 1950; French and Young, 1952). Within the plastids, the phycobilins fluoresce very little. However, if the cells are injured, the phycobilins readily diffuse out of the plastids and are then intensely fluorescent in visible light. In order to account for the absence of fluorescence in the uninjured rhodoplast, it is necessary to assume that the phycobilin molecules are arranged so that the light energy which is absorbed by them is transmitted to chlorophyll. If the spatial organization is disrupted, the light energy is not transmitted but is rather emitted in part as fluorescent energy by the tetrapyrrole prosthetic groups of the phycobilin (see Section VII).

4. *Blue-Green Chromatophores*

The characteristic color of blue-green algae is due to the accessory proteins phycocyanin and phycoerythrin. These pigments, together with chlorophyll a and carotenoids, are probably associated with dense flat lamellae, as seen in electron microscope pictures of *Phormidium uncinatum* (Fig. 18).

5. *Chromatophores of Photosynthetic Bacteria*

The reddish color of the purple sulfur bacteria (Thiorhodaceae) and nonpurple sulfur bacteria (Athiorhodaceae) is due to special carotenoids (Goodwin, 1952). The chlorophyll type pigment is bacteriochlorophyll which absorbs predominantly in the far-red. In sections of *Rhodospirillum rubrum* (Fig. 19) lamellar structures also have been found. In green sulfur bacteria (Thiochloraceae), the chlorophyll type pigment is bacterioviridin.

B. *Chromoplasts Devoid of Photosynthetic Activity*

These chromoplasts in general contain carotenoids but lack chlorophylls. They are red to yellow in color and are of highly variable shape. The colors of flowers and fruits are often due to chromoplasts. Striking color changes occur as the green fruits and flowers mature and chlorophyll disappears. For example, the yellow chromoplasts of the petals of the buttercup first develop as chloroplasts (see Section V).

The new colors that arise are often due to the formation of special carotenoids not normally found in the green part of plants, as, for example, the lycopene of tomato fruits, prolycopene in the berries of *Arum orientale,* and capsanthin of red peppers. Many of the xanthophylls of fruits and flowers occur with their hydroxyl groups esterified with fatty acids. In contrast, xanthophyll esters rarely occur in chloroplasts (Strain, 1949). In autumn leaves of *Ginkgo biloba* or *Fagus silvatica* colored droplets containing carotenoids are present. The droplets represent plastids that have degenerated so that only the fatty material containing the carotenoids remains (Möbius, 1937). Colored carotenoid droplets might also arise by solution in oil droplets of carotenoid molecules produced elsewhere, or the droplets might represent vacuoles that have concentrated the more water-soluble carotenoids.

Carotenoids also occur in fungi and in bacteria. Are these carotenoids synthesized in specific cytoplasmic bodies that represent vestiges of plastids, or do the colored droplets arise independently of a plastid-like apparatus? In fungi, acidic carotenoids similar to astaxanthin are often present; the xanthophylls characteristic of higher plants are not found in them (Goodwin, 1952; Karrer and Jucker, 1950). In *Neurospora crassa,* a complex mixture of carotenoids has been isolated by Haxo (1949); this includes lycopene, carotene, and spirilloxanthin as major components. Garton *et al.* (1951) have observed that if growth of *Phycomyces blakesleeanus* occurred in visible light, then the yield of carotenoids was doubled. In bacteria, which contain carotenoids, the carotenoids are mostly xanthophylls. Bacteria in general lack the acidic carotenoids that are characteristic of fungal carotenoids. In *Micrococcus lysodeikticus* and *Sarcina lutea* the carotenoids seem to reside in the protoplast membrane (Stanier, 1960).

As an example of a yellow chromoplast structure, one may consider the studies on the chromoplast of the carrot root. These chromoplasts develop from leucoplasts that contain starch. As the carotene increases in concentration the starch disappears. The older chromoplasts appear as large flat plates in the form of parallelograms or as needles with a typically crystalline appearance; they are birefringent and dichroic (Frey-Wyssling, 1935). Analysis of the large chromoplasts by Straus (1954) indicates that carotene may make up 20–56% of the plastid on a dry weight basis. When the carotene content of the plastid is about 30%, the other lipids amount to 30–40%, about 15% may be protein, and 0.5% may be ribonucleic acid. After extraction of the dried chromoplasts with ether, Straus observed a residue consisting of very fine rectangular fibrils or rods, which presumably represented the protein component and which appeared to have a lamellar structure (see Section V, C).

C. *Leucoplasts*

The term, leucoplast, is applied to all mature colorless plastids. (The immature colorless plastids that occur in meristematic tissues are designated proplastids and will be discussed in Section V.) When starch is predominant, as in storage organs such as the potato, the leucoplast is called an amyloplast; when oil is predominant, an elaioplast; and when protein granules or crystals are predominant, an aleurone-plast.

1. *Amyloplasts*

These are mature plastids filled with starch and are generally found in storage tubers, cotyledons, and endosperm. Starch is characteristic of the colored and colorless plastids of green algae and the higher plants; it is rarely formed in the cytoplasm apart from the plastids. In algae such as *Spirogyra* and *Chlamydomonas,* starch is deposited primarily around the pyrenoid, although under conditions such as nitrogen starvation, other portions of the plastid may contain starch. On the assumption that starch is formed only in plastids, the proplastids of meristematic tissues may be distinguished from mitochondria or other granules if starch grains can be observed in the proplastids.

The deposition of carbohydrates other than starch may not necessarily be within the plastid. For example, in red algae, floridean starch is said to be produced at the surface of the plastid. In *Euglena,* paramylum grains are considered to be produced outside of the plastids but generally in the neighborhood of the pyrenoids. In some Peridineae starch is said to be produced inside, and in others outside, the plastid.

Two types of starch grains may be distinguished, the assimilation starch and reserve starch. Assimilation starch is characteristic of actively photosynthesizing chloroplasts; the starch grains may be formed in large numbers per plastid, but they remain small because they are continually dissolving and the sugars are being transported elsewhere. In corn, the bundle sheath plastids which are pale green serve as temporary storage organs for starch during periods of rapid photosynthesis (Rhoades and Carvalho, 1944).

Reserve starch is characteristic of storage organs and is most familiar, for example, as the large starch grains of the potato tuber. A comprehensive review of the chemistry and biology of starch has been written by Badenhuizen (1959). The starch grain of the potato is generally made up of a series of concentric layers successively deposited about a center or hilum. As starch continues to be deposited around the starch grain in concentric layers, the amyloplast membrane and accompanying stroma may become greatly distended. The amyloplast may rupture and remain only on one side of the starch grain. It is noteworthy that in the latter case, new starch will be formed only where the amyloplast membrane and stroma remain in contact with the starch grain.

According to Frey-Wyssling (1953), in the starch grain the formation of a new concentric layer begins as a dense region of high refractive index, then it becomes less dense (i.e., more hydrated) outward until deposition of that particular layer ceases. When such a starch grain is examined between crossed nicols, a dark spherite cross is seen. This fact indicates that the chains of starch molecules lie more or less parallel and extend from one concentric layer outward to the next concentric layer. Starch consists of two components. A major component is the highly branched α-amylose, which stains red with iodine. The minor component (20–30%) is β-amylose, a helical chain structure that stains blue with iodine (Hassid, 1954). In a layered starch grain, α-amylose molecules appear to be deposited in the inner denser and more refractive portion of a concentric layer whereas β-amylose helixes accumulate in the outer, looser portion of the concentric layer. The concentric structure of the starch grain is due to a periodic layering connected presumably with a fluctuation in available substrates, enzymes, pH, etc. At times this periodic layering may be correlated with alternation of night and day as observed by Meyer (1920).

2. *Elaioplasts*

These may be defined as plastids that develop a preponderance of oil. In most monocotyledons, oil appears in old chloroplasts which lose their chlorophyll. In epidermal cells of Orchidaceae and Liliaceae, the oily disorganized plastids fuse to form a droplet which has also been called an "elaioplast." In many brown algae, the reserve storage product is oil instead of carbohydrate, and this oil appears to be formed in the pheoplasts (Mangenot, 1923; Sharp, 1934). Oil droplets are undoubtedly also formed independently of plastids.

3. *Aleurone-plasts*

Aleurone-plasts, or proteinoplasts, may be defined as colorless plastids that contain much protein. For example, the leucoplasts of *Phajus grandifolius* may contain a cluster of parallel needlelike protein crystals. Proteinoplasts have been studied in epidermal cells of *Helleborus corsicus* (Hartel and Thaler, 1953). Protein crystals and granules are produced in seeds of numerous plants especially in those seeds that also form large amounts of oil, as for example, in the seeds of *Ricinus* and the Brazil nut. According to Mottier (1921), protein or aleurone formation involves the activity of permanent plastid primordia; these plastid primordia are considered to aggregate in large numbers in vacuoles where their combined products unite to form aleurone grains. The synthesis of proteins within the plastids is one of the markedly active properties of the plastid. This becomes evident when it is considered that more than half of the protein of a leaf parenchyma cell may reside in the chloroplasts. However, the formation of protein crystals is not exclusively a function of plastids. Protein crystals may be formed in the cytoplasm (independent of plastid action), as for example in cells poor in starch in peripheral layers of potato, or even in nuclei as in *Lathraea squamaria* and in many Scrophulariaceae and Oleaceae (Meyer, 1920). Likewise, protein granules (aleurone grains) of Soja cotyledons do not appear to be formed within plastids (Muschik, 1953). Protein crystals in epidermal cells may also represent virus material or may be the result of virus action.

III. Structure of Photosynthetically Active Chromoplasts

By way of introduction, the structure of the chloroplast of higher plants (Fig. 3) is summarized to introduce the terminology that will be used. The chloroplast is contained within a semipermeable double membrane (each 35–50 A. thick). Within is a stroma, a proteinacious material containing tiny granules (50–250 A. diameter), variable amounts of osmiophilic droplets (50–5000 A. diameter) and starch grains. Embedded in this stroma is a series of double-membrane lamellae extending the width of the plastid and stacked one on the other. The lamellae are differentiated into less dense regions, the stroma lamellae (20–30 A. thick per membrane) and denser regions, the grana lamellae (40–60 A. thick per membrane). A dense double-membrane structure from a granum region is designated as a disk. A cylindrical pile of disks (10–100) is a granum (0.3–1.0 μ in diameter).

A. *Chloroplast Structure As Revealed by the Light Microscope*

The literature on the structure of the chloroplast, as revealed by the light microscope, has been reviewed in recent years by Weier (1938a), Rabinowitch (1945), Weier and Stocking (1952a), and Frey-Wyssling (1953). Here we summarize some of the more recent findings.

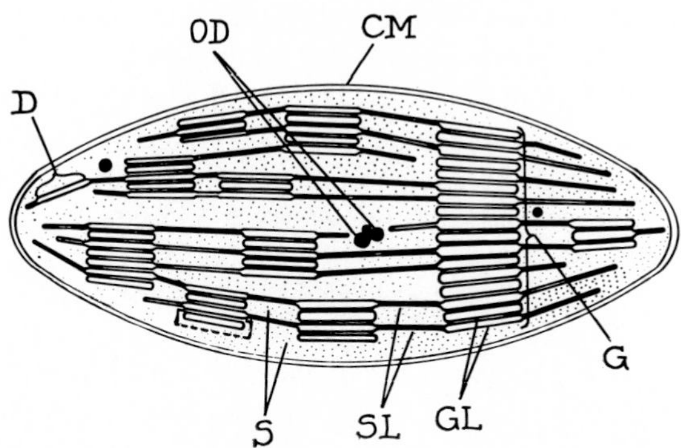

FIG. 3. Interpretation of the structure of a barley chloroplast (von Wettstein, 1958). *CM*, chloroplast double membrane (each 35–50 A. thick); S, stroma. *G*, granum (made up of a cylindrical pile of disks); *SL*, stroma lamella (each membrane is 20–30 A. thick); *GL*, granum lamella (each membrane is 40–60 A. thick); *D*, disk; *OD*, osmiophilic droplet. A molecular model of the dashed area in the lower left is presented in Fig. 41.

In intact cells of higher plants such as the leaf parenchyma cells of tomato and spinach, the chloroplast may appear homogeneous or may be seen to contain fine granules, the grana. On injury of the cell, the chloroplast may become more or less coarsely granular, swell, appear foamy, and then no distinct grana may be seen. Swelling studies suggest that the chloroplast membrane and lamellae within the chloroplast are not readily permeable to sucrose or larger molecules. From cells torn apart in an isotonic or hypertonic solution of sucrose (Knudson, 1936; Granick, 1938a) or polyethylene glycol (McClendon and Blinks, 1952), the chloroplasts float out into the solution, and may be seen to retain their smooth appearance for several minutes or longer, slowly becoming granular. In distilled water or isotonic saline, they rapidly become granular; at the same time vesicles or blebs are observed to form on their surface. The swellings are probably due to osmotic effects caused by the rapid hydrolysis of larger molecules into smaller ones. One or several blebs which develop seem to have their origin on the concave side of the chloroplast where a starch grain appears to be localized. Numerous smaller blebs which form over the surface of the chloroplast may represent the swelling of units originally separated by lamellae which have partially coalesced, as well as newly formed myelin membranes and precipitation membranes that appear on injury perhaps as a result of phosphatidase activity (Kates, 1957). Electron microscope observations on the swelling of *Nitella* chloroplasts which lack grana show that the swelling is due primarily to osmotic forces. First the lamellae spread apart; later some of the lamellae coalesce to form walls of elongated spheres, which continue to enlarge (Mercer *et al.*, 1955). In spinach and *Fucus* ultrasonic vibration will separate the double membrane layers or disks from each other, but the disks do not swell internally, thus indicating that a relatively strong binding exists between the upper and lower membranes of a disk (Thomas *et al.*, 1958).

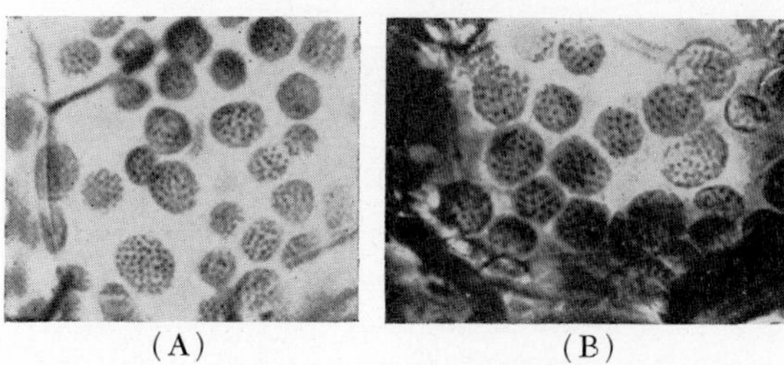

(A) (B)

FIG. 4. A. Grana in the chloroplasts of an intact cell of a water plant. B. Hand section through the leaf of a land plant examined in 6% sucrose solution. From Heitz (1936).

The granules that appear in the chloroplasts were first reported by Meyer (1883), who called them "grana." According to Schimper (1883) all chloroplasts of pteridophytes and spermatophytes contain grana. These grana are especially clearly seen in orchids like *Acanthephippium* and in many Crassulaceae. The grana were later rediscovered by Doutreligne (1935) and by Heitz (1936) (Fig. 4) in various higher plants, but not in the chloroplasts of some algae like *Spirogyra* or *Mougeotia,* a fact later confirmed by electron microscopy. Heitz observed that the grana are not spherical but disk-shaped. The grana are 0.5–2 μ in diameter, the size depending on the species. The grana of "sun" plants are smaller than those of "shade" plants. The grana are smaller in chloroplasts of the upper palisade cells and larger in chloroplasts of the lower spongy mesophyll cells. In *Elodea* the grana are larger in August and smaller in January (Beauverie, 1938). Etiolated plastids of *Agapanthus umbellatus* develop more rapidly in blue light, and the grana are more numerous but tinier than when developed in red light (Fasse-Franziskett, 1955). Weier (1938b) observed that both homogeneous and granular chloroplasts occur in healthy plant cells of higher plants.

Electron microscope studies have clarified these observations. The homogeneous appearance of the chloroplasts of some cells may be due to the closeness of packing of the grana (Fig. 10) or to the relatively high density of the stroma lamellae, whereas the granular appearance of other chloroplasts is the result of well-separated grana (Fig. 8). However, nongranular chloroplasts are found in many algae, in *Anthoceros*, and in the bundle sheath chloroplasts of *Zea mays* (Hodge *et al.*, 1956).

B. *Chloroplast Structure and Composition As Revealed by the Polarizing Microscope*

The large chloroplasts of certain algae such as *Mougeotia, Mesocarpus, Spirogyra,* and *Anthoceros* consist of parallel sheets of relatively coarse lamellae. Significant details of lamellar organization have become available from studies with the polarizing microscope on single chloroplasts that in part confirm and supplement details of electron microscopy. When viewed on edge, chloroplasts are negatively birefringent with reference to the thickness of the plastid. For example, in *Closterium moniliferum* Menke (1938) found a negative uniaxial form-birefringence in the living cell. This suggested that layers of material were present in the chloroplasts which had a refractive index higher than the medium in which the layers were embedded and that the layers were parallel to the long axis of the plastid (e.g., Fig. 13). Further evidence of a lamellar structure was presented by Menke and Koydl (1939). The fixed *Anthoceros* chloroplast, when photographed on edge in ultraviolet light, appeared to consist of very thin lamellae. Electron microscope studies later indicated that a single lamella is too thin to be distinguished in ultraviolet light. It is probable that what was observed was a partial unleafing of a series of groups of lamellae at a frayed edge of the plastid.

Suggestions about the chemical organization of the chloroplast may also be obtained from studies with polarized light. Layers of protein are present in the *Mougeotia* chloroplast, according to the studies of Frey-Wyssling and Steinmann (1948). In *Mougeotia* the rather thick platelike chloroplast is almost as wide and as long as the cell itself. The refractive index of the dense-layer component of this chloroplast may be determined by embedding the fixed chloroplast in liquid embedding mixtures of acetone and methylene iodide of gradually increasing refractive indexes (i.e., from 1.36 to 1.74). When the chloroplast was fixed in Zenker's fluid (picric acid and $HgCl_2$), the birefringence of the chloroplast changed along a hyperbolic curve as the refractive index of the embedding medium was increased. This is the behavior to be expected of a layered structure in which the thickness of the layers is small compared with the wavelength of the visible light used. At refractive index 1.58, the chloroplast became isotropic; this is the point where the layered component has the same refractive index as the embedding medium. Since acetone removed the lipids, the layered component consists of protein, a result compatible with the refractive index observed.

Evidence that the lipids are organized spatially was suggested by fixing the *Mougeotia* chloroplast with OsO_4 (Frey-Wyssling and Steinmann, 1948). With this fixative the lipids are said to become partially insoluble in organic solvents. In addition to the form birefringence due to the layers, a constant intrinsic birefringence was now found which was considered to be due to an oriented lipid component. Menke (1938) obtained evidence for a lipid component which was oriented so that the long-chain lipid molecules were in a direction perpendicular to the layers. Strugger (1936) had found that rhodamine B was a vital stain for chloroplasts. When the chloroplast of *Closterium moniliferum* was stained with rhodamine B by Menke (1938), the stained chloroplasts were optically dichroitic, a result which was interpreted to indicate that the dye molecule, which has a planar structure, becomes oriented parallel to the lipid molecules and that the long axes of the lipid molecules lie perpendicular to the protein layers. Chloroplasts in dried cells, or isolated chloroplasts that have dried, or chloroplasts in plasmolyzed cells, become positively uniaxial due to the overcompensation of the form birefringence by the intrinsic birefringence presumably of the lipid molecules; this fact also would appear to support the idea of orientation of the long axes of the lipid molecules in the direction of the thickness of the chloroplast.

Studies with polarized light have been made to determine whether chlorophyll molecules have a specific orientation in the chloroplast. Three tests have been used: optic dichroism, anomalous dispersion of birefringence, and polarization of fluorescence. The property of optic dichroism depends on the fact that the color and intensity of absorption of a planar dye molecule is due to its orientation with respect to the polarized light used (Branch and Calvin, 1941). Both the intensity and wavelength of absorption of light by the molecule are greatest when the electric vector of the polarized light vibrates parallel to the major electronic oscillation (excitation dipole) which is in the plane of the flat molecule and not perpendicular to the plane. In chlorophyll (Fig. 38, A) the major oscillation is from the dihydropyrrole ring IV across to the opposite pyrrole ring II (Platt, 1956). Crystals, containing flat dye molecules with their planes stacked parallel to each other, will absorb different wavelengths depending on the orientation of the crystal with respect to the direction of the polarized light entering the crystal. Such crystals are said to exhibit optic dichroism. Interaction of π-orbitals between dye molecules organized in a crystal may result in a new absorption band at longer wavelengths; the excitation dipole in this case may be inclined or even perpendicular to the plane of the flat molecule.

Menke (1938) found that the optic dichroism of the chloroplasts, observed in white light, was very weak in very young pale green chloroplasts; the dichroism was not detectable in older chloroplasts. With red light of 681 mμ the dichroism sometimes appeared positive. According to Goedheer (1957), the *Mougeotia* chloroplast seen in profile is dichroitic over the visible spectral region, i.e., polarized light vibrating parallel to the long edge of the chloroplast is more strongly absorbed than light vibrating perpendicularly. However, Ruch (1957) concluded that this is a "form" dichroism and not an "intrinsic" dichroism, as shown by embedding the chloroplast in glycerol medium, i.e., in a medium of higher refractive index. Therefore the dichroism observed is not due to orientation of chlorophyll molecules.

Another and more sensitive test for chlorophyll orientation is to observe the change in refractive index using polarized light in the neighborhood of the 680-mμ absorption band of chlorophyll, i.e., the anomalous dispersion of birefringence. Goedheer found a relatively high value, which he interpreted to indicate that the chlorophyll molecules were concentrated in thin lamellae, with the planes of the porphyrin rings somewhat oriented parallel to the lamellae.

Still another test for chlorophyll orientation is to excite the chloroplast with polarized light and examine whether the emitted fluorescent light at 685 mμ is polarized. In the chlorophyll molecule the excitation dipole for absorption in the plane of the chlorophyll molecule is parallel to the fluorescence dipole. Little or no polarization of emitted fluorescence could be detected (Goedheer, 1957; Ruch, 1957).

No indication of carotenoid orientation was obtained by polarized light measurements. The results of rhodamine B staining, suggested that chlorophylls and carotenoids were not in the lipid layers.

Goedheer's interpretation of his studies are that a lamella may consist of two monolayers of spherical protein molecules in juxtaposition, on the outer surfaces of which chlorophyll and carotenoid molecules are attached. The dihydroporphyrin planes would lie somewhat parallel to the protein surface, but not to each other. A layer of oriented lipid molecules would separate the lamellae.

It should be noted that polarized light studies provide information of a statistical nature. For example, if chlorophyll molecules were oriented with their dihydroporphyrin planes parallel to each other in groups of several hundred, yet if each group were oriented at random with respect to others, then no specific orientations would be observed with polarized light. Thus the possibility of orientation of chlorophyll molecules in groups cannot be ruled out as yet by the above experiments. Another explanation for the absence of polarization of light by chlorophyll is that the packing of chlorophyll would resemble that in crystals like nickel phthalocyanin the molecules of which are at 45° to each other, are isotropic, yet the crystals have conductivity (Calvin, 1958).

C. *Electron Microscope Studies of the Chloroplasts of Higher Plants*

The electron microscope has provided a powerful tool for investigating details of chloroplast structure. Osmic acid fixation together with examination of thin sections embedded in plastic have already added important data. Differential solvent extraction, enzyme digestion studies, and the use of other fixation methods may be expected to lead to information on the localization of specific compounds. In the photographs presented in this section, the unit of length, 1 μ, is represented by a bar: 1 μ = 10,000 A.; 1 mμ = 10 A.; 1 cm. = 10,000 μ or 10^8 A.

Although OsO_4 is an excellent fixative, its action does not prevent the leaching out of chlorophylls (Granick, unpublished) and possibly some other fatty materials during the dehydration through the alcohols, a step which is required for embedding in plastic. However, some details of fine structure may be more readily seen if more material has been leached out, as occurs on more prolonged action of OsO_4. Studies are desirable on the amounts and types of compounds that leach out in relation to time, temperature, pH, and concentration of OsO_4. The contrast in thin sections is believed to be due to OsO_2 in amorphous or crystalline form (Menke, 1957). According to Menke, dehydration in *Elodea densa* causes a decrease in chloroplast thickness of 43%, which means that the *in vivo* distances between stroma lamellae and between disks are greater than the measurements reported in this review.

Electron microscope studies of isolated chloroplasts fixed by vacuum drying reveal the grana clearly as dense morphological units. Here there is no loss of lipid-soluble material or pigments. In a spinach chloroplast of a mesophyll cell (Fig. 5), 40–60 grana are present (Granick and Porter, 1947). The grana appear to be dense wafer-shaped bodies some 6000 A. in diameter and 800 A. thick, embedded in a protein-containing matrix, the stroma. The diameters of the grana are rather uniform in individual chloroplasts, but may vary from one chloroplast to another. In general, the density of the dried granum is high, but some grana have a lesser density than most. Frequently, "disks" (presumably double-membrane structures) were observed which had the same diameter as that of the grana but were of low density; occasionally the "disks" were seen to spread out in a manner which suggested that a granum is made up of a pile or stack of disks (Fig. 6). The photograph by Steinmann (1952) of

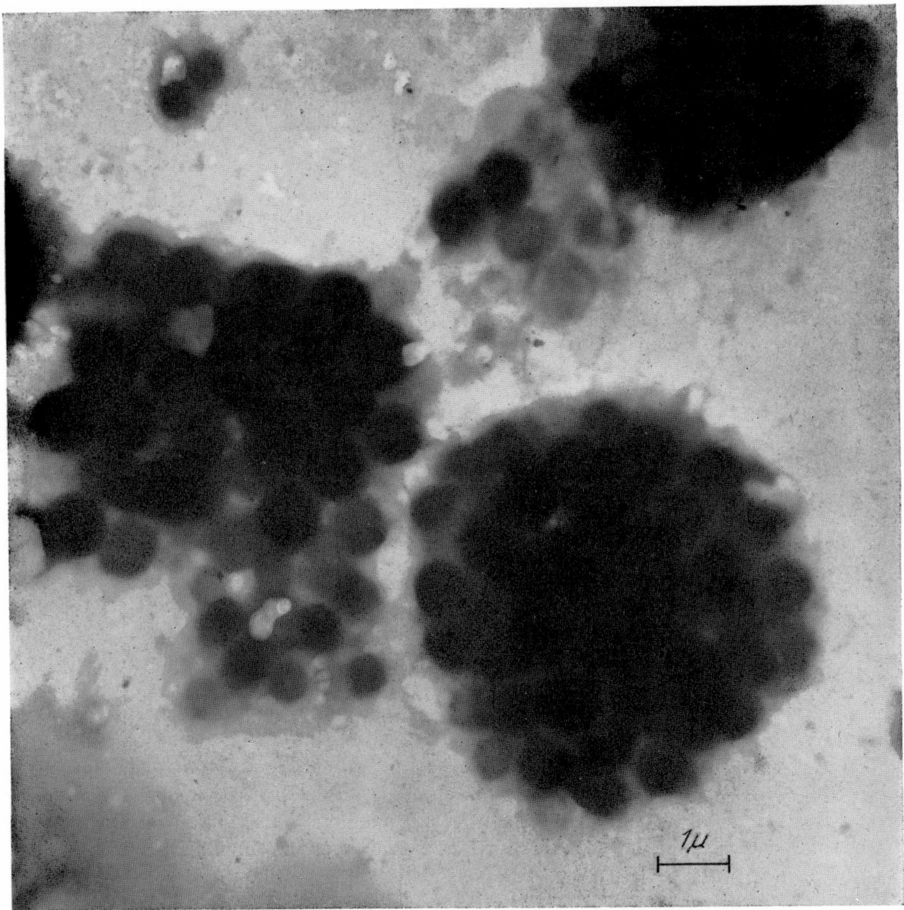

FIG. 5. Grana in isolated spinach chloroplasts not extracted with organic solvents. From Granick and Porter (1947).

the vacuum-dried cylindrical granum of *Aspidistra* clearly shows an arrangement of about 30 disks, each about 70 A. thick (Fig. 7). The grana thus appear to be discrete units, which must be held together by some material that is only poorly revealed by the thin-sectioning technique. When the grana are extracted with methanol, a residue, probably of protein, remains which appears to make up about half of the original material (Granick and Porter, 1947). Similar results have been obtained with isolated tulip chloroplasts by Algera *et al.* (1947).

Finer details of chloroplast structure have been revealed by thin-sectioning techniques, and the use of buffered osmic acid as fixative.

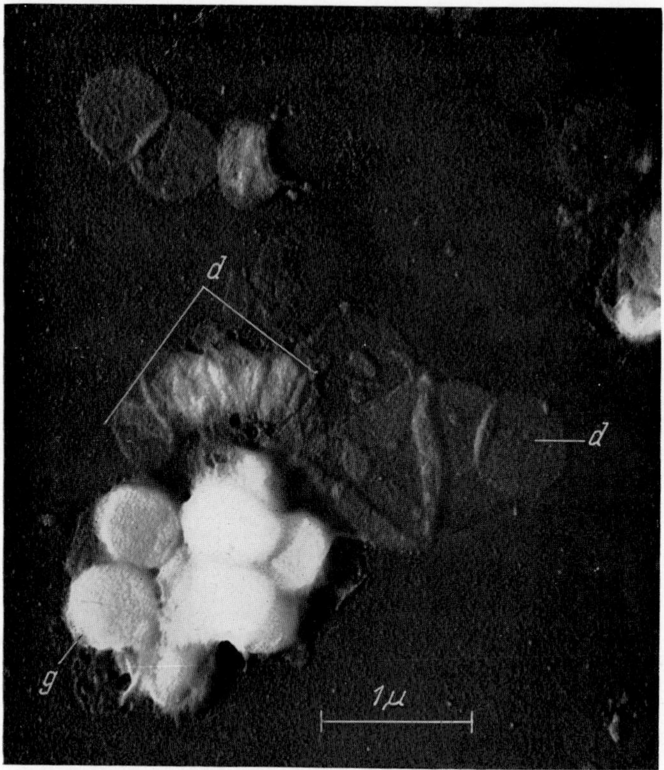

FIG. 6. Grana (*g*) and disks (*d*) of spinach chloroplasts, shadowed with gold. The disks are presumably double-membrane structures. From Granick and Porter (1947).

FIG. 7. *Aspidistra* granum, isolated in distilled water, showing disks. From Steinmann (1952).

Tobacco chloroplasts have been fixed and sectioned (Palade, 1953) at 300–500 A. thicknesses. They are made up of 40–80 grana per chloroplast (Fig. 8). The grana appear as cylinders rather than as flat wafers. The grana have not collapsed, because the embedding medium remains in the section. Each granum is 0.3–0.4 μ in diameter and appears to contain

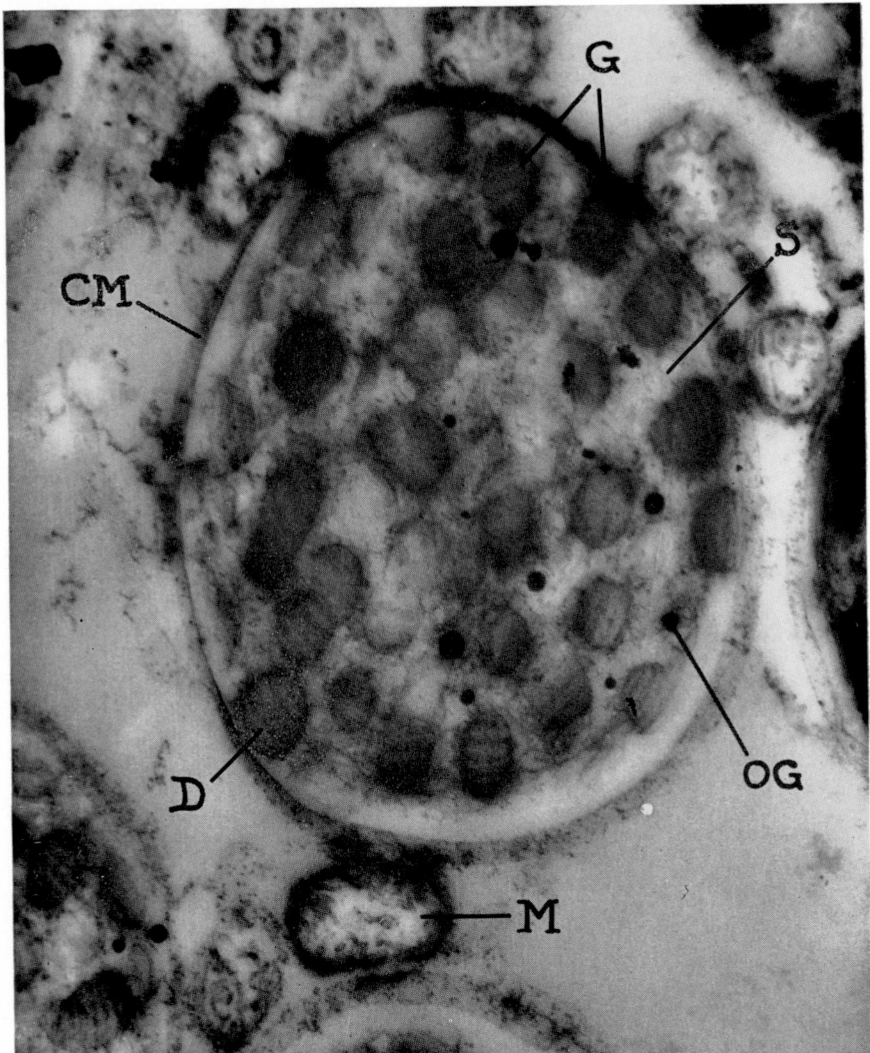

FIG. 8. Tobacco chloroplast with chloroplast membrane (*CM*), grana (*G*), disks (*D*) within the grana, the less dense stroma (*S*), osmiophilic granules (*OG*). A mitochondrion (*M*) is also shown. From Palade (1952).

10–15 disks. *Lemna* chloroplasts have somewhat larger grana (Fig. 9), each 0.7 μ in diameter and containing about 20 disks.

An interesting feature of the photograph of the tobacco chloroplast is the fact that the flat disks in the grana appear to lie parallel to the chloroplast membrane. The arrangement of the disks in *Aspidistra elatior* appears to be similar (Leyon, 1953).

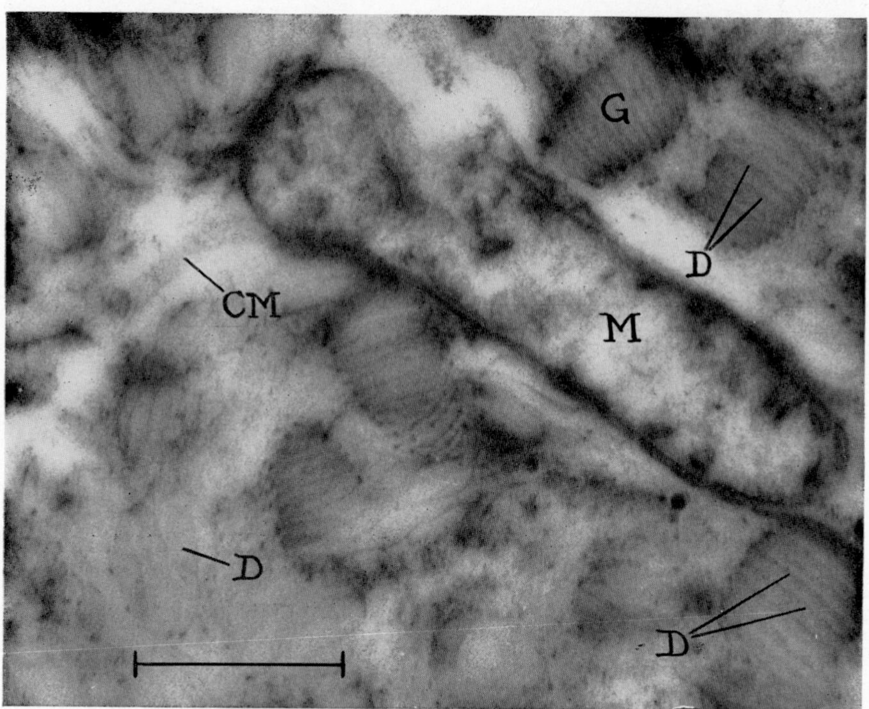

FIG. 9. *Lemna* chloroplasts showing grana (*G*), disks (*D*), chloroplast membrane (*CM*), and mitochondrion (*M*) (Palade, unpublished).

The studies of Steinmann and Sjöstrand (1955) on the chloroplasts of *Aspidistra elatior* reveal that a granum in this type of chloroplast is represented by a tall column or stack of double-membrane disks (Fig. 10). The width of the disks, and therefore the width of the column, is somewhat variable. Likewise, the height of the column is variable; a granum column may be made up of stack of 20–30, or even more than 100, disks. Two adjacent grana may be connected through a series of fine lamellae that extend through both grana. There are in general more grana lamellae than stroma lamellae. The fine stroma lamella is 30 A. thick, the dense granum lamella that makes up a membrane of a disk is

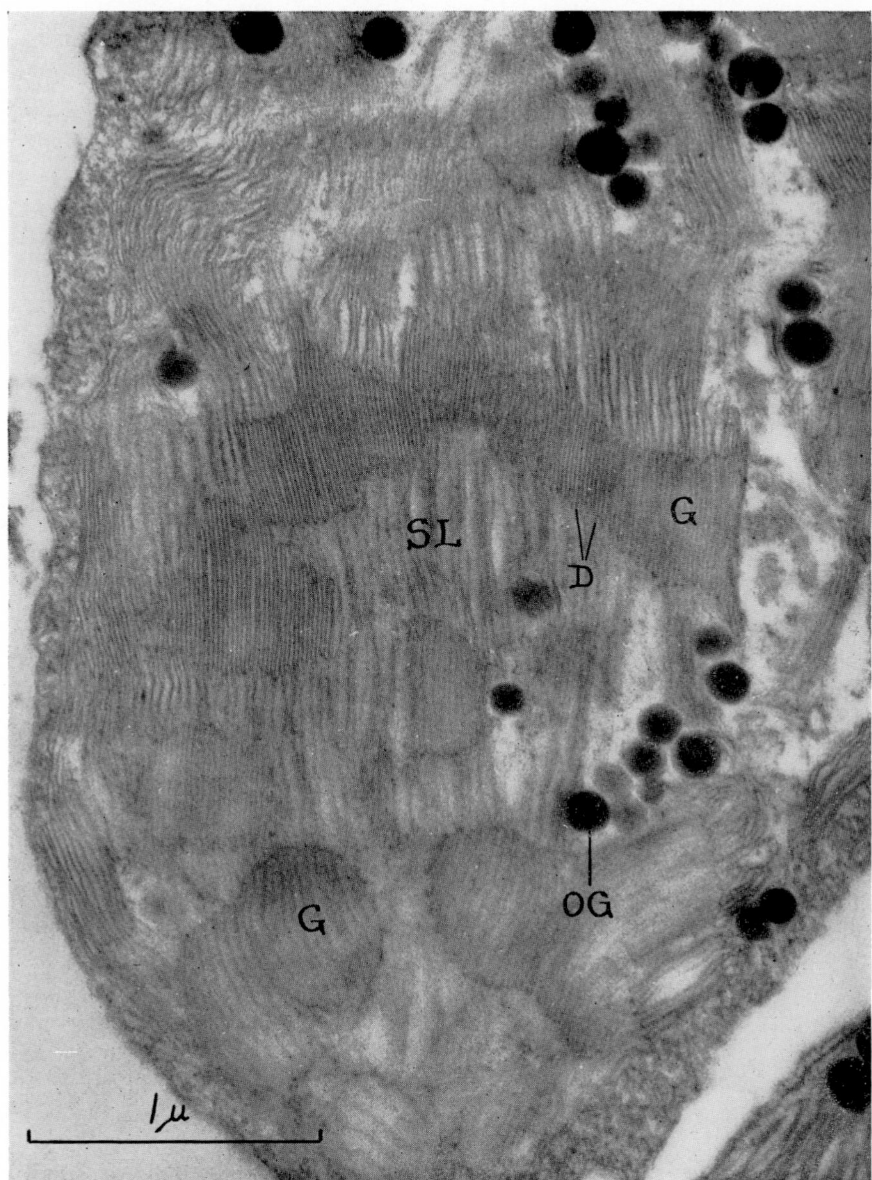

FIG. 10. *Aspidistra* chloroplast with grana (*G*), disks (*D*), stroma lamellae (*SL*), and osmiophilic granules (*OG*). From Steinmann and Sjöstrand (1955).

35 A. thick. The space enclosed by the upper and lower membranes of a disk is about 65 A. The repeat period in a granum is about 250–270 A.

von Wettstein (1958) interprets the lamellar structures as consisting of collapsed continuous double membranes (cisternae), those of the stroma regions being 20–30 A. thick and those of the grana regions 40–60 A. thick (Fig. 3). Within the double membrane is a homogeneous material of low osmium-staining density yet which does not collapse on dehydration with alcohols. At the borders of a granum the dense double-membrane structures (i.e., disks) are held together by some diffuse material. It is not known whether the disks are closed at the ends or are continuous with the stroma lamellae. Hodge *et al.* (1956) interpret their photographs on corn to indicate that a stroma lamella is not a double-membrane structure, but bifurcates to form the upper and lower membranes of a disk (Fig. 33).

Species vary in the number and dimensions of the stroma and grana lamellae and the distance separating the two membranes of a stroma lamella (Fig. 11, *Aspidistra, Hordeum*). The number of layers in a chloroplast which has just reached the mature state (see Section V) is fairly constant. For barley there are 6–12 layers, for tomato 8–12, for *Aspidistra* about 10, for *Oenothera* 7–10, and for mosses 5–10. The number of layers increases as the chloroplasts enlarge (von Wettstein, 1957a). In monocots the formation of a continuous lamellar system precedes grana formation; in dicots (*Oenothera* and tomato) grana regions appear first (von Wettstein, 1959).

In electron microscope pictures of the chloroplasts the disks of a granum column appear to adhere to each other closely so that the inter-distance between disks appears to be small compared to the intradistance between upper and lower membranes of a disk. A similar configuration is seen in algae (Fig. 12). On the basis of this short interdistance Thomas *et al.* (1958) suggest that there may be an interaction between the disks serving for the transmission of light energy. On the other hand the shrinkage of osmium-fixed chloroplasts observed by Menke (1957) may mean that the interdistance represents a dilute fluid, removal of which results in the collapse of the disks one upon the other; whereas the intra-distance represents a dense material that does not collapse readily.

A rather frequent component of chloroplasts of both algae and higher plants are osmiophilic globules 0.05–0.5 μ in diameter, which accumulate especially in inactive or degenerating plastids (Leyon, 1956).

D. *Electron Microscope Studies of Chromoplasts of Algae and Bacteria*

The chloroplasts of higher plants may be considered to have the greatest differentiation in terms of structure. Several simpler types are found in algae and bacteria. They will be discussed in the order of apparent decreasing complexity of morphological structure (Fig. 11).

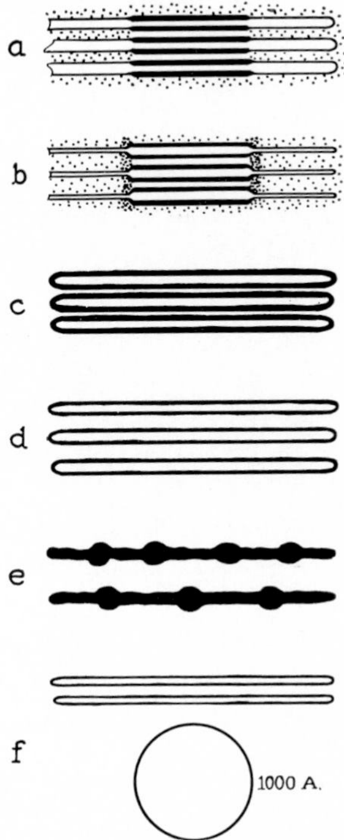

FIG. 11. Types of lamellae in photosynthetic chromatophores. a. Higher plant (*Aspidistra*). The two membranes of the cisternum are continuous and equidistant in both the granum region (dense lines) and the stroma region (lighter lines). b. Higher plant (*Hordeum*). The two membranes are closer in the stroma region than in the granum region. c. Brown alga (*Fucus*). The coarse lamella consists of a group of closely apposed double-membrane lamellae. d. Green alga (*Chlamydomonas; Spirogyra*). The double-membrane lamellae are well separated from each other. e. Blue-green alga (*Phormidium*). The lamellae contain large granules. f. Bacterium (*Rhodospirillum*). The lamellae are separate. Flat disks about 1000 A. in diameter have also been reported.

In the brown algae, the plastid consists of a pile of coarse lamellae surrounded by a semipermeable membrane. The coarse lamella is made up of four closely apposed double-membrane lamellae (Fig. 12) and extends from one end of the chloroplast to the other. The spindle-shaped pheoplast of an egg cell of *Fucus vesiculosus* (Fig. 13) measures 1.4–3.3 μ in length and contains 8–10 coarse lamellae. The outermost lamella runs parallel to the plastid membrane and appears to extend completely around the plastid. There are two optically clear "vacuoles," one at either

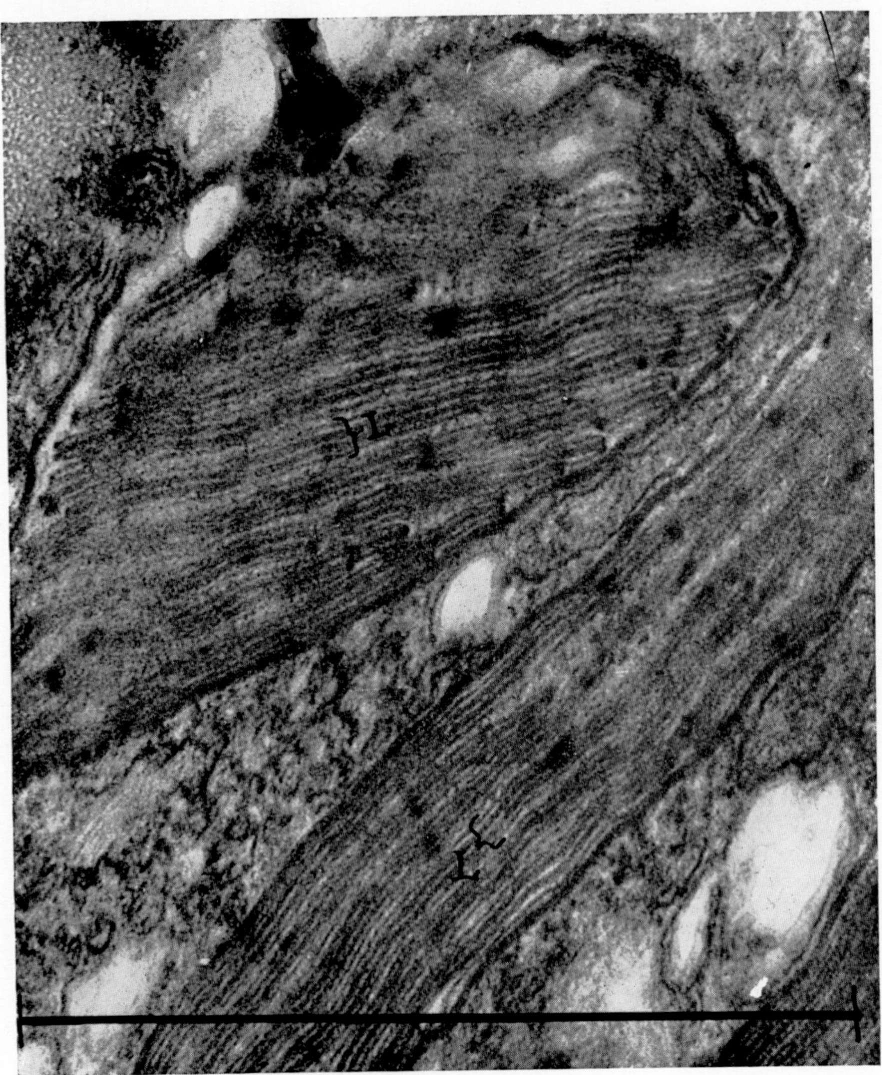

FIG. 12. Dinoflagellate, *Amphidinium elegans,* showing the coarse lamella (*L*) of a chloroplast to consist of four double membranes. From Grell and Wohlfarth-Botterman (1957).

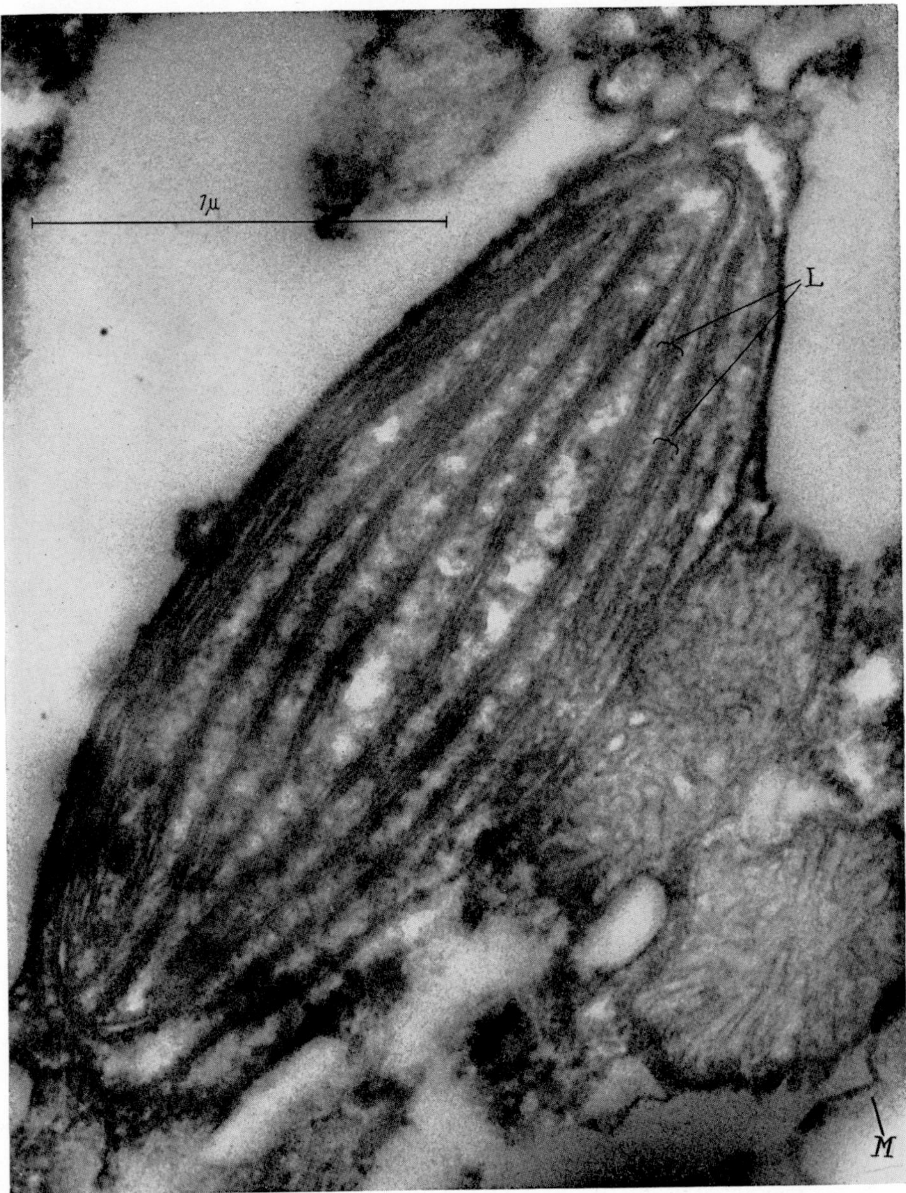

FIG. 13. Spindle-shaped brown plastid of *Fucus vesiculosus* with coarse lamellae (*L*). From Leyon and von Wettstein (1954).

end of the plastid, which lie within the boundary of the outermost lamella. A coarse lamella is about 370 ± 80 A. thick and consists of four finer lamellae, each 60 A. thick (Leyon and von Wettstein, 1954). In vegetative meristem cells the plastid contains 8–10 coarse lamellae, 280 ± 70 A. thick, each made of two fine lamellae. In old cells, the coarse lamellae have increased to 12–17 per plastid. The increase in number of plastids by a pinching-off process has been well documented by the studies of von Wettstein (1954). The constriction of the elongated pheoplast is in a plane perpendicular to the long axis; the innermost lamellae are first cut through, then the outer ones, resulting in the formation of two daughter plastids. The increase in the number of lamellae by a sheetlike splitting apart of coarse lamellae is also suggested by his studies. In *Poteriochromonas stipitata,* which is an oblong unicellular flagellate chrysomonad 7–8 μ in diameter, there are present two lens-shaped pheoplasts which flank the nucleus and are connected to each other by a membrane that lies closely apposed to the anterior part of the nucleus. The average chloroplast is 3.3 μ × 0.8 μ and contains ten dense lamellae, each about 400 A. thick. When this cell is placed in the dark, it loses its color, the plastids shrink, and the lamellae disappear (Wolken and Palade, 1953).

Among the green algae, two types of lamellae appear. One is a coarse lamella type as seen in *Euglena* and *Chlorella* which resembles that seen in brown algae. The other is a finer lamella type made up of well separated double-membrane lamellae as seen in *Chlamydomonas, Spirogyra, Mougeotia,* etc.

Euglena gracilis var. *bacillaris* is an elongate unicellular organism, about 70 × 20 μ (Pringsheim, 1956). It contains 8–12 chloroplasts 1.2 × 6.5 μ; each has about 20 coarse lamellae 240 A. thick (Fig. 14, A). In the dark on inorganic substrates, *Euglena* loses its chlorophyll and the lamellae disappear. After exposure to some 4 hours of light, lamellae become recognizable in the plastids. The lamellae at this time are relatively few in number and of low density. With longer exposure to light, the lamellae increase in number and density to a maximum by 72 hours. Mitochondria (Fig. 14, C) were not observed to be converted to plastids (Wolken and Schwertz, 1953). According to Pringsheim (1948), *Euglena* which has been grown in the dark on organic substrates contains yellowish plastids smaller than chloroplasts. Thus, small amounts of carotenoids can be synthesized in plastids developing in darkness, but not a lamellar structure. In *Chlorella pyrenoidosa* 4–8 coarse lamellae are contained in the cup-shaped chloroplast; each lamella consists of 4 fine membranes, 50 A. thick (Albertsson and Leyon, 1954).

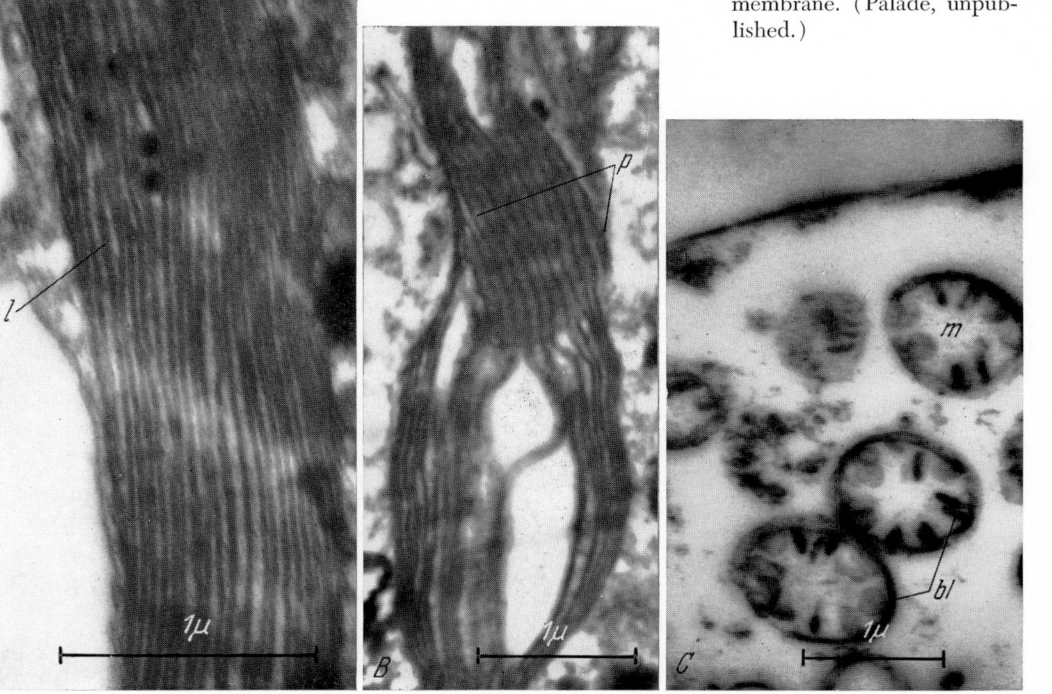

FIG. 14. *Euglena* chloroplast. A. Parallel lamellae (*l*) in a longitudinal section of a single-granum chloroplast. B. Organization of pyrenoid (*p*) with the lamellae traversing the pyrenoid (Wolken and Schwertz, 1953). C. Cross section of mitochondria (*m*) showing pocketlike bladders (*bl*) projecting into the mitochondria. At the top is seen the outer edge of the *Euglena* membrane. (Palade, unpublished.)

Leyon (1954a) reported that the chloroplasts of *Spirogyra, Mougeotia, Closterium,* and *Enteromorpha* contain less-dense, extended layers of lamellae. The lamellae have a thickness of 80 $\pm$ 20 A. (possibly representing double-membrane lamellae?). In *Closterium lunula* the sheets of lamellae, although generally layered parallel to the length of the cell, may bend upon themselves as seen in a cross section of the pyrenoid (Fig. 22, A, B).

The cup-shaped chloroplast of *Chlamydomonas reinhardi* has been examined by Sager and Palade (1954). The chloroplast is surrounded by a double membrane, as in the chloroplasts of higher plants. Parallel to the membrane are about 10 double-membrane lamellae (each membrane 50 A. thick) which are not continuous (Fig. 15, A). One closed edge of the double-membrane lamella may be attached to the chloroplast membrane (Fig. 15, B, C) and the other lies free.

In blue-green algae, a central region of the cell, the centroplasm, is present which is devoid of pigment and stains for nucleic acid; the remainder of the cell is colored and is designated chromatoplasm. No plastid membrane separates the chromatoplasm from the cytoplasm, nor is a nuclear membrane present. Studies with the electron microscope by Niklowitz and Drews (1956, 1957) reveal the chromatoplasm to be filled with lamellae (120 A. thick), which, however, do not penetrate the centroplasm (Figs. 16–18). (It is not known whether these are double-membrane structures.) Embedded in them are granules about 200 A. in diameter (phycocyanin?). In cross sections of the cells, lamellae are perpendicular to the cell wall as in Fig. 17 or may be radially arranged as in *Phormidium uncinatum.* The lamellae may extend the length of the cell as in *P. uncinatum* (Fig. 18) or *P. frigidum,* the lamellae may be more loosely arranged in islets as in *Oscillatoria limosa,* or the lamellae may be variously folded and contorted as in *Anabaena.* The number of lamellae per cell may vary from 4 in *P. frigidum* to 80 in *P. retzii.*

FIG. 15. *Chlamydomonas reinhardi,* electron micrographs. A. Chloroplast (*c*), although continuous, appears in sections owing to its undulating surface. Chloroplast membrane (*mb*) and pyrenoid (*p*) are shown. B. Anterior end of *Chlamydomonas,* showing cytoplasmic membrane (*cm*), chloroplast membrane (*mb*), eyespot region with two layers of carotenoid granules at *e* and *e*, mitochondrion in cross section (*m*) with pocketlike projection (*bl*), and lamellae (*l*) consisting of two membranes. C. Chloroplast membrane (*mb*) to which are attached the thin single membranes at (*a*). D. Pyrenoid in the lower half of the cell. The white areas (*sp*) are the regions from which starch has fallen out. Tubular elements (*t*), possibly extensions of the disks, project between the starch plates (*sp*) into the central area. From Sager and Palade (1954).

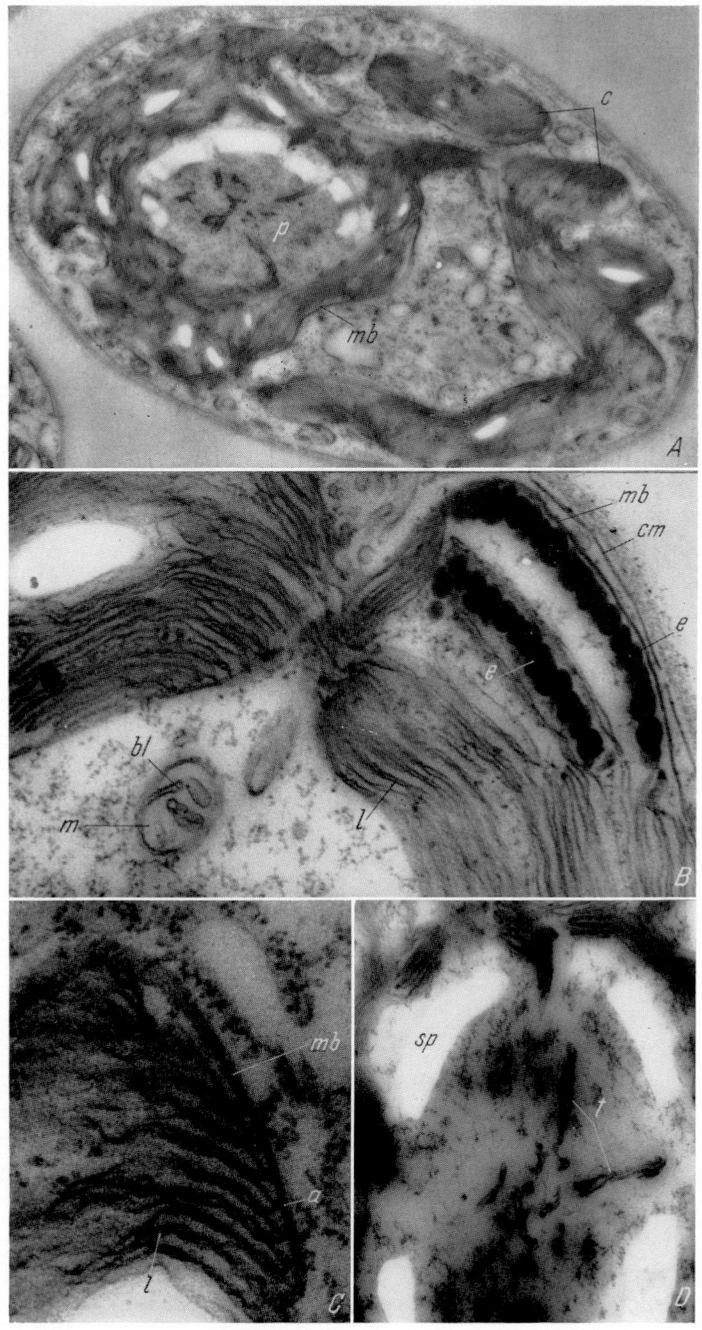
c
p
mb
A
mb
cm
e
e
bl
m
l
B
mb
a
l
C
sp
t
D

A comprehensive review of the chromatophores of bacteria and their photosynthetic properties has been written by Frenkel (1959). Photosynthetic bacteria also contain a lamellar photosynthetic apparatus as seen in sections (Fig. 19, A, B) of *Rhodospirillum rubrum* (Niklowitz and Drews, 1955). Pardee *et al.* (1952) have isolated granular particles of about 1100 A. diameter which contain bacteriochlorophyll, carotenes, proteins, and a small amount of pentose nucleic acid. Whether similar particles as seen in Fig. 20 represent lamellae or disks derived from the chromatophores is not yet known. No "disk"-like particles could be re-

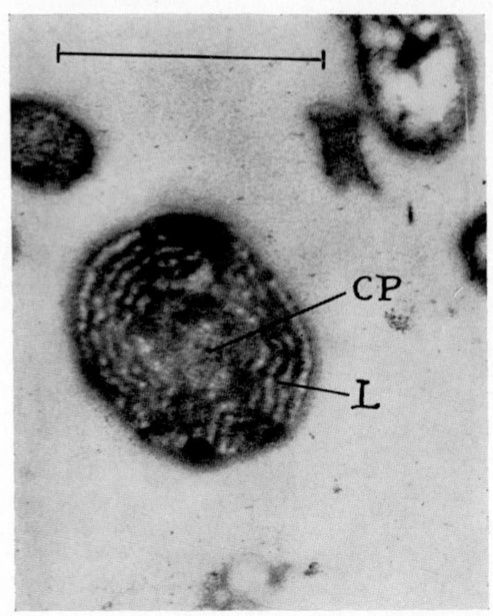

FIG. 16. *Phormidium frigidum*, a filamentous blue-green alga, in cross section showing four lamellae (*L*) arranged concentrically, and the region of the centroplasm (*CP*) free of lamellae. From Niklowitz and Drews (1956).

covered from cells that had been grown aerobically in the dark on organic substrates; under these growth conditions, the cells do not produce pigments and are colorless. Granules of 50–200 A. can carry on photosynthetic phosphorylation (Anderson and Fuller, 1958; Frenkel, 1958) but cannot fix CO_2. According to Newton and Kamen (1957) the pigmented particles of *Chromatium* contain an ethanolamine-phospholipoprotein with which are associated chlorophyll, carotenoid, and cytochrome in a molar ratio of 10:5:1. The particles also contain bound pyridine nucleotide, flavine, and large amounts of non-heme iron. There are about 200 bacteriochlorophyll molecules per chromatophore or 1 molecule per 50,000 mol. wt. protein. Bergeron has isolated 300-A.

particles from *Chromatium* and suggests that they consist of an outer rim of protein 60 A. wide, on the inner surface of which reside the pigments.

E. Summary

The basic structure of all photosynthetically active chromoplasts appears to be a flattened double membrane or cisternum structure. (In

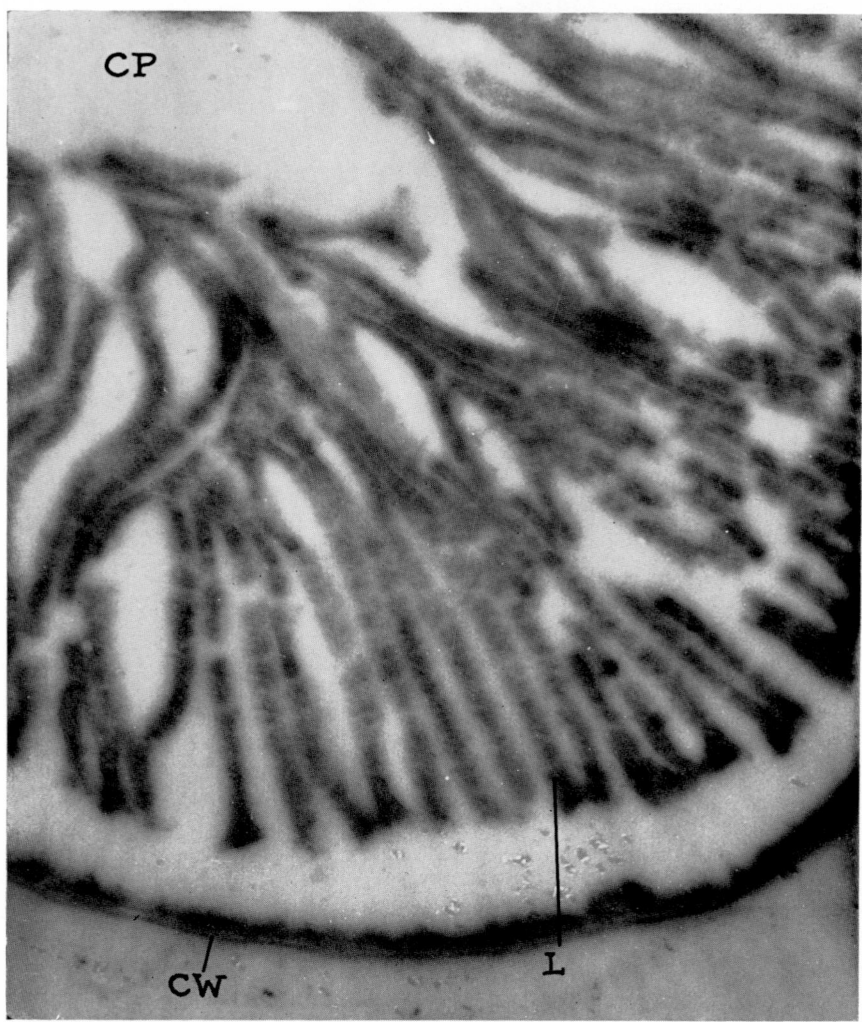

FIG. 17. *Phormidium* sp. Portion of cross section of cell to show granularity of lamella (*L*) and region of centroplasm (*CP*). The cell wall (*CW*) has torn away from the lamellae during preparation. (Granick, unpublished.)

bacteria and Cyanophyceae it may be a single membrane.) The membranes, at least their protein components, are 25–50 A. thick; the space enclosed by the two membranes is 25–75 A. wide.

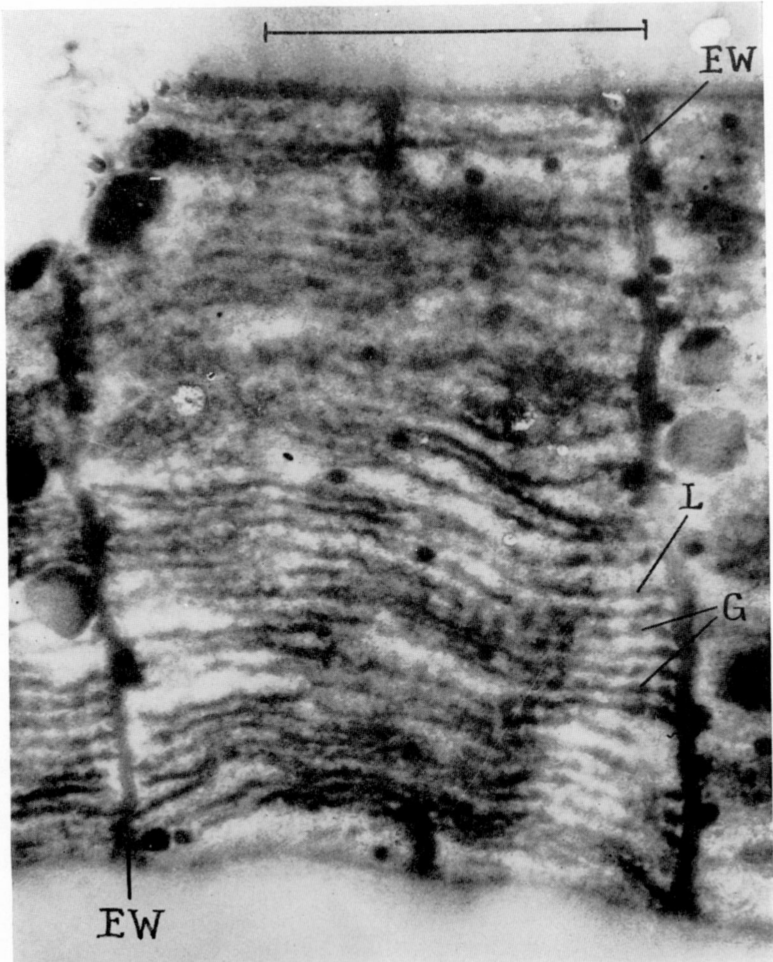

FIG. 18. *Phormidium uncinatum* in a longitudinal tangential section of a filamentous cell to show the lamellae (*L*) extending from one end cell wall (*EW*) to the next, and the large 200-A. granules (*G*) of the lamellae. From Niklowitz and Drew (1956).

In chloroplasts of higher plants the double-membrane lamellar structure is differentiated into more dense grana lamellae and less dense stroma lamellae; the lamellae are embedded in a proteinaceous stroma

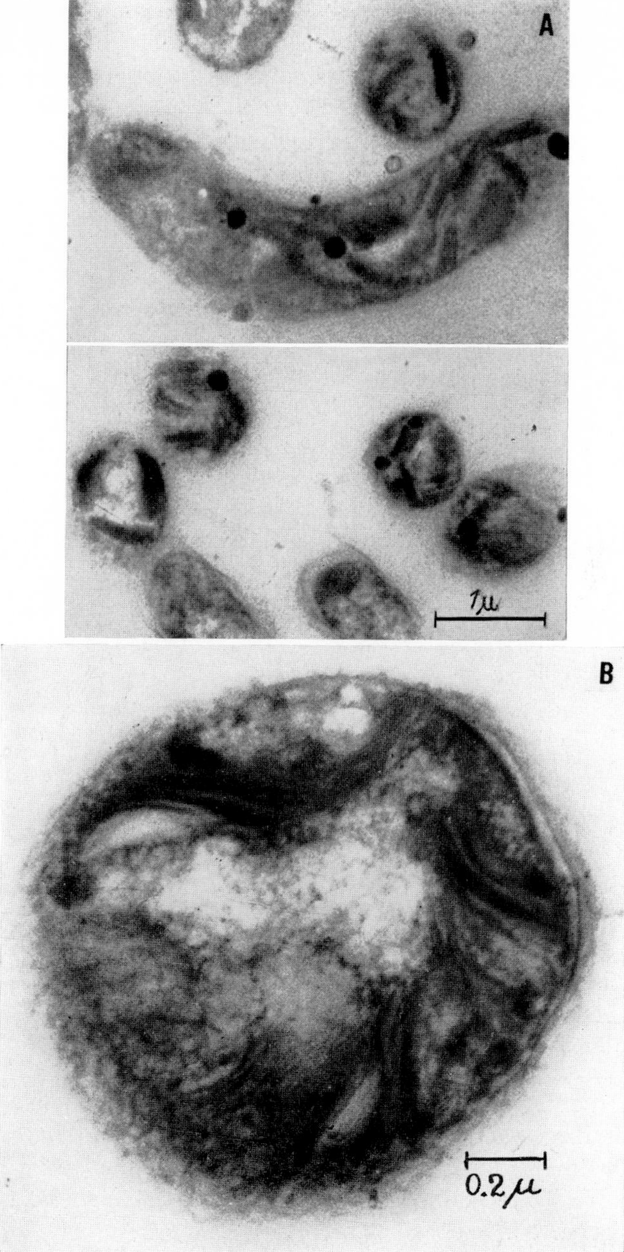

Fig. 19. A. Longitudinal and cross sections of *Rhodospirillum rubrum* fixed in Champy's, showing lamellar organization. B. The finest lamellae are less than 100 A. thick. From Niklowitz and Drews (1955).

and enclosed in a semipermeable double membrane, the chloroplast membrane (Fig. 3). In all other plastids (Fig. 11) this distinction between grana lamellae and stroma lamellae is lacking; only one uniform double-membrane lamella is seen. Certain green and brown algae contain coarse layers consisting of 2–4 tightly associated double-membrane

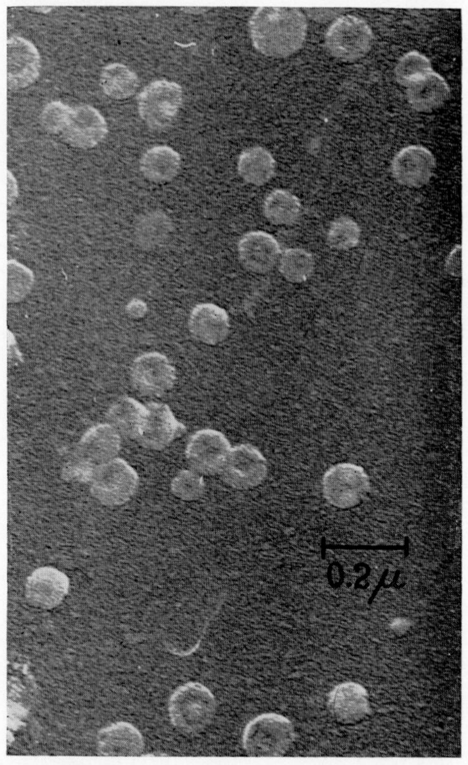

FIG. 20. Particles from some extracts of *Rhodospirillum rubrum* purified by differential centrifugation. The average diameter of the particles is about 850 A. Young, light-grown cells contain particles 200 A. in diameter that are photosynthetically active. From Frenkel (1958).

lamellae. In other green algae the double membrane lamellae are spaced equidistant from each other. In blue-green algae and bacteria no enclosing plastid membranes are present and the lamellae lie free in the cytoplasm. In bacteria fine membranes are present; spherical units have also been reported.

The chloroplast membrane in higher plants is not readily permeable to sucrose or larger molecules. On injury, the plastids swell, possibly owing to the release of hydrolytic and other enzymes that normally are structurally bound. The internal swelling suggests compartmentation by membranes originally present and by membranes newly formed on injury.

The grana are 0.5–2 μ in diameter and vary in size, depending on the species, the season, the light intensity, and the wavelength under which the plants were grown.

Evidence from polarized-light studies indicates that the lamellae contain protein as a major constituent. Chlorophyll appears to be associated with the protein layers, probably mainly in the grana. Long-chain lipid molecules lie oriented perpendicular to the protein layers. Chlorophyll and lamellar formation increase more or less together, and both disappear in darkness, suggesting a close structural association. No direct evidence for chlorophyll localization has yet been obtained with the electron microscope as the embedding techniques used at present remove chlorophyll and possibly some other fatty materials.

IV. Differentiated Regions of Some Algal Plastids

A. *Eyespot or Stigma*

An eyespot or stigma is an orange-red colored dot or streak region, consisting of carotenoid droplets, that is often formed as a portion of one of the plastids of certain motile algal species. It is usually present in the anterior portion of the cell, but in some species it may be in a median or posterior portion. An eyespot is found in motile forms of green algae, either unicellular or colonial, and also in the zoospores and gametes of vegetatively nonmotile forms, e.g., *Asterococcus*, which is considered to be a relic of a motile ancestral condition. The eyespots of the colonial *Volvox* are larger in the anterior cells of the colony and become progressively smaller or even are absent in the posterior cells (Bold, 1951).

The eyespot is said to be produced by the division of a pre-existing one or to arise *de novo*. In the Volvocales, the division of the protoplast may be accompanied by a division of the eyespot or new eyespots may be formed *de novo* in each of the protoplasts. In the filamentous green alga *Ulva*, prior to gametogenesis the chloroplast becomes yellow green. At the time of the first division, a tiny red region develops in the plastid. This region appears to grow more intensely red and to divide by repeated bipartitions each time the protoplast divides, so that each gamete swarmer ultimately receives an eyespot derived genetically from the original red region of the plastid (Schiller, 1923). In *Euglena* the eyespot region is in the anterior portion outside the chloroplast and consists of a cluster of dense orange-red spheres. During cell division this region divides and does not arise *de novo*. The colored droplets may disperse during prophase and then reaggregate during anaphase (Jahn, 1951).

In *Chlamydomonas reinhardi* (Sager and Palade, 1954) the droplets are present in a small area within the chloroplast. The electron micrographs show that the eyespot is made up of two parallel layers of dense uniform droplets packed hexagonally (Fig. 15). What factors of differentiation cause this region to accumulate the carotenoid droplets is not known. In the yellow mutant plastid of *Chlamydomonas* which lacks lamellae, the eyespot is present but not so well organized. One layer is made up of droplets of uniform size, but the second layer is represented by dense droplets of varying size which are scattered in the neighborhood of the eyespot.

Euglena is phototactic to light absorbed by the eyespot pigment, antheraxanthin. Euglenaceae that lack an eyespot are not phototactic. However, an additional light-sensitive reaction is probably present because members of the colorless families (e.g., *Peranema*) are sensitive to changes in light intensity. The complex locomotor apparatus and its morphological relations to the eyespot are described for *Chromulina,* a chrysomonad, by Rouiller and Fauré-Fremiet (1958).

B. Pyrenoids

The pyrenoids are dense spherical bodies, embedded in the chloroplast matrix of some species of algae and usually surrounded by a sheath of starch grains. The pyrenoids are present in a great majority of the green algae, the Chlorophyceae. They are also found in the chloroplasts of some diatom species, several Xanthophyceae, Dinophyceae, Bangiales, Nemalionales, and Ectocarpales. They are absent from *Microsporum,* from certain Siphonales, and from the great majority of brown and red algae. The chloroplast of the primitive moss *Anthoceros* possesses a pyrenoid. Other bryophytes and the higher plants lack pyrenoids (Czurda, 1928).

There may be one or more pyrenoids per chloroplast, depending on the species. For example, *Chlamydomonas* and *Zygnema* contain one; *Spirogyra* may contain several. Pyrenoids are said to be rich in protein because they generally stain more intensely with protein stains. Characteristic of many of the pyrenoids is the formation of a starch sheath surrounding the pyrenoid. This fact suggests that the appropriate enzymes and conditions for starch synthesis are localized in this region of the pyrenoid. Although starch may form in other parts of the cell, starch is deposited in the region of the pyrenoid first and disappears from there last. For example, in *Spirogyra* starch is formed around the pyrenoids but minute grains of stroma starch may also be formed independently of the pyrenoids.

In some cases pyrenoids are said to be duplicated by fission (Fig. 21), although this may be only apparent. In *Spirogyra,* Szejnman (1933) has followed the development of pyrenoids. In a hanging drop he was able to observe the same cell for a number of days. Growth of the ribbon chloroplast was found to attain 5% of its length per hour. The growth was partly intercalary. The most intense growth occurred during the night after a cellular division. During the morning hours pyrenoids could be seen to form *de novo.* Sometimes, doubling of the length of the chloroplast doubled the number of pyrenoids. No evidence was obtained that the pyrenoids arose from pre-existing ones by division.

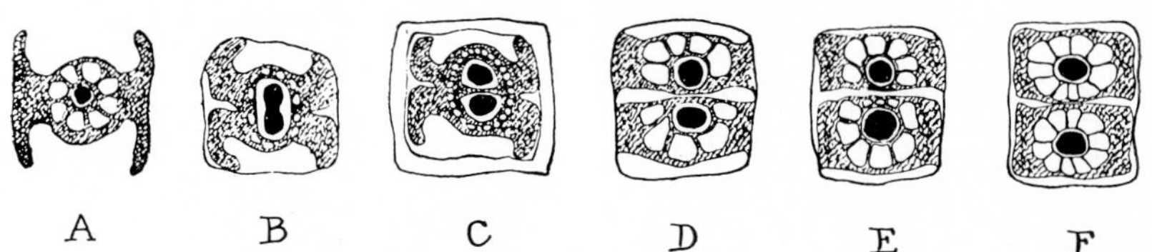

FIG. 21. Division of plastid and pyrenoid (area in black) in *Hyalotheca mucosa* with its surrounding starch masses. From Sharp (1934).

In *Euglena gracilis* (Wolken and Palade, 1953) electron photomicrographs reveal the pyrenoid (Fig. 14, B) to be a dense central region through which extend dense lamellae. The spaces between the lamellae in this region may contain a higher concentration of protein which holds the lamellae more firmly together during fixation and sectioning. No carbohydrate sheath surrounds the pyrenoid. Frequently, the pyrenoid protrudes at the surface of the chloroplast and its protrusion is adjacent to a vacuole which may or may not be connected with the formation of insoluble carbohydrate grains, paramylum, in this organism. In *Euglena fracta,* however, the chloroplast contains a pyrenoid sheathed with two paramylum grains; the pyrenoid divides at the time of fission of the chloroplast (Johnson, 1956).

In *Chlamydomonas,* a sheath of starch plates is deposited around the pyrenoid. The pyrenoid is a relatively large, nonlaminated body, 1.5–2 μ in diameter, embedded in the posterior part of the chloroplast. Tubular elements, possibly extensions of the double-membrane lamellae, project between the starch plates into the central area. These tubules are characterized by irregular thickenings along their walls (Fig. 15, A, D). In the yellow mutant no starch plates are found, but in the posterior region of the cell bits of the tubular elements with irregular thickenings are present.

Leyon (1954a) has examined the pyrenoids of *Spirogyra, Closterium, Cladophora,* and *Enteromorpha* in the electron microscope. The lamellae of the chloroplast appear to continue their parallel path through the pyrenoid as seen in *Spirogyra* or *Closterium acerosum* (Fig. 22, A); or a number of groups of parallel lamellae may fold back on themselves as in *Closterium lunula* (Fig. 22, B). Starch appears to be deposited between the lamellae adjacent to the pyrenoid. The pyrenoid of *Closterium acerosum* was dissected out of the plastid and was noted to be green.

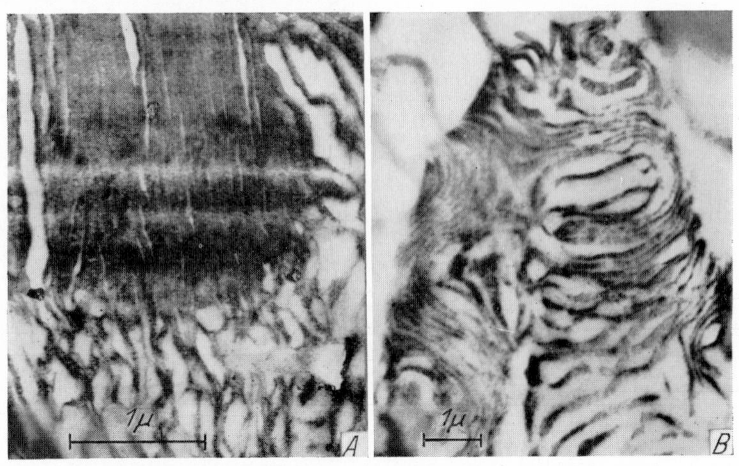

FIG. 22. A. Parallel arrangement of lamellae of the pyrenoid of *Closterium acerosum.* B. Convoluted arrangement of the pyrenoid of *Closterium lunula.* From Leyon (1954a).

The pyrenoids of *Anthoceros* divide in two before the plastid divides. The swelling of the chloroplasts by alkali shows that the pyrenoid consists of a compact cluster of tiny lens-shaped bodies. When cells divide rapidly, the number of these bodies may decrease to 6, whereas in old cells 200 may be present. An evolutionary trend is suggested by the appearance of the pyrenoids of *Anthocerotales:* homogeneous pyrenoids of one species are replaced by a compact cluster of tiny pyrenoid bodies in another species, and in still a third species these bodies are spread throughout the plastid (Kaja, 1955).

C. *Karyoids*

In *Spirogyra* the karyoids are seen in the electron microscope to consist of about ten layers of double lamellae, each 70 A. thick, extending 2.5–3.5 μ in length. The karyoids lie on the inner surface of the spiral plastid, are two to four times as frequent as pyrenoids, and stain blue with picric acid-aniline blue. Their function is not known (Butterfass, 1957).

Heitz (1958) has found in the cytoplasm structures which consist of 5–7 closely packed double lamellae in both higher and lower plants. No membrane surrounds these structures; they lack ribosomes. He considers them to represent the Golgi apparatus.

V. Origin and Development of Chloroplasts

The chloroplast is of interest not only because it represents the primary seat of photosynthesis, but also because it represents a unit of cytoplasm which, like the nucleus, can be shown in favorable examples to be self-reproducing. The evidence for the self-reproducing ability of the chloroplast is threefold: (1) In the lower plants the chloroplast can be observed to divide, and its continuity can be established from one cell to the next by direct observation. (2) In certain algae the chloroplast may disappear from a cell and then cannot be recognized in successive generations. (3) The transmission of the chloroplast through the cytoplasm of only one parent demonstrates continuity of the chloroplast and inheritance in a non-Mendelian fashion. If a plastid mutates, such a mutation is inherited in a non-Mendelian fashion (see Section VI).

A. *The Division and Continuity of the Chloroplasts of the Lower Plants*

Division of the chloroplasts may be observed directly in the cells of the lower plants, especially in those plants that contain only one or a few chloroplasts per cell (Fig. 23, H, I). After cell and chloroplast division, the flat lamellae within the chloroplast grow or extend out as broad sheets while the chloroplast and cell enlarge. In vegetative cell division of algae, the division of the chloroplast may occur at an earlier or somewhat later stage than the division of the nucleus. The pyrenoids may divide at the same time, but in many forms they disappear during cell division to reappear in each daughter cell. On the basis of direct observations of this kind Schimper (1883) and Meyer (1883) proposed that plastids never originate *de novo*, but always arise from pre-existing plastids by division.

Not only may the chloroplasts of the lower plants be followed in vegetative division, but a number of cases have been reported in which the plastid has been followed during zygospore and gamete formation. These examples also demonstrate the fact that the plastid does not arise *de novo*, but from a pre-existing plastid. For example, in *Zygnema* each vegetative cell contains one nucleus and two plastids, all of which divide at each vegetative-cell division (Fig. 23). In sexual reproduction, the entire protoplast with its nucleus and two plastids passes through the conjugating tube as a "male" gamete and unites with a similar complete protoplast ("female" gamete) of another filament. The two nuclei fuse, forming the primary nucleus of the new individual (zygospore nucleus), while the two plastids contributed by the male gamete degenerate, leaving the two furnished by the "female" gamete as plastids of the new individual. A similar process occurs in *Spirogyra*.

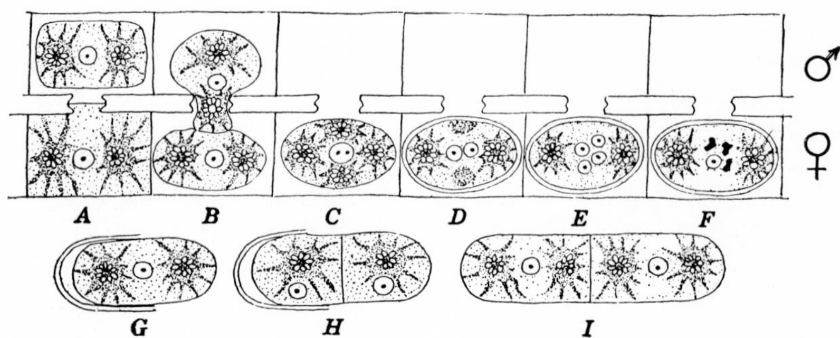

FIG. 23. A–F. Conjugation and zygospore formation in *Zygnema*. The male gamete passes through the conjugating tube and fuses with the female gamete. The two "male" chloroplasts degenerate. Thus, all the subsequent chloroplasts arise by divisions of the "female" chloroplasts i.e., maternal inheritance. In meiosis, four nuclei are formed, of which three degenerate. G–I. Germination of zygospore and vegetative division of filament showing chloroplast division at H. From G. M. Smith, "Cryptogamic Botany," Vol. I, 2nd ed. McGraw-Hill, New York, 1955.

A study now classic, of chloroplast inheritance in a species related to *Spirogyra*, was made by Chmielevsky (1890) (Fig. 24). In this alga, a *Rhynchonema* species, the vegetative cells contain one chloroplast per cell. At the time of conjugation tannins disappear from the cell, but starch and oil droplets increase. The starch and oil droplets, which are plentiful in the conjugating cells, pass into the young zygote and gradually disappear. When the cuticular wall of the zygote forms and becomes brown, the green spiral chloroplasts are still visible through the wall. The zygotes were fixed at a stage when the second and third coats of the zygote were formed. Fixation occurred in 1% OsO_4 for 5–10 seconds. The filaments containing the zygotes were then washed with water and placed in dilute glycerin; the glycerin was permitted to thicken by evaporation. The zygotes were then quite transparent. It was possible to observe by this technique that in the zygote the spiral chloroplast derived from the "male" gamete gradually became twisted, turned yellow, became thinner, and broke up into particles that later united and were transferred to the vacuole. Thus, in the successive vegetative cell divisions, only the chloroplast of the "female" gamete was inherited.

In some algae the disappearance of chloroplasts from a cell leads to a cell lineage that never again forms chloroplasts. It has been questioned whether the disappearance signifies a complete absence of plastids or whether there is an irreversible conversion to colorless plastids which cannot be recognized (Lwoff, 1950). The presence of paramylum grains in *Euglena* has been considered an indication that colorless plastids were present (compare Section IV, B).

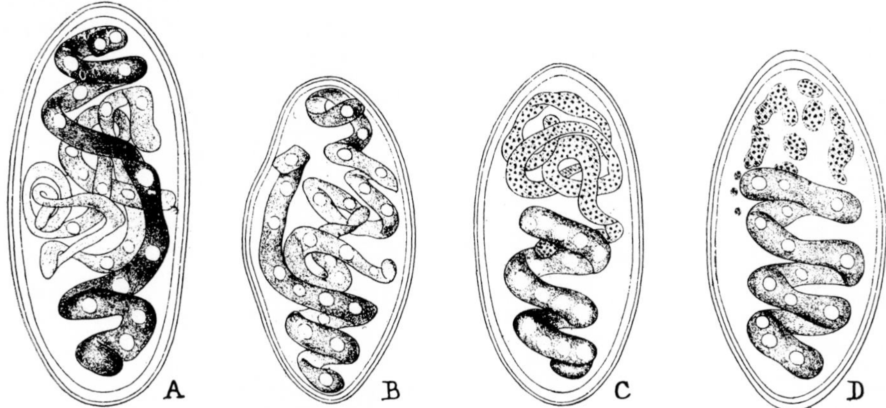

FIG. 24. A–D. Degeneration of the "male" spiral chloroplast in the zygospore of *Rhynchonema*. From Chmielevsky (1890).

Euglena gracilis when grown in the dark loses its pigments but continues to form paramylum. As yet it has not been possible to decide by light or electron microscope studies whether the resulting colorless *Euglena* cells contain colorless plastids regressed in size and derived from the original plastids or whether the colored plastids have broken down completely and proplastids are present from which new chloroplasts may arise.[1] The present data may perhaps be interpreted to indicate that the chloroplasts of *Euglena* behave like those of higher plants, i.e., arising from tiny colorless proplastids which can divide in the dark by fission and which in the presence of light enlarge, become green, and do not divide in their mature state. Also as in higher plants, streptomycin does not appear appreciably to affect the mature plastids. Gross and Vil-

[1] Studies with a fluorescence microscope reveal that in *Euglena* the plastids multiply by fission; plastids do not disappear when *Euglena* is grown in the dark; irreversibly "bleached" *Euglena* contains no fluorescent plastids (Granick, unpublished).

laire (1960) could find no evidence for the division of green *Euglena* plastids. De Deken-Grenson and Godts (1960) treated *Euglena* with suboptimal concentrations of streptomycin. They found that there was no relation between persistence of a visible green plastid in a treated cell and the transmission to its descendants of its ability to synthesize chlorophyll. Brawerman and Chargaff (1959) in a series of excellent studies on *Euglena* also studied the bleaching of *Euglena*, using elevated temperatures instead of streptomycin. When normal colorless *Euglena* cells (i.e., grown in the dark) are exposed to light at 34° in a medium that prevents cell multiplication, the chloroplasts form. When such green cells are returned to room temperature to a medium in which the cells multiply, colorless cells will develop. The authors interpret this result to indicate that elevated temperature (or streptomycin) inhibits the activity of or destroys irreversibly a "self-reproducing catalyst" required for chloroplast formation.

In nature a large number of colorless algae exist. On the basis of cellular morphology, use of similar organic substrates, and storage of identical carbohydrate products, these colorless cells are considered to be related to species of green algae. For example, species of *Astasia* are related to species of *Euglena*, *Polytoma* to *Chlamydomonas*, and *Prototheca* to *Chlorella* (Pringsheim, 1941). Have these colorless species completely lost their plastids or only lost the ability to form colored pigments and lamellae in the plastids? For example, the persistence of ribulose diphosphate carboxylase in *Astasia* suggests that a plastid remnant is still present (Fuller and Gibbs, 1959).

The problem of presence or absence of plastids is related to the problem of whether a cell can survive without plastids; i.e., are there in some species essential enzymatic reactions that are present only in plastids which are required for the life of the cell? The present evidence is difficult to assess. Older workers—Scherffel, Doflein, and Pascher—observed in certain chrysomonads which contain single plastids that if the fission of the plastid is retarded at the time of cell division then one daughter cell will contain no plastid. Although such cells which do not contain a plastid survive for some time they have not yet been isolated in culture. In *Euglena mesnili*, the plastids can form chlorophyll in the dark and can thus be followed readily. When this species is cultivated in the dark for months, a progressive decrease occurs in the number of green plastids per cell. Cells lacking both plastids and paramylum have been observed but could not be cultivated, nor could similar cells be cultivated that were obtained by streptomycin treatment (Lwoff, 1950). In *Oenothera* certain evidence suggests that plastids may be essential apart from their function of photosynthesis (see Section VI, B, 4, d).

A few examples of chloroplast division and continuity in the higher cryptogams may be cited. In the liverwort *Anthoceros* the cells of the thallus contain a single plastid, which divides with the nucleus at each cell division. The egg likewise contains a plastid, but the tiny spermatozoid has none. The zygote and the sporocyte cells which are later formed from the fertilized egg are therefore characterized, like the cells of the gametophyte, by the presence of one plastid per cell, this plastid having been derived from the division of the original egg cell plastid. Although it is difficult to demonstrate the plastid in the young sporogenous cells, every sporocyte contains one. As shown by Davis (1899), the sporocyte plastid divides twice during the prophase of the first division of the sporocyte nucleus, so that each spore of the resulting quartet receives one plastid. Upon germination, the spore produces a gametophyte with one plastid in each cell, and the cycle is complete. The plastids therefore remain as morphological entities throughout the whole life cycle, multiplying exclusively by division. The studies of Scherrer (1914) support the continuity of the chloroplasts through the life cycle of *Anthoceros*. Evidence for the continuity of chloroplasts in the liverworts and mosses has been obtained more recently by Kaja (1955). The apical cells contain relatively few chloroplasts, 1.5–2 μ in diameter and each with about 3–5 grana as seen in the light microscope. In younger moss leaves, the time required for a chloroplast division is about 8 hours (Reinhard, 1933). As the cells mature, the chloroplasts enlarge to 7 μ and contain 60–80 grana per chloroplast.

In the ferns *Selaginella* and *Isoetes*, the plastids have also been found to behave as morphological individuals multiplying exclusively by division. In such cases the plastids possess an individuality comparable to that of the nuclei, from which they differ conspicuously, however, in undergoing no fusion at the time of sexual reproduction. For example, in the sporophyte of *Isoetes* Stewart (1948) found that the meristematic cells of the root or leaf contain a single plastid per cell (as in the meristematic cells of many bryophytes and pteridophytes). The division of the single plastid precedes the division of the nucleus. Prior to cell division, the plastid of the meristematic cell becomes greatly elongated, the daughter plastids passing to the opposite poles of the spindle. At the interphase, a resting cell will contain only one plastid. As the cell matures and enlarges, the single plastid may divide to form a variable number of plastids per cell. The daughter plastids thus formed will become either starch-filled leucoplasts of storage cells or chloroplasts of leaf cells. During nuclear division these plastids show no tendency to occupy the poles of the cells, in contrast to the behavior of the plastid in a meri-

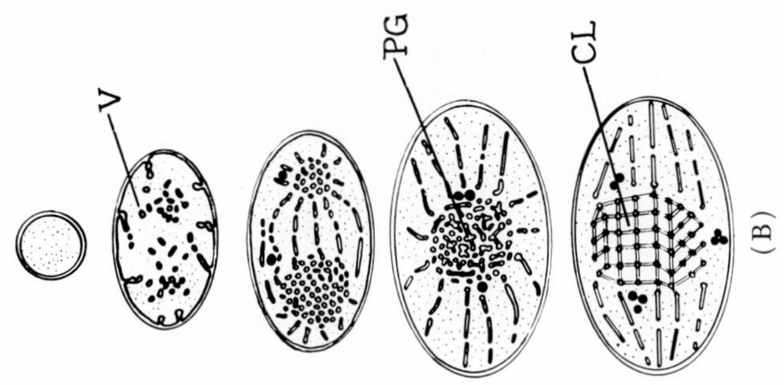
a
b
c
d
e
f
CM
V
CIS
G
Proplastid phase
Differentiation phase
Maturation phase
(A)
V
PG
CL
(B)

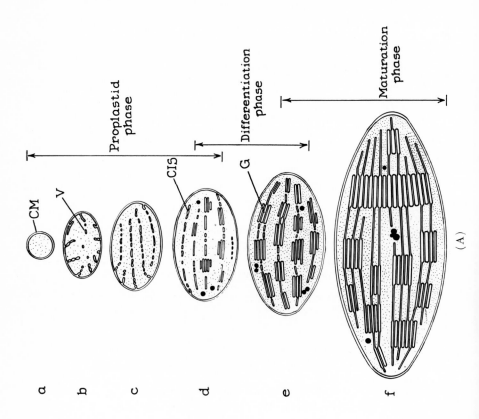

stematic cell. The plastids of such maturing cells lose their intimate association with the nucleus and lie at random within the cell.

The behavior of the chloroplasts in this higher fern may represent an evolutionary transition. In *Isoetes,* the young meristematic cell contains only a single chloroplast during cell division, but as the cell matures, it comes to have a number of chloroplasts per cell. In gymnosperms and angiosperms, however, well-defined plastids are not seen in meristematic cells, but precursors of these plastids, called proplastids, are present. If one or only a few proplastids were found to be present per meristematic cell of the phanerogams, this knowledge would be invaluable in the interpretation of variegations that suggest inheritance of mutated plastids in these higher plants. In *Agapanthus umbellatus,* about 20 proplastids are reported in meristematic cells (Fasse-Franziskett, 1955), and in *Epilobium* from 7 to 20 (Michaelis, 1958).

B. *Origin and Differentiation of Chloroplasts of Higher Plants*

Chloroplasts of higher plants develop from tiny colorless or pale yellow bodies, the proplastids. Studies of this development are in progress and the interpretations presented here should be considered tentative. The description below follows to a large extent the interpretations of von Wettstein (1958). This is a description for monocot proplastids, which, unlike dicot proplastids, enlarge in complete darkness.

The development of the chloroplast may be divided into three overlapping phases (Fig. 25): (1) The phase of proplastid multiplication and the budding off from the inner membrane of flattened tubules or vesicles. (2) The phase of differentiation and enlargement in which the vesicles increase in number, widen, and form collapsed cisternae (double-membrane lamellae); then the membranes begin to thicken in some areas and a pale green color develops. (3) The phase of maturation in which the chloroplast loses its ameboid character and well-differentiated grana lamellae and stroma lamellae are distinguished.

1. *Phase of Proplastid Multiplication*

In the rapidly dividing cells of the shoot and root tips, the metabolic activity is preponderantly that of nucleic acid synthesis. Here, per meri-

---

FIG. 25. A. Phases in the development in the light of the proplastid to the chloroplast. B. Phases in the development in the dark, showing the formation of the prolamellar body and the crystal lattice. *CM,* double chloroplast membrane; *PG,* primary granum or prolamellar body or proplastid center; *V,* vesicle; *CIS,* flattened cisternum; *C.L.,* crystal lattice; *G,* granum. Modified from von Wettstein (1958).

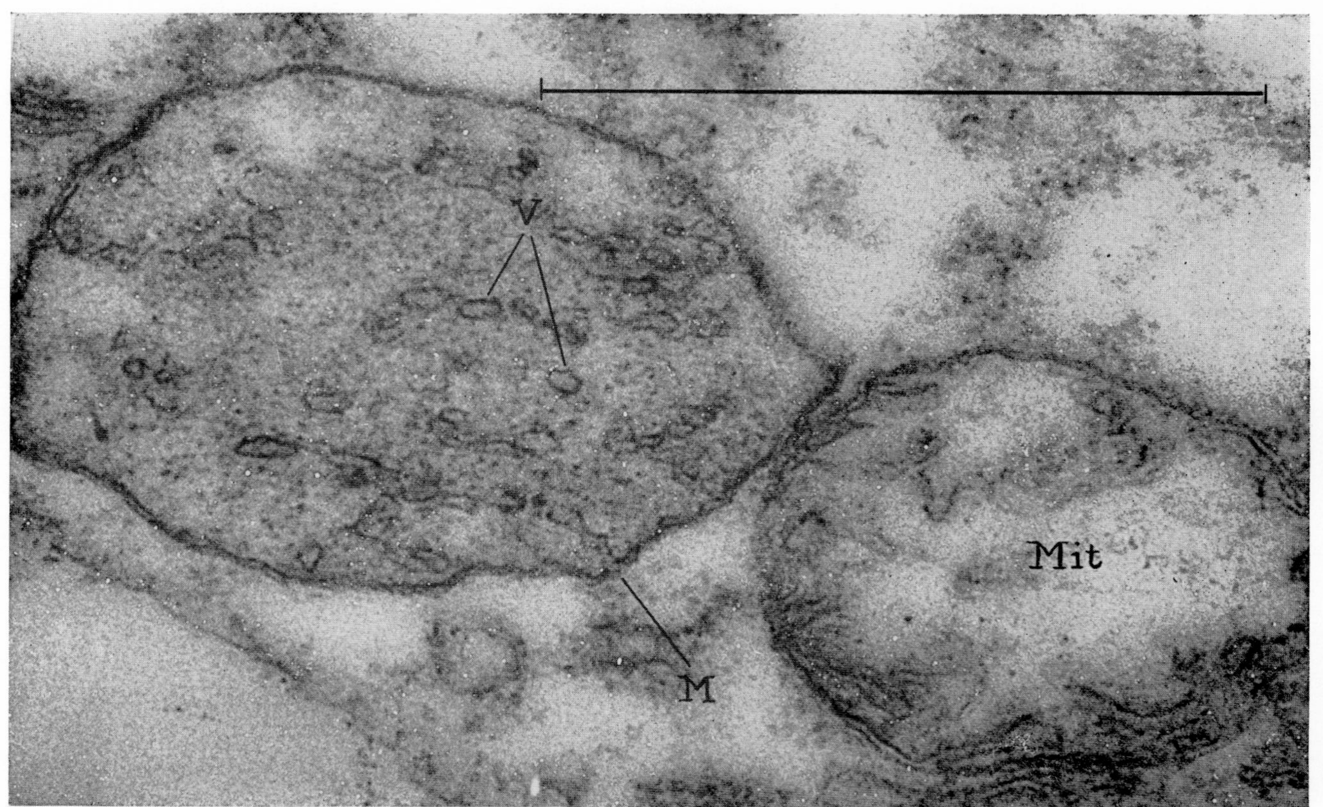

FIG. 26. Barley leaf meristem section showing proplastid (left) and mitochondrion (right). In the proplastid are vesicles (*V*); a vesicle appears to form as an invagination of the inner membrane at *M*. From von Wettstein (1958).

stematic cell 7–20 ameboid colorless bodies (Michaelis, 1958) 0.4–0.9 μ in diameter, the proplastids, are present. The proplastid is surrounded by a double membrane (each 35 A. thick). Inside is a granular background matrix or stroma. At an early stage, when the proplastid is about

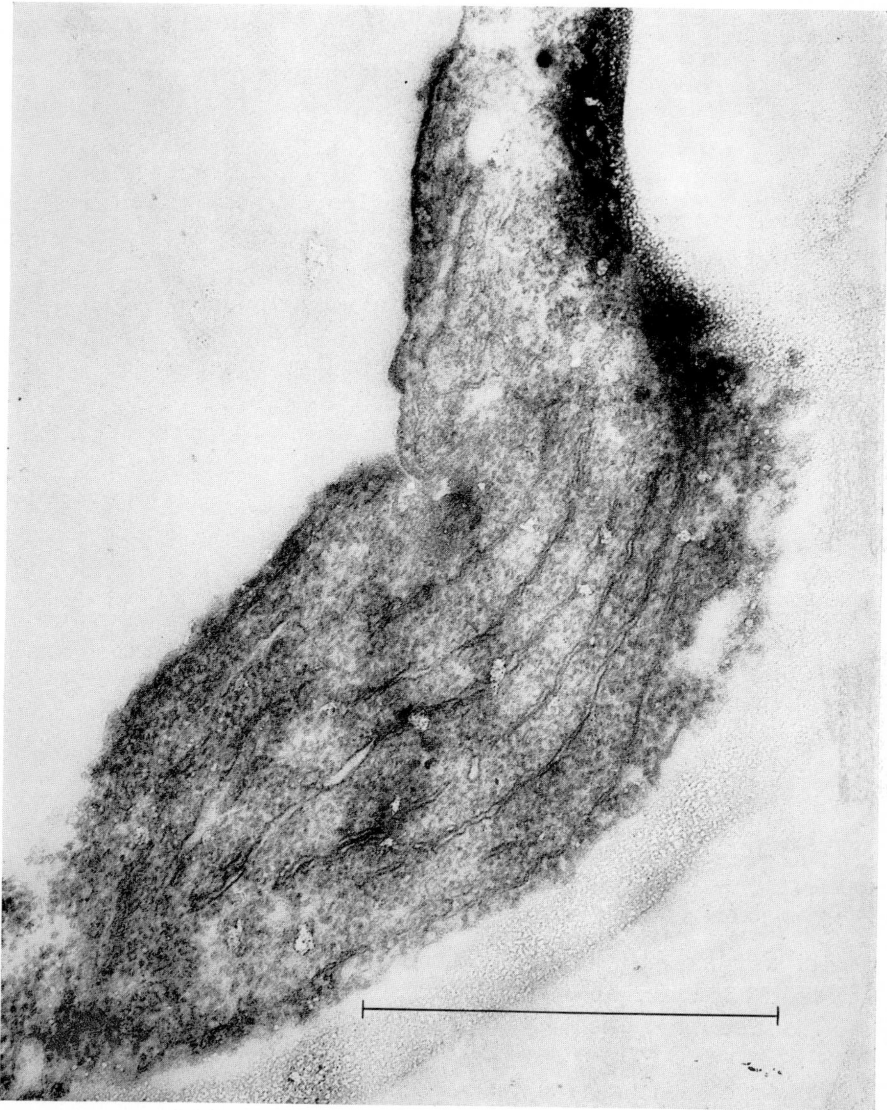

FIG. 27. Proplastid of a 10-day-old etiolated barley seedling showing the parallel alignment of vesicles and cisternae. From von Wettstein (1958).

1 μ in diameter spherical or elongate vesicles (250–400 A. diameter) bud off (Figs. 25, 26) from the inner membrane layer (Mühlethaler and Frey-Wyssling, 1959). Normally they fuse in rows to form a pile of 4–15 parallel flattened cisternae as in barley (Fig. 27). The ameboid proplastids elongate and pinch off to form two proplastids, and it is estimated that they may undergo a total of three or four such fissions. Starch grains,

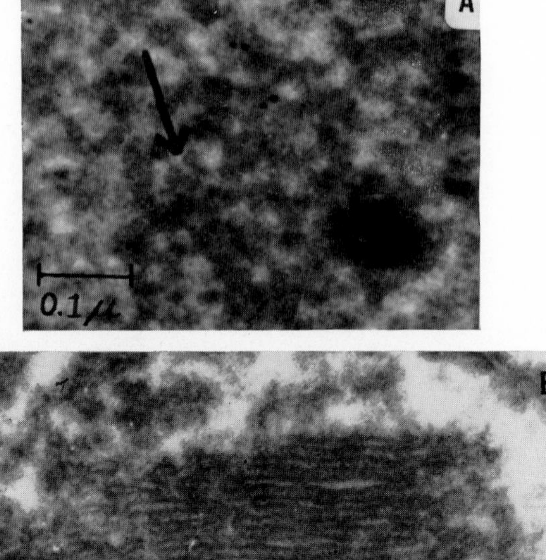

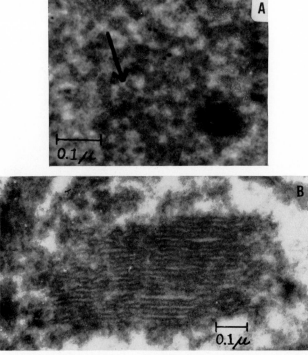

FIG. 28. A. Prolamellar body in the proplastid of etiolated maize, consisting of a cluster of vesicles, each about 250 A. in diameter. Arrow points to a vesicle. B. Section through a prolamellar body of an etiolated leaf of maize that has been exposed to light. The vesicles appear to have fused to form rows of flattened cisternae. From Hodge *et al.* (1956).

tiny oil droplets, and traces of carotenoids may be present at this stage. In contrast to mitochondria the youngest proplastids are larger in size, contain starch and often contain one or a few flat cisternae (Menke, 1960).

Details of this phase and other interpretations may now be considered. When seedlings are grown in the dark (Fig. 25, B), the rate of fusion of vesicles in proplastids to form flattened cisternae is slow (von

Wettstein, 1958) so that vesicles accumulate in a cluster (Fig. 28, A) to form a dense body variously designated as a primary granum (Strugger and Perner, 1956), prolamellar body (Hodge *et al.*, 1956), or proplastid center insoluble in fat solvents. In seedlings grown in the dark for 3–10 days the proplastids continue to enlarge and the cluster of vesicles fuses to form a "crystal lattice" structure (Fig. 29, A and B), first observed by Heitz (1954) and by Leyon (1954c). This structure is interpreted by

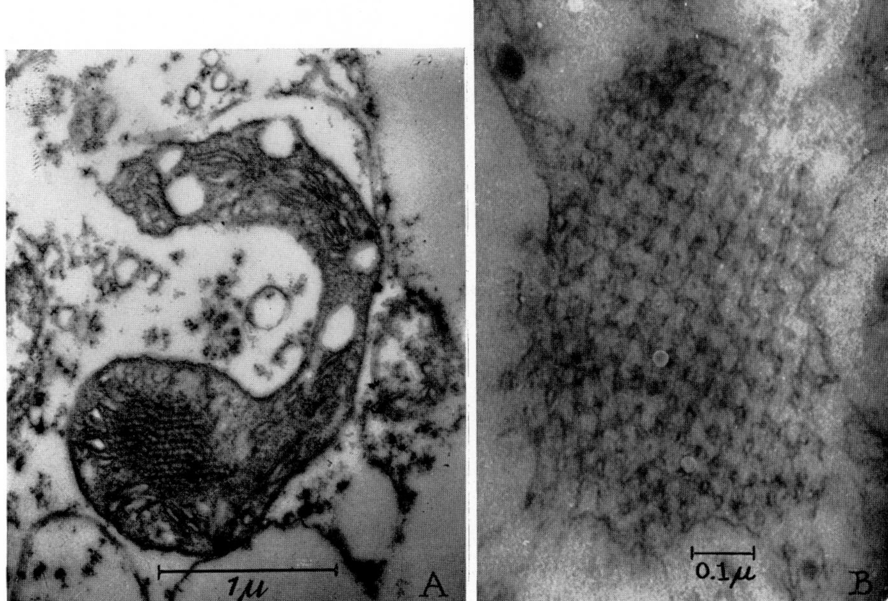

FIG. 29. A. Elongated proplastid of *Chlorophytum comosum*, possibly undergoing fission into two daughter proplastids. A lattice structure is at the lower end, and a cluster of loosely arranged flattened vesicles at the upper end. In the granular stroma are starch grains (white areas) and scattered vesicles. B. Lattice structure consisting of tubes interconnected in three dimensions. From Perner (1956).

Perner as a three-dimensional array of beaded strands, whereas von Wettstein suggests that the lattice is a tubular cubic arrangement where the bead represents a point where six tubes meet (Fig. 30). When barley seedlings are maintained still longer in darkness, concentric lamellae (reminiscent of myelin structures) arise in connection with the lattice, the lattice disappears, and finally only the concentric lamellae remain (Fig. 31). When etiolated plants are placed in light the vesicles may fuse into layers (Fig. 28, B) or vesicular outgrowths may occur from the primary granum or from the lattice structure (Fig. 32) to develop into grana and stroma lamellae. A different interpretation of the proplastid is presented by Strugger and Perner (1956), who consider that a primary granum (dense, prominent body) is always present in a proplastid and that at the time of fission a primary granum divides in two so that each daughter proplastid contains one, as is suggested by Fig. 29, A.

2. *Phase of Differentiation*

In the elongating cells behind the meristematic stem tip, the predominant syntheses are those of proteins and lipids. Here the proplastids enlarge, become lens-shaped, and gradually lose their ameboid character. The flattened cisternae (= collapsed vesicles, double-membrane lamellae) elongate. New vesicles may grow out and lie parallel in the zone

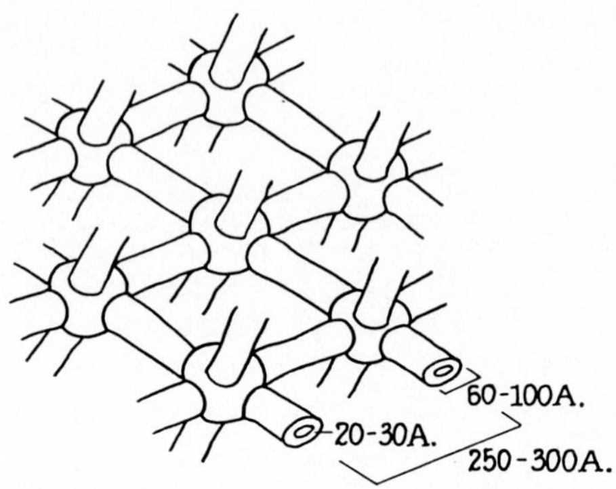

FIG. 30. Interpretation of lattice structure that develops in the proplastid of an etiolated seedling by fusion of vesicles.

just within the plastid membrane (peristromium) (Fig. 33, A, B) and fuse, so that a continuous stack of cisternae is organized (Fig. 34). Concomitantly, certain regions of the cisternal membranes increase in thickness to form the grana regions, or the grana regions may develop before complete fusion of vesicles and cisternae has occurred. A rapid increase in chlorophyll occurs at the time the grana regions differentiate.

According to Mühlethaler and Frey-Wyssling (1959) there is no evidence for reduplication of grana lamellae; the lamellae arise either from invagination of the plastid membrane, or from the prolamellar body. However, von Wettstein suggests that grana disks in barley may also arise by thickening and fission of lamellae lengthwise. A small amount of chlorophyll may form independently of lamellar formation; it may form to a small extent when no grana regions are differentiated, as in the bundle sheath chloroplasts of *Zea* (Hodge *et al.*, 1955).

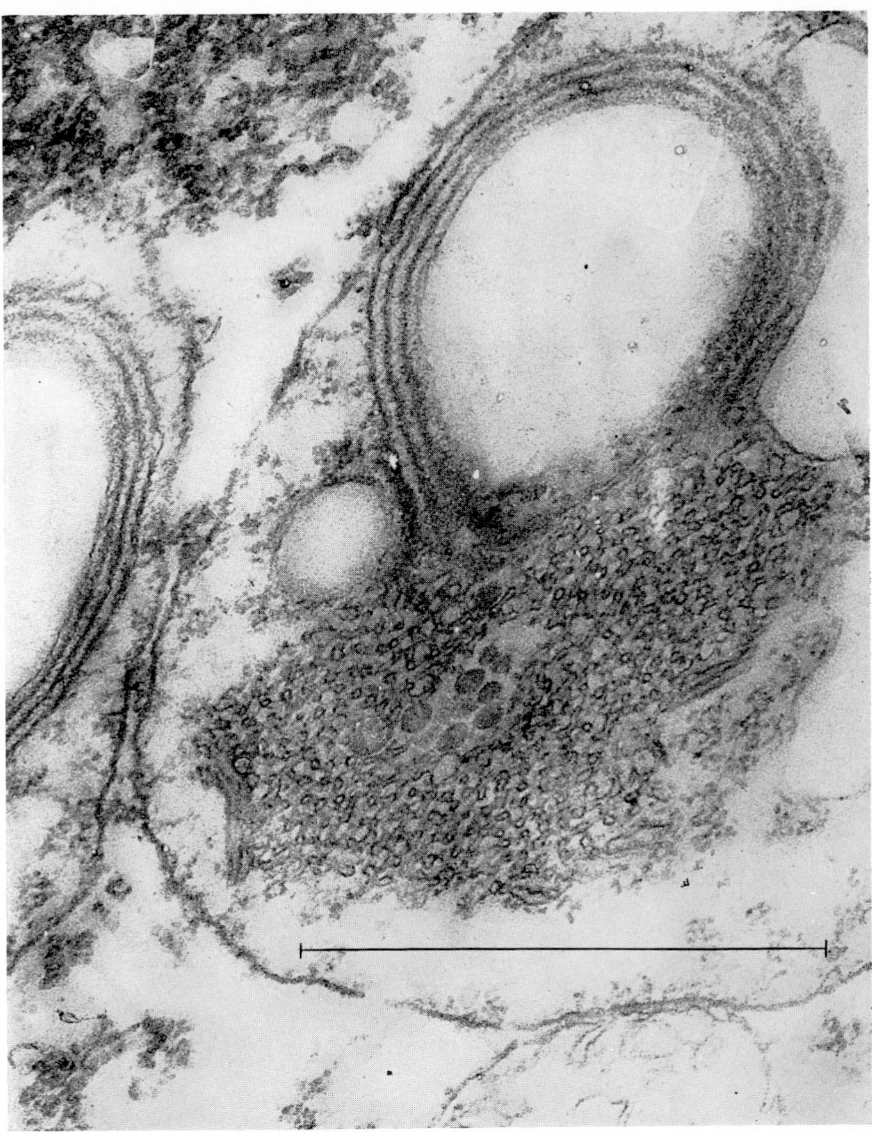

FIG. 31. Section from upper portion of a 15-cm.-long leaf from a barley seedling which was in continuous darkness for 16 days. The proplastid contains a tubular lattice structure (lower portion) from which develop the series of concentric lamellae (upper portion).

FIG. 32. Part of a young chloroplast of *Aspidistra elatior* from a section 1 mm. from the tip of the shoot. The lamellae appear to be extensions of tubules of the lattice structure. From Leyon (1954c).

3. *Phase of Maturation*

The chloroplast enlarges to final size, double-membrane lamellae increase in number, the grana become better defined, and fission ceases. The number of chloroplasts in the mature cell may be three to four times the number of proplastids originally present in the meristematic leaf cell. By the time the cells have expanded to one-third their final length the plastids cease division almost completely (Granick, 1938b).

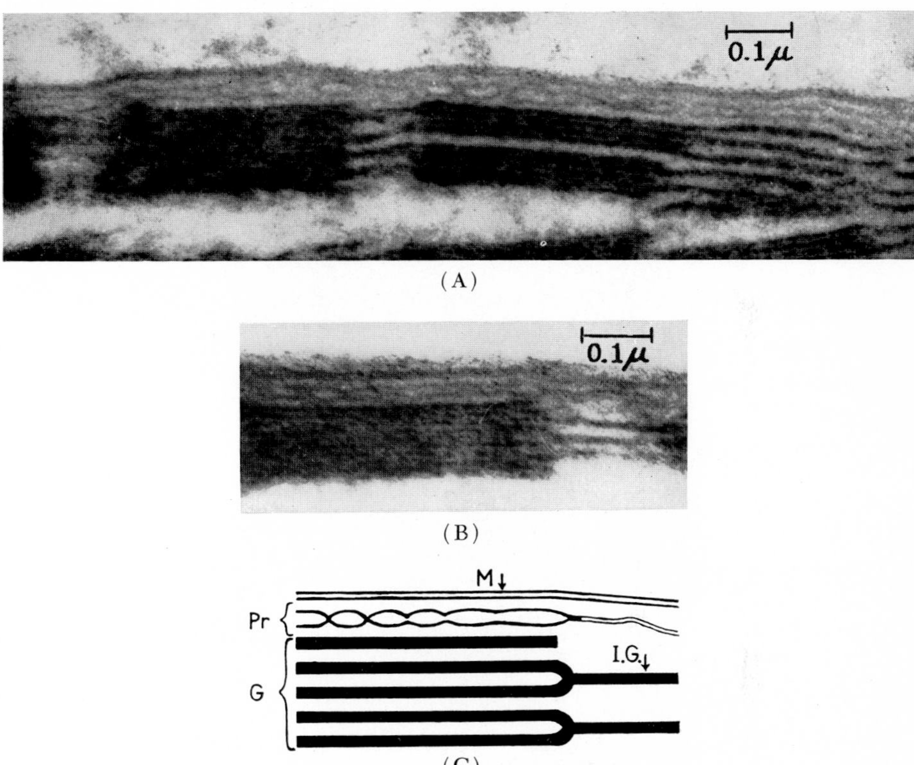

FIG. 33. A. Mesophyll chloroplast of etiolated *Zea mays* 48 hours after illumination. Note the dense grana lamellae and dense stroma lamellae. At the upper surface, immediately beneath the double chloroplast membrane are elongated vesicles which appear to fuse to form longer, flattened vesicles (Fig. 33, C, *Pr*). In the grana region a disk is interpreted to be connected to another disk by a stroma (intergranum) lamella (Fig. 33, C, *G* and *I.G.*). Compare, however, with interpretation in Fig. 3. B. Vesicles produced adjacent to the chloroplast membrane have matured into the less dense but already differentiated structures of the disk lamellae and stroma lamellae as seen in the upper portion of the figure. From Hodge *et al.* (1956).

In the larger vegetative leaf cells of *Fuchsia* and *Sedum* the division time of the maturing chloroplast is 1–2 days (Reinhard, 1933). Dangeard (1947) observed the chloroplast to divide by fission in *Elodea canadensis* and in *Mnium undulatum* even when starch grains were present. At times the fission was unequal.

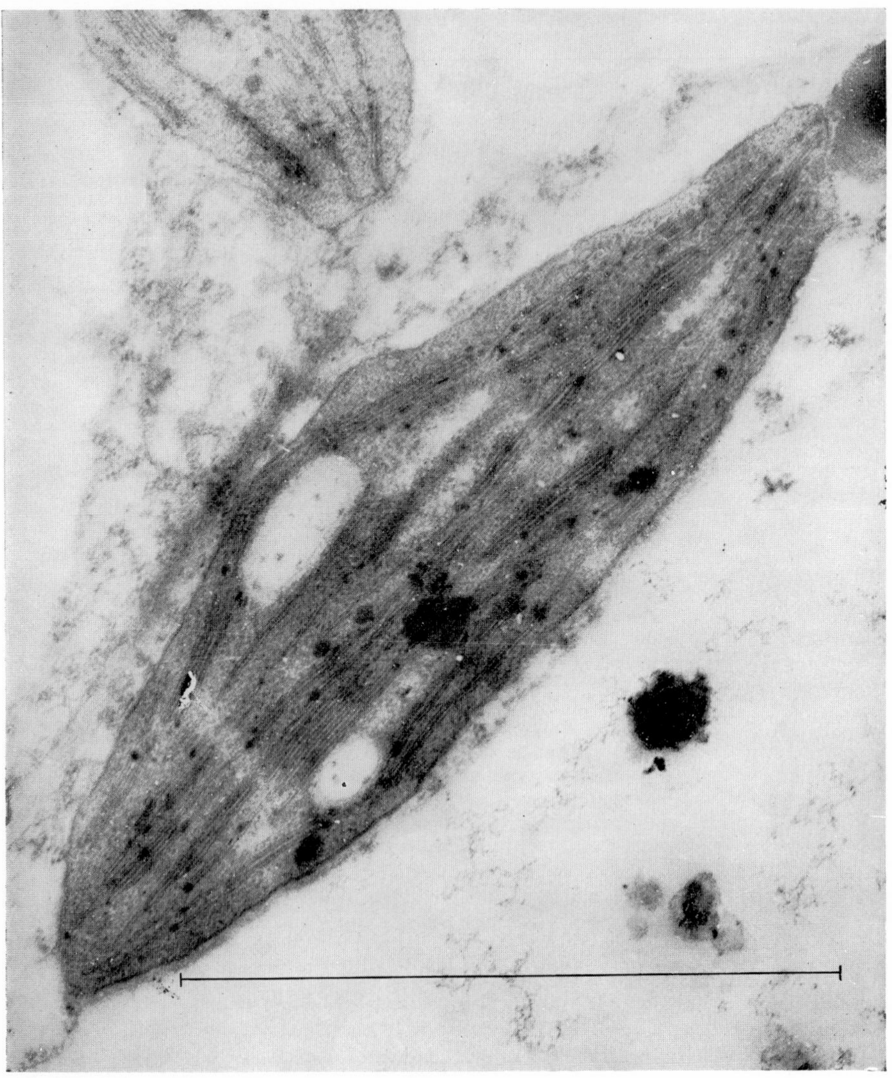

Fig. 34. Plastid of tomato leaf with numerous double-membrane lamellae extending the length of the plastid and only slightly differentiated into grana regions. From Lefort (1957).

C. Origin of Plastid Types

Other plastid types such as leucoplasts and chromoplasts (see Section II) also originate from proplastids. The simplest hypothesis to explain these plastid types is that they arise by arrest of particular phases of plastid development. This arrest, resulting from the action of internal or external factors (see Section VI), may be temporary or permanent and may include degenerative changes. Dedifferention is considered to be very limited (Frey-Wyssling and Kreutzer, 1958a). The biochemical basis of the metabolic controls involved await study.

A tentative scheme showing the relationships of plastid types is presented in Fig. 35. In roots the proplastids elongate to form leucoplasts

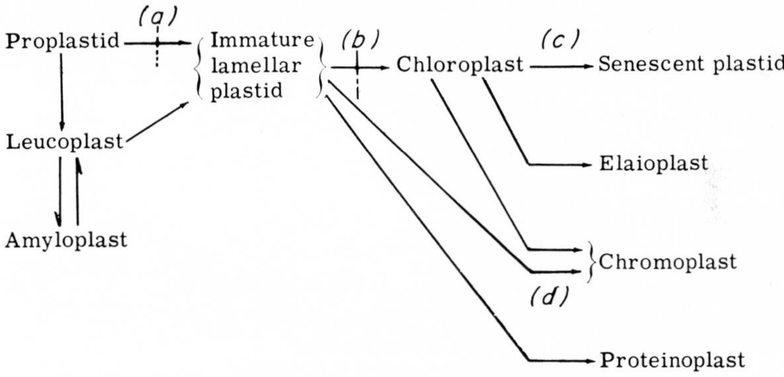

FIG. 35. Relationships of plastid types. (*a*) Arrested in normal root cells and in leaves of some barley or corn mutants. (*b*) Arrested in normal leaf epidermal cells. (*c*) Degeneration in autumn leaf cells; hypertrophy in some *Oenothera* mutants. (*d*) Chromoplasts of buttercup petals.

which do not develop further. In the root meristem of the pea the proplastid, 4 μ long by 0.25 μ wide, is surrounded by a double membrane 100 A. thick; the membrane appears as two dense lines, 30 A. (lipid?), and a 35 A. space between (protein?). A granular matrix (50 A. particles) is contained within, as well as lipid droplets, and starch grains that are not separated by membranes. Some of the starch grains show a layered structure (Sitte, 1958). In *Zea mays* or *Vicia faba* the plastid contains 10–20 round or ellipsoidal vesicles 300–400 A. in diameter lying either loosely aggregated in a few clusters or arranged in a long row; also seen are one to three single long strands which may have their origin from the fusion of a row of vesicles, and small starch grains (Heitz, 1957). Amyloplasts are characteristic of storage organs such as the potato tuber. Chlorophyll and presumably a mature chloroplast may develop from an amyloplast (Fig. 36) (Jungers and Doutreligne, 1943). Streptomycin inhibits the differentiation of proplastids to chloroplasts (see Section VI, C, 2). In epidermal cells, plastid development is arrested at the im-

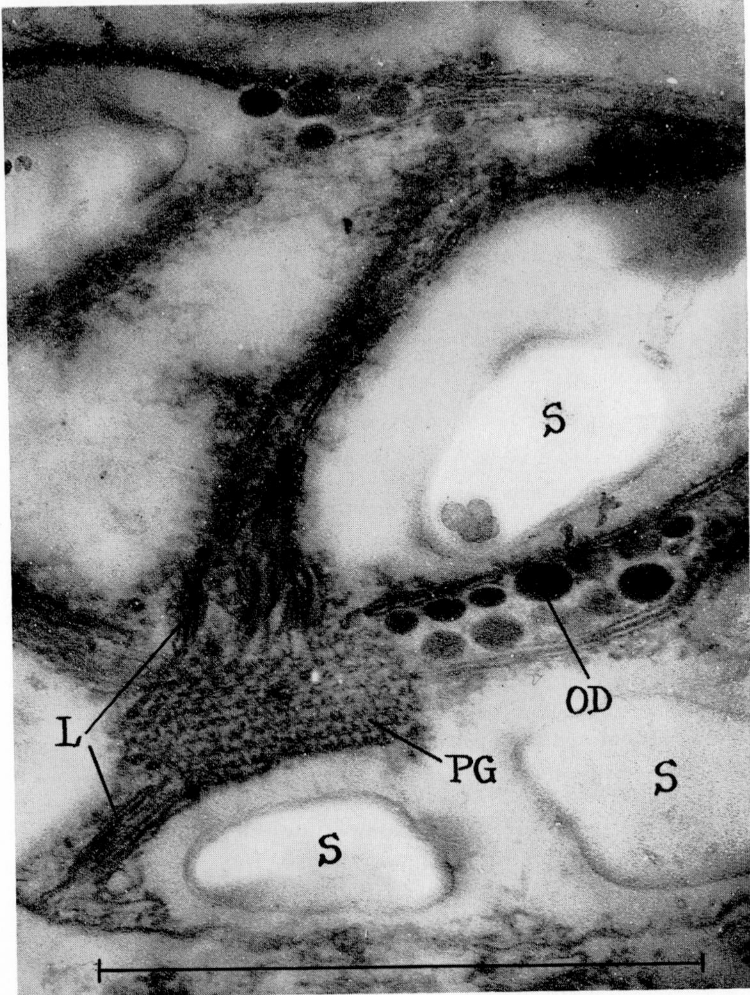

FIG. 36. Amyloplast of spruce cotyledon containing starch grains (*S*), a lattice structure (*PG*) with lamellae (*L*) extending from it, and osmiophilic droplets (*OD*). From von Wettstein (1958).

mature lamellar plastid phase. Chromoplasts may form from mature plastids or from immature pale green plastids containing only a few grana. The chromoplasts appear to represent a form of fatty plastid degeneration in which fats and carotenoids increase and accumulate in droplets, the lamellae do not multiply (suppression of protein synthesis?) but become distorted by the droplets, and finally the plastids break down to release the yellow, oily droplets into the cytoplasm (Frey-Wyssling and Kreutzer, 1958a). In carrot chromoplasts the carotenoids tend to crystallize in birefringent plates. In chromoplasts of *Capsicum annuum* spindle-shaped birefringent bundles are formed; the individual filaments of the bundle are 50 A. in diameter and consist of protein and elongated carotenoid molecules running parallel to the filaments (Frey-Wyssling and Kreutzer, 1958b).

Plastid mutations may also arrest development (von Wettstein, 1957b). Thus in Albina-20 of barley, a prolamellar body and some vesicles, but not lamellae, are formed although the plastid enlarges to the size of a mature normal plastid; pigments are absent or in low concentration; chlorophyll is broken down in light. In Xantha-3, the prolamellar body and some cisternae are formed; large amounts of oily droplets containing chlorophylls and carotenoids also form (Fig. 37) and later disappear. The interest in this mutant is that chlorophyll appears to form in the absence of lamellar development. In Xantha-10, grana regions are formed but chlorophyll is lacking. In certain *Oenothera* mutants, a normal plastid may form and then degenerate rapidly or slowly; hypertrophy with vacuolar formation occurs, the grana being most resistant to the swelling. Chemical studies on albinos are discussed in Section VIII, D.

In certain cases mature cells can give rise to meristematic tissues which later produce green cells. For example, the epidermal cells of the flax hypocotyl form buds which later form green leaves. Does this signify a dedifferentiation of immature plastids of epidermal cells to proplastids or, as is thought possible by Leyon (1954a) and von Wettstein (1958), may some proplastids reside even in mature cells?

D. *Summary*

In certain algae, chloroplasts may be seen to develop from pre-existing chloroplasts not only during the vegetative, but also through the sexual phase. In *Euglena,* once green plastids disappear from a cell by the action of streptomycin or heat, no green plastids will be found to arise *de novo;* whether colorless plastid precursors remain is not known. Examples are presented of the continuity of chloroplasts in liverworts, mosses, and ferns.

Limited plastid formation occurs in the meristem cells of the fern *Isoetes,* in which only one plastid is present per cell, and in the spermatophytes, in which the plastids are in the form of tiny proplastids. Evidence that the proplastids arise from existing ones is indirect; it is based on the behavior of mutant plastids, on plastid inheritance in interspecies crosses, and on the maternal inheritance of plastids (see Section VI). In meristematic cells the youngest proplastids can be distinguished from mitochondria by electron microscopy.

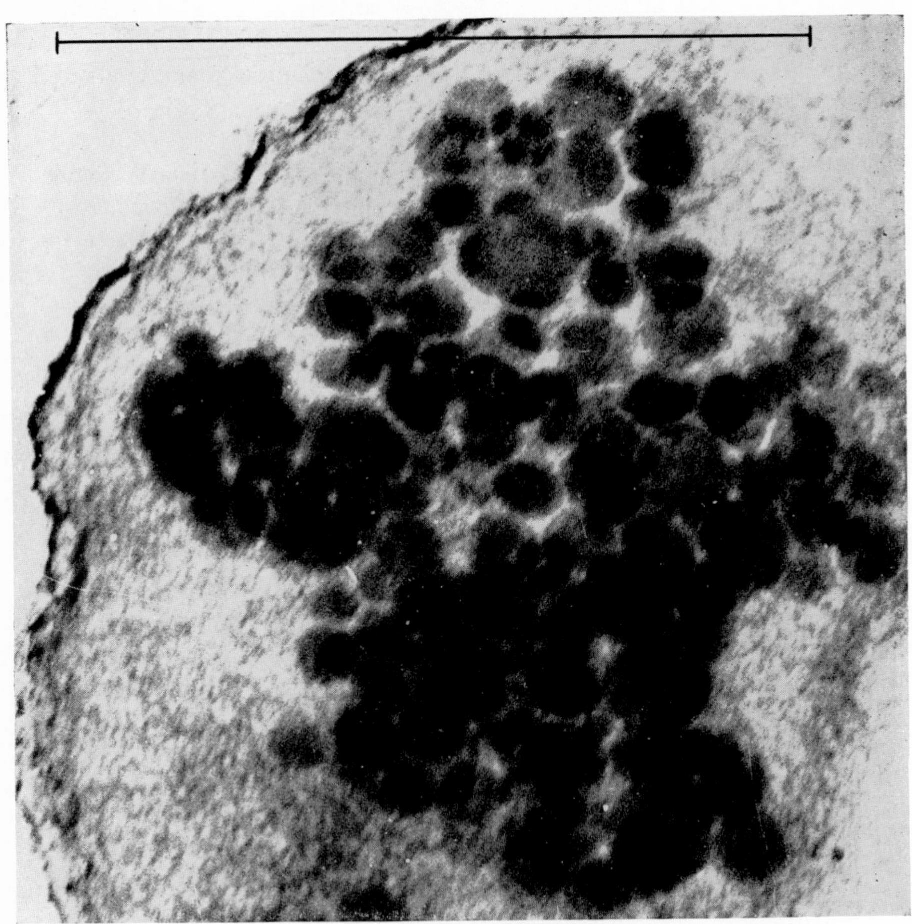

FIG. 37. Degeneration of mutant barley proplastid (Xantha-3) with the formation of osmiophilic droplets containing green and yellow pigments but no lamellae. From von Wettstein (1957a).

Data and hypotheses on the multiplication and differentiation of proplastids are discussed and an interpretation of the stages is summarized in Fig. 25.

The differentiation of plastid types in higher plants is discussed. A scheme (Fig. 35) is presented indicating the relations of the plastid types; it is based on the hypothesis of arrested development or degeneration of the plastids at specific phases of plastid development.

VI. External and Internal Factors That Influence Chloroplast Development

The chloroplast is the most convenient cytoplasmic object for the study of the effect of various factors on the differentiation and growth of a cytoplasmic particle. This is so because slight changes in the shape, size, content of starch and pigments of the chloroplast may be readily observed. The factors that affect chloroplast development may be classified into those of the external environment (i.e., the cell culture environment), such as light, temperature, the composition of the culture medium including minerals and diffusible products from neighboring and distant cells; and those of the internal environment (i.e., the culture medium of the plastids) including the factors of inheritance and differentiation.

A. *External Factors*

Only the effect of light will be discussed here because it involves the chloroplast more directly.

1. Light

There are a number of direct and indirect effects of light on the development, composition, and orientation of chloroplasts. (a) In proplastids of higher plants, light absorbed by protochlorophyll brings about its conversion to chlorophyll (see Section VII). (b) The light absorbed by chlorophyll is used in photosynthesis to make carbohydrates and to supply energy and compounds for the growth of the young plastid. (c) Light may be absorbed by accessory pigments of chloroplasts to be transmitted to chlorophyll a, thus aiding in photosynthesis. (d) Light also affects carotenoid synthesis as is suggested by the increase in colored carotenoids in fungi during illumination (see Section VII). (e) High light intensity causes destruction of chlorophyll. (f) For effects of light on grana, see Section III, A. (g) Effects of light on chloroplasts may result indirectly from the action of light on auxin. Not only does the intensity of light affect the size and number of the palisade cells as compared to the mesophyll cells, but it also affects the size and pigment content of the chloroplasts. The mechanism suggested for the destruction of auxin by light is that of a photo-oxidation of auxin resulting from the light absorbed by riboflavine and possibly also by β-carotene (Galston, 1950). (h) Light of 650 mμ causes expansion of etiolated bean leaves.

The effect on proplastid development has not been studied. Illumination with 735-mμ light can inhibit the promoting effect of the 650-mμ light (Liverman *et al.*, 1955). (i) Still another effect of light is to orient the plastids within the cells (Haupt, 1959). In a number of plants there appears to be a well-developed tendency for the chloroplasts to become arranged in the cells so that at a low intensity of light the broad surface of the plastid will face the light; and at high light intensity the edge of the plastid will face the light. Thus, at low light intensity, the chloroplast makes full use of the light for photosynthesis, and at high light intensity possibly avoids the destructive effects of excessive light. Blue light, possibly absorbed by carotenoids, but not red light, is effective in the phototaxis of the chloroplasts (Voerkel, 1933). The effect of light intensity on the orientation of the chloroplasts is clearly observed in certain algae. For example, the flat, band-shaped chloroplast of *Mesocarpus* which is suspended lengthwise in the center of the cell tends to be perpendicular to the light beam at intermediate intensities; at high light intensities the chloroplast moves so as to be parallel to the light beam. In *Vaucheria*, which is coenocytic, the chloroplasts accumulate only in the illuminated region of the filaments, but if the light intensity is too great, they move away from this region.

2. *Other External Factors*

The effects of soil nutrients including metals on chloroplast development are discussed in Section VII. Virus diseases may interfere or redirect cell metabolism so that plastid growth and development is often inhibited. Affected leaves may contain irregularly scattered pale green or whitish areas. It has been suggested that tobacco mosaic virus may develop within chloroplasts (see Section VII, A).

B. *Internal Factors*

The internal factors that affect the development of the chloroplast may be subdivided into four groups, namely: (1) the inheritable factors of the nucleus, the genome; (2) the inheritable factors of the cytoplasm exclusive of the plastids, the plasmon; (3) the inheritable factors of the plastids, the plastidome; (4) the factors that control tissue and plastid differentiation, the differentiation factors. This last category is discussed first.

1. *Differentiation Factors*

Little is known of the factors that influence the multiplication and degree of development of the proplastids in different tissues. The differentiating factors that affect the plastid may be products, or even the interaction of products, of the three groups of inheritable factors, the genome, plasmon, and plastidome. In general, in dividing cells plastids or proplastids are inhibited from maturing.

a. The effect of differentiating factors on chloroplast development in the leaf. This may be readily seen on examination of the different kinds of cells of the flat leaf of a dicotyledonous plant. The leaf is an organ the structure of which is specialized for photosynthesis to provide the maximum area in proportion to mass for the absorption of sunlight. With respect to its surfaces the leaf is a compromise between two opposing tendencies, namely the tendency to absorb CO_2 by exposing the photosynthesizing cells to the atmosphere and the tendency to protect the cells from being desiccated by the atmosphere.

The mesophytic leaf (Fig. 2) consists of one to several somewhat compact layers of cells, the palisade parenchyma near the upper surface of the leaf, and several layers of loosely arranged mesophyll cells, the spongy parenchyma, below them. These cell layers are enclosed above and below by a layer of epidermal cells that represent the outer covering of the leaf. All the parenchyma cells are arranged rather loosely so as to be in contact at some cellular surface with the air spaces of the leaf. The air spaces serve for the rapid exchange of gases, diffusion being very rapid in the gaseous state as compared with diffusion in the liquid state (Goddard, 1945). The spongy mesophyll cells, in contrast to the palisade cells, are often in fairly close association with vascular bundles and bundle ends (i.e., the leaf veins of phloem and xylem), through which substances are transported in the dissolved state to and from the leaves (Haberlandt, 1914).

Photosynthesis occurs primarily in the parenchyma cells, where the chloroplasts, containing lamellae and grana, are deep green and where a single chloroplast may absorb as much as 30–60% of the light incident on a leaf (Seybold, 1933).

In corn, cells in a single layer surrounding a vascular bundle, the bundle sheath cells, contain pale green chloroplasts which serve as temporary storage units for starch; these plastids contain lamellae but no grana (Hodge *et al.*, 1955).

The epidermal cells, except for the guard cells, in general contain small poorly developed plastids often devoid of green and yellow pigments. Light microscope studies of epidermal cells of the spinach leaf show pale chloroplasts ($2.5 \times 3.2\,\mu$) containing 10–16 grana, which Kaja (1955) interprets as being inhibited from further development. Tissue culture of epidermal cells might reveal the causes of inhibition of their plastid growth. Whether similar inhibitory factors may result in certain kinds of leaf variegations is not known.

The guard cells are specialized differentiated structures of the epidermis that serve to regulate the size of the stomatal opening through which direct gaseous exchange occurs between the intercellular mesophyll spaces and the air outside (Heath, 1959). The guard cells, in contrast to other epidermal cells, have functional green plastids, which generally contain starch, even at a time when the parenchyma cells lack starch. The guard cells are generally characterized by special thickenings of the walls that border the stomatal orifice (Haberlandt, 1914). These anatomical features cause the stomata to open when the guard cells become turgid and to close when they lose their turgidity. The osmotic value of the other epidermal cells is relatively constant and always lower than that of the guard cells. The osmotic value of the guard cells, however, is not constant. It is lowest when the stomata are closed and highest (some 2–10 atmospheres more than in the epidermal cells) when the stomata are opened. According to Scarth and Shaw (1951) a low content of CO_2 in the atmosphere results in wider opening. When the guard cells of a leaf are exposed to light in the morning, their pH increases, the osmotic pressure of their cell sap increases, their starch content decreases, and the stomata open (Miller, 1938). (In parenchyma cells, however, illumination brings about an increase in the starch content of the chloroplasts rather than a decrease.) The control of the opening and closing of the guard cells thus is a result of osmotic changes. The osmotic changes are probably primarily influenced by the sugar-starch equilibrium in the chloroplasts, but the precise step of enzyme control is not known.

b. On plastids in roots. Internal differentiating factors affect the plastids so they remain undeveloped even when exposed to light. However, roots of some species such as those of *Cucurbita, Menyanthes,* and *Cazea* may become pale green in light. In the root tip cells the proplastids are elongate, $0.5 \times 3\ \mu$, and contain tiny vesicles arranged in one or two clusters, or in one or a few chains which may represent the beginnings of a lamellar organization; tiny starch grains are also present (Heitz, 1957) (see Section V). Bartels (1955) believes that a proplastid of the root tip cells of *Vicia faba* always contains a primary granum. Examples of factors that affect plastid segregation in lower plants are discussed by Bünning (1948). Inhibition of root growth with an action spectrum of protochlorophyll has been reported by Hejnowicz (1958).

c. On the number of plastids per cell. In the cells also the number of plastids is controlled by internal factors. When the number is one to several, a marked constancy of plastids per cell is evident. For example, species of *Spirogyra* are known that contain only one or two or three spiral chloroplasts per cell. However, when the number of chloroplasts per cell becomes greater, the variation in number becomes greater. Thus, in *Spirogyra crassa,* the number may vary from 4 to 12 chloroplasts per cell.

The number of plastids per cell in certain plants is maintained constant by several devices, even through the process of sexual fusion (Sharp, 1934). For example, in *Zygnema* each gamete provides two plastids, but in the zygote the male plastids degenerate. In subsequent vegetative divisions the two plastids divide at the time the nucleus divides to give rise to two new cells, each of which contains a nucleus and two plastids (Fig. 23). In *Coleochaete,* during meiosis at one division of the zygote, the nuclei divide while the plastids do not divide; this results in the production of one chloroplast per nucleus. In *Anthoceros* the male gamete is thought to carry no plastid. Thus, the zygote has one plastid, and at each division the plastid and nucleus divide simultaneously, resulting in one plastid per cell. In the meristematic cells of *Isoetes,* there is one chloroplast per cell, and for a time at each subsequent division only one chloroplast per cell is produced. Yet, as the cells enlarge, and in mature cells, several chloroplasts may be present. What maintains the constancy of chloroplasts in these cells? One of the factors is obviously a mechanism correlated with the process of cell division. But in the vegetative cells other factors of differentiation also appear to be involved.

In the leaves of higher plants, the palisade cells contain more chloroplasts than those of the spongy mesophyll cells. "Mutant" chloroplasts of *Polypodium* may be of different sizes (Knudson, 1940; Maly, 1951). In general, the correlation is observed, that if a cell has large chloroplasts, the number of chloroplasts per cell are decreased. Since there is usually only one layer of plastids in the cytoplasm, the maximum number of plastids per cell may be related to the available cell surface. Nutritional factors have been found to affect the number and size of chloroplasts per cell. Tabentskii (1953) has observed that increase in nitrogen supplied to a sugar beet leaf culture resulted in as much as 2.5-fold increase in number of the plastids per cell, but the plastids were smaller.

d. The inheritance of plastids is maternal. In plants at sexual fusion, the plastids of the male gamete, and perhaps also the rest of the cytoplasm of the male gamete, may often be discarded more or less completely. The mature plant would then be a result of the interaction between male and female nucleus in the environment of female cytoplasm.

In animals the spermatozoan carries a cytoplasmic layer around the nuclear head, and the midpiece is derived from mitochondria. Whether these cytoplasmic elements are incorporated into the fertilized egg is not known. Perhaps tritium-labeling experiments may be useful to decide this problem.

Early evidence which indicated that male cytoplasm was discarded at zygote formation came from direct observations on lower plants that contain one or a few chloroplasts. In general, the male gamete carries no chloroplast; or the chloroplasts derived from the male gamete disintegrate before fertilization occurs; or the "male" chloroplasts disintegrate in the zygote after fertilization. A few examples may be cited. In *Anthoceros,* the spermatozoid is believed to bear no remnant of a chloroplast, so no "male" chloroplast enters the egg; the chloroplast of the egg is then the chloroplast of the zygote. In *Zygnema* and *Spirogyra* the "male" chloroplast degenerates in the zygote as if the female were a foreign and toxic environment for the male plastid. In *Pinus,* disintegration of the cytoplasmic granules (mitochondria and proplastids) of the male also occurs in the oösphere. Mangenot (1938) observed that these granules of the male were larger than those of the female. The larger granules of the pollen tube could thus be followed during fertilization as the male cytoplasm penetrated into the oösphere. After fertilization, the larger granules could still be recognized. During the development of the embryo, the granules of male origin remained in that portion of the oösphere which did not contribute to the formation of the embryo and the "male" granules disintegrated later. There did not appear to be a mixing of the granules of male and female origin. These observations suggest that not only the proplastids, but also the mitochondria and other visible granular elements of the male, are discarded on fertilization.

2. *Genome Factors*

These are inheritable factors which reside in the chromosomes and segregate according to Mendelian laws; when mutations occur the phenotypic expression of the plastid may be altered. Other chromosomal variations may also affect the expression of the plastid.

a. Simple Mendelian recessives. These are the most common types of chlorophyll mutants, indicating that the mutations are due to the action of single genes (Demerec, 1935). For example in maize, the plant in which the most extensive genetic studies of chloroplasts have been made, thirteen genes are known, each of which independently produces an albino seedling. Some twenty genes are known which produce virescent seedlings, i.e., seedlings that are white in the early stages, but develop pigment later. Several genes are known which result in pale green seedlings in which the amount of chlorophyll is reduced at certain stages of development. Some variegated seedlings have been found in which chlorophyll is lacking in spots (piebald) or in streaks (zebra). In the albinos, the plastid primordia are present and begin to develop in the seedling stage at the same rate as in the green seedlings; they are low in carotenoids; protochlorophyll is converted to chlorophyll or chlorophyllide in the light and then is rapidly bleached (Smith *et al.*, 1959). Luteus seedlings which contain carotenoids result when, in addition to the albino gene, several other mutant genes are present. The virescent types differ greatly from each other in the rate at which chlorophyll develops. At one extreme are the virescent plants which become green if grown under favorable conditions of temperature. Some virescent plants may become green at a rate that is too slow for survival. Among the pale green types there is also considerable variation; certain types do not live beyond the seedling stage; other types are pale green as seedlings and as mature plants. Whether a seedling will survive does not depend merely on its chlorophyll content, but on various other factors involved in chloroplast growth and enzyme content (see Sections V and VII). For example, Xantha-1 and Xantha-2 types are pale green viable plants that contain only 10% of the normal content of chlorophyll although they contain a normal concentration of carotenoids. A pale green type which contains 75% of the normal content of chlorophyll and a normal content of carotenoids dies at the same time as a pure albino seedling, i.e., at the three-leaf stage. Other work on the plastid pigments of mutant maize is reported by Schwartz (1949). Smith and Koski (1947) found that one of the albino strains of corn did not turn green to an appreciable extent because light caused an excessive rate of destruction of chlorophyll. Several of their virescent corn mutants were found to be partially defective in the rate with which they made protochlorophyll, but they possessed the mechanism for the continued production and conservation of chlorophyll.

Highkin (1950) has studied the pigments of barley mutants. One of his mutants, Chlorina 2, was found to possess a normal content of chlorophyll a but to be devoid of chlorophyll b. In yellow-green plastids of a barley mutant, electron microscopy showed that the plastids never fully develop; lamellae form, but no grana. In plastids of a white barley mutant not even lamellae are formed (Stubbe and von Wettstein, 1956).

The pale green plastids of a variegated leaf of *Chlorophytum comosum* contain one or two large green granules in a vacuolated stroma and represent an arrested development (Strugger and Losada-Villasante, 1955).

In fern prothallia, X-irradiation produced various kinds of morphological plastids which were maintained vegetatively (Knudson, 1940). Maly (1958a) has shown that these plastids differ because of nuclear genes.

b. Polyploidy. This may result in larger cells and larger chloroplasts.

c. Chromosomal chimeras. These chimeras may arise, resulting in sectors of plants that are pale green or white. If gene mutation or change in the number of chromosomes occurs in meristematic tissue of a very young bud, then the product of the bud, i.e., the branch, flower, or flower cluster, may have a new genotype constitution and will exhibit an appearance often strikingly different from that of the other branches or flowers of the plant (Jones, 1940). Such a change of genome may result in a chimera in which a distinct portion of the mature structure, commonly a sharply defined sector, differs constitutionally and in appearance from the other portions. For example, *Datura* plants have been studied which may bear $2n+1$, $2n-1$, and $4n$ branches (where n is the number of haploid chromosomes). The differing portions may at times also be arranged concentrically, i.e., periclinally. In spinach the plerome and dermatogen may be $2n$ and the periblem may be $4n$ or $8n$. Such sectorial and periclinal types are known as chromosomal chimeras.

d. Irregularities in the distribution of chromosomes or portions of chromosomes. Such irregularities may also result in inheritable changes that affect chloroplasts. In maize, the studies of McClintock (1939) have shown that the mosaic patterns seen in the endosperm may be the result of an unstable chromosome which breaks and re-forms with a high frequency. Other forms of unequal mitoses might arise by crossing-over or reciprocal translocation in somatic tissues (Jones, 1940). These mosaic variegations in somatic tissue must be differentiated from others, probably of virus origin. It may be noted that undoubted factor mutations occur in somatic cells of *Drosophila,* which result in mosaic individuals (Caspari, 1948).

e. A sex-linked factor that influences plastid development is postulated in the case of *Ginkgo biloba* where the female gametophyte has large chloroplasts, but the male gametophyte contains leucoplasts only (Tuleke, 1953).

3. *Plasmon Factors or Plasmagenes*

Another group of inheritable factors that may influence chloroplast development may reside in the cytoplasm outside of the chloroplast. Some twenty cases of cytoplasmic inheritance are discussed in a comprehensive review by Caspari (1948). Evidence from interspecies crosses

(Sirks, 1938) suggests that factors of the cytoplasm, apart from those of the plastids, are independently inheritable. The extensive work of Michaelis (1954) on *Epilobium* has demonstrated clearly that inheritable factors in the cytoplasm influence the size of the plant, size differences of leaves, flowers, etc. Even in combination with a homozygous foreign nucleus these cytoplasmic factors may remain relatively unchanged.

a. Mirabilis jalapa var. *albomaculata.* Certain studies on variegated plants suggest that plasmon factors are active, but proof is still largely indirect. A classic case suggesting plasmon inheritance is that first described by Correns (1909). In the variegated plant *Mirabilis jalapa* var. *albomaculata* white areas are irregularly distributed over the leaf. Intergrades are also present, varying from green leaves with a few white spots to white leaves with a few green spots. The boundary between green and white tissues is not always sharp and the intensity of whitening is very unequal. The white regions contain colorless plastids; the green regions contain normal chloroplasts. Branches may form that are pure green and bear normal flowers; or branches may form that are colorless and bear colorless flowers and fruit. When flowers from a colorless branch are fertilized with pollen from a branch bearing green plastids, offspring with colorless plastids are produced which cannot photosynthesize, and these plants die in the seedling stage. The reciprocal cross produces progeny with normal green plastids and the plant is viable. Thus:

♀ green × ♂ white → green plants
♀ white × ♂ green → white plants

These results are readily explained on the assumption that only the cytoplasm of the egg, including its plastids, is inherited, i.e., maternal inheritance occurs. The male gamete furnishes only the nucleus. The cytoplasm and plastids of the male gamete either do not enter the egg, or if they enter, they degenerate. It cannot be decided from this experiment whether the absence of chlorophyll in the plastids of the colorless regions is a result of some inheritable factor in the plasmon or plastidome. In the variegated plant the irregular splotching of the leaves appears to bear no relation to the developmental morphology of the leaves. The white areas may arise both early and late in leaf development. The assumption of a segregation of white and green plastids in the embryonic cells would not appear to explain satisfactorily the random distribution of the white leaf areas. Rhoades (1943, 1946) has suggested that there may be a ratio of normal to abnormal plasmagenes in the cytoplasm (i.e., outside the plastid) which may affect the outcome of the cell plastid type. A biochemical interpretation may be that some enzyme system is limiting in concentration. For example, depending on environmental factors, certain leaf areas may not be entirely equivalent to other leaf areas, and the substance produced by this enzyme system may be too low in concentration to result in greening of the plastids in the white areas. Or differentiation factors may be at work, of the type that cause the epidermal cells to contain colorless plastids. The irregular splotching in this plant is the kind that might be expected of a virus disease, but as yet there has appeared no evidence to implicate a virus.

b. In Humulus japonicus var. *albomaculata.* This variety was found by Winge (1919) to produce variegated progeny from variegated parents irrespective of the type of pollen that had been used in the crosses. "Selfed" seeds of these mosaics never produced pure white or pure green branches or seedlings. Winge made reciprocal crosses of mosaic plants with normal green plants of this species of hop through four generations. The inheritance of green versus mosaic plants proved to be exclusively maternal, i.e., no cytoplasm including plastids was contributed to the zygote from the pollen. The mosaic character in the hop has been thought by Winge to be due to some inhibitory factor carried in the cytoplasm, rather than to a factor in the plastids. In *Capsicum annuum* var. *albomaculata* the mosaic character was found by Ikeno (1917) to be inherited through both parents, suggesting that some cytoplasm of the pollen passed into the egg.

4. *Plastidome Factors or Plastid Genes*

Evidence has been presented that plastids can arise only from preexisting plastids or proplastids (see Section V). Here, evidence is presented that the plastid has an inheritable individuality which is highly stable even in a foreign cytoplasm. This individuality is transmitted through sexual as well as vegetative stages. Whether the individuality is connected with RNA in plastids (see Section VII) is not known.

a. In Spirogyra triformis. The inheritable individuality of the plastid may be inferred from an important chance observation of van Wisselingh (1920). In *Spirogyra triformis* he found an abnormal chloroplast together with normal chloroplasts in the same vegetative cell. This abnormal chloroplast lacked pyrenoids, and the starch grains instead of being formed as sheaths around the pyrenoids were scattered in the chloroplast. When the cell divided, this chloroplast as well as the normal ones divided and came to be present in the new cell. If a change had occurred in the nucleus or cytoplasm that had been responsible for the abnormality in this chloroplast, it should have affected all the chloroplasts of the cell to the same degree. Since in the same cell an abnormal chloroplast was present together with the normal chloroplasts, and vegetative divisions maintained this condition, it seems necessary to conclude that chloroplasts possess mechanisms for their self-duplication and factors for the maintenance of their phenotypic individualities.

b. In Pelargonium zonale var. *albomarginata.* Another important line of evidence for the inheritable individuality of plastids was supplied by the classic work of Baur (1909, 1930) on *Pelargonium zonale* var. *albomarginata.* This plant has leaves with white margins. It often produces pure white and pure green branches which can bear flowers. If a flower from a green branch is pollinated with pollen from a white branch, the resulting progeny may occasionally be mosaic seedlings, i.e., ♀ green × ♂ white → mosaic. Mosaic seedlings in the course of growth produce green, colorless, and variegated leaves and branches. Likewise the reciprocal cross of a flower from a colorless branch when pollinated with a flower of a green branch occasionally results in the same kind of mosaic seedlings, i.e., ♀ white × ♂ green → mosaic. These crosses are of special significance since they help to distinguish between plasmon and plastidome inheritance. To explain the appearance of mosaic seedlings, it is assumed that cytoplasm and plastids from the male may sometimes enter the egg; in this way, for example "green" plastids would be brought into an egg which had "colorless" plastids. If the female plastids were colorless because of factors in the female cytoplasm, then the green male plastids entering into the female cytoplasm should also become colorless, i.e., colorless plants should result. Since mosaic plants are produced, it is concluded that in this case the cytoplasms do not affect the greening or nongreening of the plastids. The factor or factors governing the greening must here reside in the plastids themselves. The plastids must be of two kinds. Either kind of plastid arises only from the division of previously existing similar plastids. The colorless or "mutated" plastid is colorless because it contains within itself some inheritable lesion.

The conclusion that a mutant plastid is present in this variety of *Pelargonium* is based on several assumptions. First, it is assumed that plastids may be transferred from the pollen to the egg so that the fertilized egg will contain both kinds of plastids. Although the general rule is that only the nuclear material of the pollen is transmitted to the egg, there is some evidence that occasionally in some plants the cytoplasm and plastids do enter the egg. For example, Ishikawa (1918) in a study of the fertilization process of *Oenothera* found that starch grains in the pollen migrated with the cytoplasm through the pollen tube and entered the embryo sac. Presumably, the starch grains were contained in plastids.

A second assumption that is made in the interpretation of the above experiment is that after fertilization the green and colorless proplastids segregate during subsequent cell divisions of the early embryo. Thus certain cells will contain only colorless plastids, and it is these cells that will give rise to the colorless areas. If such a segregation is to occur, it seems necessary to make several additional assumptions. The idea of segregation might appear more plausible if it could be established that very few proplastids were present in the fertilized egg, and if the rate of division of the plastids in the meristematic cells were to be limited to the maintenance of a small number of plastids per cell. Another explanation of segregation may be that the male cytoplasm and plastids do not mix readily with the female cytoplasm and plastids, so that the two different kinds of proplastids become segregated in the first few divisions of the fertilized egg. Relatively few instances of two types of plastids in the same cell have been reported. An assumption to explain the almost complete absence of both types of plastids in the same cell is that the colorless plastids may divide at a slower rate than the green ones, and eventually the colorless plastids would be lost. However, a number of workers have encountered green and pale plastids together in the same cell, in plants such as *Primula sinensis, Capsella bursa pastoris, Stellaria media albomaculata,* etc. Küster (1935) observed three different kinds of plastids in mesophyll cells of the above species. Maly and Wild (1956) observed cells that contained two kinds of chloroplasts that differed in size of grana in a mutant of *Antirrhinum majus* (albomaculata). It is necessary to consider in each of these cases whether these plastids in the same cell represent inheritably different kinds, or degeneration stages of a single kind.

A further assumption which is necessary to explain the results observed in Baur's *Pelargonium* is that the green and colorless cells in the embryo will give rise to cell lineages and areas that are a consequence of the mechanics of the embryonic development of the plant. Embryological studies appear to support this assumption (Jones, 1934). See, however, the critical discussion on plastid segregation by Weier and Stocking (1952).

c. In Primula sinensis albomaculata. An experiment, similar to that of Baur, was carried out by Gregory on *Primula sinensis albomaculata.* He grew pure yellow plants (i.e., containing chlorotic plastids) to flower and crossed these flowers with pollen from a normal green plant. The albomaculata character was transmitted only by the egg, and not by the pollen. In the cells of quite young leaves he found simultaneously normal and chlorotic chloroplasts. The defect therefore was considered to lie in the chlorotic plastid rather than in the cytoplasm. For other literature on this subject compilations of Jones (1934), Küster (1935), and Weier and Stocking (1952a) are recommended.

d. In Oenothera interspecies crosses. Additional support for the concept of inheritable factors in the plastids is derived from the studies on interspecies crosses of *Oenothera* which have been carried out by Renner (1936) and by Schwemmle and his students (1938). The species of *Oenothera* have arisen during evolution not only by changes in genome but also in plastidome. Renner showed that the plastids derived from one *Oenothera* species maintained their identity, even when combined with the genome of another species for fourteen generations and then recombined with the genome of its own species. The following example will show that the green plastids of one *Oenothera* species differ in inheritable properties from the green plastids of another *Oenothera* species. For example, when reciprocal crosses between certain species i.e., A and B, were made, the cross A ♀ × B ♂ was found to produce green plants with healthy plastids, whereas B ♀ × A ♂ was found to produce plants with inhibited plastids. The female parent would provide its A chromosomes, A cytoplasm, and A plastids. The male parent would provide the B chromosomes and presumably the B cytoplasm, but no plastids. Thus:

$$\text{Normal plant} = \text{A}♀ + \text{B}♂ = \left\{ \begin{array}{l} \text{A chromosomes} + \text{B chromosomes} + \text{A cytoplasm} \\ + \text{B cytoplasm} + \text{A plastids} \end{array} \right\}$$

In the reverse cross the male parent A would provide A chromosomes and A cytoplasm but not A plastids. Thus:

$$\text{Pale plant} = \text{B}♀ + \text{A}♂ = \left\{ \begin{array}{l} \text{A chromosomes} + \text{B chromosomes} + \text{A cytoplasm} \\ + \text{B cytoplasm} + \text{B plastids} \end{array} \right\}$$

Here the nuclear and the cytoplasmic components (other than plastids) are identical in the two crosses. The difference in response must lie in the differences between the inheritable properties of A plastids as compared to the B plastids.

To strengthen the idea that the cytoplasm of the male parent is incorporated, cases have been studied in which the plastids of the male parent as well as the female parent are brought into the embryo sac. It is then assumed that if the plastid gets in, the cytoplasm also will get in. The zygote then would have the following composition:

$$\begin{array}{c} \text{A chromosomes} + \text{B chromosomes} + \text{A cytoplasm} + \text{B cytoplasm} \\ + \text{A plastids} + \text{B plastids} \end{array}$$

If segregation of the plastids occurred, a variegated plant would arise. The green areas would have the composition given above for the normal plant and the pale areas the composition of the pale plant. In other words, the difference in response of the cells would reside in differences in the plastidome.

A specific example may be cited. *Oenothera odorata* has two Renner complexes or two groups of chromosomes, *v* and *I*. During reduction division the diploid number of chromosomes do not segregate at random, but rather one group, *v*, goes to one pole to form one haploid nucleus, and the other group of chromosomes, *I*, goes to the other pole to form the other haploid nucleus. Thus the female or male nucleus of *Oenothera odorata* will have either a Renner complex *v* or *I*. *Oenothera berteriana* also has two Renner complexes, namely *B* and *l*. When reciprocal crosses are made between *O. berteriana* and *O. odorata* the following data are obtained, as summarized by Weier and Stocking (1952a) in Table I. Here the egg supplied the cytoplasm of the zygote.

It may be seen from Table I that although the chromosomes, which constitute the nuclear complexes, may be the same in the diploid plants, yet the result is profoundly influenced by the composition of the cytoplasm + plastids. For example, in the diploid, with the Renner complexes (i.e., chromosomes groups) *B* + *v*, if *B* + *v* interacts with *O. berteriana* cytoplasm + plastids, then the resulting plant dies. However, if *B* + *v* interacts with *O. odorata* cytoplasm + plastids, then the resulting plant is normal.

Next it is necessary to determine whether it is the plastids or the cytoplasm that is responsible for the above results. For such an analysis it is necessary that the cytoplasm and plastids of the pollen enter the egg. This happens occasionally, and can be detected when it does happen, under the following conditions. Suppose the ♀ parent of composition:

l + *v* chromosomes + (*O. odorata* cytoplasm + plastids)

is crossed with pollen of *O. berteriana*, i.e., of

B + *l* chromosomes + (*O. berteriana* cytoplasm + plastids)

Then the following combinations will occur: *B* + *l*, which is lethal; *l* + *l*, which is lethal; *B* + *v*, which gives a normal plant; and *l* + *v*, which gives a weak, pale green plant. However, if *O. berteriana* cytoplasm + plastids of the pollen get into the egg, then a variegated white-green plant is obtained which can be readily distinguished from all the other viable plants. This variegated plant was interpreted to have the following composition in the white areas:

B + *l* chromosomes + *O. odorata* cytoplasm + *O. berteriana* cytoplasm
+ *O. odorata* plastids

TABLE I

RECIPROCAL CROSSES BETWEEN *Oenothera odorata* AND *O. berteriana*

Cross ♀ *O. berteriana* × ♂ *O. odorata*		Cross ♀ *O. odorata* × ♂ *O. berteriana*	
Appearance of hybrid with *O. berteriana* cytoplasm including chloroplasts; the specific chromosome groups or Renner complexes are indicated.		Appearance of hybrid with *O. odorata* cytoplasm including chloroplasts; the Renner complexes are indicated.	
Renner complex	Appearance	Renner complex	Appearance
$B + v$	Not formed	$B + v$	Normal
$B + I$	Normal	$B + I$	Weak and pale green
$l + v$	Lower leaves sometimes yellow, otherwise normal	$l + v$	Weak and pale green
$l + I$	Lower leaves sometimes yellow, otherwise normal	$l + I$	Nonviable; dies in embryo stage

In the green areas the composition was interpreted to be:

$B + l$ chromosomes + *O. odorata* cytoplasm + *O. berteriana* cytoplasm + *O. berteriana* plastids

The above interpretation is based on assumptions of the kind that have been discussed for the case of *Pelargonium*. For example, if the cytoplasm of the *O. berteriana* pollen tube enters the egg it is assumed that it will become thoroughly mixed with the *O. odorata* cytoplasm and no random distribution would be able to segregate the cytoplasms; whereas the plastids of *O. berteriana* could be segregated from the plastids of *O. odorata*. The presence of *O. berteriana* plastids has been tested for by crossing a flower from a green branch of the variegated plant with a standard odorata pollen. When this is done, the F_1 generation resembles that of the standard *O. berteriana* × *O. odorata*, as seen in the left half of Table I.

Thus it may be concluded that *O. odorata* plastids differ from *O. berteriana* plastids in their response to the same chromosomes and same cytoplasms and that the difference is due to an inheritable mechanism within the plastids.

A curious and important observation that remains to be explained satisfactorily is the following: In the cross of *O. odorata* × *O. berteriana*, weak pale green $B + I$ and $l + v$ plants are obtained which contain the *O. odorata* plastids. During the succeeding generations these plants become normal green and healthy. The authors believe that this change from inhibited to normal plastids cannot be due to a progressive change in the cytoplasm or plastids, but that the genome has been gradually changed so that its products no longer act to inhibit the plastids.

Another result that is suggested from these studies is that the proplastids may have a vital function independent of developing into functional chloroplasts. Binder (1938) has studied the $l + I$ complex containing *O. odorata* plastids (Table I). Fertilization occurs, but the young embryo dies. Death is therefore not a result of the inability of plastids to develop to the stage at which they carry on photosynthesis. Perhaps death may be due to the abnormal behavior of the *O. odorata* proplastids, i.e., to the inability of the *O. odorata* proplastids to supply some metabolite essential for the life of the young embryonic cell.

A new approach to studies of plastid inheritance has been developed by Michaelis (1957). When *Epilobium* plants are treated with P^{32} and S^{35}, various chloroplast changes are observed that are interpreted to be due to somatic dominant gene mutations, to labile gene mutations, and to plastid mutations of the albomaculata type that are inherited maternally. In *Antirrhinum majus* Maly (1958b) treated vegetative meristems with chemicals but could induce no plastid mutations.

C. *Agents That Cause Plastid "Mutations"*

The evidence has been presented above to show that plastids are "self-duplicating" bodies. Likewise, it has been shown that in the same internal environment of cytoplasm and nucleus, some plastids may respond differently than others. On the analogy with nuclear genes, it is logical to assume that plastid factors or plastid genes may be present in the plastids. Since specific substances like the carotenoids, chlorophylls, lipoproteins, and certain enzymes are localized in the chloroplasts, it may be assumed that the synthesis of these compounds, at least in the end steps of their biosynthetic chains, as well as the organization of the lamellae, are accomplished by enzymes in the plastids. To what extent these enzymes arise from the action of plastid genes is not known. Presumably the plastid genes are made up in part of nucleic acid.

It is difficult to observe plastid mutations because there may be a number of identical, i.e., replicate, plastid genes per cell. For example, in a chloroplast there may be only one plastid gene of each kind per chloroplast or there may be several replicates. The number of replicate plastid genes per cell is increased if, instead of the cell containing only one chloroplast, there are a number of chloroplasts per cell. The chances of finding a plastid with a mutated plastid gene in a cell containing normal plastids is very small since, if such a mutated plastid happened to arise in a meristematic cell, the mutated plastid might tend to grow more slowly and would be lost in subsequent somatic cell generations.

From such considerations it would appear that in order to bring about an inheritable observable plastid change it is necessary to change at the same time all replicates of a particular plastid gene in all the plastids of a cell. Several techniques for accomplishing this action may be considered, such as the action of a specific nuclear gene, or the action of streptomycin and possibly other chemicals, or the action of physical agents like heat, visible light, ultraviolet (UV) light, and X-rays.

1. *Action of Nuclear Gene on Plastid Genes*

If a certain recessive nuclear gene is present in homozygous condition, a change might be brought about in the cell which, for example, would prevent the plastids from forming chlorophyll. This change might be permanent. Even though the recessive gene would later be replaced by its dominant allele, the plastids would remain colorless. Three cases of this kind have been studied in detail. One occurs in maize, another in catnip, and a third in barley.

a. In maize. A chlorophyll striping or variegation of the leaves and culm is produced in maize when the recessive gene *ij* (iojap) is present in the cells in a double dose (Rhoades, 1943, 1946). The leaf pattern varies from fine white streaks to wide white bands which may occupy half or more of the leaf. Very few seedlings show the striping, since the stripes form mostly on leaves that develop later. The iojap gene is situated in a known segment of maize chromosome 7. The presence of one recessive and one dominant iojap gene ($ij + Ij$) will not cause variegation. When plants bearing the double recessive dose ($ij + ij$) of this gene are used as a female parent and crossed with pollen from an ($Ij + Ij$) plant, there result white seedlings, variegated seedlings, and green seedlings. Assuming that no cytoplasm and plastids passed from the pollen to the egg, the white seedlings may be interpreted to contain in their cells the $ij + Ij$ genes and the maternal cytoplasm with its small colorless plastids. The variegated and green seedlings are assumed to have arisen when the cytoplasm and plastids from the pollen have been able to enter the egg. In the variegated plants, at the boundary between white and green areas, cells were found that contained both the green and colorless plastids in the same cell. In further crosses it was established that if the cells had the $Ij + Ij$ gene constitution in the nucleus, and contained the cytoplasm and plastids derived from the $ij + ij$ plant, then the plant would be colorless. From this analysis it may be concluded that when no dominant *Ij* gene is present in the nucleus, a damaging change will occur in the cytoplasm (or plastids), which will be inherited; this change becomes evident as an inability of the plastids to form chlorophyll, and the defect is thenceforth inherited independently of the nuclear genes.

These results suggest that, directly or indirectly, the action of the *Ij* gene may be involved in the multiplication of some hereditary units in the cytoplasm or plastids. The inhibition of multiplication of these units might be due to various incompatibility factors, on analogy with those factors that prevent a specific virus from multiplying in different tissues of the same organism or in closely related species. One might even suppose that the *Ij* gene governed some step in nucleic acid biosynthesis in the cytoplasm or plastids. Then the lack of the dominant gene might result in a diminution in the rate of nucleic acid synthesis. As a consequence certain hereditary particles in the cytoplasm or plastids might not multiply fast enough and would be diluted out in subsequent cell divisions. Thus cells would be formed which now lacked these hereditary particles and could not regain them even if the recessive genes were replaced by dominant ones.

b. In *Nepeta cataria* (Woods and du Buy, 1951) a recessive and highly mutable gene, when present in homozygous condition, also appears to bring about inheritable changes in plastids, which are irreversible. The appearance of the plastid is thenceforth independent of the presence or absence of this recessive gene. In this case, some fifteen different plastid abnormalities have been observed, such as light green plastids, cream-colored plastids, white vacuolate plastids, and also absence of recognizable plastids. These types appear to result from changes in a number of factors that affect plastid development, i.e., degree of retention of plastid pigmentation, size of grana, plastid vacuolation, etc. There are also types in which the tissue may contain several kinds of plastids, including normal ones, all in the same cell. The fact that several stable types of plastids may occur in the same cell along with normal green plastids is evidence that the abnormalities in the plastids are caused by some inheritable defects in the plastids per se and not in the cytoplasm. An irreversible gene-induced plastid mutation has also been described for barley by Arnason and Walker (1949).

2. *Action of Chemical and Physical Agents on Plastid Genes*

A chemical agent which enters a cell may be expected to act equally on replicates of a plastidome, i.e., on the same factors which are present in all the plastids, and perhaps eliminate them simultaneously. Likewise, a physical agent like heat or UV light may also be expected to affect all the plastids similarly. Unique examples of such actions are found in *Euglena.*

Euglena gracilis var. *bacillaris* contains about eight chloroplasts per cell. It loses its green color in the dark but becomes green when placed in the light. Prolonged action of streptomycin causes an irreversible loss of green color (Provasoli *et al.,* 1948). The concentration which results in "bleaching" (40 μg./ml.) does not affect the rate of multiplication of the cell in organic media. The concentration is 50,000 times less than the concentration that will kill the cell. "Bleached" cells no longer can form green plastids. De Deken-Grenson and Godts (1960) consider that streptomycin affects some inheritable factor outside the plastids (see Section V, A).

Aureomycin (Robbins *et al.,* 1953) was also found to be effective in producing chlorotic colonies in *Euglena,* but the differential between bleaching and killing was much lower than in the case of streptomycin.

Relatively high temperatures may convert some strains of *Euglena* to permanently colorless ones (Pringsheim and Pringsheim, 1952). They may be bleached in 4–6 days at 34–35°C. Robbins *et al.* (1953) found that green *Euglena* cells, when placed at 36° in the dark for 22 days were all converted to colorless cells. Cell divisions were necessary if bleaching was to occur, which suggests that a diluting out of the plastids occurred.

Several observations on the eyespot of *Euglena* should also be mentioned (Pringsheim and Pringsheim, 1952). In strains of *Euglena* that are bleached by heat or streptomycin one of several results may be obtained: (1) The eyespot may disappear simultaneously with bleaching. (2) The eyespot may persist in the light even though the chloroplasts are bleached; however, once the bleached strain is placed in the dark, even six months after heat-bleaching, then the eyespot is lost permanently. (3) The eyespot may persist even if the bleached strain is placed in the dark. Perhaps these phenomena may be explained on the basis of differences in the rate of diluting out of an inheritable factor in the eyespot region.

Other studies on higher plants with streptomycin indicate interference with chlorophyll production, probably indirectly. Seeds germinated on filter paper moistened with streptomycin solution stronger than 2 mg./ml. developed colorless first leaves (von Euler *et al.*, 1948). The seedlings investigated were barley, lettuce, rye, spinach, and radish. With less-concentrated streptomycin solutions, only the tips of the first leaves became green. Streptomycin appeared to retard or arrest chlorophyll formation in developing leaves, but chlorophyll was not affected if it was already present in the leaves before treatment. In barley seedlings, streptomycin inhibits differentiation of the proplastid to the chloroplast without affecting heredity (De Deken-Grenson, 1955). De Ropp (1948) observed that streptomycin changed crown gall tumors from green to white and inhibited the growth of normal meristematic and tumerous tissues of sunflower. The action of streptomycin was also studied on seedlings of *Pinus jeffreyi* by Bogorad (1950). At 0.2% of streptomycin and in the dark, the cotyledons of almost all the seedlings developed without chlorophyll, although these cotyledons would normally form chlorophyll in the dark. When the cotyledons were then removed from the megagametophyte tissue which surrounds them, and placed in a medium devoid of streptomycin, they became green only in the presence of light. This behavior would suggest that one of the damaging effects is on the enzyme system (or its precursors) which serves in the dark to reduce protochlorophyll to chlorophyll in the pine cotyledons.

D'Amato (1950) observed that treatment of barley seeds with acriflavine or acridine orange or 9-aminoacridine produced mutations that led to a decrease of chlorophyll in the plastids. Whether these chemical effects are a result of changes in the plastidome is not known. Ultraviolet light has been found to inactivate the photochemical production of O_2 by chloroplasts (Holt *et al.*, 1951). Chlorophyll is not destroyed by this treatment, but probably some enzyme system related to photosynthesis is adversely affected. No studies have been carried out to find whether this effect is inherited.

D. Summary

Various factors, both external and internal, affect the multiplication of the proplastid and its differentiation into a chloroplast. Light is one of the external factors which serves not only as the energy source for photosynthesis, but also for stimulation of chlorophyll and carotenoid formation, for enlargement of plastids (in dicots) including protein synthesis, for destruction of auxin, and for orientation of plastids. Internal factors include those whose action results in the maternal inheritance of plastids and in the retardation of plastid development in roots and in epidermal cells of leaves. In lower plants factors are present that maintain a constant number of plastids per cell.

The inheritable factors that influence chloroplast development are those of the genome, the effects of which are well documented. Evidence for inheritable factors of the plasmon (i.e., cytoplasm exclusive of the plastids) is based on studies of interspecies crosses and of certain variegated plants. Evidence for inheritable factors within the plastids (plastidome) is based on the finding that in the same internal environment of cytoplasm and nucleus some plastids may respond differently than others; the evidence includes the observations on *Spirogyra* of differences of plastids within the same cell, studies on variegated *Pelargonium* var. *albomarginata,* and studies on interspecies crosses of *Oenothera.*

The presence of certain recessive genes may lead to permanent plastid changes, which are not reversed even when the recessive genes have been replaced by dominant ones. Examples of this kind are reported for corn and catnip.

The effects on plastids of streptomycin, other chemicals, heat, and UV light are discussed.

VII. Chemistry of the Developing and Mature Chloroplast of Higher Plants[2]

In higher plants growing in the light, after young cells derived from the leaf meristem have ceased dividing, the cells enlarge and the tiny colorless plastids develop through an orderly sequence of phases into

[2] Abbreviations used: ATP = adenosine triphosphate; δAL = δ-aminolevulinic acid; PBG = porphobilinogen; PN+ = pyridine nucleotides, DPN+ + TPN+;

chloroplasts (see Section V) with a concomitant increase in photosynthesis to a maximal level. [In unicellular algae there is a more or less marked decrease in the rate of photosynthesis at the time of cellular division. For the study of the phases of plastid development it is desirable to synchronize cell division, as has been done in *Chlorella* (two to nine cycles per day). In *Chlorella,* which contains a single cup-shaped chloroplast, the deep-green mother cell divides into 8–16 tiny pale-green daughter cells, each of which contains a single plastid. The plastid enlarges and becomes deeper green as the cell enlarges. Per volume of packed cells it is found that chlorophyll remains relatively constant during growth; but the rate of photosynthesis increases, becomes doubled during rapid enlargement of the cells, and then drops to a lower value during cell division (Tamiya *et al.,* 1953; Sorokin, 1957).]

The proplastids are cytoplasmic units which multiply by fission in meristem cells where nucleic acid synthesis predominates. In the enlarging cells behind the leaf meristem, protein synthesis predominates. In monocots the plastids may enlarge to some extent in complete darkness. A red-infrared system is present in dicots for leaf and plastid enlargement. Differentiation into pale-green plastids occurs in this region if light is present. Then the green and yellow pigments increase, and the lipoprotein lamellae begin to form. With the beginning of photosynthesis, energy in the form of PNH and ATP is made available for further synthesis, so the young plastid grows rapidly.

It is not known to what extent the proplastid and the young plastid can synthesize proteins, fats, RNA, etc., from small precursor molecules like amino acids, purines, sugars, etc., supplied via the cytoplasm, or to what extent large molecules from the cytoplasm may be incorporated. This problem may eventually be studied by culture methods. For starch, for pigment molecules, and for the structural lipoprotein lamellae, at least the last steps in their synthesis must occur within the plastid; this implies that the plastid contains active enzymes not found elsewhere in the cell.

During growth of the young plastid the lipoproteins and pigments increase more or less proportionately. There is evidence, however, that certain of the simultaneous reactions may be rate limiting at different

---

PNH = reduced pyridine nucleotides; FlN = flavine nucleotides; RNA = ribonucleic acid; DNA = deoxyribonucleic acid; TCA = trichloroacetic acid; —ase = enzyme; —P = phosphate; ~P = high-energy phosphate; G-1P = glucose-1-phosphate; G-3P = glyceraldehyde-3-phosphate; F-6P = fructose-6-phosphate; F-1,6diP = fructose-1,6-diphosphate; Ru-5P = ribulose-5-phosphate; Ru-1,5diP = ribulose-1,5-diphosphate; 3-PGA = phosphoglyceric acid; 1,3di-PGA = 1,3-diphosphoglyceric acid.

times. For example, etiolated barley on exposure to light does not photosynthesize immediately although protochlorophyll has been converted to chlorophyll; a period of minutes to hours either in light or dark must intervene during which chlorophyll is organized presumably into newly forming lipoprotein lamellae; then on re-exposure to light O_2 is evolved (Smith and Young, 1956). Similarly in Thatcher wheat, the etiolated plants on exposure to light do not fix $C^{14}O_2$ during the first 4–5 hours. Then phosphoglyceric acid, alanine, and sucrose-phosphate become labeled. After 12 hours of light the rate of CO_2 fixation is only 15–20% of that of a fully green plant. After 20 hours, one of the last pathways to be activated is that of glycolic acid to serine. Only after 32 hours of illumination is the plant fully green (Tolbert, 1957). The events of plastid development may be different for dicots.

Smillie and Fuller (1960) have studied the enzymes of pea leaves at different stages of growth. They classify the enzymes into (1) those that reach a peak activity in the early development, primarily the citric acid cycle enzymes and anaerobic glycolysis enzymes; (2) those that develop later and are probably localized in the chloroplast, e.g., Ru-1,5diP carboxylase, TPN-linked G-3P dehydrogenase, and photosynthetic-TPN reductase; and (3) those that participate in both photosynthesis and respiration, e.g., PGA-kinase, DPN-linked G-3P dehydrogenase, triose-P isomerase, aldolase, and phosphoglucose isomerase.

The "assimilation number," i.e., the milligrams of CO_2 fixed in light per milligram chlorophyll, is a reflection of the relative concentration and efficiency of the enzymes connected with the "light" and "dark" reactions of photosynthesis. For example, a yellow variety of *Ulmus* is ten times as efficient in photosynthesis as a normal variety when the comparison is based on an equal weight of chlorophyll (Willstätter and Stoll, 1918).

In Table II is summarized the approximate composition of a mature chloroplast, which in terms of fresh weight contains about 50% water. Table III summarizes known and postulated enzyme activities in the plastid at different phases of development; it is based on the assumption that at least the last steps in the synthesis of protein, starch, etc., occur in the plastid. Russian contributions to plastid enzymology are summarized in a review by Sissakian (1958).

A. *Nucleic Acids of the Plastid*

Evidence for the formation of hereditary units in the plastids was presented in Section VI. Only recently has experimental evidence appeared to support the hypothesis of nucleic acids in plastids. The data suggest that RNA is present in proplastids and chloroplasts; DNA may be present in proplastids but is not detectable in chloroplasts. Whether all the RNA found in plastids is produced there is not known. In this connection the studies on *Acetabularia,* a large algal cell, are of interest. When the nucleus is removed from this cell, RNA no longer is formed although growth and protein synthesis continue for 21–28 days and then cease. Chloroplast multiplication and protein synthesis in the plastids

TABLE II
APPROXIMATE ANALYSIS OF CHLOROPLASTS OF HIGHER PLANTS[a]

Constituent	% of dry weight	Components			
Proteins	35–55	About 80% is insoluble			
Lipids	20–30	[b] Fats	50%		
		Sterols	20	[c] Choline	46%
		Waxes	16	Inositol	22
		Phosphatides	2–7	Glycerol	22[e]
				Ethanolamine	8
				Serine	0.7
Carbohydrate	Variable	Starch, sugar phosphates (3–7 C.)			
Chlorophyll	~ 9	Chlorophyll a	75%		
		Chlorophyll b	25%		
Carotenoids	~ 4.5	Xanthophyll	75%		
		Carotene	25%		
Nucleic acids					
RNA	2–3				
DNA	0.5(?)				
Cytochromes					
f	~ 0.1				
b_6	—				
Vitamins					
K	0.004				
E	0.08				
Ash[d]	~ 3				
Fe	0.1				
Cu	0.01				
Mn	0.016				
Zn	0.007				
P	0.3				

[a] See respective sections for literature citations; also Rabinowitch (1945).
[b] On runner bean leaves (Eberhardt and Kates, 1957).
[c] On tobacco chloroplasts (Benson *et al.,* 1958).
[d] On broken chloroplasts (Warburg, 1949).
[e] Besides phosphatidyl glycerol there are appreciable amounts of galactosyl glycerides including a sulfate glycerol and diglycerol phosphate (Benson *et al.,* 1959).

also continue up to this time (Clauss, 1958). Although there is no net synthesis of RNA in these enucleated cells, Brachet (1960) considers that there is a true synthesis of chloroplastic RNA which occurs at the expense of the RNA of the microsomes and cytoplasm.

TABLE III
ENZYME SYSTEMS OF PLASTIDS

A. Postulated enzyme systems in duplicating proplastid used for the synthesis of
   1. RNA; DNA
   2. Protein
   3. Lipid
   4. Starch

B. Postulated enzyme systems in differentiating plastid used for the synthesis of
   1. Protein: (a) structural, (b) enzymatic
   2. Lipid
   3. Chlorophylls
   4. Carotenoids

C. Known enzymes in mature plastids that function in photosynthesis
   1. Enzymes coupled with the photodecomposition of water, reduction of TPN and DPN, oxidation of cytochrome f, formation of O_2
   2. Enzymes that form ATP by photophosphorylation
   3. Enzymes that fix CO_2 in Ru-1,5diP by the Calvin cycle using TPNH or DPNH and ATP
   4. Enzymes of the pentose-P shunt that form Ru-5P indirectly from G-3P
   5. Enzymes that form starch indirectly from G-3P.

Both analytical and histochemical techniques have been used to detect nucleic acids in the plastids of higher plants.

1. *Analytical Techniques and Methods of Isolating Chloroplasts*

Mature chloroplasts are isolated by differential centrifugation and then analyzed. The success of this method requires that there be no contamination with nuclear or other cytoplasmic materials. Unbroken chloroplasts may be isolated by gently grinding leaves against sand covered with an isotonic sucrose solution (Granick, 1938a), a method used by Arnon (1956) to obtain chloroplasts that fix CO_2 in the light. Weier and Stocking (1952b) found that under these conditions the nuclei also were unbroken and a chloroplast preparation uncontaminated with nuclear material could be obtained. Methyl green when added to the suspension of material rapidly stains the nuclei and nuclear fragments. Permanent slides may be made by fixing the material in osmic vapor and staining with 1% toluidine blue for ½ hour; the nuclei, nuclear derivatives, and disorganized chloroplasts are colored an intense blue while the intact chloroplasts appear in a delicate shade of green. McClendon (1952) found by direct analysis that chloroplasts of tobacco prepared with the use of the blendor contained DNA; but this DNA, he concluded, was derived from adsorbed nuclear DNA as seen with an aceto-orcein stain or as seen in a phase microscope. Analyses of chloroplasts isolated by nonaqueous methods (Behrens and Thalacker, 1957; Stocking, 1959) should diminish the possibility that nucleic acid may be leached out of the chloroplasts during isolation.

What are the available analytical data and the interpretations placed on them? McClendon (1952) used the method of Ogur and Rosen (1950) for the determination of RNA and DNA after a preliminary extraction of lipids and nucleotides from the chloroplasts. RNA was extracted for 18 hours at 4°C. with 1 *N* perchloric acid. (Some RNA, however, appeared to be as resistant to perchloric acid as did DNA in this treatment.) Then DNA was extracted with 1 *N* perchloric acid for 20 minutes at 70°. The chloroplasts contained on the average about 2% RNA or 0.2% RNA-P. [According to Warburg (1949) the total P of chloroplast fragments is 0.3%.] Because of contamination with nuclear fragments, it could not be decided whether DNA was present in chloroplasts. Holden (1952) found that tobacco chloroplasts contained 3–4% of RNA, but no DNA. Cooper and Loring (1957) have also found a similar amount of RNA in chloroplasts of Turkish tobacco. The chloroplasts, washed with TCA and extracted with alcohol-ether, contained per unit dry weight 3% RNA, 0.7% DNA, 9–11% N, and 0.36% P. However, Jagendorf (1955), who purified tobacco chloroplasts by differential centrifugation, found that the chloroplast fraction with the higher chlorophyll:nitrogen ratio contained negligible amounts of nucleic acid and of catalase; young chloroplasts were more dense than older ones.

Still another line of evidence suggesting the presence of nucleic acids in chloroplasts is the hypothesis of Cooper and Loring, that tobacco mosaic virus (TMV) found associated with the chloroplasts may multiply within the chloroplast. On this assumption RNA precursors might be formed within chloroplasts. Epidermal and hair cells which contain immature chloroplasts also may be filled with TMV. Perhaps TMV can multiply in plastids as well as in cytoplasm. The quantitative differences in RNA bases between chloroplasts and TMV is seen from the ratios of the bases—adenine:guanine:cytosine:uracil, in chloroplasts 1:1.38:1.09:0.95, and in TMV 1:0.87:0.71:1.03, respectively.

No study of the nucleic acid content of isolated proplastids has yet been published. The proplastids ought to contain the hereditary units uncontaminated with the lipoproteins and pigments of the photosynthetic apparatus. They should be readily separable from mitochondria, etc., because of their higher density due to their starch content.

The interesting studies of Brawerman and Chargaff (1959) indicate differences in protein and RNA in colorless and green cells of *Euglena. Euglena* cells (grown in the dark) are colorless. When such colorless normal cells are transferred to a simple medium in which no cell division can occur, the cells given light can still form chloroplasts; the newly formed chloroplasts contain 17% of the protein of the cell; a concomitant decrease of soluble cytoplasmic protein occurs, however, to provide the protein-N for the chloroplast. In this "resting" medium there is a turnover of both protein and RNA as determined with tracer leucine and adenine, respectively. The RNA of colorless etiolated cells and those of permanently "bleached" cells are alike. However, the RNA of green *Euglena* has a slightly higher proportion of adenylic and uridylic and a slightly lower proportion of cytidylic and guanylic acids.

2. *Histochemical Techniques*

Histochemical methods for the detection of DNA and RNA avoid problems arising from contamination of chloroplasts with other cytoplasmic constituents if proper fixation is used. However, for the amount of nucleotides revealed by the isolation methods, the histochemical methods are at about the limit of their sensitivity. The papers by Littau (1958) and by Spiekermann (1957) review the most recent methods. In general the methods depend on fixation in Carnoy's (ethanol-acetic) or dehydration from the frozen state (Woods and Pollister, 1955) and extraction with lipid solvents. Other extractions or enzyme digestions may follow, but finally the nucleic acids are detected by the nucleal reaction of Feulgen, and staining with methyl green-pyronine (Kurnick and Mirsky, 1950) or with azure B (Flax and Hines, 1952). Studies by McDonald and Kaufmann (1954) reveal the pitfalls that may be encountered with nuclease digestion procedures.

Proplastids in the meristematic leaf cells of *Chlorophytum* and *Helianthus tuberosus* contain both RNA and DNA localized in the primary granum (Fig. 29); in addition some RNA is detectable in the stroma (Spiekermann, 1957). Fixation of the meristematic tissue was by a freeze-drying technique. RNA was concluded to be in the primary granule because basic staining was decreased after RNAase digestion. The presence of DNA was shown by DNAase digestion after which the Feulgen reaction disappeared.

In mature chloroplasts the evidence for the presence of RNA is perhaps positive, but it is negative for DNA according to the studies of

Littau (1958) on four species of monocotyledons. Metzner (1952) could detect no DNA in *Agapanthus* with the Feulgen reaction, but the methyl green-pyronine stain appeared to indicate that DNA was present in the grana. Chiba (1951) reported a positive Feulgen reaction for DNA in chloroplasts of *Selaginella savatieri, Tradescantia fluminensis,* and *Rheo discolor.* Both Metzner and Chiba reported RNA in the grana.

Another method that may be useful for detection of nucleic acid in chloroplasts is the examination of the absorption spectrum in the ultraviolet region. The grana but not the stroma of chloroplasts of *Eucharis grandiflora* which have had the pigments extracted absorb 254-mμ light intensely according to a preliminary note by Frey-Wyssling *et al.* (1955) suggestive of the presence of nucleic acid. An absorption maximum has not yet been demonstrated, however, in this region. Still another technique that may be useful for the detection of nucleic acids in plastids is to grow the tissues in the presence of tritium-labeled thymidine or cytidine and by radioautography see if the labeling is in the plastids.

B. Proteins

The mature chloroplasts of tomato and tobacco leaves (Granick, 1938b), Sudangrass (Hanson *et al.*, 1941), and oat leaves (Galston, 1943) make up 35–45% of the total protein-N of the cell; and 35–55% of dry weight of the chloroplasts is protein. About 80% of the total protein-N of the chloroplast is in an insoluble form, probably as lipoproteins and structurally built-in enzymes. To extract the lipoproteins Menke used a solution of 60% ethanol containing 0.3% NaOH. Sissakian (1958) describes an extraction procedure with *n*-butanol at 37°, pH9, by which he isolated a nucleoprotein, a glycoprotein and two chromoproteins.

Some data are available on the increase of protein in chloroplasts as the leaves develop. In tomato and tobacco, the young meristematic cells enlarge rapidly during the interval of growth from one-third to two-thirds their final size. Within this period of several days, there is a tenfold increase of protein N in the leaves. The proplastids cease dividing by the time the meristematic cells have ceased dividing. By the time the leaves are two-thirds expanded, the chloroplasts have enlarged considerably and their protein N has increased twelvefold (Granick, 1938b). It is likely that most of this protein synthesis occurs within the chloroplast. In general the increase in chloroplast protein is proportional to the increase in total leaf protein.

There is little information on the amino acid-synthesizing abilities of chloroplasts. Glycine, alanine, and aspartate are synthesized in isolated chloroplasts (Whatley *et al.*, 1956; Rosenberg *et al.*, 1958). Glycine may be formed from active glycolaldehyde originating from pentose phosphate; alanine from phosphoglyceric acid; and aspartate from transamination of oxalacetate which arises from CO_2 addition to phosphoenol pyruvate. Can other amino acids be synthesized here or must they be supplied to the chloroplast by the cytoplasm?

Protein synthesis may still occur in isolated chloroplasts as suggested by the studies of amino acid incorporation. Stephenson and Zamecnik (1956) reported that chloroplasts of tobacco incorporated amino acids. Sissakian (1958) found that labeled glycine was incorporated into washed chloroplasts of tobacco and bean, but not into unwashed chloroplasts; the optimum pH was 8.5. Mg^{++} activated the reaction as well as addition of a mixture of other amino acids in equimolar ratio to the added glycine.

The rate of turnover of proteins in the chloroplast is unknown. The structural proteins of chloroplasts which constitute 80% of the proteins probably have negligible turnover on analogy with the behavior of other structural proteins. A breakdown of chloroplast structure is readily observed in older leaves when bean plants are placed in the dark for several days; there is then a drain of amino acids and sugars to supply the growing meristem and young leaves. Whether the protein content in the chloroplasts represents a dynamic equilibrium between synthesis and breakdown, or whether when sugars are not available the breakdown process becomes activated, is not known. An electron microscope study of this breakdown should provide useful morphological information.

C. *Lipids*

The lipid content of chloroplasts is high, from 20 to 40% of the dry weight, depending on the species (Table II). The comparatively low lipid content in the rest of the cytoplasm is indicated by the analyses of Chibnall (1939) on spinach leaves; the ratio of protein to lipid was 39:25 in the chloroplasts and 96.5:1.9 in the rest of the cytoplasm. The mitochondria are also rich in lipoproteins, but their volume in the cytoplasm of the mature cell is negligible compared to the chloroplast volume, so their contribution to the lipid content of the cell is small. Benson *et al.* (1958, 1959) have fractionated the phosphatidyl esters of tobacco chloroplasts (Table II). The phosphatidyl esters are rapidly metabolized. Galactosyl glycerol, phosphatidyl glycerol and a glycosyl glycerol are present in concentrations $\sim 0.02\,M$.

Sissakian (1958) has found that chloroplasts from cotyledons of 10-day-old sunflower plants converted 80% of the labeled acetate into higher fatty acids; this conversion did not occur in bean chloroplasts.

The mechanism for acetyl CoA formation in chloroplasts is still unknown. In brown algae (e.g., diatoms) the storage form is lipid rather than carbohydrate; the factors controlling carbohydrate versus lipid storage are unknown.

D. *Chlorophylls and Hemes of Chloroplasts*

The two major pigments of protoplasm, heme and chlorophyll, are synthesized along the same biosynthetic chain to protoporphyrin (Granick, 1950, 1951). At this step Fe is inserted to form Fe protoporphyrin or heme (Fig. 38A); and Mg is inserted to form Mg protoporphyrin which

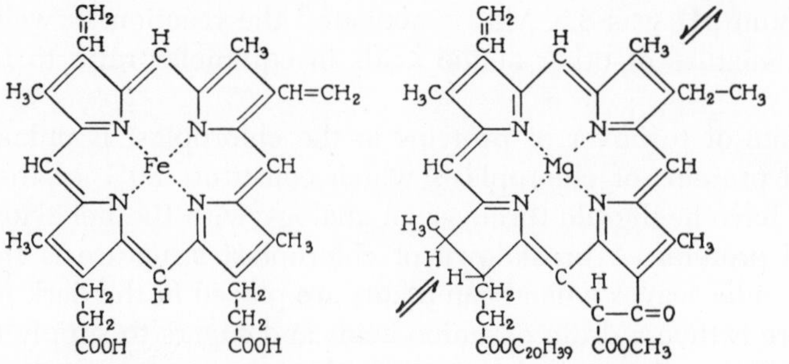

FIG. 38A. Structures of heme (left) and chlorophyll a (right). The direction of polarization of the electric vector and of fluorescence emission of chlorophyll a is indicated by the arrows.

is further converted in a series of steps to chlorophyll (Granick and Mauzerall, 1961).

1. *Iron Protoporphyrin or Modified Iron Porphyrins*

These serve as the prosthetic groups of cytochromes. The cytochromes are localized in mitochondria and chloroplasts; one cytochrome has been found in microsomes. The cytochromes of plants have been reviewed by Hartree (1955, 1957). The cytochromes b_3 ($E'_o = + 0.04$), b_6 ($E'_o = - 0.06$) and f ($E'_o = + 0.365$) are found in leaves. Cytochromes b_6 and f are restricted to the grana (James and Leech, 1960). Cytochrome f is found in the molecular ratio of 1 cytochrome f to 400 chlorophyll. The cytochrome f of parsley has a molecular weight of about 110,000 (Hill and Whittingham, 1955). Cytochrome oxidase, i.e., cytochrome a_3, is present in mitochondria but absent from chloroplasts of tobacco leaves (McClendon, 1953). Cytochrome c is also absent from chloroplasts. Cytochrome oxidase activity is said to be present in young plastids of barley leaves and roots and to decrease in older plastids (Rubin and Ladygina, 1956); the possibility of contamination with mitochondria must, however, be considered in the isolation of young plastids.

2. The Biosynthetic Chain of Heme and Chlorophyll (Fig. 38B)

The early precursors of porphyrin synthesis are glycine and succinyl CoA (Shemin, 1956). These form δ-amino levulinate (δ AL) via a cycle of glycine oxidation according to Shemin. Prepara-

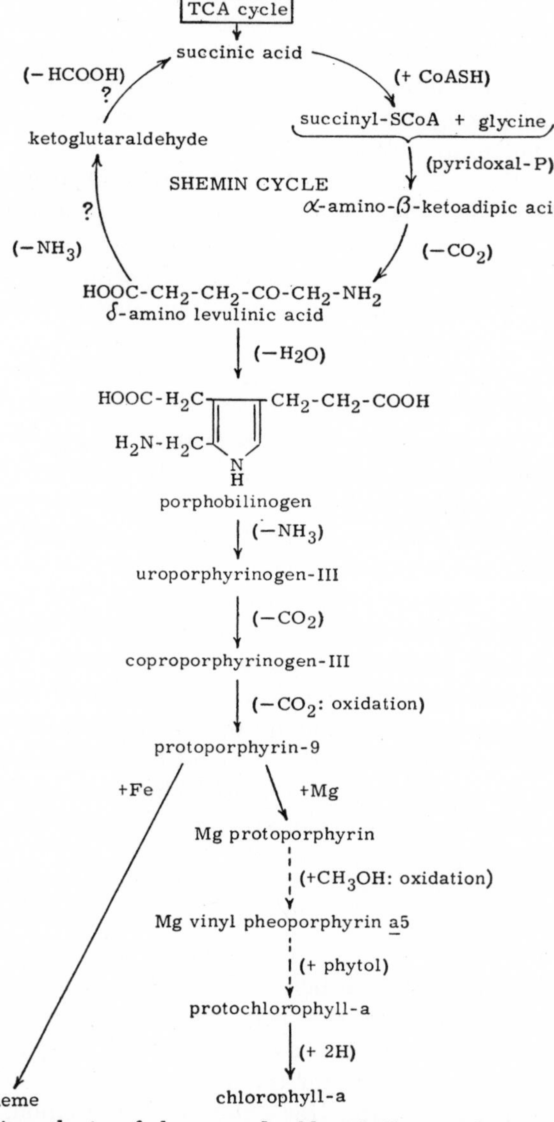

FIG. 38B. Biosynthesis of heme and chlorophyll according to Granick and Mauzerall (1961).

tions made from purple bacteria synthesize δ AL from glycine, succinyl CoA, and pyridoxal P (Kikuchi *et al.,* 1958, Gibson, 1958). Here succinyl-CoA is formed from succinate, ATP and CoA by an enzyme also found in spinach. In mitochondria of animals (also plants?) succinyl-CoA is formed as part of the citric acid cycle. Perhaps mitochondria and plastids both synthesize δ AL, but mitochondria form only Fe porphyrins whereas plastids, in addition, form Mg porphyrins.

The steps in the synthesis of porphyrins from δ AL are the condensation of 2 δ AL molecules to form the monopyrrole porphobilinogen (PBG); the condensation of 4 PBG molecules to form the colorless uroporphyrinogen III (i.e., a reduced porphyrin); the decarboxylation of the acetic side chains to methyl groups to form coproporphyrinogen III; the oxidation of two propionic acid side chains and autoxidation of the ring to protoporphyrin, which is the colored porphyrin. Metal chelates are not formed with the porphyrinogens but may be formed with the partially oxidized porphyrinogens.

Studies with *Chlorella* mutants suggest that further intermediates are Mg protoporphyrin, Mg vinylpheoporphyrin a_5, protochlorophyll a, and chlorophyll a (Granick, 1950). Phytol appears to be esterified at both the protochlorophyllide and chlorophyllide stages (Loeffler, 1954-55). The action of phytol esterase (i.e., chlorophyllase) is discussed by Mayer (1930). The possible relations of the phycobilin pigments, chlorophyll c and bacteriochlorophyll to the scheme of chlorophyll a biosynthesis have been discussed by Granick (1950). Chlorophyll b does not originate directly from chlorophyll-a (Smith and Young, 1956). Both C atoms of acetate and the β-C atom of glycine were found to be incorporated by *Chlorella* into chlorophyll (Della Rosa *et al.,* 1953) in conformity with the studies of Shemin on the labeling of heme. It should be possible to establish which of the enzymes that convert δ AL to chlorophyll are localized within the developing chloroplast. When δ AL is fed to etiolated barley leaves in the dark, protochlorophyllide is formed which is localized in the tiny plastids (Granick, 1960).

3. *Protochlorophyll*

In a few species of plants such as *Chlorella,* chlorophyll synthesis proceeds in the dark although better in the light. Therefore a full complement of enzymes are present including one for reduction of protochlorophyll to chlorophyll (i.e., pyrrole ring III is reduced). But for most plants, light is necessary for this reduction step, as determined by studies of the action spectrum for chlorophyll formation, which shows that protochlorophyll is the absorbing molecule (Frank, 1946; Smith and

Young, 1956). The sequence of changes in spectra from protochlorophyll to chlorophyll a is complex and suggests that changes in physical state (structural organization in the lamellae?) occur with time (Shibata, 1955-1956). Thus in dark-grown pole bean leaves, on exposure to light for 1 minute the protochlorophyll band maximum at 650 mμ is changed immediately to a band with a maximum at 684 mμ. The leaves are then placed in the dark and after 20 minutes the band maximum has shifted to 673 mμ; after 40–60 minutes the band maximum is at 677 mμ, which is that of photosynthetically active chlorophyll. During this time, in the dark, new protochlorophyll develops. The substances with band maxima in the plastid at 684, 673, and 677 mμ are all chlorophyll a. Short periods of irradiation in barley seedlings produce only chlorophyll a, but on continued irradiation chlorophyll b also appears, and then the two chlorophylls increase in constant proportion to each other (Smith and Young 1956). Smith *et al.* (1955-1956) isolated a protochlorophyll fraction bound to protein of molecular weight 400,000 which retains the ability to be converted to chlorophyll on illumination. Weissbach *et al.* (1956) found that the enzyme which catalyzes the carboxylation of ribulose diphosphate occurs in this fraction.

The concentration of protochlorophyll that accumulates in (etiolated) plants in the dark is usually very low. In a barley seedling there is a maximum of 0.5 μg. Earlier precursors have been detected by feeding δ AL (Granick, 1960). Light is not necessary for the steps from δ AL to protochlorophyllide.

Albino mutants of corn have been examined to see if the metabolic defects were related to the biosynthetic chain of chlorophyll. Albino mutants were found which could form protochlorophyll as effectively as the normal, but the resulting chlorophyll became bleached in light, possibly owing to a lack of carotenoids (see Section VII, E). Some albinos showed a decreased ability to add phytol; in others the normal shift from 684 mμ to 673 mμ was lacking (Smith *et al.*, 1956-1957).

E. *Carotenoids* (Goodwin, 1952; Karrer and Jucker, 1950; Stanier, 1959)

1. *Formation*

Proplastids contain small amounts of carotenoids which precede the appearance of the chlorophylls. In barley seedlings grown in the dark, the leaves are yellowish; a number of carotenols (xanthophylls) are already present in ratios similar to the ratios found in mature chloroplasts (Strain, 1938). When seedlings are exposed to light, the carotenoids increase in amount. In mature leaves of a number of higher plants, the weight ratios of chlorophylls to carotenoids is 2.4–3.0:1 and the weight ratios of carotenols to carotenes (i.e., principally xanthophyll and carotene) is 4–6:1 (Rabinowitch, 1945). In terms of molecules, the approximate ratio of chlorophyll a:b:xanthophyll:carotene is as 6:2:3:1.

2. *Chemistry and Scheme of Biosynthesis*

The carotenoids are branched C_{40} compounds containing conjugated double bonds in all-*trans* configuration, usually with an ionone ring at either end. They consist of two C_{20} units; each one is made up of four C_5 isoprene units linked head to tail; but the two C_{20} units are linked tail to tail. The carotenoids belong to the large family of compounds, the isoprenoids, which include the terpenes, sterols, and guttapercha which are all-*trans;* and rubber which is *cis* (Simonsen, 1952). A tentative scheme of synthesis is presented in Fig. 39, proceeding from the common precursors of fat synthesis, acetoacetate and acetyl CoA, to mevalonic acid pyrophosphate and to isopentenyl pyrophosphate which isomerizes and condenses with dimethyl acrylic pyrophosphate to geranyl pyrophosphate. Further condensation leads to the C_{20} and C_{40} units. Subsequent dehydrogenation with the formation of the alternating single- and double-bond structures via phytoene and phytofluene leads to the formation of lycopene. Then cyclization occurs at the ends of the chain to form the ionone rings, giving rise to carotenes. Oxidation of these rings produces the hydroxyl-containing carotenols. Mevalonic acid has also been shown to be a precursor of *Hevea* rubber (Park and Bonner, 1958) and of sterols (Tavormina *et al.,* 1956).

The genetic studies on tomatoes by Porter and Lincoln (1950) suggest that the carotenoids are formed by stepwise dehydrogenation of saturated perhydrolycopene. A *Chlorella* mutant has been examined which forms small amounts of chlorophyll (i.e., containing phytol) but no carotenoids that absorb above 300 mμ (Granick, unpublished). This result supports the previous suggestion that nonconjugated isoprenoids may be formed first. The C_{20} alcohol phytol with one double bond is an ester substituent of chlorophyll and is a component of vitamins E, K, and Q. Studies on purple bacteria have provided evidence for the later steps of oxidation and ionone ring formation (Stanier, 1959). Claes (1956) has obtained a *Chlorella* mutant that forms the more saturated carotenoids in the dark but the normal unsaturated ones in the light.

3. *Functions*

Two functions have been established for the carotenoids. One is the protection of the plastids against photo-oxidation, especially in the presence of fluorescent porphyrins or chlorophylls which catalyze photo-

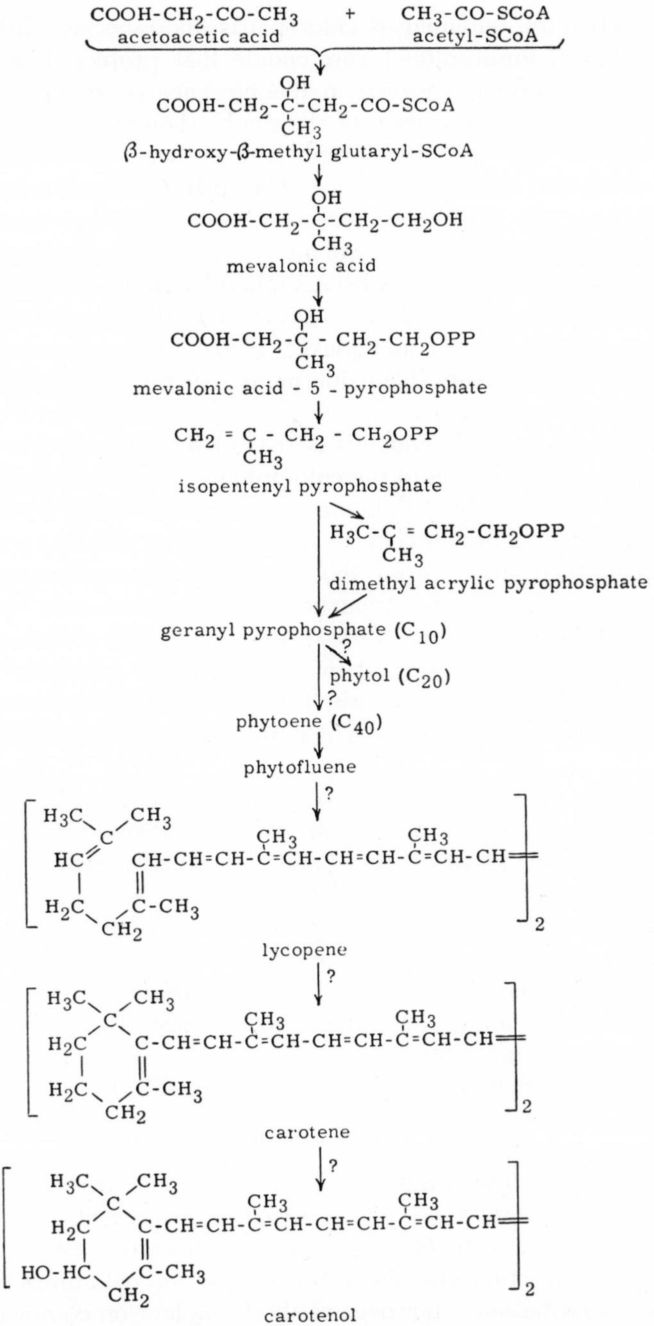

$COOH\text{-}CH_2\text{-}CO\text{-}CH_3$ + $CH_3\text{-}CO\text{-}SCoA$
acetoacetic acid acetyl-SCoA

↓

$COOH\text{-}CH_2\text{-}C(OH)(CH_3)\text{-}CH_2\text{-}CO\text{-}SCoA$
β-hydroxy-β-methyl glutaryl-SCoA

↓

$COOH\text{-}CH_2\text{-}C(OH)(CH_3)\text{-}CH_2\text{-}CH_2OH$
mevalonic acid

↓

$COOH\text{-}CH_2\text{-}C(OH)(CH_3)\text{-}CH_2\text{-}CH_2OPP$
mevalonic acid - 5 - pyrophosphate

↓

$CH_2 = C(CH_3) - CH_2 - CH_2OPP$
isopentenyl pyrophosphate

↓ ↘ $H_3C\text{-}C(CH_3) = CH_2\text{-}CH_2OPP$
dimethyl acrylic pyrophosphate ↙

geranyl pyrophosphate (C_{10})

↓ ↘? phytol (C_{20})

↓ ?

phytoene (C_{40})

↓

phytofluene

↓ ?

$[H_3C\text{-}C(CH_3)\text{-}HC{=}\ldots; H_2C\text{-}CH_2\text{-}C(CH_3){=}]\; CH\text{-}CH{=}CH\text{-}C(CH_3){=}CH\text{-}CH{=}CH\text{-}C(CH_3){=}CH\text{-}CH{=}]_2$

lycopene

↓ ?

$[H_3C, CH_3 \text{ C}; H_2C; H_2C; CH_2; C\text{-}CH_3]\; C\text{-}CH{=}CH\text{-}C(CH_3){=}CH\text{-}CH{=}CH\text{-}C(CH_3){=}CH\text{-}CH{=}]_2$

carotene

↓ ?

$[H_3C, CH_3 \text{ C}; H_2C; HO\text{-}HC; CH_2; C\text{-}CH_3]\; C\text{-}CH{=}CH\text{-}C(CH_3){=}CH\text{-}CH{=}CH\text{-}C(CH_3){=}CH\text{-}CH{=}]_2$

carotenol

FIG. 39. Tentative scheme of carotenoid biosynthesis.

oxidation. (Healthy, undamaged chloroplasts fluoresce very little.) It is the colored (i.e., unsaturated) carotenoids that protect. For example, when purple bacteria are grown in the presence of small amounts of diphenylamine, dehydrogenation of the double bonds of the polyenes is inhibited; in this case bacteriochlorophyll is destroyed in the light in the presence of O_2, but not in its absence (Cohen-Bazire and Stanier, 1958). The carotenoids may protect by accepting the activated oxygen from an unstable intermediate. This antioxidant effect in protecting unsaturated fatty acids against oxidation has been studied by Bergström and Holman (1948). Epoxide structures of some carotenoids also suggest their function as antioxidants (Karrer and Jucker, 1950). In the dark xanthophyll decreased and violaxanthin diepoxide increased; the reverse occurred in the light (Blass *et al.*, 1959).

Another function of carotenoids is that of light absorption and subsequent energy transmission to the chlorophylls (Smith and Koski, 1947). For example, in brown algae, light absorbed by fucoxanthin is transmitted to chlorophyll a. The brown color of brown algae is due to a shift of the absorption of fucoxanthin from 480 up to 560 mμ suggestive of some structural or coupling organization of the fucoxanthol with chlorophyll in the plastid lamellae. Teale (1958) has found that transfer of light energy may occur in a detergent micelle when the pigments lutein or fucoxanthol together with chlorophyll a are at 0.1 M concentration; no transfer is found to pheophytin a. The absorption oscillators of the yellow pigments are coupled with the fluorescence oscillators of chlorophyll a as shown by polarization studies. This result suggests that the long chains of carotenoid lie parallel to the plane of the dihydroporphyrin ring and in a line that connects rings II and IV of chlorophyll (Fig. 38, A).

Other functions have been postulated. It is not excluded that electron transmission from activated chlorophyll via carotenoids may occur. Colored carotenoids, however, are not essential for photophosphorylation in *Chromatium* bacteria which lack the unsaturated polyenes when grown on diphenylamine (Anderson and Fuller, 1958). In spinach chloroplasts from which β-carotene (but not xanthophyll) has been removed the Hill reaction can still occur (Bishop, 1958).

F. Some Vitamins in Chloroplasts

Besides the chlorophyll molecules which bear the phytol chain, three other substances, vitamins K, E, and Q, are also present in the chloroplasts and also contain phytol as part of their structure. Studies of Dam (1942) have shown that vitamin K develops as the chloroplasts enlarge, but vitamin K formation is not necessarily dependent on complete formation of the normal chloroplasts. For example, the content of vitamin K is approximately the same in the pale green and in the normal green leaves of *Sambucus*. It is low in pea seedlings grown in the dark, but is synthesized in the light. Seedlings of *Picea canadensis*, however, which form chlorophyll in the dark, also form vitamin K in the dark. Spinach leaves contain approximately 40 μg. vitamin K per gram dry weight. When leaves wither in the fall the vitamin K disappears only slowly. Vitamin K has been implicated in photosynthetic phosphorylation of isolated chloroplasts.

Vitamin E is localized for the most part in the chloroplasts, in which it is present to the extent of 0.08% on a dry weight basis; in the cytoplasm there is only 0.002%. Vitamin E also appears to develop in regions where photosynthesis does not occur, especially in connection with the formation of oils as in wheat germ, etc. (see Section II, C, 2). Vitamin E appears to function in oxidative phosphorylation in mitochondria (Mason, 1957) and to act as an antioxidant.

Vitamin or coenzyme Q is a benzoquinone with a C_{20}–C_{50} side chain containing four to ten isoprenoid units. It is localized in mitochondria and chloroplasts. In the chloroplasts the peak absorption is at 255 mμ, so it is referred to as Q_{255} in contrast to Q_{275} of mitochondria (Crane, 1959). The quinone probably functions as an electron transport molecule through a lipid medium.

Another chloroplast constituent, carotene, is provitamin A, i.e., a precursor which is split in the intestinal mucosa to vitamin A.

Leaves, in general, have a higher ascorbic acid (vitamin C) content than other parts of plants, except meristematic tissues. The leaf attains its maximal content of vitamin C just before flowering. Giroud suggested that the blackening of chloroplasts with silver nitrate, first observed by Molisch, was due to the presence of ascorbic acid in these bodies. Other ene-diol structures also react similarly. Weier (1938b) found that silver nitrate in the cell was reduced to elemental silver at a pH of 7, or above, but very slowly below pH 4, tending to corroborate the idea that this reducing substance was ascorbic acid. The reducing ability of the cells was lost very rapidly if they were killed by anesthetics in the presence of oxygen, but if killed in an atmosphere free from oxygen, or by procedures which inactivated the oxidative enzymes, the reducing substance was still preserved. Although silver reduction appears to be greatest in or on the chloroplasts, it cannot be assumed as proved that ascorbic acid is present exclusively in the chloroplasts. Ascorbic acid appears to function in photosynthetic phosphorylation of isolated chloroplasts. Negative Molisch reactions have been reported under conditions in which ascorbic acid was known to be present (Nagai and Ogata, 1952). In radish leaves, a compound associated with chloroplasts is described which can be hydrolyzed at 90° for 5 minutes at pH 1.5 to ascorbic acid (Fujimara and Hamaguchi, 1951). This is probably the indoleacrylic derivative of ascorbic acid "ascorbigen" isolated from cabbage by Procházka *et al.* (1957).

G. *Some Metals in Chloroplasts*

Because trace metals usually are active constituents of enzymes, the content and concentration of these metals (Table II) are suggestive of their importance in an analysis of the development and functioning of the chloroplast (Rabinowitch, 1945, 1957; Pirson, 1955, 1958). Of the metals of the sugar beet leaves, Fe and Cu are concentrated in the chloroplast but not Mn, Zn, or Mo.

In spinach leaf over 80% of the iron was found to be in the chloroplast (Liebich, 1941). Iron is required for the heme enzymes at the growing points of root and shoot and in developing leaves previous to the appearance of chlorophyll. As the chloroplasts enlarge, a further requirement for iron develops. Iron in mature chloroplasts is present in cytochrome f in the molar ratio of 1:400 chlorophyll molecules. The molar ratio of Fe to chlorophyll is 1:4 to 1:10 in most plants (Hill and Lehman, 1941). The inhibition of the Hill reaction by *o*-phenanthrolene suggests that at least one other iron compound besides cytochrome f is essential for photosynthesis. For bacteriochlorophyll production by photosynthetic bacteria, a relatively high content of iron in the nutrient medium is required (Lascelles, 1955). Iron deficiency, leading to chlorosis and decreased photosynthesis in green algae, is only very slowly reversed by addition of iron, and the return of photosynthesis is proportional to the increase in chlorophyll and protein. Iron is not transported from older leaves to meristems, so that a more or less continuous supply of iron must be obtained through the roots of growing plants or supplied to the leaves by spray. Granick (1958) has reviewed iron metabolism in plants.

The function of copper in chloroplasts is not known although it is there in a relatively high concentration. The absence of copper does not appear to affect chlorophyll or protein content. Copper enzymes have the property of acting as "mixed-function oxidases" on O_2, i.e., of adding electrons to one O atom and attaching the other O atom to the substrate (e.g., phenol oxidase) (Mason, 1957) or of transferring electrons to O_2 (Mahler, 1958).

The absence of zinc leads to a fairly uniform chlorosis. In *Chlorella,* chlorophyll content is greatly diminished and selectively restored by zinc (Pirson, 1955). Zinc is known to function in some pyridine nucleotide enzymes such as alcohol dehydrogenase and may thus be required for functioning in chloroplast growth as well as in dark reactions of photosynthesis.

Manganese deficiency leads to a lower rate of photosynthesis in both weak and strong light (Pirson, 1958). Recovery to a normal rate occurs within half an hour after addition of manganese added either in light or dark. The Hill reaction is low in manganese-deficient algae. Kessler (1957) has shown that manganese is required for the release of O_2 in photosynthesis. Manganese-deficient cells of *Chlorella* grow well in glucose in the dark but peculiarly do not appear to use glucose in the light and grow poorly. There is a very small requirement of manganese also for some dark steps in metabolism. In algae, manganese deficiency does not affect chlorophyll content but in higher plants a chlorosis develops.

A vanadium requirement for photosynthesis in strong light by *Scenedesmus obliquus* has been found (Arnon and Wessel, 1953). The chlorophyll content of the vanadium-deficient cells was 60% of normal. After addition of vanadium, recovery to normal is slow.

Magnesium deficiency leads to chlorosis. Magnesium is a coenzyme in many phosphorylation reactions, and it is a constituent of chlorophyll. It is not known how magnesium is inserted into protoporphyrin (see Section VII, D). The effects of magnesium deficiency are difficult to reverse. Smith (see Smith and Young, 1956) investigated the ether-soluble magnesium of etiolated barley seedlings and found it to increase in early stages of illumination and to be higher in content than the magnesium required for chlorophyll synthesis. After the initial stages of greening, the ether-soluble magnesium was only slightly in excess of the magnesium in chlorophyll. Albino corn seedlings contained only traces of ether-soluble magnesium, but in green corn seedlings it was proportional to the chlorophyll magnesium.

H. Carbohydrate Metabolism of the Chloroplast (Racker, 1954; Aronoff, 1957)

The immediate product of photosynthesis from the Calvin cycle (see Section VII, I) is glyceraldehyde-3P. Two products derived from G-3P are starch and ribulose-5P. In Table IV are summarized the two enzyme systems—one which converts G-3P via F-6P to starch, the main carbohydrate storage product; the other which converts G-3P plus F-6P to

TABLE IV

Some Enzyme Systems of Carbohydrate Synthesis[a]

Enzyme system	Substrate		Product
A. Conversion of glyceraldehyde-3-phosphate to starch			
(a) Triose phosphate isomerase	G-3P	⟷	DHA-3P
(b) Aldolase	G-3P + DHA-3P	⟷	F-1,6diP
(c) Fructose-1,6-diphosphatase	F-1,6diP + ADP	⟷	F-6P + ATP
(d) Hexose phosphate isomerase	F-6P	⟷	G-6P
(e) Phosphoglucomutase	G-6P	⟷	G-1P
(f) Phosphorylase forming amylose	42-116(G-1P)	⟷	(α-1,4glucosan) + 42-116 P
(g) Q enzyme forming amylopectin	α-1,4glucosan	⟷	1,4 + 1,6 branched chain
B. Conversion of glyceraldehyde-3-phosphate and fructose-6-phosphate to ribulose-5-phosphate			
(a) Transketolase	F-6P + G-3P	⟷	E-4P + Xu-5P
(b) Transaldolase	F-6P + E-4P	⟷	S-7P + G-3P
(c) Transketolase	S-7P + G-3P	⟷	R-5P + Xu-5P
(d) Pentose phosphate isomerase	R-5P	⟷	Ru-5P
(e) Epimerase	Xu-5P	⟷	Ru-5P
C. Sucrose formation (in cytoplasm)			
(a) UDPG pyrophosphorylase	G-1P + UTP	⟷	UDPG + PP
(b) UDP transglucosylase	UDPG + F-6P	⟷	Sucrose-P + UDP
(c) Sucrose phosphatase	Sucrose-P	⟶	Sucrose + P

[a] From Racker (1954) and Vishniac (1955).

Abbreviations: DHA = dihydroxyacetone; E = erythrose; S = sedoheptulose; UDPG = uridine diphosphoglucose. For other abbreviations see pages 569 and 570.

Ru-5P, the compound required for the maintenance of the Calvin cycle, thus:

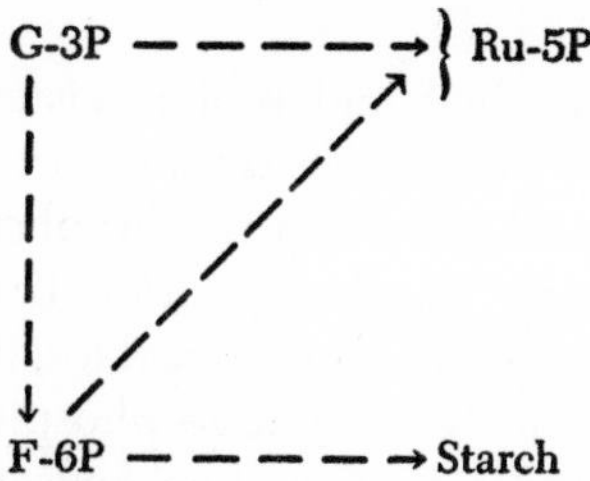

There is no evidence that sucrose phosphate (Table IV, C) is formed in the chloroplasts so it is probably formed in the cytoplasm. The enzymes converting glucose-6P through gluconate-6P to Ru-5P have been found in extracts from chloroplasts, and this system may serve to prime the Calvin cycle with pentose-P (Trebst *et al.*, 1958).

The synthesis of starch from glucose-1-PO_4 (G-1P) is catalyzed by the enzyme phosphorylase according to the reaction A(f) (Table IV). Starch is formed in the stroma of plastids (see Section II, C). Two kinds of starch molecules are generally present. The minor component, amylose, consists of long spiral polymer chains; each chain consists of about 4000 α-1,4 linked glucose units formed by the action of phosphorylase. These chains spontaneously take a helical configuration. Staining with iodine results in a blue color because I_2 molecules are held loosely within the helix (i.e., as a clathyrate compound). The major component, amylopectin, $\sim 10^7$ mol. wt., contains besides these links, also 1,6 links which are places of chain branching. The 1,6 links are formed by the action of the Q enzyme and are about twenty glucose residues in length. Amylopectin stains red with iodine. Waxy maize and waxy sorghum consist almost entirely of amylopectin. According to Yin and Sun (1949) the addition of glucose-1-P to plant tissues results in the formation of starch exclusively in the plastids, therefore phosphorylase is considered to be localized only in plastids. This effect is obtained only if the tissue dies gradually so that glucose-1P can penetrate into the plastid before phosphorylase diffuses out from the plastid (Badenhuizen, 1959).

Amylase inhibits starch formation from glucose-1P; 5×10^{-3} *M* $HgCl_2$ inhibits amylase, but not starch formation from glucose-1P. Starch may be hydrolyzed by amylase, especially during seed germination when the amylase activity is relatively high. But in leaves, amylase activity appears to be negligible. In many leaves the starch content decreases rapidly as the leaves wilt; whether this is a result of amylase or phosphorylase activity is not known. Leaves killed by chloroform or frozen and then kept sterile for a month at 30° show no decrease in starch, possibly indicating that such treatments readily diminish phosphorylase activity in the chloroplasts.

The proplastids in the leaf and root meristems often contain starch and thus presumably the phosphorylase and Q enzymes. Since sucrose is the main sugar of transport, starch in the proplastids is probably derived from sucrose. In the dark, starch formation from sucrose requires oxidative phosphorylation (Porter and Runeckles, 1956). Evidence is lacking concerning the presence of oxidative phosphorylation in proplastids, which often contain starch. In meristem cells the mitochondria are in relatively high concentration and might supply the ATP for hexose phosphorylation in proplastids (see Section V). Chloroplasts in the dark do not have an oxidative phosphorylation, nor do they contain cytochrome oxidase.

I. Enzymes of Photosynthesis (Arnon, 1956; Bassham and Calvin, 1957; Vishniac *et al.*, 1957; Aronoff, 1957)

1. Photodecomposition of Water (Tentative)

In photosynthesis, light energy brings about the photodecomposition of water to form a reductant Hx and oxidant yOH (Fig. 40). Then enzyme systems take over, to form three main products, PNH, ATP, and O_2; 2(Hx) indirectly cause the reduction of pyridine nucleotides, PN^+ to PNH; 4(yOH) are converted to O_2; ATP is formed by way of electron transfer from PNH to Fe^{+++}-cytochrome f. Possibly the photodecomposition of water and the above enzyme systems are localized in the grana. The chemical energy made available in PNH and ATP is then used in the Calvin cycle to fix CO_2 and form glyceraldehyde-3P and thence starch. Possibly the enzymes that fix CO_2 and convert it finally to starch are localized in the stroma of the chloroplast (Trebst *et al.*, 1958).

2. Enzymes That Convert 4(yOH) to O_2 (Tentative)

These enzymes include a Mn-activated enzyme (Pirson, 1958), which is required for the release of O_2. Hill found that the addition of oxidants of high potential to isolated chloroplasts in the presence of light would result in O_2 evolution (Hill and Whittingham, 1955). One interpretation of the Hill reaction is that the oxidant which is added reacts with Hx (or a reduced product) and thus the interaction of Hx with yOH is prevented (Mehler and Brown, 1952). The yOH then forms O_2. At least four photons are required to form one O_2; therefore a stepwise mechanism must be involved. It is possible that an iron enzyme closely coupled with a Mn enzyme may be the catalyst. The presence of an iron enzyme is suggested by the high iron content of chloroplasts and the inhibition of the Hill reaction by *o*-phenanthrolene; however, neither α,α′-dipyridyl nor ethylenediaminetetraacetic acid (EDTA) inhibits. The enzymes involved in the Hill reaction are not solubilized by grinding the chloroplasts. Particles of 100-A. diameter containing about 200 chlorophyll molecules are still 60% as efficient as intact chloroplasts in the Hill reaction under intense illumination (Thomas *et al.*, 1953). Chloroplasts

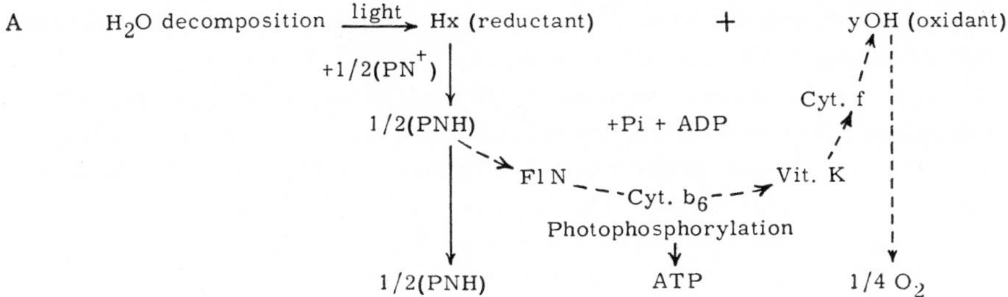

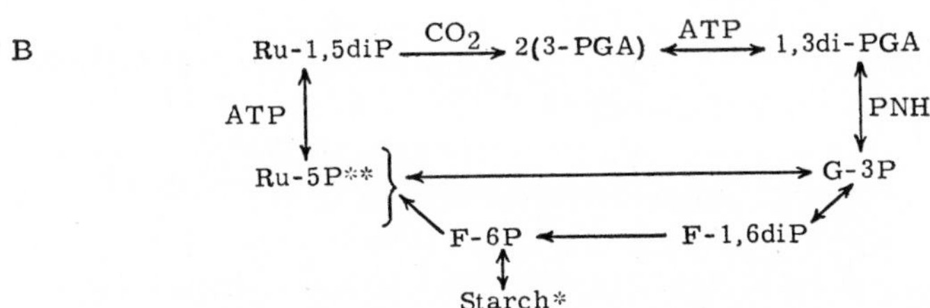

FIG. 40. Starch formation from CO_2, H_2O, and light by chloroplasts. A. Photochemical act and the production of reduced pyridine nucleotides, adenosine triphosphate, and O_2. (In grana?) B. CO_2 fixation via the Calvin cycle and starch formation, using reduced pyridine nucleotides and adenosine triphosphate. (In stroma?) See Table IV, A for starch* formation and Table IV, B for Ru-5P** formation.

that have been treated with digitonin lose the ability to photoreduce; they now can photo-oxidize; this is called the photo-oxidase activity, which can be fractionated into two heat-labile components, one containing chlorophyll (Nieman and Vennesland, 1959).

3. *Enzymes of Photophosphorylation* (*Tentative*)

These enzymes form ATP, perhaps through an electron transfer chain from PNH to ferric cytochrome f (Fig. 40). Since cytochrome oxidase is not present in chloroplasts (Jagendorf, 1955), cytochrome f may be the penultimate acceptor. A portion of the oxidant yOH may bring about the oxidation of Fe^{++}-cytochrome f. A similar system is present in plastid particles of photosynthetic, anaerobic, purple sulfur bacteria, where far-red light, absorbed by bacteriochlorophyll, is used to bring about the oxidation of the cytochrome to serve in the synthesis of high-energy phosphate bonds and also to serve for the oxidation of external H donors (Vernon and Kamen, 1954; Frenkel, 1954). In preparations of spinach chloroplasts, coenzymes and some of the more soluble proteins leach out. Ascorbic acid, vitamin K, Mg^{++}, and flavine mononucleotide enhance $\sim$ P formation when added to such preparations (Arnon, 1956). The further addition of the solubilized proteins returns the phosphorylation activity to the same order as the rate of O_2 evolution in intact leaves, i.e., 200–250 μmoles O_2 released and 200–250 μmoles CO_2 fixed per milligram chlorophyll per hour (Avron *et al.*, 1957). In the absence of CO_2, *Chlorella* in the light accumulates metaphosphate in the cells (as a storage form of $\sim$ P?) (Wintermans, 1954). Literature on this rapidly developing subject of photophosphorylation is well summarized by Jagendorf (1959).

4. *Starch Formation from* CO_2

Starch formation by chloroplasts requires a full complement of enzymes and coenzymes, including those necessary for: (a) the Hill reaction, (b) photophosphorylation, (c) the reduction of pyridine nucleotides, (d) the Calvin cycle, (e) the formation of ribulose-5P (Table IV, B), and (f) the formation of starch (Table IV, A).

CO_2 fixation is low in broken chloroplasts. However, it can be enhanced by adding substances that enhance photophosphorylation, plus pyridine nucleotides, ATP, small amounts of phosphorylated sugars, and a water extract containing enzymes from whole chloroplasts (Trebst *et al.*, 1958).

The over-all reaction of photosynthesis via the Calvin cycle from CO_2 to triose phosphate may be written:

$$3\ CO_2 + 7\ ATP + 6\ PNH \rightarrow G\text{-}3P + 6\ P + 7\ ADP + 6\ PN^+ + 7H^+$$

Assuming that 4 light quanta are sufficient for the reduction of 2 PN^+ and that an additional light quantum is required for the formation of 2–3 $\sim$ P, then the conversion of CO_2 to a reduction level of G-3P would require between 4 and 5 quanta as a minimum. Other reactions that fix CO_2 are discussed in the review by Vishniac *et al.* (1957).

5. *Molecular Model of Photosynthesizing Chloroplast*

What is the molecular anatomy of the chloroplast that permits it to function so efficiently in photosynthesis? A satisfactory model cannot be constructed with the available data. Such a model should explain not only the percentage composition and molecular distribution of the components, but also the nature of the energy sink, the pathways for PNH and ATP formation, the regions of CO_2 fixation and starch formation, etc. For discussions of various models see the reviews by Aronoff (1957), Rabinowitch (1956), Calvin (1956, 1958).

Considering the properties discussed above, the following interpretations have been advanced. In the stroma region (i.e., outside the dense grana region) there appears to be segregation of those enzymes that fix CO_2 and convert it to starch; this conclusion is based on the premise that the enzymes of the stroma region are readily dissolved out of the chloroplasts whereas those of the grana region are relatively insoluble. The compounds supplying the energy for CO_2 fixation and starch formation, i.e., PNH and ATP, are formed by photosynthesis in the disks of the grana where chlorophyll is localized; this conclusion is suggested by the fact that isolated chloroplasts can carry out the Hill reaction and photophosphorylation (i.e., ATP formation).

Chlorophyll molecules appear to be arranged in a monolayer in the disk membranes of the grana; this is suggested by the high concentration of the chlorophyll and the polarized light studies. The chlorophyll molecules need to be oriented with respect to each other so that the electron orbitals of the flat dihydroporphyrin rings overlap; this assumption would permit the energy of a photon that is absorbed by one chlorophyll molecule to be transmitted (inductive resonance transfer, exciton migration?) through several hundred chlorophyll molecules to some energy sink containing a specific chlorophyll molecule where the energy is converted directly or indirectly to chemical energy. The number of chlorophyll molecules taking part in the transmission of energy from light represents a "photosynthetic unit" for which there are several lines of evidence.

At the energy sink two different mechanisms have been postulated. One is a relatively direct decomposition by the excited chlorophyll molecule of a water molecule to xH and yOH without a spatial separation, where xH may reduce an organic molecule immediately in the neighborhood and yOH eventually will combine with another yOH in the presence of an Mn enzyme to form O_2; two light quanta are required before a chemical reaction can occur; a mechanism for the storage of the first light quantum in the excited chlorophyll is discussed by Franck (1958).

(In mitochondria various oxidations and reductions occur without apparent interference because the enzymes are highly specific for substrate.)

The other mechanism which has been postulated is one by which a spatial separation of xH and yOH occurs (e.g., Calvin, 1958); here also two quanta of light are required to shuttle into the same energy sink to

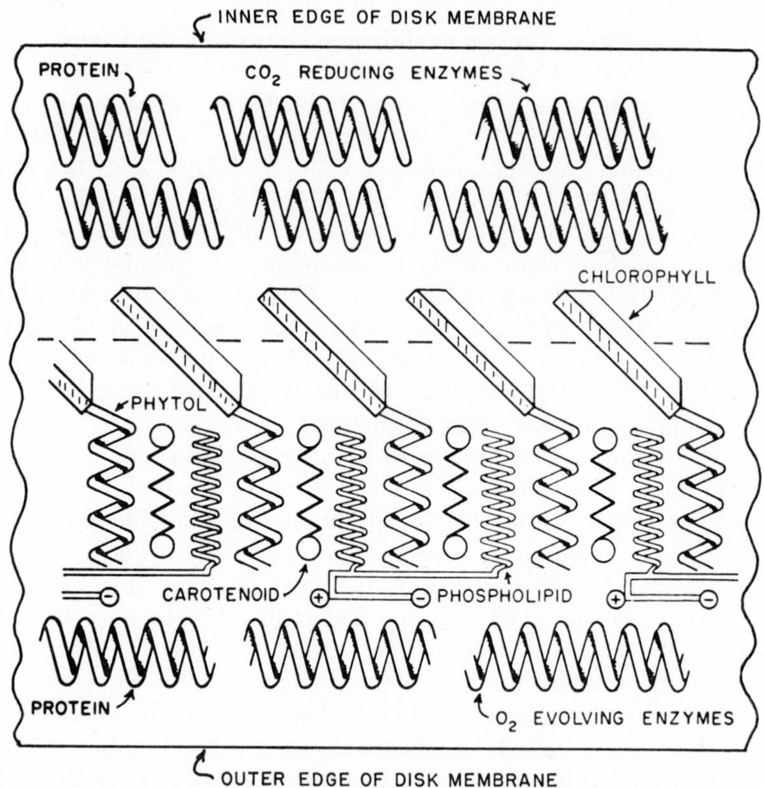

FIG. 41. Model of portion of disk membrane to show chlorophyll molecules in a fatty layer and the separation of oxidant from reductant (Calvin, 1958). (See Fig. 3.)

attain a stable event. The separation of oxidant and reductant could occur by electron transfer across a membrane via a carotenoid so that on one side (inside or outside of the disk?) O_2 would eventually be formed and on the other side PNH would be formed (Fig. 41). Present evidence suggests, however, that carotenoids are not essential for the photosynthetic steps; rather they serve for protection against photooxidation and also for transmission of light energy from carotenoid to chlorophyll as in the brown algae.

A serious effort to study chloroplast structure with the electron microscope using specific enzymes for digestion, selective solvents for proteins and lipids, and known inhibitors, may well reveal more information about this important organelle.

J. Summary

The approximate analysis of mature chloroplasts is summarized in Table II. The enzyme systems of plastids at different phases of development are summarized in Table III, heme and chlorophyll biosynthesis in Fig. 38, B; and carotenoid biosynthesis in Fig. 39.

The process of photosynthesis may be divided into two groups of reactions. The first occurs in the grana region of the chloroplast; it begins with the photochemical act and leads to the photodecomposition of water; it results in the reduction of pyridine nucleotides, in the phosphorylation of adenosine diphosphate, and in the production of O_2. The second group of reactions occurs in the stroma region of the chloroplast; it uses reduced pyridine nucleotides and adenosine triphosphate to convert CO_2 to glyceraldehyde-3-phosphate and thence to starch. These reactions are summarized in Table IV and Fig. 40. A molecular model of a portion of a disk membrane of the photosynthesizing chloroplast is presented in Fig. 41.

Acknowledgments

The author desires to acknowledge the invaluable criticisms of certain sections of this manuscript by Dr. D. von Wettstein, Dr. M. Gibbs, and Dr. D. Mauzerall.

Preparation of this manuscript was in part aided by funds from a grant of the National Institutes of Health, RG-4922.

References

Albertsson, P. A., and Leyon, H. (1954). *Exptl. Cell. Research* **7**, 288.

Algera, L., Beijer, J. J., van Iterson, W., Karstens, W. K. H., and Thung, T. H. (1947). *Biochim. et Biophys. Acta* **1**, 517.

Anderson, I. C., and Fuller, R. C. (1958). *Arch. Biochem. Biophys.* **76**, 168.

Arnason, T. J., and Walker, G. W. R. (1949). *Can. J. Research* **C27**, 172.

Arnon, D. I. (1956). *Ann. Rev. Plant Physiol.* **7**, 325.

Arnon, D. I., and Wessel, G. (1953). *Nature* **1172**, 1039.

Aronoff, S. (1957). *Botan. Rev.* **23**, 65.

Avron, M., Jagendorf, A. T., and Evans, M. (1957). *Biochim. et Biophys. Acta* **26**, 262.

Badenhuizen, N. P. (1959). *In* "Protoplasmatologia," Vol. 2,B2b δ. Springer, Berlin.

Bartels, F. (1955). *Planta* **45**, 426.

Bassham, J. A., and Calvin, M. (1957). "The Path of Carbon in Photosynthesis," 104 pp. Prentice-Hall, Englewood Cliffs, New Jersey.

Baur, E. (1909). *Z. Induktive Abstammungs- u. Vererbungslehre* **1**, 330.

Baur, E. (1930). "Einführung in die Vererbungslehre." 2 Aufl., 431 pp. Gebrüder Bornträger, Berlin.

Beauverie, J. (1938). *Rev. cytol. et cytophysiol. vegetables* **3**, 80.
Behrens, M., and Thalacker, R. (1957). *Naturwissenschaften* **44**, 621.
Benson, A. A., Wiser, W., Ferrari, R. A., and Miller, J. A. (1958). *J. Am. Chem. Soc.* 4740.
Benson, A. A., Wintermans, J. F. G. M., and Wiser, R. (1959). *Plant Physiol.* **34**, 315.
Bergeron, J. A. (1958). *Brookhaven Symposia in Biol. No.* **11**, 118.
Bergström, S., and Holman, R. T. (1948). *Advances in Enzymol.* **8**, 425.
Binder, M. (1938). *Z. Induktive Abstammungs- u. Vererbungslehre* **75**, 739.
Bishop, N. I. (1958). *Brookhaven Symposia in Biol. No.* **11**, 332.
Blass, K., Anderson, J. M., and Calvin, M. (1959). *Plant Physiol.* **34**, 329.
Bogorad, L. (1950). *Am. J. Botany* **37**, 676.
Bold, H. C. (1951). *In* "Manual of Phycology" (G. M. Smith, ed.), p. 203. Chronica Botanica, Waltham, Massachusetts.
Brachet, J. (1960). *Nature* **186**, 104.
Branch, G. E. K., and Calvin, M. (1941). "The Theory of Organic Chemistry," 523 pp. Prentice-Hall, Englewood Cliffs, New Jersey.
Brawerman, G., and Chargaff, E. (1959). *Biochim. et Biophys. Acta* **31**, 164, 172, 178.
Bünning, E. (1948). "Entwicklungs- und Bewegungsphysiologie der Pflanze." Springer, Berlin.
Butterfass, T. (1957). *Protoplasma* **48**, 368.
Calvin, M. (1956). *Proc. Intern. Congr. Biochem. 3rd Congr. Brussels 1955*, p. 211.
Calvin, M. (1958). *Brookhaven Symposia in Biol. No.* **11**, 160.
Caspari, E. (1948). *Advances in Genetics* **2**, 1.
Chiba, Y. (1951). *Cytologia (Tokyo)* **16**, 259.
Chibnall, A. C. (1939). "Protein Metabolism in the Plant," 306 pp. Yale University Press, New Haven, Connecticut.
Chmielevsky, V. (1890). *Botan. Ztg.* **48**, 773.
Claes, H. (1956). *Z. Naturforsch* **11b**, 260.
Clauss, H. (1958). *Planta* **52**, 334.
Cohen-Bazire, G., and Stanier, R. Y. (1958). *Nature* **181**, 250.
Cooper, W. D., and Loring, H. S. (1957). *J. Biol. Chem.* **228**, 813.
Correns, C. (1909). *Z. Induktive Abstammungs- u. Vererbungslehre* **1**, 291.
Crane, F. L. (1959). *Plant Physiol.* **34**, 128.
Czurda, V. (1928). *Botan. Centr. Beih.* Abt. 1, **45**, 97.
Dam, H. (1942). *Advances in Enzymol.* **2**, 285.
D'Amato, F. (1950). *Carylogia* **3**, 211.
Dangeard, P. (1947). "Cytologie végétale et cytologie générale," 611 pp. Paul Lechevalier, Paris.
Davis, B. M. (1899). *Botan. Gaz.* **28**, 89.
De Deken-Grenson, M. (1955). *Biochim. et Biophys. Acta* **17**, 35.
De Deken-Grenson, M., and Godts, A. (1960). *Exptl. Cell Research* **19**, 376.
Della Rosa, R. J., Altman, K. I., and Salomon, K. (1953). *J. Biol. Chem.* **202**, 771.
Demerec, M. (1935). *Cold Spring Harbor Symposia Quant. Biol.* **3**, 80.
de Ropp, R. S. (1948). *Nature* **162**, 459.
Doutreligne, J. (1935). *Proc. Koninkl. Akad. Wetenschap. Amsterdam.* **38**, 886.
Dutton, H. J., Manning, W. M., and Duggar, B. M. (1943). *J. Phys. Chem.* **47**, 308.
Eberhardt, F. M., and Kates, M (1957). *Can. J. Botany* **35**, 907.

Eyster, W. H. (1929). *Science* **69**, 48.
Fasse-Franzisket, U. (1955). *Protoplasma* **45**, 194.
Flax, M. H., and Himes, M. H. (1952). *Physiol. Zoöl.* **25**, 297.
Franck, J. (1957). *In* "Research in Photosynthesis" (H. Gaffron, ed.), p. 19. Interscience, New York.
Frank, S. R. (1946). *J. Gen. Physiol.* **29**, 157.
French, C. S., and Young, V. M. K. (1952). *J. Gen. Physiol.* **35**, 873.
Frenkel, A. W. (1954). *J. Am. Chem. Soc.* **76**, 5568.
Frenkel, A. W. (1958). *Brookhaven Symposia in Biol. No.* **11**, 276.
Frenkel, A. W. (1959). *Ann. Rev. Plant Physiol.* **10**, 53.
Frey-Wyssling, A. (1935). "Die Stoffausscheidung der höheren Pflanzen," 378 pp. Springer, Berlin.
Frey-Wyssling, A. (1953). "Submicroscopic Morphology of Protoplasm," 2nd ed., 411 pp. Elsevier, New York.
Frey-Wyssling, A., and Kreutzer, E. (1958a). *Planta* **51**, 104.
Frey-Wyssling, A., and Kreutzer, E. (1958b). *J. Ultrastruct. Research* **1**, 397.
Frey-Wyssling, A., and Steinmann, E. (1948). *Biochim. et Biophys. Acta* **2**, 254.
Frey-Wyssling, A., Ruch, F., and Berger, X. (1955). *Protoplasma* **45**, 97.
Fujimura, K., and Hamaguchi, Y. (1951). *Bull. Research Inst. Food Sci. Kyoto Univ. No.* **1**, 57; *Chem. Abstr.* **46**, 6169 (1952).
Fuller, R. C., and Gibbs, M. (1959). *Plant Physiol.* **34**, 324.
Galston, A. W. (1943). *Am. J. Botany* **30**, 331.
Galston, A. W. (1950). *Botan. Rev.* **16**, 361.
Garton, G. A., Goodwin, T. W., and Lijinsky, W. (1951). *Biochem. J.* **48**, 154.
Gibson, K. D. (1958). *Biochim. et Biophys. Acta* **28**, 451.
Goddard, D. R. (1945). *In* "Physical Chemistry of Cells and Tissues" (R. Höber, ed.), p. 371. Blakiston, Philadelphia, Pennsylvania.
Goedheer, J. C. (1957). "Optical properties and in vivo orientation of photosynthetic pigments," 90 pp. Janssen-Nijmegen (Proefschrift-Utrecht).
Goodwin, T. W. (1952). "The Comparative Biochemistry of the Carotenoids," 356 pp. Chapman and Hall, London.
Granick, S. (1938a). *Am. J. Botany* **25**, 558.
Granick, S. (1938b). *Am. J. Botany* **25**, 561.
Granick, S. (1950). *Harvey Lectures Ser.* **44**, 220.
Granick, S. (1951). *Ann. Rev. Plant Physiol.* **2**, 115.
Granick, S. (1958). *In* "Trace Elements" (C. A. Lamb, O. G. Bentley, and J. M. Beattie, eds.), p. 365. Academic Press, New York.
Granick, S. (1960). *Federation Proc.* **19**, 330.
Granick, S. (1961). Fifth International Congress of Biochemistry 1961 Symposium No. 6.
Granick, S., and Mauzerall, D. (1961). *In* "Metabolic Pathways" (D. M. Greenberg, ed.), Vol. 2, Chapter 20. Academic Press, New York.
Granick, S., and Porter, K. R. (1947). *Am. J. Botany* **34**, 545.
Grell, K. G., and Wohlfarth-Botterman, K. E. (1957). *Z. Zellforsch. u. mikroskop. Anat.* **47**, 7.
Gross, J. A., and Villaire, M. (1960). *Trans. Am. Microscop. Soc.* **79**, 144.
Haberlandt, G. F. J. (1914). "Physiological Plant Anatomy," 777 pp. Macmillan, New York.
Hanson, E. A., Barrien, B. S., and Wood, J. G. (1941). *Australian J. Exptl. Biol. Med. Sci.* **19**, 231.

Härtel, O., and Thaler, I. (1953). *Protoplasma* **42**, 417.
Hartree, E. F. (1955). *In* "Moderne Methoden der Pflanzenanalyse" (K. Paech and M. V. Tracey, eds.), Vol. 4, p. 197. Springer, Berlin.
Hartree, E. F. (1957). *Advances in Enzymol.* **18**, 1.
Hassid, W. Z. (1954). *In* "Chemical Pathways of Metabolism" (D. M. Greenberg, ed.) Vol. 1, Chapter 6, p. 235. Academic Press, New York.
Haupt, W. (1959). *In* "Handbuch der Pflanzen physiologie" (W. Ruhland, ed.) Vol. 17. Springer, Berlin.
Haxo, F. (1949). *Arch. Biochem.* **20**, 400.
Haxo, F., and Blinks, L. R. (1950). *J. Gen. Physiol.* **33**, 389.
Heath, O. V. S. (1959). *In* "Handbuch der Pflanzen physiologie" (W. Ruhland, ed.), Vol. 17. Springer, Berlin.
Heitz, E. (1936). *Planta* **26**, 134.
Heitz, E. (1954). *Exptl. Cell Research* **7**, 606.
Heitz, E. (1957). *Z. Naturforsch.* **12b**, 283.
Heitz, E. (1958). *Z. Naturforsch.* **13b**, 663.
Hejnowicz, Z. (1958). *Physiol. Plantarum* **11**, 878.
Highkin, H. R. (1950). *Plant Physiol.* **25**, 294.
Hill, R., and Lehmann, H. (1941). *Biochem. J.* **35**, 1190.
Hill, R., and Whittingham, C. D. (1955). "Photosynthesis," p. 165. Wiley, New York. (ed. 2, 1957, 175 p. London, Methuen).
Hodge, A. J., McLean, J. D., and Mercer, F. V. (1955). *J. Biophys. Biochem. Cytol.* **1**, 605.
Hodge, A. J., McLean, J. D., and Mercer, F. V. (1956). *J. Biophys. Biochem. Cytol.* **2**, 597.
Holden, M. (1952). *Biochem. J.* **51**, 433.
Holt, A. S., Brooks, I. A., and Arnold, W. A. (1951). *J. Gen. Physiol.* **34**, 627.
Ikeno, S. (1917). *J. Genet.* **6**, 201.
Ishikawa, M. (1918). *Ann. Botany (London)* **32**, 279.
Jagendorf, A. T. (1955). *Plant Physiol.* **30**, 138.
Jagendorf, A. T. (1959). *Federation Proc.* **18**, 974.
Jahn, T. L. (1951). *In* "Manual of Phycology" (G. M. Smith, ed.), Chapter 4, p. 69. Chronica Botanica, Waltham, Massachusetts.
James, W. O., and Leech, R. M. (1960). *Endeavor* **19**, 108.
Johnson, L. P. (1956). *Trans. Am. Microscop. Soc.* **75**, 271.
Jones, D. F. (1940). *Am. J. Botany* **27**, 149.
Jones, W. N. (1934). "Plant Chimeras and Graft Hybrids," 136 pp. Methuen, London.
Jungers, V., and Doutreligne, J. (1943). *Cellule* **49**, 407.
Kaja, H. (1955). *Protoplasma* **44**, 136.
Karrer, P., and Jucker, E. (1950). "Carotenoids," 384 pp. Elsevier, New York.
Kates, M. (1957). *Can. J. Biochem. Physiol.* **35**, 127.
Kessler, E. (1957). *In* "Research in Photosynthesis" (H. Gaffron, ed.), p. 243. Interscience, New York.
Kikuchi, G., Shemin, D., and Bachmann, B. J. (1958). *Biochim. et Biophys. Acta* **28**, 219.
Knudson, L. (1936). *Am. J. Botany* **23**, 694.
Knudson, L. (1940). *Botan. Gaz.* **101**, 721.
Kurnick, N. B., and Mirsky, A. E. (1950). *J. Gen. Physiol.* **33**, 265.
Küster, E. (1935). "Die Pflanzenzelle," 672 pp. Fischer, Jena.

Lascelles, J. (1955). *Ciba Foundation Symposium on Porphyrin Biosynthesis and Metabolism* p. 265.
Lefort, M. (1957). *Compt. rend. acad. sci.* **244**, 2957.
Lemberg, R., and Legge, J. W. (1949). "Hematin compounds and Bile Pigments," 748 pp. Interscience, New York.
Leyon, H. (1953). *Exptl. Cell Research* **4**, 371.
Leyon, H. (1954a). *Exptl. Cell Research* **6**, 497.
Leyon, H. (1954b). *Exptl. Cell Research* **7**, 265.
Leyon, H. (1954c). *Exptl. Cell Research* **7**, 609.
Leyon, H. (1956). *Svensk Kem. Tidskr.* **68**, 70.
Leyon, H., and von Wettstein, D. (1954). *Z. Naturforsch.* **9b**, 471.
Liebich, H. (1941). *Z. Botan.* **37**, 129.
Littau, V. C. (1958). *Am. J. Botany* **45**, 45.
Liverman, J. L., Johnson, M. P., and Starr, L. (1955). *Science* **121**, 440.
Loeffler, J. E. (1954-1955). *Carnegie Inst. Wash. Yearbook* **54**, 159.
Lwoff, A. (1950). *New Phytologist* **49**, 72.
McClendon, J. H. (1952). *Am. J. Botany* **39**, 275.
McClendon, J. H. (1953). *Am. J. Botany* **40**, 260.
McClendon, J. H., and Blinks, L. R. (1952). *Nature* **170**, 577.
McClintock, B. (1939). *Proc. Natl. Acad. Sci. U.S.* **25**, 405.
McDonald, M. R., and Kaufmann, B. P. (1954). *J. Histochem. and Cytochem.* **2**, 387.
Mahler, H. (1958). *In* "Trace Elements," (C. A. Lamb, O. G. Bentley, and J. M. Beattie, eds.), p. 311. Academic Press, New York.
Maly, R. (1951). *Z. Induktive Abstammungs- u. Vererbungslehre* **83**, 447.
Maly, R. (1958a). *Z. Vererbungsl.* **89**, 469.
Maly, R. (1958b). *Z. Vererbungsl.* **89**, 692.
Maly, R., and Wild, A. (1956). *Z. Induktive Abstammungs- u. Vererbungslehre* **87**, 493.
Mangenot, G. (1923). *Compt. rend. soc. biol.* **88**, 522.
Mangenot, G. (1938). *Compt. rend. acad. sci.* **206**, 364.
Mason, H. S. (1957). *Advances in Enzymol.* **19**, 79.
Mayer, H. (1930). *Planta* **11**, 294.
Mehler, A. H., and Brown, A. H. (1952). *Arch. Biochem. Biophys.* **38**, 365.
Menke, W. (1938). *Kolloid Z.* **85**, 256.
Menke, W. (1957). *Z. Naturforsch.* **12B**, 654.
Menke, W. (1960). *Z. Naturforsch.* **15B**, 800.
Menke, W., and Koydl, E. (1939). *Naturwissenchaften* **27**, 29.
Mercer, F. V., Hodge, A. J., Hope, A. B., and McLean, J. D. (1955). *Australian J. Biol. Sci.* **8**, 1.
Metzner, H. (1952). *Biol. Zentr.* **71**, 257.
Meyer, A. (1883). "Das Chlorophyllkorn in chemischer, morphologischer und biologischer Beziehung." A. Felix, Leipzig.
Meyer, A. (1917). *Ber. deut. botan. Ges.* **35**, 658.
Meyer, A. (1920). "Morphologische und physiologische Analyse der Zelle der Pflanzen und Tiere," Teil 1, 629 pp. Fischer, Jena.
Michaelis, P. (1954). *Advances in Genet.* **6**, 287.
Michaelis, P. (1957). *Protoplasma* **48**, 403.
Michaelis, P. (1958). *Planta* **51**, 600.
Miller, E. C. (1938). "Plant Physiology," 2nd ed., 1201 pp. McGraw-Hill, New York.

Möbius, M. (1920). *Ber. deut. botan. Ges.* **38**, 224.
Möbius, M. (1937). *Botan. Rev.* **3**, 351.
Mottier, D. M. (1921). *Ann. Botany (London)* **35**, 349.
Mühlethaler, K. (1955). *Protoplasma* **45**, 264.
Mühlethaler, K., and Frey-Wyssling, A. (1959). *J. Biophys. Biochem. Cytol.* **6**, 507.
Muschik, M. (1953). *Protoplasma* **42**, 43.
Nagai, S., and Ogata, E. (1952). *J. Inst. Polytech. Osaka City Univ., Ser. D* **3**, 37.
Newton, G. A., and Kamen, M. D. (1957). *Biochim. et Biophys. Acta* **25**, 462.
Nieman, R. H., and Vennesland, B. (1959). *Plant Physiol.* **34**, 255.
Niklowitz, W., and Drews, G. (1955). *Arch. Mikrobiol.* **23**, 123.
Niklowitz, W., and Drews, G. (1956). *Arch. Mikrobiol.* **24**, 134, 147.
Niklowitz, W., and Drews, G. (1957). *Arch. Mikrobiol.* **27**, 150.
Ogur, M., and Rosen, G. (1950). *Arch. Biochem.* **25**, 262.
Palade, G. E. (1952). *J. Exptl. Med.* **95**, 285.
Palade, G. E. (1953). *J. Histochem. and Cytochem.* **1**, 188.
Pardee, A. B., Schachman, H. K., and Stanier, R. Y. (1952). *Nature* **169**, 282.
Park, R. B., and Bonner, J. (1958). *J. Biol. Chem.* **233**, 340.
Perner, E. S. (1956). *Z. Naturforsch* **11b**, 567.
Pirson, A. (1955). *Ann. Rev. Plant Physiol.* **6**, 71.
Pirson, A. (1958). *In* "Trace Elements" (C. A. Lamb, O. G. Bentley, and J. M. Bettie, eds.), Chapter 5, p. 81. Academic Press, New York.
Platt, J. R. (1956). "Radiation Biology" (A. Hollaender, ed.), Vol. III, Chapter 2, p. 71. McGraw-Hill, New York.
Porter, J. W., and Lincoln, R. E. (1950). *Arch. Biochem.* **27**, 390.
Porter, H. K., and Runeckles, V. C. (1956). *Biochim. et Biophys. Acta* **20**, 100.
Pringsheim, E. G. (1941). *Biol. Revs. Cambridge Phil. Soc.* **16**, 191.
Pringsheim, E. G. (1948). *New Phytologist* **47**, 52.
Pringsheim, E. G. (1956). *Nova Acta Leopoldina* **18**, No. 125.
Pringsheim, E. G., and Pringsheim, O. (1952). *New Phytologist* **51**, 65.
Procházka, Z., Sanda, V., and Sorm, F. (1957). *Collection Czechoslov. Chem. Communs.* **22**, 654.
Provasoli, L., Hutner, S. H., and Schatz, A. (1948). *Proc. Soc. Exptl. Biol. Med.* **69**, 279.
Rabinowitch, E. I. (1945, 1951, 1957). "Photosynthesis and Related Processes," Vols. 1, 2A, 2B. Interscience, New York.
Racker, E. (1954). *Advances in Enzymol.* **15**, 141.
Reinhard, H. (1933). *Protoplasma* **19**, 541.
Renner, O. (1936). *Flora (Jena)* **30**, 218.
Rhoades, M. M. (1943). *Proc. Natl. Acad. Sci. U.S.* **29**, 327.
Rhoades, M. M. (1946). *Cold Spring Harbor Symposia Quant. Biol.* **11**, 202.
Rhoades, M. M., and Carvalho, A. (1944). *Bull. Torrey Botan. Club* **71**, 335.
Robbins, W. J., Hervey, A., and Stebbins, M. E. (1953). *Ann. N.Y. Acad. Sci.* **56**, 818.
Rosenberg, L. L., Capindale, J. B., and Whatley, F. R. (1958). *Nature* **181**, 632.
Rouiller, C., and Fauré-Fremiet, E. (1958). *Exptl. Cell Research* **14**, 47.
Rubin, B. A., and Ladygina, M. E. (1956). *Biokhimiya* **21**, 347.
Ruch, F. (1957). *Exptl. Cell Research Suppl.* **4**, 58.
Sachs, J. (1887). "Lectures on the Physiology of Plants." Oxford Univ. Press, London and New York.

Sager, R., and Palade, G. E. (1954). *Exptl. Cell Research* **7**, 584.
Scarth, G. W., and Shaw, M. (1951). *Plant Physiol.* **26**, 207.
Scherrer, A. (1914). *Flora (Jena)* **107**, 1.
Schiller, J. (1923). *Österr. Botan. Z.* **72**, 236.
Schimper, A. F. W. (1883). *Botan. Ztg.* **41**, 105, 121, 137, 153.
Schmitz, F. (1884). *Jahrb. wiss. Botan.* **15**, 2.
Schürhoff, P. N. (1924). *In* "Handbuch der Pflanzenanatomie" (K. Linsbauer, ed.), Bd. 1, Liefg. 10, 225 pp. Gebrüder Borntrager, Berlin.
Schwartz, D. (1949). *Botan. Gaz.* **111**, 123.
Schwemmle, J. (1938). *Z. Induktive Abstammungs- u. Vererbungslehre* **75**, 358.
Seybold, A. (1933). *Planta* **21**, 251.
Sharp, L. W. (1934). "An Introduction to Cytology," 3rd ed., 567 pp. McGraw-Hill, New York.
Shemin, D. (1956). *Harvey Lectures, Ser.* **50**, 258.
Shibata, K. (1955-1956). *Carnegie Inst. Wash. Yearbook* **55**, 248.
Simonsen, J. L. (1931-32, 1952). "The Terpenes," Vols. 1, 2; 2nd ed., Vol. 3. Cambridge Univ. Press, London and New York.
Sirks, M. J. (1938) *Botan. Rev.* **4**, 113.
Sissakian, N. M. (1958). *Advances in Enzymol.* **20**, 201.
Sitte, P. (1958). *Protoplasma* **49**, 447.
Smillie, R., and Fuller, R. C. (1960). *Federation Proc.* **19**, 328.
Smith, G. M. (1955). "Cryptogamic Botany" Vol. I, 2nd ed. McGraw-Hill, New York.
Smith, J. H. C., and Benitez, A. (1955). *In* "Moderne Methoden der Pflanzenanalyse" (K. Paech and M. V. Tracey, eds.), Vol. 4, Springer, Berlin.
Smith, J. H. C., and Koski, V. M. (1947-1948). *Carnegie Inst. Wash. Yearbook* **47**, 93.
Smith, J. H. C., and Young, V. M. K. (1956). *In* "Radiation Biology" (A. Hollaender, ed.), Vol. 3, Chapter 7, p. 393. McGraw-Hill, New York.
Smith, J. H. C., Kupke, D. W., and Giese, A. T. (1955-1956). *Carnegie Inst. Wash. Yearbook* **55**, 243.
Smith, J. H. C., Durham, L. J., and Wurster, C. F. (1956-1957). *Carnegie Inst. Wash. Yearbook* **56**, 279.
Smith, J. H. C., Durham, L. J., and Wurster, C. F. (1959). *Plant Physiol.* **34**, 340.
Sorokin, C. (1957). *Physiol. Plantarum* **10**, 659.
Spiekermann, R. (1957). *Protoplasma* **48**, 303.
Stanier, R. (1960). *Harvey Lectures, Ser.* **54**, 219.
Steinmann, E. (1952). *Exptl. Cell Research* **3**, 367.
Steinmann, E., and Sjöstrand, F. S. (1955). *Exptl. Cell Research* **8**, 15.
Stephenson, M. L., and Zamecnik, P. C. (1956). *Federation Proc.* **15**, 362.
Stewart, W. N. (1948). *Botan. Gaz.* **110**, 281.
Stocking, C. R. (1952). *Am. J. Botany* **39**, 283.
Stocking, C. R. (1959). *Plant Physiol.* **34**, 56.
Strain, H. H. (1938). Leaf Xanthophyll. *Carnegie Inst. Wash. Publ. No.* **490**, 147 pp.
Strain, H. H. (1949). *In* "Photosynthesis in Plants" (J. Franck and W. E. Loomis, eds.), p. 133. Iowa State College Press, Ames, Iowa.
Strain, H. H. (1951). *In* "Manual of Phycology" (C. M. Smith, ed.), Chapter 13, p. 243. Chronica Botanica, Waltham, Massachusetts.
Straus, W. (1954). *Exptl. Cell Research* **6**, 392.
Strugger, S. (1936). *Flora (Jena)* **31**, 113.
Strugger, S., and Losada-Villasante, M. (1955). *Protoplasma* **45**, 540.

Strugger, S., and Perner, E. (1956). *Protoplasma* **46**, 711.
Stubbe, W., and von Wettstein, D. (1955). *Protoplasma* **45**, 241.
Szejnman, A. (1933). *Acta Soc. Botan. Polon.* **10**, 331.
Tabentskii, A. A. (1953). *Izvest. Akad. Nauk. (S.S.S.R.) Otdel. Khim. Nauk Ser. Biol.* **1**, 71.
Tamiya, H., Iwamura, T., Shibata, K., Hase, E., and Nihei, T. (1953). *Biochim. et Biophys. Acta* **12**, 23.
Tavormina, P. A., Gibbs, M. H., and Huff, J. W. (1956). *J. Am. Chem. Soc.* **78**, 4498.
Teale, F. W. J. (1958). *Nature* **181**, 415.
Thomas, J. B. (1955). *Progr. in Biophys. and Biophys. Chem.* **5**, 109.
Thomas, J. B., Blaauw, O. H., and Duysens, L. N. M. (1953). *Biochim. et Biophys. Acta* **10**, 230.
Thomas, J. B., Daemen, F. J. M., and Schaap, A. (1958). *J. chim. phys.* **55**, 934.
Tolbert, N. E. (1957). *In* "Research in Photosynthesis" (H. Gaffron, ed.), p. 224. Interscience, New York.
Trebst, A. V., Tsujimoto, H. Y., and Arnon, D. I. (1958). *Nature* **182**, 351.
Tulecke, W. R. (1953). *Science* **117**, 599.
Vernon, L. P., and Kamen, M. D. (1954). *Arch. Biochem. and Biophys.* **51**, 122.
Vishniac, W. (1955). *Ann. Rev. Plant Physiol.* **6**, 115.
Vishniac, W., Horecker, B. L., and Ochoa, S. (1957). *Advances in Enzymol.* **19**, 1.
Voerkel, S. H. (1933). *Planta* **21**, 156.
Van Wisselingh, C. (1920). *Z. Induktive Abstammungs- u. Vererbungslehre* **22**, 65.
von Euler, H., Bracco, M., and Heller, L. (1948). *Compt. rend. acad. sci.* **227**, 16.
von Wettstein, D. (1954). *Z. Naturforsch.* **9b**, 476.
von Wettstein, D. (1957a). *Hereditas* **43**, 303.
von Wettstein, D. (1957b). *Exptl. Cell Research* **12**, 427.
von Wettstein, D. (1958). *Brookhaven Symposia in Biol. No.* **11**, 138.
von Wettstein, D. (1959). *In* "Society for the Study of Development and Growth," Symposium No. 16 (D. Rudnick, ed.), p. 123. Ronald Press, New York.
Warburg, O. H. (1949). "Heavy Metal Prosthetic Groups and Enzyme Action," 230 pp. Oxford Univ. Press, London and New York.
Weier, T. E. (1938a). *Botan. Rev.* **4**, 497.
Weier, T. E. (1938b). *Protoplasma* **31**, 346.
Weier, T. E., and Stocking, C. R. (1952a). *Botan. Rev.* **18**, 14.
Weier, T. E., and Stocking, C. R. (1952b). *Am. J. Botany* **39**, 720.
Weissbach, A., Horecker, B. L., and Hurwitz, J. (1956). *J. Biol. Chem.* **218**, 795.
Whatley, F. R., Allen, M. B., Rosenberg, L. L., Capindale, J. B., and Arnon, D. I. (1956). *Biochim. et Biophys. Acta* **20**, 462.
Willstätter, R., and Stoll, A. (1918). "Untersuchungen über die Assimilation der Kohlensäure, 448 pp. Springer, Berlin.
Winge, O. (1919). *Compt. rend. trav. lab. Carlsberg* **14**, No. 3.
Wintermans, J. F. G. M. (1954). *Koninkl. Ned. Akad. Wetenschap. Proc. Ser. C* **57**, 574.
Wolken, J. J., and Palade, G. E. (1953). *Ann. N.Y. Acad. Sci.* **56**, 873.
Wolken, J. J., and Schwertz, F. A. (1953). *J. Gen. Physiol.* **37**, 111.
Woods, M. W., and duBuy, H. G. (1951). *J. Natl. Cancer Inst.* **11**, 1105.
Woods, P. S., and Pollister, A. W. (1955). *Stain Technol.* **30**, 123.
Yin, H. C., and Sun, C. N. (1949). *Plant Physiol.* **24**, 103.

CHAPTER 8

Golgi Apparatus and Secretion Granules

By A. J. DALTON

I. Discovery and Early Background

From the first revelation of the "internal reticular apparatus" by Golgi (1898), speculation regarding its significance and even reality existed and has continued up to the present day. The earlier studies were primarily descriptive in character, in many cases being concerned with attempts to demonstrate the ubiquity of the apparatus. Thus, from the beginning and on into the early 1920's, the "apparatus" was described in a variety of animal cells. The common factor in each of these pioneering investigations was the search for an area of cytoplasm which contained reduced silver (or osmium). Its pleomorphism plus the capriciousness of the early impregnation techniques combined to cause uncertainty and skepticism as to the basic importance and significance of the Golgi apparatus. The later modifications of the silver impregnation method by Ramon y Cajal (1914) and da Fano (1920), of the Kopsch OsO_4 impregnation method by Ludford (1925a), and the development of the Kolatchew-Nassonov method (Nassonov, 1923) removed some of the technical difficulties, but the problem of variability in form remained. These differences applied not only to details of structure, but to location within the cell, amount of reduced metal (size), and gross differences in form; e.g., vertebrates versus invertebrates. The original description by Golgi (Fig. 1) of a concentric network about the nucleus of Purkinje cells of the cerebellar cortex was found to be by no means characteristic of vertebrate cells generally, and furthermore the spherical or concave disk-shaped form of the dictyosomes of invertebrates appeared to have nothing in common structurally with the vertebrate "apparatus." In spite of these difficulties and in spite of the fact that observers occasionally described structures under the name "Golgi apparatus" which in retrospect were incorrectly identified, one conclusion can be drawn from these

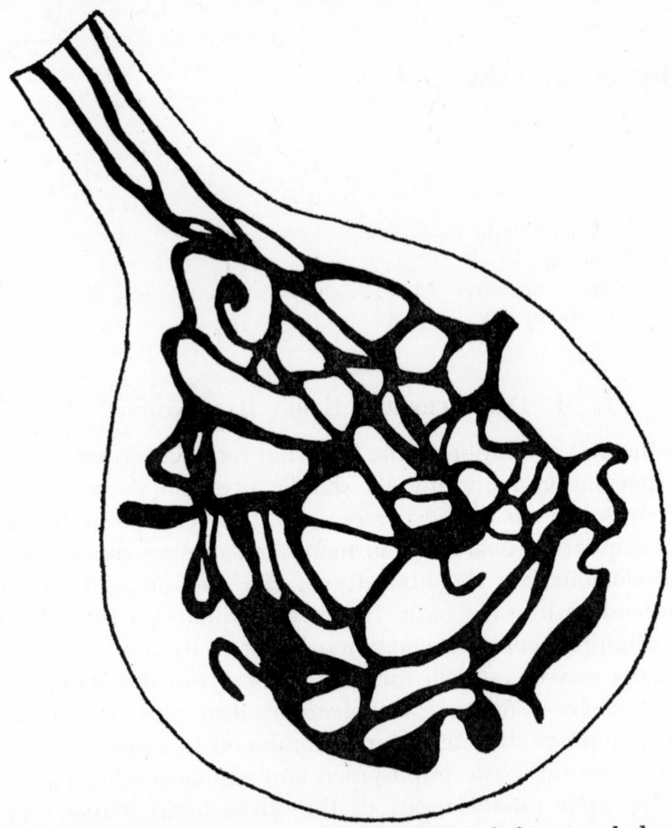

FIG. 1. Diagram of the Golgi apparatus—after Golgi's original drawing of a Purkinje cell of the cerebellum of the barn owl.

studies, and that is that in the overwhelming majority of cell types studied, vertebrates and invertebrates including protozoa, a cytoplasmic area capable of reducing silver nitrate and/or osmium tetroxide could be demonstrated. What this area represented in terms of its relationship to other cell components, what its significance might be in regard to the economy of the cell as a whole, and whether it existed as a distinct entity in the living cell had not been determined.

Beginning in the early 1920's and continuing on into the 1930's, however, many descriptive and experimental studies were carried out with the aim of gaining some insight into the possible function of the Golgi apparatus. As a result of the early efforts of Gatenby (1917), Hirschler (1918), and others it became clear that although form and size difference existed, there was good reason to consider the Golgi apparatus of vertebrates and the dictyosomes of invertebrate cells as basically similar structures and that any resolution of the problem of function would have to take this evidence into consideration.

Nassonov (1924b) called attention to the fact that the area immediately surrounding the contractile vacuole of certain protozoa was capable of reducing osmium tetroxide in a manner similar to the Golgi apparatus of vertebrate cells, and he suggested that a homology exists between these two structures. Tuzet (1932), Duboscq and Tuzet (1937), Gatenby and Singh (1938), Jepps (1947), and others later extended this evidence. Implied in this homology is the likelihood that the Golgi apparatus is involved in the control of water balance.

It was Bowen's brilliant work (Bowen, 1929) demonstrating a clear-cut relationship between the Golgi apparatus and the developing acrosome of spermatids which drew attention from Nassonov's suggestion toward the possibility of the direct intervention of the Golgi apparatus in the secretory functions of the cell. Actually these two concepts are not necessarily mutually incompatible, but from the time of Bowen's publications on, the major effort was expended in attempting to determine whether or not a correlation between the apparatus and secretory activity existed. Beams (1930), Hirsch (1939), Ludford (1925b), and Nassonov (1924a), as well as many other investigators, presented evidence supporting this correlation. Beams and King (1933), for example, showed a reversal in position of the apparatus at the onset of secretory function in ameloblasts. Much of the evidence during this interval involved the demonstration of changes in amount and distribution of the apparatus in relation to changes in secretory activity of glandular cells. As a result of all this evidence, there developed a more or less generally accepted thesis that the Golgi apparatus was involved in some direct way in secretory activity.

II. The Vacuome

The interest in and effective exploitation of this area of cytologic investigation was hampered somewhat by the tendency of investigators to introduce new terms, either to describe processes of change in form of the apparatus in any given cell or to call attention to differences in its form in different cells (Hirsch, 1939). More important, however, was the fact that much of the evidence depended upon observation of a cell component, parts of which, we now know, are so minute as to be well beyond the capabilities of the light microscope to resolve. It was this fact that made possible the acceptance by some of the views of Walker and Allen (1927) that the Golgi apparatus in fixed material is an artifact and also gave credence to the position held by many that inferences concerning cellular activity based on the study of fixed tissue were of no value. It probably also contributed indirectly to the enunciation by Parat and Painlevé (1924a, b, c) of the vacuome hypothesis. It was Parat's view that vacuoles stainable supravitally with neutral red were the living representation of the Golgi reticulum of classic cytology. Parat did not imply in any of his writings that the vacuome (Golgi apparatus) was of no significance. Actually he envisaged it as the center of synthetic activity. The practical result of the acceptance of his views, was, however, to discredit to an unwarranted degree the observations made and conclusions drawn from studies using classic impregnation methods. The effect of accepting Parat's views can be estimated by comparing the general tenor of the review by Kirkman and Severinghaus (1938a, b) with that of Hibbard (1945). The first of these shows little evidence of being affected by Parat's views whereas the latter is considerably influenced by them. Parat later modified his position somewhat by introducing the concept of the "chondriome actif." This modification became necessary because the "vacuome" alone could not account for all structures, particularly the dictyosomes seen in the Golgi zone in living as well as in fixed material (Gatenby, 1931).

III. The Golgi Controversy

The controversy involving the question of the oneness or separateness of the vacuome and the Golgi apparatus continued with the publication of papers by Worley (1943a, b), Baker (1944, 1949), and Palade and Claude (1949a, b). While varying in detail, the conclusions drawn from the results of this series of studies constituted a return to the basic concepts of Parat. In common, they considered droplets stainable with neutral red and/or methylene blue as the precursor in the living cell of the reticulum of the Golgi apparatus in fixed cells. They differed primarily in regard to the importance to be given to these droplets in the economy of the cell and as to the mechanism by which they were transformed into the Golgi reticulum. More recent work of Baker and coworkers tended to emphasize the variability in chemical constitution of the "lipochondria" (Baker, 1957), although additional evidence was presented suggesting that the Golgi reticulum was produced by the piling up of reduced osmium on the surfaces of mitochondria (Thomas, 1948). At some variance with this position was that of Hirsch (1939), who carried out careful studies on living and fixed cells and developed his concept of Golgi "presubstance."

The majority of investigators who considered the Golgi reticulum an artifact accepted the existence of the dictyosomes of germ cells as real structures since they are visible in the living state (Brice *et al.*, 1946; Oettlé, 1948; Gresson, 1950). Dictyosomes, however, were considered to be structures distinct from the Golgi apparatus, partly because they are readily visible in fresh material, partly because they are stainable with a variety of dyes in fixed tissue, and partly because spermatocytes may be considered as a very special cell type.

Those who continued to support the concepts of Parat ignored or disregarded certain facts. These were the evidence that in somatic cells certain vacuoles or canals distinct from lipid droplets (lipochondria) and considered to be associated with the Golgi reticulum of impregnated material had been observed consistently and over a period of years by very competent observers (Bensley, 1911; Beams, 1930; O'Leary, 1930; Moussa, 1952). In addition there was the frequently repeated demonstration that lipid droplets, mitochondria, and the Golgi complex could be visualized side by side in the same cell (Beams, 1930, 1931; Dalton and Felix, 1953; Lacy and Challice, 1957). Furthermore, the Golgi apparatus was demonstrated in frozen-dried material and without impregnation as distinct from lipid droplets (Gersh, 1949). For a concise critique of the status of the controversy before the advent of electron microscopy see Bensley (1951).

IV. Electron Microscopy

As indicated earlier, this controversy could not be completely resolved to the satisfaction of all parties involved because of the inadequacy of instrumentation. The advent of the electron microscope and the development of technical methods that made possible the exploitation of its high resolving power changed all this. It was soon shown that the complex identified as the Golgi apparatus in fresh material and in fixed, impregnated cells possessed a characteristic fine structure when viewed with the electron microscope (Dalton, and Felix, 1954). In highly differentiated, functional cells the complex was shown to consist of three components: first a group of large vacuoles, which are considered to be responsible for the so-called "negative image" of the Golgi apparatus readily visualized in ordinary histologic preparations in principal cells of the small intestine, proximal tubule cells of the kidney, cells of the islets of Langerhans, etc.; second, a system of flattened sacs that marginate the vacuoles and on dilation may become vacuoles; third, clusters of small vesicles derived by budding from the periphery of the flattened sacs (Fig. 2). Examination of a series of impregnated (postosmicated) cell types indicated that all three components may react to reduce osmium tetroxide and thus contribute to the form of the classic Golgi reticulum (Dalton and Felix, 1956). However, it is felt that osmium tetroxide reduced in relation to the flattened sacs is primarily responsible for the development of the reticular and lamellar pattern seen with the light microscope (Fig. 3) (Pollister and Pollister, 1957). The fact that in anaplastic tumor cells the Golgi complex may be reduced to a small cluster of vesicles and only occasional flattened sacs (Howatson and Ham, 1955) is probably correlated with the frequent failure to demonstrate a Golgi reticulum in such cells with impregnation techniques. Actually in some neoplastic cells the Golgi complex may be highly developed (Haguenau and Bernhard, 1955), but in nonfunctional cells, in many normal and neoplastic cells in tissue culture, and in ascites tumor cells, the writer has found the complex to be generally poorly developed (Fig. 4).

In addition to the delineation of the fine structure of the Golgi complex, observations with the electron microscope have removed some of the confusion that would otherwise develop in attempting to relate the work of Baker and co-workers to earlier studies on the Golgi apparatus. This recent work demonstrated considerable variation in the types of lipid present in the Golgi zone of certain cells, particularly in principal cells of the small intestine, and lead to the suggestion that it is inappropriate to give the name Golgi apparatus to a component which is so variable in chemical composition from cell to cell and from time to time (Baker, 1957). Electron microscopic analysis has shown, in principal cells of the small intestine at least, that Baker and his group determined the types of lipid absorbed into the vacuoles of the Golgi complex. Figure 5 shows the fine structure of the Golgi complex of a principal cell of the

---

FIG. 2. An electron micrograph of a portion of the cytoplasm of two adjacent epithelial cells of the epididymis of the mouse. Plasma membranes (*PM*) of the two cells extend vertically in close proximity upward along the mid-line of the figure. Adjacent to them and also to the extreme right and left of the figure are cisternae of the ergastoplasm (basophilic substance) (*E*). Two rows of vertically oriented Golgi membranes (*GM*) may be seen to dilate into Golgi vacuoles (*GV*) at various points (arrows). Clusters of Golgi vesicles (*V*) are also evident. In this and subsequent figures, the line at the lower margin represents 1 μ. Chrome-osmic fixation, pH 7.4. Magnification: approximately × 57,000.

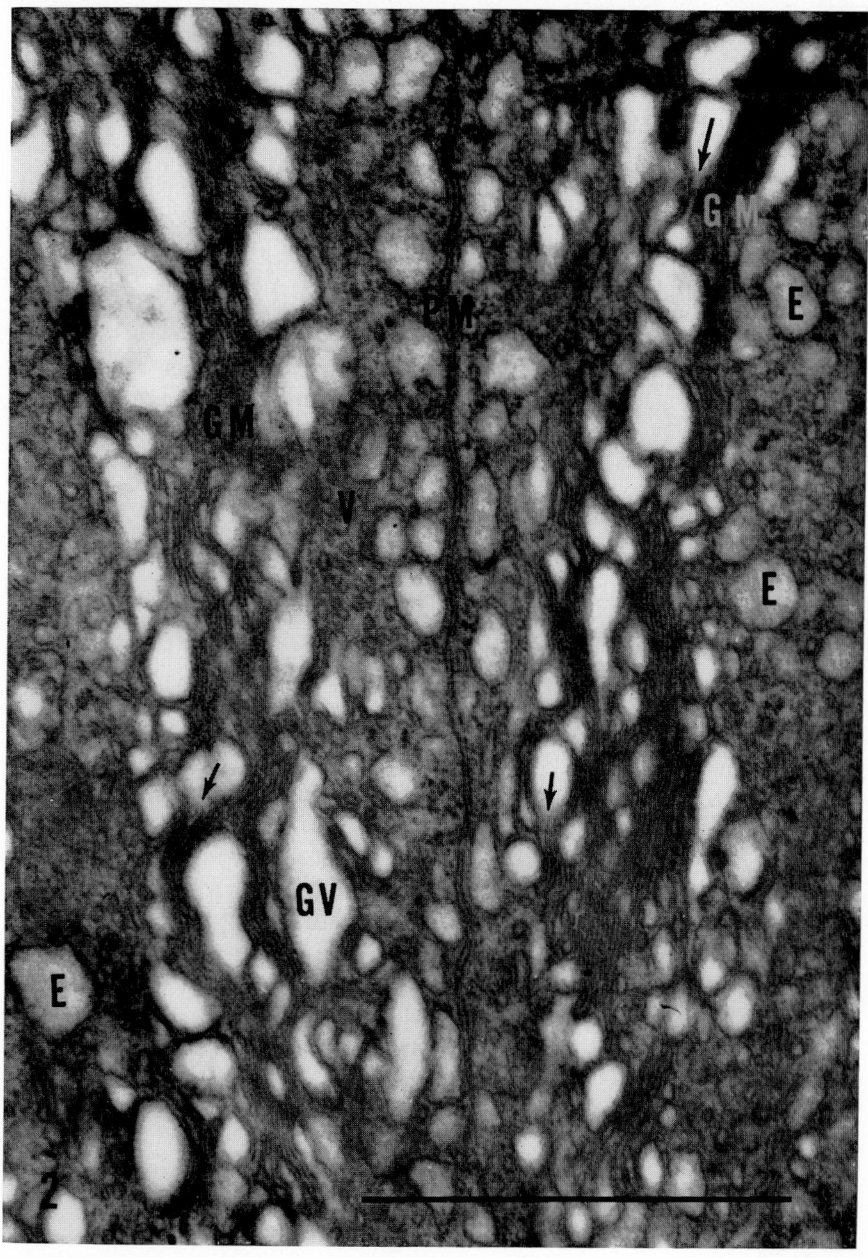
GM
E
PM
GM
V
E
GV
E
2

small intestine of the mouse after a 24-hour fast, and Figs. 6 and 7 show the absorbed lipid present in the vacuoles of the Golgi complex of similar cells 40 minutes after feeding corn oil. In Fig. 5 the contents of the vacuoles of the Golgi complex is either electron lucent or has been removed during processing; in Figs. 6 and 7 the absorbed lipid has been taken into the vacuoles, where it appears to have been emulsified into

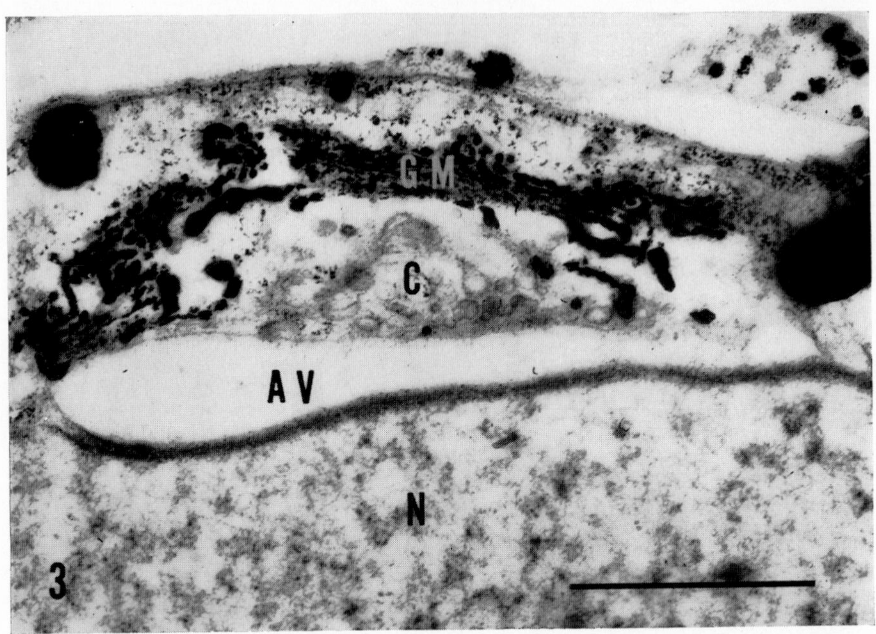

FIG. 3. An electron micrograph of the dictyosome (Golgi body) of a mouse spermatid showing reduced OsO_4 associated with the piled-up layers of flattened sacs (*GM*). The metal appears to be deposited primarily within the sacs, not between them. The chromophobic component (*C*) is marginated by the flattened sacs above and the acrosome vesicle (*AV*) below. The nucleus of the spermatid (*N*) is under the acrosome vesicle. Veronal-acetate-buffered OsO_4 fixation (pH 7.2) followed by a wash in running tap water for one-half hour followed by immersion in 2% OsO_4 at 37°C. for 44 hours. Magnification: approximately × 32,000.

smaller droplets. After an undetermined sojourn in these vacuoles the lipid is released through the plasma membranes at the lateral and proximal boundaries of the cells. These results, together with those of Palay and Karlin (1956) substantiate the views of Krehl (1890) and Weiner (1926), who much earlier indicated that the area of the cytoplasm in which the Golgi complex is located is directly involved in the storage of lipid during and after its absorption.

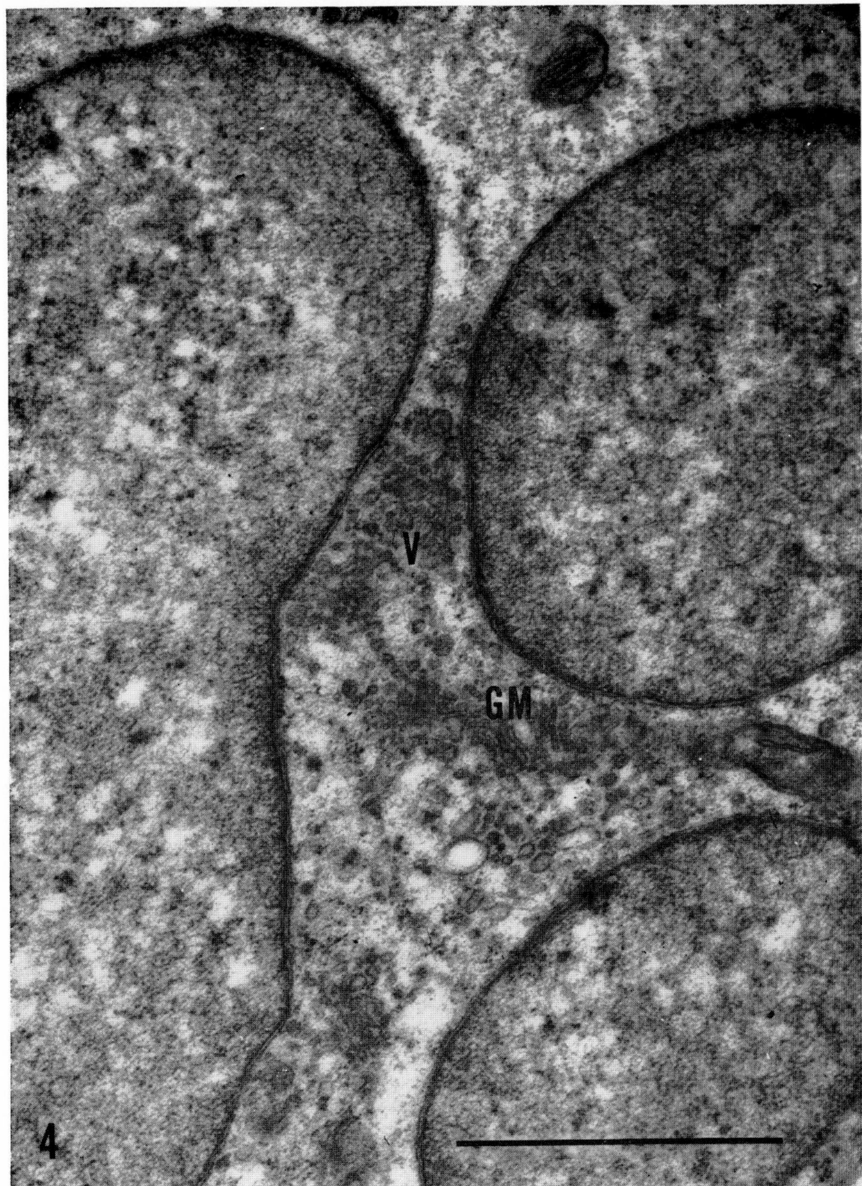

FIG. 4. An electron micrograph of the Golgi zone of a lymphoma cell grown in the peritoneal fluid of a mouse. The Golgi complex in this instance consists of but a few flattened sacs (*GM*) and a cluster of vesicles (*V*) situated in the "hoff" of the lobulated nucleus. Chrome-osmic fixation (pH 7.4). Magnification: approximately $\times$ 43,000.

Chou and Meek (1958) have reported electron microscopic observations on the neurons of the snail *Helix aspersa* which they and Baker (1959) have interpreted as indicating that phospholipid droplets of these cells are transformed into the Golgi complex during fixation. Similar studies on *Helix pomatia* neurons have not confirmed these findings (Dalton, 1960).

Other electron microscopic studies have confirmed and extended earlier work which at the time was interpreted as demonstrating the intervention of the Golgi complex in secretory activity. Burgos and Fawcett (1955) confirmed in excellent detail the observations of Bowen that the acrosome is formed in close association with the Golgi complex of developing sperm. Ferreira (1957) has shown that secretory granules of embryonic beta cells of the islets of Langerhans originate within the Golgi complex of these cells. Haguenau and Bernhard (1955) presented similar evidence in the case of cells of adenomas of the hypophysis. Sjöstrand and Hanzon (1954) and Farquhar and Wellings (1957) interpreted their results as indicating a relationship between the Golgi complex and the formation of zymogen granules in exocrine cells of the pancreas. Most recently, Palay (1958) has demonstrated a close relationship between the Golgi complex and the formation of lipid in cells of the sebaceous gland. On the other hand, Weiss (1953) in the rat and Palade (1956) in the guinea pig have shown that zymogen granule formation occurs within the membrane system of the ergastoplasm (basophilic substance) of exocrine cells of the pancreas. Also Hendler *et al.* (1957) have shown that the ergastoplasm, not the Golgi complex, is directly involved in the synthesis of precursors of albumin in the oviducal glands of the hen.

It would appear then that the Golgi complex is involved in storage and possibly modification of some substances, particularly lipids following their absorption. However, whether or not it is directly concerned in the synthesis of secretory products remains uncertain, since, as in-

---

FIG. 5. An electron micrograph of the Golgi zone of a principal cell of the duodenum of a mouse sacrificed at the end of a 24-hour fast. The three elements of the Golgi component, flattened sacs (*GM*), vacuoles (*GV*) and vesicles (*V*), are evident to the left of the nucleus (*N*). The vacuoles are electron lucent. Veronal-acetate buffered OsO_4 fixation (pH 7.2). Magnification: approximately × 25,000.

FIG. 6. An electron micrograph of a portion of a cell similar to that shown in Fig. 5 but from a mouse fed corn oil 40 minutes previously. Golgi membranes (*GM*) and vesicles (*V*) are evident above and to the left of the nucleus (*N*). The Golgi vacuoles (*GV*) contain many small lipid droplets. Veronal-acetate buffered OsO_4 fixation (pH 7.2). Magnification: approximately × 25,000.

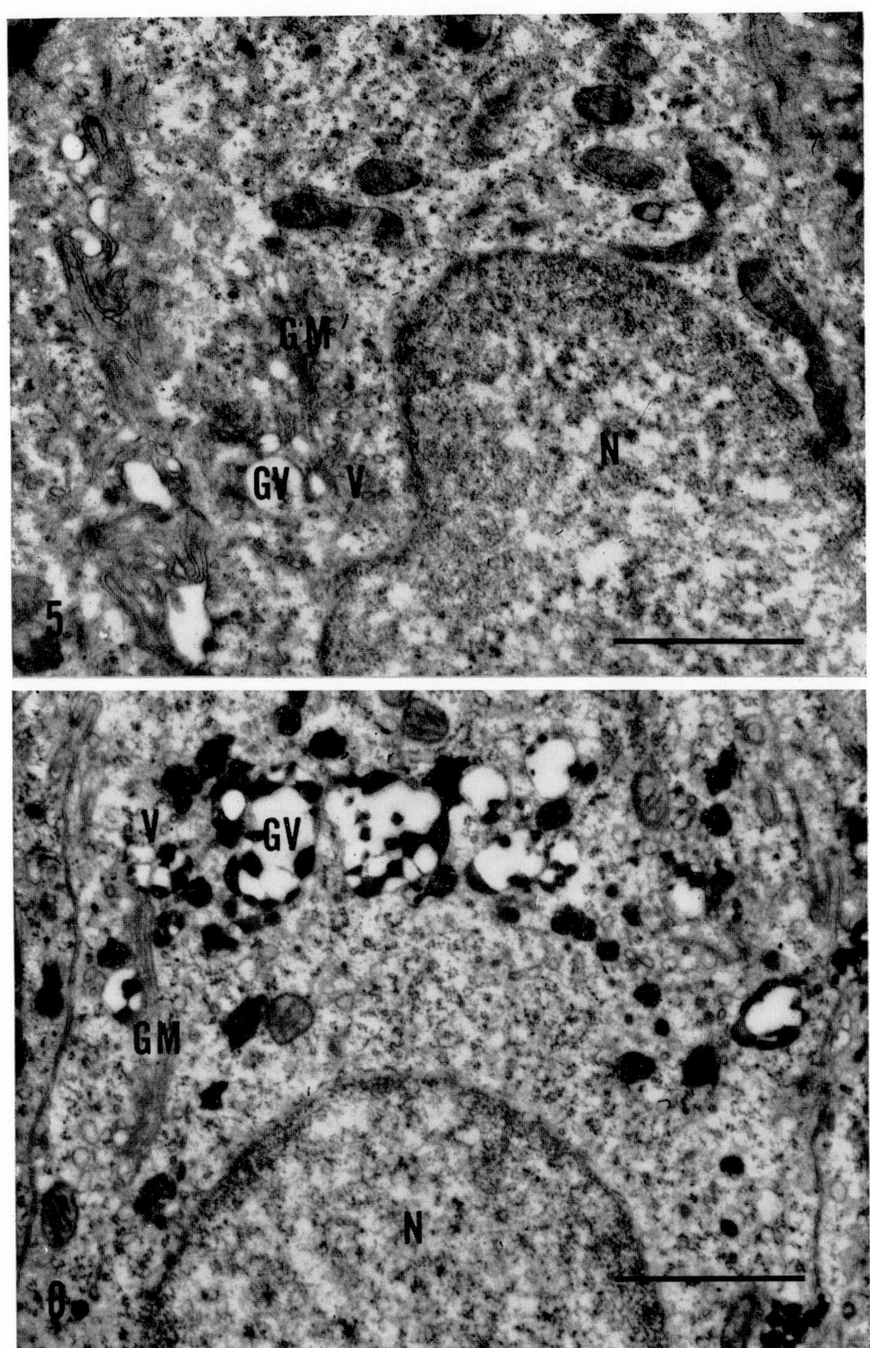
GM
GV
V
N
5
V
GV
GM
N
6

dicated above, in some instances at least it definitely is not. The possibility must not be overlooked that Nassonov's original suggestion of a homology between the Golgi complex and the contractile vacuoles of Protozoa and Porifera has some basis in fact (Gatenby *et al.*, 1955). If further study confirms this hypothesis, then the possibility that the function of the Golgi complex is concerned with the control of cellular water balance must receive serious consideration. The presence of secretory products within the vesicles of the Golgi complex of gland cells is not necessarily evidence against such a hypothesis. In fact the view that such products, synthesized elsewhere, are matured and concentrated within the Golgi complex by the removal of water to form definitive secretory droplets would be but a slight modification or amplification of the views of Bowen (1929) and Palay (1958).

In other studies, analysis of the fine structure of the dictyosomes of invertebrate spermatids amply supports the early views of Gatenby (1917) and Hirschler (1918) by showing a clear homology between them and the vertebrate Golgi apparatus (Grassé *et al.*, 1956; Dalton and Felix, 1956; and Beams *et al.*, 1957). Examination of electron micrographs of this material (Fig. 8) reveals the piling up, layer on layer, of flattened sacs which *in toto* would become readily visible with phase microscopy while the relatively few sacs present in the vertebrate complex would make such observations difficult. Membrane systems similar to those characteristic of animal cells have recently been demonstrated in plant cells (Buvat, 1957), emphasizing the ubiquity of the Golgi complex.

Progress has been made in the isolation of the Golgi complex by the technique of homogenization and ultracentrifugation using a layering technique controlled by inspection of aliquots with the electron microscope. The earliest report on this subject demonstrated a high concentration of phospholipid, ribonucleic acid, and alkaline and acid phosphatase in the layer showing a concentration of Golgi bodies (Schneider and Kuff, 1954). Refinements in technique, with the layering of the homogenate at the bottom rather than at the top of the layered density gradient,

---

FIG. 7. An electron micrograph of a portion of a principal cell from the same duodenum as that from which Fig. 6 was taken. In this case the tissue was secondarily treated with OsO_4. The Golgi complex is situated between the nucleus (*N*) and the lateral cell borders (*PM*). The Golgi vacuoles contain dense particles of reduced OsO_4 in addition to lipid droplets (*L*), which on the average are considerably smaller than those present in the more distal cytoplasm. Mitochondria (*M*) are numerous but poorly defined. Veronal-acetate buffered OsO_4 fixation followed by a one-half hour wash in running tap water followed by immersion in 2% OsO_4 (pH 6.5) at 37°C. for 40 hours. Magnification: approximately × 13,500.

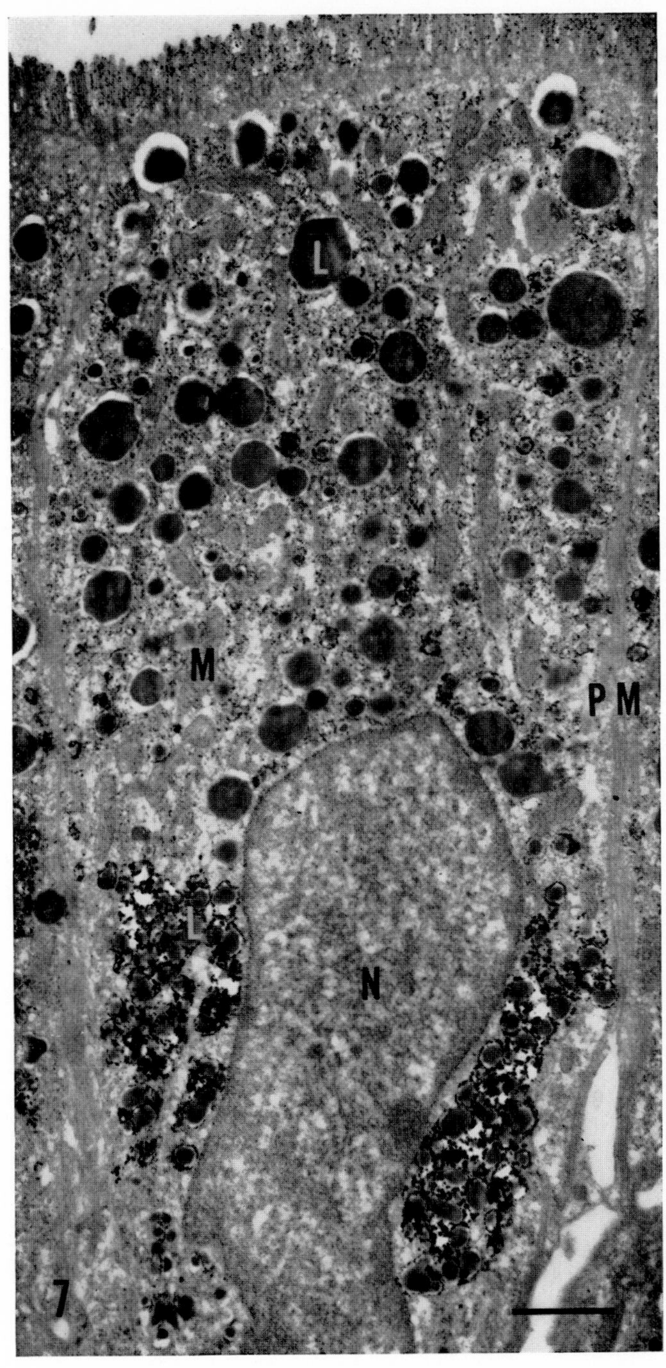
L
M
PM
L
N
7

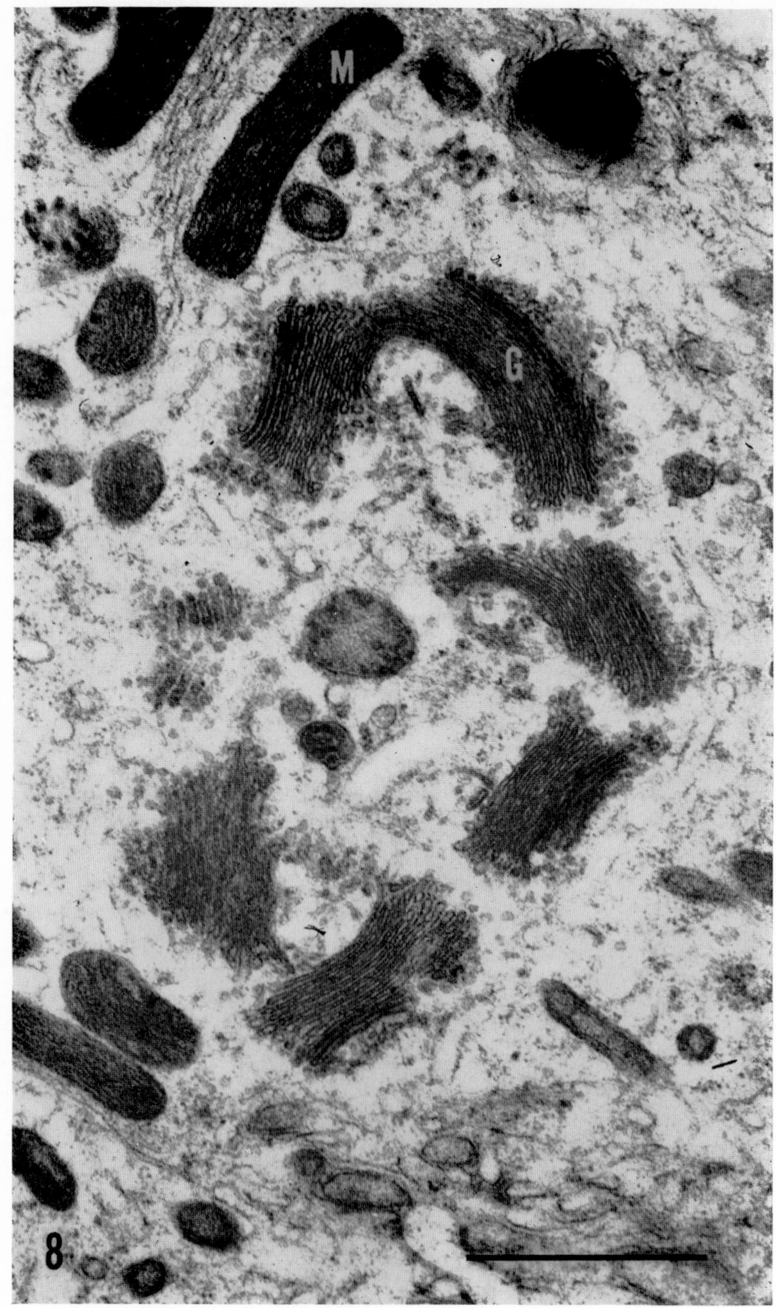
M
G
8

resulted in the isolation of what appeared to be the morphologically intact Golgi component relatively free of contaminating elements (Fig. 9). In these more recent studies (Kuff and Dalton, 1960) an almost complete absence of ribonucleic acid was found together with a great reduction in the concentration of alkaline phosphatase but with high specific concentrations of acid phosphatase and phospholipid.

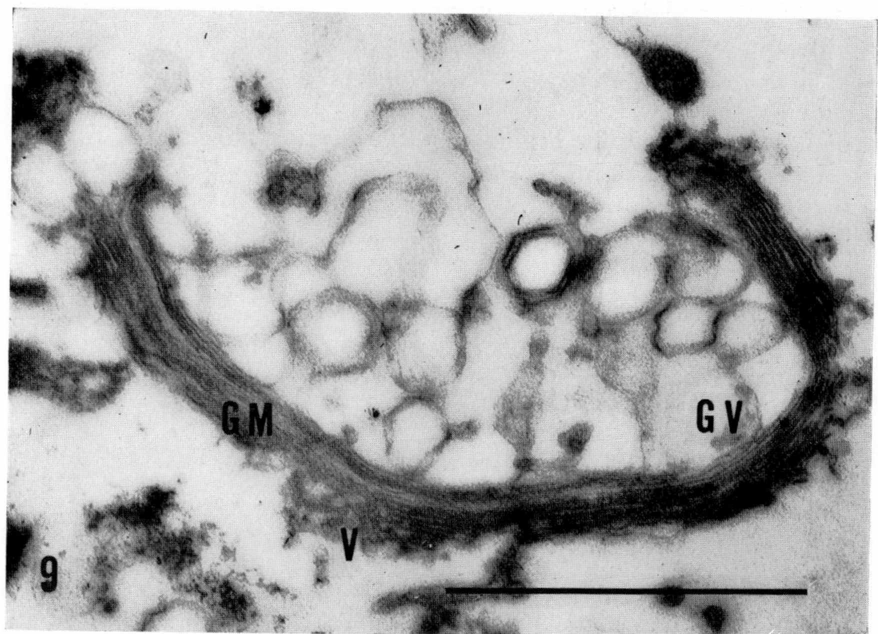

FIG. 9. An electron micrograph of isolated Golgi substance obtained by centrifugation with a layering technique (Kuff and Dalton, 1960). Characteristic layers of flattened sacs (*GM*) can be seen to delaminate into vacuoles (*GV*). Small vesicles (*V*) are also present. Fixation in isotonic sucrose containing 1% OsO_4. Magnification: approximately × 48,000.

In a more general sense, evidence is accumulating in support of the view that the Golgi complex as well as the ergastoplasm actually form parts of a membrane system extending throughout the cytoplasm, possibly including both the plasma and nuclear membranes of the cell

---

FIG. 8. An electron micrograph of dictyosomes (*G*) of a spermatid of *Helix aspersa*. Five separate bodies are visible, the upper one of which appears to be in the process of division. A sixth is visible cut tangentially at the left of center. The typical layering of flattened sacs with budding off of vesicles at their periphery is visible in each body. Mitochondria (*M*) with longitudinally arranged cristae are also evident. Chrome-osmic fixation (pH 7.2). Magnification: approximately × 32,000.

(Palade 1955). This cytoplasmic membrane system, the endoplasmic reticulum, is considered in detail in Chapter 9.

Thus, it may be said that the form of the classic Golgi reticulum is correlated directly with the fine structure of the complex as revealed by electron microscopy, that electron microscopic analysis has demonstrated a homology between the vetebrate Golgi apparatus and the dictyosomes of vertebrate and invertebrate germ cells and has given some rather equivocal answers concerning the function of the complex. Beyond this, with the electron microscope serving as a check on contamination with other cellular components, the Golgi complex has been isolated in pure enough form to allow the beginning of biochemical analysis.

References

Baker, J. R. (1944). *Quart. J. Microscop. Sci.* **85**, 1.

Baker, J. R. (1949). *Quart. J. Microscop. Sci.* **90**, 293.

Baker, J. R. (1957). "Mitochondria and Other Cell Inclusions," p. 1. Cambridge Univ. Press, London and New York.

Baker, J. R. (1959). *J. Roy. Microscop. Soc.* **77**, 116.

Beams, H. W. (1930). *Anat. Record* **45**, 137.

Beams, H. W. (1931). *Anat. Record* **49**, 309.

Beams, H. W., and King, R. L. (1933). *Anat. Record* **57**, 29.

Beams, H. W., Tahmisian, T. N., Devine, R. D., and Anderson, E. (1956). *J. Roy. Microscop. Soc.* **76** (Part 3), 98.

Bensley, R. R. (1911). *Am. J. Anat.* **12**, 37.

Bensley, R. R. (1951). *Exptl. Cell Research* **2**, 1.

Bowen, R. H. (1929). *Quart. Rev. Biol.* **4**, 299, 484.

Brice, A. L., Jones, R., and Smyth, J. D. (1946). *Nature* **157**, 553.

Burgos, M., and Fawcett, D. W. (1955). *J. Biophys. Biochem. Cytol.* **1**, 287.

Buvat, R. (1957). *Compt. rend. acad. sci.* **244**, 1401.

Chou, J. T. Y., and Meek, G. A. (1958). *Quart. J. Microscop. Sci.* **99**, 279.

da Fano, C. (1920). *J. Physiol. (London)* **53**, 92.

Dalton, A. J. (1960). "Symposium on Cell Physiology of Neoplasia," 14th Annual Symposium on Fundamental Cancer Research, p. 161. U. of Texas, M. D. Anderson Hospital and Tumor Institute. U. of Texas Press, Austin.

Dalton, A. J., and Felix, M. D. (1953). *Am. J. Anat.* **92**, 277.

Dalton, A. J., and Felix, M. D. (1954). *Am. J. Anat.* **94**, 171.

Dalton, A. J., and Felix, M. D. (1956). *J. Biophys. Biochem. Cytol. Suppl.* **2**, 79.

Duboscq, P., and Tuzet, O. (1937). *Arch. zöol. exptl. et gen.* **79**, 157.

Farquhar, M. G., and Wellings, S. R. (1957). *J. Biophys. Biochem. Cytol.* **3**, 319.

Ferreira, D. (1957). *J. Ultrastruct. Research* **1**, 14.

Gatenby, J. B. (1917). *Quart. J. Microscop. Sci.* **62**, 216.

Gatenby, J. B. (1931). *Am. J. Anat.* **48**, 421.

Gatenby, J. B., and Sigh, B. N. (1938). *Quart. J. Microscop. Sci.* **80**, 567.

Gatenby, J. B., Dalton, A. J., and Felix, M. D. (1955). *Nature* **176**, 301.

Gersch, I. (1949). *A.M.A. Arch. Pathol.* **47**, 99.

Golgi, C. (1898). *Arch. ital. biol.* **30**, 60.

Grassé, P.-P., Carasso, N., and Favard, P. (1956). *Ann. sci. nat. Zool. et biol. animale* **18**, 339.
Gresson, R. A. R. (1950). *Quart. J. Microscop. Sci.* **91**, 73.
Haguenau, F., and Bernhard, W. (1955). *Arch. anat. microscop. morphol. exptl.* **44**, 27.
Hendler, R. W., Dalton, A. J., and Glenner, G. G. (1957). *J. Biophys. Biochem. Cytol.* **3**, 325.
Hibbard, H. (1945). *Quart. Rev. Biol.* **20**, 1.
Hirsch, G. C. (1939). "Form- und Stoffwechsel der Golgi Körper." Gebrüder Borntraeger, Berlin.
Hirschler, J. (1918). *Arch. mikroskop. Anat. u. Entwicklungsmech.* **91**, 140.
Howatson, A. F., and Ham, A. W. (1955). *Cancer Research* **15**, 62.
Jepps, M. (1947). *Proc. Roy. Soc.* **B134**, 408.
Kirkman, H., and Severinghaus, A. E. (1938a). *Anat. Record* **70**, 413.
Kirkman, H., and Severinghaus, A. E. (1938b). *Anat. Record* **71**, 557.
Krehl, L. (1890). *Arch. Anat. u. Entwicklungsges.* p. 97.
Kuff, E. L., and Dalton, A. J. (1960). "Subcellular Particles," p. 114. Ronald Press, New York.
Lacy, D., and Challice, C. E. (1957). "Mitochondria and Other Cell Inclusions," p. 62. Cambridge Univ. Press, London and New York.
Ludford, R. J. (1925a). *J. Roy. Microscop. Soc.* **31**, 6.
Ludford, R. J. (1925b). *Proc. Roy. Soc.* **B98**, 354.
Moussa, T. A. (1952). *Am. J. Anat.* **90**, 379.
Nassonov, D. N. (1923). *Arch. mikroskop. Anat. u. Entwicklungsmech.* **97**, 136.
Nassonov, D. N. (1924a). *Arch. mikroskop. Anat. u. Entwicklungsmech.* **100**, 433.
Nassonov, D. N. (1924b). *Arch. mikroskop. Anat. u. Entwicklungsmech.* **103**, 437.
Oettlé, A. G. (1948). *Nature* **162**, 76.
O'Leary, J. L. (1930). *Anat. Record* **45**, 27.
Palade, G. E. (1955). *J. Biophys. Biochem. Cytol.* **1**, 567.
Palade, G. E. (1956). *J. Biophys. Biochem. Cytol.* **2**, 417.
Palade, G. E., and Claude, C. (1949a). *J. Morphol.* **85**, 35.
Palade, G. E., and Claude, C. (1949b). *J. Morphol.* **85**, 71.
Palay, S. L. (1958). "Frontiers in Cytology," p. 305. Yale Univ. Press, New Haven, Connecticut.
Palay, S. L., and Karlin, L. J. (1956). *Anat. Record* **124**, 343.
Parat, M., and Painlevé, J. (1924a). *Compt. rend. acad. sci.* **179**, 543.
Parat, M., and Painlevé, J. (1924b). *Compt. rend. acad. sci.* **179**, 612.
Parat, M., and Painlevé, J. (1924c). *Compt. rend. acad. sci.* **179**, 844.
Pollister, A. W., and Pollister, P. F. (1957). *Intern. Rev. Cytol.* **6**, 85.
Ramon y Cajal, S. (1914). *Trabajos Lab. invest. biol. (Madrid)* **11**.
Schneider, W. C., and Kuff, E. L. (1954). *Am. J. Anat.* **94**, 209.
Sjöstrand, F. S., and Hanzon, V. (1954). *Exptl. Cell. Research* **7**, 415.
Thomas, O. L. (1948). *Quart. J. Microscop. Sci.* **89**, 333.
Tuzet, O. (1932). *Arch. zool. exptl. et gen.* **10**, 74.
Walker, C., and Allen, M. (1927). *Proc. Roy. Soc.* **B101**, 468.
Weiner, P. (1926). *Arch. Rus. anat., hist., et embryol.* **5**, 10.
Weiss, J. M. (1953). *J. Exptl. Med.* **98**, 607.
Worley, L. G. (1943a). *Proc. Natl. Acad. Sci. U.S.* **29**, 225.
Worley, L. G. (1943b). *Proc. Natl. Acad. Sci. U.S.* **29**, 228.

CHAPTER 9

The Ground Substance; Observations from Electron Microscopy

By KEITH R. PORTER

I. Current General Concepts

Present-day concepts of the cytoplasmic ground substance are greatly influenced by the images derived from electron microscopy. This can be regarded as appropriate even though the material studied has for the most part been fixed with osmium tetroxide and thence dehydrated and embedded in methacrylate or some other resin. Though the resulting image is truly an image of artifacts, there is good evidence that these images are closely descriptive of the native content. For example, certain of the larger elements, though having dimensions below the theoretical resolving power of the light microscope, can be seen as shadows in phase contrast microscopy of living cells. These and other structures are present in patterns of organization that repeat from cell to cell even after the use of a variety of preparative procedures, including those that dry the cell from the frozen state. And finally, a number of the submicroscopic structures have been found to retain, after isolation from the living cell, the same form they show in the intact cell after fixation. It can be concluded with full justification that the evidence of fine structure provided by electron microscopy of the osmium-fixed specimen deserves serious consideration and that interpretations of far-reaching significance relative to the living cell may be based on this kind of information.

In general the "submicroscopic" ground substance examined at electron microscope resolutions presents a picture of a highly complex, polyphasic mixture. There are innumerable line-limited units which are reasonably interpreted as profiles of membrane-bounded units (Figs. 1–20). In any one image derived from a thin section of a cell, these seem to represent unconnected vesicles of many shapes and sizes. But, as varied as they are, it is evident that they fall into classes and may properly be thought of as representing systems, which can in turn be identified with the more classic microscopic elements of the light microscope image. Thus some are recognized as forming, in a characteristic association, part of the ergastoplasm; others the Golgi component; and still others as belonging to the centrosphere. There are in addition other membrane-limited units that find no clear equivalent in classic cytology.

Regardless of their detailed differences, these membranous elements are surrounded by or suspended in a matrix or continuous phase of the cytoplasm. This same matrix can be said, on the basis of electron microscopy, also to support in many types of cells a population of re-

---

FIG. 1. Electron micrograph of part of an epidermal cell from larval form of *Ambystoma punctatum*. The margin of the cell is at the upper right and includes a desmosome at (*d*). The nucleus (*N*) is at the left. The micrograph is presented to show the conglomeration of structural elements recognized now as part of the "ground substance." The Golgi component (*G*) near the nucleus is made up of relatively small, membrane-limited elements in close array. Profiles through vesicles and cisternae of the endoplasmic reticulum (*er*) are relatively large. Many of these are encrusted with small dense particles (seen to better advantage in Fig. 12) suggesting that the system in this cell is active in protein synthesis. This is appropriate since these cells are secretory. A partly exploded secretory granule is shown at *sg*. The ground substance otherwise is rich in extremely fine fibrils of keratin, characteristic of epidermal cells. They associate in bundles that sweep around among the other components of the cytoplasm but concentrate chiefly in the cortical zone of the cell. A mitochondrion is at (*m*). Magnification: × 25,000.

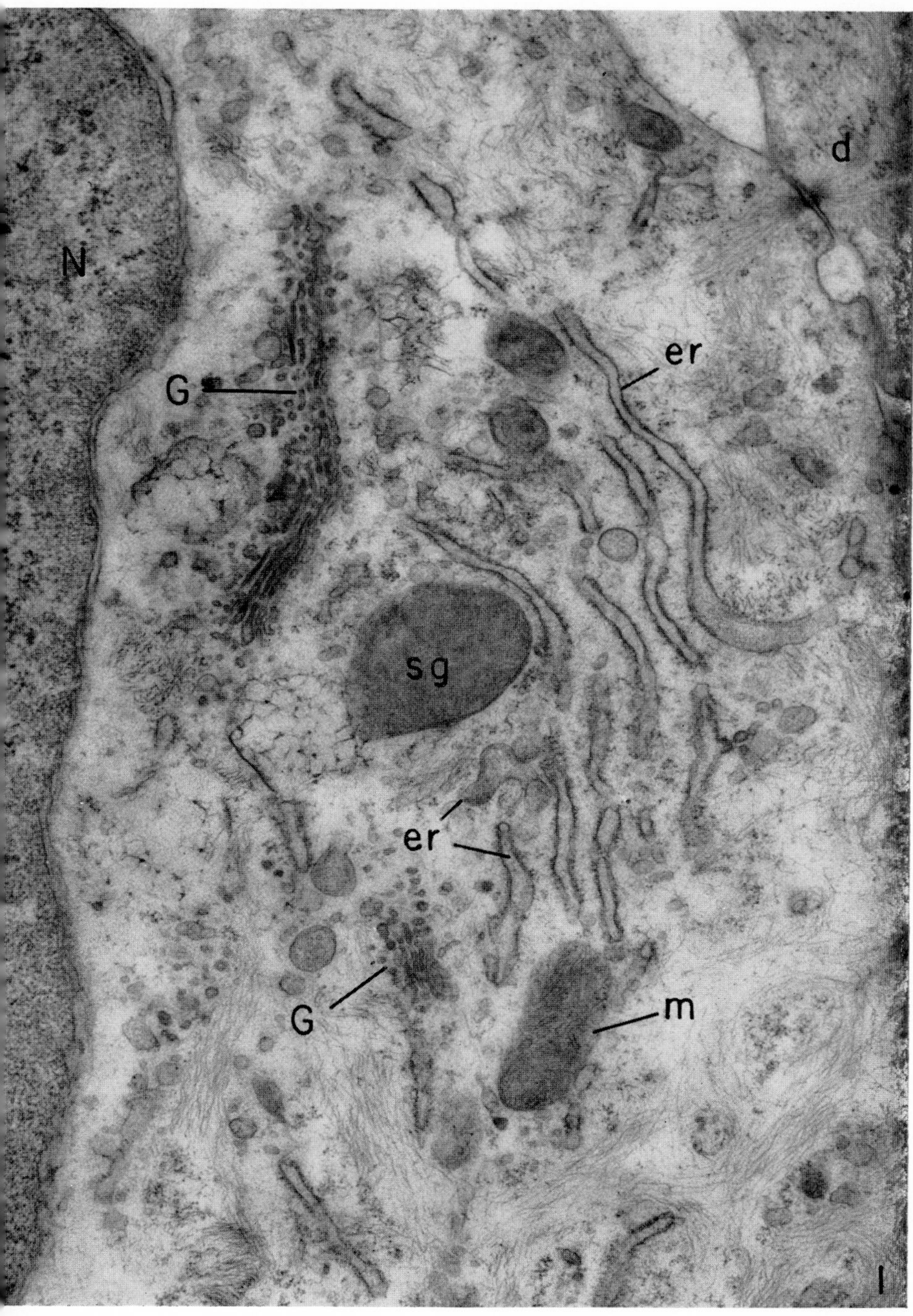
d
N
er
G
sg
er
G
m
I

solvable dense particles which fall within a fairly narrow size range of 15–20 mμ and which have been shown to be extraordinarily rich in ribonucleic acid (RNA) (Fig. 12) (Palade, 1955a). The matrix itself is without ordered structure except where it gives way to arrays of fine fibrous elements (~ 10 mμ diameter). Such well-known fibrous elements as myofilaments of muscle fibers (Fig. 11), the rootlet fibers of ciliated cells (Fawcett and Porter, 1954), the keratin filaments (tonofibrils) of epithelial cells (Fig. 1), and fibrous material forming such cortical differentiations as the "terminal web" (Leblond and Clermont, 1960) fall within this broad category. But just as properly are included here the more transient fibers of the mitotic spindle (Porter, 1954b; Favard, 1959) and the fibroglial fibers of motile cells, which doubtless represent temporary polymerizations of other-time globular macromolecular units of the ground substance (Porter, 1953).

The matrix therefore occupies, in present-day concepts of the cytoplasm, the same position as that held by the ground substance or hyaloplasm of the earlier light microscope image. It is the "structureless" medium in which are suspended all the resolvable elements of the cytoplasm, including of course the larger elements such as mitochondria, plastids, lipid droplets, and vacuoles. There is no reason to believe that with improved methods of microscopy and specimen preparation this part of the cytoplasm will not in time be shown to contain complex organizations of macromolecules.

This newer image of the cytoplasmic ground substance obviously provides a basis for many new hypotheses regarding cells and cell phenomena. The membrane-limited units probably represent various and varying packets of metabolites and enzymes—not, incidentally, in their movement conforming to the usual laws of diffusion. The relatively large surfaces possessed by these intracellular membranes may provide for the spatial

---

FIG. 2. This micrograph depicts the adjacent halves of two parenchymal cells of the rat liver. Nuclei (*N*), Golgi components (*G*), and mitochondria (*m*) are all indicated. Lysosomes (*e*) hover around a bile canaliculus (*bc*). The endoplasmic reticulum appears in two forms. In the one the profiles are long, arranged more or less parallel in groups and coated with small particles (*erg*); this is the ergastoplasmic form of the ER. In the other, which is continuous with the first, the profiles are short and agranular or smooth (*ers*). These represent sections through the tubular elements of an irregular lattice. The rough and smooth distinction cannot be made at this magnification but is apparent in Figs. 6 and 7. Apart from these elements the cytoplasmic matrix shows a more or less homogeneous to finely fibrous composition. Frank bundles of fibrous material are only rarely encountered in liver cells. Magnification: × 12,500.

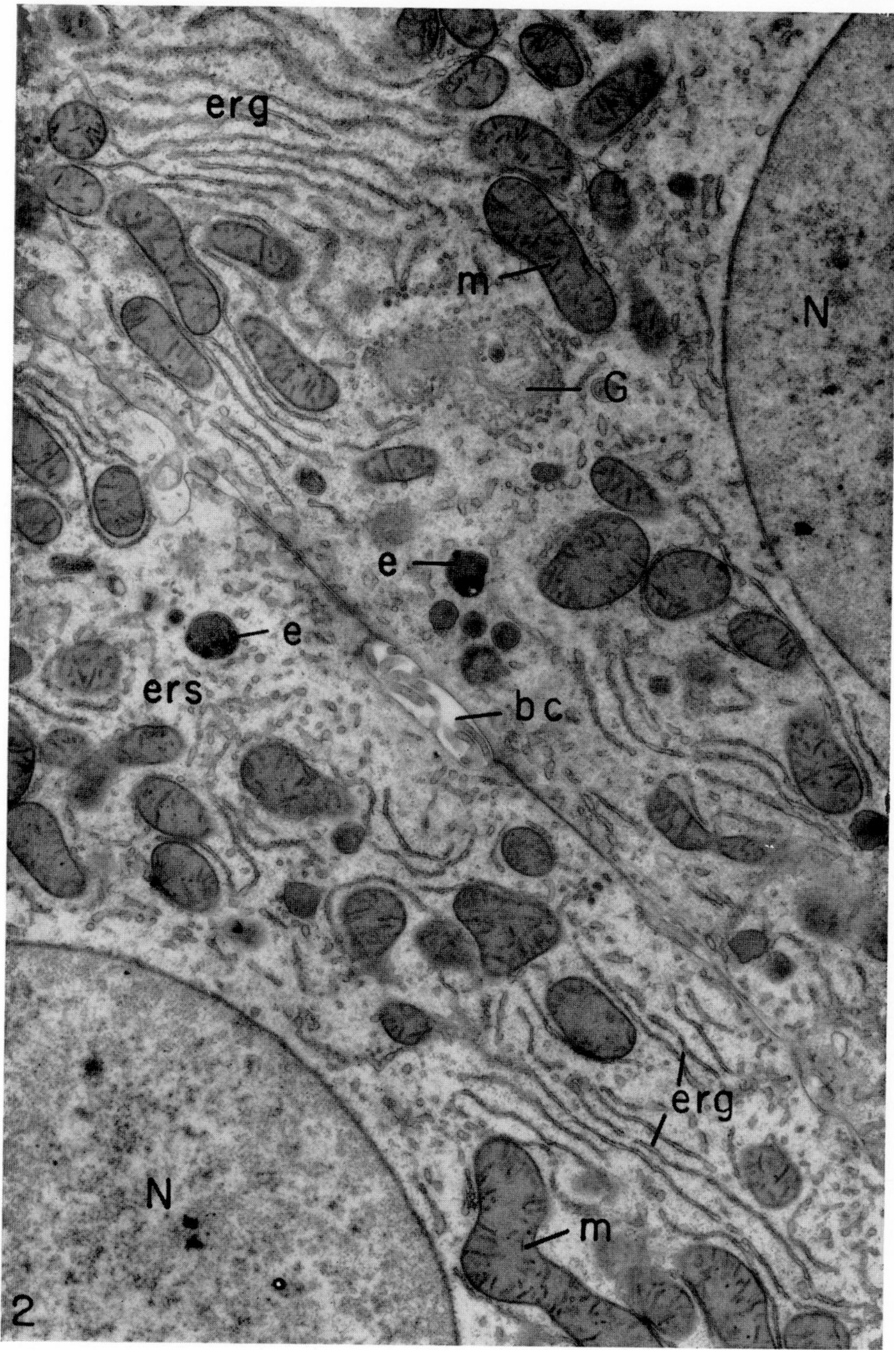
erg
m
N
G
e
e
ers
bc
erg
N
m
2

distribution of enzymes and substrates in the cell. A finely divided, membrane-enclosed phase within the cytoplasm would provide for the development of electrical membrane potentials of possibly great significance in life processes. The nucleoprotein particles of uniform size and character could represent packets of genetic information with specific roles in protein synthesis. The fibrils, which may be transient or relatively permanent structures, are bound to be associated in one's thinking with cell motion and sol-gel transformations. These are only a few of the ideas that have developed around these new images of cell structure.

An attempt is made in what follows to describe and examine in some detail the properties and current concepts regarding certain of these newly resolved components of the ground substance. Because of time and space limitations a complete coverage is not achieved, for even though exploration of these levels is a relatively recent development in the history of cell biology, the observations have already mounted to dimensions which make an exhaustive review a forbidding task. Therefore some valuable observations are sure to have been overlooked.

II. The Endoplasmic Reticulum

A. *Definition*

It was noted just above that one of the newly resolved components of the cytoplasm is represented in electron micrographs of thin sections by line-limited units having the form expected of sections through vesicles and tubules. In some types of cells these are relatively few, but in other types the cytoplasm is literally packed with them. From such displays of membranes there has developed the concept that the cytoplasm is extraordinarily rich in membranes, and one finds in the recent literature frequent references to membrane systems of the cytoplasm. Actually the picture is more meaningful than implied by such references because these membranes, in the majority of instances, enclose spaces and thus create in the cytoplasm an internal phase separated by a membrane from an outer or continuous phase which is the cytoplasmic matrix. After fixation and other preparatory procedures, these two phases are observed to differ in density and morphology.

The experienced observer of cell fine structure readily detects that

---

Fig. 3. Micrograph of thin section through malarial parasite, *Plasmodium berghei*, contained in a rat erythrocyte. The limits of the protozoan are indicated by arrows. It shows a nucleus at *N* with a double-membraned envelope (*ne*); ER vesicles in the cytoplasmic ground substance (*er*) and a scattering of small dense particles. Even in these retrograde forms, the nuclear envelope and ER are represented as in other cells. Magnification: × 20,000. Micrograph by courtesy of Maria Rudzinska.

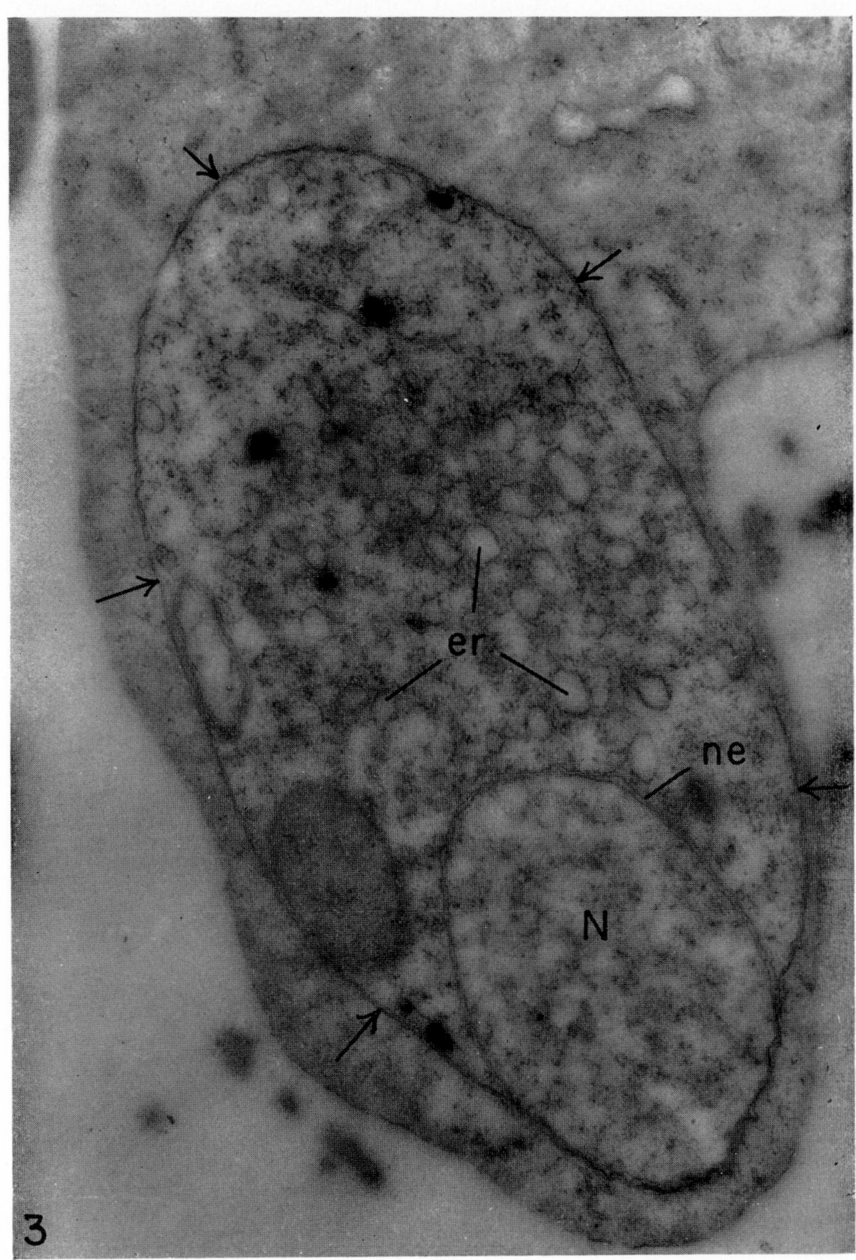
er
ne
N
3

these membrane-limited structures, or profiles, fall into classes defined in terms of the form, dimensions, and character of association of the constituent units. Thus some are readily recognized as part of the Golgi complex. Others are relatively constant components of the ergastoplasm, where this is found. And these associations are recognized as one part or one differentiation only of a larger class or system of vesicles and tubules which pervades almost the entire cytoplasm. To this system, by virtue of its form and distribution, the name endoplasmic reticulum has been applied (Porter and Kallman, 1952; Porter, 1953; Palade and Porter, 1954). It will, hereinafter, be referred to also as the ER.

It would be misleading to suggest that identification of all elements and parts of this newly recognized system is as easy and therefore as definite as is now the identification of mitochondria. Nevertheless, using criteria of form, dimensions, disposition, and association, it is possible to distinguish elements of the endoplasmic reticulum from most other membrane-limited elements where any possibility of confusion exists at all. In part, this is made possible by the peculiar character and restricted disposition of vesicular units comprising the other major systems. Thus, the units identified with the Golgi complex have smaller dimensions, are associated in a characteristic fashion, and tend to be confined to special regions of cells (Fig. 1). Likewise, vesicular elements of the centrosphere area are distinctive in being there, in being spherical, and in never being connected into anything like a continuous system.

The form displayed by the component elements of the ER is highly variable. Thus the profiles may appear round and in this form range in size from approximately 25 to 500 mμ (Figs. 3, 9, 12); they may be oblong—representing sections cut at various degrees of obliquity through

---

FIG. 4. Micrograph of ergastoplasm in rat liver cell. This is the typical image of the granular or rough forms of the endoplasmic reticulum (*er*) in these cells. The long slender profiles are made up of two parallel lines separated by a narrow space and represent vertical sections through vesicles of a flat or lamellar form. The outer surfaces are studded with small dense particles—the RNP particles or ribosomes. A few similar particles can be seen in the cytoplasmic matrix which lies between the lamellar units. This element of the liver cell is the chief component of the microsomal fraction. Mitochondria are at *m*. Magnification: × 40,000.

FIG. 5. Another part of a rat liver cell showing the nuclear envelope and associated elements of the cytoplasmic ground substance. The double-membrane structure of the envelope is apparent as is also its similarity to the profiles of the ER (*er*) in the neighboring cytoplasm. The outer surface of the envelope is coated with ribosomes like the cisternae of the ER. At points indicated by arrows (and at other places), there are profiles of pores through the envelope at which points the cytoplasmic matrix and the nucleoplasm are in continuity. Magnification: × 25,000.

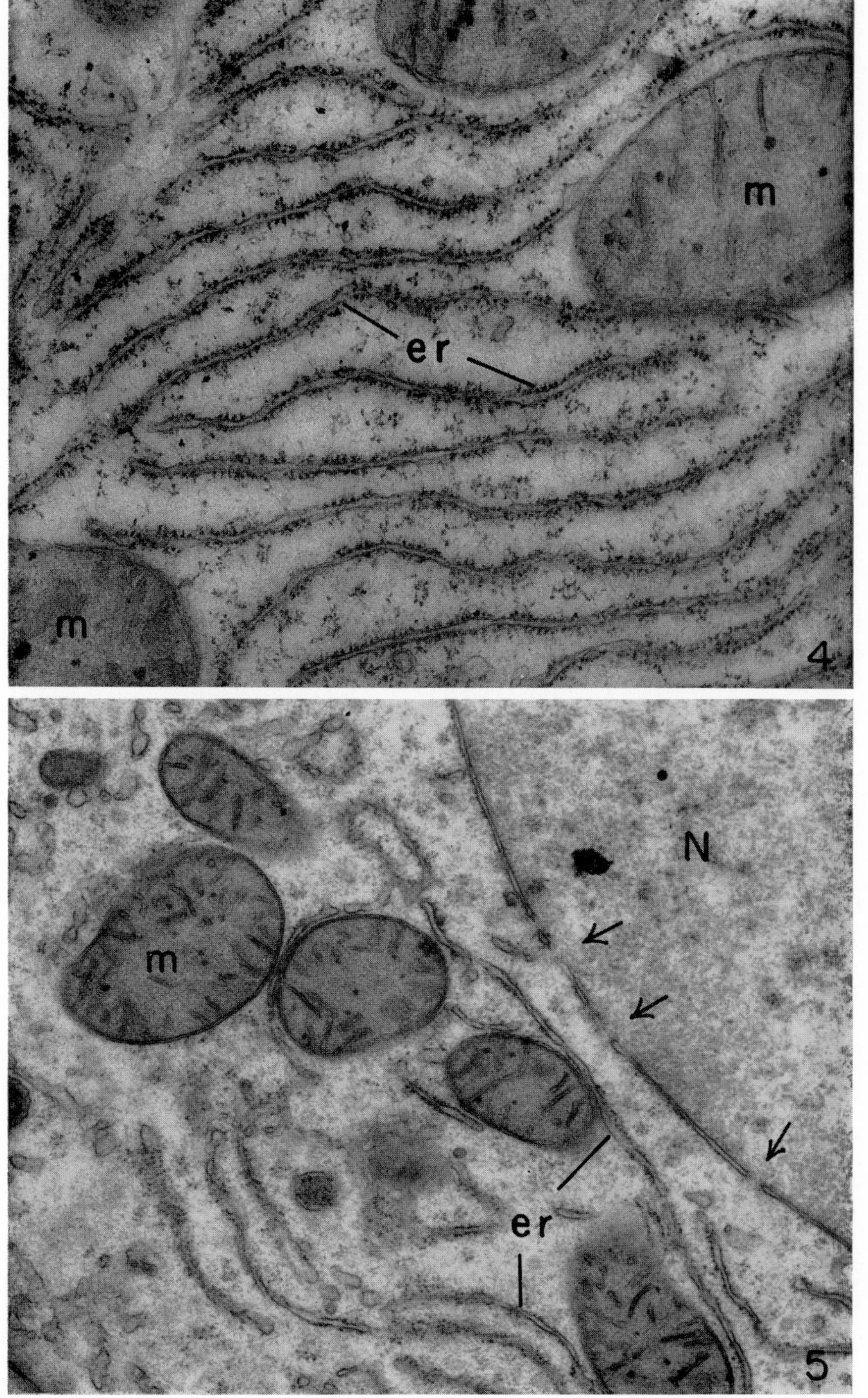

m
er
m
4
N
m
er
5

tubules (Fig. 6); or they may be long and slender—obviously representing sections through flat, lamellar vesicles or cisternae (Palade and Porter, 1954) the thickness of which is frequently in the range of 40–50 mμ (Figs. 1, 2, and 4). The dimensions are in general greater than those of elements comprising the Golgi or centrosphere zones (Fig. 1). The limiting membrane is of the order of 5 mμ in thickness and shows within the cell only a single-line structure (see Robertson, 1959; Karrer, 1960). The enclosed space usually shows no evidence of structure, though occasionally, in terms of density to the electron beam, there is evidence of content (Figs. 12, 13). The presence or absence of content is referable, in some instances, to the fixative used, but it also probably relates to the functioning of the system. This profile—a membrane-limited space—may be thought of as the "unit structure" of the ER as far as images of thin sections are concerned.

Profiles thus described are found in all parts of the cytoplasm except in the region occupied by elements of the Golgi complex (Fig. 2) and in the region of the centrosphere, if and when this is evident. Where the cortex or exoplasm of the cell is especially differentiated, as in intestinal epithelia or in epidermal cells (possibly with fibrous materials), these vesicles of the endoplasmic reticulum are confined to the more central zones of the cytoplasm, or endoplasm (Fig. 1). The ER is, however, usually widespread in its distribution and may occupy the greater part of the cytoplasmic ground substance.

It is common also for the elements to be associated in such a way as to form a continuum (see, e.g., Figs. 14 and 15). This is not universally the case nor is it always recognizable in a single section, and it can frequently be discerned only by examining a series of adjacent sections (Porter and Blum, 1953).

Fig. 6. This micrograph is designed to show the presence in liver cells of a second form of the ER, this one without attached particles and therefore referred to as agranular or smooth (*ers*). It is frequently found along the cell margins—as here—but appears elsewhere and is preferentially associated with glycogen. Its component tubules seen in section profile make up an irregular lattice. A bile canaliculus is shown at *bc* and is observed to be formed by slight depressions in the contiguous surfaces of two cells. The cells are joined at desmosomes (*d*) at opposite margins of bile canaliculus. Lysosomes (*e*) are evident as usual in this region of the cell. Magnification: × 20,000.

Fig. 7. An area of a rat liver cell in which glycogen has been preserved. It appears as clusters or rosettes of granular material (*gl*). Intermingled with these there are numerous profiles of the smooth ER (*ers*). Since similar condensations of glycogen are not found among the profiles of the ergastoplasm (*erg*) or among the Golgi elements, the association depicted is regarded as preferential. Magnification: × 28,000.

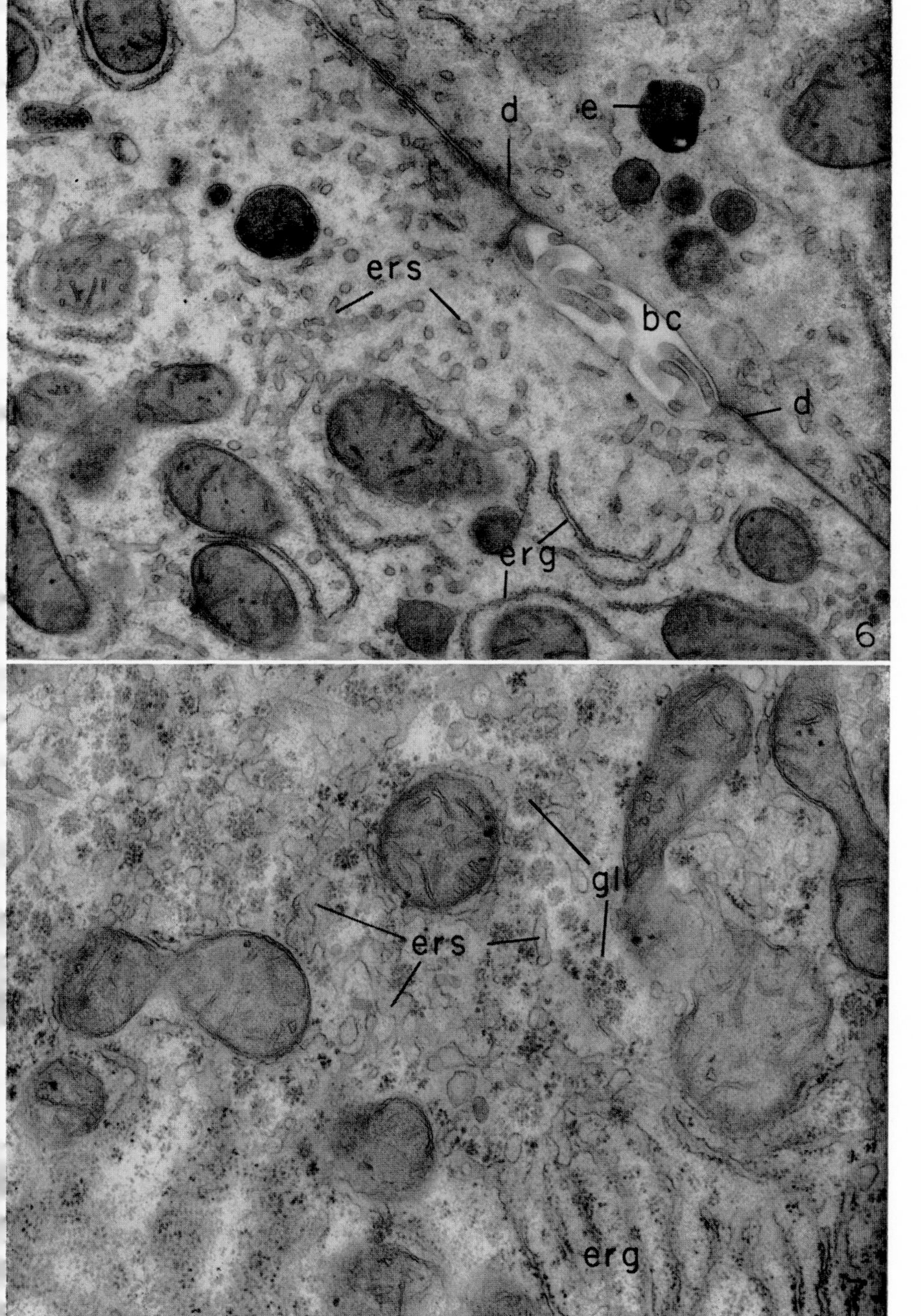

d
e
ers
bc
d
erg
6
gl
ers
erg
7

One of the best criteria for defining the ER is its association with the nuclear membrane, or envelope. As noted by Hartmann (1953) and later by Watson (1955) in a variety of cells, the nuclear membrane of classic cytology really consists of two membranes and an intervening, perinuclear space (see also Bahr and Beerman, 1954). In structure and dimensions, it appears identical in profile to a section through a flattened cisterna of the ER (Figs. 5 and 8). The outer of the two membranes has been shown, in many instances now, to be continuous with the membrane limiting adjacent elements of the ER (Watson, 1955, and others). The association is indeed a constant one, and it is this striking fact which makes it proper to regard the nuclear envelope as part of the endoplasmic reticulum. For conceptual purposes, then, in a situation where valid simplifications are welcome, it is reasonable to regard the nuclear envelope as *the constant part* of the system and to think of the cytoplasmic parts as derivatives or extensions of the envelope. In other words, the ER becomes a unit system of the cell based on the nuclear envelope (Porter and Machado, 1960a).

B. *Occurrence*

With the system thus defined, it is desirable to examine its occurrence among cells and so seek a basis for forming some opinion of its significance. For the convenience of the hurried reader, it can be stated at the outset that the available evidence describes the system (nuclear envelope and cytoplasmic extensions) as universally present in the cells of the higher animals and plants. It has been reported in all nucleated cells of vertebrate animals thus far studied. This applies to the germ cells as well as all others and leaves only the erythrocyte (mammalian) totally free of any. The same is true for the cells of metazoan animals in general, including such lower forms as *Hydra* (Chapman and Tilney, 1959; Slautterback and Fawcett, 1959). Likewise in the protozoans studied, a distinct nuclear envelope is present plus cytoplasmic elements having characteristics of the ER. In *Amoeba proteus,* as pointed out by Pappas (1956), the nuclear envelope takes on a bizarre form, but it is nevertheless possible to identify as part of it the double membrane structure and pores analogous to those in other cells and other protozoans including the smaller amebae (Mercer, 1959) and the sporozoan *Plasmodium berghei* (Rudzinska and Trager, 1959) (Fig. 3). The vesicles in the cytoplasm, identified by these authors with the endoplasmic reticulum, vary greatly in size and in association with particles. It is rare if ever in those amebae that one sees any continuous reticular system—a fact which, as Mercer (1959) points out, is not too surprising in view of the flowing motion evident in the cytoplasm. Similar observations have been made on *Paramecium* (Sedar and Porter, 1955).

In plants, the observations available similarly point to the ubiquity of the reticulum. From the higher forms (Buvat and Carasso, 1957; Porter, 1957; Whaley *et al.*, 1959; Porter and Machado, 1960a) on down and including the slime molds (Mercer and Shaffer, 1960) and the yeasts (Hashimoto *et al.*, 1960), there is evidence of a typical nuclear envelope, and characteristic elements of the ER in the cytoplasm. Only a few representatives of the algae have been examined, and among these *Porphyridium cruentum* (Brody and Vatter, 1959) among the red algae and *Chlamydomonas* (Sager and Palade, 1957) from among the green algae have been shown also to have a nuclear envelope and typical ER elements. It is apparently not until one reaches the blue-green algae that the nuclear envelope is lost, and even here there is evidence of a membrane-limited system in the cytoplasm which resembles that in animal cells (Bradfield, 1956; Elbers *et al.*, 1957). Recently this system has been shown to contain the chlorophyll and carotenoids of these cells and has been described as functional in photosynthesis (Shatkin, 1960). It was felt for a time that the bacteria were unique in possessing nothing even remotely resembling the ER, but more recent observations on *Streptomyces coelicolor* (Glauert and Hopwood, 1959, 1960; Moore and Chapman, 1959) and on other forms, such as *Bacillus subtilis* (Ryter *et al.*, 1958) and *Escherichia coli* (Kellenberger *et al.*, 1958), make it evident that the cytoplasms of bacteria may contain complex membrane systems—sometimes showing continuity with the plasma membrane and morphological association with crosswall formation (Fauré-Fremiet and Rouiller, 1958; Van Iterson, 1960). In these forms, however, such membranes show no preference for the nucleoloid (for review see Murray, 1960). It is difficult to judge at this time whether the membranes of the bacterial cells should be regarded as analagous to the ER of higher forms or whether, as suggested by their continuity with the external cytoplasmic membrane, they may not be more appropriately related to the multiple infoldings encountered in other cells (Pease, 1956). It would appear that in units having the small dimensions of the bacterial cell and shorter distances for diffusion, an internal vacuolar system is not important.

C. *Morphological Subdivisions of the Reticulum*

The general picture one may form of the ER is that of a complex, finely divided vacuolar system extending from the nucleus throughout the cytoplasm to the margins of the cell. In the fully differentiated cell the system shows a number of subdivisions—what may be termed special differentiations—which, one may guess, are designed to perform certain separate and distinct functions. The most constant of these subdivisions is the nuclear envelope.

1. The Nuclear Envelope

It was mentioned earlier that the nuclear envelope appears in section as a large lamellar or cisternal unit of the endoplasmic reticulum enclosing the nucleus. The two membranes which limit the structure are separated by a space (the perinuclear space) 20–40 mμ in width; it constitutes a sort of moat around the nucleus. At certain points the continuity of this moat is interrupted by openings or pores in the envelope through which the nucleoplasm and cytoplasmic matrix are continuous (Figs. 5 and 8). Callan and Tomlin (1950) early recognized the double structure of the envelope, and these other features of its morphology have been added by many observers including Bahr and Beerman (1954), Hartmann (1953), and Watson (1955, 1959).

Of the two membranes that make up the envelope, the inner one is in intimate contact morphologically with the peripheral or surface chromatin of the nucleus. This seems to be more than a random association for one seldom sees the inner membrane without dense granular material adjacent to it. Possibly because of this association the inner membrane frequently appears thicker than the outer, which in its 50-A. thickness resembles other membranes of the ER. At the pores, the inner and outer membranes join in a ring formation—the circumference of the pore. As noted by Watson (1955), and several other observers since, the outer membrane turns away from the envelope at many points to continue as the membrane limiting cytoplasmic elements of the ER (Porter, 1960).

FIG. 8. This micrograph depicts the nuclear and adjacent cytoplasm of another rat liver cell. The image here includes several instances of structural continuity between the outer membrane of the nuclear envelope and elements of the rough ER (arrows, cytoplasmic). Thus the perinuclear space between the membranes is continuous with the space phase of the ER and so a channel is provided for transport from one to the other. Several pores can be seen in the envelope (arrows, nuclear), in most instances opposite regions of low density within the nucleus. Nucleus (*N*) and mitochondria (*m*) are indicated. Magnification: × 30,000.

FIG. 9. Micrograph of portion of rat spermatid showing a group (stack) of annulate lamellae. It is to be noted that each has a double-membrane structure which is interrupted at points (arrows) by pores. These repeat approximately the image of pores in Fig. 8. As is usual, the assembly of lamellae is surrounded by numerous large vesicles. Magnification: × 52,000. From Palade (1955b).

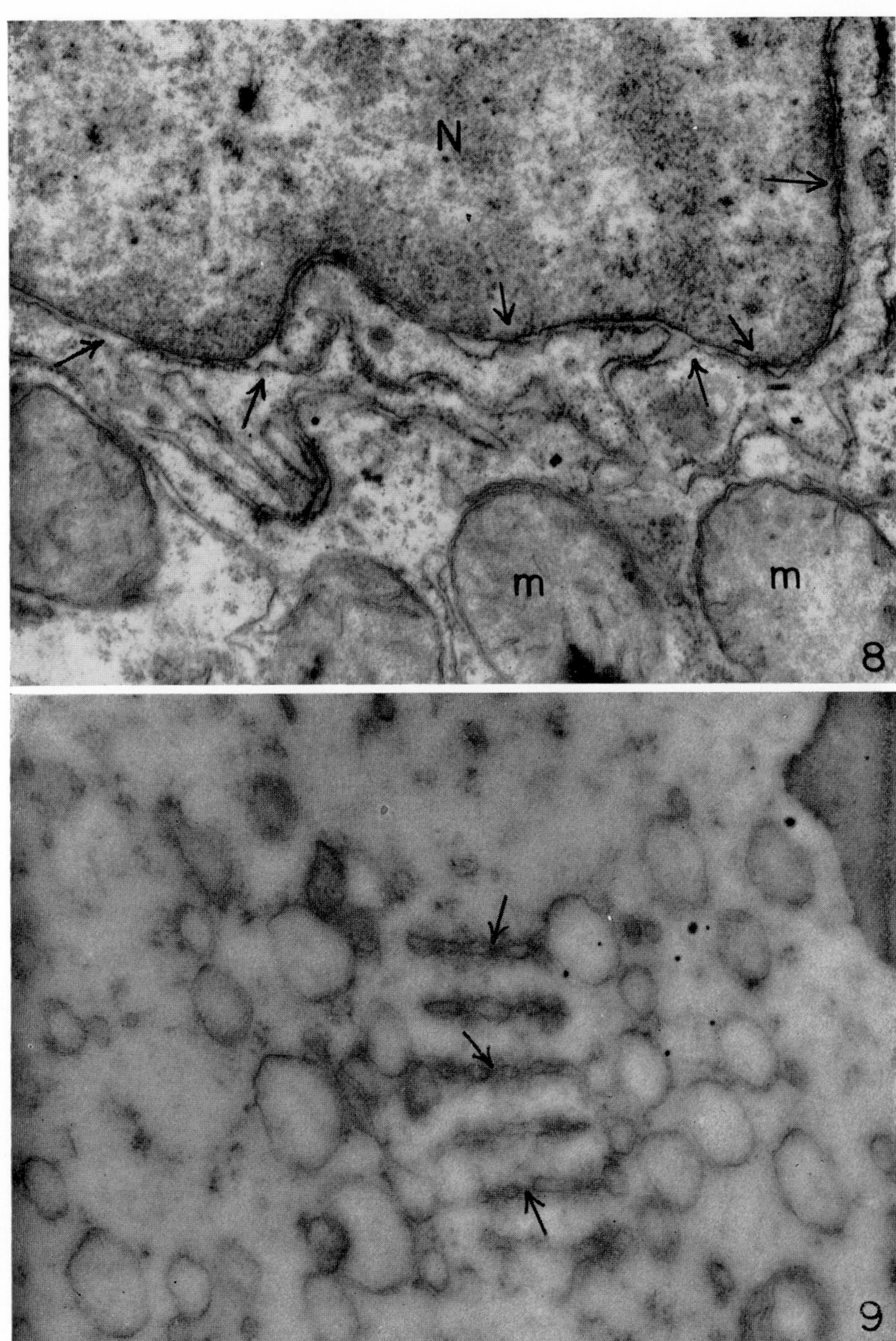
N
m
m
8
9

The pores, which vary in diameter from 50 to 100 mμ [and larger, see Gall (1959)], also vary in number in different cell types. Their arrangement in the envelope is not highly ordered; they seem instead to coincide with places where the nucleoplasm extends through the chromatin to the nuclear surface. The positioning of the pore may therefore be determined by the peripheral distribution of chromatin. Obviously the existence of the pores makes the envelope, otherwise continuous except during mitosis, a reticular structure.

Materials of the cytoplasmic matrix and/or nucleoplasm tend to varying degrees to condense in ring formation around the channel provided by the pore, and it is these condensations that appear as annuli (Afzelius, 1955; Wischnitzer, 1958; Watson, 1959). In *Amoeba proteus* this annulate structure is especially prominent on the infranuclear side and leads to a honeycomb-like development about 250 A. thick (Pappas, 1956).

The pore itself has been variously described as patent or closed by a membrane, but current evidence suggests that in most instances it is an open structure providing for the migration, between nucleus and cytoplasm, of relatively large particles (Watson, 1959). Such a high degree of continuity makes it reasonable to question the classic concept of separation between these two major divisions of the cell. It would be more reasonable in the light of this newer information to regard the nuclear envelope as a special part of the ER which for some functional reason is associated intimately with the interphase chromosomes. The pore, on the other hand, coincides in position with the low-density nucleoplasm. Evidence of migration of materials from the nucleus to the cytoplasm has been reported from electron micrographs by Pollister *et al.* (1954) and by Anderson and Beams (1956).

2. *The Cytoplasmic ER*

The reticular form and continuity of the cytoplasmic part of the endoplasmic reticulum are neither so constant nor so readily demonstrable as those of the nuclear part of the system. Here in the cytoplasm the variability is much greater and here the system is apparently more labile within the limits of normal cell functioning. Thus a system of tubules and vesicles that is continuous at one time in the life of the cell may be represented by a system of discontinuous vesicles at another. To what extent a return from this latter state is possible has not been determined, and it may well be that in any apparent recovery the typical reticular form is a product of regeneration from the nuclear envelope rather than a recoalescence of erstwhile separate vesicles. In a few cell types there is scant evidence of continuity, and the name for the system is hardly appropriate except as it applies to the nuclear envelope and a structural state that may have existed prior to the general vesiculation encountered.

The picture of continuity and reticular form for the cytoplasmic part of the ER has been accorded close scrutiny and skeptical comment by a number of observers and so deserves some consideration here.

a. Its "reticular" form. The earliest observations on this system—observations in which the concept of a continuous unit system were initially based—were made on the margins of whole, cultured cells (Porter *et al.*, 1945; Porter, 1953). This was fortunate in the sense that one obtained an image of the whole structure, albeit somewhat distorted by flattening, rather than an image of apparently unconnected profiles provided by thin sections. From these early observations there evolved then the concept of a reticular system made up of tubular strands or trabeculae varying somewhat in diameter and form. The trabeculae were vesiculated in some cells and so appeared as strings of vesicles; in others they showed a relatively smooth outline. The strands evident in the majority of instances were unquestionably interconnected to form a continuous system. These early images were readily related to the membrane-limited profiles apparent in the thin sections, and it was possible to relate the shapes of the profiles seen in thin sections to the forms encountered earlier in cultured cells (Palade and Porter, 1954). These included the large, flat cisternal units commonly seen in sections as well as vesicular and tubular units. It was recognized as well from the cultured cell studies that the continuity and form of the system is not fixed or rigid but is quite labile and that under conditions of cytolysis the trabeculae break up into separate vesicles.

As suggested just above, these concepts of this newly resolved component of the cytoplasmic ground substance are not so readily developed from a study of thin sections. The section shows only fragments, and these are frequently not joined to form a continuum, even over small areas of the field. Partly on this basis, the interpretation of the system as reticular has been criticized (Sjöstrand, 1956). As more and more cell types have been examined, however, the interpretation has been borne out, supported especially by studies of serial sections (Porter and Blum, 1953; Bang and Bang, 1957; Andersson-Cedergren, 1959).

Despite the validity of these statements, it must be admitted that the term reticular has to be stretched rather far to accommodate all the forms encountered. Thus some of the lamellar units characteristic of pancreas and protein-synthesizing systems are frequently continuous and reticular only at their margins where they join the noncisternal units of the system. The concentric arrangements (*Nebenkern*) of cisternal elements which are occasionally seen around lipid or protein granules are neither fenestrated nor connected, yet have other properties of ergastoplasm (see Favard, 1958, for good example). Here, obviously, reticular is not appropriate. Such formations, of which there are others, may come to be recognized as special differentiations of a system for which the broad term, endoplasmic reticulum (ER) may continue to be appropriate. The term, while usefully descriptive at this time, should not be regarded as more precise than it really is. In the main it applies to a complex membrane-limited system of the cytoplasm about which very little is known. Thus the limits of application of the term remain vague and will continue so until a better understanding of the whole complex is achieved. It may then properly be replaced by something more accurately descriptive.

b. Patterns of structure and organization. The nuclear envelope (perinuclear ER) as indicated earlier, is similar in the grosser aspects of its morphology from one cell type to another. It appears that the number, size, and distribution of pores vary somewhat; it is also evident that the pores show varying degrees of morphological complexity, but other major differences are not apparent. The cytoplasmic part of the system is, on the other hand, much more variable and difficult to analyze. Despite this, however, there is a degree of order in the picture, and when the many observations now available are compared a few generalizations emerge.

It requires, for example, the examination of only a few mature cells of any single type to reach the conclusion that the ER for that particular cell type possesses a characteristic pattern of structure. Thus it is typical for rat liver cells to show groups of eight or ten slender profiles in parallel array, representing thin sections through lamellar vesicles or cisternae arranged in a pile or stack (Figs. 2 and 4). There are obvious variations in this picture where mitochondria intrude in the stacks or other disturbances are introduced by dynamic events in the cell. But such groups of profiles, which coincide with the basophilic bodies or masses of the stained preparation, may be described as the expected form of the ER in the rat liver cell (Fawcett, 1955; Porter and Bruni, 1960a).

FIGS. 10 and 10A. These two pictures show electron micrographs of thin sections through similar parts of two notochord cells in a larval stage of *Ambystoma*. The sections cut transversely an arm of the cell. The bottom surface in each rests on the chord sheath (*cs*). The two micrographs depict the similarity in structural pattern of the ER in two separate cells of the same tissue. Magnification: × 25,000.

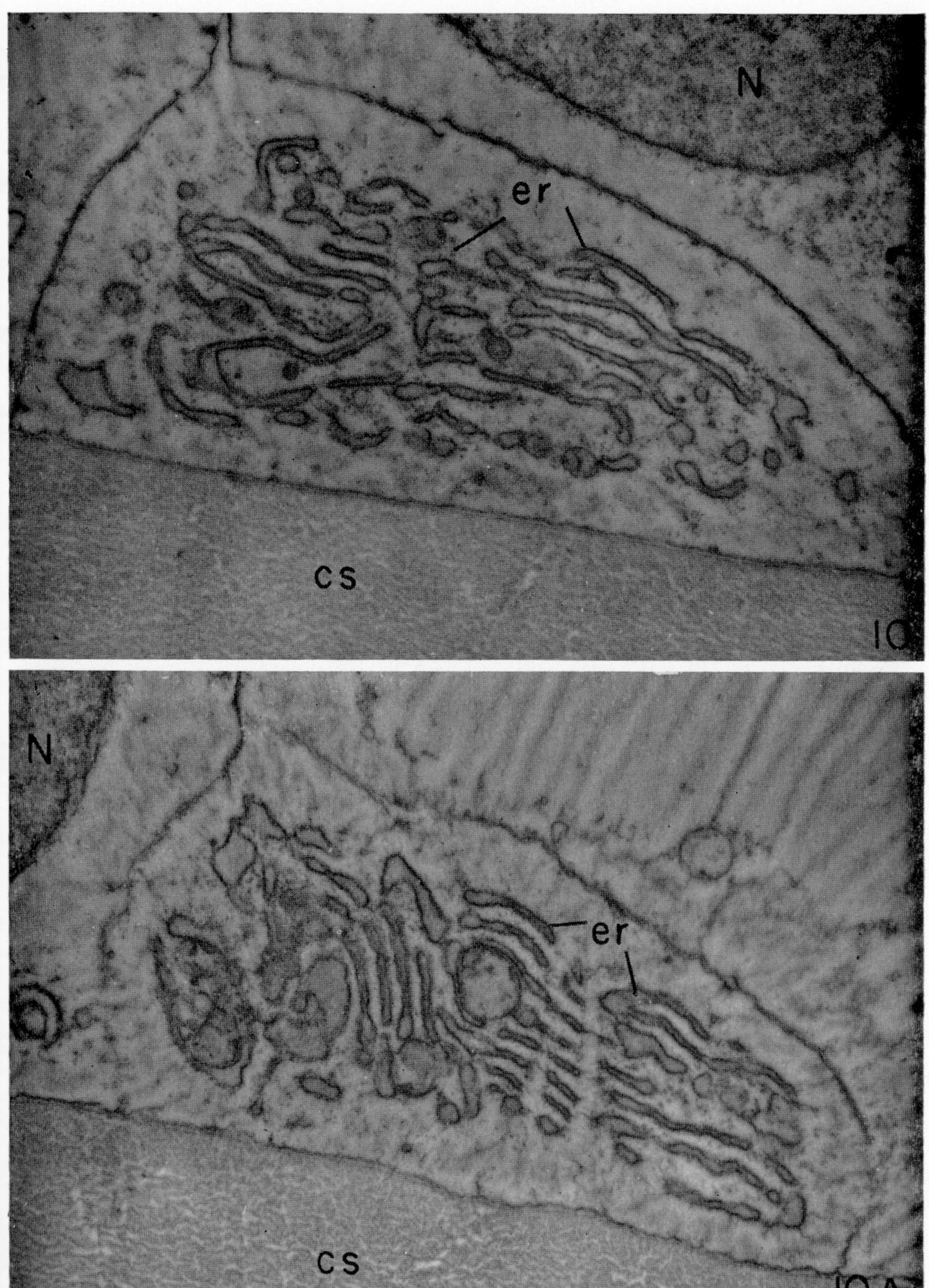
N
er
cs
10
N
er
cs
10 A

The notochord cells of *Ambystoma* larvae likewise show repeating patterns of cisternae (Figs. 10 and 10A). Both the arrangement and the dimensions of the component units are distinctly different from those encountered in liver cells. From cell to cell, however, within the notochord, the picture is the same. Many other examples of this phenomenon could be cited. In striated muscle cells the organization achieved with respect to the fibrils is remarkable (Fig. 11), and here again the pattern repeats in each fiber or cell of a given muscle type. Where, however, the muscle has somewhat different physiologic properties even though in the same animal, the organization is different. Thus the ER pattern varies subtly with the functional characteristics of different muscle cells (Peachey, 1960).

To some degree these structural patterns repeat as well in cells of the same type in different species of animals. As is well known, it is characteristic for the pancreas cell, wherever taken, to show closely packed cisternae in the basal half of the cell. Similarly one can recognize a plasma cell wherever encountered on the basis of the large, inflated cisternae and the slender intervening layers of cytoplasmic matrix (Fig. 13). There are recognizable differences in design in each instance, but the similarities are sufficiently pronounced in most instances to provide for identification. In frog liver, the profiles of the cisternae in the typical basophilic body are shorter than in the rat and more of them seem to be collected in any one group (Fawcett, 1955). Such differences are recognized as minor compared with those that exist between most cells of different type.

Even among different cell types derived from different tissues of the same animal or from different species, similarities in the ER may be noted if the cells are performing similar functions. Thus, as will be developed more fully below, the ER shows similar structural features in all cells engaged in synthesizing protein for export (Figs. 4, 12, and 13), or similar features in cells producing steroids for secretion (Fig. 15). It is possible therefore to conclude that the form of the ER in any one cell is a rather precise mark of differentiation. It is differentiated for a few tasks one can recognize, and doubtless for others we have yet to learn.

FIG. 11. Portion of a muscle fiber from caudal somite of an *Ambystoma* larva. Myofibrils run in longitudinal section from lower left to upper right. The sarcolemma (*sl*) is included in the field at the upper left. Between the myofibrils the endoplasmic reticulum, in this case smooth and called the sarcoplasmic reticulum (*sr*), comprises a lacework of tubules. The pattern of this lacework repeats in each sarcomere, the limits of which are marked by Z-lines (Z). Magnification: × 42,000. From Porter (1957).

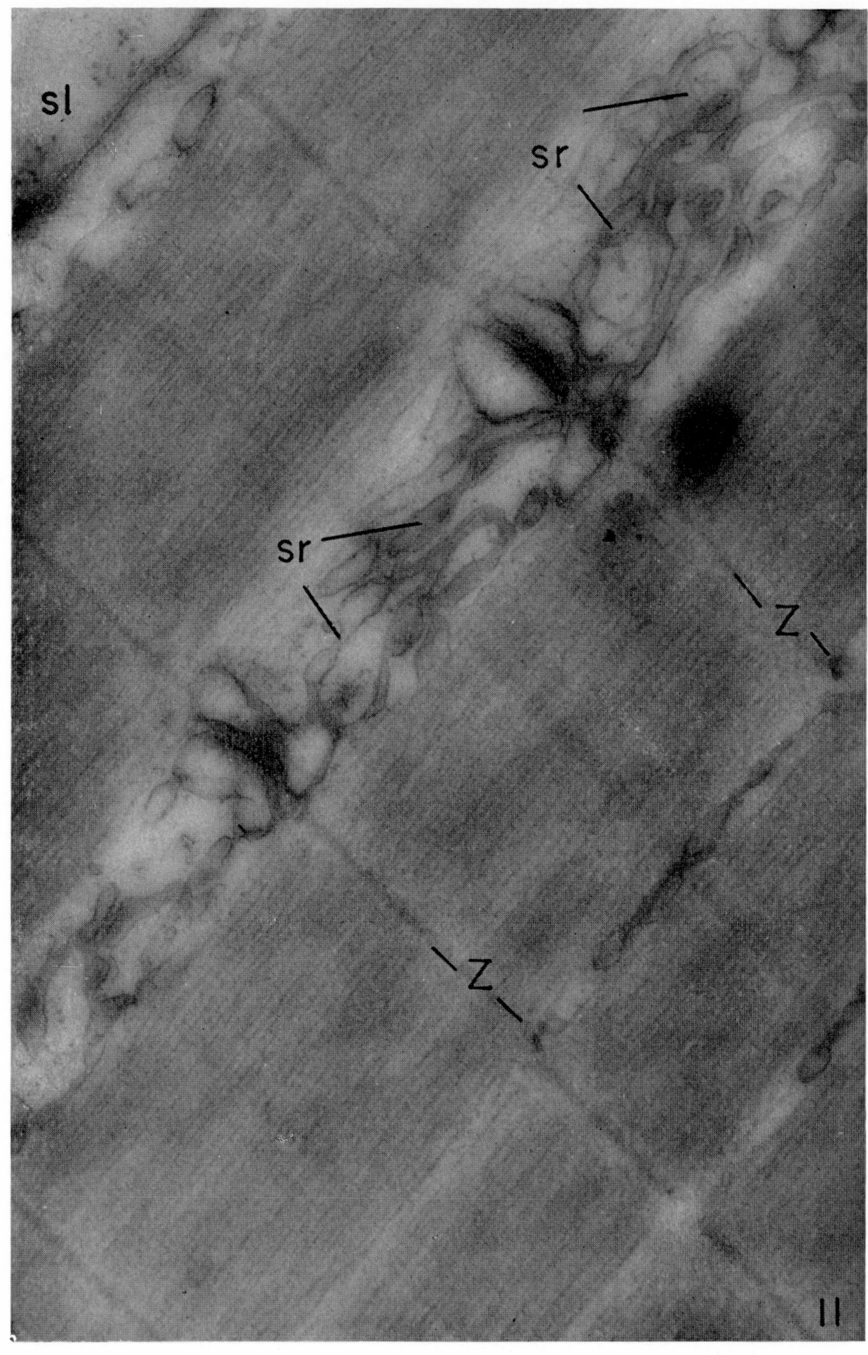
sl
sr
sr
Z
Z
11

As we have already indicated, these patterns reflect variations in the form of unit components, the number and proximity per unit volume of cytoplasm, and the way they are arranged. The similarities, evident in the two-dimensional image derived from a thin section, would probably be more striking in the three-dimensional image, if such were available. These patterns are of course neither rigid nor particularly stabile, but represent the state of a dynamic system at the time of fixation. Thus limited departures from the typical form for a cell occur, but it is suggested that the system within any differentiated cell "inclines" to adopt a form characteristic for that cell. As a structural component, then, the ER is essentially elastic and probably attempts, under normal physiologic conditions, to recover from any distortion. What factors control this form is impossible to say, but apparently they reside in part in the molecular texture of the membranes themselves and in forces operating through the matrix between the vesicular units of the system.

In addition to the configurations of the ER as a whole there exist a number of local structural divisions of unique design within the cytoplasm of any one cell. These represent subdivisions of the system and are thought of as special or local differentiations. The existence of continuity between these morphologically dissimilar parts provides the major excuse for grouping them together under the umbrella—endoplasmic reticulum.

c. The granular or rough form. One of the easiest of these to recognize is that which has associated with it a uniform particulate component of the ground substance found in great abundance in growing cells and in others engaged in protein synthesis (Palade, 1953, 1955a). These particles are always on the outer surfaces of the ER elements, i.e., on the surface of the limiting membrane facing the continuous phase, the matrix of the cytoplasm (Figs. 4 and 12). Since it is not the property of all ER membranes to associate with these particles, those parts which do are distinguished by the association. Besides possessing this property, it

---

FIG. 12. Micrograph of basal part of exocrine cell, guinea pig pancreas. The lighter, membrane-limited areas represent inflated cisternae (*er*) of ER. They contain large (200–300 mμ) dense granules, known as intracisternal granules (*icg*). These are similar in density and composition to the secretory granules ordinarily found in apical pole Golgi zone of cell. Their presence here describes the segregation of the product of synthesis in cavities of ER. In most other instances (see Fig. 13) the segregated material does not condense into granules. The small dense particles on the surfaces of the cisternae and scattered throughout the matrix are ribosomes. The cell rests on a basement membrane (*bm*) which in this field is underlain by a capillary (*c*). Magnification: × 30,000. Micrograph by courtesy of G. E. Palade.

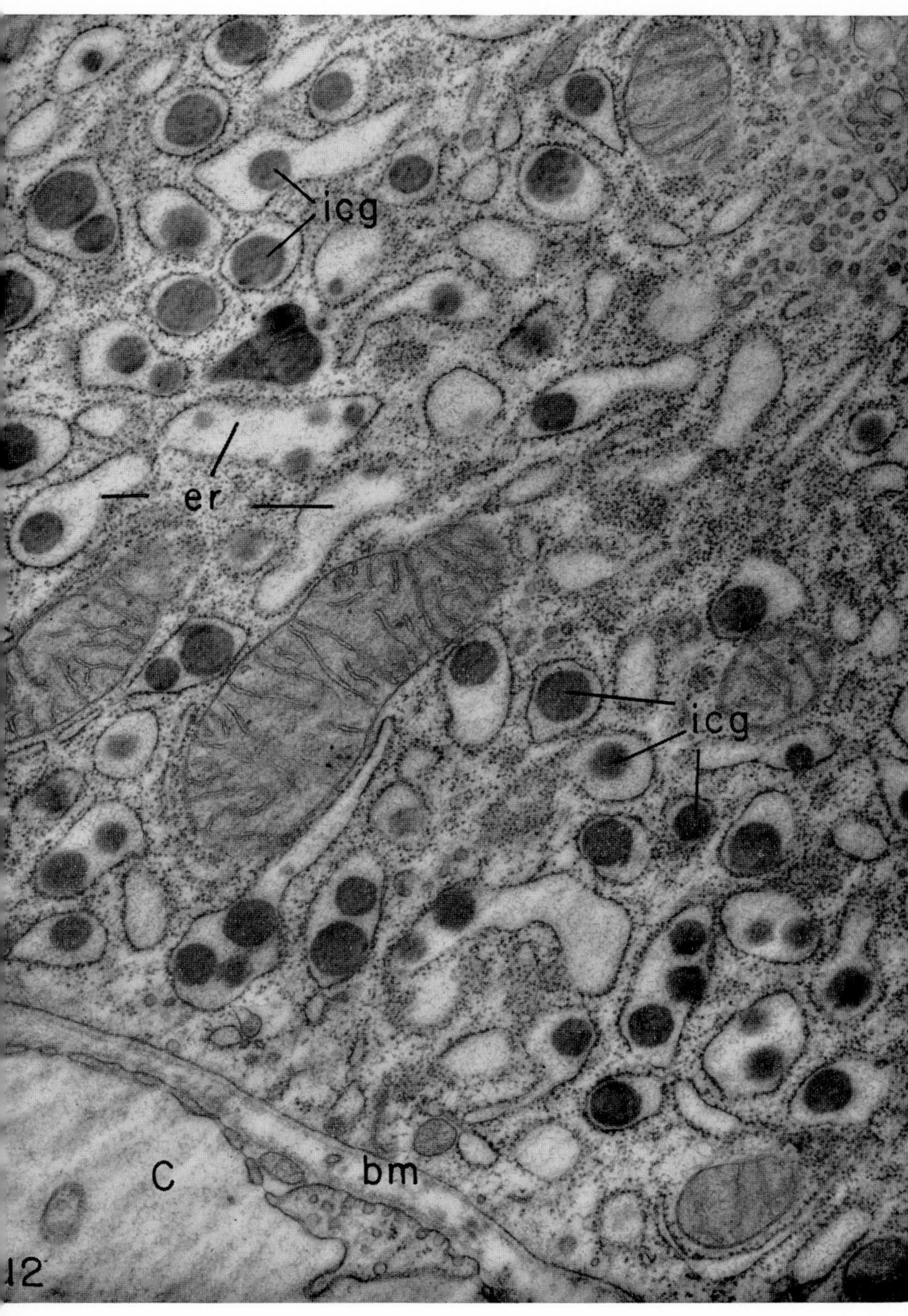
icg
er
icg
c
bm
12

may be said that elements of the ER showing this association have a fairly characteristic shape which is, in general, that of the lamellar or flat cisternal vesicle. These are sometimes inflated or otherwise distorted by an intracisternal accumulation of material (Figs. 12 and 13), but the usual form is the large lamellar cisterna (Fig. 4). This has been called the rough or granular form of the ER. The number of particles varies from a mere scattering (as in plant meristem cells) to a fairly close packing (as in acinar cells of the pancreas) (Sjöstrand and Hanzon, 1954). In these latter instances, the particles are arranged in patterns: circles, rosettes, spirals, etc. The particles may be induced to leave the surfaces of the cisternae without the cisternae losing their form, or probably their capacity to associate again with the particles (Porter and Bruni, 1960a). This form of the ER is perhaps the easiest to recognize, and for some observers it is the typical one. It was one of the first to be noted in thin sections of liver and pancreas cells (Dalton *et al.*, 1950; Bernhard *et al.*, 1951; Palade, 1952; Palade and Porter, 1952; Sjöstrand, 1953; Porter, 1954a), where it was recognized as the fine-structure equivalent of Garnier's ergastoplasm. Subsequently it was found to be the form of the ER always in coincidence with the basophilic regions of the cytoplasm [see Haguenau (1958) for an excellent review of ER-ergastoplasm relationship]. It is the part of the endoplasmic reticulum which, as will be discussed below, is involved in protein synthesis.

d. The agranular or smooth form. The other large division of the system owes its identity in part to the absence of particles and is therefore commonly referred to as the smooth or agranular form.

Like the rough ER (the ergastoplasm), the agranular or smooth shows a characteristic morphology which is tubular rather than cisternal (Figs. 2, 6, 14, and 15) (exceptions noted below). The resulting structure is a complex lattice of tubules having diameters of 50-100 mμ. These appear in varying degrees of compaction. They are common in cells that have a form to maintain (crown cells, sustentacular cells) (Porter, 1956). They are found in cells engaged in the synthesis of steroids (Christensen and Fawcett, 1960). They are particularly striking in pigment epithelial cells of the retina (Porter and Yamada, 1960). A form of similar character

---

FIG. 13. Part of a plasma cell encountered in circulation of *Ambystoma* larva. The form of the ER (*er*) repeats that encountered in plasma cells in other animals and less closely that of the ER in other cells synthesizing protein for export. The cisternae contain a relatively dense material, presumably accumulated globulins. Ribosomes are abundant and preferentially attached to cisternal surfaces. The plasma membrane (*pm*) and nucleus (*N*) are indicated. Magnification: × 26,000.

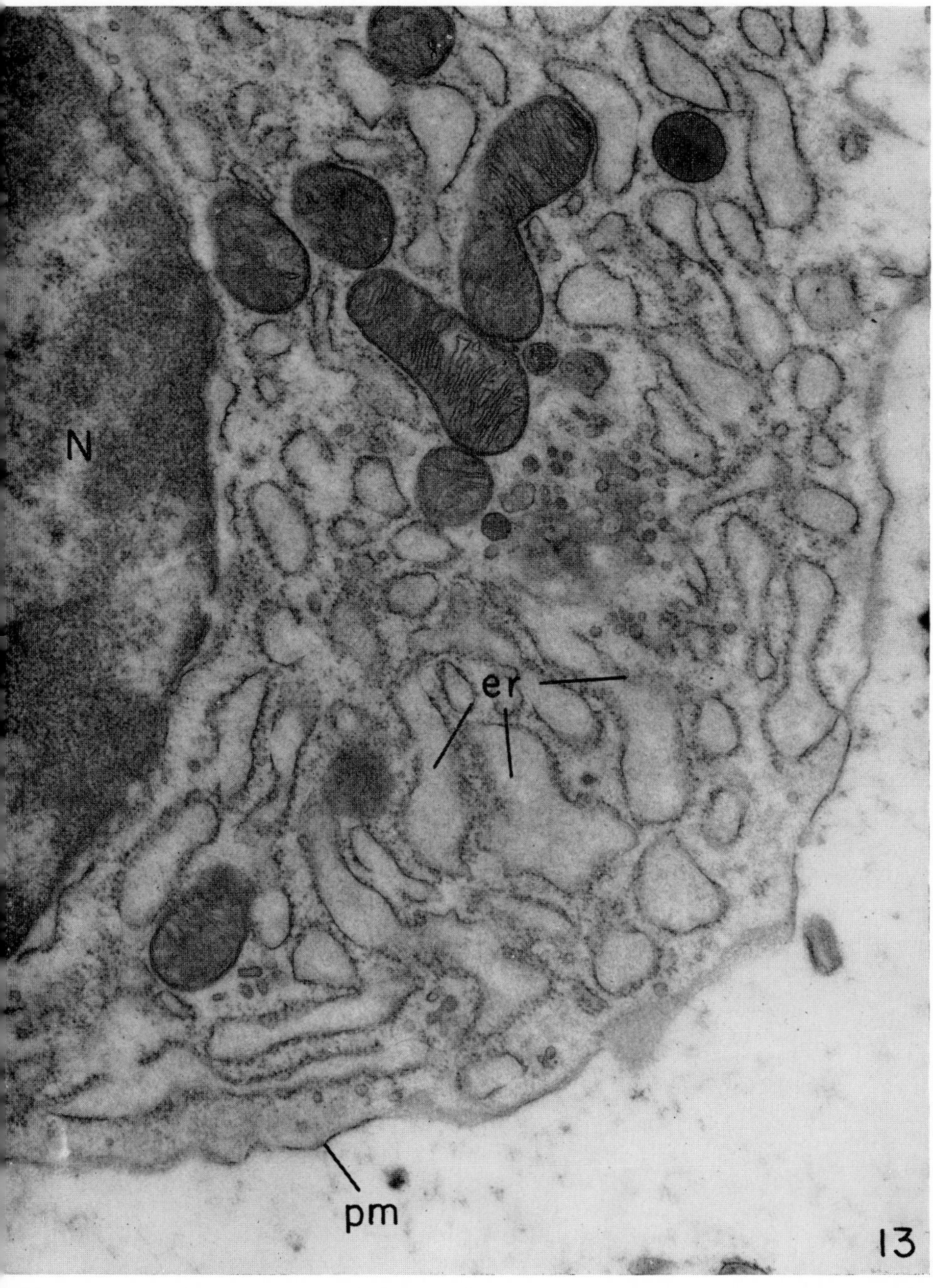
N
er
pm
13

occurs in liver cells in association with glycogen stores (Porter and Bruni, 1960b) and along the surfaces of plant cells where cell walls are being formed (Porter and Machado, 1960b). It is the dominant form that occurs widely in striated muscle cells (Porter and Palade, 1957; Andersson-Cedergren, 1959).

Continuity between this smooth form of the ER and the rough has been repeatedly demonstrated. It has been suggested, moreover, that one grows from the other, but which from which is unsettled. This form of the ER, like the rough, has the capacity to form patterns, and indeed appears to possess greater versatility in this regard than the rough. The repeating design in muscle, phased to the sarcomeres of the fibrils, is an outstanding example of this. The capacity to develop complex local differentiation seems also inherent.

e. The myeloid body. One of the best examples of a local differentiation of the endoplasmic reticulum is provided by the myeloid bodies found in pigment epithelial cells of the retina. The reticulum in these cells comprises a compact lattice of tubules and vesicles without attached particles (Fig. 14). At points within this reticulum there are remarkable bodies constructed of paired membranes in close parallel array. The whole body has the shape of a biconvex lens—a double cone (Fig. 14). At their margins the paired membranes or lamellar vesicles develop fenestrae, and this creates a reticular structure which is found to be continuous with that of the general ER of the cell. There is no mistaking these as true differentiations of the ER based on continuity of structure (Porter, 1957; Porter and Yamada, 1960). The role of these structures is unknown, but the similarity they show in fine structure to photoreceptors, such as the outer segments of rod cells, suggests that they may depend on light stimulation for functional activation. Similar if not identical components have been reported in the pigment epithelial cells from a variety of vertebrates (Yamada *et al.*, 1958; Yamada, 1958).

f. The Golgi component. The Golgi component (apparatus or complex), when examined for knowledge of its fine structure, is seen to be made up of membrane-limited vesicles and cisternae which differ chiefly

---

FIG. 14. This micrograph depicts a part of a pigment epithelial cell in retina of the frog. The ground substance in these cells is characterized by a dense development of the smooth ER (*er*). At scattered points in the basal halves of these cells, the reticular form of the ER gives way to the differentiation of paired membranes which are organized in tightly packed arrays called myeloid bodies (*mb*). These have the shape of a biconvex lens. Their fine structure suggests that these are light-sensitive organelles. The basement membrane on which these cells rest is shown at *bm*. Magnification: × 20,000. From Porter and Yamada (1960).

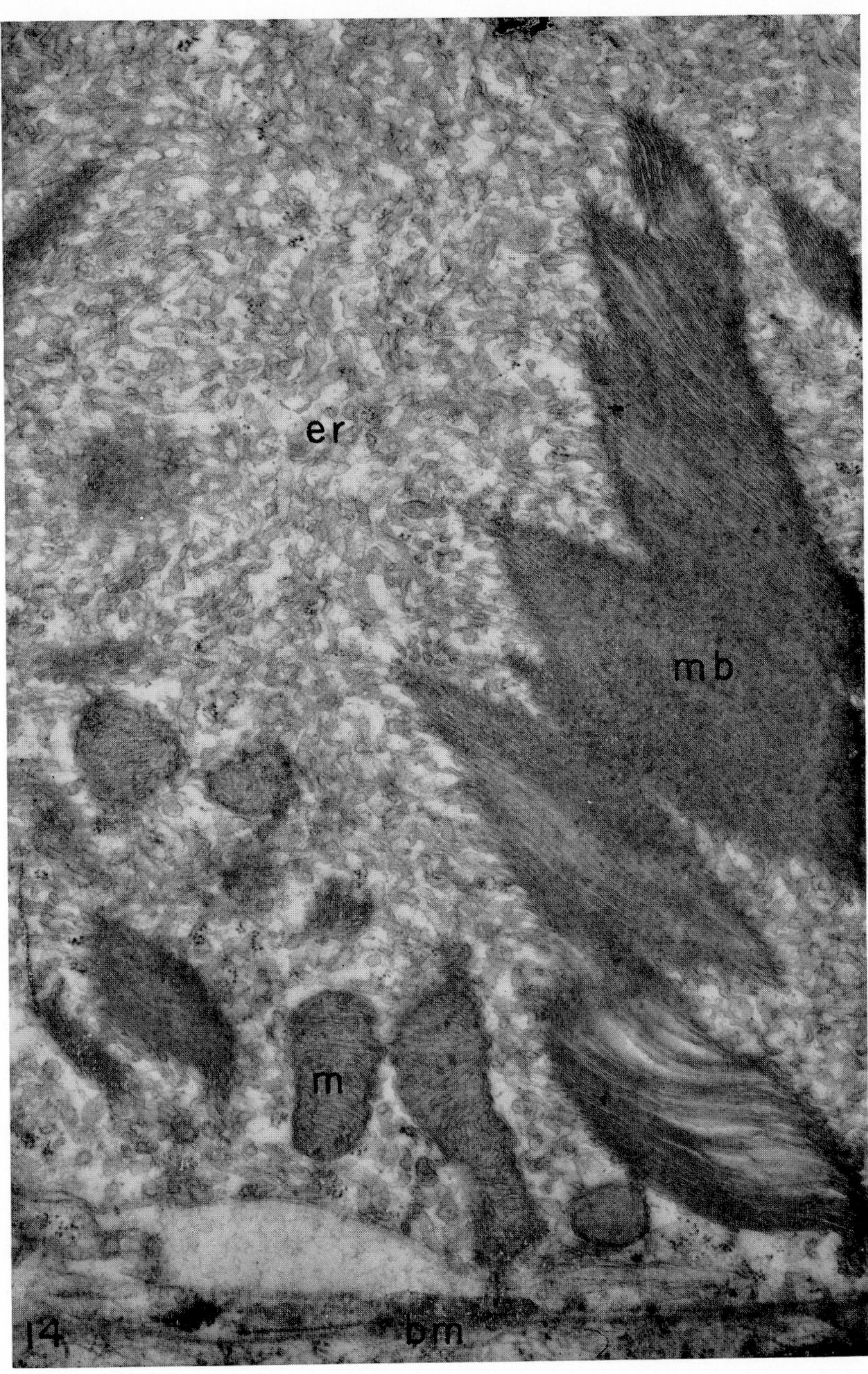
er
mb
m
bm
14

from those of the ER in being smaller and entirely free of granules and in achieving a more intimate association within the complex (Fig. 1). Thus it has a distinctive form which makes it one of the most easily identified components of the cytoplasm. These and other features of the Golgi component are considered elsewhere in this volume (Chapter 8).

A question of interest here is whether the Golgi is related in any way to the cytoplasmic-nuclear complex of membrane-enclosed spaces, which comprises the endoplasmic reticulum. Are the two systems closely enough related to be regarded as one, with the Golgi a special differentiation of the reticulum? There is no simple and direct answer to this. The Golgi component has morphological features that are similar to those encountered in parts of the ER and in no real sense more remarkable in terms of complexity than such differentiations of the ER as the myeloid bodies. In lieu of any good criteria except morphological ones for assessing relationship, the best evidence would come from a demonstration of continuity between the two systems. Palade (1955b) noted a suggestion of this in rat spermatids, i.e., between elements of the idiosome and adjacent vesicles identified with the ER, and proposed that intermittent continuity might be established. Subsequent to this, it has become clear that in some instances at least, notably in the exocrine pancreas cell, the content of the ER finds its way into the Golgi and that it does this during intervals, at least, of continuity between the two systems rather than by diffusion across the cytoplasmic matrix (Palade, 1960). Such connections may be established only during the building phase in the production of secretory granules. Whether this is adequate excuse for relating the two systems remains to be determined. Studies on plant cells have in general failed to provide any clear evidence of continuity between dictyosomes (the Golgi equivalent) and the ER. Possibly a later study of the comparative ontogeny of the reticulum and Golgi complex will yield clues to their relationship, if any.

g. Annulate lamellae. There is, as an additional component of the cytoplasmic ground substance, a remarkable class of structures that has come to be referred to as annulate lamellae. They were noted early by Afzelius (1955) in *Arbacia* eggs and interpreted by him as fragments of

---

FIG. 15. This represents a portion of the cytoplasm of an interstitial cell from the testis of the opossum. The cytoplasm here is strongly acidophilic. As can be seen in the micrograph, the ground substance is filled with a tangle of smooth (agranular) tubules of the ER. Similar formations are found in testicular interstitial cells of man and rat, and there is cytochemical evidence that derived microsomes carry enzymes which mediate the synthesis of steroids. Magnification: × 30,000. From Christensen and Fawcett (1960).

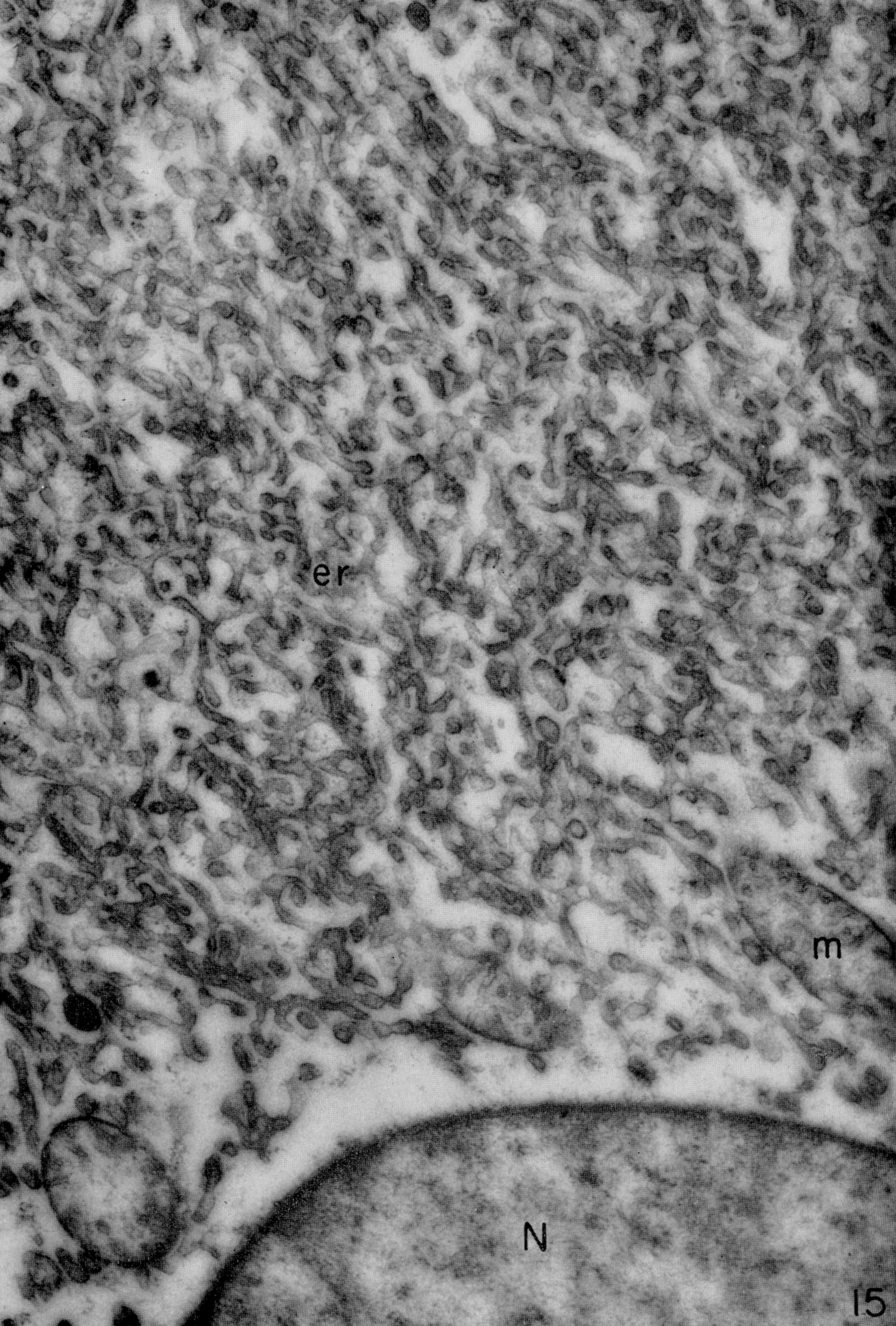
er
m
N
15

the nuclear envelope. Subsequent to this they were identified in a wide variety of cells by Swift (1956).

An individual annulate lamella consists of two parallel membranes separated by distances of 20–40 mμ. At their edges these two membranes fuse, thus giving the component the character of a thin lamellar vesicle. The characteristic which distinguishes these lamellar units from similar units of the ER is, however, the presence of perforations or pores identical with those appearing in the nuclear envelope. It is as though each individual annulate lamella were a replica of a small patch of the nuclear envelope. As in the envelope, each pore shows around its margins a condensation of dense material—an annulus. The dimensions of the lamellae, apart from their thickness, vary greatly.

It is not unusual for these structures to be arranged parallel to each other in piles or stacks and for these in turn to be adjacent or parallel to the nuclear envelope. From this relationship there has developed the idea that this component of the cytoplasm represents a delamination from the nuclear envelope (Rhebun, 1956). It must be noted, however, that individual lamella appear elsewhere in the cytoplasm more or less randomly arranged (Wischnitzer, 1960). In all these occurrences no clear evidence of continuity with what is more properly regarded as endoplasmic reticulum has been encountered. Hence the chief excuse for including these annulate lamellae within this category of cell structure (the ER) resides in the structural similarity they show to the nuclear envelope.

When arranged in large bundles these special lamellae make up a structure that can easily be seen by light microscopy and examined for staining properties. It is found to be strongly basophilic and to owe this to the presence of RNA (Rebhun, 1956). One might expect from this that lamellae would be encrusted with RNP particles, but actually there are few if any when the lamellae are near the nuclear surface (Merriam, 1959). Instead there are only the annuli around the pores and a diffuse dense material between the lamellae. Hence, in this case, RNA is associated with something other than particles.

The origin and role of these striking structures is unknown and a

---

FIG. 16. This shows a portion of the margin of a liver cell fixed 2 hours after refeeding following a 5-day fast. The cell margin is at *X*, the nucleus at *N*. Mitochondria are readily identified. The ground substance of the cytoplasm is filled with small clumps or masses of glycogen (*gl*) interspersed with profiles of smooth-surfaced vesicles. Though not well represented in this micrograph, instances of continuity between the smooth and rough forms are frequently encountered in these cells. Magnification: × 26,000.

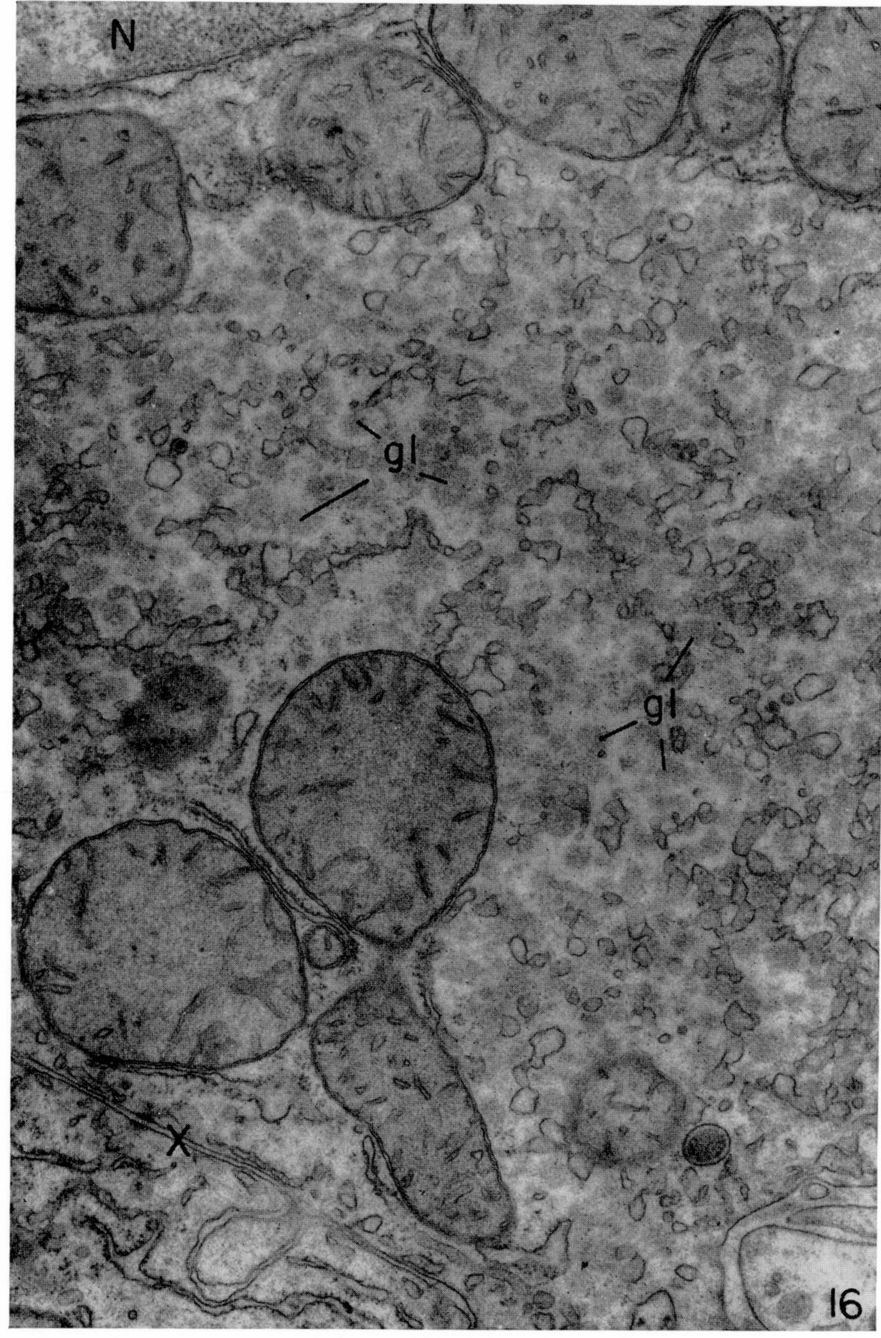
N
gl
gl
X
16

comparative study of their occurrence among cells gives as yet no solid clues. They are apparently common in immature, young egg cells, but have been reported in rat spermatids (Palade, 1955b), in exocrine pancreas cells of *Ambystoma* larvae (Swift, 1956), and in the cells of chicken tumors (Bingelli, 1959).

Their relation, if any, to the phenomenon of nuclear blebbing noted by Gay (1956b) in salivary gland cells of *Drosophila* is in doubt. A greater similarity is seen in the nuclear membrane duplication noted by Barer *et al.* (1960) in the spermatocytes of the locust. It may be pertinent to observe that annulate lamellae seem to be most prominent in cells that are actively growing and multiplying. But apart from this there is no suggestion of how they increase, and contemplating their origins augments the general impression that we know almost nothing about the growth of membranous structures of the cytoplasm.

D. The Endoplasmic Reticulum in Living Cells

The endoplasmic reticulum of the electron micrograph has come to be accepted as closely equivalent to a native cytoplasmic component. Early attempts to see the structure in living cells (Porter, 1953) were less than convincing and did little to diminish doubts then prevalent regarding its reality in such material. When later, however, it was found that the parallel linear striations evident in the Nissl areas of living neurons (Palay and Wissig, 1953) are repeated in electron micrographs of thin sections of such areas as cisternal units of the ER (Palay and Palade, 1955), the existence of the system was given greater credence. Belief in it was further strengthened by evidence of birefringence in the ergastoplasm of living pancreas cells (Sjöstrand, 1953; Munger, 1958), where parallel and closely arrayed elements of the reticulum make up the bulk of the cytoplasm.

Certainly the best images to date of the system in the living cell appear in a report by Fawcett and Ito (1958). Their phase contrast micrographs not only provide a clear view of cisternal elements, but also demonstrate the possibility of following their behavior over a limited period of time. Though the structures involved are usually smaller in one dimension than the limits of light resolution, an optical section of a cisternal element may include enough in depth of the curved structure to provide a resolvable shadow. The nuclear envelope is thought to be visible for similar reasons. In this regard, moreover, it is important to recall the point made by Françon (1954) that structures having sufficient density or refractility will cast a visible shadow in phase contrast microscopy even though their dimensions are theoretically too small for light resolution. Phase contrast images of similar clarity depict the ER in acinar cells of the mouse pancreas after 30 minutes' isolation of the cells in saline (Haguenau, 1958).

Most recently, Rose and Pomerat (1960) have taken cinematographic pictures of cultured cells which show, in phase contrast, structures having the characteristic form of ER elements. Their studies do not, unfortunately, include parallel electron micrographs, as do Fawcett and Ito's (1958) observations on guinea pig spermatocytes; but the conditions which provide the images are those which also permit long-term observation, so much more can be expected of these procedures.

Thus far these observations on living cells have been valuable mostly for demonstrating what is technically possible. In an early study the lability of the system was stressed. It was noted, for example, that the system appeared in a variety of forms in identical cells after simultaneous fixation in vapors of OsO_4 (Porter, 1953). Out of this grew the suggestion that the system is labile and sensitive to changes in the physiology of the cell and that it is prone to disintegrate in cytolysis into a suspension of separate and unconnected vesicles. This conclusion finds support as well in dark-field observations on cultured cells, where the cytoplasm in cytolysis shows an increase in vesicular and refractile structures.

While this is probably one behavior of the system, it is apparently only one, and the less interesting because it represents events associated with cell death. It would obviously be more valuable to follow its behavior during phases of normal function. Again here the observations of Fawcett and Ito (1958) are of special interest. When first isolated and photographed with phase contrast, spermatids show a configuration of ER (cisternal) elements easily referable to the electron microscope image. With the passage of time under *in vitro* conditions, these linear shadows increase in clarity (probably through a change in hydration of the cytoplasmic matrix) and increase also in length. A change in distribution is noted, and they tend to associate in close parallel arrangements. This process continues, the authors suggest, through a confluence of all the ER elements in the cell. Finally, if the cell is kept under such conditions for several hours, the lamellae break up into vesicles—repeating the behavior noted in cultured cells. Thus, one gains an impression of greater stability on the part of the lamellar units than was anticipated from the earlier studies. The organization of the ER characteristic for the cell disappears first, but the cisternal form persists.

This stability of the cisternal elements, i.e., the lamellar form of them, is demonstrated also in their behavior upon isolation from the cell where in the microsome fraction they still appear long and thin (Palade and Siekevitz, 1956). This is true where the fractionation is done in the presence of 0.88 *M* sucrose. When, however, they are isolated in media with a lower tonicity, these cisternal fragments vesiculate before fixation. The isolated units therefore behave as small osmometers and the limiting membrane can be regarded as semipermeable. It can be concluded that the lamellar form is retained in environments where water and electrolytes are kept in low concentration or under conditions, as in the spermatids, where the plasma membrane of the cell is intact and presumably is controlling the electrolyte balance and the character of the cytoplasmic matrix. Once these conditions are no longer regulated, the system fragments and vesiculates.

E. Behavior of the Endoplasmic Reticulum in Mitosis

The significance of the close morphological association between the inner membrane of the nuclear envelope and the peripheral chromatin is at this time impossible to assess, as are also the various observations suggesting that the cytoplasmic part of the system is derived from the envelope by outgrowth, by replication as in the formation of annulate lamellae (Swift, 1956), or by the formation of blebs which detach into the cytosome (Gay, 1956a, 1956b). Though their meaning is unclear, these observations have prompted the somewhat vague proposal that the ER through its cytoplasmic patterns and pathway serves as a vehicle for the transfer of genetic information to the cytosome (Kaufmann and Gay, 1958; Porter and Machado, 1960a). To establish the validity of such proposals will obviously require a lot more information than is currently available and, among other things, some knowledge of the behavior of the nuclear envelope and the rest of the ER during mitosis. Do they for example persist, and if so are they parceled out to the daughter cells according to some plan? These and many other important questions naturally present themselves.

Unfortunately not many reports are as yet available on the fine structure of cells in division. The reasons are several, but outstanding among the difficulties is that of obtaining with thin sections an adequate sampling of the specimen to give an accurate and thorough picture of events.

The story that emerges from the available published reports is as follows. When the nuclear envelope breaks down (as it does in most cases studied), it gives rise to a number of recognizable fragments that retain at first a lamellar form, but later disintegrate into small vesicles (Moses, 1960; Barer *et al.*, 1960). Such remnants persist in the cytoplasm around the spindle and some may be recognized within the body of the spindle. At late anaphase and telophase, similar if not identical vesicular elements associate with the karyomere surfaces and then apparently proliferate or fuse in sufficient number to cover the whole nucleus. Some of these vesicles and parts of the early envelope may be carried into the interior of the nucleus; others persist in the interzonal region of the telophase spindle.

The behavior of the cytosomal ER is also poorly documented. In some cells it appears to fragment and to remain only as isolated vesicles (Porter, 1956). In others it persists in a more characteristically interphase form outside the spindle zone and appears not to participate morphologically in mitosis, though eventually it is distributed more or less evenly to the daughter cells (Porter and Machado, 1960a). In this and other respects, however, the variation in different cells and organisms seems to be great. For example, Ito (1960), in a study of mitosis in spermatogenic cells of *Drosophila virilis,* finds evidence that the ER contributes to the formation of extensive arrays of lamellar elements (appearing as paired membranes) which are applied closely to the surface of the spindle and integrated into the structure of the asters. They persist through mitosis.

With recent improvements in techniques for the preservation and sectioning of plant cells it has been possible to use profitably a classic material—the growing root tip (Whaley *et al.*, 1959; Porter and Machado, 1960a). Just how broadly the observations made here apply among dividing cells remains to be seen, but the material offers considerable promise for a better understanding of these phenomena.

The behavior of the envelope and cytoplasmic portions of the ER are perhaps better depicted here than in any other material thus far examined. It is evident, for example, that after the initial breakdown the envelope persists in relatively large lamellar units within the margins of the spindle. The ER elements belonging to the cytoplasm of the parent cell remain outside the spindle. There is some indication that as mitosis enters metaphase the remnants of the nuclear envelope move to the poles of the spindle. Whatever their origin, certain elements in this region begin a proliferation which sends extensions of the ER into the spindle and into regions between the chromosomes (Fig. 18). Simultaneously or just prior to this, the ER and nuclear-envelope remnants of the parent cell are distributed more or less evenly between the two poles of the cell to form two daughter systems. Thus proliferation occurs simultaneously with separation. Though these developments bring ER elements (possibly all developed from the old nuclear envelope) into the interior of the spindle, it has not been possible to recognize any evidence of an active participation of this component in the events of mitosis. This invasion into the spindle, partly from the poles and partly from the sides, places the ER into a favorable position to form the new envelope. Thus as the karyomeres move to the poles of the spindle they are close to potential membrane elements, and it is possible to see associations as early as late anaphase (Fig. 18). By late telophase the new envelope is complete (Fig. 17). Occasionally pieces of the ER are seen within the daughter nuclei as though accidentally trapped by envelope formation. This observation suggests to this reviewer that the association of envelope elements and chromosomes is a haphazard process and apparently wasteful. It is impossible, however, from evidence available and the thin sections one has to study to see the full sequence of events and to determine to what extent, if any, these are ordered.

Elements of the spindle ER that do not participate in envelope formation extend down into the interzone and approach the equator of the spindle. At this level the advancing margins reticulate to form a lattice of microtubules contributed by both of the daughter systems. Within this lattice the cell plate first appears (Fig. 17). This is an event peculiar to plant cells and one of great interest. Obviously the close morphological relation of the ER to the plate suggests functional responsibilities, but what these are is in doubt.

The observations thus far reported do not, unfortunately, tell us very much about the division of the parent envelope, i.e., whether it is in any sense equal. Likewise, one cannot be sure to what extent the daughter systems proliferate from fragments of the old envelope. One gets the impression, however, that there is a good deal of randomness in the events depicted at this juncture.

F. The Endoplasmic Reticulum and Cell Differentiation

One of the most interesting questions relative to the endoplasmic reticulum is that of its relation to differentiation. This review has already referred to the system in the fully differentiated cell as an expression of differentiation. Certainly the ergastoplasm of the pancreatic acinar cell

---

FIG. 17. Two cells of an onion root tip just after mitosis. The cell plate (*cp*) has almost completed its lateral growth to the cell margins. The nuclei (*N*) are in late telophase. The dense meandering lines in the cytoplasm represent sections through lamellar units of the endoplasmic reticulum (*er*). At higher magnifications they appear, like the nuclear envelope, as two dense lines with intervening light space. Continuity with nuclear envelope is evident at arrows. At the level of the cell plate the profiles of the ER represent sections through tubules, for at this level the ER transforms into a lattice of tubular elements. Proplastids are indicated at *pp*, dictyosomes (Golgi component) at *d*, and mitochondria at *m*. The cell is surrounded by a thick wall (*cw*). Magnification: × 7000.

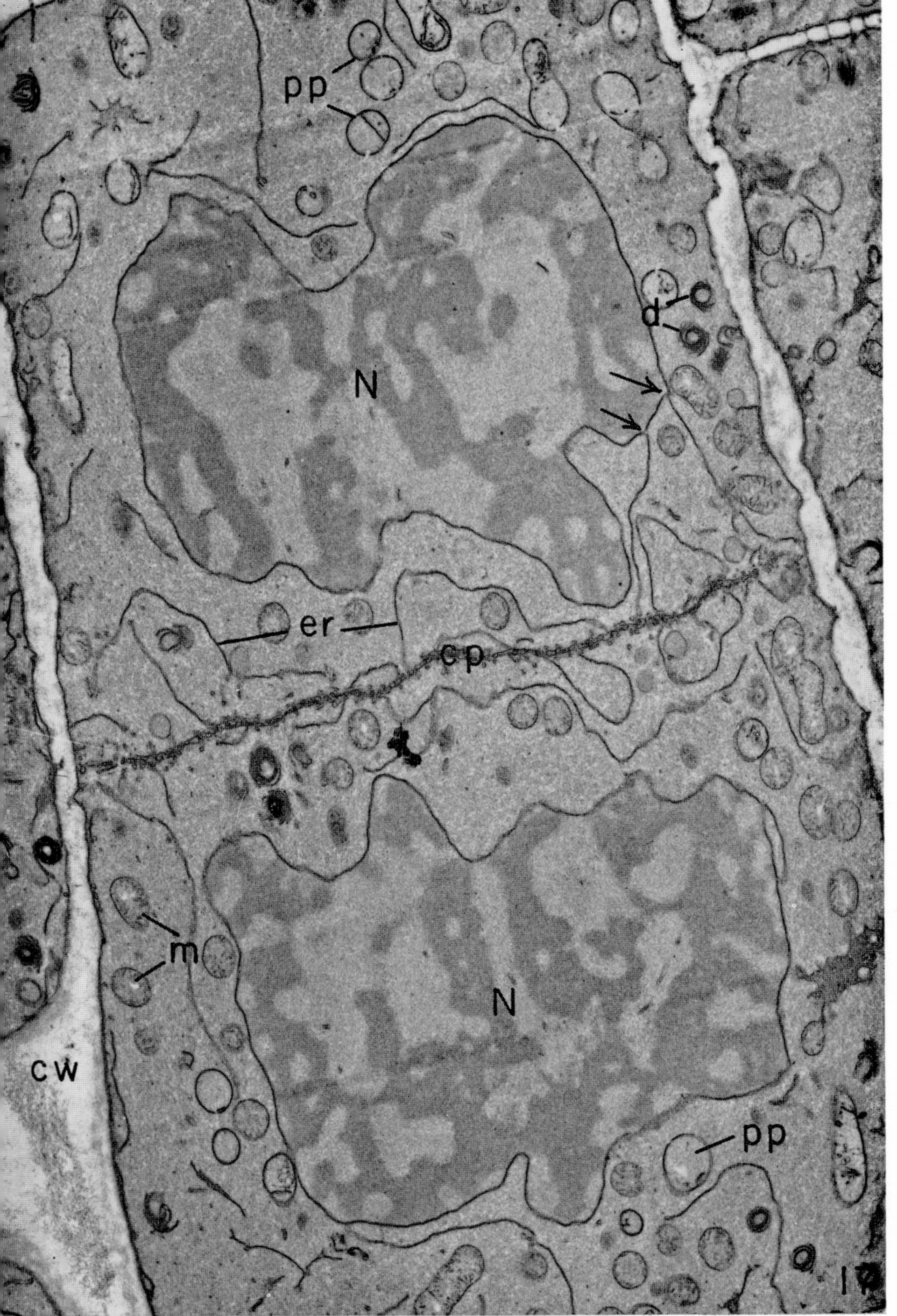
pp
N
d
er
cp
m
N
cw
pp
17

should be so regarded, for it is a highly specialized structure with a distinct function to perform. The same conclusion is probably justified in other instances. In a discussion of patterns in the ER and their role in the cell (see below), it is suggested that cell form, surface configurations, and even the distribution within the cell of formed structures may be greatly influenced by patterns in the ER. This obviously implies that the ER, with dependence on the nucleus alone, may exert an important influence on the development of what are commonly accepted as expressions of cellular differentiation.

It is interesting, therefore, to look into its behavior during the process and at least summarize the fairly meager information available.

First, it should be noted that as a rule the endoplasmic reticulum is not a prominent structure in the relatively undifferentiated cell. This has been repeatedly observed. It was noted relative to the cells in the crypts of intestinal epithelium of the rat (Palade, 1955a). The cytoplasmic matrix of such cells is always dense with RNP particles, but there are only few profiles of the ER except for the nuclear envelope. These features are so constant for rapidly proliferating, undifferentiated cells as to be useful for identification (Porter, 1954a). Palay notes a scarcity of ER profiles in the marginal (basal) cells of the sebaceous gland, but an elaborate development of the system in the more central and mature cells. In a study of the *Hydra,* Slautterback and Fawcett (1959) described the interstitial cell from which the cnidoblast develops as intensely basophilic, presumably related to a dense population of RNP particles, and almost entirely devoid of elements identifiable with the ER (Fig. 19). Only later, after a series of divisions, when the cell begins its differentiation does the system become prominent (Fig. 20). Another interesting observation to come from these studies is that the ER appears to regress from its highly organized state as the nematocyst matures. The undifferentiated cells of the blastema have been noted by Hay (1958) to possess early in differentiation only the simplest expression of a reticulum. And in normal embryonic growth and differentiation of the nervous sys-

---

FIG. 18. Onion root tip cell in anaphase of mitosis. The chromosomes (*ch*) have moved to the poles of the spindle. Long profiles (*er*) of the ER, typical for these plant cells, have invaded the spindle from the poles and sides (between the chromosomes) by proliferation, presumably from fragments of the nuclear envelope. At one or two places (arrows) elements of this spindle ER have returned to the chromosome surfaces as the first stage in construction of new envelope. The complex of ER elements at the equator of the spindle differentiates into a lattice work of tubules in which the new cell plate and wall are formed. Magnification: × 7000. From Porter and Machado (1960a).

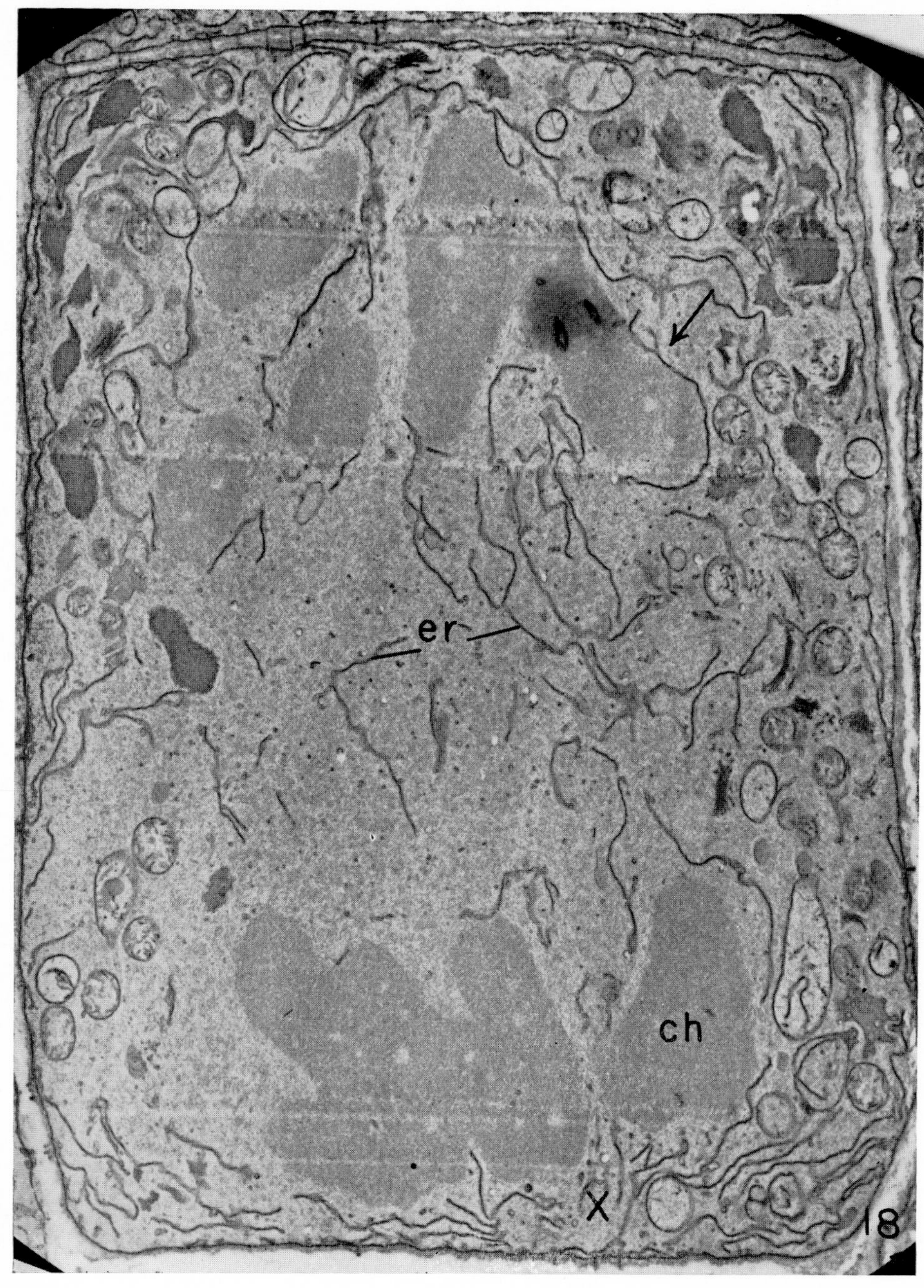
er
ch
X
18

tem of chick embryos, Bellairs (1959) has quantitated the development of the ER and notes a three- to fourfold increase in its amount in the progression from embryonic stage 4 to stage 36.

Similar examples could be cited from other sources, including differentiating plant tissues. In all instances, the ER seems initially to be represented by the nuclear envelope, scattered vesicles in the cytoplasm, or a tracery of fine tubules possibly extending from the nuclear envelope. The form of the meagerly developed ER in these early cells needs further exploration.

In this connection it is interesting to examine the character of the ER in the earliest stages of development after fertilization. An examination of two-cell stage in the rat ovum (Sotelo and Porter, 1959) found the system represented (though not strikingly so) whereas in the mature oöcyte it is difficult to identify. Even the Golgi component is said by Odor (1960) to be absent from the egg during the first maturation division and just prior to fertilization. One can conclude, therefore, that the mature ovum and the cells of the earliest embryonic stages are similar to basal cells and other undifferentiated, proliferating units in possessing remarkably little that can be called an endoplasmic reticulum, except for the nuclear envelope. Obviously, electron microscopy of embryonic tissue offers many possibilities for achieving a better understanding of the role of cytoplasmic organelles in the processes of differentiation.

On the basis of available evidence it would appear that the ER does not attempt an elaborate development during the proliferative phases of any cell line. The possibility is suggested that with each division, after the nuclear envelope is rebuilt, a new ER is formed which survives and performs a function only during that generation interphase, to be replaced by a slightly different ER in the successive generation, until the

---

FIG. 19. Micrograph of undifferentiated interstitial cell or early cnidoblast from ectoderm of *Hydra*. The limits of the cell are represented by the lines which cross the corners at the upper left, lower left, and lower right. The nucleus (*N*) is surrounded by a typical, double-membraned envelope. Otherwise the ER is meagerly represented (a few profiles of vesicles at arrows) and the ground substance appears mostly as a suspension of small dense particles (probably ribosomes). These features are characteristic of embryonic (undifferentiated) cells. Magnification: × 12,000. From D. B. Slautterback and D. W. Fawcett (1959).

FIG. 20. A cnidoblast of *Hydra* in an advanced stage of differentiation. Here, when the cell is most actively engaged in synthesizing nematocyst materials (*nm*), the endoplasmic reticulum is as highly developed as in an exocrine pancreas cell. The change in differentiation from the image in Fig. 17 is striking. Magnification: × 13,000. Micrograph by courtesy of D. B. Slautterback and D. W. Fawcett. "Developmental Cytology" (D. Rudnick, ed.). Ronald Press, New York, 1959.

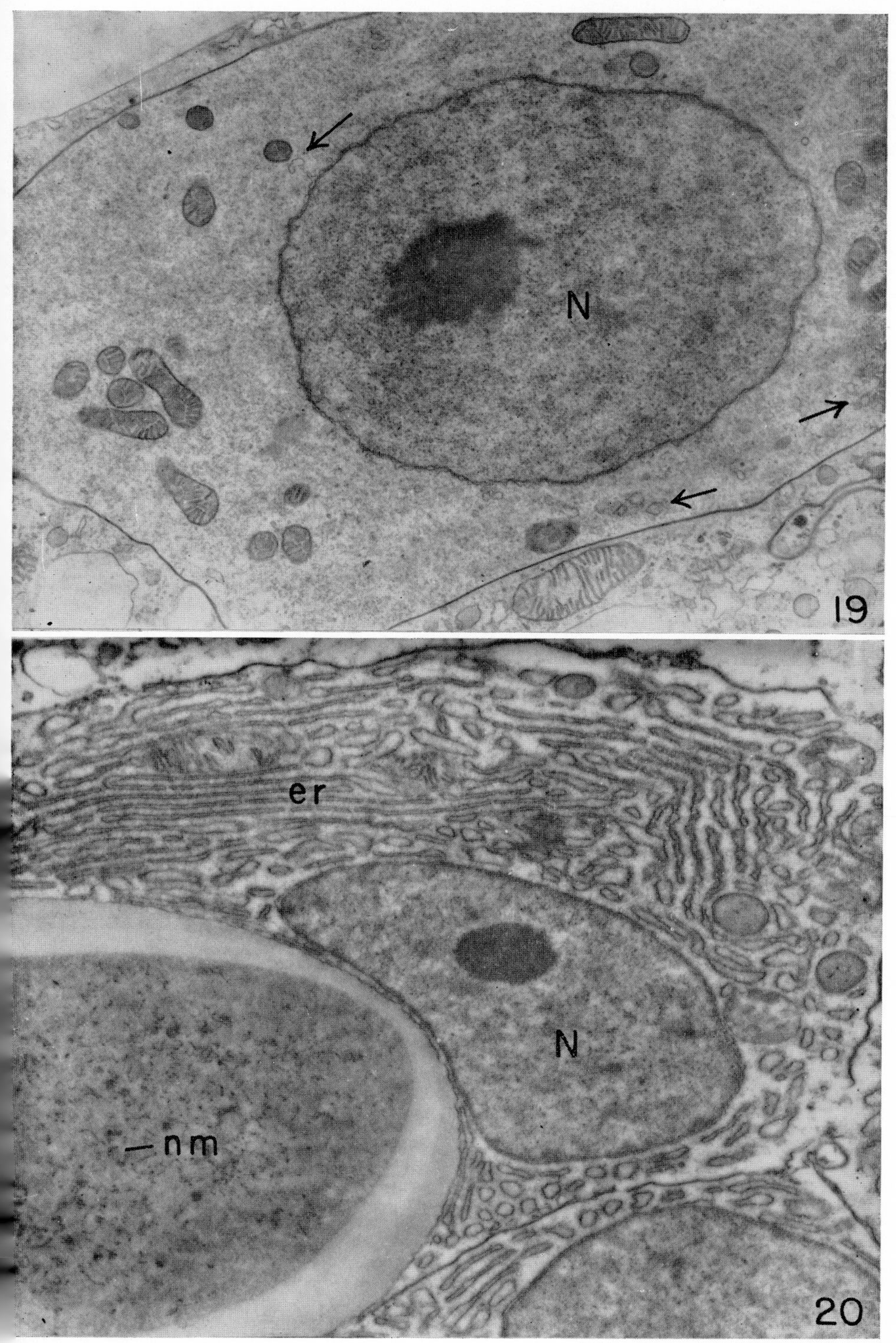
N
19
er
N
— nm
20

final one, when the form characteristic of the fully differentiated unit appears.

G. *Functions of the ER*

The roles of the endoplasmic reticulum in the cell are poorly understood, and the problem of defining them is one that will probably occupy cell biologists for some time to come. There is good and constantly improving evidence of its involvement in the synthesis and segregation of proteins for export from the cell. There is suggestive evidence that certain forms of the system (broadly defined) play a role in glycogen storage and release and in the synthesis of steroids for secretion. It seems to function also in intracellular transport. But beyond the data supporting these statements the available information regarding functions is extremely limited and the discussion highly speculative.

1. *Protein Synthesis*

The recent biological literature contains a number of good reviews on the apparent involvement of the endoplasmic reticulum or its derivative, the microsome, in protein synthesis (see Loftfield, 1957; Hogeboom *et al.*, 1957; Haguenau, 1958; Palade, 1958; and Siekevitz, 1959). These not only report on the current status of knowledge, but also review the development of this knowledge. In this account, therefore, we shall be more general and simply summarize what is currently recognized as valuable and established information.

It has already been pointed out that this complex, membrane-limited structure in the cytoplasm exists in at least two forms—one rough, with particles on its surface, the other without particles, in other words smooth or agranular. The former has now repeatedly been shown to coincide with what appears in the optical image as discrete masses of (filamentous) cytoplasm having an affinity for basic dyes and containing a relatively large amount of RNA—masses that have long been recognized as common in cells engaged in synthesizing proteins for secretion. The rough form of the ER is therefore known to represent the ergastoplasm of Garnier (1897) and even without the available biochemical evidence it would doubtless be described as functional in protein synthesis.

It is, however, the definitive demonstration of the origin of the liver microsome (Claude, 1941, 1943) in large part from the ergastoplasm, or rough ER, that provides most of our knowledge of the function of at least this division of the system. This correlation, early suggested by the electron microscope observations of Slautterback (1953) and later by Chauveau *et al.* (1955) was firmly established by Palade and Siekevitz (1955, 1956). Whereas formerly the microsome had been a fraction of the cell defined largely by the technique of centrifugation used to isolate it (Claude, 1941), it now became a part of a well-characterized intracellular system and all the data on liver microsomes became applicable to the endoplasmic reticulum. Thus it could be stated that the rough form of the ER is capable, *in vivo* (or *in vitro* as microsomes), of incorporating amino acids into proteins [beginning with the pioneer efforts of Hultin (1950); Borsook *et al.* (1952)]. It is equally valid to speak of the ER (derived from smooth as well as rough ER) as possessing glucose-6-phosphatase activity (Hers *et al.*, 1951) and as containing high concentrations of DPN-cytochrome c reductase (Hogeboom, 1949) and 40–50% of the RNA in the whole tissue (liver) from which they are derived. These are the now well-known properties of microsomes and, with little doubt, the endoplasmic reticulum from which they are derived.

The distribution of function among these components of the ground substance has now been even more finely differentiated. As noted above, the rough form of the ER owes this quality to the presence on the membrane surfaces of particles. Thus there are at least three parts to the microsome or fragment of the ER: the particles, the membrane, and the content. With deoxycholate (DOC) the particles can be separated from the membranes and by electron microscopy are shown to be relatively free of contamination (Palade and Siekevitz, 1956). Thus their properties, distinct from those of the intact microsomes, have been investigated, and it is now known that the particle as isolated is extraordinarily rich in RNA. Here again isotope incorporation studies have described the particle as having the capacity to combine free, labeled amino acids into proteins (Littlefield *et al.*, 1955). In spite of these clear-cut results the "ribosome" story requires further clarification, for there is evidence of RNA in membranes of the so-called smooth ER (Kuff *et al.*, 1956; Moulé *et al.*, 1960). The fact remains, however, that the particle is an important site of synthesis and would appear to be responsible for determining the character of the protein synthesized. This conclusion finds strong support in the recent studies of Siekevitz and Palade (1960a, b) on the relation of the RNP particles to the synthesis of pancreatic enzymes.

The rest of the microsome, which is to say the *in situ* cisterna of the rough ER, is apparently involved in the segregation of the product of synthesis from the site of synthesis. The swollen ergastoplasmic sacs were interpreted by Weiss (1953) as containing zymogen, but the observations of Palade (1956) that, in guinea pig pancreas, relatively large granules having the density of zymogen granules develop within the cisternae of the ER are far more definitive (Fig. 12). These granules have now been isolated and analyzed and found to be similar in enzymatic content to the zymogen granules (Siekevitz and Palade, 1958a, b, c). Thus the story of protein synthesis in this case has achieved remarkable completeness. There is, in addition, some evidence that a morphological continuity between the rough ER and the Golgi component provides for the movement of prozymogen from the ergastoplasm to the Golgi, where it is apparently packaged for export from the cell (Palade, 1960).

While these studies provide us with perhaps the most striking and completely analyzed case of protein synthesis in the ergastoplasm where structure and biochemistry are correlated, it is not the only one. In another somewhat parallel investigation, the albumen-producing cells of the hen's oviduct have been fractionated and the product of synthesis recognized as the content of the cisternae (Handler *et al.*, 1957). No definite granule was evident here in the intact cell; the content of the cisternae appeared instead to be homogeneous. On the basis of this latter observation, one may assume that other instances of expanded cisternae depict the accumulation of products synthesized by the system. Thus the cisternae in the thyroid cell have been recognized as being filled with colloid (Dempsey and Peterson, 1955; Wissig, 1960). The cells of the coagulating gland of the mouse prostate show enormously expanded cisternae that occupy the greater part of the cytoplasm (Brandes and Portela, 1960). The same may be noted in plasma cells, though no one has identified the content as rich in immune bodies (Fig. 13). And finally, the ergastoplasm of fibroblasts and chondrocytes active in collagen and cartilage production is highly developed and frequently shows within the cisternae an accumulation of dense material (Porter and Pappas, 1959; Godman and Porter, 1960).

2. *Intracellular Transport*

The point was made earlier that the reticulum by its basic structure and distribution provides the cytoplasm with an internal phase distinct in character from that of the cytoplasmic matrix. Into this internal phase the products of biosynthesis may be segregated from the rest of the continuous phase, the cytoplasmic matrix (the "new" ground substance). Assuming that the dimension and content of the channels would permit it, these products of biosynthesis and other selected metabolites might reasonably diffuse from the perinuclear space to all parts of the cytoplasm or vice versa. This potential functional property of the ER has been suggested before (Porter, 1956), but without evidence to support it. In some respects the idea is difficult to accept. The dimensions of the space within the channels of the system are usually small (75–100 mμ) and would presumably permit little flow in the usual sense. One must, however, recognize that, without knowledge of the behavior of the system in the living cell and the effects of the "jostling" it receives from the minor contractions and sol-gel transformations in the surrounding matrix, any assumptions on what the system can or cannot permit are highly speculative. At very least, it would seem that transport of smaller molecules by channeled diffusion is likely. There is also the possibility, as suggested by Bennett (1956), of membrane flow from sites of membrane synthesis to sites of breakdown. Presumably such membrane flow would carry with it the adjacent content of the ER cavities as well as the matrix component (such as RNP particles) attached to the external surface. With so little information available concerning the behavior of the system in the living cell it is difficult to suggest how transport may actually be effected. That the content does move seems, however, to be supported by a number of observations.

One of the most valuable observations in this regard is described in a series of papers by Palade (1956, 1959) and Siekevitz and Palade (1958a, b, c). As noted earlier, in the guinea pig pancreas the new-synthesized proteins (enzymes) make their way across the cisternal membranes from the ribosomes and then condense out as large intracisternal granules. From this initial site of condensation, then, they must be transported (as granules or dissolved) to the Golgi complex, where they are again condensed into granules. Since the Golgi complex seems a stabile structure distinct in many respects from the particle-studded elements of the ergastoplasm, it is not likely that the membrane of one is "flowing" into the other. Instead the intracisternal "zymogen" would appear to be transported through the channels of one system into the other. Subsequently the required continuity of channels was found (Palade, 1960). Although such granules have been encountered thus far only in the cisternae of guinea pig pancreas, it is probable, as noted above, that the dense material reported in other forms represents the product of synthesis and that the same phenomenon of transport from ergastoplasmic cisternae to condensation in a Golgi vesicle occurs in a majority of instances where the cell product is packaged before secretion (Slautterback, Fawcett, 1959; Palay, 1958).

There are, however, other instances where the contents of the expanded cisternae seemingly are not packaged into granules within the Golgi complex but instead are contributed directly to the cell's environment by focal externalization of the cortical cytoplasm at multiple points. It would appear that plasma cells must employ this method in discharge of antibody globulins. There is evidence that in chondrocytes, engaged in producing cartilage ground substance, large sections of the cytoplasm including ER elements are "pooled" and then discharge, leaving excavations in the cell surface (Godman and Porter, 1960). It is assumed that these losses are repaired through regeneration to the degree that an aging cell repairs. An analogous situation is encountered in plant cells when they are engaged in producing a cell wall. The plasma membrane disappears and the cell cortex, with terminal elements of the ER, is shed to become part of the new layer of wall (Porter and Machado, 1960b). These alternate methods of secretion, which are not as yet well defined, do not detract from the basic conclusion that the contents of the endoplasmic reticulum do move within the spaces of the system and independently of the limiting membrane.

The phenomenon of transport within the ER is supported also by the studies of Palay and Karlin (1959) on quite another material. They followed the absorption of fat by epithelial cells of the intestinal villus and observed that small fat droplets ($\sim$ 50 mμ in diameter), which appeared to pass through the terminal web via pinocytotic vesicles, were taken up by the ER and transported within its channels to the lateral margins and thence through intermittent connections into the intercellular spaces. Particle-studded elements of the ER and even the nuclear envelope were observed to contain the droplets within 30 minutes after intragastric instillation of corn oil (Palay, 1960). Besides demonstrating transport within the ER, the observations are descriptive of continuity within the system. Palay and Karlin suggest, therefore, that the fat particles which appear in the apical cisternae of the ER move in continuous channels through the cell. They propose further that since the fat is picked up by what they interpret as pinocytic activity and passed directly into the lumina of the ER (which are intermittently continuous with extracellular space) the entire passage through the epithelium "may be considered as extracellular."

This interpretation of events in fat absorption is discussed further here because it relates to speculation that the ER represents essentially an elaborate infolding of the cell's surface (Robertson, 1959) and further that materials taken up in small bites (quanta) at the cell's surface are discharged helter-skelter into the lumina of the ER. Thus a space in which the products of synthesis accumulate becomes also a space into which all manner of metabolites and nonmetabolites from the environment are dumped. These ideas have grown out of observations on pinocytic activity at the cell surface (Palade, 1956) and out of observations descriptive of occasional and intermittent continuity between membranes of the ER and the cell surface (Epstein, 1957). The idea of direct continuity with the outside in the uptake of materials is still open to question.

Apart from the Palay-Karlin observations (see alternate interpretation below) there is no real evidence to suggest that materials taken up by pinocytosis are discharged directly into the ER. In two instances, where the question was approached experimentally (Hampton, 1958, and Karrer, 1960) in a study of the uptake of particulate materials by the rat liver cell and alveolar macrophages, evidence of continuity was *not* provided. It appeared rather that in the normal course of events, pinocytized materials remain in vacuoles,[1] essentially digestive vacuoles derived from the cell surface, where they are presumably hydrolyzed if possible and made available to the cell by absorption. A barrier of two membranes (plasma and ER membrane) and a layer of the cytoplasmic matrix would seem always to intervene on the uptake side between the extracellular space and the cavity of the ER.

Even the Palay-Karlin observations do not refute this concept. As the authors admit in their paper, clear evidence of continuity between pinocytic vesicles and ER is not available and it is difficult to account for all the rapid fat uptake in a given time by this process. Furthermore, the triglycerides in the absorbed fat, observed as chylomicrons, are not the same as those placed in the animal's stomach (Peterson, 1960); they have been altered chemically. It seems therefore more proper to interpret the granules Palay and Karlin see in the pinocytotic vesicles as the residue of incomplete hydrolysis and to believe that only the products of triglyceride hydrolysis pass the plasma membrane and enter the cytoplasm. The lipid droplets that appear in the cavities of the ER could represent triglycerides newly synthesized through the activity of enzymes resident in the membrane that limits the smooth ER vesicles and tubules. Thus fatty acids, mono- or diglycerides which diffuse through the matrix of the cytoplasm from the apical pole of the cell could be incorporated into fats again at all levels in the ER and segregated in the internal phase of the system in much the same manner as amino acids are put into the zymogenic proteins of the pancreas. The requirement for rapid uptake would be satisfied by this process as would also the evident differences in chemistry between the fat absorbed and that found in the cisternae.

It is justifiable to conclude that transport within the ER does occur, though the mechanism needs clarification. On the basis of available observation it would seem appropriate to regard the ER as a special differentiation or organelle of the nonbacterial cell, designed, among other things, for the intracellular transport of metabolites and the widespread (but patterned) distribution of enzyme-rich surfaces or membranes.

[1] Karrer (1960) makes the important point that membranes limiting the phagocytic vesicles show the "triple-layered" structure of the plasma membrane.

Thus a structural device is provided for some degree of equilibration among all parts of the cell and, via the nuclear envelope, the genetic material of the nucleus.

3. *The Endoplasmic Reticulum and Other Intracellular Synthetic Mechanisms*

Among the various differentiations shown by the endoplasmic reticulum, the most prominent and common one is that characterized by its association with the RNP particles, or ribosomes. As noted earlier, the typical unit structure in this form of the ER is a relatively large, lamellar cisterna that frequently appears in parallel orientation with other cisternae to form arrays or stacks.

Within the other large division of the reticulum, the particle-free, or agranular, form, there exist a number of configurations and also evidence of morphologic association with several metabolic processes. For example, in the cell engaged in the synthesis of lipid materials for export, it is common to find the cytoplasm filled with smooth-surfaced tubules. Thus Palay (1958) has reported this form in cells of the sebaceous glands (Meibomian gland of the rat). The secretory (lipid) droplets form initially as unusual, ordered structures in continuity with the tubular elements of the smooth ER and later becomes sudanophilic and soluble in lipid solvents. For present purposes the clear association of an agranular form of the ER with this kind of synthesis is simply noted.

A similar development of the smooth ER has been reported in cells that produce steroid hormones. Christensen and Fawcett (1960), for example, have described a tangled network of interconnected tubules in the interstitial cells of the opossum testis (Fig. 15). This ER is entirely devoid of dense particles and the cells are intensely acidophilic. This finding is of special interest because a biochemical study on similar material by Lynn and Brown (1958) has provided evidence that enzymes involved in the production of testosterone from progesterone are associated with microsomes and, while inactivated by lipases, are stable to the action of ribonuclease at concentrations which normally inhibit protein synthesis in liver homogenates. It would appear that the steroid enzymes in question are associated with the RNA-free microsomal membranes. Fetal zone cells of the human adrenal provide still another example of a richly developed smooth ER in association with steroid production (Ross *et al.*, 1958).

Prominent differentiations of the smooth ER have been encountered in other cells in which the probable function is not so apparent (Porter, 1956). In the cells of the pigment epithelium of the retina, the cytoplasm is packed with vesicles and tubules in a compact lattice. It is conceivable that the system here plays an important role in the metabolism of vitamin A and its conversion to retinene for use in the construction of visual pigment. If so, we can record another instance where a tremendous development of the smooth ER is found in cells active in the metabolism of unsaponifiable lipids.

Another, though less striking, development of the smooth or agranular endoplasmic reticulum is found in liver cells. Its presence has been repeatedly noted, and variations in its appearance with fasting and refeeding have been examined by Fawcett (1955). In the normal cell it is prominent along the cell margins and to varying degrees in regions of the cytoplasm that contain glycogen. In the presence of a toxin, 3′-methyl-4-dimethylaminoazobenzene, which destroys the capacity of the cell to store glycogen, the system undergoes a tremendous hypertrophy (Porter and Bruni, 1960a). This led to a study of the relation of the system to glycogen metabolism. Thus in fasted (5 days) animals the smooth vesicles and tubules of the ER were found concentrated around the residual glycogen, and at the margins of these masses numerous examples of continuity between smooth and rough ER were observed. Upon refeeding there was an apparent increase in the amount of smooth ER, and this retained a specific association with the accumulating glycogen (Fig. 16). The morphological association suggests a correlated function, but whether in glycogen storage or lysis is not clearly indicated and will not be known without suitable biochemical and physiological experiments. In this same connection, a number of additional observations are important. In the paraboloid of the turtle retina, for example, Yamada and Porter (1961) report an association of a smooth form of the ER with glycogen storage similar to that observed in liver cells. Then, also in plant cells there is an elaborate development of the smooth ER along surfaces where cellulose walls are being formed (Porter and Machado, 1960b) and a clear morphological intermingling at the site of polymerization. In none of these instances has the study progressed far enough to define the role of the reticular elements, but the obvious implication is that the smooth ER is carrying enzymes and metabolites important in physiological events taking place within the localized regions where it is found.

4. *Intracellular Impulse Conduction*

The existence of the endoplasmic reticulum in the cytoplasmic ground substance creates as already noted a continuous and elaborate two-phase system in the cell separated by a membrane. If the membrane possesses properties of permeability characteristic of other biological membranes, it is probable that the two phases will differ in ionic species and concentrations and a membrane potential will be established. This hypothesis provides, then, conditions for depolarization and impulse conduction between all parts of the continuous system. The full significance of this for the cell is difficult to imagine.

This line of thought relative to the endoplasmic reticulum was initiated by electron microscope observations on the character of the ER in striated muscle cells of vertebrates (Bennett and Porter, 1953; Bennett, 1955; Porter, 1956; Porter and Palade, 1957; Edwards *et al.*, 1956; Peachey and Porter, 1959). The salient features of the ER in these muscle fibers are (1) that it is clearly organized with respect to the sarcomeres of the myofibrils, as though intimately associated functionally with the contractile process, and (2) that it is continuous across the fiber from close contact with the sarcolemma to the deepest fibrils. The system was recognized at once by thoughtful physiologists as a structure suitable for the lateral conduction within the fiber of the excitatory impulse. Thus interpreted, the existence of the system solved the mystery of how the central fibrils in a fiber are excited to contract almost simultaneously with those at the periphery, even though removed from the sarcolemma by distances as great as 50 microns. Whether or not the entire membrane system, referred to in muscle as the sarcoplasmic reticulum, is involved in this assumed impulse transmission remains to be determined, but that some part (such as the special differentiations opposite the I band) of it is, seems to be indicated by the experiments of Huxley and Taylor (1958) which tested with microelectrodes the relative responsiveness of different levels in the sarcomere (for details, see Chapter 7, Vol. IV).

If, on the basis of these observations, one is permitted to assign electrical properties to the phase-separating membranes in muscle fibers, it is appropriate to make the assignment general wherever a continuous ER exists in cells. Thus most cells, at intervals at least, are provided with a signal transmitting system which could influence a wide variety of enzyme-controlled processes. Though ignorance regarding all this is at the moment profound, it seems important to include the hypothesis as descriptive of one of the several possible ER functions.

III. Concluding Comments

The observations reviewed here represent of course only a fraction of the new information becoming available at this time on the ground substance and its components. Even the membranous system on which the major attention has here been focused deserves a larger discussion. The phenomena represented by the patterned organization of the system are in themselves deserving of an extended examination. It is apparent that in the fully differentiated cell the ER should be regarded as one expression of differentiation in the same sense that myofibrils are so regarded. In each type of cell it shows a characteristic or preferred form by which, indeed, the cell can be recognized. This pattern is labile and shows several changes or modulations with function. A good example of this is found in the response of the system to hormonal stimulation. In epithelial secretory cells of the mouse seminal vesicle it is normal for the ER (ergastoplasm) to be essentially as prominent as in the exocrine pancreas. Under the influence of exogenous androgen, the cells increase slightly in height and the cytoplasm becomes abnormally packed with particle-covered cisternae. The system obviously responds to hyperstimulation by an increase in total surface of ER membranes. Following castration, on the other hand, the cells lose height and total volume and the ER is greatly reduced in prominence (Deane and Porter, 1960).

In this example of form modulation with function, as in other instances, the general pattern of the ER seems to be retained though the degree of development varies. In the case of the castrate animal, the system is reminiscent of an early stage in development.

The significance of these configurations in the ER is poorly understood at the present time. Obviously if enzymes are built into the membranes and these membranes are distributed in patterns and local differentiations, then the product of synthesis, if destined to accumulate, will be located relative to the ER pattern. This is particularly well exemplified by the apparent correlation between the organization of the ER in developing plant cells and the distribution of secondary thickenings during cell wall formation (Porter and Machado, 1960b). To the degree that it controls the form of the cell and the distribution of cell components whose formation it influences, the ER becomes a kind of cytoskeleton. In the same context, it represents the formerly invisible purveyor of genetic information from nucleus to cytoplasm.

We have considered *in extenso* only one major component of the ground substance which before the introduction of electron microscopy was almost unknown. Through its newness, striking characteristics, and isolatability it has captured attention to the partial neglect of several other elements of equal interest concerning which only fragmentary observations are available. The fibrous systems of the ground substance are particularly representative in this regard. They are present almost universally among cells and may challenge the membrane systems for the distinction of performing as the cell's skeletal framework.

References

Afzelius, B. A. (1955). *Exptl. Cell Research* **8**, 147.

Anderson, E., and Beams, H. W. (1956). *J. Biophys. Biochem. Cytol.* **2**, *Suppl.* 439.

Andersson-Cedergren, E. (1959). *J. Ultrastruct. Research Suppl.* **1**.

Bahr, G. F., and Beermann, W. (1954). *Exptl. Cell Research* **6**, 519.

Bang, B. G., and Bang, F. B. (1957). *J. Ultrastruct. Research* **1**, 138.

Barer, R., Joseph, S., and Meek, G. A. (1960). *In* "Proceedings Fourth International Conference on Electron Microscopy," p. 233. Springer, Berlin.

Bellairs, R. (1959). *J. Embryol. Exptl. Morphol.* **7**, 94.

Bennett, H. S. (1955). *Am. J. Phys. Med.* **34**, 46.

Bennett, H. S. (1956). *J. Biophys. Biochem. Cytol.* **2**, *Suppl.* 99.

Bennett, H. S., and Porter, K. R. (1953). *Am. J. Anat.* **93**, 61.

Bernhard, W., Gautier, A., and Oberling, Ch. (1951). *Compt. rend. soc. biol.* **145**, 566.

Bingelli, M. F. (1959). *J. Biophys. Biochem. Cytol.* **5**, 143.

Borsook, H., Deasy, C. L., Haagen-Smit, A. J., Keighley, G., and Lowy, P.H. (1950). *J. Biol. Chem.* **187**, 839.

Bradfield, J. R. G. (1956). *Symposium Soc. Gen. Microbiol.* **6**, 296.

Brandes, D., and Portela, A. (1960). *J. Biophys. Biochem. Cytol.* **7**, 505.

Brody, M., and Vatter, A. E. (1959). *J. Biophys. Biochem. Cytol.* **5**, 289.

Buvat, R. M., and Carasso, N. (1957). *Compt. rend. acad. sci.* **244**, 1532.

Callan, H. G., and Tomlin, S. G. (1950). *Proc. Roy. Soc.* **B137**, 367.

Chapman, G. B., and Tilney, L. G. (1959). *J. Biophys. Biochem. Cytol.* **5**, 69.

Chauveau, J., Gautier, A., Moulé, Y., and Rouiller, Ch. (1955). *Compt. rend. acad. sci.* **241**, 337.

Christensen, A. K., and Fawcett, D. W. (1960). *Anat. Record* **136**, 333.

Claude, A. (1941). *Cold Spring Harbor Symposium Quant. Biol.* **9**, 263.

Claude, A. (1943). *Biol. Symposia* **10**, 111.

Dalton, A. J., Kahler, H., Streibich, M. J., and Lloyd, B. (1950). *J. Natl. Cancer Inst.* **11**, 439.

Deane, H. W., and Porter, K. R. (1960). *Z. Zellforsch. u. Mikroskop. Anat.* **52**, 47.

Dempsey, E. W., and Peterson, R. R. (1955). *Endocrinology* **56**, 46.

Edwards, G. A., Ruska, H., Souza-Santos, P., and Vallejo-Freire, A. (1956). *J. Biophys. Biochem. Cytol.* **2**, *Suppl.* 143.

Elbers, P. F., Minnaert, K., and Thomas, J. B. (1957). *Acta. Botan. Neerl.* **6**, 345.

Epstein, M. A. (1957). *J. Biophys. Biochem. Cytol.* **3**, 851.

Fauré-Fremiet, E., and Rouiller, C. (1958). *Exptl. Cell Research* **14**, 29.

Favard, P. (1958). *Compt. rend. acad. sci.* **247**, 531.

Favard, P. (1959). *Compt. rend. acad. sci.* **248**, 3344.

Fawcett, D. W. (1955). *J. Natl. Cancer Inst.* **15**, 147.

Fawcett, D. W., and Ito, S. (1958). *J. Biophys. Biochem. Cytol.* **4**, 135.

Fawcett, D. W., and Porter, K. R. (1954). *J. Morphol.* **94**, 221.

Françon, M. (1954). "Le microscope à contraste de phase et le microscope interférentiel." Centre National de la Recherche Scientifique, Paris.

Gall, J. G. (1959). *J. Biophys. Biochem. Cytol.* **6**, 115.

Garnier, Ch. (1897). *Bibliographie Anat.* **5**, 278.

Gay, H. (1956a). *J. Biophys. Biochem. Cytol.* **2**, *Suppl.* 407.

Gay, H. (1956b). *Cold Spring Harbor Symposium Quant. Biol.* **21**, 257.

Glauert, A. M., and Hopwood, D. A. (1959). *J. Biophys. Biochem. Cytol.* **6**, 515.

Glauert, A. M., and Hopwood, D. A. (1960). *J. Biophys. Biochem. Cytol.* **7**, 479.
Godman, G., and Porter, K. R. (1960). *J. Biophys. Biochem. Cytol.* **8**, 719.
Haguenau, F. (1958). *Intern. Rev. Cytol.* **7**, 425.
Hampton, J. C. (1958). *Acta Anat.* **32**, 262.
Handler, R. W., Dalton, A. J., and Glenner, G. G. (1957). *J. Biophys. Biochem. Cytol.* **3**, 325.
Hartmann, J. F. (1953). *J. Comp. Neurol.* **99**, 201.
Hashimoto, T., Gerhardt, P., Conti, S. F., and Naylor, H. B. (1960). *J. Biophys. Biochem. Cytol.* **7**, 305.
Hay, E. (1958). *J. Biophys. Biochem. Cytol.* **4**, 583.
Hers, H. G., Berthet, J., Berthet, L., and de Duve, C. (1951). *Bull. soc. chim. biol.* **33**, 21.
Hogeboom, G. H. (1949). *J. Biol. Chem.* **177**, 847.
Hogeboom, G. H., Kuff, E. L., and Schneider, W. C. (1957). *Intern. Rev. Cytol.* **6**, 425.
Hultin, T. (1950). *Exptl. Cell Research* **1**, 376.
Huxley, A. F., and Taylor, R. E. (1958). *J. Physiol. (London)* **144**, 426.
Ito, S. (1960). *J. Biophys. Biochem. Cytol.* **7**, 433.
Karrer, H. E. (1960). *J. Biophys. Biochem. Cytol.* **7**, 357.
Kaufmann, B. P., and Gay, H. (1958). *The Nucleus,* **1**, 57.
Kellenberger, E., Ryter, A., and Séchaud, J. (1958). *J. Biophys. Biochem. Cytol.* **4**, 671.
Kuff, E. L., Hogeboom, G. H., and Dalton, A. J. (1956). *J. Biophys. Biochem. Cytol.* **2**, 33.
Leblond, C. P., and Clermont, Y. (1960). *Anat. Record* **136**, 230.
Littlefield, J. W., Keller, E. B., Gross, J., and Zamecnik, P. C. (1955). *J. Biol. Chem.* **217**, 111.
Loftfield, R. B. (1957). *Progr. in Biophys. and Biophys. Chem.* **8**, 347.
Lynn, W. S., Jr., and Brown, R. H. (1958). *J. Biol. Chem.* **232**, 1015.
Mercer, E. H. (1959). *Proc. Roy. Soc.* **B150**, 216.
Mercer, E. H., and Shaffer, B. M. (1960). *J. Biophys. Biochem. Cytol.* **7**, 353.
Merriam, R. W. (1959). *J. Biophys. Biochem. Cytol.* **5**, 117.
Moore, R. T., and Chapman, G. B. (1959). *J. Bacteriol.* **78**, 878.
Moses, M. J. (1960). *In* "Proceedings Fourth International Conference on Electron Microscopy," Part II, p. 230. Springer, Berlin.
Moulé, Y., Rouiller, C., and Chauveau, J. (1960). *J. Biophys. Biochem. Cytol.* **7**, 547.
Munger, B. L. (1958). *J. Biophys. Biochem. Cytol.* **4**, 177.
Murray, R. G. E. (1960). *In* "The Bacteria" (I. C. Gunsalus and R. Y. Stanier, eds.), Vol. 1, Chapter 2. Academic Press, New York.
Odor, D. L. (1960). *J. Biophys. Biochem. Cytol.* **7**, 567.
Palade, G. E. (1952). *J. Exptl. Med.* **95**, 285.
Palade, G. E. (1953). *J. Appl. Phys.* **24**, 1419.
Palade, G. E. (1955a). *J. Biophys. Biochem. Cytol.* **1**, 59.
Palade, G. E. (1955b). *J. Biophys. Biochem. Cytol.* **1**, 567.
Palade, G. E. (1956). *J. Biophys. Biochem. Cytol.* **2**, 417.
Palade, G. E. (1958). First Symposium Biophysical Society. "Microsomal Particles and Protein Synthesis" (R. B. Roberts, ed.) p. 36. Pergamon Press, New York.
Palade, G. E. (1959). *In* "Subcellular Particles," p. 64. American Physiological Society, Washington, D. C.

Palade, G. E. (1960). *In* "Symposium on Electron Microscopy," p. 176. British Anat. Assoc. Arnold Press, London.
Palade, G. E., and Porter, K. R. (1952). *Anat. Record* **112**, 370.
Palade, G. E., and Porter, K. R. (1954). *J. Exptl. Med.* **100**, 641.
Palade, G. E., and Siekevitz, P. (1955). *Federation Proc.* **14**, 262.
Palade, G. E., and Siekevitz, P. (1956). *J. Biophys. Biochem. Cytol.* **2**, 171.
Palay, S. L. (1958). *In* "Frontiers in Cytology" (S. L. Palay, ed.), p. 305. Yale Univ. Press, New Haven, Connecticut.
Palay, S. L. (1960). *J. Biophys. Biochem. Cytol.* **7**, 391.
Palay, S. L., and Karlin, L. J. (1959). *J. Biophys. Biochem. Cytol.* **5**, 373.
Palay, S. L., and Palade, G. E. (1955). *J. Biophys. Biochem. Cytol.* **1**, 69.
Palay, S. L., and Wissig, S. L. (1953). *Anat. Record* **116**, 301.
Pappas, G. D. (1956). *J. Biophys. Biochem. Cytol.* **2**, *Suppl.* 431.
Peachey, L. D. (1960). *Federation Proc.* **19**, 257.
Peachey, L. D., and Porter, K. R. (1959). *Science* **129**, 721.
Pease, D. C. (1956). *J. Biophys. Biochem. Cytol.* **2**, *Suppl.* 203.
Peterson, M. L. (1960). The Transport of Fat in Man: A Study of Chylomicrons. Thesis submitted for the Ph.D. degree at The Rockefeller Institute.
Pollister, A. W., Gettner, M., and Ward, R. (1954). *Science* **120**, 17.
Porter, K. R. (1953). *J. Exptl. Med.* **97**, 727.
Porter, K. R. (1954a). *J. Histochem. and Cytochem.* **2**, 346.
Porter, K. R. (1954b). *In* "Symposium on the Fine Structure of Cells," p. 236. Interscience, New York.
Porter, K. R. (1956). *J. Biophys. Biochem. Cytol.* **2**, *Suppl.* 163.
Porter, K. R. (1957). *Harvey Lectures Ser.* **51**, 175.
Porter, K. R. (1960). "Proceedings Fourth International Conference on Electron Microscopy," Part II, p. 186. Springer, Berlin.
Porter, K. R., and Blum, J. (1953). *Anat. Record* **4**, 685.
Porter, K. R., and Bruni, C. (1960a). *Cancer Research* **19**, 997.
Porter, K. R., and Bruni, C. (1960b). *Anat. Record* **136**, 260.
Porter, K. R., and Kallman, F. (1952). *Ann. N.Y. Acad. Sci.* **54**, 882.
Porter, K. R., and Machado, R. (1960a). *J. Biophys. Biochem. Cytol.* **7**, 167.
Porter, K. R., and Machado, R. (1960b). European Regional Conference on Electron Microscopy, Delft (in press).
Porter, K. R., and Palade, G. E. (1957). *J. Biophys. Biochem. Cytol.* **3**, 269.
Porter, K. R., and Pappas, G. D. (1959). *J. Biophys. Biochem. Cytol.* **5**, 153.
Porter, K. R., and Yamada, E. (1960). *J. Biophys. Biochem. Cytol.* **8**, 181.
Porter, K. R., Claude, A., and Fullam, E. (1945). *J. Exptl. Med.* **81**, 233.
Rebhun, L. I. (1956). *J. Biophys. Biochem. Cytol.* **2**, 93.
Robertson, J. D. (1959). *Biochem. Soc. Symposium* **16**, 3.
Rose, G. G., and Pomerat, C. M. (1960). *J. Biophys. Biochem. Cytol.* **8**, 423.
Ross, M. H., Pappas, G. D., Lanman, J. T., and Lind, J. (1958). *J. Biophys. Biochem. Cytol.* **4**, 659.
Rudzinska, M. A., and Trager, W. (1959). *J. Biophys. Biochem. Cytol.* **6**, 103.
Ryter, A., Kellenberger, E., Birch-Andersen, A., and Maaløe, O. (1958). *Z. Naturforsch.* **13b**, 597.
Sager, R., and Palade, G. E. (1957). *J. Biophys. Biochem. Cytol.* **3**, 463.
Sedar, A. W., and Porter, K. R. (1955). *J. Biophys. Biochem. Cytol.* **1**, 583.
Shatkin, A. J. (1960). *J. Biophys. Biochem. Cytol.* **7**, 583.
Siekevitz, P. (1959). *Exptl. Cell Research, Suppl.* **7**, 90.

Siekevitz, P., and Palade, G. E. (1958a). *J. Biophys. Biochem. Cytol.* **4**, 203.
Siekevitz, P., and Palade, G. E. (1958b). *J. Biophys. Biochem. Cytol.* **4**, 309.
Siekevitz, P., and Palade, G. E. (1958c). *J. Biophys. Biochem. Cytol.* **4**, 557.
Siekevitz, P., and Palade, G. E. (1960a). *J. Biophys. Biochem. Cytol.* **7**, 619.
Siekevitz, P., and Palade, G. E. (1960b). *J. Biophys. Biochem. Cytol.* **7**, 631.
Sjöstrand, F. S. (1953). *Nature* **171**, 31.
Sjöstrand, F. S. (1956). *Intern. Rev. Cytol.* **5**, 455.
Sjöstrand, F. S., and Hanzon, V. (1954). *Exptl. Cell Research* **7**, 393.
Slautterback, D. B. (1953). *Exptl. Cell Research* **5**, 173.
Slautterback, D. B., and Fawcett, D. W. (1959). *J. Biophys. Biochem. Cytol.* **5**, 441.
Sotelo, J. R., and Porter, K. R. (1959). *J. Biophys. Biochem. Cytol.* **5**, 327.
Swift, H. (1956). *J. Biophys. Biochem. Cytol.* **2**, *Suppl.* 415.
Van Iterson, W. (1960). *J. Biophys. Biochem. Cytol.* **9**, p. 183.
Watson, M. L. (1955). *J. Biophys. Biochem. Cytol.* **1**, 257.
Watson, M. L. (1959). *J. Biophys. Biochem. Cytol.* **6**, 147.
Weiss, J. M. (1953). *J. Exptl. Med.* **98**, 607.
Whaley, W. G., Mollenhauer, H. H., and Kephart, J. E. (1959). *J. Biophys. Biochem. Cytol.* **5**, 501.
Wischnitzer, S. (1958). *J. Ultrastruct. Research* **1**, 201.
Wischnitzer, S. (1960). *J. Biophys. Biochem. Cytol.* **8**, 558.
Wissig, S. L. (1960). *J. Biophys. Biochem. Cytol.* **7**, 419.
Yamada, E. (1958). *J. Biophys. Biochem. Cytol.* **4**, 329.
Yamada, E., and Porter, K. R. (1961). *J. Biophys. Biochem. Cytol.* **10** (in press).
Yamada, E., Tokuyasu, K., and Iwaki, S. (1958). *J. Electron microscopy (Chiba)* **6**, 42.

CHAPTER 10

The Interphase Nucleus

By ALFRED E. MIRSKY and SYOZO OSAWA

INTRODUCTORY

The nucleus of a cell when it is not visibly engaged in division is said to be "resting," "metabolic," or in "interphase." The first term, "resting," is misleading because it implies a low level of activity. The second term, "metabolic," recognizes the activity of the nucleus of the nondividing cells but implies that nuclear components may not be metabolically active in the course of cell division. Such an implication is at present unwarranted. The third term, "interphase," being purely descriptive, has been chosen for the title of this chapter.

The chromosomes of a cell are contained within its nucleus during interphase. Since the Mendelian factors, or genes, are located in chromosomes, it is expected that the nucleus would play a central role in activities of the interphase cell. This expectation is borne out by experiments on unicellular organisms in which enucleation leads to impairment of vital activities. Such experiments on unicellular organisms are discussed in Chapter 11 of this volume. In this chapter the nucleus is considered with respect to its morphology, chemical composition, and metabolic activity, always bearing in mind that the nucleus is loaded with the germinal material of the cell.

Nuclei of various cell types differ from each other, and often these differences are very striking. Such differences will be referred to frequently in the course of this chapter. The most obvious variations are in size and shape. Nuclei of many cells (hepatic cells, for example) are round; others (those of muscle fibers, for example) are elongated; and occasionally they have irregular shapes, as in polymorph leucocytes. Of all nuclei, probably those with the most irregular forms are the branched nuclei in the silk-spinning cells of insect larvae. The unusual form of these nuclei appears to be related to their intense activity. In certain insects (not in others, however) the silk-synthesizing cells develop in the Malpighian tubule. In the tubule there is an obvious transition from ordinary cells with round nuclei to secreting cells with branched nuclei (Fig. 1). Branching results in an increased nuclear surface, which would be significant when interactions between nucleus and cytoplasm are intensified.

Another type of variation appears as soon as even a preliminary survey of different nuclei is made. Some compact nuclei contain little more than the germinal material; in others there is so much other material that the germinal material is almost as difficult to find as is the needle in a haystack. Differences in concentration of the germinal material can be expressed quantitatively with reference to an essential component of the germinal material—deoxyribonucleic acid. DNA can be detected microscopically after Feulgen staining, and it can also be determined quantitatively in dried preparations of isolated nuclei. Sperm nuclei have relatively high concentrations of DNA, that of the trout, for example, having 60%. Nuclei of somatic cells have a lower DNA concentration, the figure for calf liver being 15.6%. Egg nuclei tend to have the lowest DNA concentrations. These nuclei are often exceedingly large and yet there is evidence that each contains the same amount of DNA as does the compact sperm nucleus of the same species. The concentration of DNA in the egg nucleus of the sea urchin is so low that it is just detectable by Feulgen staining, and then only because it is relatively concentrated at the periphery of the nucleus.

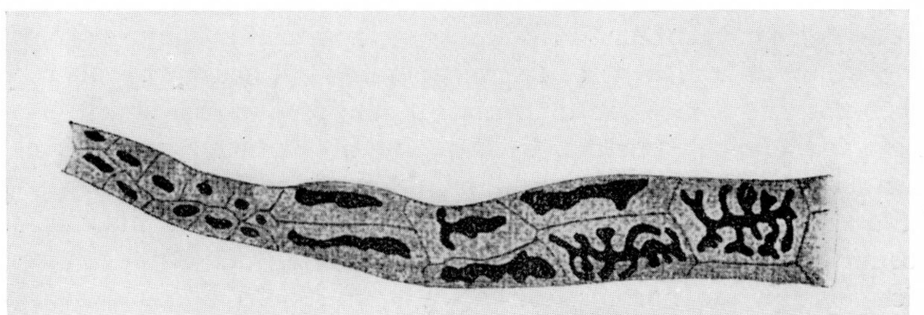

FIG. 1. A Malpighian tubule of *Myrmeleon,* the ant lion, which has become a silk-spinning gland. The ordinary tubule cells are on the left and those that have changed into silk-spinning cells are on the right. From Lozinski (1911).

The significance of the marked differences in DNA concentration of various nuclei is understood if we consider that the nucleus performs two distinct, though related, functions: one function is to transmit germinal material from one cell generation to the next; the other is concerned with the way in which germinal material influences the metabolism of the cell in which it is located. Since sperm cells are highly specialized for the transmission of germinal material, we would expect to find little more than germinal material in the sperm nucleus, and accordingly there should be in it a high concentration of DNA. Furthermore, we would expect to find an exceedingly low level of metabolic activity in the sperm nucleus. Experiments do, in fact, show that whereas the proteins in nuclei of liver and other cells rapidly incorporate labeled amino acids, only the slightest activity is observed in the sperm nucleus.

I. THE NUCLEAR MEMBRANE

A. *Permeability*

The membrane of the nucleus, which is receiving a great deal of attention today, was mentioned as far back as 1833 by Robert Brown, in one of the first papers on the nucleus.

The importance of the nuclear membrane is clearly shown by microdissection experiments. It is well known that the cell membrane can be punctured and that injuries so produced are rapidly repaired, so that quite extensive microsurgery can be performed on cell cytoplasm. The situation with respect to the nucleus is entirely different (Chambers and Fell, 1931). The nucleus is very sensitive to injury; in every case in which the nuclear membrane is even slightly punctured, once the needle penetrates, the nucleus slowly collapses into a shriveled mass. The cell sooner or later dies. These experiments were done with a variety of cells in tissue culture. In a binucleate, cell puncture of one nucleus results in its destruction, but the cell usually survives.

The nuclear membrane makes possible a special intranuclear environment, different from that of the surrounding cytoplasm; the membrane makes possible a measure of control of the medium in which chromosomes and nucleoli operate. Considering the known properties of the *cell* membrane, it is probable that much of this control is due to activity of the nuclear membrane itself. Certainly much material must pass between nucleus and cytoplasm, so that if membrane activity regulates conditions within the nucleus, this activity must also be concerned with interactions between nucleus and cytoplasm.

Some signs that nuclear activity may be especially intense at its periphery are: (1) certain pieces of chromatin may be located just within the membrane, and in some cases most or all of the chromatin is close to the periphery, as in the nucleus of the sea urchin egg (Fig. 2; and see the discussion on page 703) or in the association between chromatin and nuclear membrane outpocketings (Gay, 1956; Wischnitzer, 1958); (2) in the amphibian oöcyte, when hundreds of ribonucleoprotein globules, often considered to be nucleoli, are present, they are mostly at the periphery and indeed in many appear to be stuck to the nuclear membrane (Gall, 1954).

Observations on the passage of substances across nuclear membranes have been made both on isolated nuclei and on nuclei within cells. The isolated nuclei on which such observations have been made are those of amphibian oöcytes or of mammalian tissues. Oöcyte nuclei are large enough so that individual nuclei have been dissected out and immersed in various media. It was found that small molecules, sugars and polypeptides, readily pass into isolated amphibian oöcyte nuclei, whereas proteins such as ovalbumin and serum albumin do not readily enter. Thymus and liver nuclei can be isolated in masses. Ribonuclease and deoxyribonuclease readily penetrate into these nuclei as shown by decomposition of RNA and DNA within the nuclei. Considering how drastically liver nuclei are altered in the course of isolation, it is doubtful whether observations on isolated liver nuclei have any relevance to an understanding of the nuclear membrane within a liver cell; and the same comment can be made concerning observations on the permeability of other isolated nuclei.

Hundreds of *in vivo* experiments with isotopically labeled amino acids and other small molecules show that they readily pass through the nuclear membrane. There are also a few observations concerning pene-

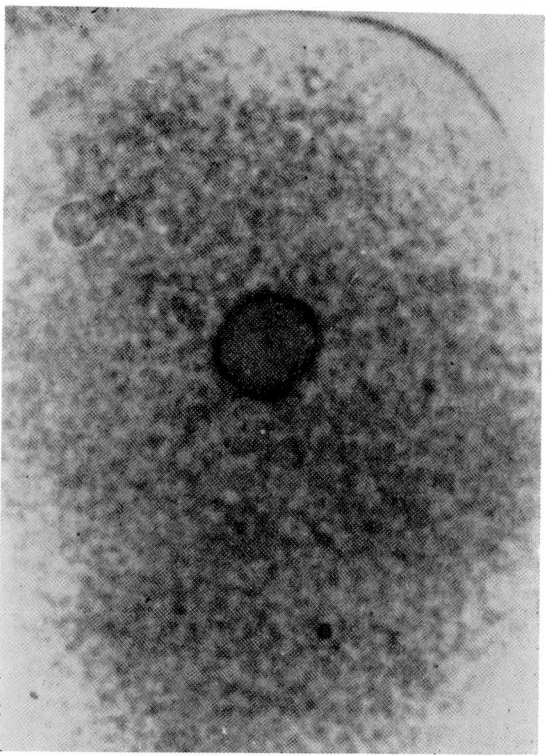

FIG. 2. A mature unfertilized sea urchin egg (*Arbacia punctulata*) showing an accumulation of Feulgen reactive material on the inner aspect of the nuclear membrane. Magnification: × 1800. From Burgos (1955).

tration by proteins. Some of the smaller protein molecules are able to penetrate into cells and then into the nuclei. That protamines and histones penetrate into the cell nucleus was shown by their effect on nuclear staining. It is well known, however, that these proteins have serious toxic and indeed lethal effects on cells, and they surely change the cells into which they penetrate so that penetration may actually have been made possible by the damage caused (Becker and Green, 1960). Ribonuclease, a relatively small molecule (molecular weight 13,000) has been found to penetrate into the cell nucleus of many cells. If penetration was due to digestion of some RNA-containing constituent of the nuclear membrane, all that can be said is that this hypothetical change in the membrane is readily reversible, for the cells recovered from the effects of ribonuclease.

There are several experiments showing that proteins which penetrate into the cytoplasm of a cell are unable to pass into the nucleus. In one experiment serum albumin labeled with fluorescein was injected into rats. The albumin was taken up by the Kupffer cells of the liver. Microscopic examination showed that although the cytoplasm of the Kupffer cells was

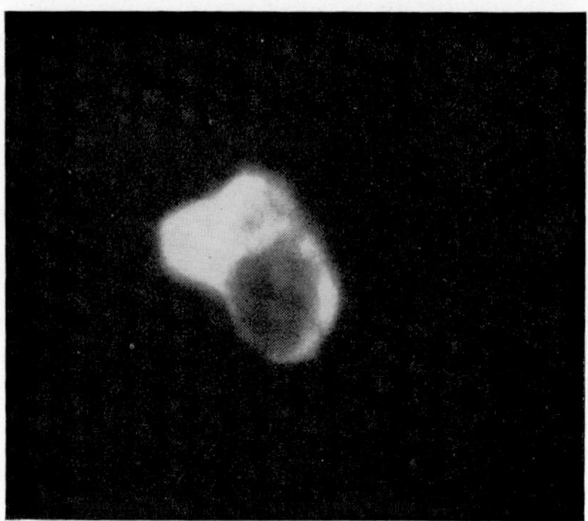

FIG. 3. A plasma cell from a lymph node to which had been added a fluorescein labeled antibody to γ-globulin. The fluorescence localization, due to the label, sharply defines the cytoplasmic margins. The nucleus is unstained. Magnification: × 1500. From Ortega and Mellors (1957).

loaded with fluorescein, none penetrated into the nuclei (Schiller *et al.*, 1952). In another set of experiments an antibody labeled with fluorescein passed into the cytoplasm of the plasma cells taken from a lymph node, but failed to penetrate into the nuclei (Fig. 3) (Ortega and Mellors, 1957; Neil and Dixon, 1959). The most systematic experiments concerning the ability of proteins to diffuse into the nucleus are those by Harding and Feldherr (1959) and those by Feldherr and Feldherr (1960). In the former, bovine serum albumin (molecular weight 65,000) in varying concentrations was injected into the cytoplasm of the frog oöcyte and measurements of nuclear size were made. Isotonic solutions of albumin caused no change in nuclear volume whereas more concentrated solutions caused a decrease in volume, indicating that the albumin did not penetrate from the cytoplasm into the nucleus. In experiments by the Feldherrs γ-globulin labeled with fluorescein was injected into the cytoplasm of immature *Cecropia* oöcytes. As shown in Fig. 4, the labeled protein did not penetrate into the nucleus. Apparently the nuclear membrane is, at least under certain conditions, impermeable to large protein molecules. These experiments should be borne in mind when looking at the "holes" and "pores" on the electron micrographs of nuclear membranes, for the most

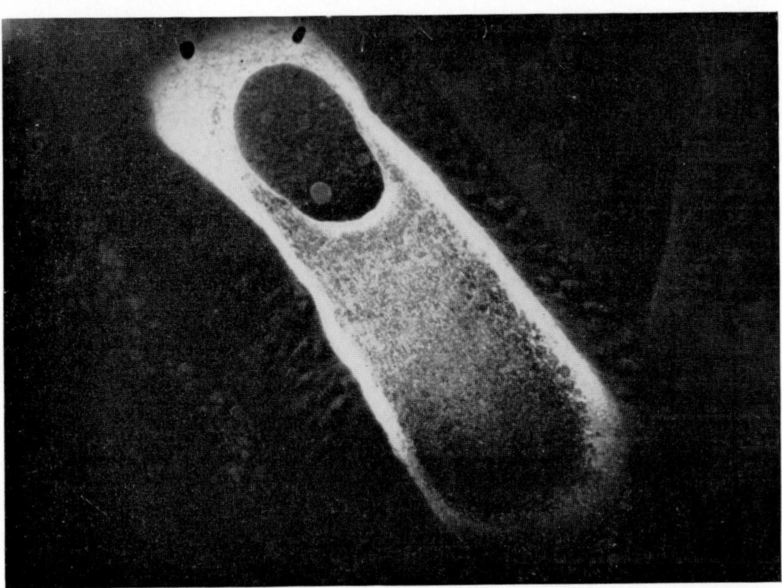

FIG. 4. A section through an injected oöcyte of a *Cecropia* moth showing the absence of fluorescein in the nucleus. Under the fluorescence microscope the nucleus appeared blue and the entire cytoplasm was yellow. Due to a high concentration of granules, the fluorescence in the end of the cell away from the nucleus did not show up with black-and-white photography. From Feldherr and Feldherr (1960).

direct physiological experiments now available show that large protein molecules even though much smaller than the so-called pores in the nuclear membrane do not simply diffuse through the membrane. Before coming to morphological studies of the nuclear membrane the present situation may be summarized in general terms: a clear distinction between nucleus and cytoplasm is indicated by such experiments as puncturing of the nuclear membrane by a microneedle and injection into the cytoplasm of labeled proteins; on the other hand, intimate interactions between nucleus and cytoplasm are demonstrated by a host of genetic, physiological, and cytochemical observations; interactions that involve the passage of macromolecules between nucleus and cytoplasm probably require a more active process than simple diffusion.

B. Morphology

Our present picture of the nuclear membrane has been made possible by the electron microscope. Numerous skillful and ingenious microscopists have shown that the nuclear membrane is part of what can now be seen to be a complex cell membrane system. The nuclear membrane must be considered in relation to the whole system. In the cytoplasm there is a complex of membranes (usually referred to as ergoplasmic lamellae or endoplasmic reticulum, two rather cumbersome terms which no longer seem necessary), which are continuous both with the external membrane of the cell and with the nuclear membrane. In most bacteria there are no internal membranes and no nuclear membrane. In mycobacteria there are internal membranes continuous with the outer cell membrane, but there is no nuclear membrane (Fauré-Fremiet and Rouiller, 1958). The situation is the same in an actinomycete, *Streptomyces coelicolor* (Glauert and Hopwood, 1959). In some fungi, at least, as in the cells of higher organisms, there is a nuclear membrane as well as a general cell membrane system (Turian and Kellenberger, 1956; O'Hern and Henry, 1956).

The membrane surrounding the nucleus is double (Hartmann, 1953) with a space between the two membranes in the range of 100–300 A. The outer of the two membranes is sometimes continuous with membranes of the general cytoplasmic membrane system (Watson, 1955). This means that the space between the nuclear membranes is continuous with the channels lying between the internal cytoplasmic membranes. Since in some cells (notably macrophages and kidney tubule cells) these channels ultimately open out into the extracellular space, it means that there is an open pathway, however tortuous, and perhaps only intermittently open, from the space between the double nuclear membrane through the spaces between the internal cytoplasmic membranes out to the exterior of such a cell (see the diagram, Fig. 5). A micrograph was published by Epstein (1957) showing this open approach from the cell surface to the nucleus. More recently, it has been found that fat droplets, 60 to 200 mμ in diameter, absorbed from the lumen of the small intestine appear in the space between the nuclear membranes in epithelial cells of the intestine (Palay, 1960). The movement of the droplets demonstrates that there is a pathway, open perhaps only intermittently, from the cell surface, through the channels between the cytoplasmic membranes right up to the nucleus, indeed into the space between the nuclear membranes (Fig. 6, a and b).

That the pathways to the nucleus from the exterior of the cell through the spaces between the cytoplasmic membranes should be regarded as constantly in flux is shown by moving pictures of living cells and especially by those in which rotation of the cell nucleus is observed (Pomerat, 1953). Moving pictures of tissue cultures show that nuclei rotate at an

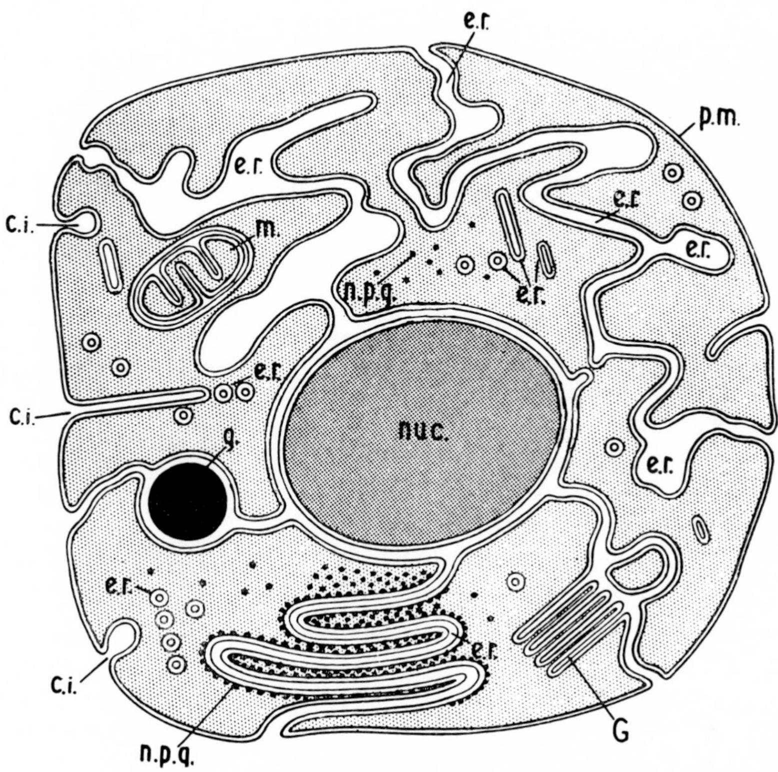

FIG. 5. Diagram of a hypothetical cell illustrating relationships of the cell membrane to various cell organelles; *e.r.* (endoplasmic reticulum, the cytoplasmic membrane system), *G* (the Golgi apparatus), *n.p.g.* (nucleoprotein granules), *m.* (mitochondrion), and *nuc.* (the nucleus). Modified from Robertson (1959).

average of 280 seconds per revolution. Sometimes a complete revolution may be made in 75 seconds. Changes in the direction of movement are frequent.

The fact that there are connections between the nuclear membrane and the cytoplasmic membranes does not necessarily mean that the nuclear membrane is derived from those of the cytoplasm. Observations on the formation of the nuclear membrane at telophase do, however, strongly suggest that the nuclear membrane is actually a specialized cytoplasmic structure. A micrograph made by Yasuzumi (1959) of a Yoshida sarcoma cell in telophase (Fig. 7) shows the nuclear membrane in formation and strongly suggests that it has its origin in the cytoplasmic membrane system. Quite another picture of nuclear membrane formation at telophase is given by observations on insect spermatocytes (Barer *et al.*, 1959). In these cells, vesicles, which are indistinguishable from the cytoplasmic membrane system, surround the chromosomes at telophase and finally fuse to form a complete nuclear membrane. Membranes around the telophase chromosomes are also visible in electron micrographs of other cells (Lafontaine, 1958) (Fig. 8).

If the nuclear membrane is actually a specialized part of the cytoplasm, this should be considered in the interpretation of nuclear transplantation experiments. Indeed it is doubtful whether nuclear transplantation, strictly speaking, is a possibility.

The fine structure of the nuclear membrane has revealed special features not found in other of the many membranes of the cell. Presence of "pores" in one of the nuclear membranes was first observed by Callan and Tomlin (1950) in amphibian oöcyte nuclei. Since then these structures have been seen in the nuclear membranes of a great variety of cells. An excellent detailed study of the nuclear membrane of the sea urchin oöcyte was made by Afzelius (1955). The details mentioned by him have also been found in mammalian tissue cells by Watson (1959), among others.

FIG. 6. a. Portion of an intestinal epithelial cell from a rat 30 minutes after intragastric instillation of 1.5 ml. of corn oil. The nucleus (*N*) lies at the left margin of the figure and the highly folded, interdigitating lateral cell membranes of this cell and its neighbor course diagonally across the figures at the right. Crowds of fat droplets occupy the cisternae and vesicles of the endoplasmic reticulum (*er*), that is, the channels between the cytoplasmic membranes. Two extracellular fat droplets lie between the two epithelial cells. Near the left upper corner of the figure a small fat droplet is located within the perinuclear cisterna, that is, the space between the two nuclear membranes (arrow). Magnification: × 42,000. From Palay (1960).

b. Portions of three intestinal epithelial cells from a rat 3.5 hours after intragastric instillation of 1.5 ml. of corn oil. The nuclei (*N*) of these cells are visible in the left upper, left lower, and right lower corners of the figure. The intercellular junction (*ic*) courses diagonally across the middle of the picture. Numerous fat droplets appear in this intercellular space. Other fat droplets are intracellular, enclosed within the endoplasmic reticulum (*er*), in the channels between the cytoplasmic membranes. The nuclear envelopes of the nuclei at the left upper and right lower corners of the figure contain fat droplets (arrows). Magnification: × 42,000. From Palay (1960).

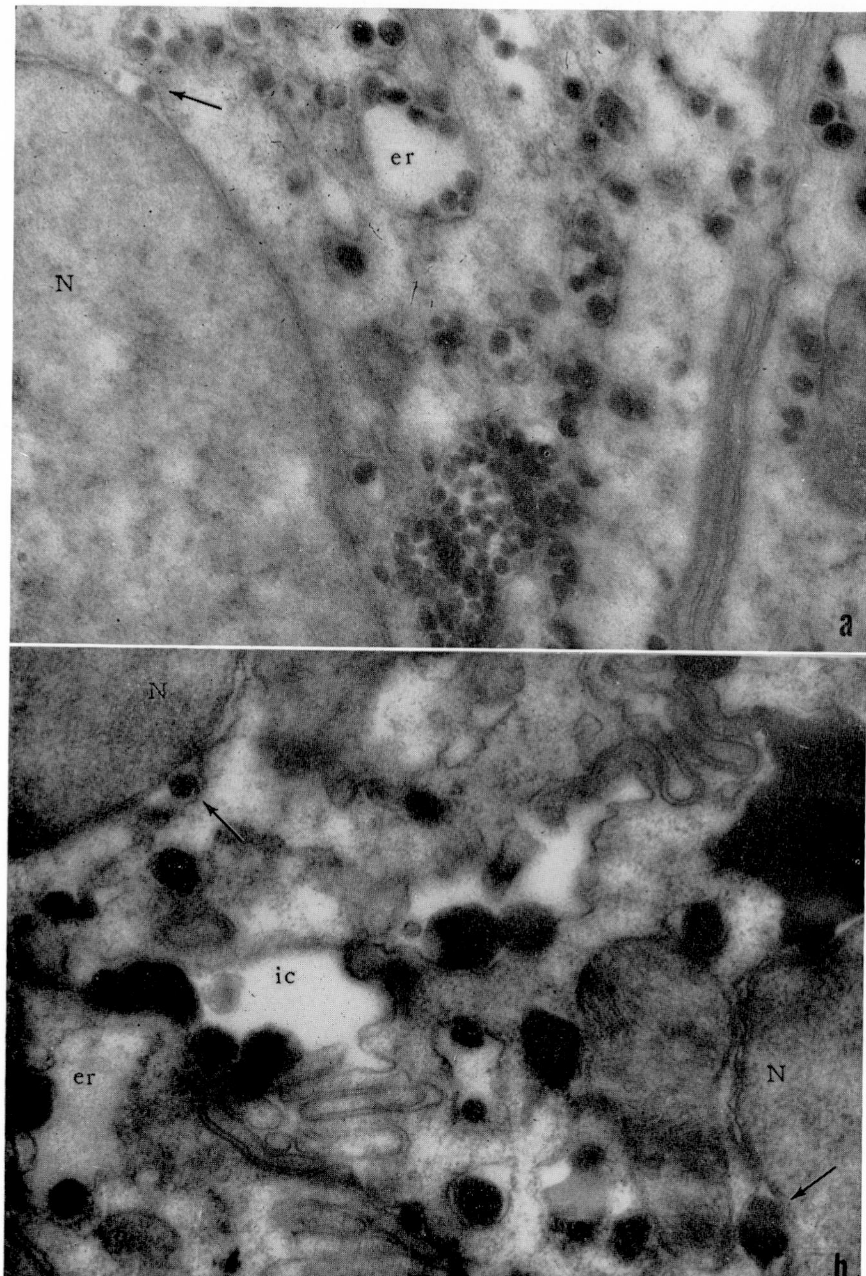
er
N
a
N
ic
er
N
b

In the inner and outer nuclear membranes, which are more or less parallel, there are discontinuities at which the two membranes join on either side of a gap of somewhat variable width. The gap is occupied by

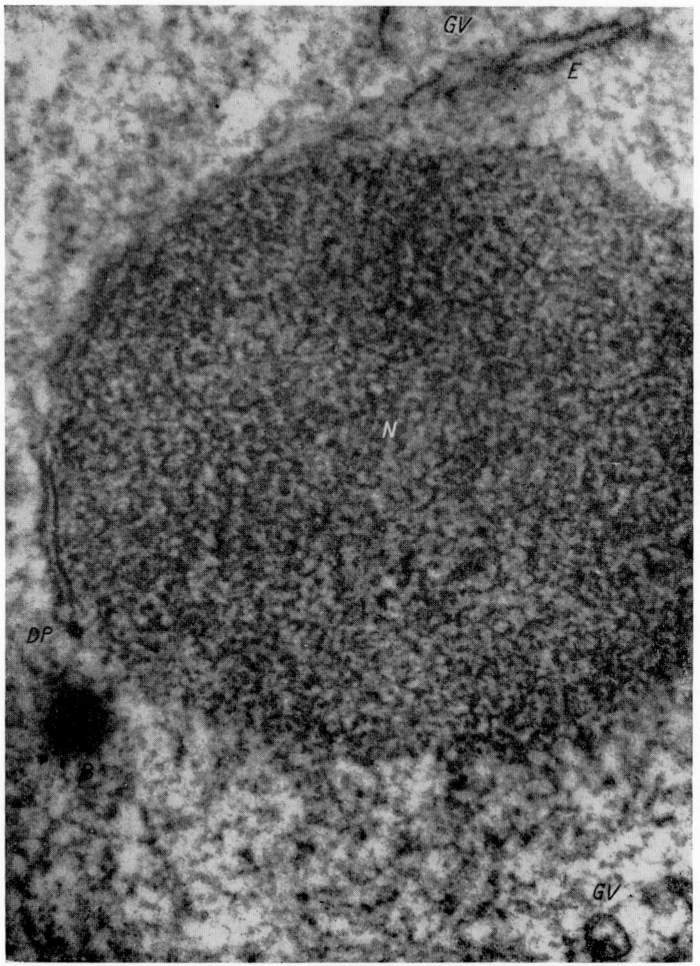

FIG. 7. A micrograph showing development of the nuclear membrane at telophase in a Yoshida sarcoma cell. The double membrane of the nucleus (*N*) is continuous with the cytoplasmic membrane system (*E*). Magnification: × 66,600. From Yasuzumi (1959).

a cylindrical structure extending from just within and extruding slightly into the cytoplasm, as seen in the schematic diagram (Fig. 9) and in the micrographs in Fig. 12. The cylindrical structures, or "pore complexes," when seen in sections cut tangentially, obliquely, and transversely to the nuclear surface appear as "pores" (Figs. 10–12). Experiments that have been referred to on page 683 show that the pore complexes do not convert the nuclear membrane into a sieve through which macromolecules readily diffuse. There are, however, details of structure which suggest that the

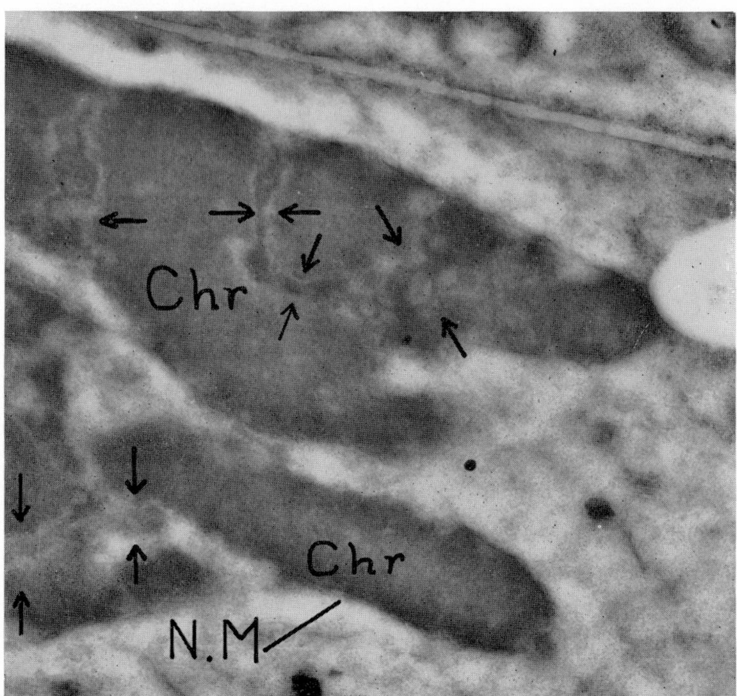

FIG. 8. An early telophase nucleus in *Vicia faba* showing prenucleolar material (arrows) between the chromosomes (*Chr*). Note how closely the nuclear membrane (*N.M.*) follows the contours of the chromosomes. From Lafontaine (1958).

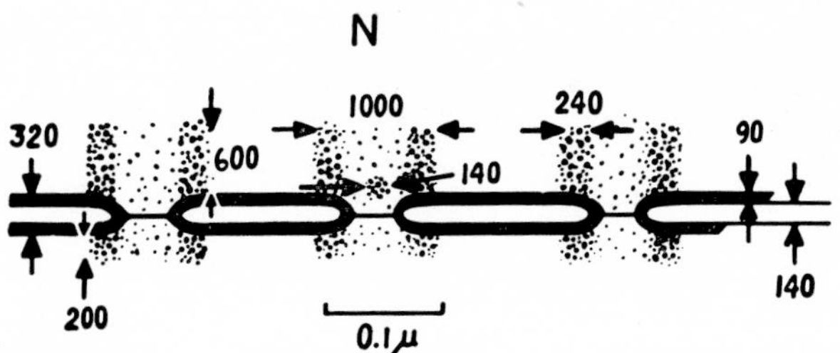

FIG. 9. Scheme of the nuclear membrane as seen in a transverse section. The dimensions are in angstrom units. *N:* nuclear side. Magnification: × 110,000. From Afzelius (1955).

pore complexes are concerned with the exchange of material between nucleus and cytoplasm. The cylindrical pore complexes can be traced some distance into the nucleus and there are structures in the perinuclear

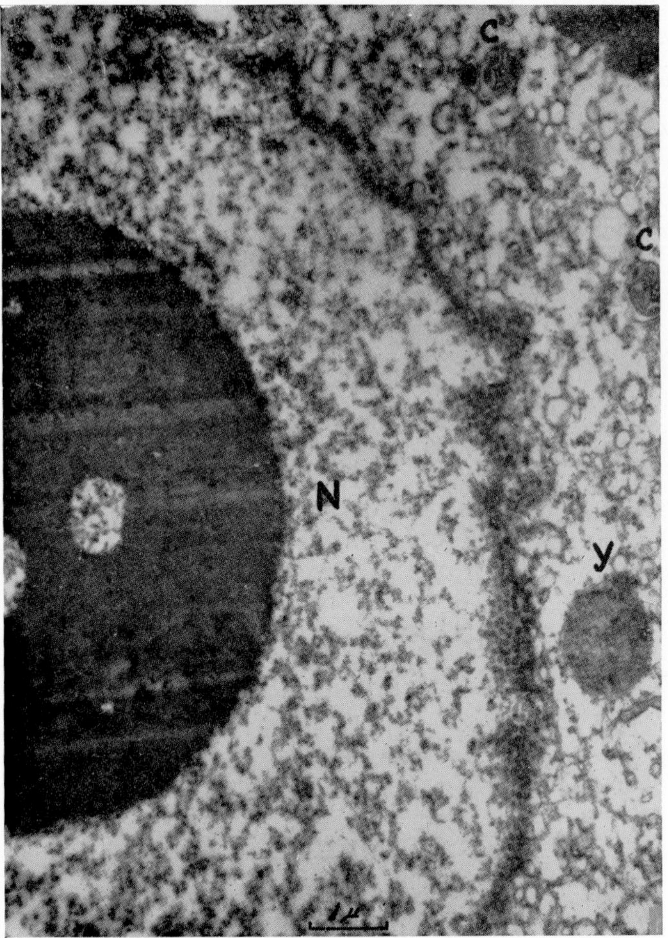

FIG. 10. *Echinus esculentus* oöcyte. The nucleus (*N*) with the dark nucleolus to the left. In the obliquely cut nuclear membrane annuli are to be seen. *C:* cortical granule, *Y:* yolk granule. Magnification: × 1500. From Afzelius (1955).

cytoplasm which may possibly be derived from material of the pore complexes (Watson, 1959).

The fine structure of the nuclear membrane is so complex that one readily supposes it is more than a nuclear envelope, even more than a semipermeable envelope, and that the membrane plays an active role in nucleocytoplasmic interactions. Along these lines the nuclear membrane has been thought of "as a region in which materials from both nucleus and cytoplasm may be gathered and there organized into definite lamellate structures. It is proposed that by some process these structures

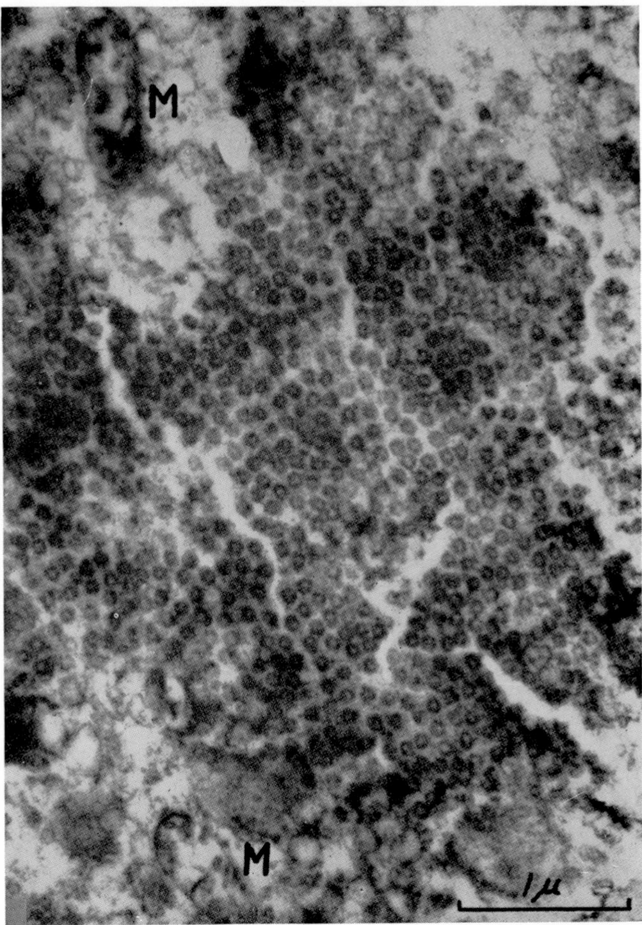

FIG. 11. *Brissopsis lyrifera* oöcyte. The nuclear membrane tangentially sectioned. *M*: mitochondrion. Magnification: × 32,000. From Afzelius (1955).

migrate into the cytoplasm from this region" (Rebhun, 1956). An excellent opportunity to test this hypothesis is provided by the regeneration of cytoplasmic nucleoprotein in liver on refeeding after starvation. Bernhard and Rouiller (1956) made an electronmicroscopic study of regenerating ergastoplasmic structures (which are rich in ribonucleoprotein) of liver after starvation. They found that the ergastoplasmic membranes reappear *simultaneously* at the periphery of the cell and along the nuclear envelope; no density gradient from nucleus to cell surface was detectable.

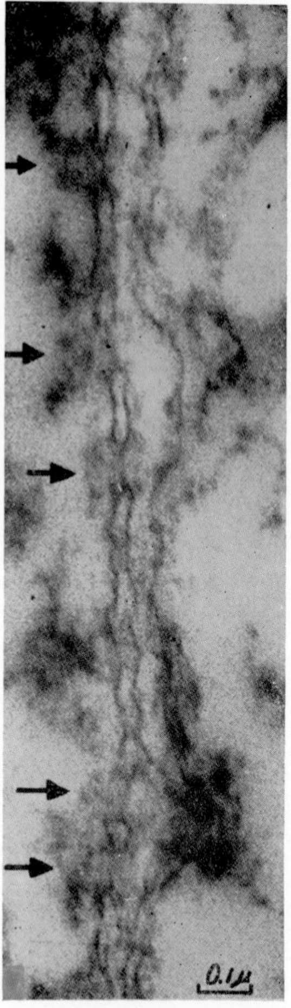

FIG. 12. *Echinus esculentus* oöcyte. The nuclear membrane transversely sectioned. The arrows in the nuclear side show five annuli. From Afzelius (1955).

Observations on the erythrocyte nucleus provide clear evidence that there is a correlation between nuclear activity and the fine structure of the nuclear membrane. Experiments with isotopically labeled amino acids show that protein synthesis in mature erythrocyte nuclei is very slight indeed compared with the generality of nuclei (Allfrey and Mirsky,

1952). Electron micrographs (Fig. 13) of mature nucleated erythrocytes show the double membrane characteristic of nuclei but that very few, if any, "pores" are present.

Interaction between nucleus and cytoplasm has always been considered a necessary step in communication between the nucleus and the

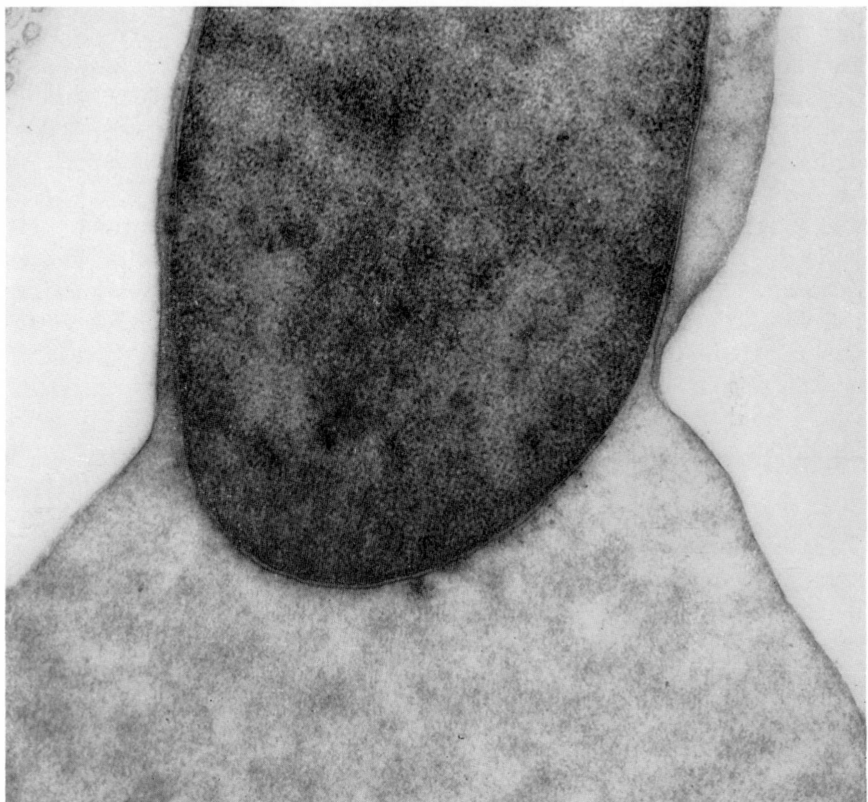

FIG. 13. Part of nucleus in erythrocyte of toadfish *Opsanus tau*. Magnification: approximately × 24,000. From Fawcett (1959).

region outside of the cell. It has always been supposed that a molecule in the medium surrounding the cell must first pass through the cytoplasm, if it is to enter the nucleus. We now know that this need not be the case. There is a direct pathway from the exterior of the cell to the nucleus. Furthermore, the nuclear membrane is a specialized part of the cell membrane system so that we must consider the possibility that some substances present outside the cell will penetrate the nuclear membrane more readily than the other membranes. It would, accordingly, not be surprising if a substance in the medium surrounding the cell were to get into the nucleus without at the same time appearing in the surrounding cytoplasm. Are there any observations indicating that molecules do actually penetrate directly into the nucleus?

The acridine dyes provide a possible example of penetration directly into the nucleus from the exterior of the cell. De Bruyn and his colleagues have shown (1950, 1952) that a number of diamino acridine dyes stain the chromatin and nucleolus of living mammalian tissue cells without at the same time staining the cytoplasm of these cells. Nuclear staining occurs without any gross injury to the cells if they are not exposed to light, for these are fluorescent dyes. Absence of gross injury is evidenced by the fact that although their nuclei are stained, these cells can pass through mitosis. When the cells are killed the picture changes, for these basic dyes then stain basophilic materials of the cytoplasm as well as those of the nucleus. In a living cell, even one such as a pancreatic acinar cell with an exceedingly basophilic cytoplasm, only the basophilic components of the nucleus are stained. It has not hitherto been possible to explain this curious behavior of the acridine dyes. A simple explanation would be that the dye penetrates directly into the nucleus without passing through the cytoplasm.

II. Distribution of Chromatin

It is commonly supposed that the chromosomes are distributed at random in the chromatin network of an interphase nucleus. There are, however, some indications that chromosomes are located in definite and recurrent sites even in the interphase nucleus. A cursory glance at the location of the so-called female sex chromatin in the nuclei of mammalian somatic cells suggests that at least one piece of chromatin is not situated at random. This suggestion becomes much firmer when all the observations of Barr and his colleagues (Barr *et al.*, 1950; Graham and Barr, 1952; Graham, 1954) are considered, for it is then recognized that the sex chromatin may be localized in different positions in nuclei of different cell types of the same organism but that the position is definite for each cell type. Furthermore it becomes apparent that the localization changes in an orderly way during embryonic development and during certain physiological episodes. Thus in the cat, sex chromatin is visible adjoining the nucleolus in neurons, whereas in other somatic cells of the cat it is located at the nuclear membrane. In *all* embryonic cells of the cat in which there is sex chromatin, it is located at the nuclear membrane, but at an early stage it moves up to the nucleolus of the neurons. At the earliest stages of embryonic development, while nuclei are still very large, sex chromatin is not seen (Austin and Amoroso, 1957). When Nissl material is depleted the sex chromatin moves away from the nucleolus toward the nuclear membrane (Lindsay and Barr, 1955). In man the sex chromatin is found adherent to the inner surface of the nuclear membrane in all cells, including neurons.

Since "sex chromatin" occurs mainly in female mammals, it was regarded by Barr and his associates (Prince *et al.*, 1955) as being the heterochromatic portion of 2X chromosomes. In birds where the males are the 2X sex, the remarkable fact is that "sex chromatin" appears also primarily in females (Kosin and Ishizaki, 1959), so that "sex chromatin" can no longer be regarded as the heterochromatic portion of the 2X chromosomes.

The remarkable observations on "sex chromatin" make it seem likely that chromosomes other than the "sex chromatin" occupy definite positions in the interphase nucleus. Localization of chromosomal activity would surely have a special significance depending on the positions of adjoining chromosomes and on proximity to the nucleolus or a particular region of the nuclear membrane. If chromosomes are located at definite sites in the nucleus, nuclear rotation would mean that a given chromosome is constantly being exposed to different regions of cytoplasm.

In the living interphase nucleus there is frequently nothing visible except nucleoli. The lampbrush chromosomes of certain oöcytes, which have been carefully studied in an unfixed state, are among the few interphase chromosomes which are readily visible without fixation. After fixation, however, in most interphase nuclei a great variety of images are produced—finely or coarsely granular structures or a network of fibers with smaller or larger clumps of chromatin. Even a mild "fixative" such as citric acid produces the network of chromatin fibers seen in isolated nuclei. When chromosomes are isolated from interphase nuclei and compared with the chromatin strands of nuclei isolated in citric acid it is seen that many of the chromosomes have forms that can be recognized in the chromatin strands. It was discovered by Lewis (1923) and van Herwerden (1924) that almost any interference with the living cell causes the appearance of visible structures in the interphase nucleus, but that in many cases the cell recovers from the injury, and when it does so the chromatin structures in the nucleus disappear. Why can these chromosomes not be seen ordinarily in the living nucleus and what change occurs during fixation in citric acid?

In experiments on isolated nuclei and chromosomes it has been shown by a number of investigators that interphase chromosomes can be in an *extended state* or *condensed state* (Cohen, 1937; Ris and Mirsky,

1949a; Anderson and Wilbur, 1952; Pasteels, 1954; Philpot and Stanier, 1956). By staining for DNA it was found that the state of the chromosomes depends upon how diffusely this chromosomal component is distributed. For further understanding of the state of chromosomes in the ordinary interphase nucleus the higher resolution of electron microscopy was needed.

For years, however, the contribution of electron microscopy fell short of what was needed. Many of the procedures for fixing and staining used in electron microscopy are essentially refinements of classical cytological techniques. In light microscopy elegant procedures for visualizing DNA have been devised, but these could not be carried over to electron-microscopy. Quite recently however novel techniques have been introduced and they promise to give a new understanding of the chromatin material of the interphase nucleus (Stäubli, 1960; Leduc and Bernhard, 1960).

III. Chemical Composition

A. *DNA*

1. *Distribution*

The two types of nucleic acid, RNA and DNA, differ strikingly in their distribution within the cell: RNA is widely distributed, being present in sites in the cytoplasm, in the nucleolus, and in the chromosomes; DNA, on the other hand, is with very few exceptions, found only in the chromosomes. Indeed, one of the best ways of staining chromosomes, Feulgen staining, is actually a spot-test for DNA discovered by Feulgen (Feulgen and Rossenbeck, 1924; Feulgen *et al.*, 1937) and later applied cytologically by him. Until recently almost everything known about the general occurrence of DNA in chromosomes and the limitation of DNA to chromosomes was due to this remarkable staining procedure. In the past few years staining with fluorescent dyes has been used, especially in studies of virus synthesis, for cytological identification of DNA (De Bruyn *et al.*, 1952; Armstrong, 1956).

Although in the generality of cases all the DNA of a cell is confined to its nucleus, a small number of interesting exceptions to this rule are now known. An extranuclear structure, the kinetoplast, which occurs in trypanosomes, has been repeatedly described as Feulgen positive (Breslau and Scremin, 1924; Reichenow, 1928; Lwoff and Lwoff, 1931). The kinetoplast in trypanosomes is also stained by methyl green, a dye considered to be selective for polymerized DNA (Sen-Gupta *et al.*, 1953).

Steinert's recent work (Steinert, 1960) shows that the kinetoplast is an even more interesting body than had previously been supposed. In one sense it can be considered to be another nucleus, for it divides at the same time as the main nucleus. Electron microscopic study reveals a most surprising difference between the kinetoplast and the nucleus. The kinetoplast is seen (Fig. 14) to be a composite body, consisting of a

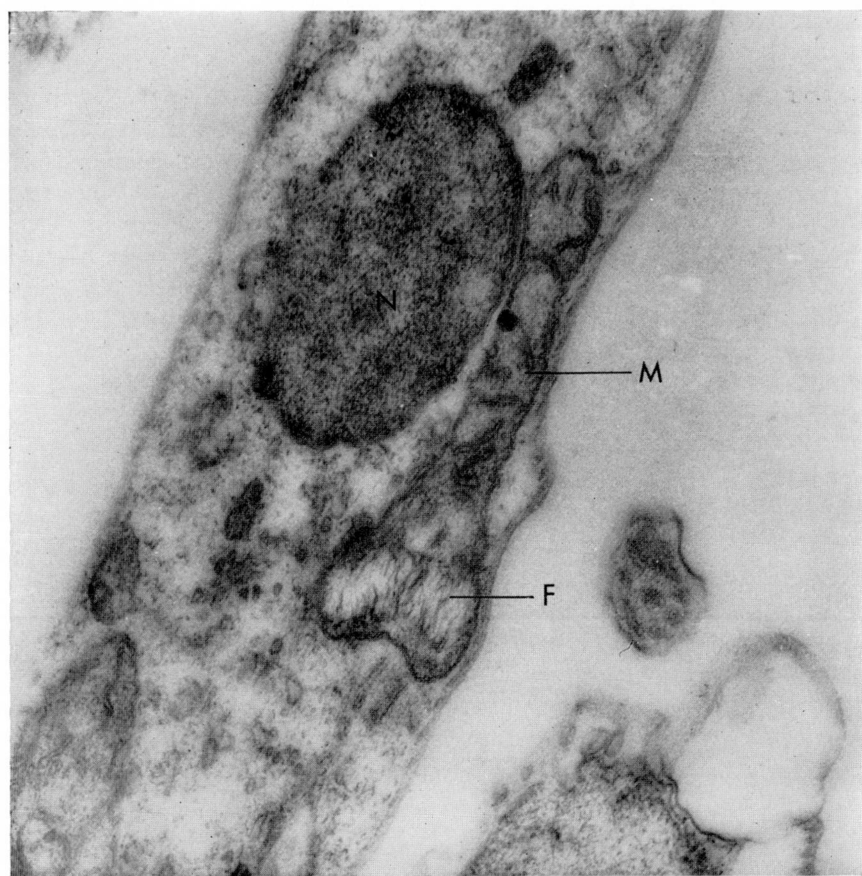

FIG. 14. *Trypanosoma mega. Crithidia* form grown *in vitro. N:* nucleus, *M:* mitochondrial part of kinetoplast, *F:* Feulgen-positive part of kinetoplast. Magnification: × 35,000. From Steinert (1960).

Feulgen-positive (DNA-containing) moiety and another part that is mitochondrial in nature. In the course of time the mitochondrial part changes in shape, enlarges, and becomes detached from the Feulgen-positive body.

The relationship between a mitochondrion and DNA seen in the kinetoplast may have a general significance, if one considers Steinert's observations in relation to the rather surprising experiments of Chèvremont and his colleagues (1959). In these experiments living fibroblasts in tissue culture were treated with an acid DNAase prepared from the thymus and by no means a pure enzyme. (The effect of pancreatic DNAase was entirely different.) This treatment had no effect on the DNA content of the cell nucleus; but it had an unexpected effect on the mitochondria. These bodies now became Feulgen positive and capable of incorporating tritiated thymidine. Apparently the DNA of the mitochondria did not come from the nucleus. The total quantity of DNA in the mitochondria of a cell could amount to nine-tenths that of the nucleus. If one considers that synthesis of DNA requires to be primed by some DNA, it seems likely that mitochondria originally contain small quantities of DNA, not detectable by Feulgen staining. The effect of acid DNAase would be to induce a large increase in the amount of DNA normally present. This interpretation of Chèvremont's experiments (of which a confirmation would be highly desirable) would bring them in line with Steinert's observations on the kinetoplast, and would throw new light on a very old problem—the genetic continuity of mitochondria.

There are several reports of the existence of "Feulgen-positive" material in the cytoplasm of certain reproductive cells, e.g., oöcytes (Konopacki, 1936) and microsporocytes (Sparrow and Hammond, 1947). In addition to the evidence of staining reactions there exist direct chemical demonstrations of extranuclear DNA or of related compounds: (1) the observation that avidin, a protein isolated from egg white, contains bound deoxyribonucleic acid (Fraenkel-Conrat *et al.*, 1952); (2) the demonstration of deoxyribonucleosides in the cytoplasm of frog oöcytes (Hoff-Jørgensen, 1954); and (3) the demonstration of deoxyribonucleosides in the eggs of *Drosophila* (Schultz, 1956). It is an interesting feature of these exceptions to the rule of DNA localization in the nucleus that closer investigation of the exceptions further substantiates the rule. Thus, the DNA in the cytoplasm of reproductive cells can sometimes be shown to originate in the nuclei of surrounding nurse cells, and the process of the transfer has been studied in detail in Ascidians (Konopacki, 1936). A related phenomenon has been observed in plants where chromatin is transferred from the nuclei of pollen mother cells to the sporocyte nucleus (Cooper, 1952). In oöcytes the amount of cytoplasmic DNA is far greater than that which exists in the oöcyte nucleus, and it is probable that it represents storage material for the synthesis of the nucleic acids in the thousands of cells which arise soon after fertilization of the ovum. The experiments of Grant (1955, 1956) show that deoxyribonucleotides of the cytoplasmic reserve provide the material for most of the DNA synthesis required by the nuclei formed in the early cleavage stages, but that even at this time there is also some *de novo* synthesis of nucleotides (see also B. C. Moore, 1957).

Oöcytes also provide an exception to the rule that all of the DNA of the nucleus is in the chromosomes. In the toad's egg Painter and Taylor (1942) showed that there are Feulgen-positive granules, clearly detached from the chromosomes and close to the nuclear membrane. This DNA-containing material is probably a precursor of the DNA of the egg cytoplasm. If DNA is being synthesized in the egg nucleus for storage in the cytoplasm, there should be an uptake of P^{32} into nuclear DNA, something that in general does not occur in the interphase nucleus, excepting in a cell which will soon divide. Uptake of P^{32} by nuclear DNA of the frog egg has in fact been observed by Finamore and Volkin (1958).

2. *Measurement of the DNA per Nucleus*

How much DNA is present in the nucleus? The most unequivocal answers to this question have been obtained by analyzing isolated nuclei and cell suspensions. In such preparations, the nuclei can be counted under the microscope and analyzed for DNA by routine microchemical procedures. This permits calculation of the average amount of DNA per nucleus, and was first done, independently of each other, by Boivin and the Vendrelys (Boivin *et al.*, 1948) and by Mirsky and Ris (1949). Their results (and those of other workers) reveal a striking constancy in the amounts of DNA per nucleus in different somatic tissues of the organism (Table I) (Mirsky and Ris, 1949; Ris and Mirsky, 1949b; Leuchtenberger *et al.*, 1951; Thomson *et al.*, 1953; Davidson *et al.*, 1950; Mirsky and Ris, 1951a; Vendrely and Vendrely, 1948, 1949; Mandel *et al.*, 1951; Davidson *et al.*, 1951). The average lymphocyte nucleus, for example, has the same DNA content as a kidney nucleus in the same animal; and the nucleated erythrocyte of the fowl has the same DNA per nucleus as the fowl liver cell. Even more striking was the observation that sperm cell nuclei contain only one-half the amount of DNA found in the nuclei of the various somatic cells. Since the sperm cell contains only one set of chromosomes and the lymphocyte or kidney cell has two sets, it became evident that the DNA content of the nucleus is a measure of the number of chromosome sets that the nucleus contains. In support of this conclusion, Ris and Mirsky found that the nuclei of tissues containing polyploid cells (i.e., cells having more than the usual two sets of chromosomes) have correspondingly high average DNA contents per nucleus. The effect of polyploidy in raising the average DNA per nucleus is particularly evident in mammalian liver tissues (data in Ris and Mirsky, 1949b).

TABLE I
AVERAGE DNA CONTENT PER NUCLEUS IN VARIOUS ANIMAL TISSUES[a]

Animal	Sperm	Erythro-cyte	Leuco-cyte	Liver	Kidney	Spleen	Heart	Pancreas	Thymus	References
Fowl	1.26	—	—	—	—	—	—	—	—	Mirsky and Ris, 1949
Fowl	—	2.58	—	2.65	2.28	2.63	2.54	2.70	—	Davidson *et al.*, 1950
Calf	—	—	—	6.22	6.25	—	—	—	7.15	Mirsky and Ris, 1949
Calf	—	—	6.93	—	—	—	—	—	—	Mandel *et al.*, 1951
										Vendrely and Vendrely, 1949
Beef	3.42	—	—	7.05	6.63	7.26	—	7.15	—	Leuchtenberger *et al.*, 1951
Beef	2.82	—	—	—	—	—	—	—	—	Mirsky and Ris, 1949
Beef	—	—	6.98	—	—	—	—	—	—	Mandel *et al.*, 1951
Human	3.25	—	7.30	10.36	8.6	—	—	—	—	Davidson *et al.*, 1951
Rat	—	—	—	9.47	6.75	6.55	6.50	7.38	7.44	Thomson *et al.*, 1953
Rat	—	—	6.64	—	—	—	—	—	—	Mandel *et al.*, 1951
Pig	—	—	—	5.0	5.2	—	—	—	—	Vendrely and Vendrely, 1949
Dog	—	—	—	5.5	5.3	—	—	—	—	Vendrely and Vendrely, 1949
Carp	1.64	3.49	—	3.33	3.5	—	—	—	—	Mirsky and Ris, 1949
Shad	0.91	1.97	—	2.01	—	—	—	—	—	Mirsky and Ris, 1949
Lungfish (*Protopterus*)	—	100.00	—	—	—	—	—	—	—	Mirsky and Ris, 1951
Toad	3.70	7.33	—	—	—	—	—	—	—	Mirsky and Ris, 1951
Frog	—	15.0	—	15.7	—	—	—	—	—	Mirsky and Ris, 1951
Jellyfish	0.33	—	—	—	—	—	—	—	—	Mirsky and Ris, 1951
Sea Urchin	0.98	—	—	—	—	—	—	—	—	Mirsky and Ris, 1951
Neroid worm	1.45	—	—	—	—	—	—	—	—	Mirsky and Ris, 1951
Squid	4.5	—	—	—	—	—	—	—	—	Mirsky and Ris, 1951

[a] Expressed in milligrams $\times 10^{-9}$.

Since these experiments were performed, a mass of data has accumulated in support of the concept that the average amount of DNA per chromosome set is fixed and characteristic of the species. For example, the analysis of haploid and diploid *Aspergillus* conidia shows that the latter have twice as much DNA as the former (Heagy and Roper, 1952). A parallelism between DNA and number of chromosome sets has also been demonstrated for haploid through tetraploid yeasts (Ogur *et al.*, 1952). Similarly, it has been found that the average DNA per cell in Ehrlich ascites tumors, which are known to contain four sets of chromosomes (Hauschka, 1952), is just twice that observed in normal diploid cells of the Krebs-Rask-Nielsen sarcoma (Klein and Klein, 1950).

Another factor of considerable interest is the wealth of evidence which shows that the amount of DNA per cell nucleus is not only constant in the different somatic tissues of an organism, but is also essentially invariant despite drastic changes in the physiological state of the organism. It remains constant in prolonged fasting (Thomson *et al.*, 1953; Mirsky and Kurnick, unpublished experiments), and is independent of changes in dietary regimen (Thomson *et al.*, 1953; Ely and Ross, 1951; Campbell and Kosterlitz, 1952; Rose and Schweigert, 1952). No significant differences are observed in the amount of DNA per cell of normal and leukemic leucocytes (Mandel and Metais, 1950; Davidson *et al.*, 1953), nor do precancerous or tumor cells produced by feeding carcinogens differ in this respect from normal liver cells (Cunningham *et al.*, 1950). The remarkable constancy in the amount of DNA per cell nucleus has made it possible to calculate the number of cells in a given tissue sample from its DNA content and to express the analytical results for protein, ribonucleic acid, or lipid composition on a per cell basis. This procedure has been used to study changes in cell composition during development (Davidson *et al.*, 1950; Mills *et al.*, 1950; Leslie and Davidson, 1951), as well as dietary effects on the protein and nucleic acid composition of the cell (Allfrey *et al.*, 1952; Thomson *et al.*, 1953).

An apparent exception to the constancy of DNA is found when cortisone is administered to rats (Lowe and Rand, 1956b; Lowe *et al.*, 1959). The rule of constancy holds in so many cases that what seems to be an exception is worthy of careful scrutiny. Cortisone produces changes in the DNA of liver cells so that analysis of DNA per nucleus shows a drop of 25%. This occurs after administration of cortisone for 5 days and in the following 3 days the analytical value returns to normal. The drop in the quantity of analytically determined DNA is not due to presence of necrotic cells, for microscopic examination does not show that they are present, and the number needed to account for the change in DNA would easily be detected.

The most likely explanations of the *apparent* loss of DNA per nucleus are: (1) that less DNA per nucleus is in fact present; or (2) that some of the DNA has been changed in such a way that it is lost in the course of analysis. If there is actually less DNA per nucleus, then in the 3 days following cortisone administration, when the DNA value returns to normal, there should be a marked synthesis of new DNA. DNA in the liver is metabolically so stable that the rate of incorporation of P^{32} is very low, and a synthesis of DNA would be revealed by a marked incorporation of P^{32}. This does not happen in the 3 days during which a "recovery" of DNA occurs. The second explanation, therefore, seems to be the more likely of the two mentioned. In the course of DNA analysis, acid-soluble nucleotides are discarded and it may be that in cortisone-treated animals 25% of the DNA has been changed (perhaps partly depolymerized) so that it is discarded with the acid-soluble fraction. This change would be reversed in the 3 days following cortisone administration. Somewhat similar changes in the cytoplasmic RNA of rat liver have been observed after cortisone administration (Lowe and Rand, 1956a). There may even be some doubt as to how profound an alteration in nuclear metabolism is brought about by the change in state of DNA due to cortisone, for it has been found that when the state of cytoplasmic RNA is changed by cortisone there is no accompanying change in the incorporation of N^{15}-labeled glycine by liver proteins (Mirsky *et al.*, 1954).

The amount of DNA per nucleus obtained by the analysis of isolated cell nuclei is of necessity an average value, and if variations exist from nucleus to nucleus they would escape detection by the method of measurement used. This is a problem which has been approached, but not yet solved, by the application of cytophotometric techniques to single nuclei. The most widely used of these cytochemical procedures makes use of the Feulgen color reaction for DNA, and the color intensity in a single nucleus is measured using the microscope as a spectrophotometer (Ris and Mirsky, 1949b; Swift, 1950, and many subsequent papers). It was demonstrated by Ris and Mirsky that, when the experimental conditions are carefully controlled and when one standardizes against nuclei of known DNA content, the microspectrophotometric method permits the comparison of individual nuclei for their relative DNA contents. By and large, the results of many such comparisons agree with the conclusions obtained by the analysis of isolated nuclei, namely, that the amount of DNA per chromosome set is a constant, and that when variations in the amount of DNA per nucleus are observed as, for example, in the various classes of liver nuclei, they are the expected results of varying degrees of ploidy.

Measurement of the DNA in the nucleus of the ovum was for some time an elusive undertaking, but it has been accomplished by Feulgen staining. The special difficulties of dealing with egg nuclei have been due partly to their large size but also to the considerable quantities of DNA-like storage material present in the abundant cytoplasm of eggs. Considering the size of many egg nuclei, it would be expected that if the quantity of DNA present is the same as other nuclei of the organism, Feulgen staining of the egg nucleus might be exceedingly faint. It has in fact been stated that the egg nucleus of the sea urchin is Feulgen negative and that it contains no DNA (Marshak and Marshak, 1955; Immers, 1957). Careful attention to cytological technique has, however, yielded Feulgen-positive preparations, as shown in Fig. 2 (Burgos, 1955; Brachet and Ficq, 1956). Because of difficulties with eggs, such as those of echinoderms, which have large nuclei, observations were made by Mirsky and Ris on the egg of *Ascaris megalocephala*, which has a small Feulgen-positive nucleus. Following van Beneden's classical observations, the egg nucleus can best be compared with the sperm nucleus after fertilization when the sperm nucleus has already penetrated into the egg and has enlarged, just before fusing with the egg nucleus, at a time when the two nuclei are of the same size and also have the same structure. A Feulgen preparation made at this time shows that the two nuclei are indistinguishable from one another; the two haploid nuclei, therefore, have identical quantities of DNA (Mirsky and Ris, 1951a).

Those oöcyte nuclei in which DNA is being synthesized in the nucleus and stored in the cytoplasm (and in some of which Feulgen-positive granules detached from chromosomes are visible close to the nuclear membrane) would be expected to contain more DNA per nucleus than is present in an ordinary somatic cell nucleus. Determinations of DNA content per nucleus have recently been made in frog oöcyte nuclei. Such determinations are exceedingly difficult and the procedure followed is subject to criticism. It would seem that the frog oöcyte nucleus contains 800 times as much DNA as is found in a frog liver cell and this amount, large as it is, is only one-sixth to one-tenth that of the total DNA of the oöcyte (Finamore *et al.*, 1960).

Cytophotometric determinations of DNA on Feulgen-stained preparations have in many cases given results in conformity with DNA constancy per set of chromosomes, but this procedure has also yielded results indicating a variable DNA content (Moore, 1952; Leuchtenberger *et al.*, 1952; Pasteels and Lison, 1950; Lison and Pasteels, 1951; Leuchtenberger *et al.*, 1954; Fautrez and Laquerriere, 1957; Leeman, 1959). There are many possible causes of variability, and of course one of these may be that under certain conditions the quantity of DNA per set of chromosomes is not constant. Before this can be established, however, it is necessary to exclude other sources of variability.

An important *biological* variable to be considered is that doubling of DNA, which occurs during interphase, may be in process at the time the cells are fixed for subsequent Feulgen staining. This variable has been avoided in a recent investigation by Richards *et al.* (1956) by making their measurements in cells in metaphase, at which time no DNA doubling occurs. The difficulties involved in making measurements on material as heterogeneous as a field of metaphase chromosomes were overcome by a special crushing and scanning technique. Feulgen stain was measured on counted numbers of chromosomes and a direct relationship was found between DNA Feulgen and chromosome number, as shown in Fig. 15. These measurements show that cytophotometric work by other investigators in which variable DNA content is found should be scrutinized to see whether due attention has been given to chromosome number, process of DNA duplication, and optical technique.

There are also other factors to be considered, as can be seen from the work of La Cour *et al.* (1956) on the DNA content measured by Feulgen staining of cold-treated plant cells. Exposure to 0° for a few days seems to decrease the DNA per nucleus by 20 to 23%. Recovery of DNA occurs in 48 hours. Since these measurements were made in the same laboratory as were those that have just been described relating chromosome number to DNA content, there is every reason to have confidence in the optical technique used. A question which should be raised concerning the changes observed as a result of cold treatment is whether there was a variation in the amount of DNA or merely in the *state* of the DNA which might influence the Feulgen staining. This is the problem that Lowe and Rand considered in their work on the effect of cortisone in causing an apparent decrease in DNA content of liver nuclei. During the "recovery" of DNA, when plants are brought out of the cold, is there a synthesis of DNA as shown by incorporation of P^{32}? If there is not, then it may well be supposed that cold treatment caused an alteration such as "depolymerization" of some of the DNA so that it was lost during the partial hydrolysis that occurs in the Feulgen process. The decrease in nuclear DNA of the liver due to cortisone is detected by Feulgen staining as well as by conventional chemical determination (Lowe *et al.*, 1959).

A large apparent decrease in the DNA content per cell has been found to occur in bovine spermatozoa during *in vitro* storage (Birge *et al.*, 1960). It is common veterinary practice to store bovine semen in a yolk-citrate medium at 5°C. Fertility of such sperm drops somewhat during storage for 5 days, but even after this time a relatively high fertility is retained. Feulgen absorption cytophotometry indicates a drop in DNA content of 30% and, furthermore, this change is quite uniform for all the cells of any particular semen sample. It would be of interest to know whether the change in the Feulgen cytophotometric value is due to a partial depolymerization of the DNA or to an over-all loss of DNA from

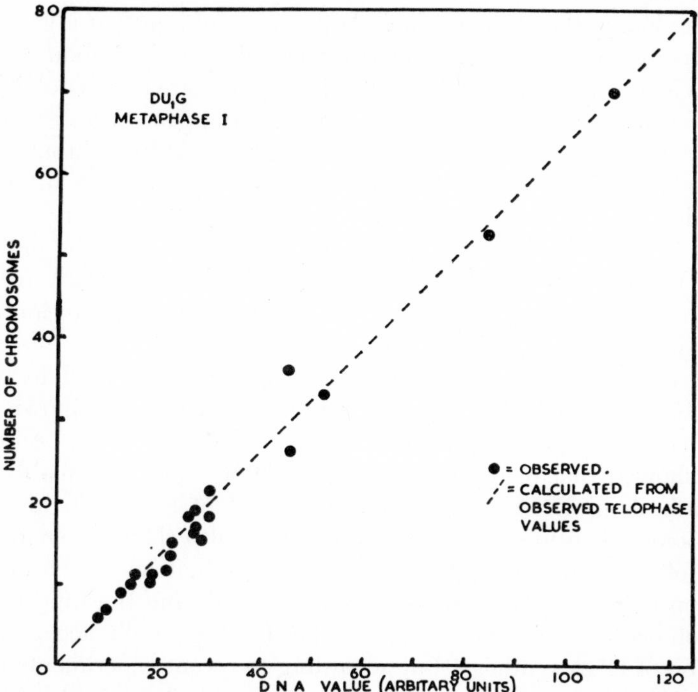

FIG. 15. Plot of DNA amount against number of chromosomes in Krebs ascites tumor. Broken line connects the origin with the mean DNA value (125) of a large telophase sample, assuming that these contain 80 chromosomes. From Richards *et al.* (1956).

the sperm. Chemical determinations could show whether there has been an absolute loss in deoxynucleotides or only a decrease in the quantity of DNA that remains in the sperm after the treatment with hot, concentrated HCl, a first step in the Feulgen procedure.

The observations of Breuer and Pavan (1955) on the giant polytene chromosomes of the sciarid *Rhynchosciara angelae* show that variations in the quantity of DNA in a chromosome do in fact occur under normal physiological conditions. The variations, present in certain bands only, are of such magnitude that they are readily observed; and yet they would not be detected by either conventional chemical methods for determining DNA or by Feulgen staining on interphase nuclei for the variations are found at a given time in only a few out of a very large number of bands. These polytene chromosomes are in the salivary glands, Malpighian tubules, and intestines of *Rhynchosciara* larvae.

After the larvae have grown to full size, pufflike developments form at definite loci in the chromosomes. Appearance of a puff at a given locus is a reproducible event, occurring at a particular locus at about the same time in each cell type. The locus at which puff formation is observed is characteristic of the cell type; thus a puff will appear at a certain locus in a salivary gland chromosome and at another locus in a chromosome of a Malpighian tubule. A puff remains for only a limited time, after which it subsides. There are three types of puffs: in one type there is first an increase in DNA, as shown by Feulgen staining, and then the puff forms (Fig. 16); in another type, puff formation and increase of DNA are simultaneous events; and in a third type, the puff forms without an increase in DNA (Pavan, 1958; Rudkin, 1959).

That the increase in Feulgen staining material observed in phase f of Fig. 16 is probably due to newly synthesized DNA was shown by experiments with tritiated thymidine (Ficq and Pavan, 1957). Autoradiographs showed that incorporation of thymidine (and presumably therefore synthesis of DNA) was much more evident in certain loci of the chromosomes than in others. Autoradiographs also show a general background of thymidine incorporation suggesting that an increasing polytenization is in progress in these chromosomes during the time that puff formation occurs. Uptake of thymidine has been generally accepted as an indication of DNA synthesis, but according to recent experiments by Pelc (1958b; 1959a) thymidine may be taken up under certain conditions as part of an exchange reaction with DNA and not as synthesis of DNA. The uptake of tritiated thymidine in giant chromosomes may therefore not indicate synthesis of DNA. In the present writer's opinion the best evidence for DNA synthesis in certain puffs is observation of the marked change in intensity of the Feulgen preparations. The giant chromosomes of other Diptera have been studied with results essentially like those of Pavan and his colleagues (references in Rudkin and Woods, 1959).

At present it is uncertain what implications the observations on these chromosomes have for an understanding of the generality of interphase chromosomes. It is possible that the puff formation is related to the polytene nature of these giant chromosomes and to the fact that they are in cells which will not divide but in which polytenization is taking place.

Cells with giant chromosomes are examples of growth accompanied by massive increase of DNA but without mitosis and its accompanying nucleolar reconstitutions. Puff formation may well be related to what in other cells would be nucleolar phenomena.

To sum up concerning the distribution of DNA in the cell and its

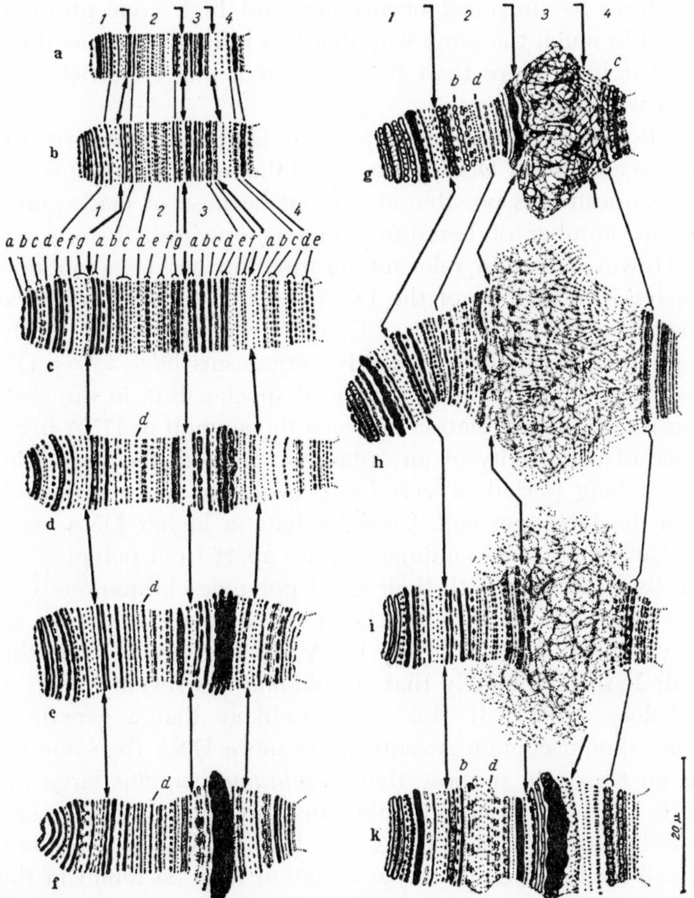

FIG. 16. Camera lucida drawings of the distal end of chromosome c of the salivary gland of *R. angelae,* in different stages of larval development. Phase c is from a chromosome of a full grown larva. Phase a is from a chromosome 16 days younger than the one in phases h and i. Phase b is from a chromosome 8 days older than the one in phase a. Phases c to k represent 8 days of larval development. The arrows and italic letters indicate corresponding bands. The arabic numbers indicate sections into which this part of the chromosome was arbitrarily divided. From Breuer and Pavan (1955-56).

content in the nucleus: The occurrence of DNA in the chromosomes, together with the parallelism between DNA content and ploidy, label the DNA as part of the genetic machinery of the cell. This conclusion draws further support from the finding that the transforming principle in bacteria is also a deoxyribonucleic acid. It should also be said that the special significance of DNA in hereditary mechanisms might have been surmised from its universal occurrence and high concentration in the sperm head in much the same way that this inference is now drawn from the fact that little more than the DNA of a phage particle enters the bacterium which it infects.

A question immediately suggests itself: Does a simple correlation exist between the amount of DNA per cell and the number of genes? No direct test of this question is possible at present because of inadequate knowledge of the number of hereditary factors operating in a complex organism. However certain relevant information has been obtained from an extensive investigation of the DNA per nucleus of a wide variety of animal species (Mirsky and Ris, 1951a). It was shown that, in the invertebrate phylum, very simple primitive organisms have lower DNA contents than is observed in more advanced species. But, in the vertebrates, there was no simple correlation between the amount of DNA per cell and the degree of complexity of an organism. Indeed, the results indicated that over a long period of vertebrate evolution there was probably a decline in the DNA per cell. Urodeles have a higher DNA content per nucleus than do any other animals (quite apart from polyploidy and the special cells of Diptera with their giant polytene chromosomes). *Amphiuma* cells have 168×10^{-9} mg. DNA per nucleus. Birds, on the other hand, have only 2.4×10^{-9} mg. DNA per nucleus. How is this to be explained? It seems unlikely that *Amphiuma* has seventy times as many genes as does a bird. It also seems unlikely that a genetic locus in *Amphiuma* should contain seventy times more DNA than one in a bird. There is no reason to suppose that *Amphiuma* contains large masses of nongenetic DNA, for in both amphibians and birds, spermatozoa contain the haploid quantity of DNA, i.e., one-half the quantity present in the somatic cell nuclei. Perhaps the explanation is to be found in the multistranded or polynemic nature of vertebrate chromosomes (Kaufmann, 1960). This would mean that the number of strands in the chromosomes of *Amphiuma* is some seventy times greater than in those of a bird. Such an explanation would imply that, when a mutation occurs, it extends across all the strands.

The relationship between the quantity of DNA in a nucleus and the number of genes is puzzling, but an interesting relationship has been found to exist between the DNA content and the size of the cell (Mirsky and Ris, 1951a). In general, when homologous cells of different species are compared, the greater the DNA content, the larger the cell. For example, in the nucleated red cells of vertebrates, a series of homologous cells, there is an approximately direct relationship between cell mass and DNA content (Table II). In all these cells the nuclei are diploid. It is of

TABLE II
DNA CONTENT AND MASS OF ERYTHROCYTES OF VARIOUS VERTEBRATES[a,b]

Animal	DNA	Mass
Dipnoan		
African lungfish, *Protopterus*	100	161
Amphibians		
Amphiuma	168	368
Necturus	48.4	40.5
Frog	15.0	27
Toad	7.33	13.7
Reptiles		
Green turtle	5.27	18.4
Woodturtle	4.92	14.1
Snapping turtle	4.97	
Alligator	4.98	14.9
Water snake	5.02	13.7
Pilot snake	4.28	13.3
Black racer snake	2.85	10.2
Birds		
Domestic fowl	2.34	4.39
Guinea hen	2.27	4.58
Duck	2.65	5.44
Goose	2.92	7.37

[a] Allfrey *et al.* (1955).

[b] DNA expressed as milligrams $\times 10^{-9}$ per cell and mass as milligrams $\times 10^{-8}$ per cell.

course well known that polyploidy and polyteny are associated with marked increases in cell size. The significance of this observation is discussed later in this chapter.

3. *Heterogeneity of DNA*

The evidence that the deoxyribonucleic acids are concerned in the maintenance and transmission of genetic factors has been brought forward to support the viewpoint that DNA preparations probably consist of large numbers of molecular species, chemically related but individually distinct and specific. The chemical evidence for the heterogeneity of the deoxyribonucleic acids will now be considered. A test of DNA heterogeneity is a successful resolution of a single DNA preparation into fractions of different chemical composition. A number of such fractionations have been accomplished. Most of them depend upon the use of varying concentrations of sodium chloride to release DNA from its combination with histone and other proteins. Concentrated sodium chloride is commonly used to extract nucleohistone from the nucleus, and it is known that an important factor in the procedure is the tendency of the salt to dissociate DNA from its combination with histone (Mirsky and Pollister, 1946). One form of fractionation, therefore, is simply to extract calf thymus nucleohistone (prepared by aqueous extraction) with salt solutions of varying concentrations (Chargaff *et al.*, 1953). In another fractionation procedure an ingenious chromatographic method has been devised by preparing a column of kieselguhr coated with calf thymus histone. To the column is added a deproteinized calf thymus DNA preparation and this is eluted by sodium chloride in stepwise increasing concentrations (Brown and Watson, 1953). Both these methods of fractionation give quite a number of DNA fractions which differ in their base composition.

DNA has also been fractionated on a histone column, using for elution a concentration gradient of NaCl instead of stepwise increases in concentration (Brown and Martin, 1955). The elution diagram showed presence of two distinct, though not well separated, fractions and this was borne out by purine and pyrimidine analyses. Considering the number of distinct species of DNA that are generally supposed to be present in a DNA preparation, separation into two distinct species would be no more than a beginning of fractionation.

Elution of DNA by a combination of concentration gradients and stepwise changes of eluant produces a more complex elution diagram than does a concentration gradient alone. In this way, and using a substituted cellulose column instead of one containing histone, Bendich and his colleagues (1956) have obtained "profiles" of DNA's prepared from many different sources. By a "profile" is meant the resolution of the DNA as extracted from a tissue into a number of components and the relative quantities of the components so obtained—what is usually called a chromatographic pattern. Bendich claimed that such profiles are different from one tissue to another even in a single mammalian species. Such a pronounced differentiation of DNA during ontogenesis, though somewhat surprising, would be of great importance. A careful repetition of Bendich's experiments by Kondo and Osawa (1959) and also by Kit and Gross (1959) and Kit (1960) failed to confirm Bendich's results; no significant differences were found in the DNA profiles of preparations made from various tissues of the same animal. There might well be subtle differences in the secondary structure of DNA as it occurs in various tissues of an organism, but they have not yet been found.

DNA preparations from microorganisms and from tissues have been fractionated, but it is generally supposed that these preparations are far more heterogeneous than is indicated by the fractionation so far accomplished. The relative degree of heterogeneity of DNA preparations can be estimated by two procedures used by Doty and his colleagues (Marmur and Doty, 1959; Sueoka *et al.*, 1959). In one of these the form of the heat-denaturation curve of DNA is related to its heterogeneity. DNA is kept in its native configuration by hydrogen bonds which hold together the double helix. DNA is denatured when the hydrogen bonds are broken. The bonds between guanine and cytosine are broken at a higher temperature than the bonds between adenine and thymine. There is in fact a linear relation between the guanine-cytosine content of a DNA and its denaturation temperature. The steepness of the denaturation-temperature curve depends on heterogeneity. The steepest curve is that given by the synthetic adenine-thymine DNA which has no heterogeneity due to composition. The denaturation-temperature transition is just as sharp for T4r bacteriophage DNA and hence it is supposed that the DNA molecules in this case have identical composition. DNA preparations from some bacteria have much less steep denaturation-temperature curves and that for thymus DNA is still broader. In thymus DNA there is evidence that heterogeneity is at two levels: there are different kinds of DNA molecules and also there is heterogeneity on a scale very much smaller than that of whole molecules.

Heterogeneity of DNA is also disclosed by centrifugation in a density gradient, as first used for DNA by Meselson and his associates (1957), who showed that in an analytical ultracentrifuge DNA sediments in a well-defined band, rate of sedimentation depending on the density of the DNA. It was subsequently found (Sueoka *et al.*, 1959) that there is a linear relation between the guanosine-cytosine content of a DNA and its density and also that the width of the band in the ultracentrifuge is an expression of the heterogeneity of a DNA sample.

B. Proteins

One of the most interesting problems in the chemistry of the nucleus is concerned with the nature of its protein constitution, and with the relationship between this constitution and nuclear activity.

When nuclei of different tissues are compared one observes striking differences as well as similarities. The remarkable similarity in DNA content per nucleus has already been mentioned. One of the most striking differences is the variation in the total amount of protein in different nuclei.

Some information on the total protein contents of a diversity of nuclei is summarized in Table III. This information was obtained by the chemical analysis of nuclei isolated in nonaqueous media (Allfrey *et al.*, 1952)

TABLE III
DNA-P, RNA-P, AND PROTEIN CONCENTRATIONS OF NUCLEI ISOLATED IN NONAQUEOUS MEDIA[a]

Source of nucleus	DNA-P (%)	DNA (%)	RNA-P (%)	Ratio: DNA-P / RNA-P	Protein (%)	Ratio: Protein / DNA
Calf						
Thymus	2.56	26.5	0.17	15	71.8	2.7
Heart	1.90	19.7	0.11	17	79.2	4.0
Kidney	1.71	17.7	0.16	11	80.7	4.6
Liver	1.50	15.6	0.15	9.7	82.9	5.3
Intestinal mucosa	1.96	20.3	1.04	14	78.3	3.9
Bone marrow	2.34	24.3	0.09	26	74.8	3.1
Pancreas	1.70	17.6	0.11	15	81.3	4.6
Beef						
Heart	1.65	17.1	0.22	7.6	80.7	4.8
Pancreas	1.71	17.7	0.13	13	81.0	4.6
Horse						
Pancreas	2.20	22.8	0.20	11	75.2	3.3
Liver, normal	1.20	12.4	0.16	7.5	86.0	7.0
Liver, fasted	1.77	18.3	0.08	21	80.9	4.4
Chicken						
Kidney	1.45	15.0	0.20	7.3	83.0	5.5
Erythrocytes	2.55	26.4	0.01	255	73.5	2.9

[a] Allfrey *et al.* (1955).

—the only isolation procedure now available which precludes a loss of nuclear proteins or an absorption of proteins from the cytoplasm. It can be seen that the nuclei of different tissues vary considerably in the amounts of protein which they contain. The variability is also observed in nuclei of the same tissue under different physiological conditions. The nucleus of the liver cell in starvation, for example, has far less protein than that found in the liver nuclei of fed animals. Furthermore, enzyme studies which will be discussed below make it evident that the type of protein in the cell nucleus also varies in starvation.

Thus, both the nature and the total amount of protein in the nucleus vary from tissue to tissue, and, in a given organ, they vary with changes in physiological state. This variability in the protein composition of nuclei is one of the factors underlying the differences in appearance and stainability of nuclei which have long been familiar to histologists. Histologists, however, have sometimes tended to confuse the problem by interpreting the differences in intensity of nuclear staining in terms of varying "nucleic acid charge," a viewpoint which is rendered untenable by the chemical demonstration of DNA constancy, at least within certain limits, in the nucleus.

The variability in the nature and amount of nuclear proteins suggests that the nucleus is responsive to and concerned with the metabolic activity in the rest of the cell. This is a picture of the nucleus that is in accord with enzyme studies and with tracer experiments that will be discussed later in this chapter.

1. Protamines and Histones

The protamines and histones are basic proteins found combined with DNA in many nuclei. Protamines, discovered by Miescher in salmon sperm (Miescher, 1897), were soon found in the sperm of many other fish (Kossel, 1928) and more recently gallin, a protamine-like protein was isolated from rooster sperm (Daly *et al.*, 1951). No basic protein has yet been isolated from mammalian sperm. There is, however, evidence that protamine-like substance is present in mammalian sperm: in sperms known to contain protamines there is in the sperm head a high arginine:DNA ratio, much higher than is found in the generality of nuclei; and such a ratio has been found in mammalian sperm (Vendrely and Vendrely, 1953).

Histones were discovered in the goose erythrocyte (Kossel, 1884) and soon thereafter were found in the thymus gland (Lilienfeld, 1894). Kossel (1928) considered histones to be present in the nuclei of only "certain kinds of tissues." More recently the extraction of nucleohistones with 1 *M* NaCl showed that histones are present in the nuclei of many tissues (Mirsky and Pollister, 1942), and since then they have been found in the somatic nuclei of the vertebrates wherever looked for. Histones have also been found in certain plant cell nuclei, notably wheat and cedar nut embryos (Mirsky and Pollister, 1946; Belozerskii and Uryson, 1958). It is still uncertain whether histones are present in bacteria. According to Vendrely and his colleagues (Palmade *et al.*, 1958), nucleohistone can be prepared from bacterial protoplasts by extraction with 1 *M* NaCl just as from animal tissues. On the other hand, Zubay and Watson (1959) report that in bacteria the protein combined with DNA is not a histone.

Recently basic proteins quite similar to histones have been found in the microsomes of cytoplasm (Butler *et al.,* 1960).

The distinguishing characteristics of protamines and histones, apart from the fact that they are found in combination with DNA, are their basicity and their simplicity compared with the generality of proteins. In both of these characteristics the protamines are the more extreme; they are both more basic and simpler than histones. In clupeine and salmine, the two most thoroughly investigated protamines, there are some 85 gm. of arginine per 100 gm. of protein. Glutamic and aspartic acids are absent. Histones have far less arginine and considerable quantities of glutamic and aspartic acids. Clupeine and salmine are lacking in all the aromatic amino acids and in quite a number of other amino acids. Except for cystine and tryptophan, which are absent from histones, the usual amino acids are found in most histones. Salmine has a molecular weight of about 4000 and readily passes through cellophane membranes. Histones have molecular weights several times larger and are held back by cellophane membranes.

There are basic proteins which have been isolated from sperm and which have chemical properties intermediate between the typical protamines, clupeine and salmine, and thymus histone. Gallin of rooster sperm is such a protein. Its amino acid composition, for example, lies in between that of salmine on the one hand and that of thymus histone on the other (Daly *et al.,* 1951). On chemical grounds there could be some doubt as to whether gallin should be classified as a protamine or histone; and yet on biological grounds there can be little doubt that gallin should be grouped with the protamines, the basic proteins of sperm.

The significant point here is that the basic proteins of the various somatic nuclei of an organism are very much alike, if not identical, and they are distinctly different from the basic protein present in the sperm of the same organism. Gallin, for example, is unmistakably different from the histone of fowl erythrocytes. Although the difference is less than that between salmine and the histone of salmon erythrocytes, the properties of both gallin and salmine are removed in the same direction from histones present in the somatic nuclei of the fowl and salmon; in this sense both sperm proteins are protamines. Replacement of histone by protamine, or a protamine-like protein, seems to be a general phenomenon in the vertebrates. It would be of interest to know what the situation is in other organisms. What we know in this field has its origin in Miescher's work on salmon sperm (Miescher, 1897); the contributions of the next generation are summarized in Kossel's monograph (Kossel, 1928). Miescher showed that salmine appears only as the sperm mature. This was the first demonstration of a chemical difference in the chromosomes of the various cell types of the same organism. That the difference involves the replacement of histone by protamine became clear only when it was shown that histones are present in all somatic nuclei including those of the salmon (Mirsky and Pollister, 1943). The final step was the demonstration that this replacement occurs at an advanced stage in spermatogenesis. This was shown in salmon testis by staining with fast green (Alfert, 1956); it has also been possible to show the replacement of histone in bull testis by using 1 *M* NaCl to extract nucleohistone (Vendrely *et al.*, 1957).

2. *Fractionation of Protamines and Histones*

Clupeine preparations have been fractionated by Felix and his colleagues (Felix *et al.*, 1956, an excellent summary). Countercurrent distribution gave results indicating the presence of at least six components, but it is likely that there are even more. The fractions differ slightly in amino acid composition and molecular weight. Clupeine has also been separated into three distinct fractions by paper chromatography. An uncertainty that should be mentioned concerns the heterogeneous nature of the cell population in the testes from which the clupeine was prepared, for the clupeine was extracted from mature testes and not mature sperm. In the course of spermatogenesis histone is replaced by protamine. The whole testis may be expected to contain cells at all stages in this process. Therefore, a certain amount of protamine heterogeneity might be expected on biological grounds. It would be better to obtain mature sperm from fish by stripping and then extract protamine. This can be done easily with salmon and trout. Recently an investigation of salmine prepared in this way has been begun, and the first results indicate that the salmine is heterogeneous (Callanan *et al.*, 1957).

Histones have been fractionated by partial extraction, precipitation, chromatography and electrophoresis (Davison and Butler, 1954; Crampton *et al.*, 1955; Daly and Mirsky, 1955; and references in Moore, 1959). The first clear fact to emerge from fractionation is that there are in each tissue examined two types of histone: lysine-rich histone, in which lysine accounts for 40% of the nitrogen, alanine for 20%, and arginine for 5%; and arginine-rich histone in which arginine accounts for 28% of the nitrogen, alanine for 7%, and lysine for 16%. To what extent each of these fractions can be further fractionated, is still uncertain. The uncertainty is due to the fact that autolysis proceeds rapidly in histone extracts, so that it is frequently difficult to decide whether a histone fraction preexisted as such in the nucleus or is an artifact produced by autolysis.

Proteinases have been found in preparations of histone (Phillips and Johns, 1959). Their action can be partly inhibited by diisopropyl fluorophosphate. When histones are prepared under conditions minimizing proteolysis, the proline and alanine *N*-terminal end groups account for up to 96% of all end groups. These two end groups pre-existed in the deoxynucleohistone and were not produced by the acid extraction. Presence of more end groups in a histone preparation is evidence of proteolysis. The lysine-rich histone fraction contains predominantly NH_2-terminal proline.

Chromatography on the carboxylic resin IRC-50 (Crampton *et al.*, 1957) and on carboxymethylcellulose (Davison, 1957b) is effective in the fractionation of histones. Starch gel electrophoresis, which has great resolving power, has also been used (Neelin and Connell, 1959), but not under conditions minimizing proteolysis.

An interesting problem associated with histone fractionation is whether or not these proteins differ in various somatic tissues of the same organism. The first claim for a difference came from Stedman and Stedman (1943) when they reported that tumor cells have a low histone content. It was then found that the histone content of normal and malignant leucocytes is the same (Allfrey *et al.*, 1955), and this was also noted by Davison (1957a). Indeed, in relation to DNA content the same quantity of histone was extracted from different tissues (Daly *et al.*, 1951). Furthermore, when histones were fractionated into lysine-rich and arginine-rich components, both were found in histones of all somatic tissues examined.

With the chromatographic separation of histones the means were vastly improved for answering the question of tissue specificity (Crampton *et al.*, 1957). The fractions that appear to be most homogeneous have had their amino acid composition examined and also the chromatographic patterns of their tryptic hydrolyzates. By all the criteria employed for comparison, the fractions examined from calf thymus, liver, and kidney appear to be remarkably similar, and those from guinea pig testis barely detectably different. Essentially similar results have been obtained by Davison (1957a). Of course in such work the failure to find differences does not imply that no differences exist.

Tissue-specific differences in histones, especially from malignant cells, have however been reported by Stedman and his colleagues (Cruft *et al.*, 1954, 1957b). The differences observed are in electrophoretic mobility and in amino acid composition. They are both open to question: electrophoretic mobility is much affected by presence of impurities and by aggregation; differences in amino acid composition are readily brought about by autolysis in histone extracts (Davison, 1957b; Crampton *et al.*, 1957). As a matter of fact, the electrophoretic experiments of Davison do not show the differences claimed by Cruft between the histones of normal tissues and tumors; nor did he find differences by chromatographic procedures. The conclusion to which we come at present is that if there are somatic tissue-specific differences between histones, they have not yet been demonstrated.

Both protamines and histones are attached to DNA by saltlike linkages. For protamines this has been shown by dissolving the nucleoprotamine of trout sperm in 1 *M* NaCl and then dialyzing against 1 *M* NaCl. All the protamine passes through the membrane, leaving an essentially protein-free DNA inside the membrane (Pollister and Mirsky, 1946). Histone also is to some extent dissociated from DNA in salt solutions of high ionic strength. Dissociation in this case is shown not by dialysis, but by high speed centrifugation. Centrifugation of nucleohistone in 1 *M* NaCl leaves a layer at the top of practically DNA-free histone. Also the study of isolated chromosomes has shown that histone is bound to DNA by saltlike linkages: most of the histone in a suspension of chromosomes in isotonic electrolyte can be displaced by protamine, a more basic protein (Mirsky and Ris, 1951b).

The function of histones and protamines in the metabolism of the chromosome is still not clear. In considering this problem it is necessary to distinguish between the functions of histones and protamines, for, in the course of spermatogenesis, protamines take the place of histones. This is a clear indication of some difference in function of these two proteins. Further evidence of functional difference comes from tracer experiments which show that the histones of many somatic cells incorporate N^{15}-glycine, whereas the basic proteins of sperm (measurements having been made on *Arbacia* sperm) are practically inert in this respect (Daly *et al.*, 1952). The tentative conclusion at present is that, if the basic protein of sperm is considered to be essentially a passive agent combined with the acid groups of DNA, which is itself metabolically inactive in sperm, this can hardly be the case for the histone of somatic cells. The DNA of somatic cells is metabolically active and, as will be mentioned later, there is evidence that histone may control this activity.

3. *Nucleohistone*

In the nucleus, histones are combined with DNA and it is possible to extract DNA and histone as a complex, as nucleohistone, in which many of the relatively small histone molecules are attached to a single DNA molecule. Nucleohistone is most readily prepared from the thymus. From this tissue nucleohistone is readily extracted in water by keeping the ionic strength low. Thymus nucleohistone was prepared by Huiskamp as far back as 1901. It was carefully studied electrophoretically by Hall (1941), who found that nucleohistone migrates as a single component. In a recent investigation (Zubay and Doty, 1959) nucleohistone was prepared from isolated thymus chromosomes (Mirsky and Ris, 1951b) and electron micrographs were made. The electron micrographs of nucleohistone are similar to those of DNA, that is, they show long, only slightly coiled threads, those of DNA being about 20 A. in diameter and those of nucleohistone about 30 A.

4. *Nonhistone Proteins of the Nucleus*

a. Proteins soluble in dilute salt solutions. Some of the nonhistone proteins are present in the nucleoplasm. This is evident especially in large nuclei such as amphibian germinal vesicles, but very little is known about proteins in these nuclei. In smaller nuclei, such as those of the thymus, liver, kidney, and pancreas, there is much less protein in the nuclear sap; but again practically nothing is known about it. The fact that a nuclear protein is extracted by an isotonic salt solution has frequently been taken to mean that this protein is extrachromosomal. This need not, however, be true. Thymus nuclei isolated in isotonic sucrose contain a protein fraction that is extracted in neutral, isotonic saline. Experiments with labeled amino acids show that uptake of amino acids by this protein fraction is exceedingly rapid and furthermore that uptake stops when the DNA of the nucleus is removed by DNAase. This property would at least raise the question of whether such a soluble protein fraction is a chromosomal component (Allfrey *et al.*, 1957).

There are other indications that the nuclear proteins soluble in isotonic saline are not retained in the nucleus by the nuclear membrane but are bound to the deoxyribonucleoprotein structures in a manner similar to that observed with ion-exchange resins (Barton, 1960). The experiments on which this suggestion is based were done with liver nuclei isolated in sucrose media. Such nuclei do not liberate soluble protein when they are chopped to pieces in sucrose solution by means of a Waring blendor, although they do release a considerable amount of soluble protein when they are gently washed with isotonic saline, in which medium the nuclei retain their form, as seen under the microscope. When such nuclei are returned to a sucrose medium and exposed to the protein previously released in saline, they combine with the protein. Nuclear proteins extracted in neutral isotonic saline from liver nuclei have been studied electrophoretically (Barton, 1960).

Proteins have been extracted in a 0.1 *M* neutral buffer from isolated thymus nuclei and were then fractionated by high speed centrifugation (Frenster *et al.*, 1960). More than one-half of the extracted material was sedimented by centrifugation at 40,000 r.p.m. for 16 hours. The sedimented material was a ribonucleoprotein while that remaining in solution consisted of protein and but little RNA. No DNA was present in either fraction. The sedimentable fraction was examined in the electron microscope and was found to consist of granules similar to those present in the nuclei before extraction. Granules resembling these extracted from thymus nuclei have been observed in various nuclei before—on the lateral loops of lampbrush chromosomes (Gall, 1956), on the chromosomal rings of Balbiani (Beermann and Bahr, 1954), in the blebs (Gay, 1956), in pore annuli of the nuclear membrane (Wischnitzer, 1958; Watson, 1959), and in the nuclear sap (Callan, 1956).

The nucleoproteins and proteins extracted in neutral buffer, when prepared from thymus nuclei which previously had been incubated with alanine-*1*-C^{14}, or other labeled amino acids, were found to have incorporated more of the labeled amino acid than did the bulk of the nuclear protein. The fact that the extracted proteins were so "hot" was of value in demonstrating that the extracted protein was actually of nuclear origin, and not a cytoplasmic impurity present in the isolated nuclei. Extracted protein was only about 10% of the nuclear mass, so that without the evidence which will now be presented, there could be some doubt as to whether it was actually of nuclear origin. When nuclei were incubated in a medium containing potassium ions (rather than sodium ions) or when the nuclei had previously been treated with DNAase, very little of the labeled amino acid was present in the extracted protein. These are definite signs, as will be seen later in this chapter, that the extracted protein was of nuclear origin. Considering the presence in the annuli of the nuclear membrane of granules similar to those extracted, it is tempting to suppose that the extracted material was a nuclear product destined for export to the cytoplasm.

Ribonucleoprotein particles isolated from thymus nuclei are able actively to incorporate amino acids in the presence of certain metabolites and also of the pH 5 amino acid activating enzyme (Frenster, *et al.*, 1960).

b. Nonhistone proteins that are insoluble in dilute salt solutions. After extraction of nuclei with a neutral isotonic buffer they still contain a large quantity of nonhistone protein. From liver or kidney nuclei isolated in sucrose (Chauveau *et al.*, 1956) it is possible to extract both histone and DNA (after the proteins soluble in isotonic neutral buffer have already been removed), leaving behind well-formed nuclei consisting primarily of nonhistone protein combined with some RNA. Such a preparation is made by first extracting histone and then DNA. Histone is extracted by 1 *M* NaCl at pH 3.8, and DNA is then readily removed by the action of pancreatic DNAase. The residue of nonhistone proteins remaining after removal of histone and DNA has been called the residual protein fraction. In a preparation of isolated liver nuclei containing 19.5% DNA there is 28.6% of residual protein. The amino acid composition of residual protein and its properties when freed of DNA show it to be entirely different from the histones.

Different as residual protein is from histone, it too is combined with the phosphoric acid groups of DNA (Mirsky and Ris, 1951b). This is shown by staining experiments with a basic dye such as crystal violet. Both histone and residual protein tend to prevent crystal violet from combining with the phosphoric acid groups of DNA. Further evidence for the combination of DNA with residual protein is that nuclei from which all histone has been extracted can be placed in isotonic saline at neutrality and still retain their DNA. Since DNA is readily soluble under these conditions, it is apparent that some, at least, of the nonhistone protein is combined with DNA. Once this DNA has been removed, it is not possible to attach DNA again to the nonhistone protein.

By the procedures that have just been described histone and DNA can be removed from the isolated chromosomes prepared from interphase nuclei of liver and kidney. It is then found that the form of the chromosome as seen under the microscope is due to DNA and residual protein; both are needed if chromosomes are to remain intact. Histone can be removed leaving the appearance of chromosomes unchanged, but if the residual protein of histone-less chromosomes is broken down by trypsin, the chromosomes disappear, leaving a DNA gel. On the other hand, if the histone-free chromosomes are treated with deoxyribonuclease, nothing remains of the chromosomal structure but a mass of minute protein threads. Thus, the morphological configuration of the chromosome, as seen under the microscope, is due to the combination of DNA with residual protein, and once these components are separated, neither the combination nor the configuration can be restored (Mirsky and Ris, 1951b).

The relative amounts of DNA and residual protein vary considerably in nuclei isolated from different tissues. Since it has been shown that the amount of DNA per set of chromosomes is a constant for different cells of the same organism, it follows that the quantity of residual protein varies in different tissues. There must be a lower limit, for without some residual protein there would be no chromosome structure, and this minimum quantity may also have some genetic significance.

A significant correlation is observed when one considers the relationship between the compositions of isolated nuclei and the metabolism of the cells from which they are derived. For example, liver and kidney cells, which have an abundant, metabolically active cytoplasm, have a relatively large amount of residual protein in their nuclei. On the other hand, a cell such as the lymphocyte, with only a scanty layer of cytoplasm around its nucleus, has less residual protein; and the metabolically sluggish red cell has even less. Thus, the amount of "active" cytoplasm and of residual protein run roughly parallel. The correlation suggests that the residual protein of the nucleus has an active role in cell metabolism. The nature of the role is not known. The activity of the residual protein, however, is made evident by tracer experiments which show that this fraction of the nucleus incorporates N^{15}-glycine at a rate comparable to that of the cytoplasmic proteins of the cell, and much higher than that of most histones (Daly *et al.*, 1952).

With what we know about histone and residual protein, the staining of chromosomes by basic dyes can be explained. Since *both* of these proteins are combined with the phosphoric acid groups of DNA, they *both* tend to prevent basic dyes from combining with DNA. This has been shown by experiments with crystal violet and isolated nuclei and chromosomes in test tubes (Mirsky and Ris, 1951a). The same type of experiment has been done with nuclei on microscope slides using methyl green, a more suitable dye, for Brachet has shown that it combines specifically with DNA, so that de-staining is not needed (Bloch and Godman, 1955). In addition they compared the interphase nuclei which have doubled their DNA and histone in the period before mitosis with interphase nuclei after telophase. They found that the posttelophase nuclei bound less dye than did the premitotic nuclei. This was interpreted to mean that more residual protein is present in interphase nuclei after telophase. These nuclei are acquiring more residual protein, a fraction active in protein synthesis. It may well be that the amount and character of the residual protein acquired is influenced by the surrounding cytoplasm, which would mean that the state of differentiation of the cytoplasm is affecting the variable component of the chromosomes.

The residual protein (and by this is meant the nonhistone protein remaining in the nucleus after extraction of protein in neutral saline) has been fractionated to some extent. One form of fractionation is to treat nuclei with neutral *M* NaCl. In this way nucleohistone is removed and along with it more than half of the residual protein. Just how much can be calculated for liver nuclei by subtracting from the total residual protein (28.6% of the nucleus) the quantity of protein (8.6% of the nucleus) remaining after extraction with neutral *M* NaCl, showing that 70% of the residual protein is extracted at the same time that nucleohistone is removed. The material remaining after neutral *M* NaCl extraction consists largely of the nucleolus and nuclear membrane and contains much RNA (Bessis, 1954; Allfrey and Mirsky, 1957b; Zbarsky and Georgiev, 1959). The *M* NaCl extract is saturated with NaCl at pH 11. The precipitate obtained contains much of the histone and a gelatinous nonhistone protein. Still attached to the DNA, which remains in solution, is the rest of the histone and a nonhistone protein quite firmly attached to DNA. DNA is removed from this protein by treatment with pancreatic DNAase (Allfrey *et al.*, 1957). Residual protein which is firmly attached to DNA has been investigated by Kirby and Frearson (1960). Both of the nonhistone proteins that are associated with nucleohistone much more actively incorporate labeled amino acids (when experiments are carried out with intact thymus nuclei) than do the histones.

C. *The Nucleolus and RNA*

Since it is difficult to discuss the RNA of the nucleus without constantly referring to the nucleolus it will be convenient at this point to consider both the nucleolus and nuclear RNA.

The nucleolus is readily seen in the interphase nucleus. Indeed in a living cell no other object in the nucleus may be visible. With the electron microscope it has become clear that the nucleolus is not surrounded by a membrane (Fig. 17). Interactions between the nucleolus and other nuclear components are not limited by the properties of a membrane, a factor of much significance for interactions between the nucleus and cytoplasm or between the nucleus and the medium surrounding the cell.

The somatic cell nucleus ordinarily contains one nucleolus for each set of chromosomes. There are nuclei, notably those of amphibian oöcytes, which contain hundreds of nucleoli unattached to chromosomes. It has always been, and it still is, questionable whether these bodies are in the same category as the nucleoli of somatic cells. Gall's observations suggest however that they are homologous (Gall, 1955). He has reported that the locus of attached nucleoli in oöcyte chromosomes may correspond to the nucleolar organizing region of the mitotic chromosomes. If the extrachromosomal nucleoli in the oöcyte also form at this site, followed by their continuous release, then these hundreds of nucleoli are indeed homologous to the somatic nuclei in the sense that they originate at a common locus.

The nucleolus, visible during interphase, disappears in the course of prophase and reappears at telophase. This cycle raises questions concerning the continuity of the nucleolus during mitosis, its formation at telophase, and its function during interphase.

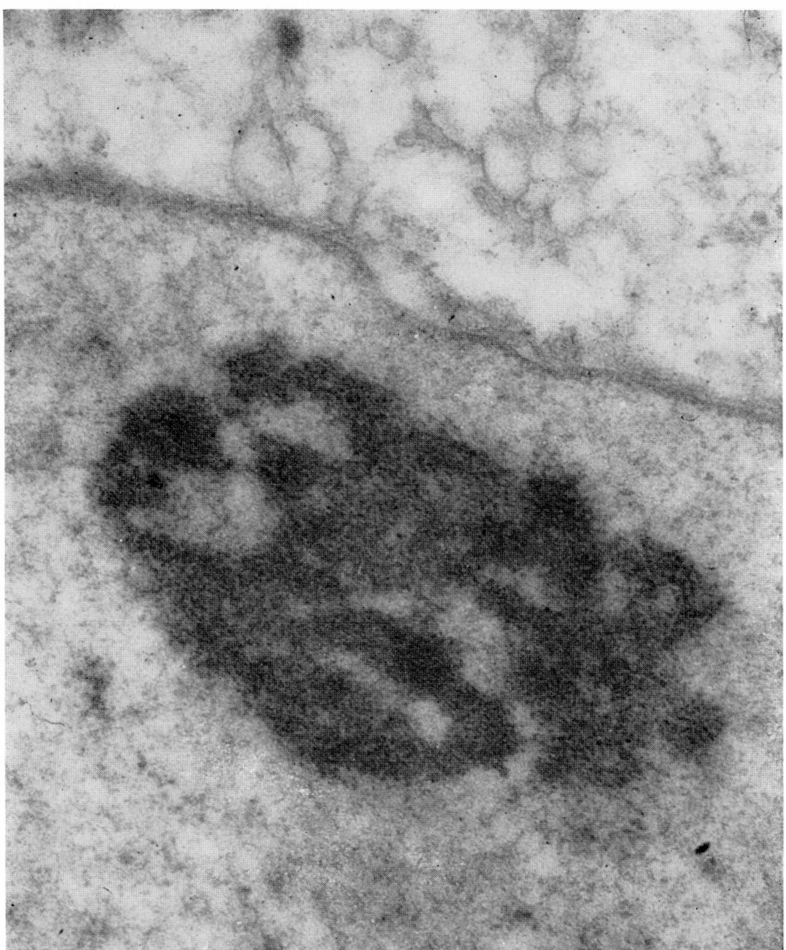

FIG. 17. Nucleolus, with the nuclear membrane above it, of a normal rat liver cell. Magnification: × 42,000. From Bernhard and Byczkowska (1960).

1. *Formation of the Nucleolus*

In a classical paper Heitz (1931a) showed that a nucleolus is formed at a definite region at one chromosome in each haploid set of chromosomes. He insisted that the nucleolus is formed *at* this point on the chromosome and not *from* the chromosomal material at that place. Later this particular chromosomal element was called by McClintock the nucleolar-organizing element (Fig. 18). She added to the evidence that the nucleolus develops through the activity of this body by obtaining a reciprocal translocation which broke this body into two parts. Both interchanged chromosomes possessed a section; and nucleoli developed at *each* of these two segments.

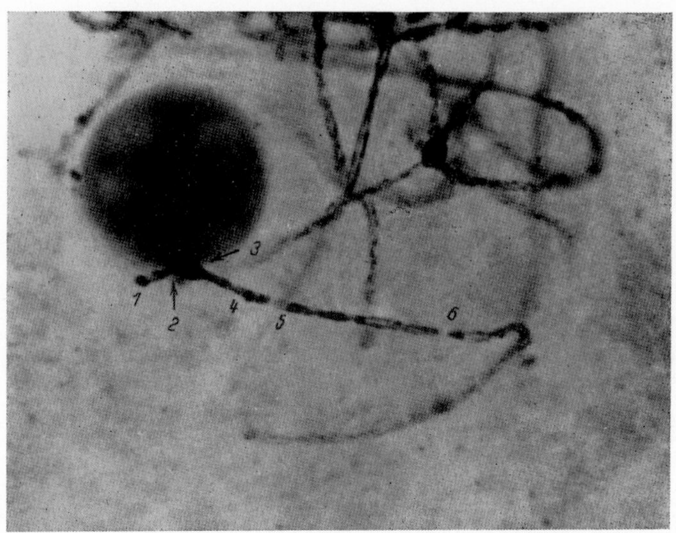

FIG. 18. Photomicrograph of two homologously associated chromosomes 6 (satellited chromosome) in the mid-prophase of meiosis showing the relationship to the nucleolus. The parts of the chromosome have been numbered: 1, the satellite proper; 2, the region joining the satellite to, 3, the deep-staining nucleolar-organizing body; 4, the chromatic thread of the short arm; 5, the spindle fiber attachment region; 6, the chromatic thread of the long arm. Magnification: × 1400. From McClintock (1934).

Evidence suggesting that nucleolar substance is produced by every chromosome is provided by McClintock's observations (1934) on chromosomes in which the activity of the nucleolar-organizing element was hindered by certain genomic deficiencies. When this occurs many small nucleolar-like droplets appear along all the chromosomes. The function of the nucleolar-organizing element, it would seem, is to organize the nucleolar substance present in each chromosome into a definite body, the nucleolus.

There are other observations which show that the potentiality for formation of nucleolar material exists along the chromosomes in places other than the locus of the nucleolar organizer. In certain plants micronuclei containing a single chromosome are formed next to the nucleus which contains the other chromosomes. Heitz (1931b) noticed that in *Vicia faba* and *Vicia monanthos* a nucleolus may appear in a micronucleus even when both chromosomes with nucleolar organizers are in the main nucleus. In chironomids Bauer (1936) found under certain conditions that, in addition to the main nucleolus, many accessory nucleoli formed in numerous places along the giant polytene chromosomes. In the onion, treatment with slow neutrons produces cells containing micronuclei, some of which contain nucleoli, while others lack nucleoli. On the whole the fragmented nuclei contain more nucleoli per cell than did the controls (Rasch, 1951).

The appearance of nucleoli in micronuclei which contain chromosomes that had not previously been found to possess a nucleolar organizer shows that the organizer usually present may suppress potential organizers. There are many instances of the occurrence of micronuclei (McLeish, 1954; Crosby, 1957) in which a nucleolus appears only when a chromosome containing the nucleolar organizing locus is present. In such chromosomes there is no sign of the potentiality for formation of a nucleolar organizer. Failure of a nucleolus to form in a micronucleus lacking an organizer is in accord with the Heitz-McClintock theory that at telophase the nucleolar organizer forms a nucleolus from material produced by all the chromosomes. The best evidence for the theory until recently was the original observation by McClintock of the appearance in her preparations of nucleolar-like droplets along chromosomes in the absence of a nucleolar organizer. It must be said that this evidence is no more than suggestive.

Recent electron microscopic studies on the formation of the nucleolus at telophase tend to support the view that the nucleolus is formed by the organization of materials that are widely dispersed and closely associated with the chromosomes (Lafontaine, 1958). Granules of a certain size and density are characteristic of the nucleolus. Such granules can be followed during the mitotic cycle. They are seen in loose clusters associated with the chromosomes at late anaphase, they become less scattered in early telophase (Fig. 8), and during late telophase these "prenucleolar bodies" fuse at special loci on chromosomes to form the mature interphase nucleoli. What is lacking in these studies is a satisfactory criterion for identification of "prenucleolar bodies." Size and density of granules are not sufficiently precise characteristics, especially when it is likely that these characteristics are dependent on methods of fixation. Even so, these observations seem to show that the nucleolus is organized at a chromosomal locus by coalescence of scattered materials.

The nucleolus as a formed body is lacking in continuity; it is formed anew at telophase of each mitosis. But do the smaller units of which it is formed possess continuity from one mitotic cycle to the next? There have been claims that one portion of the nucleolus becomes attached to the chromosomes at prophase, when the nucleolus disintegrates, and that this material (the *nucleolonema*) passes from the chromosomes to the new nucleolus at telophase (Estable and Sotelo, 1954; discussion in Lafontaine, 1958).

When it is suggested that materials pass from the nucleolus back to the chromosomes, in the course of the mitotic cycle, it should be mentioned that there are cases, as in grass, where the nucleolus does not disappear in prophase and indeed persists to metaphase or later. Furthermore such persistent nucleoli are not incorporated into the newly formed nuclei at the end of telophase but are discarded (Brown and Emery, 1957).

Once the nucleolus is formed at telophase, the nucleolar organizer seems to have some control over its composition. Materials derived from the chromosomes may still come to the nucleolus during interphase, and it has been suggested that this occurs (Goldstein and Micou, 1959b). Concerning the influence of the nucleolar organizer our knowledge comes mainly from Lin's experiments (Lin, 1955) on the same material, the microsporocytes of maize, on which McClintock's work was done. Chromosome morphology has been very well studied in maize and a large number of chromosome rearrangements are available so that it is an excellent material for investigation of the relationship between the nucleolus and the chromosomes.

In maize it is possible to produce cells with nuclei containing as many as five extra of the chromosomes which have the nucleolar organizer and also cells in which there is no increase in the number of nucleolar organizers but a considerable increase in the other chromosomes. In all these cells there is a single nucleolus. The RNA contents of nucleoli in the various nuclei were measured by ultraviolet absorption. Figure 19 shows that there is a linear relation between the RNA content of the nucleolus and the number of extra nucleolar organizers. The extra organizers do not seem to change the RNA:protein ratio of the nucleus, judging by the appearance of the ultraviolet absorption curve. The whole of the chromosome in which the nucleolar organizer lies, influences the RNA content of the nucleolus, but to a much smaller extent than does the organizer itself. Increasing the number of other chromosomes has no influence on the RNA content of the nucleolus.

The simplest explanation of Lin's measurements is that the nucleolar organizer controls the quantity of RNA in the nucleolus and that the other chromosomes have little, if any, effect on it. Since it is known from autoradiographs that there is a rapid turnover of nucleolar RNA, one conclusion, though surely not the only one, would seem to be that the organizer determines the rate of RNA metabolism in the nucleolus. This conclusion is supported by the observation that the nucleolar RNA which lies closest to the organizer has the greatest rate of labeling (Sirlin, 1960).

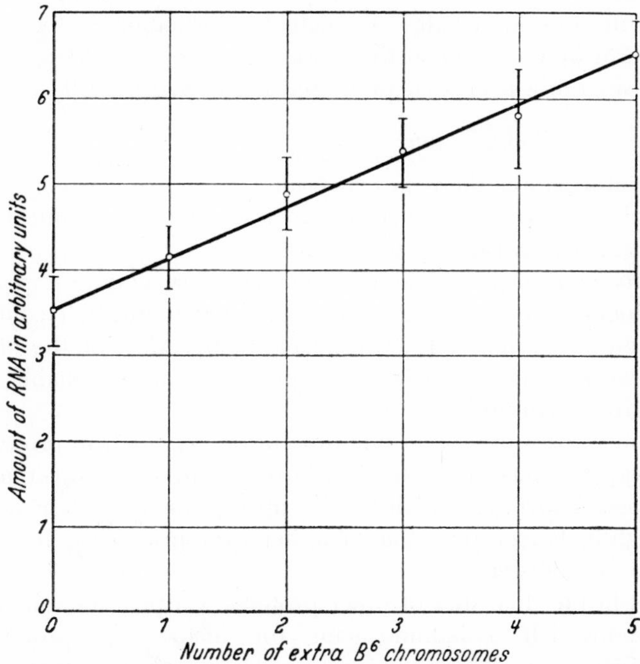

FIG. 19. Relationship between RNA content in the nucleolus and the number of extra nucleolar organizers. Vertical lines represent standard errors. From Lin (1955).

The association between the nucleolus and chromatin strands of the organizer is very intimate. Mulnard (1956) has shown that the Feulgen-positive material of the chromatin is actually within the nucleolus, for it is visible in sections thin enough to cut right through the nucleolus; and in Lafontaine's electron micrographs one also sees the deep penetration of the nucleolus by chromatin threads.

The relationship between chromosomes and the nucleolus may be described quite tentatively as follows: the nucleolus is formed at telophase by materials derived from all the chromosomes; once formed, the rate of syntheses in the nucleolus is primarily under control of the organizer. If this picture of nucleolar formation is correct, it may be supposed that the nucleolus is "seeded" by ribonucleoproteins from many different chromosomal loci and this assembly of ribonucleic acids continues to be synthesized in the nucleolus, the rate of synthesis being determined by the nucleolar organizer. The nature of the ribonucleic acids present in the nucleolus, according to this hypothesis, would not be characteristic of the chromosomal locus at the organizer but of those chromosomal loci that were active at the time of nucleolar formation. [At this time our knowledge of nucleolar formation is too sketchy to permit a firm opinion and so the reader may be referred to a view differing considerably from that presented here (Swift, 1959).]

2. *Chemical Composition*

What information there is concerning the chemical composition of the nucleolus has been obtained in two ways: by cytochemical tests and by investigations on isolated nucleoli.

a. Cytochemical tests. The main contribution, and an important one, of the cytochemical tests has been to show that nucleoli contain RNA (Caspersson and Schultz, 1939; Brachet, 1940). Most of the other cytochemical work done on nucleoli has proven to be unreliable, and two examples are worth noting.

Nucleoli were supposed at one time to be very rich in proteins of the histone type (Caspersson, 1950). The basis for this claim was that histones were supposed to have an ultraviolet absorption maximum at a wavelength slightly longer than that of other proteins. When it was shown (Mirsky and Pollister, 1942) that the ultraviolet absorption curve of histones is in fact like that of other proteins, Caspersson substituted the term "proteins rich in diamino acids" for "histone-type proteins" (Caspersson, 1950, p. 63). There is in fact no experimental evidence that "proteins rich in diamino acids" have the absorption spectrum that Caspersson attributes to them. Furthermore no histones have been found by Vincent in isolated nucleoli (1952). It has been noted that certain color reactions for basic groups of proteins, examples being the Sakaguchi reaction (Serra and Lopes, 1944) and the reaction of proteins with an acidic dye (Horn and Ward, 1957), are especially marked in the nucleolus, and these have been taken as indicating the presence of basic proteins in the nucleolus. The intensity of such reactions should, however, be related to the concentration of protein present in the test object. If this is not done, no conclusion can be drawn from the tests, for the concentration of protein in the nucleolus is exceedingly high.

It has been claimed by numerous investigators that cytochemical tests demonstrate a high concentration of the enzyme alkaline phosphatase in the nucleolus. Vincent (1957a) found none in his isolated nucleoli, but it could be said that the enzyme had been extracted during isolation. The following experiment (Osawa, 1951; many others could be cited), shows that the cytochemical "demonstration" is actually an artifact due to diffusion. The nucleolus together with the follicle layer of the growing amphibian oöcyte can be visualized as most active sites of alkaline phosphatase activity by the Gomori-Takamatsu procedure. However, when the follicle layer is removed before incubation, the nucleolus manifests no activity. Incubation of the follicle layer by itself on the other hand shows intense activity. Furthermore, when a heat-inactivated section of an oöcyte is incubated face to face with the active section, a strongly positive reaction is found in the nucleoli of the inactivated section.

b. Isolated nucleoli. So far satisfactory mass isolations have been made from the mature oöcytes of the starfish (Vincent, 1952; Baltus, 1954; Vincent, 1957a) and from ungerminated pea embryos (Johnston *et al.*, 1958). Nucleoli have also been isolated by microdissection from spider oöcytes (Edström, 1960).

In the mature oöcyte there is usually one big, well-defined nucleolus in the germinal vesicle. Gray had observed many years ago (Gray, 1931) that when oöcytes are centrifuged their nucleoli are so dense that they settle at the bottom of the nucleus. The method of isolation consists essentially in rupture of the oöcytes followed by differential centrifugation in aqueous medium, or preferably in sucrose. The isolated nucleoli show but little contamination when examined microscopically. Interesting as these preparations of isolated nucleoli are, it must be recognized that they have probably lost materials by extraction or gained them by adsorption in the course of isolation.

Little is known about nucleolar proteins except that they are present in exceedingly high concentration. The dry matter of isolated starfish nucleoli 4 μ in diameter was determined by microinterferometry (Vincent and Huxley, 1954) and average 85% (calculated as protein). With growth of the nucleolus to 20 μ in diameter the dry matter decreased to 40%, but this was due to the formation of vacuoles which contain little dry matter and occupy 50% of the total volume. The dry matter of neuron nucleoli was reported to be 70% (Nurnberger *et al.*, 1952) by use of an X-ray absorption technique, but this value is higher than that found for fresh, unfixed cells (see Hydén, Chapter 5 in Volume IV). Nucleoli isolated from starfish oöcytes contain nucleoside phosphorylase and the DPN-synthesizing enzyme. Both of these enzymes are present in high concentration in nuclei, as has already been mentioned in this chapter. It would be of great interest to know how much of the intranuclear content of these enzymes is nucleolar.

Coming now to the RNA of the nucleolus, it will be considered in relation to the RNA of the whole nucleus. It has been known since Brachet's early work (1942) that RNA occurs both in the nucleolus and chromosomes. The total ribonucleic acid of the nucleus can be measured directly by analyzing preparations of isolated nuclei. The nuclei considered were prepared in nonaqueous media under conditions that preclude a loss of ribonucleic acid (Volume I, Chapter 7). It is found that the amount of RNA per cell nucleus varies considerably in different organs of the same species. This latter finding is in sharp contrast to the constancy previously described for nuclear DNA content. It has also been found that the RNA per nucleus may vary with changes in the physiological state of a tissue. Prolonged fasting, for example, results in a decrease in the ribonucleic acid content of liver nuclei. This variability is again in contrast with DNA stability under the same conditions. Of the major nuclear constituents DNA and histone are constant, RNA and nonhistone protein variable. This variability in RNA content of the nuclei of different tissues and of the same nuclei in different states suggests that the RNA of the nucleus is closely associated with metabolic activities in the rest of the cell.

Several distinct RNA fractions have been prepared from isolated nuclei (Logan and Davidson, 1957; Osawa *et al.*, 1957; Allfrey and Mirsky, 1957b). In the fractionation carried out by Allfrey and Mirsky the nuclei are first treated with an isotonic neutral buffer, which extracts the ribonucleoprotein particles later described by Frenster *et al.* (1960). These particles seem to be associated with the chromosomes. The RNA of these particles may be designated RNA I. The nuclei are subsequently extracted with 1 *M* NaCl, which removes all the DNA, much protein, and a small amount of RNA. This is RNA II and it too is probably associated with the chromosomes. The material that remains consists of nothing but particles which stain like nucleoli and which are of the same size as the nucleoli as seen in nuclei. These nucleolar-like particles are rich in RNA. This is RNA III. Bearing in mind that RNA contents of nuclei are highly variable, the following are some analytical values obtained: In calf thymus nuclei RNA I is 0.3% of the mass of the dry, lipid-free nucleus, RNA II is 0.2%, and RNA III is 0.8%; in calf liver nuclei, isolated in 2.3 *M* sucrose, RNA I is 0.5% of the nuclear mass, RNA II 0.5%, and RNA III 1.3%; in calf kidney nuclei, isolated in 2.3 *M* sucrose, RNA I is 0.7% of the nuclear mass, RNA II is 0.4%, and RNA III is 0.3%. It will be shown later that incorporation experiments on thymus nuclei with labeled RNA precursors show that RNA I and RNA III differ markedly from each other in specific activity. Part of the RNA I fraction of thymus nuclei is a carrier RNA. It combines with "activated" amino acids and in this way resembles the s-RNA of cytoplasm. In thymus nuclei, carrier RNA appears not to be in the nucleolar fraction.

Isolated starfish nucleoli contain RNA. Vincent, in an elegant investigation on the incorporation of P^{32}, has found that there are two types of RNA in these nucleoli (Vincent, 1957b; Vincent and Baltus, 1960). In one of these P^{32} is added to terminal groups in a manner analogous to that in which nucleotides are added to the end of a pre-existing chain in the s-RNA of cytoplasm. In the other RNA the pattern of P^{32} incorporation demonstrates the presence of newly synthesized RNA.

The presence of RNA in starfish nucleoli analagous to the s-RNA of cytoplasm is in keeping with the observation (Ficq, 1955a, b) that in both the nucleoli of starfish and frog oöcytes there is a rapid synthesis of protein. In oöcyte nucleoli autoradiographs show rapid incorporation into both RNA and protein. In this respect the nucleoli of interphase nuclei in mammalian tissues seem to be strikingly different. Autoradiographic studies show almost no uptake of labeled amino acids into the proteins of the nucleoli of neurons and liver cells (Carneiro and Leblond, 1959; Schultze *et al.*, 1959). In mammalian tissue nucleoli there is rapid incorporation into RNA and only a slight incorporation into protein. Perhaps in line with this slight incorporation of amino acids is the finding that carrier RNA does not seem to be present in thymus nucleoli.

When RNA is extracted from masses of isolated nuclei it is possible to make a precise study of the biochemical properties of each fraction, but the cytological localization of each fraction within the nucleus is lacking in precision. When, on the other hand, autoradiographic techniques are used, each fraction can be located precisely, but the biochemical properties of each fraction may remain obscure. Thus in the salivary glands of the chironomid *Smittia* sp. Sirlin (1960) could observe the grains formed by RNA labeling over the sites of various regions of the chromosomes and in different parts of the nucleolus. The intensity of labeling was found to be greater in the nucleolus than in the chromosomes. Similar observations were made by Harris (1959) on connective tissue cells in tissue culture, and in these nuclei too the heaviest labeling was observed over the nucleoli. Heavier labeling over the nucleoli may indicate either that the nucleolar RNA is incorporating the precursor more rapidly or merely that there is a greater concentration of RNA at the nucleolus than elsewhere in the nucleus. What is needed is a measurement of turnover in relation to concentration of RNA. The relative turnover ratio can be determined by measuring the rate at which labeled precursor is *released* from RNA at different sites. It was then found that the turnover of nucleolar RNA does not differ greatly, if at all, from the rate of RNA turnover in the rest of the nucleus (Harris, 1959).

And yet significant steps in synthesis may elude turnover studies. There are observations which suggest that RNA synthesized in the chromosomes of an interphase nucleus may then move over to the nucleolus (Goldstein and Micou, 1959a).

3. *Nucleolar Function*

It can be said with assurance that the nucleolus is concerned with interactions between nucleus and cytoplasm. This was recognized a very long time ago, as can be seen from the following statement made in a review written at the end of the nineteenth century (Montgomery, 1898): "Where there is a close physiological *rapport,* in regard to processes of nutrition, between the nucleus and the cell body a relatively large amount of nucleolar substance occurs in the former."

There are many examples of such rapport between nucleus and cytoplasm in which there is marked enlargement of the nucleolus. Some of the most interesting examples are given in a paper by Hermann Fischer (1934) on the cells of leaves. If a leaf is injured, the cells in the neighborhood of the wound greatly increase in size and take on the appearance of embryonic cells. The cells as well as the nuclei and nucleoli within them increase in size, but the largest increase is in the nucleoli. Within 42 hours after the injury the average diameter of the nucleus in a group of cells increases from 14.4 μ to 18.3 μ but at the same time the nucleolar diameter increases from 3.6 μ to 7.1, so that the nucleolar volume goes from 1.6% to 6.0% of the nuclear volume. Essentially similar experiments had previously been done by Heitz (1925).

In another series of experiments Fischer observed a diurnal variation in nucleolar size. Measurements of nuclear and nucleolar size were made at 5 P.M. after a day of illumination and at 6 A.M. after a similar period in the dark. In one experiment at 5 P.M. the nuclear volume was 2290 μ^3 and the nucleolar volume 65 μ^3, whereas the next morning, although nuclear volume remained practically constant at 2270 μ^3, the volume of the nucleolus had shrunk to one-half, 32 μ^3. Here again similar experiments on the effect of light on nucleolar size had been made by plant physiologists for a period of some twenty years before Fischer's paper appeared (references in his paper).

Changes in the nucleoli of leaf cells caused by light are of special interest because there can be little doubt that we are here dealing with an influence of the cytoplasm, with its chloroplasts, on the nucleolus. Recently similar observations have been made on *Acetabularia* (Stich, 1955). Other experiments on *Acetabularia* by Stich (1956) in which nuclear and nucleolar changes were observed were also cited by him as examples of cytoplasmic influence on the nucleus. He observed that dinitrophenol tends to diminish nuclear and nucleolar size in *Acetabularia.* This was explained as a suppression of energy-rich polyphosphates in the cytoplasm, which, now being changed, proceeds to influence the state of the nucleus and nucleolus. The same observations concerning the effect of dinitrophenol on *Acetabularia* have been made by Brachet *et al.* (1955), and the same explanation was advanced. There is in fact no need to suppose that these effects of dinitrophenol are examples of cytoplasmic influence on the nucleus, for it is now known that polyphosphates are also present in the nucleus and that phosphorylation within the nucleus is inhibited *directly* by dinitrophenol. [This work by Osawa *et al.* (1957) will be described later in the chapter.]

In animal, as well as in plant cells there are many examples of rapport between nucleus and cytoplasm which influences the state of the nucleolus. Embryonic cells of animals, like those of plants, have prominent nucleoli. Embryonic muscle cells, for example, have large nucleoli, whereas those of mature muscle are hardly visible. A striking instance of correlation between nucleolar state and development of the cytoplasm is found in the growing oöcyte of Amphibia (already referred to on page 721), in which an active accumulation of egg yolk occurs in the cytoplasm. There are about a thousand nucleoli in the germinal vesicle, most of which are located at the inner surface of the nuclear membrane of a growing oöcyte. When the oöcyte reaches maturity the nucleoli become shrunken and are now found only in the central portion of the nucleus, and during the early cleavages after fertilization, when little protein is being synthesized, nucleoli are not visible.

In mature cells rapport between nucleus and cytoplasm is seen in a rudimentary way during starvation and the recovery therefrom (and it may be noted that maintaining a green plant cell in the dark is a form of starvation). A great many studies have been made on changes in the liver during starvation and refeeding, for the size of the liver readily responds to these conditions (see Lagerstedt, 1949—where an excellent account of the earlier literature is given). Within 24 hours of the onset of starvation the surface of liver nucleoli is reduced by 50%, and no further change takes place for the next 4 to 5 days, during which time a progressive decline occurs in the amount of cytoplasmic nucleoprotein. On refeeding, within 3 hours the nucleoli regain their normal appearance and this is the first change to be observed; cytoplasmic nucleoprotein reappears much more slowly.

Some of the most instructive examples of a correlation between the nucleolus and cytoplasmic activity are found in gland cells, for in many of them the function of the cytoplasm is to synthesize protein (Gland Cells, Volume V, Chapter 1).

An important step toward an understanding of the role of the nucleolus in the rapport between nucleus and cytoplasm was made when Caspersson found that there is RNA in the nucleolus. At the same time he recognized that there is a connection between RNA and protein synthesis; and furthermore he realized that there is a correlation between a well-developed nucleolus and the intensity of cytoplasmic protein synthesis. Caspersson's theory (given in some detail in Chapter 11 of this volume) holds that the nucleolus plays a central role in cytoplasmic protein synthesis. This role of the nucleolus, he believed, is made possible by its rich store of "diamino-acid-rich protein." We have seen, however, that there is actually no evidence for the presence of such material in the nucleolus; and so this part of the theory may well be discarded. There are indeed grounds for supposing that in some cells, neurons for example, there is little or no correlation between protein metabolism in the nucleolus and cytoplasm, for there is an intense synthesis of protein in the cytoplasm of a neuron and very little in the nucleolus of a neuron.

And yet it is important to retain the background of cytological information which connects the nucleolus with cytoplasmic protein synthesis. A firm indication that there is a rapport between the nucleolus and cytoplasmic protein synthesis is that in supraoptic neurons there is a linear relation between nucleolar volume and the amount of RNA (microchemically determined) in the cell body (Edström and Eichner, 1958b) (Fig. 20). A further impressive indication of rapport is that in spider oöcytes whereas the base composition of nucleolar RNA is distinctly different from that of the remaining RNA of the nucleus, no difference can be found in base composition of nucleolar RNA and cytoplasmic RNA (Edström, 1960). In these cells the cytoplasm contains about 99% of the total RNA, the nucleoplasm about 1%, and the nucleolus not more than 0.3%, although the highest concentration is found in the latter structures. The materials for these determinations were isolated by micromanipulation.

A useful new technique for investigation of the nucleolus is to irradiate it with a microbeam (2.2 μ diameter) of ultraviolet light (Uretz and

Perry, 1957; Perry and Errera, 1960; Perry, 1960). The procedure is to irradiate the nucleolus to such an extent that subsequent incorporation of cytidine into it is inhibited by more than 90% and then to observe the incorporation of cytidine into the RNA of other parts of the cell. It is found that damage to the nucleolus inhibits incorporation into nuclear RNA, apart from the nucleolus, by 30% and that into cytoplasmic RNA by 60–70%.

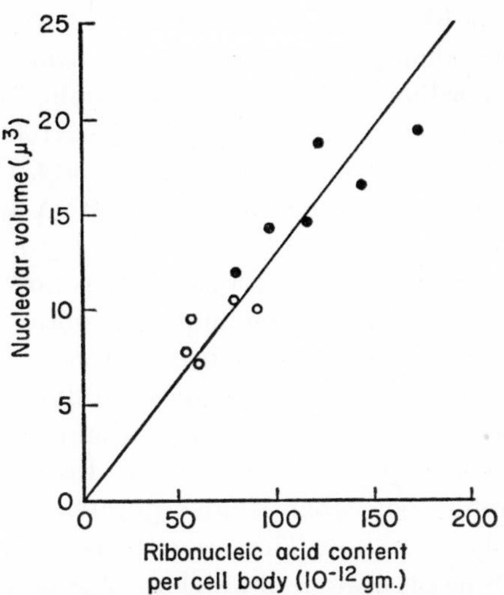

FIG. 20. Values for nucleolar volumes plotted against values for ribonucleic acid content in cell bodies of supraoptic neurons. Mean values per animal are given. Open circles designate controls, filled circles experimental animals. From Edström and Eichner (1958b).

The dependence of cytoplasmic RNA upon nuclear, and especially nucleolar, RNA has become increasingly evident ever since it was found with tracer experiments that the rate of incorporation into the RNA of the nucleus may be some ten times greater than into RNA of the cytoplasm (Marshak and Calvet, 1949; Jeener and Szafarz, 1950; Potter *et al.*, 1951; Smellie *et al.*, 1953). More recently evidence along the same lines has focused attention on the nucleolus, for in a wide variety of material it has been found that RNA metabolism proceeds rapidly in the nucleolus (references in Sirlin, 1960). The high activity of nuclear RNA has prompted the suggestion that the nucleus, and in particular the nucleolus, is one of the main sites of RNA synthesis for the cell as a whole. This theory is discussed in detail in Chapter 11, and it will be considered only briefly here.

At the present time the theory of the nuclear origin of much of the cytoplasmic RNA appeals strongly to biologists. A clear demonstration has, however, not been achieved. Enucleation experiments on microorganisms have demonstrated that, under certain conditions, synthesis of cytoplasmic RNA is dependent upon the nucleus, and it has been claimed that cytoplasmic RNA is actually synthesized in the nucleus. Some of the experiments described in the preceding pages have perhaps come closer to showing that some cytoplasmic RNA is actually synthesized in the nucleus, but nothing like proof of this has been accomplished.

It is occasionally claimed that *all* the cytoplasmic RNA is synthesized in the nucleus. According to Zalokar, for example, "at least 99% of the cellular RNA" is made in the nucleus (Zalokar, 1960). Enough is now known to believe that this claim is too far reaching. During the past few years it has been found that a certain kind of RNA synthesis takes place in the cytoplasm of liver cells and there are good grounds for considering this to be a general property of cytoplasm. It is now known that in the cytoplasm there is a system for attaching nucleotides to preformed RNA molecules, s-RNA, in a specific sequence (references in Canellakis and Herbert, 1960). In some s-RNA molecules cytidylic and adenylic acids are added to the end of the polynucleotide chain; in other s-RNA molecules guanylic and uridylic acids are added. In each instance it is the ribonucleoside triphosphate that is incorporated into s-RNA and this reaction is probably reversible. This knowledge has implications for experiments in which incorporation of labeled RNA precursors is studied in the cytoplasm of a cell after enucleation: If no incorporation is detected after enucleation it is possible that the s-RNA terminal syntheses, though occurring in the cytoplasm of an intact cell, are nuclear dependent, and this would make one suspect that some other types of RNA synthesis that fail to occur after enucleation may also be merely nuclear dependent. The most favorable experimental material would, accordingly, be a cell in which it is possible to demonstrate s-RNA terminal syntheses in the cytoplasm before enucleation, for only in such a cell could the result of enucleation for other types of RNA synthesis be properly evaluated.

Much of the investigation on the significance of the nucleus for cytoplasmic RNA synthesis has been done with autoradiographic techniques, which have the great advantage of precisely locating RNA. The autoradiographic procedure suffers, however, from certain grave disadvantages. One is that the RNA studied is not characterized with respect to composition, as it is, for example in the superb experiments of Edström (1960) in which minute quantities of RNA are isolated from clearly defined sites by micromanipulation and then characterized microchemically. In the absence of such knowledge it may too readily be assumed when a labeled component of RNA moves from nucleus to cytoplasm that the whole polynucleotides of which it forms a part has also moved.

Another defect in most autoradiographic procedures is that although the labeled material is considered to move from one part of the cell to another, a quantitative kinetic experiment is only rarely set up. An excellent example of quantitative experiments by autoradiographic methods on the relative turnover rates of the RNA's in different parts of cells is given by Harris in a recent paper (1959). Harris failed to find evidence for the movement of RNA from nucleus to cytoplasm, but this may have been due in part to his choice of material and experimental conditions.

D. Calcium and Magnesium

For many years there has been a continuing interest in the calcium and magnesium of the nucleus, and this despite a failure until recently to obtain definite information concerning these two elements in the nucleus (references in Steffensen and Bergeron, 1959). The persistent interest has been sustained by the knowledge that calcium and magnesium seem to have some special significance for nucleoproteins: thus certain nucleoproteins, nucleohistone for example, are precipitated by addition of traces of calcium or magnesium; in the isolation of nuclei it is usually necessary to have traces of calcium or magnesium ions in the medium; and these cations are required for the activation of many enzymes concerned with nucleotides or polynucleotides. The activity of pancreatic DNAase, for example, depends upon the presence of magnesium and it has been shown that magnesium combines with the substrate DNA (Kunitz, 1950).

A solid fact, quite different from the previous conjectures, concerning the calcium of the nucleus has emerged from an autoradiographic study using Ca^{45} (Steffensen and Bergeron, 1959). Autoradiographs of pollen tube nuclei with Ca^{45} have shown that all of the nuclei, namely the tube nucleus and the two sperm nuclei, retain Ca^{45} during growth and development of the pollen tube. In this process there was a mitotic division of the generative nucleus, and it was therefore inferred that the calcium is bound in a continuing nuclear component, the chromosomes. Steffensen and Bergeron were of the opinion that the site of attachment of calcium is DNA. Experiments, presently to be described, made it seem more likely that the firmly held calcium is combined with a protein of the chromosomes.

Experiments on isolated thymus nuclei have provided over-all analytical values for calcium and magnesium and also some information concerning the sites of attachment of these elements (Naora *et al.*, 1960).

Some of the nuclei used in these experiments were isolated in nonaqueous media; others were isolated in sucrose solutions containing either calcium or magnesium ions. Although presence of calcium in the isolation medium causes addition of calcium to the nuclei, such preparations can be used for analysis of magnesium. This is feasible because it was found, using nonaqueous nuclei as a standard of reference, that sucrose nuclei made in the presence of calcium contain very nearly the normal amount of magnesium, and so far the investigation has been concerned primarily, but not entirely, with magnesium.

Calf thymus nuclei contain 0.115 mg. magnesium and 0.024 mg. calcium per cent of lipid-free dry matter. Calf thymus tissue, and likewise liver, pancreas, and kidney tissues, also contain more magnesium than calcium. In thymus nuclei magnesium and calcium are held at different sites: most of the magnesium is bound to DNA; calcium is not bound to DNA and the firmly held calcium is probably combined with protein. The difference in binding sites for calcium and magnesium is in line with what is known in a general way about the physiological properties of these two elements, for they have entirely different physiological properties. When calcium or magnesium are present in the sucrose solution used for isolation of nuclei, the extra amounts of these elements found attached to the nuclei are held at different sites from those which they occupy naturally. Thus much of the "artifact" calcium seems to be associated with DNA, while none of the "artifact" magnesium is bound to DNA.

Most of the naturally occurring nuclear magnesium is attached to DNA. There is evidence that it is also combined with the mononucleotides of the nucleus, individual magnesium atoms being combined with both DNA and mononucleotide, so that magnesium atoms link the nucleotides to DNA. The point of attachment of magnesium with DNA is probably at the phosphate groups. This however is a moot point because in experiments with isolated DNA there are those (Zubay and Doty, 1958; Zubay, 1959) who claim that magnesium combines with the purines of DNA and others (Shack *et al.*, 1953; Shack and Bynum, 1959; Felsenfeld and Huang, 1959) who maintain that combination is with the phosphate groups. The weight of evidence seems to favor the latter. Considering the relative amounts of DNA and magnesium in thymus nuclei, this would mean that there is one magnesium for about every twenty phosphate groups. The remaining phosphate groups are combined mainly with histones, but also with other proteins. When an excess of magnesium is added to sucrose nuclei, very little of it combines with DNA (about one-tenth of that already present), the phosphate groups of DNA being blocked by histones (Naora *et al.*, 1960).

Later in this chapter it will be seen that the metabolic activity of the interphase nucleus is dependent upon the phosphoric acid groups of DNA, also on the presence of mononucleotides, and that this activity is inhibited by histone. According to this view, the active sites along the DNA chain are those few occupied by magnesium.

E. *Enzymes*

It goes without saying that one of the main currents in the history of biochemistry is the demonstration that cell function can be explained and interpreted in terms of enzyme composition. It is for this reason that particular interest attaches to the enzymatic composition of the nucleus, for nuclear function must eventually be understood in terms of its enzyme content.

An approach to this problem can be made by investigating the enzymatic activities of isolated cell nuclei. The methods of preparing nuclei and their limitations are discussed in Volume I, Chapter 7, which should be consulted by a reader of this section. The results about to be described were, for the most part, obtained in studies of cell nuclei isolated in nonaqueous media. This is the most reliable procedure now known for the investigation of the water-soluble components of the cell nucleus.

The enzymes studied have been grouped into three classes to facilitate the presentation and interpretation of the data (Stern *et al.*, 1952). The first group considered contains those components that characteristically reflect the differentiation of their tissue of origin, e.g., arginase of mammalian liver and fowl kidney; lipase, amylase, and deoxyribonuclease of the pancreas; alkaline phosphatase of the intestinal mucosa; hemoglobin of the erythrocyte; and myoglobin of the cardiac muscle cell. The intracellular distributions of the "special" components are summarized in Table IV. It should be emphasized that the data given in the tables actually represent the measurable enzyme "activities" expressed as units per milligram of tissue and that the enzymes were not isolated and weighed. In all cases nuclear enzyme "concentrations" (in activity units per milligram) were compared with the concentrations measured in tissue "controls," i.e., tissues treated with the same solvents for the same duration as the isolated nuclei. The column 100 N/C (or 100 N/T) gives the nuclear enzyme concentration relative to that of the cytoplasm (or whole tissue), and is a useful indicator of the intracellular enzyme distribution.

Several conclusions can be drawn from the measured distributions of these "special" components. First, the nucleus need not reflect the characteristic enzymatic composition of its tissue of origin. This is illustrated in several ways: Hemoglobin was found to be present and was isolated in crystalline form from fowl and goose erythrocyte nuclei, yet the analogous heme protein, myoglobin, does not occur in heart muscle nuclei. Similarly, arginase activity, which is high in mammalian liver and fowl kidney tissues, is found in liver but not in kidney nuclei. The specialized enzymes of the pancreas, amylase, lipase, and deoxyribonuclease, do not occur in the nucleus in significant concentrations. Thus, there is no simple conformity in enzyme pattern between cytoplasm and

TABLE IV
INTRACELLULAR DISTRIBUTION OF "SPECIAL" ENZYMES

Enzyme	Tissue	Units of activity per mg. in: Cytoplasm	Nuclei	100 N/C
Arginase	Calf liver	2.88	1.55	54
	Horse liver	1.21	0.6	57
	Fowl liver	0.04	0.025	63
	Fowl kidney	0.62	0.04	7
	Calf kidney	0.137	0.152	110
Catalase	Horse liver	41	29	71
	Calf liver	9.3	3.0	32
	Calf kidney	25	0	0
	Fowl kidney	33.5	0	0
Uricase	Horse liver	2.14	0.11	5
	Calf liver	0.41	0	0
	Calf kidney	2.9	0	0
Lipase	Horse pancreas	0.323	0.005	1.5
	Beef pancreas	0.655	0.055	8
Amylase	Horse pancreas	14.30	0.356	2.5
	Beef pancreas	36.05	2.51	7.1
Deoxyribonuclease I	Beef pancreas	330	2.5	0.76
Alkaline phosphatase	Calf intestinal mucosa	344	12	3.5
Adenosine deaminase	Calf intestinal mucosa	127	24	19

nucleus. Nevertheless, it should be pointed out that the specialized components which distinguish a differentiated cell may appear in its nucleus. This is best illustrated by the occurrence of some hemoglobin in the avian erythrocyte nucleus, but in no other. It was surprising to find hemoglobin in the erythrocyte nucleus, but experiments with "nonaqueous" nuclei clearly demonstrated its presence. This conclusion has been confirmed by a microspectrographic technique on unfixed erythroblasts demonstrating the presence of intranuclear heme granules through various stages of maturation of these cells (Carvalho and Wilkins, 1954).

Further information about the nucleus can be obtained from the distributions of the second group of enzymes considered. This group comprises a number of enzymes that are widely distributed throughout the different tissues of the body: esterases, nucleotide-specific phosphatases, β-glucuronidase, etc. Some data for the intracellular localization of these enzymes in different tissues are presented in Tables V and VI. The results can be considered in three parts. First, there is a group of enzymes which actually occur in relatively low concentrations in many nuclei, although it has frequently been supposed that some of them are especially concentrated in nuclei. Typical of this group are alkaline phosphatase, some of the nucleotide-specific phosphatases, β-glucuronidase, and "acid" deoxyribonuclease (Allfrey *et al.*, 1952). A second type of intracellular enzyme distribution is exemplified by esterase, which seems to occur in all nuclei in widely varying concentration depending upon the tissue examined. Finally, there are those enzymes (nucleoside phosphorylase, adenosine deaminase, and guanase) which are present in high proportions in many nuclei. The nuclear concentration of nucleoside phosphorylase in heart nuclei, for example, is more than four times its concentration in the cytoplasm, and comparable activity ratios exist for adenosine deaminase in both heart and liver. Furthermore assays of the coenzyme diphosphopyridine nucleotide (DPN) in the calf tissues of heart, pancreas, and liver all show concentrations in the nucleus somewhat above that in the cytoplasm (Stern *et al.*, 1952). There is evidence that all of the DPN-synthesizing enzyme is in the cell nucleus (Hogeboom and Schneider, 1952).

Nucleoside phosphorylase and the DPN-synthesizing enzyme, it should be noted, are two enzymes which have been found in high concentration in isolated echinoderm nucleoli (Baltus, 1954). It would be of great interest to know what fractions of these enzymes in the nuclei of heart and liver are in the nucleoli.

The response of nuclei to changes in physiological state was demonstrated by experiments on starvation (Table VII). In these experiments the enzymatic activities of normal liver nuclei were compared with the corresponding activities of liver nuclei prepared from an animal subjected to a prolonged fast. It was found that the catalase activity of both nucleus and cytoplasm decreased on starvation, whereas β-glucuronidase per cell and per nucleus remained essentially constant. Many of the other enzymes studied—arginase, esterase, alkaline phosphatase, nucleoside phosphorylase, and adenosine deaminase—increased in the cytoplasm and decreased in the nucleus.

The third group of enzymes considered was selected to shed some light on the problem of the energy-yielding reactions in nuclear metabolism (Stern and Mirsky, 1952). This group includes phosphoglyceral-

TABLE V

INTRACELLULAR DISTRIBUTION OF COMMONLY OCCURRING ENZYMES IN ADULT TISSUES

Enzyme	Calf							Horse liver
	Liver	Kidney	Kidney cortex	Thymus	Heart	Intestinal mucosa	Pancreas	
Esterase								
C[a]	0.161	0.061	0.078	0.075	0.011	0.336	0.398	0.160
N[a]	0.122	0.006	0.018	0.026	0.014	0.029	0.128	0.128
100 N/C	76	10	23	35	127	9	32	80
β-Glucuronidase								
C	1.88	0.50	—	0.655	Trace	0.410	Trace	0.433
N	0.324	0.06	—	0.033	Trace	0.102	Trace	0.073
100 N/C	17	12	—	5	—	25	—	17
Adenosine deaminase								
C	52	Trace	Trace	80	7	127	0.81	—
N	108	Trace	Trace	53	42	24	1.32	—
100 N/C	208	—	—	66	600	19	163	—
Nucleoside phosphorylase								
C	52	15	17	19	18.5	19	28	15
N	52	19	17	7	82	4.5	21	41
100 N/C	100	120	100	37	440	24	75	274

[a] C: activity in cytoplasm, N: activity in nuclei.

TABLE VI
INTRACELLULAR DISTRIBUTION OF PHOSPHATASES

Enzyme	Calf							Horse
	Intestinal mucosa	Thymus	Kidney	Liver	Heart	Pancreas	Spleen	Horse liver
Alkaline phenolphthalein phosphatase								
C[a]	344	37	20.4	4.32	—	—	—	17.4
N[a]	12	0.17	1.3	0.5	—	—	—	4.9
100 N/C	3.5	0.5	6	12	—	—	—	28
Adenylic-5-phosphatase								
C	50	7.33	—	17.2	<0.2	—	—	—
N	3	0.66	—	3.1	0	—	—	—
100 N/C	6	9	—	18	0	—	—	—
Adenylic-3-phosphatase								
C	41	4.75	—	4.4	<0.2	—	—	—
N	1.7	0.51	—	0.45	0	—	—	—
100 N/C	4	11	—	10	0	—	—	—
DNAase II								
C	36	4.6	1.6	3.5	—	2.1	6.5	7.3
N	1.3	0.18	0.18	2.0	—	0.25	0.8	3.9
100 N/C	3.6	4	11	57	—	12	12	54

[a] C: activity in cytoplasm, N: activity in nuclei.

dehyde dehydrogenase, aldolase, enolase, and pyruvate kinase. Some results of these investigations are presented in Table VIII. It is clear from the data in the tables that nuclei of wheat germ cells are well endowed with the energy-yielding systems of glycolysis. Similarly, calf thymus and liver nuclei have been shown to contain glucose 6-phosphate dehydrogenase, and the nuclei of pancreas, heart, and liver cells have high concentrations of diphosphopyridine nucleotide. These are indications

TABLE VII
EFFECT OF FASTING ON INTRACELLULAR ACTIVITY[a] OF HORSE LIVER ENZYMES

Enzyme	Cytoplasm		Nuclei	
	Normal	Fasted	Normal	Fasted
Nucleoside phosphorylase	100	150	274	203
Esterase	100	124	82	21
Arginase	100	132	60	21
β-Glucuronidase	100	122	17	16
Uricase	100	139	6	3
Catalase	100	12	73	0
Alkaline phosphatase	100	113	28	13

[a] Activities expressed in relative values taking that of normal liver as 100. Values for fasted liver are corrected for changes in protein content, so that comparisons are effectively per cell and per nucleus.

TABLE VIII
INTRACELLULAR DISTRIBUTION OF GLYCOLYTIC ENZYMES IN WHEAT GERM

Enzyme	Units of activity per mg. in		100 N/T
	Tissue	Nuclei	
Aldolase	1.97	3.04	155
Glyceraldehyde phosphate dehydrogenase	1.16	1.14	99.4
Enolase	27.2	43.2	160
Pyruvate kinase	3.47	4.93	142

that glycolytic activity may be a common characteristic of many cell nuclei. The nuclear enzymes concerned with glycolysis have recently been further studied by Siebert, but no more than an introduction to this work has been published (1960).

Glycolysis provides a possible way for adenosine triphosphate (ATP) formation in the nucleus, a process which will be discussed in detail later in this chapter. At present it should be mentioned that the presence of adenylate kinase in the nucleus suggests still other possibilities for ATP synthesis, for this enzyme mediates the conversion of two molecules of ADP into one each of AMP and ATP (Miller and Goldfeder, 1960).

Some enzymes of importance in energy-yielding reactions, notably succinic dehydrogenase and cytochrome c oxidase, are absent from nuclei (Hogeboom *et al.,* 1952). Cytochrome c and cytochrome c oxidase are limited to mitochondria. Elsewhere in the cytoplasm other cytochromes and enzymes associated with them have been found. It is possible that cytochromes, and enzymes associated with them, which are different from those found either in mitochondria or microsomes are present in nuclei.

A conclusion of general biological interest emerges from the enzyme studies on nonaqueous nuclei. The observed differences in nuclear activity in the enzyme systems studied frequently exceed the differences measured between the tissues themselves. This fact, together with the observation that specialized cellular components can occur in the nucleus, leads to the conclusion that differentiation is a nuclear as well as a cytoplasmic process. Based upon the observations of early cytologists that nuclei vary morphologically in the course of glandular activity (Heidenhain, 1875), the idea of a variable nucleus has been commonly accepted. A chemical expression of nuclear variability is seen in the data collected in Tables IV, V, and VI. Enzyme studies emphasize not only nuclear variability, but equally, nuclear differentiation. The chromosomes of various cells are thus placed in different enzymatic environments; surely such environmental differences must have important effects on the biochemical activities of the chromosomes. At what stage in development biochemical differentiation of nuclei appears is not known.

IV. Metabolism

A. *DNA*

Is DNA in dynamic equilibrium in an interphase nucleus? The question arose when it was discovered that the incorporation of P^{32} into the DNA of nondividing cells is negligible. This observation has at times been taken to mean that under these conditions DNA is "metabolically inert." In a discussion of the metabolic activity of DNA, or of other components of the cell, a distinction must be made between an activity which requires a replacement of active groups, and one which does not require a continuous breakdown and renewal of the molecule. In the case of ATP, "activity" involves a continuous phosphate exchange, and this lability of ATP phosphate is evident in tracer experiments involving P^{32}. On the other hand, there is the example of hemoglobin which functions continuously in oxygen transport without requiring any "turnover" of either porphyrin or protein components of the molecule. The "inertness" of the hemoglobin molecule does not imply an absence of function. The DNA of some interphase nuclei does indeed seem to be "inert" with respect to "turnover," but this does not imply that DNA is otherwise inactive in interphase nuclei.

One approach to the metabolic stability of DNA is concerned with the retention of an isotope after it has been incorporated. There are two sets of such experiments which show that the DNA of the interphase rat liver nucleus is remarkably stable (Fresco *et al.*, 1955; Hecht and Potter, 1956). In one experiment orotic acid-*6-C^{14}* was administered to partially hepatectomized rats and in the other adenine-*1,3-N^{15}* and glycine-*2-C^{14}* simultaneously. After the livers had regenerated, the disappearance of C^{14} and N^{15} from the liver DNA was followed over a three-month period. The apparent turnover of DNA during this period is very slow and can be accounted for by a small gain in liver weight. Furthermore the slight decreases in specific activity of the purines labeled with C^{14} and N^{15} were identical and the C^{14} activities of all the bases decreased at about the same rate. In contrast to DNA the nuclear RNA fractions contained very small amounts of isotope after three months.

This extraordinary stability of DNA has also been observed in dividing cells. Even in dividing cells there are many reports which seem to show that once an isotope has been taken up by DNA, it is not released. In a recent carefully thought-out experiment, Thomson *et al.* (1957) have shown that there is actually a slow breakdown of DNA in the course of cell division in a tissue culture. The point of departure in their experiments was to consider that in a rapidly growing culture, any breakdown products are likely to be reincorporated into newly synthesized materials. They accordingly studied retention of isotope in cells that were grown rather slowly; and there then was a very slow decline in isotope content of the DNA, whereas at the same time there was a far more rapid breakdown of RNA. Even in slowly growing culture it is possible that if there is any breakdown of DNA, the labeled breakdown products might be reutilized in the synthesis of new DNA. To meet this difficulty, after the cells were labeled they were grown in nonradioactive thymidine, which would compete with the hypothetical labeled breakdown product for incorporation into newly synthesized DNA. The result was a considerably increased loss of isotope from the DNA. The breakdown of DNA observed (which might be masked in the absence of nonradioactive thymidine) was about 10% per generation.

Although the metabolic stability of DNA during interphase in many types of cells is now readily shown in aurotradiographs by the *non*-incorporation of labeled thymidine, there are some types of cells in which uptake of thymidine into DNA is observed during interphase and not as part of over-all DNA synthesis. In a notable series of studies by Pelc it has been shown that cells in the seminal vesicle of the mouse and in cells beginning to elongate in roots of *Vicia faba* there is a lively uptake of tritiated thymidine into DNA, apparently without concomitant DNA synthesis. Perhaps this is an exchange reaction. The incorporation into the epithelial cells of the seminal vesicle of adult mice is at a rate equivalent to a renewal of approximately 10% of the DNA per day. The rate of incorporation appears to be correlated with the biological activity of the cell (Pelc, 1958a; Pelc and La Cour, 1959; Pelc, 1959a, b). A contrary point of view has recently been expressed by Gall and Johnson (1960).

B. Protein Synthesis

Synthesis of protein, RNA, DNA, ATP, and other triphosphates occur in the interphase nucleus. The syntheses have been studied in intact tissues *in vivo* and in isolated nuclei *in vitro*. Each type of investigation has its advantages. The final goal is to understand what happens in the intact cell, but in reaching this goal the only effective way at present to uncover the inner dynamics of synthetic processes is to investigate isolated cell components.

1. *In Vivo*

First nuclear protein synthesis *in vivo*. In these experiments N^{15}-labeled amino acids are injected into animals and then after various time intervals organs are removed; nuclei are isolated by the citric acid procedure (Volume I, Chapter 7) and the histones and nonhistone proteins are then separated from each other (Bergstrand *et al.*, 1948; Daly *et al.*, 1952; Smellie *et al.*, 1953). The first point to be noted is that in each nucleus examined, uptake of amino acids into nonhistone protein is far greater than into histone. The next point is that the rate of amino acid uptake into nuclear protein is in general correlated with uptake into the combined cytoplasmic protein, which varies from one tissue to another. In each tissue, uptake into the nonhistone nuclear protein is slightly less than into the total cytoplasmic protein.

The relation between activity in the cytoplasm and nucleus was further explored in experiments on animals in which a period of fasting was followed by feeding (Allfrey *et al.*, 1955). After a prolonged fast some mice were divided into two groups; one was fed and then after 30 minutes given labeled amino acid; the other group was given the labeled amino acid without prior feeding. Feeding immediately induced pancreatic activity, as shown by a much more rapid uptake of amino acid by the total protein of the pancreas. The difference in pancreatic activity between fed and fasted mice became apparent in 30 minutes and the difference in activity persisted for several hours. No immediate difference in amino acid uptake was found in the total protein of the livers and kidneys of fasted and fed mice.

All the tissues were fractionated, a number of cytoplasmic components being separated and from the isolated nuclei the histones and residual (nonhistone) proteins were prepared. In the pancreas all cytoplasmic fractions showed greater amino acid uptake after feeding; and in the pancreas both nuclear fractions examined, histone and residual protein, showed a marked increase of uptake within 30 minutes after feeding. In the liver and kidney there was no immediate effect of feeding on amino acid uptake into the proteins of either cytoplasm or nucleus.

In experiments that lasted not for several hours but for 3–4 days the retention of labeled amino acid by tissue proteins as affected by fasting and feeding was studied. Over the longer time interval of these experiments feeding influenced the liver and kidney as well as the pancreas. The effect of feeding was a marked decrease in retention of labeled amino acid by *both* cytoplasmic proteins and the histone and residual protein of chromosomes. Decreased retention of labeled amino acid, loss of the labeled amino acid, can be attributed to an increased rate of synthesis of unlabeled protein and to a more rapid loss of the originally labeled material.

The experiments that have been described show clearly that under certain conditions uptake and retention of N^{15} by cytoplasmic and chromosomal proteins are correlated. These experiments do not disclose just how the correlation is accomplished. In a general way, however, it may be said that feeding mice brings about changes first of all in the cytoplasm of pancreas, liver, and kidney cells and that the changed condition of the cytoplasm in some way produces a change in the nucleus. Modifications in chromosomal proteins may, therefore, be regarded as a response to an altered cytoplasm. Since it is well known that chromosomes influence activities in the cytoplasm, it may be supposed that modifications in chromosomal proteins, especially those combined with DNA, will affect the way in which the chromosomes influence the cytoplasm. There can be little doubt that we are here dealing in a fragmentary way with interactions between cytoplasm and chromosomes, and that the pattern of interaction here considered—the cytoplasm, changed by external conditions, producing modification in the chromosomes and these modifications reacting upon the cytoplasm—holds for physiological changes in nondividing cells and also for the differentiation that occurs in the course of development. An example of decisive cytoplasmic influence on the state of the nucleus is the observation by Karl

Sax (1935) on the pollen grains of *Tradescantia.* Here, in the first pollen-grain mitosis, the orientation of the mitotic spindle places the daughter nuclei in different regions of the cytoplasm, and it is this difference that causes one daughter nucleus to be vegetative and the other generative. Similar examples of nuclear differentiation have been recently observed in protozoa (Volume IV, Chapter 3).

2. *In Isolated Nuclei*

Experiments on isolated nuclei make possible an analysis of the factors within the nucleus affecting amino acid incorporation by nuclear proteins. The selection of thymus nuclei as the material of choice for the study of nuclear protein synthesis dates back to earlier experiments in which it was found that calf thymus nuclei can survive an isolation in isotonic sucrose solutions containing a small amount of $CaCl_2$. Nuclei so prepared retained their soluble proteins and nucleoproteins, including a variety of soluble enzymes (Stern and Mirsky, 1953), and even low-molecular weight compounds, such as nucleotides and free amino acids were retained (Osawa *et al.*, 1957). Nuclei isolated from other tissues, liver for example, are far less active in protein synthesis (Logan *et al.*, 1959). The difference between thymus and other nuclei is probably due to the fact that, taking nonaqueous nuclei as a standard for comparison, there is a loss of enzymes, nucleotides, and other substance from liver and other nuclei but not from thymus nuclei in the course of isolation in sucrose.

A number of tests indicate that the thymus nuclear fractions are free of appreciable cytoplasmic or whole-cell contamination. Under the light microscope the thymus nuclear preparations appear remarkably clean; there is a small amount of visible contamination, consisting of a few whole cells, occasional red cells, and small blocs of cytoplasm that adhere to some of the nuclei. Recent visual estimates of whole cell contamination in stained nuclear preparations by Ficq and Errera (1958) placed the number of intact cells at 3%. This figure is slightly lower than that based on whole cell counts made with the electron microscope, which ranged from 2.9 to 7.7% (Allfrey and Mirsky, 1955; Allfrey *et al.*, 1957). When necessary, most of the few remaining cells can be removed by centrifugation through a sucrose or Ficoll density gradient. This purification procedure was used as an added precaution in many of the enzyme preparative procedures described below. The amount of over-all cytoplasmic contamination has been estimated from nucleic acid analyses of the nuclei and the whole tissue, and by testing for the activity of enzymes known to be localized in the cytoplasm; these tests show less than 5% contamination, even prior to further purification in a sucrose gradient.

a. Nuclear function as a test for nuclear purity. Isolated thymocyte nuclei are capable of at least three classes of synthetic reactions: (1) the synthesis of adenosine triphosphate (ATP) and other energy-rich phosphate bonds (Osawa *et al.*, 1957); (2) the uptake of amino acids into nuclear proteins (Allfrey, 1954); and (3) the incorporation of many purine and pyrimidine precursors into ribo- and deoxyribonucleic acids (Allfrey *et al.*, 1957; Friedkin and Wood, 1956; Allfrey and Mirsky, 1957b, 1959; Breitman and Webster, 1958).

Many of these reactions have been found to be interdependent. Some of the relationships that cast some light on the mechanism of nuclear protein synthesis will be described below, but in connection with the problem of nuclear purity, it should be stressed that there is convincing evidence that all the reactions described are indeed localized in the nucleus. Thus it has been shown that nuclear ATP synthesis is DNA dependent (Allfrey and Mirsky, 1957a), as is amino acid incorporation into protein (Allfrey, 1954), and adenosine uptake into nuclear RNA (Allfrey *et al.*, 1957). This dependence is shown by the loss of synthetic activity which follows treatment of the nuclei with deoxyribonuclease, and by the restoration of function when DNA is added to the enzyme-treated nuclei. A second test for nuclear localization is based on the specific sodium ion requirement for nuclear protein synthesis. The reason for this requirement is discussed below. The sodium ion dependence of the nucleus stands in sharp contrast to the general observation that amino acid uptake in cytoplasmic systems is potassium dependent (e.g., Sachs, 1958).

b. Amino acid incorporation into nuclear proteins. If a suspension of isolated nuclei is incubated aerobically at 37°C., in a buffered sucrose medium in the presence of C^{14}-labeled amino acids, there is an appreciable incorporation of the isotope into the proteins of the nucleus. Figure 21 shows the time course of incorporation of three amino acids: lysine-*2-*C^{14}, alanine-*1-*C^{14}, and glycine-*1-*C^{14}. Similar curves have been obtained with radioactive leucine, valine, tryptophan, and phenylalanine. In the figure the specific activity of the total mixed proteins of the nucleus is plotted against the time of incubation. One observes a characteristic lag phase of 5–15 minutes, followed by a rapid approximately linear incorporation of amino acid into nuclear proteins which lasts over an hour. After that time the rate of uptake begins to fall off as some of the nuclei undergo autolysis.

The incorporation has been shown to involve peptide bond formation and it is essentially irreversible. If all the uptake of amino acid is assumed to represent protein synthesis, then it can be calculated that 30 mg. (dry weight) of nuclei (or 10^9 nuclei) can synthesize about 8 µg. of protein per hour. Although it should be stressed that not all the nuclei are incorporating amino acid at the same rate (differences in uptake are evident in autoradiographs), it is interesting to calculate that the average isolated nucleus is synthesizing 22 molecules of protein every second (assuming an average molecular weight of the proteins equal to 50,000).

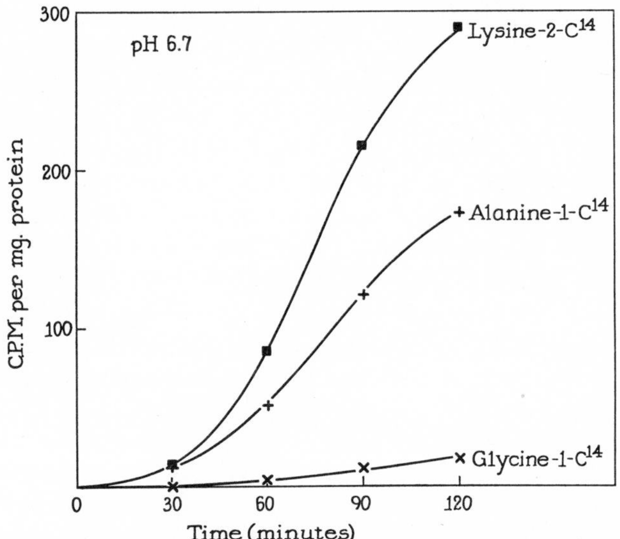

FIG. 21. Time course of C^{14}-amino acid incorporation. From Allfrey *et al.* (1957).

C. Nuclear Phosphorylation

The incorporation of amino acids into the proteins of the nucleus takes place only under aerobic conditions. The reason for this oxygen dependence became clear when other evidence accumulated relating nuclear protein synthesis to a second oxygen-dependent nuclear function, namely, the synthesis of ATP (Fig. 22).

Suspensions of isolated thymus nuclei are capable of ATP synthesis by an aerobic process, accompanied by the uptake of oxygen. Nuclear phosphorylation is inhibited by cyanide, azide dinitrophenol, and antimycin A. Nuclear ATP synthesis, however, differs in several important respects from the process of oxidative phosphorylation observed in suspensions of isolated mitochondria. For example, it is not affected by a number of inhibitors of mitochondrial phosphorylation, including Dicumarol, methylene blue, Ca^{++} ions, or an atmosphere of 95% carbon monoxide with only 5% oxygen (Osawa *et al.,* 1957). A calculation by

Mr. Bruce McEwen shows that ATP synthesis in isolated nuclei is about three times as rapid as it is in isolated mitochondria.

Experimentally, the rate of ATP synthesis in the nucleus can be followed by chromatographic isolation of the different nucleotides as illustrated in Fig. 22 by indirect measurement of the amount of inorganic

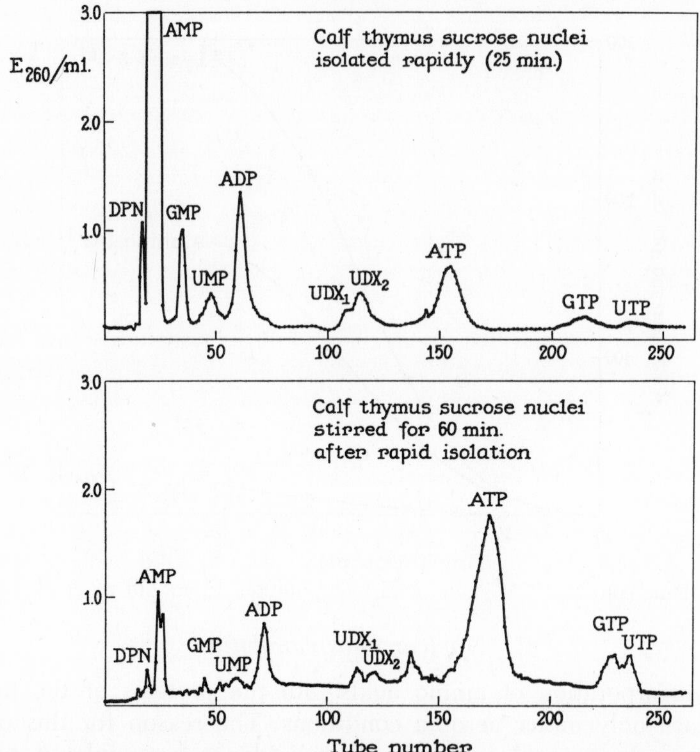

FIG. 22. Phosphorylation in isolated calf thymus sucrose nuclei. Acid-soluble nucleotides of sucrose nuclei isolated rapidly (in 25 minutes) at 2°C. (above), and of the same nuclei stirred gently in the cold for an additional 60 minutes (below). Note the increase in amount of triphosphates after 60 minutes. From Osawa *et al.* (1957).

phosphate esterified, or by studying the incorporation of P^{32}-orthophosphate into the terminal phosphate groups of the nucleotides.

The ATP Requirement for Nuclear Protein Synthesis. A comparison of the inhibitory effects of different compounds on nuclear protein and ATP syntheses suggested that these processes were related: all agents which blocked the generation of ATP also blocked amino acid incorporation into nuclear proteins.

Additional evidence that amino acid uptake is ATP dependent came from other experiments in which the nucleotides were selectively extracted from nuclei by treatment with acetate buffers at pH values below 5 (Osawa *et al.*, 1957). A clear parallelism was found between the loss in nucleotides and the decrease in capacity for amino acid incorporation. (Using citrate, succinate or other buffer systems which do not extract nuclear ATP as "controls," it could be shown that nuclei could recover from the low pH sustained in the acetate method of nucleotide removal.)

The direct demonstration of the role of ATP in nuclear protein synthesis is given below in connection with amino acid activation reactions.

An important characteristic of ATP synthesis in the nucleus is its DNA dependence (Allfrey and Mirsky, 1958). The removal of DNA (using crystalline pancreatic deoxyribonuclease) blocks subsequent ATP synthesis; but restoring the DNA or substituting other polyanions for it allows nuclear phosphorylation to resume. These findings stress the importance of the negative electrical charge on DNA in the regulation of nuclear metabolism (Allfrey and Mirsky, 1958). After removal of DNA from the nucleus, addition of such large polyanions as polyethylene sulfonate and polyacrylic acid are just as effective in restoring nuclear phosphorylation as is DNA itself. Since negative charge is the only common factor in all these molecules, it is clear that the negative charge provided by the phosphate groups of DNA is essential for nuclear phosphorylation.

Replacement of DNA by a polyanion is effective only if the latter is large; small polyanions, even if they are polynucleotides, are ineffective. Although replacement must be by a large molecule, negative charge is essential. Thus polyacrylic acid is an effective substitute for DNA, whereas polyacrylamide is not. The significance of negative charge is also shown by the effects of polycations (Mirsky and Allfrey, 1960). Addition of polycations, such as polylysine, histone, and protamine, do not restore function to nuclei depleted of their DNA. On the contrary, addition of polylysine and histone greatly inhibit the activity of nuclei which still retain their DNA (Fig. 23).

The nuclear activities dependent on DNA and its negative electrical charge, it should be stressed, are incorporation of amino acids into proteins and of precursors such as adenosine into polynucleotides, as well as the synthesis of ATP. The synthesis of ATP and other triphosphates is a necessary condition for the other activities.

The role of DNA as a cofactor in an oxidative phosphorylation within the nucleus is paralleled by a ribopolynucleotide which acts as a cofactor in the bacterium *Alcaligenes faecalis* linking the formation of ATP to the oxidation of DPNH (Pinchot, 1959). One of the components in this system is a cytochrome-containing particle. In this system, as in the DNA system, synthetic polynucleotides containing a single purine can replace the natural ribopolynucleotide, so that one suspects that other polyanions would do as well as polynucleotides in this system, as in the nucleus. Another example of a polynucleotide closely linked to an oxidative sys-

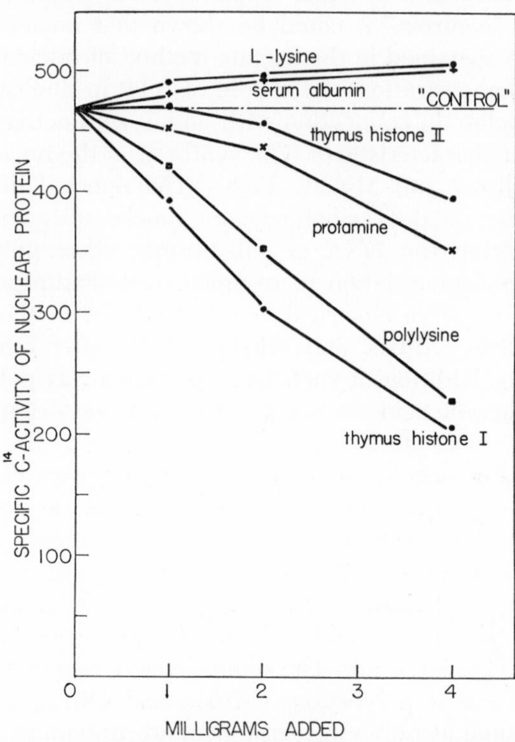

FIG. 23. The effects of added polycations on amino acid incorporation by isolated thymus nuclei. The specific C^{14}-activity of the nuclear protein after one hour's incubation is plotted against the concentration of added protein. The arginine-rich histone fraction I is a more basic protein than the lysine-containing histone fraction II, and is a more effective inhibitor of nuclear syntheses. From Mirsky and Allfrey (1960).

tem is one in which the polynucleotide is DNA. Appleby and Morton (1960) have recently crystallized the lactic dehydrogenase of bakers' yeast. The crystalline enzyme contains protein with both heme (cytochrome b_2) and flavine prosthetic groups and a deoxyribose polynucleotide which is intimately associated with the protein. Here we have wrapped up together in one complex molecule the components which may well account for the role of DNA in nuclear phosphorylation.

D. *Sodium Ions and Amino Acid Transport into Nuclei*

Amino acid incorporation into nuclear proteins requires the presence of sodium ions (Allfrey *et al.*, 1957). The magnitude of the sodium effect is very strikingly demonstrated in Fig. 24 which compares the uptake of 1-C^{14}-alanine in a sodium containing medium with that observed in the presence of equivalent amounts of potassium. The incorporation of

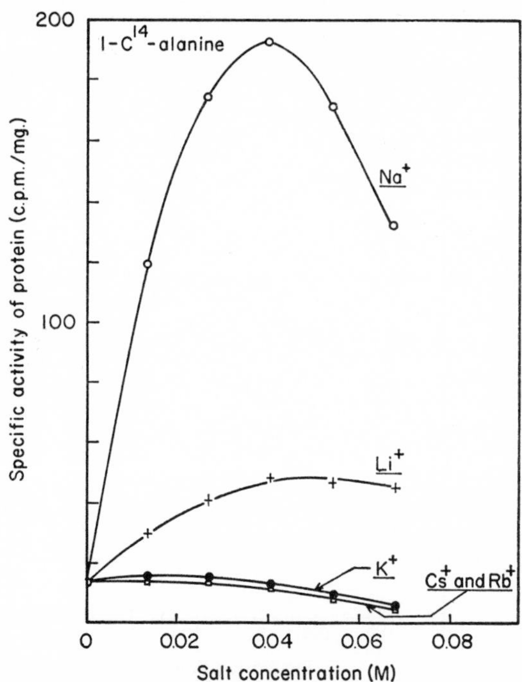

FIG. 24. The effect of adding different monovalent cations (as chlorides) on the incorporation of 1-C^{14}-alanine into the proteins of isolated calf thymus nuclei. The specific activity of the nuclear protein after 60 minutes' incubation is plotted against the salt concentration of the medium. From Allfrey *et al.* (1960).

alanine is exceptionally responsive to the addition of sodium ions, but the sodium requirement has also been observed for other amino acids as well: for most of them the level of uptake into protein is doubled by adding Na^+ (as the chloride) to the incubation medium.

The reason why amino acid incorporation into nuclear proteins increases when sodium ions are added to the incubation medium has been made clear only recently; it has been found that the transport of amino acids into the nucleus is strongly sodium dependent. This dependence can be demonstrated by experiments in which the ionic environment of the nucleus is varied and the penetration of free amino acids into the nuclear "pool" is measured. This is done conveniently by exposing nuclei to C^{14}-labeled amino acids, centrifuging them down after different periods of incubation, and extracting the acid-soluble "pool" in cold 2% $HClO_4$. The radioactivity in the extract is plotted against the time of incubation in the presence of 1-C^{14}-alanine in Fig. 25. The transport of the amino

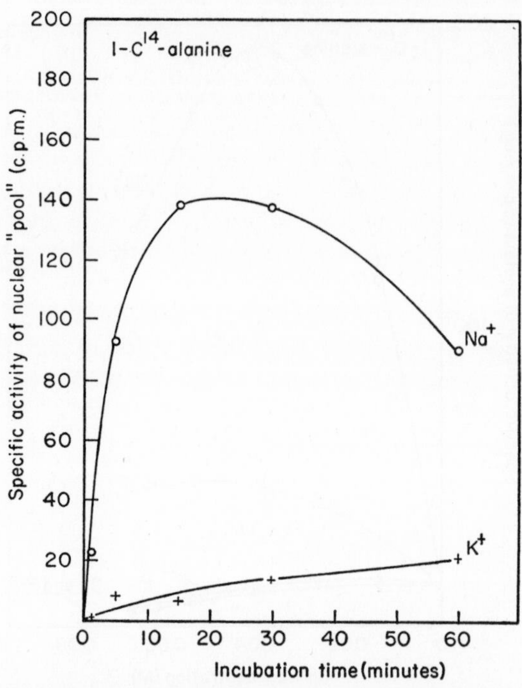

FIG. 25. The specific effect of sodium ions in promoting the transport of free amino acids into the isolated cell nucleus. The specific activity of the 1-C^{14}-alanine in the acid-soluble nuclear extract is plotted against time of incubation at 37°C. From Allfrey *et al.* (1960).

acid into the acid-soluble pool is evidently sodium dependent and does not occur to any appreciable extent when equivalent amounts of potassium replace the sodium.

The reaction is temperature dependent; it proceeds very slowly at 0°, but at 10° and higher amino acid transport is rapid. A study of alanine transport at different temperatures has shown that the rate of the reaction is doubled when the temperature is raised from 10° to 20°C. The doubling of reaction rate with a 10° rise in temperature would be expected if amino acid transport involves the activity of an enzyme.

There is some evidence which suggests that the transport of amino acids into the nucleus is an active process that requires the participation of ATP or some similar energy source. For example, the addition of 0.002 *M* cyanide or dinitrophenol (both of which suppress nuclear ATP synthesis) causes some inhibition of alanine transport. Similarly, the removal of nucleotides from the nuclei by a rapid extraction with acetate buffer at pH 4.8 destroys their capacity for subsequent amino acid accumulation (tested under neutral conditions). Yet a similar extraction process with succinate buffers at the same pH value does not extract nuclear nucleotides, nor does it affect alanine transport. Attempts to restore the activity of acetate-extracted nuclei with added ATP have not been successful, and to date, a direct demonstration that ATP is necessary for amino acid transport into the nucleus has not been achieved.

It has already been pointed out that, in contrast to nuclei, cytoplasmic systems require potassium ions for protein synthesis. It is also known that the active transport of amino acids into or across cells is potassium-, and not sodium-dependent (Christensen and Riggs, 1956). Sodium dependence therefore, becomes a good test for the nuclear localization of many synthetic reactions. The range of this dependence is discussed in more detail below.

Other experiments have made it clear that protein synthesis in the nucleus is sodium dependent mainly because sodium ions are required to get amino acids to the site of synthesis. For example, it has been found that thymus nuclei *will* synthesize C^{14}-labeled protein in a *potassium*-containing medium, provided they are first exposed to a sodium-rich medium containing the radioactive amino acid. Once transport has occurred, the continued presence of sodium ions in the medium is not necessary. This finding is in accord with later observations that the action of the nuclear amino acid activating enzymes is not sodium dependent.

Other synthetic processes in the nucleus may or may not show a demonstrable sodium ion requirement. For the incorporation of 2-C^{14}-thymidine into DNA, the requirement is clear. For the uptake of 8-C^{14}-adenine or adenosine into nuclear RNA, the need is less marked, but still definite. In both cases the results can be interpreted in terms of a specific stimulation by sodium ions of the transport of both adenosine and thymidine into the nucleus. A contrasting effect is seen in the labeling of RNA pyrimidines; no specific ionic requirement is observed for 6-C^{14}-orotic acid incorporation into the uridylic acid of nuclear RNA. This finding fits in with observations showing that orotic acid penetration into nuclei is not specifically promoted by sodium ions in the medium.

ATP synthesis in the nucleus does not appear to be sodium dependent. A study of P^{32}-orthophosphate incorporation into the terminal phosphate groups of the acid-soluble nucleotides did not reveal significant differences in labeling between nuclei incubated in sodium- or potassium-containing media.

It is now clear that many of the synthetic processes which occur in the nucleus, and protein synthesis in particular, are sodium dependent. The evidence obtained in isolated nuclei for the special and essential role of sodium ions in nuclear metabolism is supported by studies of sodium distribution within living cells. This has been studied using autoradiographic techniques in which large cells were exposed to media containing Na^{24} (Abelson and Duryee, 1949) or Na^{22} (Naora *et al.*, 1960). For example, the exposure of frog oöcytes to a medium containing radioactive sodium leads to a marked accumulation of sodium ions within the nucleus. The magnitude of the effect is strikingly demonstrated in Fig. 26, in which the high grain density over the nucleus and the low grain density over the cytoplasm testify to the specific nuclear localization of sodium ions. Considering the accessibility of the nucleus to direct contact with the medium surrounding the cell (as described on page 684), it is possible that sodium ions of the medium penetrate the nucleus without first passing through the cytoplasm.

E. Amino Acid Activation and Transfer

Some of the procedures involved in nuclear protein synthesis are peculiar to the nucleus; the role of the phosphate groups of DNA as negatively charged groups essential for the phosphorylation of nucleotides and the role of sodium ions in promoting the penetration of amino acids into the nucleus are presently known only in the nucleus. After an amino acid penetrates into the nucleus it becomes "activated" and "transferred" to RNA. These processes occur in the nucleus essentially as they have been found to occur in cytoplasm (Hopkins, 1959).

Beginning with the work of Hoagland (1955), evidence has been obtained in many laboratories to indicate that the first step in the sequence of protein-synthetic reactions involves the enzymatic activation of the carboxyl groups of the amino acids according to the reaction:

$$\text{amino acid} + \text{ATP} + \text{activating enzyme} \rightleftharpoons \text{enzyme-AMP-amino acid} + \text{PP}$$

The reaction requires ATP, pyrophosphate is split off, and the amino acyladenylate compounds which are formed remain bound to the enzyme (Hoagland *et al.*, 1956). This reaction was first shown to occur in the cytoplasmic fraction of tissue homogenates. The amino acid activating enzymes also occur in nuclei (Hopkins, 1959; Webster, 1960). The best evidence that these enzymes occur in the nucleus, as well as in the cyto-

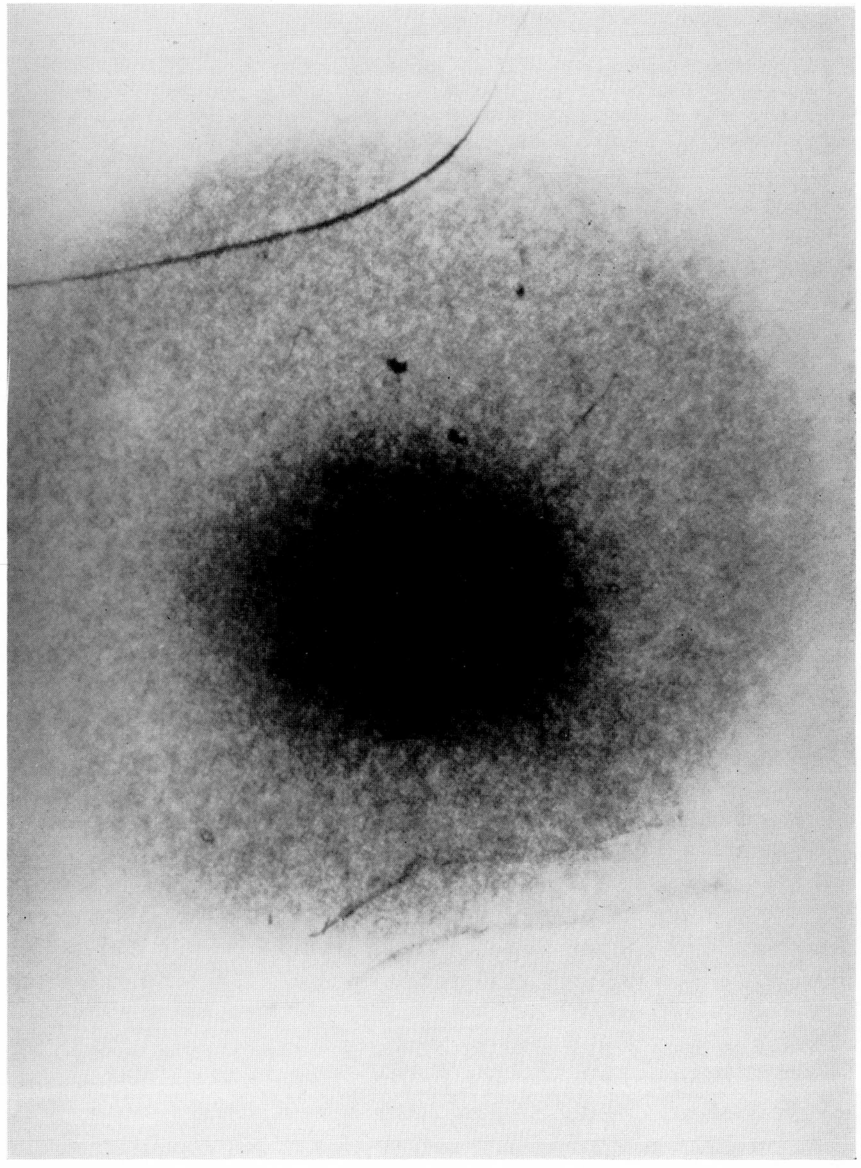

FIG. 26. Autoradiograph showing penetration of Na^{22} into an ovarian frog oöcyte. Magnification: × 88. From Naora *et al.* (1960).

plasm, is that they are present in nuclei prepared by the nonaqueous procedure (Volume I, Chapter 7).

In many cytoplasmic systems it has been shown (Hoagland *et al.*, 1957; Ogata and Nohara, 1957; Holley, 1957; Berg and Ofengand, 1958; Weiss *et al.*, 1958) that the sequel to the activation of amino acids is their transfer to "soluble" nucleic acids, according to the reaction:

$$\text{enzyme-aminoacyl AMP} + \text{s-RNA} \rightleftharpoons \text{aminoacyl-s-RNA} + \text{enzyme} + \text{AMP}$$

A similar reaction occurs in isolated thymus nuclei and in nuclei prepared from liver tissue (Hopkins, 1959). The labeling of nuclear RNA's with radioactive amino acids can be carried out either in the intact nucleus or in an isolated system containing nuclear activating enzymes, RNA, ATP, and the C^{14}-amino acid in suitable concentrations (Hopkins, 1959). The RNA of the nucleus that takes part in this reaction seems to be different from the "soluble" RNA of the cytoplasm for the corresponding RNA of the nucleus is part of the extractable ribonucleoprotein particles mentioned on page 719.

The RNA which received activated amino acids, carrier RNA, is but a small part of the total nuclear RNA. Activated amino acids do not combine with DNA or with nucleolar RNA. When it comes to experiments on incorporation of purines and pyrimidines, nucleolar RNA is about ten times as active as is the fraction of RNA in which carrier RNA is found (Allfrey and Mirsky, 1957b) (Fig. 27).

There has been much discussion about which substances that are synthesized in the nucleus pass out into the cytoplasm. In this chapter the evidence that nuclear RNA moves into the cytoplasm has been considered, and this problem is considered in much detail in Chapter 11 of this volume.

In recent years there has been less discussion about whether proteins synthesized in the nucleus pass into the cytoplasm. It is tempting to suppose that the ribonucleoproteins and proteins soluble in neutral isotonic buffer move from the nucleus into the cytoplasm. There is, however, no strong evidence for this.

There is at present an instance in which observations with microcinematography provide evidence for the passage into the cytoplasm of protein synthesized in the nucleus. In neurons it is said (Geiger, 1958) that the formation of neurofibrils can be seen to begin at the nucleolus when adrenaline is added to mammalian cortical neurons in tissue culture. "Almost immediately after the addition of adrenaline to the cultures, the optical density of various nucleolar and nuclear areas started to change. Fibrils appeared first at the nucleoli, often as a beaded chain, and from there extended gradually into the nucleoplasm and cytoplasm. They wound first around the nucleus, then spread into the remainder of the cytoplasm, the dendrites, and axon." At present this observation stands

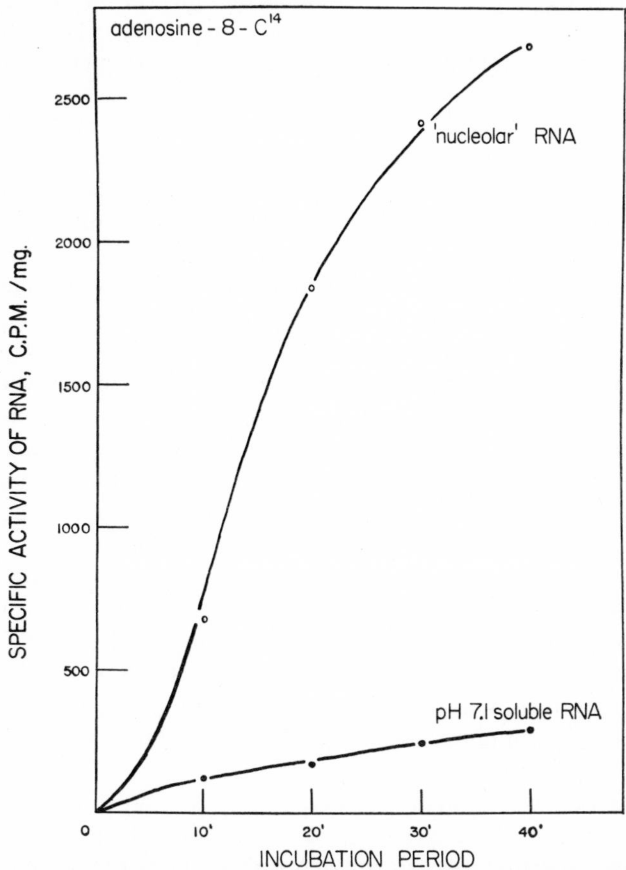

FIG. 27. The time course of adenosine-8-C^{14} incorporation into different RNA fractions of isolated thymus nuclei. The specific activity of the RNA is plotted against the time of incubation. The incubation medium contained 0.1 mg. (0.16 microcurie) of adenosine per 40 mg. of nuclei. From Allfrey and Mirsky (1957b).

almost by itself, although a suggestion that keratin is synthesized in the nucleus should be mentioned (Pelc, 1958b).

CONCLUSION

The chromosomes are the most significant components of the nucleus, and, of the substances found in them, our attention is first of all focused on DNA. We cannot help noticing DNA, for all methods of staining chromosomes are really methods for staining DNA. Our reason for dwelling on DNA, however, is that it is an essential part of the germinal material. The other substances found in chromosomes vary in amount, often considerably, per set of chromosomes in various cells of an organism and in a given cell under varying physiological conditions. The relative constancy of DNA per set of chromosomes places it in a special category.

Taking the nucleus as a whole, it shows almost as much variation in the cells of an organism as do the cytoplasmic moieties of these cells. Nuclei differ from each other morphologically, in the distribution of their chromatin, in the size of their nucleoli, in their RNA content, their protein content, their enzymatic equipment, and their metabolic activity.

The variability of chromosomes is one of their main characteristics. The lampbrush chromosomes of frog oöcyte are vastly different in appearance from the chromosomes of the frog's somatic cells, as are the giant larval polytene chromosomes of the Diptera from the ordinary chromosomes of these insects. The differences in chromosomal activity are just as striking. A lampbrush chromosome shows signs of activity throughout its length (Gall, 1958), whereas a polytene chromosome of *Rhynchosciara* or *Chironomus* (Breuer and Pavan, 1955; Beermann, 1956) show puffs at very few of its loci.

Underlying this variability there is a thread of continuity and constancy. This can be visualized in *Drosophila*. Despite the unusual appearance of the giant polytene chromosomes and their presence in highly differentiated somatic cells, it is well known that the banding seen in these chromosomes is an excellent guide to the genetic maps of *Drosophila* egg and sperm chromosomes. This brings us back to DNA, for DNA is the chemical basis for chromosomal constancy.

What is the biochemical function of DNA? One function is to make possible the synthesis of more DNA. What DNA does otherwise is best known from experiments with isolated thymus nuclei in which it is found that DNA is a cofactor needed for ATP syntheses. For this function other polyanions can replace DNA. It is, therefore, as a polyanion, as a polyphosphate, that DNA functions as a cofactor in ATP synthesis. At present this is the only known function of DNA in the interphase nucleus. It is, moreover, an essential function. Experiments on isolated nuclei have indeed shown that the presence of DNA also makes possible the synthesis of protein and RNA. It is not yet known whether in these syntheses under the conditions studied DNA does any more than provide for a necessary supply of ATP. There are of course good grounds for supposing that DNA plays a more specific role in protein and RNA synthesis. For the present, however, what is definitely established for the interphase nucleus is that the phosphate groups of DNA are needed for ATP synthesis.

Only a small part of the DNA-phosphate groups in thymus nuclei are active. Most of these groups are blocked by combination with histone. The active groups, less than one-tenth of the total, are combined with magnesium atoms. When an excess of magnesium chloride is added to isolated nuclei only a few of these magnesium ions can combine with DNA (approximately one-tenth of the magnesium already combined with DNA), for its phosphate groups are blocked by histone. Indeed when extra histone is added to isolated nuclei their activity is considerably reduced. Magnesium and histone compete for the phosphate groups of isolated DNA and of DNA within a nucleus in much the same way that basic dyes compete with histone (Mirsky and Ris, 1951b).

In many types of cells, as in thymus lymphocytes, chromosomes have relatively few of their DNA-phosphate groups occupied by magnesium, most groups being blocked by histones. The chromosomes of the various differentiated cells of an organism probably differ with respect to which patches of DNA-phosphate groups are combined with magnesium, a factor determining which genetic loci are active. The giant polytene chromosomes of the *Diptera* are among those having a small number of active loci, and in these chromosomes also activity may depend upon the condition of DNA-phosphate groups.

We are beginning to have some insight into what determines which patches of the long DNA chains are inactive and which are active. It is known that in the background cytoplasmic factors are important influences on the state of the nucleus, but the chemistry of cytoplasmic control of chromosomal activity is entirely unknown.

References

Abelson, P. H., and Duryee, W. R. (1949). *Biol. Bull.* **96**, 205.
Afzelius, B. A. (1955). *Exptl. Cell Research* **8**, 147.
Alfert, M. (1956). *J. Biophys. Biochem. Cytol.* **2**, 109.
Allfrey, V. G. (1954). *Proc. Natl. Acad. Sci. U.S.* **40**, 881.
Allfrey, V., and Mirsky, A. E. (1952). *J. Gen. Physiol.* **35**, 841.
Allfrey, V. G., and Mirsky, A. E. (1955). *Science* **121**, 879.
Allfrey, V. G., and Mirsky, A. E. (1957a). *Proc. Natl. Acad. Sci. U.S.* **43**, 589.
Allfrey, V. G., and Mirsky, A. E. (1957b). *Proc. Natl. Acad. Sci. U.S.* **43**, 821.
Allfrey, V. G., and Mirsky, A. E. (1958). *Proc. Natl. Acad. Sci. U.S.* **44**, 981.
Allfrey, V. G., and Mirsky, A. E. (1959). *In* "Subcellular Particles," p. 186. American Physiological Society, Washington, D.C.
Allfrey, V., Stern, H., Mirsky, A. E., and Saetren, H. (1952). *J. Gen. Physiol.* **35**, 529.

Allfrey, V. G., Daly, M. M., and Mirsky, A. E. (1955). *J. Gen. Physiol.* **38**, 415.

Allfrey, V. G., Mirsky, A. E., and Osawa, S. (1957). *J. Gen. Physiol.* **40**, 451.

Allfrey, V. G., Hopkins, J. W., Frenster, J. H., and Mirsky, A. E. (1960). *Ann. N. Y. Acad. Sci.* **88**, 722.

Anderson, N. G., and Wilbur, K. M. (1952). *J. Gen. Physiol.* **35**, 781.

Appleby, C. A., and Morton, R. K. (1960). *Biochem. J.* **75**, 258.

Armstrong, J. A. (1956). *Exptl. Cell Research* **11**, 640.

Austin, C. R., and Amoroso, E. C. (1957). *Exptl. Cell Research* **13**, 419.

Baltus, E. (1954). *Biochim. et Biophys. Acta* **15**, 263.

Barer, R., Joseph, S., and Meek, G. A. (1959). *Exptl. Cell Research* **18**, 179.

Barr, M. L., Bertram, L. F., and Lindsay, H. A. (1950). *Anat. Record* **107**, 283.

Barton, A. D. (1960). *In* "The Cell Nucleus (J. S. Mitchell, ed.), 142. Butterworths, London.

Bauer, H. (1936). *Zool. Jahrb. Abt. Allgem. Zool. Physiol. Tiere* **56**, 239.

Becker, F. F., and Green, H. (1960). *Exptl. Cell Research* **19**, 361.

Beermann, W., and Bahr, G. F. (1954). *Exptl. Cell Research* **6**, 195.

Beermann, W. (1956). *Cold Spring Harbor Symposia Quant. Biol.* **21**, 217.

Belozerskii, A. N., and Uryson, S. O. (1958). *Biokhimiya* **23**, 532.

Bendich, A., Pahl, H. B., and Beiser, S. M. (1956). *Cold Spring Harbor Symposia Quant. Biol.* **21**, 31.

Berg, P., and Ofengand, E. J. (1958). *Proc. Natl. Acad. Sci. U.S.* **44**, 78.

Bergstrand, A., Eliasson, N. A., Hammarsten, E., Norberg, B., Reichard, P., and von Ubisch, H. (1948). *Cold Spring Harbor Symposia Quant. Biol.* **13**, 22.

Bernhard, W., and Byczkowska. (1960). Unpublished.

Bernhard, W., and Rouiller, C. (1956). *J. Biophys. Biochem. Cytol.* **2**, Suppl., 73.

Bernhard, W., Bauer, A., Gropp, A., Haguenau, F., and Oberling, C. (1955). *Exptl. Cell Research* **9**, 88.

Bessis, M. (1954). "Traité de Cytologie Sanguine," p. 83. Masson, Paris.

Birge, W. J., Salisbury, G. W., de la Torre, L., and Lodge, J. R. (1960). *Anat. Record* **137**, 475.

Bloch, D., and Godman, G. (1955). *J. Biophys. Biochem. Cytol.* **1**, 17.

Boivin, A. R., Vendrely, R., and Vendrely, C. (1948). *Compt. rend. acad. sci.* **226**, 1061.

Brachet, J. (1940). *Compt. rend. soc. biol.* **133**, 88.

Brachet, J. (1942). *Arch. biol. (Liège)* **53**, 207.

Brachet, J., and Ficq, A. (1956). *Arch. Biol. (Liège)* **67**, 431.

Brachet, J., Chantrenne, H., and Vanderhaeghe, F. (1955). *Biochim. et Biophys. Acta* **18**, 544.

Breitman, T., and Webster, G. C. (1958). *Biochim. et Biophys. Acta* **27**, 408.

Breslau, E., and Scremin, L. (1924). *Arch. Protistenk.* **48**, 509.

Breuer, M. E., and Pavan, C. (1955-56). *Chromosoma* **7**, 371.

Brown, G. L., and Martin, A. V. (1955). *Nature* **176**, 971.

Brown, G. L., and Watson, M. (1953). *Nature* **172**, 339.

Brown, R. (1833). *Trans. Linnean Soc. London* **16**, 685.

Brown, W. V., and Emery, W. H. P. (1957). *Am. J. Botany* **44**, 585.

Burgos, M. (1955). *Exptl. Cell Research* **9**, 360.

Butler, J. A. V., Cohn, P., and Simson, P. (1960). *Biochim. et Biophys. Acta* **38**, 386.

Callan, H. G. (1956). *Symposium on the Fine Structure of Cells (Leiden)* **B21**, 89.

Callan, H. G., and Tomlin, S. G. (1950). *Proc. Roy. Soc.* **B137**, 367.

Callanan, M. J., Carrol, W. R., and Mitchell, E. R. (1957). *J. Biol. Chem.* **229**, 279.
Campbell, R. M., and Kosterlitz, H. W. (1952). *Science* **115**, 84.
Canellakis, E. S., and Herbert, E. (1960). *Proc. Natl. Acad. Sci. U.S.* **46**, 170.
Carneiro, J., and Leblond, C. P. (1959). *Science* **129**, 391.
Carvalho, S. de, and Wilkins, M. F. H. (1954). *Proc. 4th Intern. Congr. of Intern. Soc. Hematol.* p. 119.
Caspersson, T. (1950). "Cell Growth and Cell Function." Norton, New York.
Caspersson, T., and Schultz, J. (1939). *Nature* **143**, 602, 609.
Chambers, R., and Fell, H. B. (1931). *Proc. Roy. Soc.* **B109**, 380.
Chargaff, E., Crampton, C. F., and Lipschitz, R. (1953). *Nature* **172**, 289.
Chauveau, J., Moule, Y., and Rouiller, Ch. (1956). *Exptl. Cell Research* **11**, 317.
Chèvremont, M., Chèvremont-Comhaire, S., and Baeckeland, E. (1959). *Arch. biol. (Liège)* **70**, 811.
Christensen, H. N., and Riggs, T. R. (1956). *J. Biol. Chem.* **220**, 265.
Cohen, I. (1936-37). *Protoplasma* **27**, 484.
Cooper, D. C. (1952). *Am. Naturalist* **86**, 219.
Crampton, C. F., Moore, S., and Stein, W. H. (1955). *J. Biol. Chem.* **215**, 787.
Crampton, C. F., Stein, W. H., and Moore, S. (1957). *J. Biol. Chem.* **225**, 363.
Crosby, A. R. (1957). *Am. J. Botany* **44**, 813.
Cruft, H. J., Mauritzen, C. M., and Stedman, E. (1954). *Nature* **174**, 580.
Cruft, H. J., Hindley, J., Mauritzen, C. M., and Stedman, E. (1957a). *Nature* **180**, 1107.
Cruft, H. J., Mauritzen, C. M., and Stedman, E. (1957b). *Phil. Trans. Roy. Soc. London* **B241**, 93.
Cunningham, L., Griffin, A. C., and Luck, J. M. (1950). *J. Gen. Physiol.* **34**, 59.
Daly, M. M., and Mirsky, A. E. (1955). *J. Gen. Physiol.* **38**, 405.
Daly, M. M., Mirsky, A. E., and Ris, H. (1951). *J. Gen. Physiol.* **34**, 439.
Daly, M. M., Allfrey, V. G., and Mirsky, A. E. (1952). *J. Gen. Physiol.* **36**, 173.
Davidson, J. N., Leslie, I., Smellie, R. M. S., and Thomson, R. Y. (1950). *Biochem. J.* **46**, xl.
Davidson, J. N., Leslie, I., and White, J. C. (1951). *Lancet* **i**, 1287.
Davidson, J. N., Leslie, I., and White, J. C. (1953). *J. Pathol. Bacteriol.* **63**, 471.
Davison, P. F. (1957a). *Biochem. J.* **66**, 703.
Davison, P. F. (1957b). *Biochem. J.* **66**, 708.
Davison, P. F., and Butler, J. A. V. (1954). *Biochim. et Biophys. Acta* **15**, 439.
De Bruyn, P. P. H., Robertson, R. C., and Farr, R. S. (1950). *Anat. Record* **108**, 279.
De Bruyn, P. P. H., Farr, R. S., Banks, H., and Morthland, F. W. (1952). *Exptl. Cell Research* **4**, 174.
Denues, A. R. T., and Mottram, F. C. (1955). *J. Biophys. Biochem. Cytol.* **1**, 185.
Edström, J. E. (1960). *J. Biophys. Biochem. Cytol.* **8**, 47.
Edström, J. E., and Eichner, D. (1958a). *Nature* **181**, 619.
Edström, J. E., and Eichner, D. (1958b). *Z. Zellforsch. u. mikroskop. Anat.* **48**, 187.
Ely, J. O., and Ross, M. H. (1951). *Science* **114**, 70.
Epstein, M. A. (1957). *J. Biophys. Biochem. Cytol.* **3**, 567, 851.
Estable, C., and Sotelo, J. R. (1954). *Symposium VIIIth Congr. Cell Biol.*, Leyden.
Fauré-Fremiet, E., and Rouiller, Ch. (1958). *Exptl. Cell Research* **14**, 29.
Fautrez, J., and Laquerriere, R. (1957). *Exptl. Cell Research* **13**, 403.
Fawcett, D. (1959). Personal communication.
Feldherr, C. M., and Feldherr, A. B. (1960). *Nature* **185**, 250.

Felix, K., Fischer, H., and Kreckels, A. (1956). *Progr. in Biophys. and Biophys. Chem.* **6**, 1.
Felsenfeld, G., and Huang, S. (1959). *Biochim. et Biophys. Acta* **34**, 234.
Feulgen, R., and Rossenbeck, H. (1924). *Z. physiol. Chem.* **135**, 203.
Feulgen, R., Behrens, M., and Mahdihassan, S. (1937). *Z. physiol. Chem.* **246**, 203.
Ficq, A. (1955a). *Arch. biol. (Liège)* **66**, 509.
Ficq, A. (1955b). *Exptl. Cell Research* **9**, 286.
Ficq, A., and Errera, M. (1958). *Exptl. Cell Research* **14**, 182.
Ficq, A., and Pavan, C. (1957). *Nature* **180**, 983.
Finamore, F. J., and Volkin, E. (1958). *Exptl. Cell Research* **15**, 405.
Finamore, F. J., Thomas, D. J., Crouse, G. T., and Lloyd, B. (1960). *Arch. Biochem. Biophys.* **88**, 10.
Fischer, H. (1934). *Planta* **22**, 767.
Fraenkel-Conrat, H., Snell, N. S., and Ducay, E. D. (1952). *Arch. Biochem. Biophys.* **39**, 80, 97.
Frenster, J. H., Allfrey, V. G., and Mirsky, A. E. (1960). *Proc. Natl. Acad. Sci. U.S.* **46**, 432.
Fresco, J. R., Bendich, A., and Russell, P. J. (1955). *Federation Proc.* **14**, 214.
Friedkin, M., and Wood, H. (1956). *J. Biol. Chem.* **220**, 639.
Gall, J. G. (1954). *J. Morphol.* **94**, 283.
Gall, J. G. (1955). *Symposia Soc. Exptl. Biol. No.* **9**, p. 358.
Gall, J. G. (1956). *J. Biophys. Biochem. Cytol.* **2**, Suppl., 393.
Gall, J. G. (1958). *In* "A Symposium on the Chemical Basis of Development" (William D. McElroy and Bentley Glass, eds.) p. 103. Johns Hopkins Press, Baltimore, Maryland.
Gall, J. G., and Johnson, W. W. (1960). *J. Biophys. Biochem. Cytol.* **7**, 657.
Gay, H. (1956). *Cold Spring Harbor Symposia Quant. Biol.* **21**, 257.
Geiger, R. S. (1958). *Nature* **182**, 1674.
Glauert, A., and Hopwood, D. A. (1959). *J. Biophys. Biochem. Cytol.* **6**, 515.
Goldstein, L., and Micou, J. (1959a). *J. Biophys. Biochem. Cytol.* **6**, 1.
Goldstein, L., and Micou, J. (1959b). *J. Biophys. Biochem. Cytol.* **6**, 301.
Graham, M. A. (1954). *Anat. Record* **119**, 469.
Graham, M. A., and Barr, M. L. (1952). *Anat. Record* **112**, 709.
Grant, P. (1955). *Biol. Bull.* **109**, 343.
Grant, P. (1956). *Anat. Record* **125**, 623.
Gray, J. (1931). "Experimental Cytology." Macmillan, New York.
Hämmerling, J., and Stich, H. (1956). *Z. Naturforsch.* **11b**, 158.
Hall, J. L. (1941). *J. Am. Chem. Soc.* **63**, 794.
Harding, C. V., and Feldherr, C. (1959). *J. Gen. Physiol.* **42**, 1155.
Harris, H. (1959). *Biochem. J.* **73**, 362.
Hartmann, J. F. (1953). *J. Comp. Neurol.* **99**, 201.
Hauschka, T. S. (1952). *Cancer Research* **12**, 269.
Heagy, F. C., and Roper, J. A. (1952). *Nature* **170**, 713.
Hecht, L. I., and Potter, V. R. (1956). *Cancer Research* **162**, Suppl. 4, 988.
Heidenhain, R. (1875). *Arch. ges. Physiol. Pflüger's* **10**, 557.
Heitz, E. (1925). *Z. Zellforsch. u mikroskop. Anat.* **2**, 69.
Heitz, E. (1931a). *Planta* **12**, 775.
Heitz, E. (1931b). *Planta* **15**, 495.
Heitz, E., and Bauer, H. (1933). *Z. Zellforsch. u. mikroskop. Anat.* **17**, 67.

Hoagland, M. B. (1955). *Biochim. et Biophys. Acta* **16**, 288.
Hoagland, M. B., Keller, E. B., and Zamecnik, P. C. (1956). *J. Biol. Chem.* **218**, 345.
Hoagland, M. B., Zamecnik, P. C., and Stephenson, M. L. (1957). *Biochim. et Biophys. Acta* **24**, 215.
Hoff-Jørgensen, E. (1954). *In* "Recent Developments in Cell Physiology" (J. A. Kitching, ed.), p. 70. Butterworths, London.
Hogeboom, G. H., and Schneider, W. C. (1952). *J. Biol. Chem.* **204**, 233.
Hogeboom, G. H., Schneider, W. C., and Striebich, M. J. (1952). *J. Biol. Chem.* **196**, 111.
Holley, R. W. (1957). *J. Am. Chem. Soc.* **79**, 658.
Hopkins, J. W. (1959). *Proc. Natl. Acad. Sci. U.S.* **45**, 1461.
Horn, E. C., and Ward, C. L. (1957). *Proc. Natl. Acad. Sci. U.S.* **43**, 776.
Huiskamp, W. (1901). *Z. physiol. Chem.* **32**, 145.
Immers, J. (1957). *Exptl. Cell Research* **12**, 150.
Jeener, R., and Szafarz, D. (1950). *Arch. Biochem.* **26**, 54.
Johnston, F. B., Setterfield, G., and Stern, H. (1958). *J. Biophys. Biochem. Cytol.* **6**, 53.
Kaufmann, B. P. (1960). *In* "The Cell Nucleus" (J. S. Mitchell, ed.), 251. Butterworths, London.
Kirby, K. S., and Frearson, P. M. (1960). *In* "The Cell Nucleus" (J. S. Mitchell, ed.), 211. Butterworths, London.
Kit, S. (1960). *Arch. Biochem. Biophys.* **87**, 318, 330.
Kit, S., and Gross, A. (1959). *Federation Proc.* **18**, 262.
Klein, E., and Klein, G. (1950). *Nature* **166**, 832.
Kondo, N., and Osawa, S. (1959). *Nature* **183**, 1602.
Konopacki, M. (1936). *Compt. rend. soc. biol.* **122**, 139.
Kosin, I. L., and Ishizaki, H. (1959). *Science* **130**, 43.
Kossel, A. (1884). *Z. physiol. Chem.* **8**, 511.
Kossel, A. (1928). "The Protamines and Histones." Longmans, Green, London.
Kunitz, M. (1950). *J. Gen. Physiol.* **33**, 363.
La Cour, L. F., Deeley, E. M., and Chayen, J. (1956). *Nature* **177**, 272.
Lafontaine, J. G. (1958). *J. Biophys. Biochem. Cytol.* **4**, 777.
Lagerstedt, S. (1949). "Cytological Studies on the Protein Metabolism of the Liver in Rat." Hakan Ohlssons Boktrykeri, Lund.
Leduc, E., and Bernhard, W. (1960). *Compt. rend. acad. sci.* **250**, 2948.
Leeman, L. (1959). *Nature* **183**, 1188.
Leslie, I., and Davidson, J. N. (1951). *Biochim. et Biophys. Acta* **7**, 415.
Leuchtenberger, C., Vendrely, R., and Vendrely, C. (1951). *Proc. Natl. Acad. Sci. U.S.* **37**, 33.
Leuchtenberger, C., Vendrely, R., and Vendrely, C. (1952). *Exptl. Cell Research* **3**, 240.
Leuchtenberger, C., Leuchtenberger, R., and Davis, A. M. (1954). *Am. J. Pathol.* **30**, 65.
Lewis, M. R. (1923). *Bull. Johns Hopkins Hosp.* **34**, 373.
Lilienfeld, L. (1894). *Z. physiol. Chem.* **18**, 473.
Lin, M. (1955). *Chromosoma* **7**, 340.
Lindsay, H. A., and Barr, M. L. (1955). *J. Anat.* **89**, 47.
Lison, L., and Pasteels, J. (1951). *Arch. biol. (Liège)* **62**, 1.
Logan, R., and Davidson, J. N. (1957). *Biochim. et Biophys. Acta* **124**, 196.

Logan, R., Ficq, A., and Errera, M. (1959). *Biochim. et Biophys. Acta* **31**, 402.
Lowe, C. U., and Rand, R. N. (1956a). *J. Biophys. Biochem. Cytol.* **2**, 331.
Lowe, C. U., and Rand, R. N. (1956b). *J. Biophys. Biochem. Cytol.* **2**, 711.
Lowe, C. U., Box, H., Venkataraman, P. R., and Sarkaria, D. S. (1959). *J. Biophys. Biochem. Cytol.* **5**, 251.
Lozinski, P. (1911). *Zool. Anz.* **38**, 401.
Lwoff, A., and Lwoff, M. (1931). *Bull. biol. France et Belg.* **65**, 170.
McClintock, B. (1934). *Z. Zellforsch. u. mikroskop. Anat.* **21**, 294.
McLeish, J. (1954). *Heredity* **8**, 385.
Mandel, P., and Metais, P. (1950). *Bull. acad. natl. méd. (Paris)* **134**, 449.
Mandel, P., Metais, P., and Cuny, S. (1951). *Compt. rend. acad. sci.* **231**, 1172.
Marmur, J., and Doty, P. (1959). *Nature* **183**, 1427.
Marshak, A., and Calvet, F. (1949). *J. Cellular Comp. Physiol.* **34**, 451.
Marshak, A., and Marshak, C. (1955). *Exptl. Cell Research* **8**, 126.
Meselson, M., Stahl, F. W., and Vinograd, J. (1957). *Proc. Natl. Acad. Sci. U.S.* **43**, 581.
Miescher, F. (1897). "Die histochemischen und physiologischen Arbeiten." Vogel, Leipzig.
Miller, L. A., and Goldfeder, A. (1961). *Exptl. Cell Research.* In press.
Mills, G. T., Smith, E. E. B., Stary, B., and Leslie, I. (1950). *Biochem. J.* **47**, xlviii.
Mirsky, A. E. (1947). *Cold Spring Harbor Symp. on Quant. Biol.* **12**, 143.
Mirsky, A. E., and Allfrey, V. (1960). *Diseases of Nervous System,* Monograph Suppl. XXI.
Mirsky, A. E., and Kurnick, N. B. (1949). Unpublished experiments.
Mirsky, A. E., and Pollister, A. W. (1942). *Proc. Natl. Acad. Sci. U.S.* **28**, 344.
Mirsky, A. E., and Pollister, A. W. (1943). *Trans. N.Y. Acad. Sci.* [2] **5**, 190.
Mirsky, A. E., and Pollister, A. W. (1946). *J. Gen. Physiol.* **30**, 117.
Mirsky, A. E., and Ris, H. (1949). *Nature* **163**, 666.
Mirsky, A. E., and Ris, H. (1951a). *J. Gen. Physiol.* **34**, 451.
Mirsky, A. E., and Ris, H. (1951b). *J. Gen. Physiol.* **34**, 475.
Mirsky, A. E., Allfrey, V. G., and Daly, M. M. (1954). *J. Histochem. and Cytochem.* **2**, 376.
Montgomery, T. H. (1898). *J. Morphol.* **15**, 265.
Moore, B. C. (1952). *Chromosoma* **4**, 563.
Moore, B. C. (1957). *J. Morphol.* **101**, 227.
Moore, S. (1959). "On the Constitution of Histones." *XI Conseil de Chimie* (tenu à l'Université Libre de Bruxelles).
Mulnard, J. (1956). *Arch. biol. (Liège)* **67**, 485.
Naora, H., Naora, H., Mirsky, A. E., and Allfrey, V. G. (1960). Unpublished experiments.
Neelin, J. M., and Connell, G. E. (1959). *Biochim. et Biophys. Acta* **31**, 539.
Neil, A. L., and Dixon, F. J. (1959). *A.M.A. Arch. Pathol.* **67**, 643.
Nurnberger, J., Engström, A., and Lindström, B. (1952). *J. Cellular Comp. Physiol.* **39**, 215.
Odeblad, E., and Magnusson, G. (1954). *Acta Endocrinol.* **17**, 290.
Ogata, K., and Nohara, H. (1957). *Biochim. et Biophys. Acta* **25**, 660.
Ogur, M., Minckler, S., Lindegren, G., and Lindegren, C. C. (1952). *Arch. Biochem. Biophys.* **40**, 175.
O'Hern, E. M., and Henry, B. S. (1956). *J. Bacteriol.* **72**, 632.
Ortega, L. G., and Mellors, R. C. (1957). *J. Exptl. Med.* **106**, 627.

Osawa, S. (1951). *Embryologia* **2**, 1.
Osawa, S., Allfrey, V., and Mirsky, A. E. (1957). *J. Gen. Physiol.* **40**, 491.
Painter, T. S., and Taylor, A. N. (1942). *Proc. Natl. Acad. Sci. U.S.* **28**, 311.
Palay, S. L. (1960). *J. Biophys. Biochem. Cytol.* **7**, 391.
Palmade, C., Chevallier, M. R., Knobloch, A., and Vendrely, R. (1958). *Compt. rend. acad. sci.* **246**, 2534.
Pasteels, J. J. (1954). *Symposia VIIIth Congr. Cell Biol. Leyden*, p. 195.
Pasteels, J. J., and Lison, L. (1950). *Arch. biol. (Liège)* **61**, 445.
Pavan, C. (1958). Personal communication.
Pelc, S. R. (1958a). *Exptl. Cell Research* **14**, 301.
Pelc, S. R. (1958b). *Exptl. Cell Research* **6**, Suppl. 97.
Pelc, S. R. (1959a). *Nature* **183**, 335.
Pelc, S. R. (1959b). *Nature* **184**, 1414.
Pelc, S. R., and La Cour, L. F. (1959). *Experientia* **15**, 131.
Perry, R. P. (1960). *Exptl. Cell Research* **20**, 216.
Perry, R. P., and Errera, M. (1960). *In* "The Cell Nucleus" (J. S. Mitchell, ed.), p. 24. Butterworths, London.
Phillips, D. M. P., and Johns, E. W. (1959). *Biochem. J.* **72**, 538.
Philpot, J. St. L., and Stanier, J. E. (1956). *Biochem. J.* **63**, 214.
Pinchot, G. B. (1959). *Biochem. Biophys. Research Commun.* **1**, 17.
Pollister, A. W., and Mirsky, A. E. (1946). *J. Gen. Physiol.* **30**, 101.
Pomerat, C. M. (1953). *Exptl. Cell Research* **5**, 191.
Potter, V. R., Recknagel, R. O., and Hurlbert, B. (1951). *Federation Proc.* **10**, 646.
Prince, R. H., Graham, M. A., and Barr, M. L. (1955). *Anat. Record* **122**, 153.
Rasch, E. M. (1951). *Botan. Gaz.* **112**, 331.
Rebhun, L. I. (1956). *J. Biophys. Biochem. Cytol.* **2**, 93.
Reichenow, E. (1928). *Arch. Protistenk.* **61**, 144.
Richards, B. M., Walker, P. M. B., and Deeley, E. M. (1956). *Ann. N.Y. Acad. Sci.* **63**, 831.
Ris, H., and Mirsky, A. E. (1949a). *J. Gen. Physiol.* **32**, 489.
Ris, H., and Mirsky, A. E. (1949b). *J. Gen. Physiol.* **33**, 125.
Robertson, J. D. (1959). *Biochem. Soc. Symposia No.* **16**, 3.
Rose, J. A., and Schweigert, B. S. (1952). *Proc. Soc. Exptl. Biol. Med.* **79**, 541.
Rudkin, G. T. (1959). *In* "Records of the Genetics Society of America," p. 91. (Published by the Secretary, W. L. Russell, Oak Ridge National Laboratory, Oak Ridge, Tennessee.)
Rudkin, G. T., and Woods, P. S. (1959). *Proc. Natl. Acad. Sci. U.S.* **45**, 997.
Sachs, H. (1958). *J. Biol. Chem.* **233**, 643.
Sax, K. (1935). *J. Arnold Arboretum (Harvard Univ.)* **16**, 301.
Schiller, A. A., Schayer, R. W., and Hess, E. L. (1952). *J. Gen. Physiol.* **36**, 489.
Schultz, J. (1956). *Cold Spring Harbor Symposia Quant. Biol.* **21**, 307.
Schultze, B., Oehlert, W., and Maurer, W. (1959). *Beitr. pathol. Anat. u. allgem. Pathol.* **120**, 58.
Sen-Gupta, P. C., Bhattacharjee, B., and Ray, H. N. (1953). *J. Indian Med. Assoc.* **22**, 305.
Serra, J. A., and Lopes, Q. (1944). *Chromosoma* **2**, 576.
Shack, J., and Bynum, B. S. (1959). *Nature* **184**, 635.
Shack, J., Jenkins, R. J., and Thompsett, J. M. (1953). *J. Biol. Chem.* **203**, 373.
Siebert, G. (1960). *In* "The Cell Nucleus" (J. S. Mitchell, ed.), p. 176. Butterworths, London.

Sirlin, J. L. (1960). *In* "The Cell Nucleus" (J. S. Mitchell, ed.), p. 35. Butterworths, London.
Smellie, R. M. S., McIndoe, W. M., Logan, R., Davidson, J. N., and Dawson, I. M. (1953). *Biochem. J.* **54**, 280.
Sparrow, A. H., and Hammond, M. R. (1947). *Am. J. Botany* **34**, 439.
Stäubli, W. (1960). *Compt. rend. acad. sci.* **250**, 1137.
Stedman, E., and Stedman, E. (1943). *Nature* **152**, 556.
Stedman, E., and Stedman, E. (1947). *Symposium Soc. Exptl. Biol.* **1**, 232.
Steffensen, D., and Bergeron, J. A. (1959). *J. Biophys. Biochem. Cytol.* **6**, 339.
Steinert, M. (1960). Personal communication.
Stern, H., and Mirsky, A. E. (1952). *J. Gen. Physiol.* **36**, 181.
Stern, H., and Mirsky, A. E. (1953). *J. Gen. Physiol.* **37**, 177.
Stern, H., Allfrey, V., Mirsky, A. E., and Saetren, H. (1952). *J. Gen. Physiol.* **35**, 559.
Stich, H. (1955). *Z. Naturforsch.* **10b**, 281.
Stich, H. (1956). *Chromosoma* **7**, 693.
Stich, H., and Hämmerling, J. (1953). *Z. Naturforsch.* **8b**, 329.
Sueoka, N., Marmur, J., and Doty, P. (1959). *Nature* **183**, 1429.
Swift, H. (1950). *Physiol. Zoöl.* **23**, 169.
Swift, H. (1959). *Symposium on Mol. Biol. Univ. Chicago,* p. 266.
Thomson, R. Y., Heagy, F. C., Hutchinson, W. C., and Davidson, J. N. (1953). *Biochem. J.* **53**, 460.
Thomson, R. Y., Paul, J., and Davidson, J. N. (1957). *Biochem. J.* **69**, 553.
Turian, G., and Kellenberger, E. (1956). *Exptl. Cell Research* **11**, 417.
Uretz, R. B., and Perry, R. P. (1957). *Rev. Sci. Instr.* **28**, 861.
van Herwerden, M. A. (1924). *Biol. Zentr.* **44**, 579.
Vendrely, R., and Vendrely, C. (1948). *Experientia* **4**, 434.
Vendrely, R., and Vendrely, C. (1949). *Experientia* **5**, 327.
Vendrely, R., and Vendrely, C. (1953). *Nature* **172**, 30.
Vendrely, R., Knobloch, A., and Vendrely, C. (1957). *Exptl. Cell Research Suppl.* **4**, 279.
Vincent, W. S. (1952). *Proc. Natl. Acad. Sci. U.S.* **38**, 139.
Vincent, W. S. (1957a). *In* "The Beginnings of Embryonic Development." American Association for the Advancement of Science, Washington, D.C.
Vincent, W. S. (1957b). *Science* **126**, 306.
Vincent, W. S., and Baltus, E. (1960). *In* "The Cell Nucleus" (J. S. Mitchell, ed.), p. 18. Butterworths, London.
Vincent, W. S., and Huxley, A. H. (1954). *Biol. Bull.* **107**, 290.
Watson, M. (1955). *J. Biophys. Biochem. Cytol.* **1**, 257.
Watson, M. (1959). *J. Biophys. Biochem. Cytol.* **6**, 147.
Webster, G. C. (1960). *Biochem. Biophys. Research Communs.* **2**, 56.
Weiss, S. B., Acs, G., and Lipmann, F. (1958). *Proc. Natl. Acad. Sci. U.S.* **44**, 189.
Wischnitzer, S. (1958). *J. Ultrastruct. Research* **1**, 201.
Yasuzumi, G. (1959). *Z. Zellforsch. u. mikroskop. Anat.* **50**, 110.
Zalokar, M. (1960). *Exptl. Cell Research* **19**, 559.
Zbarsky, I. B., and Georgiev, G. P. (1959). *Biochim. et Biophys. Acta* **32**, 301.
Zubay, G. (1959). *Biochim. et Biophys. Acta* **32**, 233.
Zubay, G., and Doty, P. (1958). *Biochim. et Biophys. Acta* **29**, 47.
Zubay, G., and Doty, P. (1959). *J. Mol. Biol.* **1**, 1.
Zubay, G., and Watson, M. R. (1959). *J. Biophys. Biochem. Cytol.* **5**, 51.

CHAPTER 11

Nucleocytoplasmic Interactions in Unicellular Organisms[1]

By J. BRACHET

I. Introductory Remarks

In preceding chapters of this treatise, the chemical composition of the main cellular constituents obtainable by differential centrifugation of homogenates has been examined in detail.

It is well known that the homogenate method suffers from several drawbacks, such as possible loss or adsorption of enzymes (e.g., alkaline phosphatase) during isolation. In the absence of adequate electron microscopy observations, it is difficult to be sure that the isolated fractions are really identical with the pre-existing intracellular granules:

[1] This chapter is largely inspired from Chapter VII of "Biochemical Cytology" (Academic Press, 1957) by the same author. The literature has been brought up to date and a few errors have been corrected.

rupture of nuclei or mitochondria, as well as aggregation of smaller cytoplasmic particles, are difficult to avoid completely. Even after very careful work, it is therefore impossible to be certain that the fractions obtained are cytologically homogeneous.

But the great limitations of the homogenate method become apparent when, as in this chapter, the interest shifts from a description of the chemical composition of the several types of cell organelles to another question: What is the nature of the *interactions* that occur between the various constituents of the *intact* cell, between the nucleus and the cytoplasmic particles in particular? It is perfectly legitimate for a biochemist to mix together the various fractions obtained by differential centrifugation of homogenates. A system consisting of microsomes and mitochondria has been very useful to Siekevitz (1952) in his studies on the *in vitro* incorporation of amino acids into proteins. Vishniac and Ochoa (1952) have studied, with advantage, the biochemical events which occur when chloroplasts are mixed with mitochondria isolated from animal tissues. Potter *et al.* (1951) and Johnson and Ackermann (1953) have followed, in an effort toward an understanding of the biochemical role of the nucleus, the effects of the addition of nuclei on oxidative phosphorylations in mitochondria. They claimed that, although the nuclei themselves lack respiratory enzymes, they stimulate phosphorylation reactions. Experiments of that type, which have definite biochemical interest and value, are useless when the nature of the interaction between nuclei, mitochondria, and ergastoplasm in the *intact, living* cell is our main objective. For instance, stimulation of mitochondrial oxidative phosphorylations upon the addition of nuclei may simply be caused by an autolytic release of enzymes or cofactors from the nuclei; such a phenomenon might never occur in the nucleus of a living cell. In fact, the indications given by this type of homogenate experiment may be totally misleading if the importance of structural integrity inside the cell is neglected.

There is another reason for doubting, in the case of the nuclei at least, the value of the experiments performed on mixed fractions recovered by differential centrifugation of homogenates. If we take as a test of survival for isolated nuclei the capacity to divide when they are reintroduced into cytoplasm, there is no doubt that isolated nuclei are very quickly inactivated by contact with the outside medium. In experiments by Comandon and de Fonbrune (1939) on the transplantation of a nucleus in an anucleate *Amoeba* fragment, very close contact between the cell surfaces of the donor and the receptor amebae is indispensable; the slightest contact of the nucleus with the outside medium results in "death" and elimination of the transplanted nucleus. The same is true for the embryonic nuclei transplanted by Briggs and King (1953) into anucleate unfertilized frog eggs: the nucleus must be protected against the medium by the crushed cytoplasm of the cell out of which it is taken, if the nuclear transplantation is to be successful. None of the many media tried by Briggs and King (1953) has proved satisfactory so far. Until a medium is found in which nuclei can be kept a long time without losing their biological activities, experiments made on nuclei isolated from homogenates will have limited biological interest only. Such a medium will probably be difficult to find, since we know, from the work of Cutter *et al.* (1955), that the nuclei which swim freely in coconut liquid endosperm are capable only of degenerative amitosis.

In order to eliminate or reduce these difficulties, experiments have been made in our laboratory on nucleate and anucleate halves of two different organisms, *Amoeba proteus* and the unicellular alga *Acetabularia mediterranea.* Their purpose was, of course, to get a better understanding of the biochemical role exerted by the cell nucleus. Simultaneously, similar studies have been made on *Acetabularia* by Hämmerling and his group (Stich, Beth, and others) and, on amebae, by Mazia and by Danielli and their co-workers. The development of accurate cytochemical and quantitative micromethods has permitted fast progress in a field which has always fascinated many biologists. The results obtained in these studies will be the main subject of the present chapter.

In the following pages, we shall first deal with the results of a number of biological experiments in which the nucleus or part thereof has been removed (merotomy experiments); we shall then consider the most popular theories concerning the biochemical role of the nucleus; finally the main facts will be presented and discussed.

II. Results of Merotomy Experiments

The importance of the cell nucleus for regeneration and growth has been known since the classical experiments of merotomy made more than sixty years ago by Balbiani (1888), Klebs (1889), Verworn (1892), Townsend (1897), and others. A good review of this early work will be found in Wilson's (1925) classical treatise "The Cell in Development and Heredity." More recently, Comandon and de Fonbrune (1939), followed by Lorch and Danielli (1950), succeeded in grafting a nucleus in an anucleate *Amoeba* half. Important work on regeneration of *Stentor,* including the role of the nucleus, has been published by Weisz (1948) and by Tartar (1953, 1956). Of particular importance is the fundamental discovery by Hämmerling (1934) that regeneration in *Acetabularia* proceeds for some time in the absence of the nucleus. Since in most of the biochemical work *Amoeba proteus* and *Acetabularia* have been used, more details will be given about the effects of enucleation in these two species.

In *Amoeba proteus*, removal of the nucleus is quickly followed by loss of motility: even after a few minutes, pseudopod formation becomes sporadic and the anucleate halves soon become spherical. They are unable to feed on living prey, in contrast to normal amebae or nucleate halves. But they survive almost as long as the latter: if both are kept fasting, anucleate halves may survive as long as 2 weeks, whereas nucleate ones die after 3 weeks.

In another species of amebae (*A. sphaeronucleus*), the effects of enucleation are similar. The interesting and delicate experiments of Comandon and de Fonbrune (1939) have shown that motility (i.e., pseudopod formation) is resumed in a truly dramatic manner when a nucleus is reintroduced into a cytoplasmic fragment that was severed from the nucleus 2 or 3 days previously; if, however, the anucleate half comes from an ameba operated on 6 days before, the graft of a nucleus no longer has a favorable effect. Apparently, irreversible changes occur, sooner or later, in nonnucleated cytoplasm.

Very interesting also are the nuclear transfer experiments of Lorch and Danielli (1950) and Danielli *et al.* (1955) (see also Danielli, 1955, 1959a, b). These workers have succeeded in exchanging nuclei between two distinct, but closely related species of amebae, *A. proteus* and *A. discoides;* the result is the production of "hybrids" in organisms in which no sexual reproduction is known to take place. They found that the nucleus of either species is capable of restoring locomotion activity in nonnucleated cytoplasm. The morphological characters of the "hybrids" (e.g., shape, size of the nucleus) show strong cytoplasmic dominance. The shape and locomotion are intermediary between those of the "parent" strains. As we shall see later, the nucleus exerts, however, specific effects on the production of certain proteins. It is the belief of Danielli (1955, 1959a, b) that the nucleus determines the specific character of the macromolecules, while the cytoplasm is more important in the organization of these macromolecules into functional units. One thing is clearly proved by these elegant experiments: there exist *mutual* interactions and exchanges between nucleus and cytoplasm. We shall soon find other instances in which the nucleus lies under cytoplasmic control; it would thus obviously be a mistake to believe that there is only one sort of control in the cell, that exerted by the nucleus on the cytoplasm.

In ciliates, anucleate cytoplasm does not survive as long as in amebae; however, ciliary movement, in spite of its complexity, remains active and mating reactions are still normal 2 days after enucleation in *Paramecium* (Tartar and Chen, 1941). As pointed out correctly by Mazia (1952) in a recent review: "There is not a single case where an activity has not continued in an enucleated cell."

A favorite material for the study of regeneration in protozoa has long been *Stentor* (Fig. 1), because many interesting problems of morphogenesis can be studied in this organism. Among recent papers, special mention should be made of the work of Tartar (1953, 1956) and of

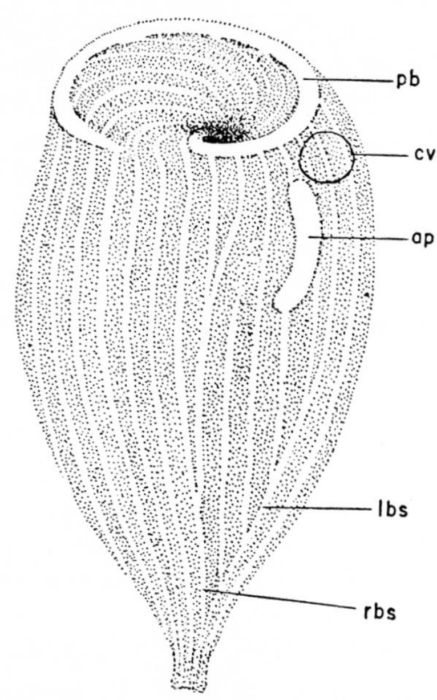

FIG. 1. Diagram of *Stentor*: *pb*, peristome band; *cv*, contractile vacuole; *ap*, position of adoral zone in physiological regeneration; *lbs* and *rbs*, left and right boundary stripes of ramifying zone. From Weisz (1954).

Weisz (1948, 1950, 1954-1956). Tartar (1953), for instance, has studied the effects of nuclear transplantations and graft combinations between two species of *Stentor*. His general conclusion is that completely successful regeneration follows when there is a preponderance of the nucleus of one species in a preponderance of its own specific cytoplasm. The nucleus supplies something essential to cell differentiation; this substance of nuclear origin, in *Stentor*, is used up at once and not stored. The result is that the *continuous* presence of the nucleus is required for growth and regeneration, in agreement with earlier observations of Balamuth (1940). The experiments of Weisz (1949) show that the nodes present in the macronuclear chain determine the differentiation activity in a specific way: while the posterior nodes fail to mediate the differentiation of a peristome, midnodes support only partial regeneration; nodes in the anterior half of the macronucleus support complete and normal regeneration. Further cytochemical work by Weisz (1950) showed that the nodes which have great morphogenetic activity also have great affinity for methyl green, and vice versa. In order to explain his results, Weisz (1950) suggested that "different degrees of morphogenetic activity are correlated with different concentration levels of highly polymerized DNA." But he has since (1954) expressed doubts about the causal significance of this curious parallelism. Another interesting and recent observation of Weisz (1956) is that grafting an actively regenerating *Stentor* can induce regenerative reorganization in an intact, parabiotically connected organism; the process is irreversible from the start. Since regeneration in *Stentor* is inhibited by acriflavine, it is likely that nucleic acids or some of their derivatives play an important part in the process (Weisz, 1955).

According to a recent and extensive study of Tartar (1956), it is only the development of the oral primordium that requires the continuous presence of the nucleus: anucleate fragments of *Stentor* survive as long as starving nucleate pieces and they retain ciliary activity and contractility of the myonemata. Grafting of a nucleus from a different species into an anucleate fragment may permit the formation of an oral primordium, but does not allow the continued life of the cell.

In the case of the human HeLa cells, removal of the nucleus has no effect on cytoplasmic activity for 40 hours; shrinkage and loss of motility are observed afterward, according to a brief report of Crocker *et al.* (1956).

For plant cells, by far the most important experimental work is that of Hämmerling and his co-workers (review by Hämmerling in 1953) on the unicellular alga *Acetabularia mediterranea.* Its life cycle is depicted in Fig. 2. The alga is made of a chloroplast-containing stalk and of rhizoids during most of its life. A single nucleus, with an extraordinarily developed and basophilic nucleolus, is located in one of the rhizoids (Fig. 3a). Later, the tip of the stalk forms a "cap" or umbrella, which serves for the reproduction of the alga. When the umbrella is almost completely formed, the nucleus breaks down and small daughter nuclei spread through the whole alga (Schulze, 1939), including the cap, where resistant spores (the "cysts") are formed. These cysts contain a few nuclei and are surrounded by a thick capsule. After a maturation period, they can be stimulated to germination by short-time exposure to distilled water: the nuclei multiply and the cytoplasm divides; flagellated gametes are produced and escape out of the broken cyst. Copulation of the gametes is quickly followed by the growth of the zygote, which soon differentiates into rhizoid and stalk. The whole life cycle takes about 5 months under laboratory conditions, and one year in nature.

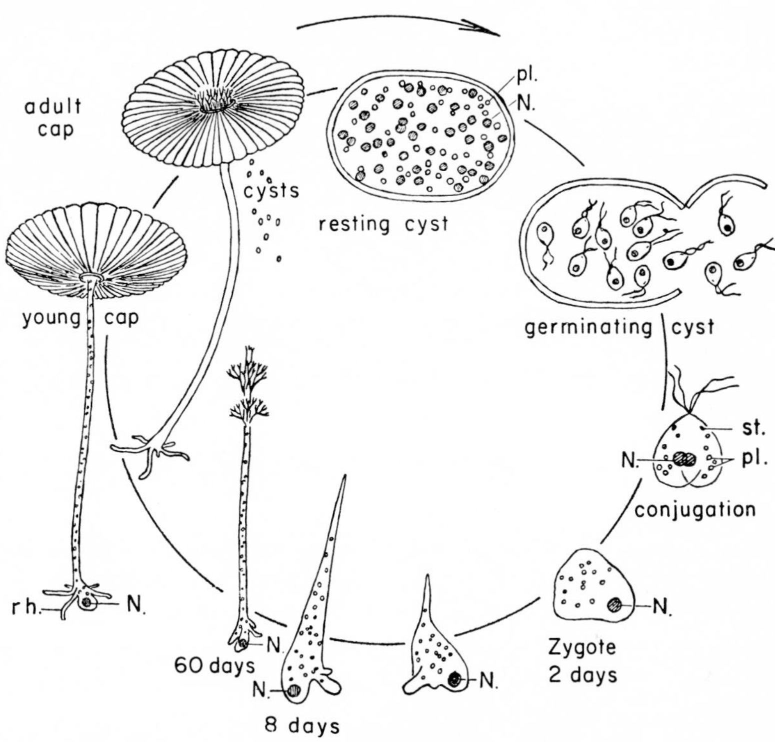

FIG. 2. Life cycle of *Acetabularia mediterranea*. *N.*, nucleus; *pl.*, chloroplasts; *st.*, stigma; *rh.*, rhizoid. From Brachet (1957).

One of the main results of Hämmerling (1934) is the demonstration that nonnucleated stalks survive for a very long time (several months); still more important is the fact that these anucleate parts are capable of important regeneration, including the formation of large-sized caps (Fig. 4). It was concluded by Hämmerling, as early as 1934, that the morphogenetic capacity of an anucleate part is determined by the amount of nucleus-dependent morphogenetic substances stored in it; these substances are distributed along an anterior-posterior concentration gradient.

Further analysis of the problem by Hämmerling (1943, 1946) and his colleagues (Beth, 1943; Maschlanka, 1946) includes very interesting ex-

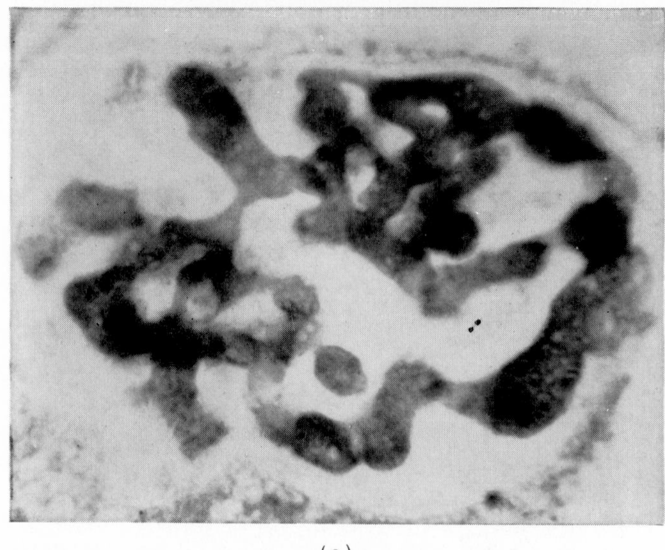

(a)

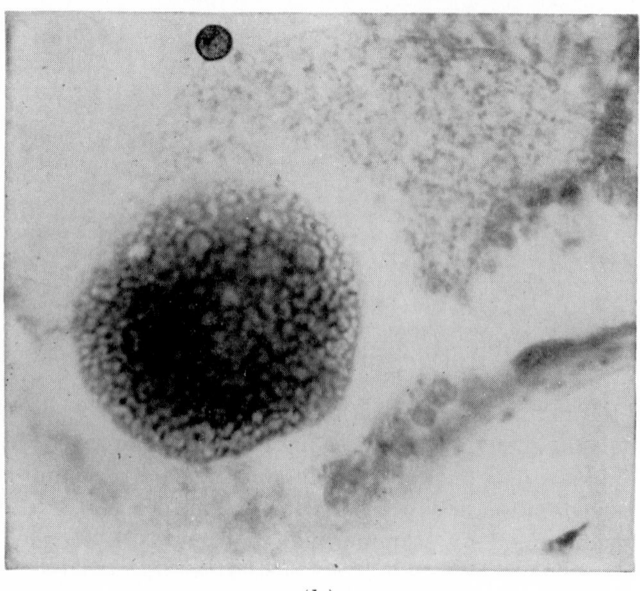

(b)

Fig. 3. a. Nucleus and nucleolus of normal *Acetabularia*. Unna staining. b. Nucleus and nucleolus of *Acetabularia* treated with dinitrophenol. Unna staining. From Brachet *et al.* (1955).

periments on interspecific grafts. For instance, binucleate grafts containing one *A. mediterranea* (*med*) and one *A. crenulata* (*cren*) nucleus form "intermediate" caps (Fig. 5); trinucleate grafts containing two *cren* and one *med* nucleus give, as would be expected, caps which more resemble normal *cren*. If now an *anucleate cren* stalk is grafted with a

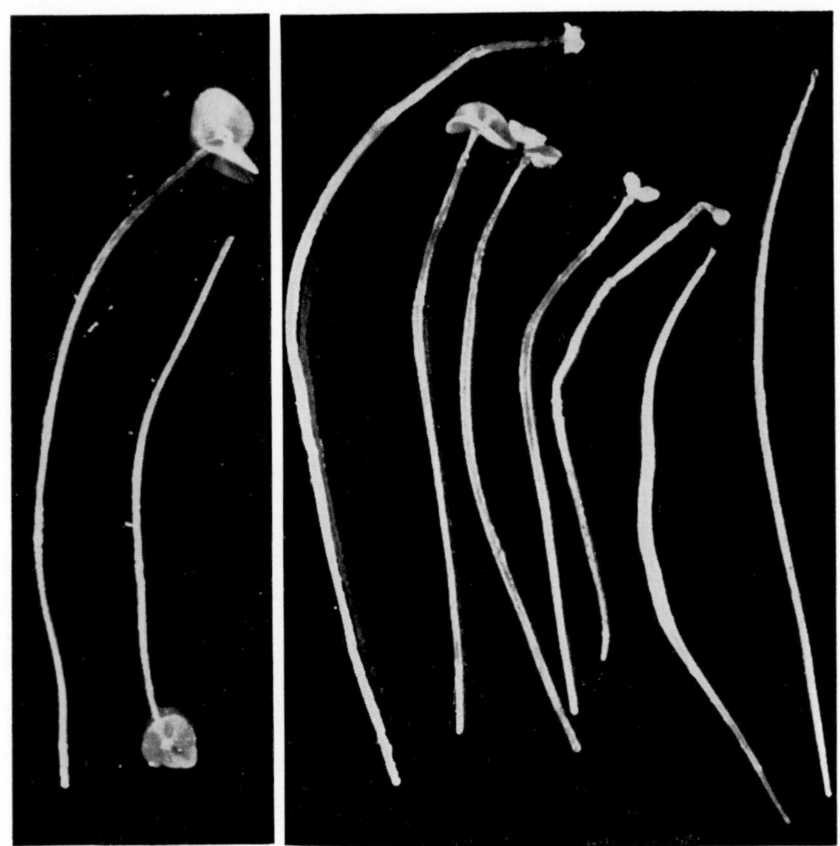

FIG. 4. Regeneration (cap formation) of anucleate stalks of *Acetabularia*. From Brachet *et al.* (1955).

med nucleate rhizoid, intermediate caps of various types are formed; if this first intermediate cap is removed, the new cap that forms will always be a typical *med* cap. These results have led Hämmerling (1953) to the following important conclusions: the nucleus-controlled morphogenetic substances show species specificity; in binucleate grafts, distinct substances will be produced by the two nuclei and intermediate caps of constant types will be formed. In uninucleate grafts, the anucleate cytoplasm contains a store of morphogenetic substances of its own species; if this store is large enough, an "intergrade" cap will be formed as a result of the competition between the substances stored in the anucleate piece and the substances produced by the grafted nucleus. Since the structure of the cap is obviously a hereditary character, it can be con-

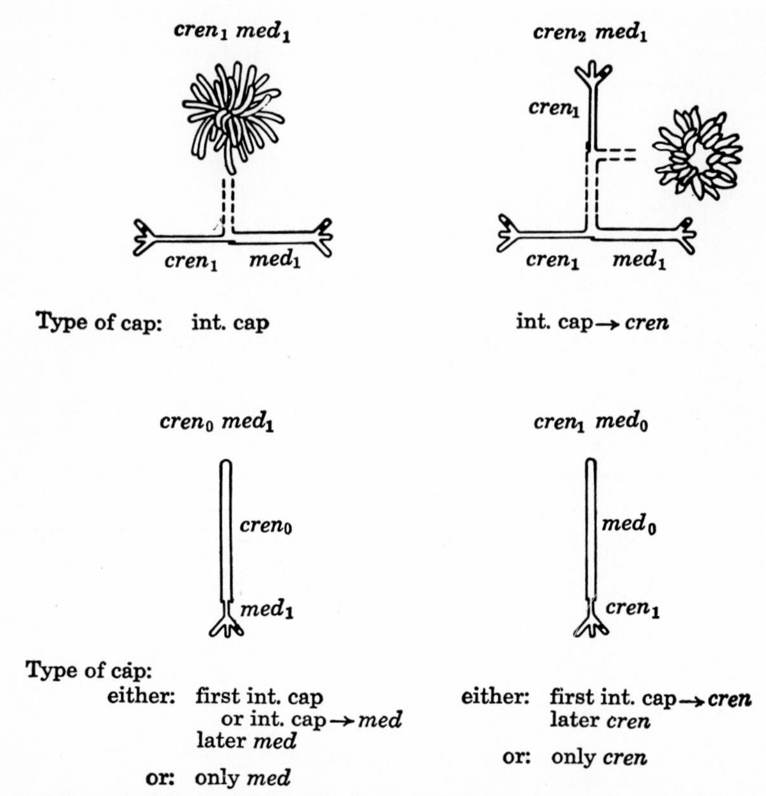

FIG. 5. Intermediate caps formed in binucleate grafts *Acetabularia mediterranea* × *A. crenulata*. From Hämmerling (1953).

cluded, with Hämmerling (1953), that the substances produced under the influence of the nucleus are "products of gene action, which stand between gene and character." It should, however, be borne in mind that, in these experiments, it is not the nucleus *alone* which is transplanted: cytoplasm surrounding the nucleus is also grafted, and it cannot be entirely excluded that cytoplasmic particles endowed with genetic continuity are also involved in the specific regenerative processes.

We have seen that, in amebae, Danielli's (1955) experiments strongly suggest that there are *mutual* interactions between nucleus and cytoplasm. Such interactions have been known since A. Brachet (1922), studying fertilization of unripe sea urchin eggs, discovered a phenomenon he called "la mise à l'unisson." In short, the sperm nuclei in these polyspermic eggs always take the same morphology as that of the egg nucleus; if the latter is dividing, the sperm heads remain condensed and are surrounded with asters. If the egg nucleus is in a swollen stage, the sperm nuclei also swell quickly. The morphology of the nuclei, in these eggs, is thus controlled by cytoplasmic events (Fig. 6). Comparable observations can easily be made on *Acetabularia* also: Stich (1951a, 1956b)

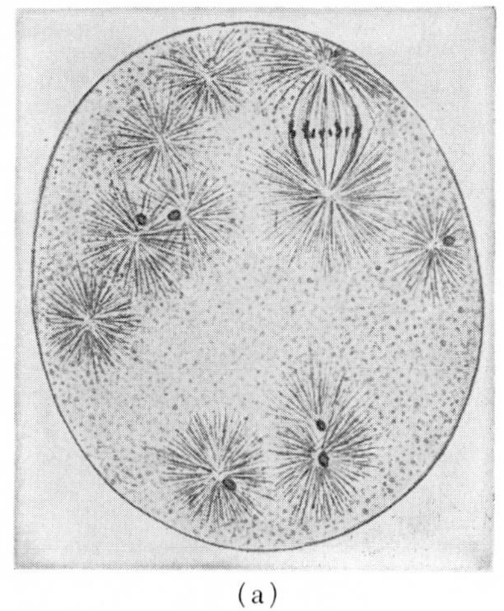
(a)

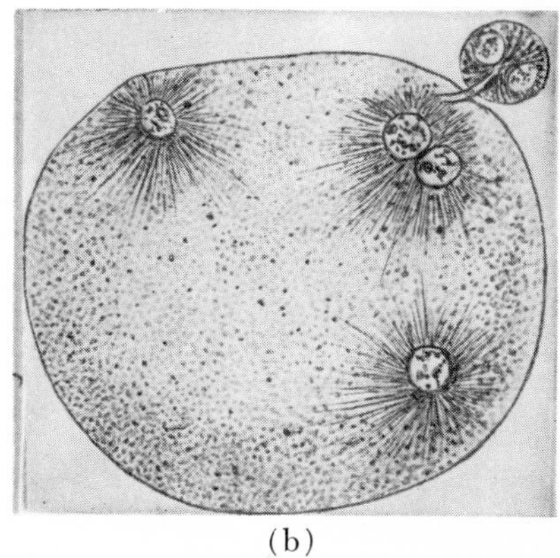
(b)

FIG. 6. Polyspermic sea urchin egg. a. Condensed sperm heads when the egg nucleus is dividing. b. Swollen sperm heads when the egg nucleus is in a swollen stage. From A. Brachet (1922).

found that the size, the shape, and the basophilia of the nucleoli are deeply modified when the algae are kept in the dark for 10 days (Fig. 3, b); the nuclear volume also markedly decreases under these experimental conditions. The phenomenon is entirely reversible in the light, and the whole experiment can be repeated several times. Similar observations have been made by the author (1952), who used poisons of oxidative phosphorylations (dinitrophenol, usnic acid) and anaerobiosis. Reduction of the energy production in the cytoplasm thus can have deep effects on the morphology and the chemical composition of the nucleus.

Elegant experiments of Hämmerling (1939, 1953) have also shown that the division of the nucleus depends on cytoplasmic conditions. If a large cap is removed just before the nucleus is going to divide, nuclear division is postponed until a new cap is formed. If the operation is repeated, mitosis can be delayed indefinitely. Conversely, if a young nucleate part (rhizoid) is grafted on a plant containing a large cap, division may begin as early as 2 weeks after grafting, instead of at the normal time of about 2 months.

There is some reason to believe that RNA is involved in the regeneration of anucleate parts of *Acetabularia.* For instance, trypaflavine, which strongly combines with nucleic acids, inhibits regeneration in both nucleate and anucleate pieces of *Acetabularia* (Stich, 1951a; Brachet *et al.*, 1955; Werz, 1957a). Regeneration of anucleate halves is very susceptible to ultraviolet (UV) irradiation at 2537 A., which is strongly absorbed by nucleic acids; X-rays, on the contrary, toward which RNA is much more stable, have little effect on regeneration in anucleate pieces (Brachet *et al.*, 1955; Bacq *et al.*, 1955, 1957; Hämmerling, 1956; Six, 1956; Errera and Vanderhaeghe, 1957; Skreb-Guilcher and Errera, 1957). More direct biochemical evidence pointing toward the same direction will be presented later.

Although the case of *Acetabularia* is somewhat exceptional, it should be pointed out that, some fifty years ago, Van Wysselingh (1908) insisted upon the importance of the biological activities persisting in nonnucleated fragments of *Spirogyra:* not only do they survive for several weeks, but photosynthesis, plastid formation, fat and tannic acid production, protoplasmic streaming, and even increase in cell length are said to continue. These old observations of Van Wysselingh (1908) have been substantiated recently by Yoshida (1956): working with plasmolyzed *Elodea* cells, he observed synthesis of chlorophyll and starch in anucleate fragments. Recently, Hämmerling (1959) also has emphasized the interest of these old, almost forgotten observations of Van Wysselingh (1908), and he has pointed out the many similarities existing between anucleate fragments of *Spirogyra* and *Acetabularia.*

Such a considerable resistance to the removal of the nucleus was, however, not found in experiments of Waris (1951) on the desmid *Micrasterias.* This organism, which possesses a complex and characteristic pattern of symmetry, can be cut into two parts by centrifugation; the anucleate half survives only 1 day under normal conditions and its cytoplasm soon shows signs of degeneration. In the presence of calcium at pH 4, the survival time can, however, be prolonged up to 9–10 days. The permeability remains normal and cyclosis movements are accelerated. The most significant finding of Waris (1951) is that the anucleate half of the cell can build a new membrane, which contains little cellulose and which is rich in a viscous, metachromatic, pectic substance. Several substances, the most active of them being DNA, increase the size of the lobes of the cell and thus have "formative" effects. Since there is no growth and since amino acids have no effect on survival time, Waris (1951) concludes that synthesis of proteins is quickly stopped; cellulose synthesis is also inhibited, and Waris (1951) suggests that the anucleate half-cell utilizes amino acids for the production of pectic substances. The remarkable effect of calcium might be the result of the hardening of the newly formed pectic membrane in the presence of this ion.

The biological experiments which have just been reviewed clearly show that great variations exist when different materials are compared; biological activities are never immediately or entirely suppressed when the nucleus is removed. In certain cases, anucleated cytoplasm is still capable of growth, for a certain time at any rate. We shall try to find out whether growth, in these cases, is really a corollary of syntheses, especially of protein synthesis, continuing in the absence of the nucleus. But, before we go into that, a short discussion of the hypotheses which have been proposed to explain the biochemical role of the cell nucleus might be useful.

III. Hypotheses on the Biochemical Role of the Nucleus

One of the oldest theories ever proposed to explain the importance of the nucleus in regeneration is that of Loeb (1899). He thought that the nucleus might be the main *center of cellular oxidations.* The fact that the indophenol oxidase reaction is often positive around the nucleus has been taken as an argument for Loeb's (1899) theory; it should be pointed out, however, that this reaction is never given by the nucleus itself.

Discussing the facts known in 1925, Wilson concluded that "the nucleus may be a storehouse of enzymes, or of substances that activate the cytoplasmic enzymes and that these substances may be concerned with synthesis as well as with destructive processes." Stated in more modern and biochemical terms, it might be said that, in Wilson's (1925) opinion, the nucleus is a site for *enzyme and coenzyme synthesis or accumulation.*

Another view concerning the biochemical role of the nucleus is that already held in 1892 by Verworn: the nucleus is the center of the synthetic activities of the cell, of its anabolism. Verworn's (1892) idea has been made more precise by Caspersson (1941, 1950), who considers the nucleus as a cell organelle especially organized as the *main center for the formation of proteins.* Caspersson's (1941, 1950) theory, which will be discussed in some detail at the end of this chapter, is based especially on his own cytochemical observations: they established that the nucleolus is well developed and rich in RNA in all cells which are the site of intensive protein synthesis. This finding has since been abundantly confirmed. According to Caspersson (1941, 1950), cytoplasmic RNA accumulates around the nuclear membrane; that such a perinuclear accumulation of RNA is a universal phenomenon appears somewhat doubtful now.

More recently (1952), Mazia proposed the new hypothesis that the nucleus is concerned with the *replacement* of the activities of the cell. Mazia's (1952) hypothesis is based on the essentially correct observation that removal of the nucleus is not followed by *immediate* effects on cellular activities; the latter decrease, however, sooner or later. For Mazia (1952), two alternatives are possible. In the first, the nucleus is the site of enzyme synthesis, as in Wilson's (1925) hypothesis. Removal of the nucleus should then lead to a continuous drop of the enzyme content in the cytoplasm. Such a drop would not necessarily occur at the same rate for all cytoplasmic enzymes, since the latter might be synthesized in different parts of the nucleus. It is also possible, according to Mazia, that enzymes which are functionally related decline together, if their replacement mechanisms in the nucleus are somewhat related. Still another possibility is that the replacement of coenzymes is a function of the nucleus.

In the second of Mazia's alternatives, the product of nuclear activity might be a cytoplasmic unit. The latter, which would be comparable to a plasmagene, would play a role in cytoplasmic synthesis; but it would require continuous replacement by the nucleus for its maintenance in the cytoplasm. A similar idea had been expressed before by Wright (1945) and by Marshak (1948); one of its consequences is that the losses of activity, after removal of the nucleus, would be discontinuous instead of gradual. The discontinuities would occur at the time when the synthetic units were exhausted.

It is possible to test such ideas, to a certain extent, by comparative work done on the biochemical activities of nucleate and anucleate fragments of *Amoeba* and *Acetabularia.* Another valuable material is the egg of either the sea urchin or the frog: as shown first by Harvey (1932), fast centrifugation of unfertilized sea urchin eggs, when they are suspended in a medium which has the same density as their own ("isopycnotic" medium), results in the sedimentation of the cytoplasmic particles present in the eggs. Ultimately, the eggs break into two parts (Fig. 7): the light, clear half contains the nucleus, the fat droplets, and most of the hyaloplasm with its basophilic granules; the heavy half is anucleate, containing most of the mitochondria and all of the yolk and pigment granules.

In amphibian eggs, enucleation is easy to perform by ligation or pricking of the maturation spindle.

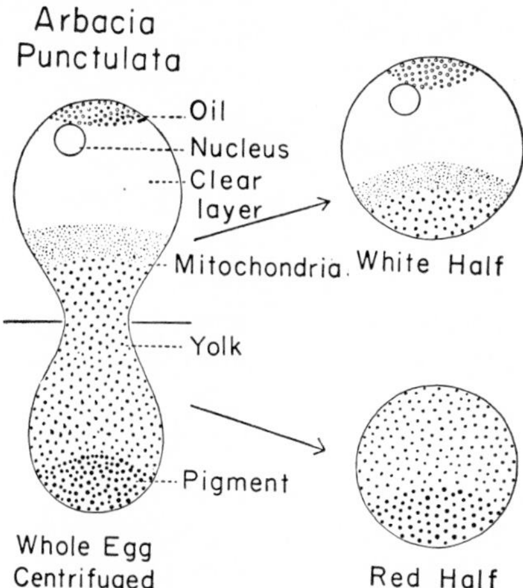

FIG. 7. Ultracentrifugation of unfertilized sea urchin egg. From Harvey (1932).

IV. THE NUCLEUS AND CELLULAR OXIDATIONS

The first experiments made in order to measure directly the oxygen consumption of nucleate and anucleate fragments of cells are those of Shapiro (1935). Working on unfertilized sea urchin eggs separated into two halves by Harvey's (1932) method, he found that the oxygen consumption is much higher in the anucleate heavy fragments than in the nucleate ones (Fig. 8). The experiments of Shapiro (1935) have been extended by Ballentine (1939), who studied the distribution of the dehydrogenases in the same material. He found that the anucleate part possessed about 70% more dehydrogenase activity than the nucleate fragments of equal volume. A high proportion of the dehydrogenases is thus bound to granules displaced by high speed centrifugation, and it is evident that the nucleus by itself can contain only small amounts. These results are enough to cast considerable doubt on Loeb's (1899) theory and they stand in excellent agreement with the results obtained with homogenates: essential respiratory enzymes are bound to large cytoplasmic particles, presumably the mitochondria.

Similar observations were made by the author (1939) on frog oöcytes: enucleation by pricking of the oöcyte produced only negligible decrease in the elimination of CO_2 and the oxygen uptake.

The oxygen consumption of nucleate and anucleate fragments of

Amoeba proteus has been studied in detail (Brachet, 1955a), because earlier results of Clark (1942) indicated that the nucleus plays a very important part in the respiration of these organisms. According to Clark (1942), who used a very sensitive—but not very reliable—method, removal of the nucleus produces a marked and rapid decrease of the oxygen consumption; in particular, oxidations sensitive to cyanide (i.e., those mediated by the cytochrome-cytochrome oxidase system) would be quickly inhibited in the absence of the nucleus. Our own experiments,

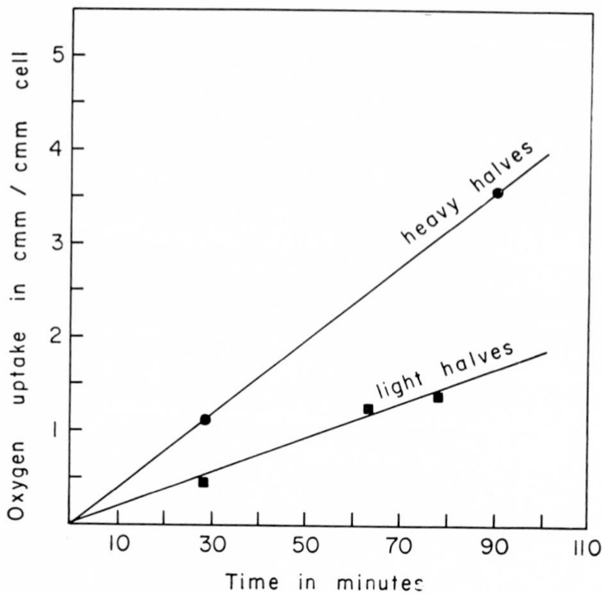

FIG. 8. Oxygen consumption of light and heavy halves of unfertilized sea urchin eggs. After Shapiro (1935).

which were performed with the Cartesian diver technique, showed that, on the contrary, removal of the nucleus exerts very little effect on the respiratory rate, for 1 week at least (Fig. 9). By that time, some of the anucleate fragments began to cytolyze and the differences between the two halves became more conspicuous.

The same results have been obtained with *Acetabularia* (Brachet *et al.*, 1955): the oxygen consumption *increases* with time in the anucleate as well as in the nucleate fragments, even after 3 weeks. In the latter, where growth and regeneration are more active, the increase is, however, larger than in the anucleate stalks (Figs. 10). Trypaflavine which, as already seen, inhibits the regeneration of both types of fragments

(Stich, 1951a), has no effect on respiration at first; after a longer treatment (4 days), a 60% inhibition of the oxygen consumption is obtained in both nucleate and anucleate halves, the effect being largely reversible on subsequent washing (Brachet *et al.*, 1955; Werz, 1957a). There is, however, a difference between the two halves when their oxygen con-

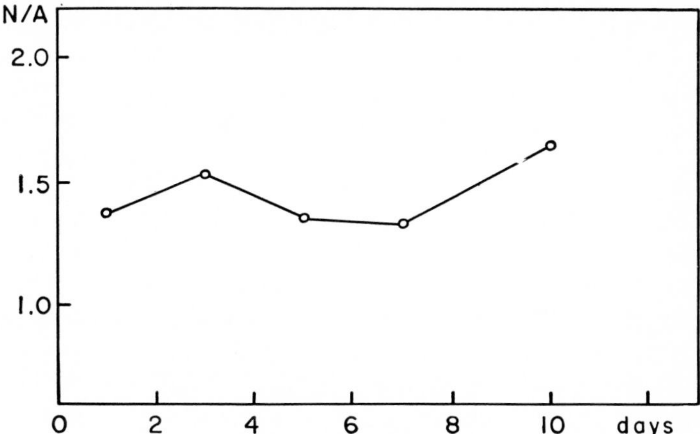

FIG. 9. Ratio between nucleate and anucleate fragments (*N*:*A*) for the oxygen consumption of amebae. Changes with time. From Brachet (1955a).

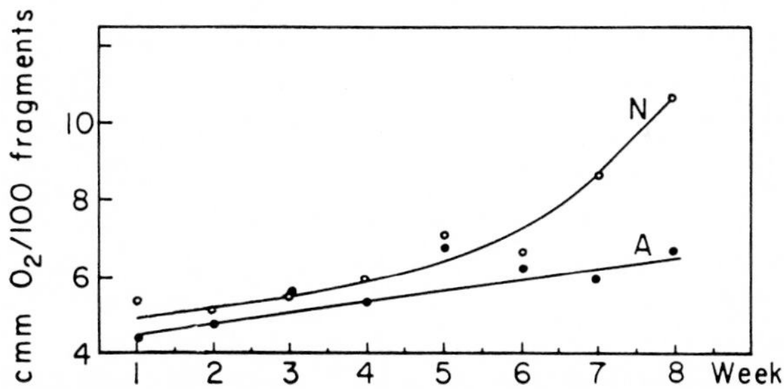

FIG. 10. Oxygen consumption of nucleate (*N*) and anucleate (*A*) halves of *Acetabularia*. From Brachet *et al.* (1955).

sumption is measured in the presence of sodium bicarbonate: a marked inhibition is found, in the anucleate pieces only, provided the concentration becomes higher than 10^{-2} *M*. The reasons for this difference are still unexplained (Brachet *et al.*, 1955).

According to a brief and recent report of Whiteley (1956), the situation in *Stentor* might be somewhat different from that found in amebae: Whiteley (1956) observed a drop in the oxygen consumption of the anucleate fragments, which did not regenerate. On the other hand, nucleate fragments regenerated after 1 day and showed an increase in the oxygen consumption during that initial period. The respiratory rate came back to normal afterward. Whiteley (1956) concluded from these experimental results that stimulation of the macronucleus tends to form the respiratory machinery of the whole organism.

The experimental results obtained with amebae and *Acetabularia* leave no doubt that Loeb's (1899) old theory is incorrect: the nucleus cannot be the main center of cellular oxidations. Furthermore, the fact that the oxygen consumption remains essentially normal for a long time after removal of the nucleus (10 days for amebae, 3 months for *Acetabularia*) clearly shows that the nucleus does not exert any appreciable control on the activity of the respiratory enzymes: it looks as if the mitochondrial enzymes were remarkably independent of nuclear control.

It is needless to say that these conclusions are in complete agreement with all that is known about the chemical composition of isolated nuclei and mitochondria. The results obtained on fragments of unicellular organisms confirm the view that nuclear metabolism is largely anaerobic (Stern and Mirsky, 1952; Stern and Timonen, 1954; Stern, 1955); on the other hand, they do not stand against the more recent observations of Allfrey and Mirsky (1957) showing that isolated thymus nuclei are capable of oxidative phosphorylations (provided their nucleic acids remain intact), since these energy-producing reactions are of quantitatively small importance in the nuclei. The observations reported earlier in this chapter on the dependence of nuclear morphology and chemical composition on the maintenance of a normal energy level in the cytoplasm also support the same view. As a matter of fact, one may wonder whether the attachment which, according to Frederic (1954), often occurs between mitochondria and nuclear membrane might not represent a mechanism for energy transfer (presumably in the form of ATP) from the mitochondria to the nucleus.

If the maintenance of mitochondrial activity in the absence of the nucleus is a general phenomenon in unicellular organisms and in eggs, it should be said, however, that the situation is different when the nucleus *normally* disappears at a stage of cellular differentiation. We refer to the well-known maturation of the red blood cells in mammals, where the immature reticulocytes have lost their nucleus, but retained both mitochondria and ergastoplasm. In the mature red blood cell, mitochondria and the basophilic "reticulofilamentous" substance have both disappeared (Chalfin, 1956). It is not surprising therefore that, as found by Warburg in his 1909 classical experiments, reticulocytes respire much more than the mature red blood cells. Later studies by Rapoport and Hofmann

(1955) have shown that, as would be expected, reticulocytes possess, in contrast to mature red blood cells, an active tricarboxylic cycle; cytochrome oxidase and succinic oxidase are also much more active than in the adult red blood cells, which have a high glycolytic activity (Rubinstein *et al.*, 1956).

But it is doubtful that these changes depend directly upon the disappearance of the nucleus; it is more probable that they are a result of the highly specialized function of the erythrocytes, because similar changes also occur in the *nucleate* red blood cells of the birds and amphibians: the chondriome strongly diminishes and may finally disappear completely, while the content of hemoglobin increases in the nucleate amphibian erythrocytes (Rojas and De Robertis, 1936). As a possible result of this particular evolution of the mitochondria, respiratory enzymes become concentrated in the nucleus or in the cell membrane, according to the chemical studies of Rubinstein and Denstedt (1955) and the cytochemical observations of Defendi and Pearson (1955) on mature bird erythrocytes.

We shall have to refer again, in this chapter, to the reticulocytes; it is good to remember that they are very different, because of their exceptional specialization for hemoglobin synthesis, from eggs or unicellular organisms which have been experimentally deprived of their nucleus.

V. Qualitative Changes in the Metabolism of Nonnucleated *Amoeba* Fragments

It is only in amebae that enough is known about the "general metabolism" of anucleate fragments to allow a fruitful discussion; work on other organisms is badly needed before any generalization can be attempted.

Cytochemical observations on the distribution of glycogen, lipids, and proteins (the latter being studied with the Millon reaction for tyrosine and the Serra reaction for arginine) have shown that considerable individual differences exist in one and the same culture (Brachet, 1955a). There are, however, a few general trends: the utilization of glycogen and fat reserves begins normally in the fasting fragments; but, after a few days, it stops in the anucleate halves (Fig. 11). After a week's time, almost all the anucleate fragments are still very rich in carbohydrate and fat reserves, while the majority of the nucleate halves have nearly exhausted these reserve stores. In regard to the proteins, no striking differences can be found between the two types of fragments, except for the fact that the arginine reaction is much stronger in the

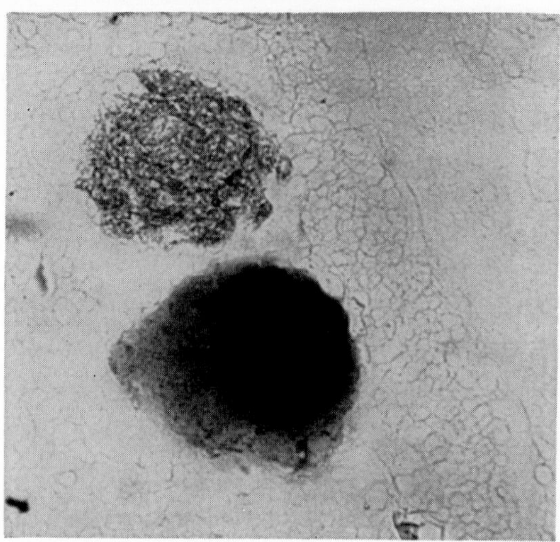

FIG. 11. Lack of utilization of the glycogen reserves in the starving anucleate half (below). From Brachet (1955a).

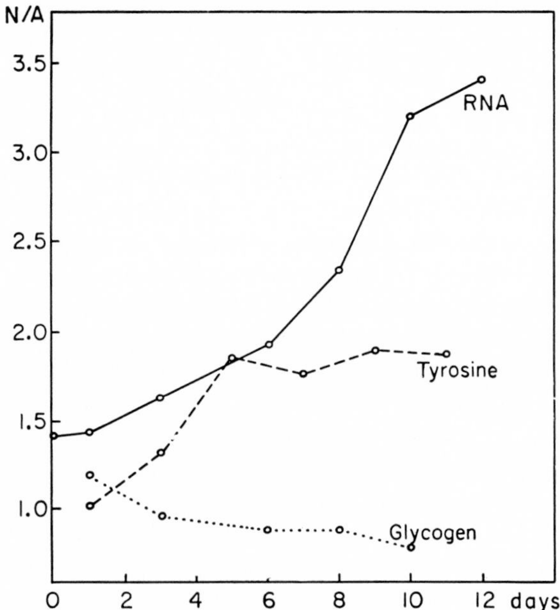

FIG. 12. Changes in RNA, protein (tyrosine), and glycogen in nucleate and anucleate ameba halves. $N:A$ ratio as in Fig. 9. From Brachet (1955a).

nucleus (both in the central chromatin and in the peripheric ring of nucleoli) than in the cytoplasm.

Quantitative estimations of the glycogen and total protein content (Brachet, 1955a) confirm the cytochemical observations. For glycogen (Fig. 12), it was found that the utilization is normal in the anucleate fragments for 3 days; utilization then stops completely, but the glycogen content of the nucleate halves still slowly decreases. In *Stentor* also, utilization of the carbohydrate reserves is considerably (50%) inhibited in anucleate fragments (Tartar, 1956). On the other hand, the protein content does not decrease much (15%) in an 11-day fasting experiment with amebae; but, in the anucleate fragments, a marked decrease (almost 50%) occurs between the third and the fifth day after removal of the nucleus (Fig. 12).

It is to be concluded that, while the over-all metabolism (measured by the oxygen consumption) remains essentially normal, removal of the nucleus induces in amebae deep qualitative metabolic changes: increased utilization of proteins coincides with the disappearance of glycogenolysis. The anucleate half, when it becomes unable to utilize its normal reserve stores, is the site of a kind of "autophagy." But, in either case, nuclear control is not immediate: a lag period of 3 days is necessary before carbohydrate and protein metabolisms are modified.

VI. Phosphate and Coenzyme Metabolism

We owe to Mazia and Hirshfield (1950) a very significant observation: the uptake of radioactive phosphate is very quickly decreased in the anucleate halves. Even a few hours after the operation, this uptake is reduced to 25–30% of normal. The main interest of Mazia and Hirshfield's (1950) results lies in the fact that this change of phosphate metabolism occurs so quickly after removal of the nucleus. The inhibition of the phosphate uptake in the anucleate halves is perhaps correlated with the loss of pseudopod formation and the change in shape. However, it is the opinion of Mazia and Hirshfield (1950) that the phenomenon is not a result of a nonspecific decrease of permeability in the anucleate halves; nor can it be explained on the basis of an unusually high uptake of phosphate by the nucleus. One cannot exclude, however, that the penetration of phosphate in amebae is an active process, linked to the metabolism of the membrane which might be altered when the nucleus has been removed.

Mazia and Hirshfield's (1950) results have been entirely confirmed by D. Thomason in this laboratory (see Brachet, 1955a). When amebae are placed in radioactive phosphate, washed, and cut into halves, there is no impressive preferential accumulation of P^{32} in the nucleate part; it can be concluded that the nucleus does not markedly accumulate the labeled phosphate. On the other hand, when freshly cut amebae are placed in a radioactive phosphate-containing medium, the uptake is two to three times higher in the nucleate than in the anucleate halves: the difference between the two types of fragments becomes six times after 3 days and more than thirty times after 6 days. There is thus no doubt that, in confirmation of Mazia and Hirshfield's results and conclusions, removal of the nucleus deeply upsets phosphate metabolism.

In some respects, anucleate cytoplasm resembles cells which have been treated with poisons of oxidative phosphorylations, with dinitrophenol for instance. In both cases, uptake of phosphate is strongly inhibited while respiration remains normal or increases. This situation has led to the hypothesis (Brachet, 1951) that removal of the nucleus might uncouple oxidations from phosphorylations; the nucleus might play a part in this coupling, by the production of a coenzyme, for instance.

If this hypothesis were correct, one would expect a marked drop in the ATP content of the anucleate *Amoeba* halves. Experiments designed to test this prediction have shown that, on the contrary, the ATP content is somewhat *higher* in the anucleate halves than in their nucleate counterparts. This increase probably reflects an imperfect utilization of ATP in the nonnucleated cytoplasm, where synthetic processes are limited (Brachet, 1955a).

Experiments on the ATP content of *Amoeba* fragments placed in anaerobiosis disclosed an interesting fact: as shown in Fig. 13, the drop in the ATP content is much larger and much faster in the anucleate halves than in the others. Removal of the nucleus thus has as a consequence a marked decrease in the capacity to maintain ATP in its phosphorylated form, in anaerobic conditions. The exact meaning of these observations is not yet clear; but, since glycolysis is usually the main ATP-producing reaction in anaerobiosis, it is likely that deficient glycolysis is responsible for the inferiority of the anucleate cytoplasm in the maintenance of the anaerobic ATP level. Such a conclusion is in obvious agreement with the aforementioned reduced utilization of glycogen in amebae that have been deprived of their nuclei.

We do not know yet why the carbohydrate metabolism of anucleate amebae becomes abnormal; it might be owing to the loss of one of the glycolytic enzymes or of a coenzyme. It might even be caused by insufficient amounts of a simple cofactor, inorganic phosphate for instance.

Too little is known about the pathway of carbohydrate metabolism in amebae to warrant drawing any definite conclusion from the scanty experimental data at our disposal; nothing is known, in amebae, about the end products of glycolysis. Hexokinase and phosphorylase could not be detected in *Amoeba proteus*, even with sensitive methods. Enolase, on the other hand, is present and has been studied in amebae fragments (Brachet, 1955a).[2] In contrast with Stern and Mirsky's (1952) observations on isolated nuclei of vegetal cells, this enzyme is not accumulated in the nucleus in *Amoeba*. Furthermore, there is no appreciable drop in the enolase content of the anucleate fragments, even 11 days after the operation. The inhibition of carbohydrate metabolism in anucleate halves therefore cannot be explained on the ground of an enolase deficiency. A

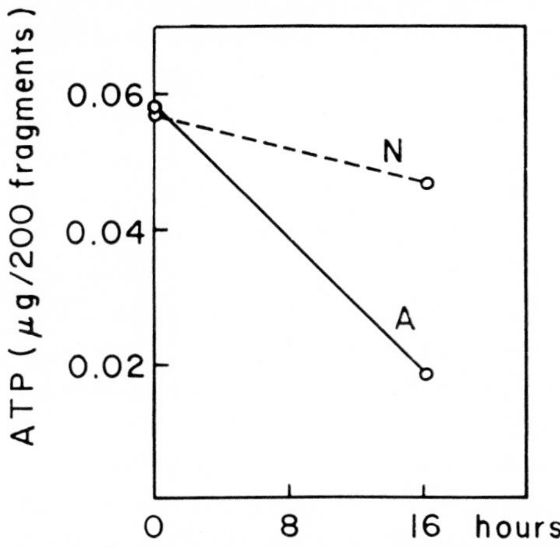

FIG. 13. Drop in the ATP content of nucleate (*N*) and anucleate (*A*) *Amoeba* halves when placed in anaerobiosis. From Brachet (1955a).

deficiency in amylase is also unlikely: the enzyme is abundantly present in amebae (Holter and Doyle, 1938); but, as shown by Urbani (1952), nucleate and anucleate halves retain equal amounts of amylase, even in 11-day experiments. There is, however, a peculiarity, which will be mentioned again later, in the behavior of amylase in the anucleate halves: the enzymatic activity markedly *increases* during the 3 days that follow the section, and then goes back to normal (Urbani, 1952).

A more likely explanation for the deficiency of carbohydrate metab-

[2] Recently, Borner and Mattenheimer (1959) have demonstrated the presence, in the large multicellular ameba *Chaos chaos*, of a great number of enzymes involved in carbohydrate metabolism (hexokinase, glucose-6-phosphate dehydrogenase, hexosephosphate isomerase, aldolase, glyceraldehydephosphate dehydrogenase, enolase, pyruvate kinase, isocitric dehydrogenase, fumarase, malic dehydrogenase). It would be of great interest, if these enzymes are also present in *Amoeba proteus*, to follow their fate after removal of the nucleus.

olism in nonnucleated *Amoeba* cytoplasm might be found in a suppression of diphosphopyridine nucleotide (DPN) formation. We know from the work of Hogeboom and Schneider (1952) that this very important coenzyme is synthesized in the nuclei (in liver); nonaqueous liver nuclei contain large amounts of DPN (Stern *et al.*, 1952). According to Baltus (1954), the enzyme which synthesizes DPN is highly concentrated in the nucleoli of starfish oöcytes. Furthermore, it is an interesting fact that this enzyme is present in the nucleate red blood cells of birds, but not in the anucleate reticulocytes and adult red blood cells of mammals (Malkin and Denstedt, 1956). If the synthesis of DPN is a function of the nucleus in amebae also, one would expect enucleation to produce a drop in the DPN content of the cytoplasm. This might lead to a decrease in the glycolytic rate. A DPN deficiency in anucleate cytoplasm would not necessarily produce a decrease in respiration, provided we assume that free DPN can be broken down by hydrolytic enzymes (DPNase) present in the cytoplasm, while DPN bound to mitochondria would be protected.

Experiments made to test this hypothesis have led to conflicting results: first, the DPN-synthesizing enzyme could not be detected with present methods in amebae, so that its intracellular localization unfortunately could not be ascertained. The next thing to do was to follow the DPN content of nucleate and anucleate *Amoeba* fragments. Using the fluorescence method of Burch *et al.* (1955), Baltus (1956) found a marked decrease in the DPN content of the anucleate halves during the 3 days following the section. The decrease was 22% after 25 hours, 38% after 50 hours, and 56% after 75 hours. But entirely different results have been simultaneously reported by A. I. Cohen (1956), who used another method. He found that almost all the DPN is present in the oxidized form in amebae, and that there is no change in the DPN level in the nonnucleated fragments for 6 days. No conclusion can be drawn until the reasons for these discrepancies, which are probably due to differences in the methods used, are made clear.

It should, finally, be added that the fate of glycogen is the same in normal nucleate fragments and in anucleate fragments which had been enriched in DPN by cultivation in a mixture of plasma albumin and DPN (Baltus, 1959a): under these conditions, the DPN content approximately doubles. Nevertheless, the glycogen decrease is of the same order of magnitude in the two types of fragments: either DPN which has been ingested by pinocytosis is useless for the amebae, or the reason for the loss of glycogen in anucleate *Amoeba* halves must be looked for elsewhere.

The observations made on carbohydrate and phosphate metabolism in fragments of *Acetabularia* give a very different picture from that obtained with amebae: this is not surprising, since the metabolism of this alga is of course dominated by photosynthesis. As shown by the work of Arnon and his colleagues (1956), isolated chloroplasts are capable of photosynthetic phosphorylation: in the absence of mitochondria and in the light (but not in the dark), they vigorously synthesize ATP from inorganic phosphate and adenosine monophosphate. Experimental conditions, especially intensity of illumination, will thus be of extreme importance for the interpretation of the results obtained on phosphorus and carbohydrate metabolism in *Acetabularia*.

A first fact is that photosynthetic activity remains entirely normal in anucleate pieces of *Acetabularia* for at least 4 weeks (Brachet *et al.*, 1955): in agreement with this finding, experiments of Vanderhaeghe (1957) indicate that the metabolism of starch is essentially the same in nucleate and anucleate stems, whatever the experimental conditions adopted (light, darkness, presence or absence of oxygen).

In regard to radioactive phosphate uptake, experiments by Brachet *et al.* (1955) have shown that, in algae operated on a considerable time before (3 months), striking differences between the two types of fragments can be found only in young algae; only relatively small differences (60%) in favor of the nucleate halves were found when older, actively regenerating algae were used. More extensive experiments by Hämmerling and Stich (1954) have shown that, as might be expected, the P^{32} uptake is linked to photosynthetic activity; no differences were found between nucleate and anucleate fragments which had been separated for less than 4 weeks. An important point, which will be examined in more detail later, is that a very strong incorporation of P^{32} occurs in the nucleolus of the rhizoid, especially in its RNA (Stich and Hämmerling, 1953; Hämmerling and Stich, 1956). This process is dependent on the energy production in the cytoplasm, since it is inhibited in the dark (Hämmerling and Stich, 1956).

A peculiarity of phosphate metabolism in *Acetabularia* is the important role played by polyphosphates (metaphosphates). Polyphosphate granules were identified in *Acetabularia* in studies by Stich (1953) on the ground of their cytochemical reactions. He found that the synthesis of polyphosphates, as one could expect, is dependent on energy production by photosynthesis: the number and size of the polyphosphate granules decreases in the dark. They increase, on the other hand, when growth of nucleate or anucleate pieces is inhibited with trypaflavine. In a more recent paper, Stich (1955) has confirmed his earlier observations and shown that the incorporation of P^{32} in the polyphosphates is dependent on photosynthetic and oxidative phosphorylation reactions; darkness, cyanide, and dinitrophenol quickly inhibit the process. The turnover of the polyphosphates is very rapid and these substances certainly play an important role in *Acetabularia.* Stich (1955) suggests that, since they contain energy-rich phosphate bonds, they may play a role similar to that of ATP; they could, for instance, be used for the phosphorylation of nucleotides. The fact that polyphosphates accumulate when regeneration is stopped, as a result of enucleation or trypaflavine treatment, suggests that they might be utilized for synthetic processes (Stich, 1955). One cannot exclude the possibility, however, that polyphosphates accumulate when suitable phosphate acceptors, ADP for instance, are lacking. A study of the acid-soluble nucleotide pool in *Acetabularia* is obviously required before any definite conclusion can be reached.

This comparative survey of phosphate and carbohydrate metabolism in *Amoeba* and *Acetabularia* clearly shows how different the two materials are. Several reactions of the cytoplasm to the removal of the nucleus are entirely different in the ameba and in the alga. In particular, Mazia and Hirshfield's (1950) findings on the reduced uptake of P^{32} in nonnucleated fragments of *Amoeba* cannot be generalized; the same is true for our own observations on the decrease in glycogen utilization in anucleate amebae. Such a conclusion is perhaps discouraging for those who hope to find some simple mechanism for nuclear control. But these comparative studies give us an important warning: no claim for a simple and unique mechanism of nuclear control can be made unless it is shown to be present in organisms as different as an alga and an ameba. We shall now see that these two organisms also react in very different ways to the removal of the nucleus when RNA and protein metabolisms are considered.

VII. The Role of the Nucleus in RNA and Protein Metabolisms

In view of the close links existing between RNA and protein synthesis (Caspersson, 1941; Brachet, 1941), RNA and protein metabolisms in parts of unicellular organisms will be studied together.

A. *Amoeba proteus*

Cytochemical observations with Unna staining show that the basophilia of the anucleate fragments begins to decrease a couple of days after the section (Brachet, 1955a). Even by the fifth day, the staining differences between the nucleate and anucleate halves are very conspicuous; the anucleate fragments become almost colorless in experiments of longer duration (Fig. 14). The cytological structure of these anucleate

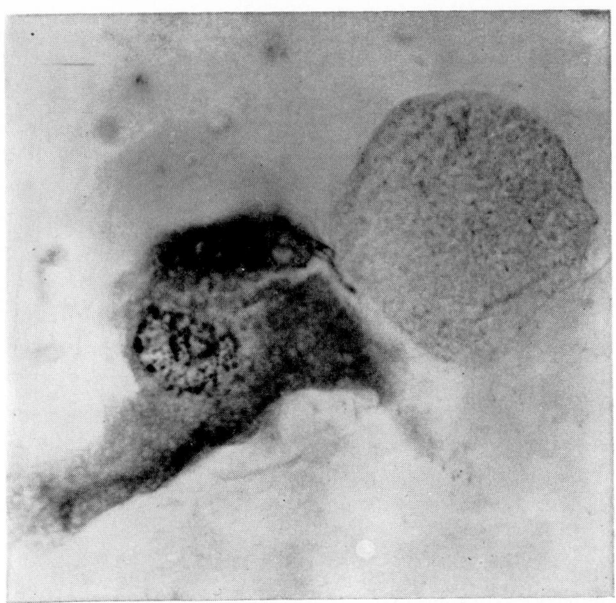

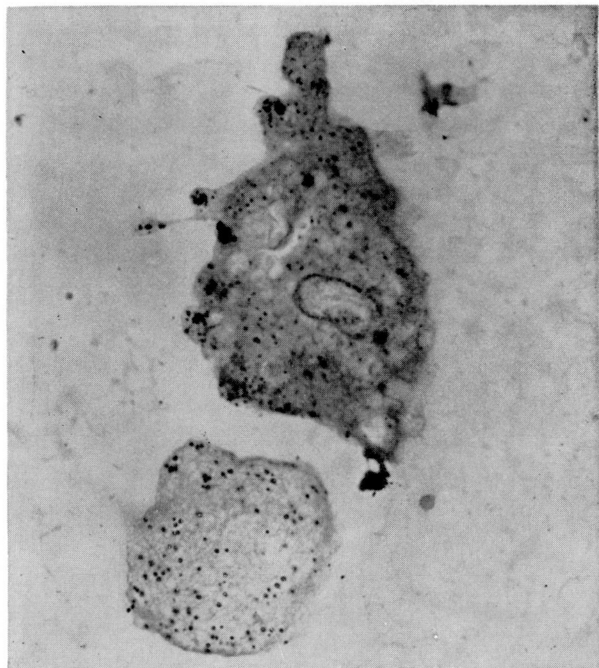

FIG. 14. Drop in basophilia occurring in anucleate fragments of amebae. From Brachet (1955a).

fragments is also changed; it is now finely granular rather than fibrillar after acid fixation. It is therefore likely that the removal of the nucleus exerts profound effects on the structure of the ergastoplasmic "small granules"; the more granular aspect of the ground cytoplasm in the anucleate halves might represent distortion or breakdown of the ergastoplasmic lamellae. Unfortunately, the effects of enucleation on the ultrafine structure of the cytoplasm have not yet been studied in detail with the electron microscope. Preliminary observations made with the electron microscope in Dr. Bernhard's laboratory indicate that removal of the nucleus has usually no visible effects on the structure of the mitochondria, although swelling of the latter can be found; the absence of a well-defined ergastoplasm in amebae makes the interpretation of the results concerning the RNA-containing structures very difficult (Brachet, 1959a).

Quantitative estimations of the RNA content (Brachet, 1955a) completely confirm the cytochemical observations. As shown in Fig. 12, the nucleate halves keep their RNA content constant, even after 12 days of fasting; in the anucleate cytoplasm, on the contrary, there is a steady and marked decrease in the RNA content: it drops by 60% within 10 days. These results have been confirmed by Prescott and Mazia (1954) and by James (1954). The latter claims, however, that the RNA content also drops when intact amebae are kept fasting. Therefore, the loss of RNA might not be a direct consequence of enucleation; but we could show (Brachet, 1955a) that the RNA content of fasting whole amebae remains constant when they are kept under such experimental conditions that they do not markedly decrease in volume.

These experiments lead to the conclusion that, in amebae, the nucleus exerts an important control on the maintenance of cytoplasmic RNA; since, in amebae, almost all the RNA is localized in the microsome fraction, it can even be said that the small RNA-rich granules of the ergastoplasm lie under close nuclear control; however, definite proof of this conclusion with the electron microscope has not yet been given.

Our experimental data on the RNA content of nucleate and anucleate *Amoeba* halves are obviously compatible with the idea that cytoplasmic RNA originates from nuclear RNA; they do not, however, prove that it is so, since RNA might disappear from nonnucleated cytoplasm for a number of conceivable reasons other than a nuclear origin. More direct experiments have been designed by Goldstein and Plaut (1955) to prove the nuclear origin of cytoplasmic RNA. They have strongly labeled amebae with P^{32} by feeding them with *Tetrahymena* cultivated in the presence of radiophosphate. The nucleus of the tagged amebae was then removed and grafted into normal unlabeled amebae or into anucleate halves. Autoradiographs showed clearly that the cytoplasm of the grafted amebae becomes radioactive after 12 or more hours. Utilization of the ribonuclease test further showed that, under the conditions adopted for the autoradiography (fixation in 45% acetic acid), all the autoradiographically detectable P^{32} in both nucleus and cytoplasm is in the form of RNA. When the tagged nucleus is grafted into a whole ameba, so as to produce a binucleate cell, it is found that the originally unlabeled nucleus does not acquire any significant amount of radioactivity. This last experiment shows that the cytoplasm does not supply RNA to the nucleus; the latter, therefore, synthesizes its own RNA. Once synthesized in the nucleus, RNA can be transferred to the cytoplasm and the transfer proceeds in that direction only.

These experiments of Goldstein and Plaut (1955) are of far-reaching importance for our understanding of the interactions between nuclear and cytoplasmic RNA. They certainly deserve a short critical discussion. There is no doubt that Goldstein and Plaut (1955) have good evidence for the view that RNA is synthesized in the nucleus and transmitted therefrom to the cytoplasm. However, as pointed out by the authors themselves, they have not proved that the labeled material migrating to the cytoplasm is the RNA as it actually existed in the nucleus: it might be a precursor, of a more or less complex nature. Goldstein and Plaut's (1955) demonstration that the labeled RNA (or its precursor) which has been passed from the nucleus into the cytoplasm cannot be used for further nuclear RNA synthesis does not seem entirely convincing. Proof that the RNA transfer can proceed from the nucleus only in the cytoplasm direction would be more complete if it had been shown that a nonlabeled nucleus grafted in a *strongly* labeled anucleate cytoplasm never becomes radioactive. In Goldstein and Plaut's (1955) experiments on binucleate amebae, the radioactivity of the cytoplasm seems to be rather weak and no unequivocal answer can be obtained from their observations.

An important additional remark is made by Goldstein and Plaut (1955): the possibility of the complete synthesis of *some* RNA in the cytoplasm is not ruled out by their data. In other words, it is possible that, besides a transfer of nuclear RNA to the cytoplasm, independent synthesis occurs in the latter. This eventuality seems to be a probable one in view of autoradiography experiments performed in this laboratory by Skreb-Guilcher (1956). Studying the incorporation of labeled adenine in nucleate and anucleate *Amoeba* halves, she obtained a measurable incorporation into the RNA of nonnucleated fragments separated for 1 and 3 days. The activity of these fragments was, however, about four times less than that of the nucleate halves. The incorporation in anucleate cytoplasm became negligible when, 8 days after the section, its RNA content had dropped markedly. Similar results have been reported by Plaut and Rustad (1956), who studied the uptake of adenine into *Amoeba* fragments and found that it is an effective RNA precursor in this organism. Their experiments show that the presence of the nucleus is not essential for the uptake of adenine; the nucleus is important, however, for this uptake during early incubation period, since the ratio between the nucleate and anucleate halves is 2.1:1.

More recently, Plaut and Rustad (1957, 1959) and Plaut (1959a, b) have reported results that agree perfectly well with those of Skreb-Guilcher (1956): using radioactive adenine as a label, they find that there is a cytoplasmic incorporation mechanism that can operate in the absence of the nucleus.

Similar results have been obtained, and the same conclusions have been drawn by Brachet (1959a, b), who studied the incorporation of $C^{14}O_2$ and of adenine into the RNA of nucleate and anucleate halves; both autoradiography and counting techniques were used in these studies, which showed that removal of the nucleus quickly reduces the incorporation process to a very low level; however, anucleate fragments always display some RNA synthetic activity, even when the nucleus has been removed 8 days before the incorporation experiment is performed.

However, recent autoradiography experiments of Prescott (1957) lead to different results and conclusions: according to him, there is no incorporation whatsoever of labeled uracil in anucleate halves: the whole of *cytoplasmic RNA* would thus, in amebae, *originate from the nucleus.*

Furthermore, new experiments by Prescott (1960) on another *Amoeba* species, *Acanthamoeba,* failed to demonstrate *any* incorporation of uracil, orotic acid, or adenine into the RNA of anucleate halves. The contradictory results obtained in *Amoeba proteus* would be due, according to Prescott (1960), to the presence of undigested, still living bacteria in the cytoplasm. The fact that, according to Plaut and Sagan (1958), thymidine can be incorporated into DNA in anucleate ameba *cytoplasm* is a warning that Prescott's (1960) objection is a serious one; future experiments of this type should obviously be carried on under strictly sterile conditions.

The main conclusion is that, in amebae, the nucleus is of considerable importance for RNA metabolism; in its absence, the RNA content of the cytoplasm drops markedly. There is good evidence, since Goldstein and Plaut's work, that the nucleus actively synthesizes RNA and that nuclear RNA is transferred to the cytoplasm. But the cytoplasm is probably not entirely inactive, and limited synthesis or turnover of RNA might continue in strictly anucleate cytoplasm, although this last point certainly deserves more study before it can be accepted. Such a picture is, of course, in very good agreement with Mazia's (1952) replacement hypothesis.

If removal of the nucleus produces, in *Amoeba,* a marked weakening of RNA metabolism, one would expect a parallel inhibition of protein anabolism, if both are really closely linked together. We have already seen that, in accordance with this expectation, the total protein content drops faster in anucleate halves than in nucleate fragments. A more precise and detailed analysis of protein metabolism has been carried out by Mazia and Prescott (1955), who studied the incorporation of S^{35}-labeled methionine into the proteins of both halves.

Their experiments show that the percentage of incorporation is lower by a factor of 2.5 in the anucleate half immediately after the cell has been cut into two; this percentage of incorporation (i.e., the relation between the total amount of methionine taken up and the amount incorporated into trichloroacetic acid-insoluble material in a given time) does not change in either half for 3 days after cutting. On the other hand, the amount of methionine which the anucleate half can incorporate into its proteins declines with time; this is attributed to the uptake mechanism and not to the incorporation mechanism. Mazia and Prescott (1955) conclude that the nucleus is either the seat of a considerable proportion of the protein synthesis in amebae or that this nucleus-linked synthesis is very closely coupled with processes that are localized in the nucleus. But the experiments also show that the nucleus is not an *exclusive* center of protein synthesis in the cell.

Autoradiography experiments of Ficq (1956) and Brachet and Ficq (1956), who used C^{14}-phenylalanine as a precursor instead of S^{35}-methionine, have essentially confirmed Mazia and Prescott's (1955) results. The differences observed were, however, less striking since the ratios between nucleate and anucleate halves were in the neighborhood of 2 (instead of 6–20) from 1 to 10 days, at the time when the RNA content of the anucleate cytoplasm has dropped considerably. Obviously, removal of the nucleus does not immediately stop protein metabolism in the cytoplasm, but it does appreciably reduce it. It is important to add that, in Brachet and Ficq's (1956) autoradiography experiments, incorporation into the nucleus itself is, per surface unit, higher than in the cytoplasm, a fact which is in agreement with some of Mazia and Prescott's (1955) expectations.

Observations on the incorporation of $C^{14}O_2$ into nucleate and anucleate halves of amebae, using both autoradiographic and counting procedures, have fully confirmed the view that RNA lies under much closer nuclear control than do proteins (Brachet, 1959a): in fact, even 8 days after removal of the nucleus, the inhibition is limited to only 40% in the case of protein synthesis. Finally, one should mention very interesting experiments of Goldstein (1958), who grafted the nucleus of a methionine-labeled ameba into a normal ameba; he observed a transfer of radioactivity from the labeled nucleus to the nonlabeled one within 4–8 hours. This observation suggests the existence of a nuclear protein that can move from the nucleus to the cytoplasm and, after a while, migrate back to the nucleus; such a protein might, in Goldstein's (1958) opinion, be associated with nuclear RNA. Finally, Goldstein's (1958)

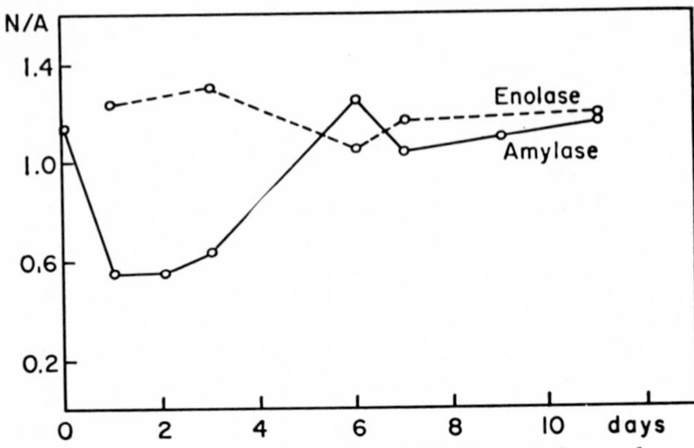

FIG. 15. Changes in the nucleate:anucleate (N:A) ratio for enolase and amylase. From Brachet (1955a).

experiments agree with the view that, in amebae, the nucleus is not quantitatively a major site of protein synthesis.

Thus we come to the conclusion that the nucleus cannot be the exclusive or even the major center of protein synthesis in amebae; amino acid incorporation into cytoplasmic proteins is maintained at a nonnegligible rate in the absence of the nucleus as long as the RNA content of the anucleate fragments remains essentially unchanged. But there is no doubt that, in amebae, the nucleus exerts a control on cytoplasmic protein metabolism. This presents us with a new question: are all the cytoplasmic proteins of the *Amoeba* equally dependent on the nucleus?

This was studied (Brachet, 1955a) by following, in the course of time, the changes of various enzymes (hence of as many specific proteins) in both types of fragments. The experiments showed that the removal of the nucleus results in widely different effects in the case of the various enzymes. Some of them, like protease, aldolase, and adenosinetriphosphatase, remain practically unchanged after removal of the nucleus; amylase (Urbani, 1952) behaves in much the same way, except for the unusual initial increase in activity in the nonnucleated halves, which has already been referred to. Dipeptidase, on the other hand, shows an initial decrease and then remains essentially constant. Acid phosphatase and esterase, finally, have practically disappeared from the nonnucleated cytoplasm after a few days (Figs. 15–17). These experiments establish

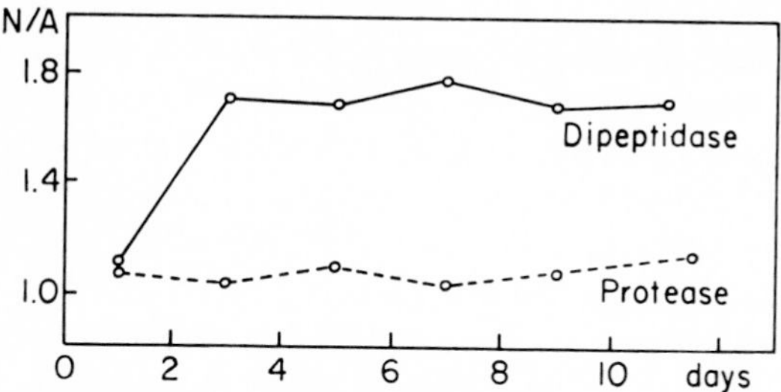

FIG. 16. Changes in the $N:A$ ratio for dipeptidase and protease. From Brachet (1955a).

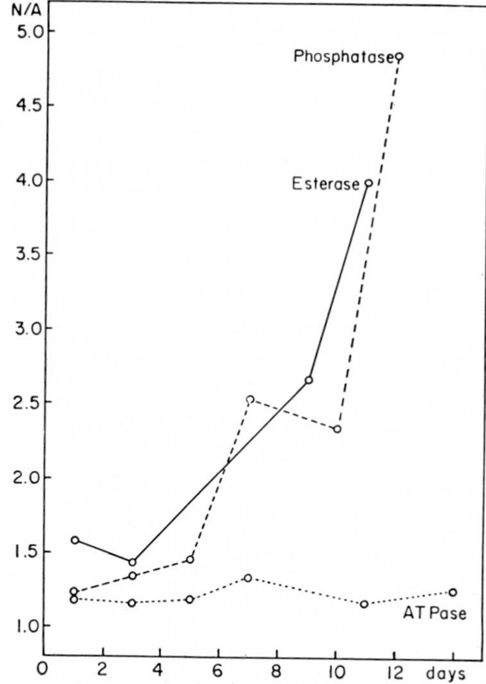

FIG. 17. Changes in the $N:A$ ratio for adenosinetriphosphatase (ATPase), esterase, and phosphatase. From Brachet (1955a).

beyond a doubt that different enzymes are placed, to different extents, under nuclear control and that this postulated "control" of the nucleus is much more complex than might have been expected. It should be noted that none of the enzymes studied ever showed a predominant nuclear localization, a finding which disposes of Wilson's (1925) theory of nuclear storage or synthesis of enzymes.

The reason for this strikingly different behavior of the various enzymes we studied remains uncertain. It is tempting to speculate that the cellular localization of the different enzymes is of importance in this respect. As shown by Holter (1954, 1955), amylase and protease are bound to large granules in the *Amoeba* [mitochondria or the lysosomes of de Duve *et al.* (1955)]. This might imply that, as indicated by the experiments made on the effects of enucleation on the respiration of amebae, mitochondria largely escape nuclear control. The different behavior of dipeptidase is not surprising since, according to Holter (1954), this ubiquitous enzyme is probably in solution in the hyaloplasm. Acid phosphatase and esterase, which behave like RNA, might be bound to microsomes. But this interpretation of the findings is weakened by the fact that according to Holter (1955), Holter and Lowy (1959), and Quertier and Brachet (1959), acid phosphatase behaves like protease: it is normally bound to large granules and is easily released in solution on homogenization. Protease, amylase, and acid phosphatase might well be present in lysosomes, these baglike particles which, according to de Duve *et al.* (1955), apparently contain enzymes in solution. Unequal resistance of the lysosomes to autolytic processes induced by the removal of the nucleus might explain all the results; such a hypothesis would also help in explaining the temporary increase in activity of amylase in the anucleate halves. Nothing more definite than that can be said until more is known about the cellular distribution of several hydrolytic enzymes in amebae.

The recent work of Danielli *et al.* (1955) is also of importance for our understanding of the control exerted by the nucleus on protein synthesis. We have seen that they succeeded in making "hybrids" between two different species of *Amoeba,* using the method of nuclear transfers. They have now prepared antibodies against the two species used for the crosses and found that the reaction (lysis) of the "hybrids" to the antisera is determined by the nucleus. Danielli (1955) and his co-workers (1955) conclude that the determination of the antigenic specific characters lies under nuclear dominance, while the morphological and physiological characters of the "hybrids" are under cytoplasmic dominance.

More recently, E. Wilson (1959) has studied in great detail the transient resistance of *Amoeba discoïdes* to shock doses of its antiserum and investigated nuclear and cytoplasmic relations in the light of this character: it appears, from this analysis, to be dependent on the presence in these amebae of an *Amoeba discoïdes* nucleus. The theoretical implications of these and other experiments have been the subject of interesting discussions by Danielli (1959a, b).

After this extensive discussion of RNA and protein metabolism in fragments of *Amoeba*, the time has now come to study the evidence obtained with *Acetabularia.*

B. *Acetabularia mediterranea*

Many technical difficulties have to be overcome for a correct estimation of RNA in *Acetabularia.* Thanks to the utilization of an adenine

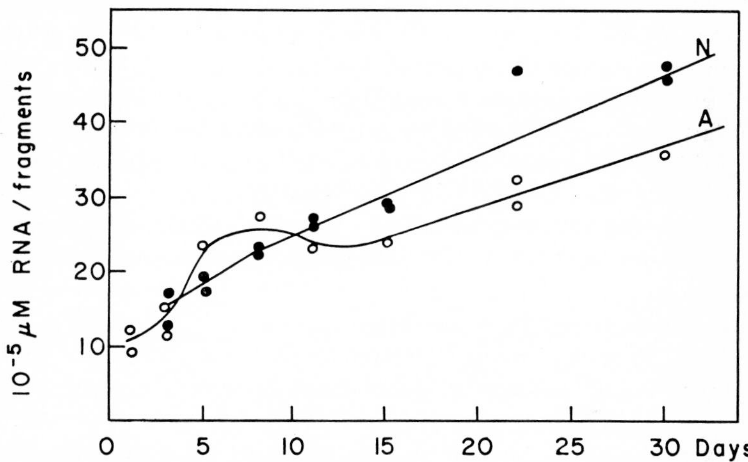

FIG. 18. Synthesis of RNA in nucleate (*N*) and anucleate (*A*) fragments of *Acetabularia.* From Brachet *et al.* (1955).

isotope dilution method, it has been possible to follow the RNA content of regenerating nucleate and anucleate pieces. As shown in Fig. 18, quite unexpected results were first obtained (Brachet *et al.*, 1955): during the first week after removal of the nucleus, there is a *net* RNA synthesis, and its rate is *faster* in the anucleate than in the nucleate halves. However, the rate of this RNA synthesis slows down considerably afterward.

Regenerating anucleate *Acetabularia* stalks and anucleate amebae thus behave in an exactly opposite way. Instead of a drop in the RNA content, enucleation produces, in *Acetabularia,* an initial, but temporary stimulation of RNA synthesis.

However, different results have been obtained by Richter (1957, 1959a), in Hämmerling's laboratory: using a more conventional method for RNA estimation, Richter (1957, 1959a, b) has come to the conclusion that, while the RNA content of the nucleate stems steadily increases, that of the anucleate fragments remains essentially constant. Richter's results obviously confirm the view that, in contrast to what is found in amebae, anucleate cytoplasm can retain its RNA content in *Acetabularia*.

Furthermore, they don't exclude the possibility of independent cytoplasmic RNA synthesis, but on a small scale only. Richter's (1957, 1959a, b) results and conclusions have been essentially confirmed by a subsequent study, carried on in our own laboratory, by Naora *et al.* (1958): using both the isotope dilution method and spectrophotometric procedures, they came to the conclusion that there is no appreciable net synthesis of RNA in anucleate fragments of *Acetabularia*. But later experiments of Naora *et al.* (1960) have gone a large step forward in trying to elucidate what remained a puzzling situation: they found that, in anucleate halves of *Acetabularia*, there is a net synthesis of *chloroplastic* RNA and a corresponding decrease in the other (microsomal and soluble) RNA fractions. If this balance between the various RNA fractions is upset in favor of chloroplastic RNA [and it is known from Hämmerling's (1934) early work that the number of chloroplasts increases in anucleate halves], net RNA synthesis might be found; it is likely that such conditions prevailed in the first experiments of Brachet *et al.* (1955). But a much more important conclusion can be drawn from the latest experiments of Naora *et al.* (1960): synthesis of the RNA which is present in the chloroplasts (which may be considered as self-duplicating bodies) is largely independent of nuclear control; in this respect, chloroplasts behave very much like mitochondria in amebae. On the other hand, nonchloroplastic (microsomal + soluble) RNA lies under strict nuclear control, just as it does in amebae. In other words, the only striking difference between amebae and *Acetabularia*, in respect of the RNA content of anucleate halves, lies in the presence of chloroplasts in the latter only. It would be of obvious interest to find out whether, as one would predict from the above considerations, total RNA markedly drops when anucleate *Acetabularia* fragments are maintained in darkness.

Experiments on the incorporation of orotic acid, a labeled precursor of RNA, into the nucleate and anucleate halves of *Acetabularia* confirm that intensive RNA metabolism continues in the absence of the nucleus: the incorporation is, from the beginning, about 50% higher in the nucleate rhizoids than in the anucleate stalks. But there is no striking fall in the latter's activity, even 70 days after the operation. The incorporation of orotic acid cannot be explained on the ground of net RNA synthesis, since it still goes on normally at a time when there is no such synthesis. It is therefore likely that the turnover of RNA is fast in *Acetabularia* and that it is almost unaffected by removal of the nucleus. The latter is probably especially active in RNA metabolism, however, as evidenced by the fact that incorporation of orotic acid is 50% higher in the rhizoid than in the stalk.

These conclusions have been confirmed by later experiments in which various labeled precursors ($C^{14}O_2$, C^{14}-adenine) have been used (Brachet, 1959a; Naora *et al.*, 1960). It is a striking fact that, even when such an unspecific RNA precursor as $C^{14}O_2$ is used (in the light), extensive incorporation into RNA takes place: one must therefore conclude that the complicated biosynthetic reactions which lead to the formation of ribose and nucleic acid bases proceed for a long time and with considerable efficiency in the absence of the nucleus. However, incorporation of all the precursors studied so far is higher in nucleate halves than in anucleate ones, a fact which suggests once more that the nucleus itself is especially active in RNA metabolism.

Such a suggestion is in perfect agreement with already mentioned observations of Stich and Hämmerling (1953) and Hämmerling and Stich (1956). They studied, with autoradiography and quantitative methods, the incorporation of P^{32} into the RNA of nucleoli which were dissected out of the algae: the incorporation of P^{32} in nucleolar RNA was found to be exceedingly fast. Autoradiography experiments of Vanderhaeghe (1957) on the incorporation of labeled adenine and orotic acid into the RNA of whole *Acetabularia* algae confirm the conclusion that RNA metabolism is very active in the nucleolus: Fig. 19 demonstrates the very strong incorporation of labeled adenine into the nucleolar RNA.

The experimental results described above thus suggest that, in *Acetabularia* as in amebae, a good deal of the RNA which is localized in the microsomes and the cell sap might be produced by the nucleus itself. There is a good deal of circumstantial evidence, but no definite proof yet, for such an assumption: for instance, it has been found that a 1-week treatment with ribonuclease irreversibly inhibits growth and cap formation in anucleate halves; in the case of nucleate halves, there is a resumption of morphogenesis after a lag period (Stich and Plaut, 1958). These observations, which have been confirmed in our laboratory by J. Quertier and F. de Vitry (unpublished) and extended by Puiseux-Dao (1958) to another unicellular alga, *Batophora,* suggest that the RNA, which is present in anucleate halves and which might be of nuclear origin, is required for regeneration; in the presence of ribonuclease, it would be destroyed and could not be resynthesized. In nucleate fragments, on the other hand, the nucleus would be able, after a while, to synthesize RNA again and would yield it to the surrounding cytoplasm.

Similar conclusions can be drawn from work done on the effects of the X-band of X-rays or UV radiations (Bacq *et al.*, 1957; Errera and Vanderhaeghe, 1957; Six, 1958, Errera *et al.*, 1959) and of specific chemical inhibitors of RNA synthesis (Brachet, 1959a); again, anucleate halves are more sensitive to these agents than nucleate ones, and the effects are usually irreversible in the case of the former. Furthermore, it has been

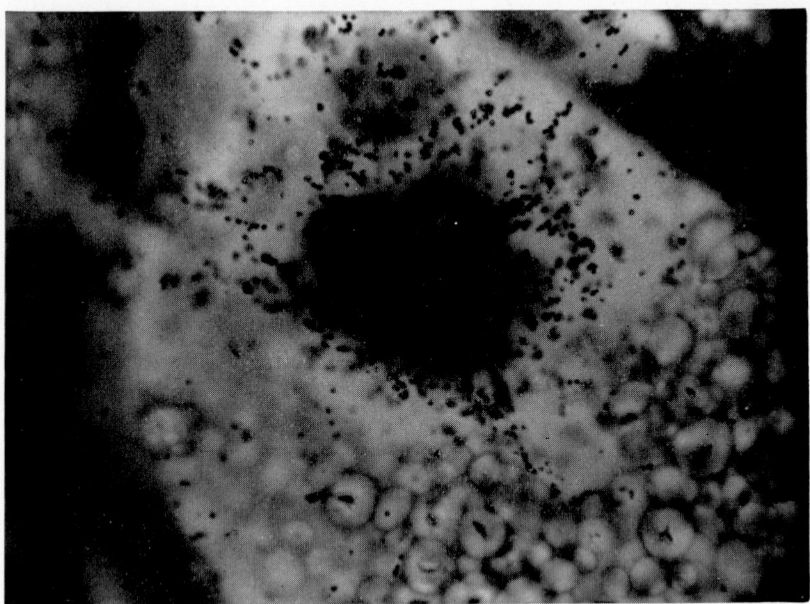

FIG. 19. Autoradiograph showing very strong incorporation of C^{14}-adenine in the nucleolus of *Acetabularia*. Courtesy of F. Vanderhaeghe; from Brachet (1957).

shown by Richter (1959b) that, whereas UV irradiation just inhibits RNA synthesis in nucleate halves, it produces a drop in the RNA content of the anucleate ones: UV radiation and ribonuclease thus act in a very similar way, and the results can be best explained by assuming that a part of the cytoplasmic RNA, which might somehow be involved in morphogenesis, originates from the nucleus.

In conclusion, two opposite processes have been detected in the RNA metabolism of *Acetabularia:* removal of the nucleus does not reduce the RNA content of the cytoplasm; but the nucleus, especially the nucleolus, is more active than the cytoplasm in RNA metabolism and it might be the origin of a part of the cytoplasmic RNA.

Can similar conclusions be obtained in the case of protein metabolism in *Acetabularia*? The answer to that question is yes, and the evidence will now be presented.

In the course of regeneration in *Acetabularia*, growth of the anucleate fragment is paralleled by increases in wet weight and in protein nitrogen (Vanderhaeghe, 1954; Brachet *et al.*, 1955). If regeneration occurs under *suboptimal* conditions, in which the stalks increase in length, but form no, or few, caps, the rate of protein synthesis is the same in the nucleate and in the anucleate pieces for 1–2 weeks (Fig. 20). Protein synthesis then stops altogether, and alterations of the chloroplasts begin: they

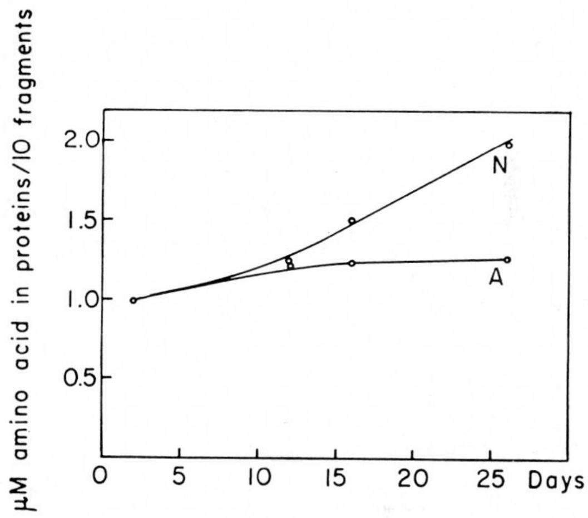

FIG. 20. Protein synthesis in nucleate (*N*) and anucleate (*A*) halves of *Acetabularia* under suboptimal conditions. From Brachet *et al.* (1955).

retain their chlorophyll, but their proteins are partially degraded. The proteins of the small granules (microsomes), on the contrary, remain quantitatively unaffected during this period (Vanderhaeghe, 1954, Clauss, 1959).

If, as in the experiments of Brachet *et al.* (1955), the algae are operated on just before the formation of the caps and if the fragments are placed under *optimal* culture conditions, the nonnucleated pieces will form a high proportion of caps: as a result, net protein synthesis will be definitely *faster* in the anucleate halves than in the nucleate rhizoids (Fig. 21). These experiments, which have been confirmed by Richter (1957), clearly show that the presence of the nucleus is not necessary for protein synthesis, although it is required for *prolonged* protein synthesis: the latter stops completely after 2–3 weeks, that is, when the growth of the cap has ceased. Therefore, in *Acetabularia,* the rate of protein synthesis is initially increased by the removal of the nucleus.

Of great interest regarding the role of the nucleus in protein synthesis are more recent observations by Werz (1957b): he has studied the effects of adding nuclei from either the *A. mediterranea* (*med*) or the *A. crenulata* (*cren*) species in nuclear transplant experiments. The main conclusion is that the addition of a homologous nucleus has no appreciable effect on protein synthesis: the rate of the latter is the same whether the alga contains two or a single nucleus from the same species. In the heterologous *med cren* combination, the *cren* nucleus speeds up protein synthesis, a fact that is in agreement with the observation that protein synthesis is faster in *Acetabularia crenulata* than it is in *Acetabularia mediterranea.*

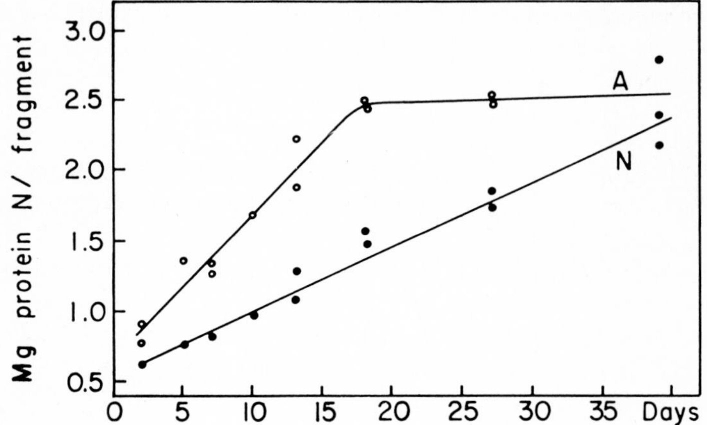

FIG. 21. Protein synthesis in nucleate (N) and anucleate (A) halves of *Acetabularia* under optimal conditions. From Brachet *et al.* (1955).

Of interest, in this respect, is also an observation of Richter (1959c): he found that, if a nucleate part is grafted on an anucleate fragment, both protein and RNA synthesis are resumed. These observations, as well as the preceding ones, might suggest that there is a close and direct link between protein synthesis and morphogenesis in *Acetabularia.* However, as pointed out by Werz and Hämmerling (1959) and Hämmerling *et al.* (1959), protein synthesis and growth processes are not necessarily tightly linked together and the two phenomena can, under given conditions, be dissociated one from the other. For instance, anucleate pieces of the stem which have been cut in the vicinity of the nucleus never form a cap; nevertheless, they are the site of active protein synthesis.

The experimental results just described are in good agreement with the observations of Beth (1953a), who showed, in *Acetabularia,* that the presence of the nucleus exerts an inhibitory effect on cap formation; this process is initially speeded up when the stalk is severed from the rhizoid just before the formation of the cap.

Extensive experiments by Beth (1953b, 1955) have disclosed another interesting fact: cap production markedly depends upon the amount of light received by the algae. Intense illumination produces algae that have a short stalk and a large cap; insufficient light supply results in the formation of very long algae with small caps. The phenomenon studied by Beth (1953b, 1955) also occurs in nature: algae collected in the Mediterranean at a depth of a few feet are short and have large caps in July; those obtained by dredging or deep-diving have long stalks and smaller caps.

It is worth mentioning that it is possible to duplicate Beth's (1953b, 1955) results by merely shifting the —SH- —SS— equilibrium in the medium which surrounds the algae (Brachet, 1959a, c): reducing substances, mercaptoethanol for instance, completely inhibit cap formation in nucleate and anucleate halves; on the contrary —SS— -containing substances, such as dithiodiglycol, stimulate cap formation. It looks as though a sulfur-containing protein, localized at the tip of the stem, plays an important role in morphogenesis in *Acetabularia* and might have something to do with cap formation according to its degree of oxidation or reduction. Further evidence for the view that sulfur-containing proteins are concerned with cap formation in *Acetabularia* and might be part of the still mysterious "morphogenetic substances" can be found in very recent observations of Olszewska and Brachet (1960): they studied, by autoradiography, the incorporation of S^{35}-methionine in anucleate fragments or whole algae and found a very distinct apicobasal gradient of incorporation into the proteins. Finally, it should be added that, according to Thimann and Beth (1959), auxin stimulates growth and cap formation in whole *Acetabularia* algae but has no effect on the increase in length of anucleate fragments: indoleacetic acid can, therefore, not be the "true" morphogenetic substance, the chemical nature of which still remains unknown.

The observations of Beth (1953b) have led Brachet *et al.* (1955) to a study of the regenerative capacities of anucleate fragments which had been left in the dark for increasing lengths of time prior to exposure to light in order to induce regeneration (Figs. 4 and 22). The experiments showed that the same percentage of caps is obtained with the stalks that had been kept 2 weeks in the dark as with those that had been immediately illuminated. But the regenerative potencies of the algae which are kept in the dark for periods longer than 2 weeks soon decrease: they disappear after 4 weeks. It may be concluded that the substances of

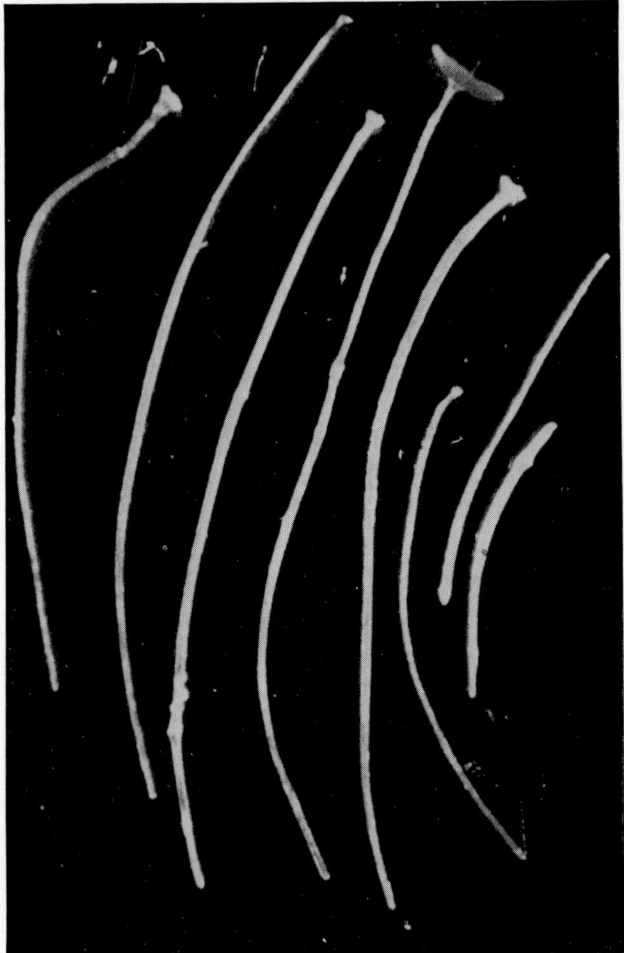

FIG. 22. Decreased regenerative potencies of anucleate parts of *Acetabularia* that have been kept in the dark for 3 weeks. Compare with Fig. 4, where the fragments were cultivated in the light soon after section. From Brachet *et al.* (1955).

nuclear origin which are required for regeneration disappear at the same rate in the light and in the dark; their maintenance is apparently not linked with the energy supply in the cytoplasm. Similar results have since been obtained by J. and C. Hämmerling (1959).

In order to get a better insight into the mechanisms of protein synthesis in *Acetabularia,* Brachet *et al.* (1955) studied the incorporation of $C^{14}O_2$ in the proteins of nucleate and anucleate halves; it was found that the incorporation reaction proceeds at the same rate in both types of fragments for 2 weeks. At that time, incorporation becomes progressively less active in the anucleate halves than in the others; after 7 weeks, the incorporation in the proteins of anucleate stalks is 2.4 times less than in the nucleate rhizoids (Fig. 23).

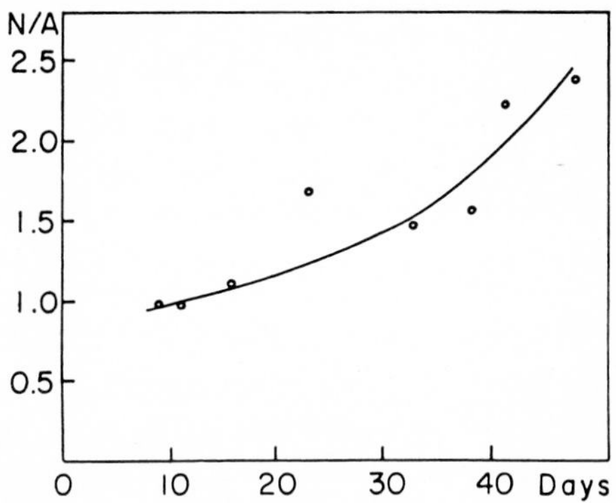

FIG. 23. Changes in the nucleate:anucleate (*N*:*A*) ratio for incorporation of $C^{14}O_2$ into protein. From Brachet *et al.* (1955).

It can be concluded that the incorporation experiments entirely confirm the results obtained for net protein synthesis; but they further show that, even when net protein synthesis has ceased in the anucleate pieces, protein turnover continues for several weeks in the absence of the nucleus.

The incorporation of $C^{14}O_2$ into the proteins of *Acetabularia,* as one would expect, requires light; it becomes negligible in the dark. Further indications that the process is closely linked to photosynthesis are found in the fact that the specific activity of chloroplastic proteins is two to three times higher than that of the other proteins.

A different situation is found when C^{14}-glycine is used as a precursor; its incorporation has, unfortunately, not been studied in fragments. But an interesting fact is that the glycine uptake and incorporation are not dependent on the presence or absence of light; the incorporation, in contrast with the results obtained with $C^{14}O_2$, is stronger in the microsomes than in the chloroplast fraction. The comparison of the results obtained with $C^{14}O_2$ and with glycine suggests that *Acetabularia* possesses two biochemically different mechanisms for protein synthesis: one of them requires CO_2, light, and chloroplasts; in the second, in which an amino acid is the precursor, microsomes are more important than chloroplasts. Such a conclusion stands in perfect agreement with the work of Beth (1953b, 1955), who, as we have seen, demonstrated that growth of the stalks requires less light than cap formation. As we have already seen, many experiments of the author strongly suggest that sulfur-containing amino acids also play a very important role in cap formation.

A few more observations pertaining to the protein metabolism in *Acetabularia* deserve mention. For instance, incorporation of C^{14}-glycine into the proteins is 45% inhibited when the whole algae are placed for 2 weeks in the dark; we know, from Stich's (1951b) work, that the nucleolus becomes spherical and loses much of its RNA under these experimental conditions.[3] Another significant fact is that treatment of the algae with 10^{-3} *M* thiouracil—which acts as an inhibitor of RNA metabolism—also inhibits (30%) the incorporation of glycine into the proteins of whole algae. These observations strongly suggest that RNA is involved, as usual, in the protein metabolism of *Acetabularia,* a conclusion that is shared by Werz (1957a), in his studies on the effects of trypaflavine on regeneration in *Acetabularia.*

We have seen that, in *Amoeba,* different enzymes are unequally placed under nuclear control; there has been no parallel systematic investigation, unfortunately, in *Acetabularia.* All that we know (Brachet *et al.,* 1955) is that, as in *Amoeba,* the enzymes which play a part in nucleotide metabolism and which are said to be accumulated in the liver nuclei (DPN-synthesizing enzyme, adenosine deaminase, nucleoside phosphorylase, guanase) are in concentrations too low to be detected. But it is interesting that aldolase is concentrated in the nucleate half; however, it is not dependent, in *Acetabularia,* on the presence of the nucleus for its synthesis (Baltus, 1959b).

The synthesis of a number of other enzymes has been studied by Hämmerling's group (Hämmerling *et al.,* 1959; Clauss, 1959; Keck and Clauss, 1958): They found a net (and often considerable) synthesis of phosphorylase, invertase, and fructosidase in anucleate fragment. All these enzymes, as is the case for aldolase, are involved in carbohydrate metabolism and are presumably largely bound to chloroplasts. It is of particular interest that, according to Keck and Clauss (1958), acid phosphatase follows an entirely different course: it remains essentially constant in anucleate fragments for several weeks. This particular behavior

[3] It is perhaps for this reason that, as found by Werz (1960), the protein content of both nucleate and anucleate halves drops to the same extent when the algae are kept in the dark.

of acid phosphatase is similar to that of RNA, and it is worth recalling that, in amebae also, RNA and acid phosphatase follow each other in dropping markedly after removal of the nucleus. It looks as if acid phosphatase, as well as RNA, might be synthesized mainly in the nucleus.

The anucleate cytoplasm is thus still capable of synthesizing enzymes, that is, specific proteins. One would very much like to know whether induced enzyme synthesis is also possible in the absence of the nucleus; experiments designed to test that possibility (Brachet *et al.*, 1955) have not, very unfortunately, led to clear-cut results; although an increase in catalase activity was found many times when anucleate pieces of *Acetabularia* were cultivated in the presence of hydrogen peroxide, it was impossible to repeat the results in later experiments. The reasons for this lack of uniformity in the results remain unknown; it might be due to bacterial contamination in the first series of experiments.

We have suggested before that the nucleus might successfully compete with the cytoplasm for RNA precursors. The same possibility exists for protein synthesis, and there is even some evidence in favor of such a possibility. It has been reported by Giardina (1954) and confirmed by Brachet *et al.* (1955) that the proportion of acid-soluble nitrogenous compounds, as compared to protein nitrogen, increases much more in anucleate than in nucleate fragments. This increased synthesis of soluble nitrogenous compounds might correspond to an increase in the pool of the precursors for RNA and protein synthesis, when the competitive influence of the nucleus has been experimentally removed. A study of the chemical nature of this pool, which probably is largely composed of amino acids, peptides, and purine derivatives, might be rewarding; the analysis should be relatively easy with chromatographic methods.

Autoradiography observations on the incorporation of labeled amino acids into nuclear proteins of *Acetabularia* (Vanderhaeghe, 1957) confirm the idea that considerable uptake occurs in the nucleus itself; the proteins of the nucleolus certainly become labeled to a measurable extent before those of the cytoplasm. But, as in amebae, the difference in the incorporation activity between the nucleus and the cytoplasm is much less striking for the incorporation of amino acids into proteins than for that of adenine into RNA.

This conclusion, however, is valid only for certain amino acids, such as glycine or phenylalanine, for which the incorporation into nucleolar proteins is about twice that found for the cytoplasm (per surface unit). The situation is very different for a sulfur-containing amino acid, methionine (Olszewska and Brachet, 1960): the incorporation is extremely high in the nucleolus and, if algae are transferred into a medium containing unlabeled methionine, the cytoplasm within a few hours becomes more intensively labeled than the nucleus. These observations are in agreement with the view that sulfur-containing proteins play a special role in morphogenesis in *Acetabularia;* they are also in agreement with the already mentioned experiments of Goldstein (1958), which suggest that a methionine-containing protein can, in amebae, leave the nucleus for the cytoplasm and be later incorporated again in the nucleus.

To summarize: the experiments made on *Acetabularia* clearly show that anucleate cytoplasm can retain its RNA content and that it is capable of protein synthesis; given favorable experimental conditions, this synthesis can be even more rapid in anucleate than in nucleate fragments. Whereas, in amebae, RNA and protein synthesis both drop together, in *Acetabularia,* the rate of protein synthesis first increases; when, after a 2-week period, synthetic activities come to a standstill in anucleate pieces of *Acetabularia,* RNA and protein still run parallel. These observations are in good agreement with the hypothesis of a close link between RNA and protein synthesis. Finally the experiments show that, although the nucleus does not exert a close and immediate control on protein and RNA synthesis, it does control these syntheses in a more remote way: removal of the nucleus ultimately leads to the arrest of protein synthesis, even though the respiratory and photosynthetic mechanisms remain intact for a very long time.

We shall now see whether these ideas and conclusions are valid for other cells, reticulocytes and eggs for instance.

C. RNA and Protein Synthesis in Other Cells in the Absence of the Nucleus

The reticulocytes (immature red blood cells that have lost their nucleus and retained their RNA in the form of a basophilic network) have been extensively studied in Borsook's laboratory. It is now a well-established fact (Borsook *et al.,* 1952; Koritz and Chantrenne, 1954; Holloway and Ripley, 1952) that the reticulocytes are capable of incorporating amino acids into their proteins, including hemoglobin, despite the loss of their nucleus. The adult red blood cells, on the contrary, have practically lost the ability to incorporate amino acids into their proteins and they contain traces only of RNA. According to Holloway and Ripley (1952), development of reticulocytosis is accompanied by a substantial increase in the RNA content, which is closely paralleled by the amount of radioactive leucine incorporated into the proteins. The authors point out that their results are compatible with the view that RNA is closely associated with amino acid incorporation into proteins. It should be added that they are not compatible with the opinion that the cell nucleus is the most important center of protein synthesis.

Slightly different results have, however, been reported by Koritz and Chantrenne (1954), who found that the maximal rate of incorporation of labeled glycine precedes the RNA maximum by 2 or 3 days: the RNA peak coincides with the maximal content of the red blood cells in hemoglobin, dipeptidase, and carbonic anhydrase. Since proteases, peptidases, and phosphatases also increase during reticulocytosis (Ellis *et al.*, 1956), it is quite possible that the nonnucleated reticulocytes are capable of specific enzyme synthesis. In fact, experimental results of Nizet and Lambert (1953), Schweet *et al.* (1958), Rabinovitz and Olson (1959), Allen and Schweet (1960), Kruh *et al.* (1960), afford good demonstration that actual synthesis of hemoglobin occurs in reticulocytes and that the small ribonucleoprotein particles (ribosomes) are directly involved in this process. In fact, autoradiography observations by Gavosto and Rechenmann (1954) had already shown that, in reticulocytes, an excellent correlation exists between RNA content and amino acid incorporation into proteins: during the ripening of the erythrocytes, loss of basophilia and decrease in the incorporation go hand in hand.

Less is known about possible RNA synthesis in reticulocytes. Isotope experiments of Kruh and Borsook (1955) indicate, however, that they do incorporate radioactive glycine into their RNA.

The results obtained on reticulocytes are thus in full agreement with the data drawn from *Acetabularia:* removal or spontaneous elimination of the nucleus does not necessarily lead to a rapid block of the mechanisms for RNA and protein synthesis.

Similar conclusions can be drawn from the few data we possess for amphibian and sea urchin eggs, where the analysis has not yet been pushed very far: in *Triton* eggs, Tiedemann and Tiedemann (1954) studied the incorporation of radioactive CO_2 into various chemical constituents, especially proteins and RNA. Working with eggs separated into two by ligation, they found no significant difference between the nucleate and the anucleate halves. In unfertilized eggs separated into "light" and "heavy" halves by E. B. Harvey's (1932) centrifugation method, Malkin (1954) observed a stronger incorporation of radioactive glycine into the RNA of the anucleate heavy halves than in the others. The difference was not so striking when Malkin (1954) studied the incorporation of the same precursor into the proteins. But the important fact remains that, in Malkin's experiments, considerable incorporation occurred in both RNA and proteins of the anucleate egg fragments. It should be added, however, that these experiments of Tiedemann and Tiedemann (1954) and of Malkin (1954) can hardly be taken to prove that RNA and protein *synthesis* occurred in the absence of the nucleus. It is likely that, in unfertilized eggs, net syntheses of protein and RNA are negligible and that we are dealing, in fact, with a *turnover*. The latter obviously remains at its normal level in nonnucleated egg cytoplasm.

A recent autoradiography study of Abd-el-Wahab and Pantelouris (1957) leads, however, to somewhat different results than those that have just been reported for sea urchin and newt eggs: they find that the isolated polar lobes of the mussel *Mytilus* (i.e., anucleate cytoplasm) show a considerable and fast decrease in the incorporation of amino acids and adenine into the insoluble materials. But no certain conclusion can be drawn from these experiments because the authors have not studied the uptake and concentration of the soluble precursors. It might well be that the permeability of the polar lobe [which is known to differ strongly from the rest of the cytoplasm by its viscosity, low oxygen consumption (Berg and Kutsky, 1951), and low phosphate uptake (Berg and Prescott, 1958)] to amino acids and purines is abnormally low.

Taken together, present evidence shows that there are quantitative, rather than qualitative, differences in the effects produced by nucleus removal on RNA and protein metabolism. In *Acetabularia*, when optimal conditions are chosen for regeneration, protein and probably RNA synthesis can be stimulated in anucleate fragments; a synthesis of specific proteins can apparently occur in the nonnucleated reticulocytes; turnover of RNA and proteins is not affected by the removal of the egg nucleus. In amebae, RNA and protein syntheses decrease markedly in anucleate fragments. But, in all cases, RNA and proteins show a striking parallelism; in all cases also, the long-term effects of nucleus removal are inhibitory for both protein and RNA anabolism.

VIII. RNA and Protein Metabolism in the Nucleus of the Intact Cell

The evidence relative to the role of the nucleus in RNA and protein metabolism comes from two different sources: work done with isolated nuclei and autoradiography techniques. We shall mainly concentrate here on the results obtained by autoradiography, since it is the only method which can give information on what happens *in situ* at the cytological level. It should be borne in mind, however, that present autoradiography techniques still have serious limitations. The main trouble does not lie in the only semiquantitative character of the techniques; the experience obtained in our laboratory has shown that autoradiography and counter measurements are usually in substantial agreement. The real difficulty is that specific activities cannot be calculated. If we study, for instance, the incorporation of adenine into the RNA of the nucleolus, we usually ignore both the RNA and the free adenine concentrations in this part of the cell. Autoradiography is also unable to tell us what happened to the precursor, which might be altered or even completely oxidized and transformed into $C^{14}O_2$ before it is incorporated; all that we can see is the *localization* of the labeled carbon atoms in the different parts of the cell. With these reservations in mind, we shall briefly review the evidence obtained with autoradiography methods and compare it with the results yielded by the other techniques.

All the autoradiography experiments published so far confirm the observations, repeatedly made on homogenates, of the great metabolic lability of nuclear RNA as compared to that of cytoplasmic RNA. It is known that, since Marshak's (1948) and Jeener and Szafarz's (1950) early experiments, many workers have confirmed that, whatever the chosen precursor, incorporation always proceeds much faster into nuclear RNA than into cytoplasmic RNA. The autoradiography observations of Ficq (1955a, b) on starfish and frog oöcytes have brought forward an additional precision: fast incorporation of labeled adenine or orotic acid into RNA is especially a characteristic of the *nucleolus*. In the germinal vesicle of amphibian oöcytes, however, the incorporation of radioactive adenine also proceeds very quickly in the RNA-containing loops of the lampbrush chromosomes (Brachet and Ficq, 1956). We have seen that, in amebae and in *Acetabularia*, autoradiography observations also indicate a very rapid incorporation of adenine into the nucleoli.

Similar observations have been made, with radioactive phosphate as label, by Vincent (1954) for starfish oöcytes, by Odeblad and Magnusson (1954) for mouse oöcytes, by Stich and Hämmerling (1953) for *Acetabularia*, and by Taylor (1953, 1954) and Taylor and McMaster (1955) for insect glands.

Two more recent observations deserve special mention, since their authors cleverly combined autoradiography studies with experimental procedures: first, Zalokar (1959) studied the incorporation of H^3-uridine into the RNA of *in vivo* centrifuged *Neurospora* hyphae; his experiments clearly showed that RNA synthesis first occurs in the nuclei and that the ergastoplasmic RNA does not become labeled in short-time experiments. Experiments in which the labeled hyphae were transferred to nonradioactive medium clearly suggest that nuclear RNA migrates toward the cytoplasm. In the ovary of *Drosophila* also, according to Zalokar (1960), RNA would be synthesized in the nucleus, while the cytoplasm would be the main site of protein synthesis.

The second set of experiments is due to Perry and Errera (1960), who irradiated with a UV microbeam the nucleolus of *in vitro* cultivated cells: they found that, as a result, RNA synthesis is inhibited not only in the nucleolus and the surrounding chromatin, but also in the cytoplasm. Again, their results are compatible with a nuclear origin of cytoplasmic RNA hypothesis.

Going back to straight autoradiography work, the observations of Goldstein and Micou (1959) on amnion human cells and those of Woods (1959) suggest that RNA synthesis occurs first in the chromosomes and that chromosomal RNA is transferred to the nucleolus; however, according to Sirlin and Elsdale (1959), who worked on a very different material (myoblasts of amphibian embryos), RNA synthesis would start first at the site of the nuclear membrane; ordinary chromatin RNA would be less strongly labeled than RNA present in the "nucleolar apparatus," i.e., the nucleolus plus chromatin associated to the nucleolus. These authors also favor the view that cytoplasmic RNA is largely or entirely of nuclear origin; however, Sirlin's (1960) more recent results on the salivary glands of larvae of *Smittia*, are compatible with the nucleus to cytoplasm RNA transfer hypothesis, but without excluding the possibility of independent cytoplasmic RNA synthesis.

A similar view has been expressed by Woodard (1958) who worked on *Tradescantia* pollen grains; Woodard (1958) is, however, more careful in his conclusions than most of the other workers in the same field: he concludes that the nucleus is not the *only* source for cytoplasmic RNA and that two independent mechanisms for nuclear and cytoplasmic RNA synthesis must exist.

Finally, Harris (1959), who studied the incorporation of adenosine and cytidine in macrophages and connective tissue cells, goes one step further: while admitting a fast turnover of nuclear RNA, he does *not* believe in a nuclear origin of cytoplasmic RNA.

Obviously the situation, as we have already seen in the case of the comparison of nucleate and anucleate fragments of unicellular organisms, remains in a fluid state: while there is no doubt that nuclear RNA always displays unusual synthetic activity, it is not yet proved that all the cytoplasmic RNA originates from the nucleus, and the existence of independent RNA synthesis remains controversial for the time being.

The situation is a little more difficult to appreciate in the case of the metabolism of the nuclear proteins. Most of the work done with homogenates indicates that the most active proteins of the nuclei (the residual proteins) incorporate labeled amino acids at a rate comparable to that found for mixed cytoplasmic proteins (Daly *et al.*, 1952; Allfrey *et al.*,

1955b, 1957; Smellie *et al.*, 1953, and others). This means that the residual proteins must be less active than some of the cytoplasmic proteins, in particular those of the microsomes. But most of this work has been done with nuclei isolated in sucrose or citric acid media, which do not prevent the loss of a large proportion of the nuclear proteins. In more recent experiments of Kay (1956), who isolated the nuclei by a nonaqueous procedure, it was found that residual proteins incorporate labeled amino acids at a very fast rate in short-term (2 hours) experiments.

The experiments performed with homogenates do not, of course, give us any information about the intranuclear localization of the labile proteins; residual proteins are supposed to be constituents of the chromatin, but a nucleolar localization of metabolically labile proteins cannot be definitely excluded either.

It is in such cases that autoradiography methods are of special value, for large cells such as the oöcytes at any rate: in starfish and amphibian oöcytes, nucleoli incorporate precursors (glycine or phenylalanine) at a faster rate than does either nuclear sap or cytoplasm, according to Ficq (1955b). But the differences are never as marked as in the case of the incorporation of purines or pyrimidines into RNA. The same remark can be made for centrifuged amphibian oöcytes: while the incorporation of adenine is, as we have just seen, very intense in the loops of the lampbrush chromosomes, the differences between the loops, the rest of the nuclear sap, and the cytoplasm is much less conspicuous for proteins. In amebae and in *Acetabularia,* as we already know, the nucleus seems to be somewhat more active than the cytoplasm for the incorporation of amino acids into proteins: but, again, the difference between the nucleus and the cytoplasm is usually not as strong as it is for RNA anabolism.

Mention should, however, be made here again of the fact that methionine, in contrast to phenylalanine, shows an unusually high rate of incorporation in the nucleolus of *Acetabularia* and that the proteins into which this amino acid is incorporated behave exactly like RNA: in particular, the cytoplasm of algae which, after a short incubation with S^{35}-methionine, are cultivated in the presence of nonradioactive methionine, after a few hours becomes more strongly labeled than the nucleus. These results are compatible, as we have seen, with the view that methionine-containing proteins might be synthesized in the nucleus and transferred to the cytoplasm (Olszewska and Brachet, 1960).

In higher organisms, liver cells show a much faster incorporation of amino acids into nuclear proteins than into cytoplasmic proteins (Ficq and Errera, 1955b; Moyson, 1955). But liver cells are the exception rather than the rule, since Ficq and Brachet (1956) did not find any conspicuous accumulation of labeled phenylalanine in the nuclei of pancreas, intestine, lung, heart, muscle, spleen, and uterus cells of the mouse. Labeling of nuclear proteins, methionine being used as a precursor, has recently been reported by Pelc (1956) also: as in *Acetabularia,* the nuclei show appreciable incorporation of the precursor, even in tissues where mitotic activity is very small (thyroid, seminal vesicle, epididymis, nerve cells). But it is not stated by Pelc (1956) whether the radioactivity of the nuclei is higher than that of the cytoplasm or not. Finally, it should be added that, according to Carneiro and Leblond (1959), incorporation of several tritiated amino acids (leucine, methionine, and glycine) is high in the chromatin material, but not in the nucleoli of the nuclei of adult mice. Their results suggest that protein synthesis occurs continuously within nuclear chromatin.

There is one instance, however, in which the nuclei always show much higher radioactivity than the cytoplasm after treatment with a labeled amino acid: it is during the early development of amphibian and avian embryos (Brachet and Ledoux, 1955; Waddington and Sirlin, 1954; Sirlin, 1955; Quertier and Brachet, 1959). This is not surprising, since in these embryos we are dealing with very actively dividing cells: extensive synthesis of nuclear proteins is only to be expected under such circumstances.

It is interesting to note, however, that, according to recent autoradiography studies by Sirlin and Waddington (1956) and Sirlin and Elsdale (1959), incorporation of amino acids into proteins of chick and amphibian embryos is higher in the nuclear membrane, the nucleoli, and the nucleolus-associated chromatin than in the cytoplasm. Such conclusions are in obvious agreement with Caspersson's (1941, 1950) ideas.

In conclusion, present autoradiography observations indicate that a very active RNA metabolism in the nucleus is the general rule; protein metabolism, on the other hand, is not necessarily more active in the nucleus than in the cytoplasm, except in fast-dividing cells.

IX. Discussion

The work of the last years has certainly helped toward an understanding of the biochemical functions of the nucleus as a whole. The question is, however, far from settled and new experimental approaches are obviously required in order to make more definite progress.

One of the positive aspects of the experiments made on unicellular organisms is the elimination of unfounded theories. For instance, Loeb's (1899) idea of the "nucleus as the center of cellular oxidations" can be dismissed safely, except for the special case of the red blood cells, which requires further investigation. It is also certain that Wilson's (1925) idea of the nucleus as a storehouse of enzymes is incorrect: very few enzymes, all of them related to nucleotide or anaerobic carbohydrate metabolisms, have been found to be accumulated in the nucleus. And it is far from certain that these enzymes always have a preferential nuclear localization.

One of the possible functions of the nucleus is the production of coenzymes; this might, of course, be only one of the many biochemical activities of the nucleus. But it is tempting to speculate that the activity of some of the cytoplasmic enzymes might be regulated by the production and release of coenzymes from the nucleus. Such a mechanism would provide, as already pointed out, a simple explanation for the frequent attachment of mitochondria to the nucleus observed by Frédéric (1954). Much more experimental work is required before we can say whether coenzyme production is a general function of the cell nucleus; nor do we know whether coenzymes other than DPN are produced in the nucleus.

But, even if the nucleus provides coenzymes to the cytoplasm, the fact remains that mitochondria, as evidenced by the absence of any important effects of nucleus removal on cellular oxidations, largely escape nuclear control. As pointed out before, it is likely that it is rather the other way round, and that the nucleus is dependent on the mitochondria for energy supply.

We are faced with a different situation in the case of the ergastoplasm. The latter is probably deeply modified after the nucleus has been removed, as evidenced by the decrease of basophilia and the finely granular structure of anucleate amebae, even though electron microscopy has so far failed in showing finer changes (Brachet, 1959a). That the nucleus exerts a marked control on the ergastoplasm is also indicated by the fact that mature mammalian red blood cells, which lack a nucleus, are among the few cells in which no ergastoplasmic structures can be detected with the electron microscope (Porter, 1955; Pease, 1956). The nature of the "control" exerted by the nucleus on the ergastoplasm would become very clear if, as suggested by Weiss (1953), by Gay (1955, 1956), and by Rebhun (1956), the nuclear membrane gave rise to the ergastoplasmic lamellae.

We now come to a more important and complex question, that of the role of the nucleus in nucleic acid and protein synthesis. The idea, that the nucleus is the main center for the formation of proteins, has been put forward by Caspersson (1941, 1947, 1950) as a part of a more general theory on the role played by nucleic acids in the cell. Since this theory remains useful in many respects, a short survey of Caspersson's ideas is required here.

Caspersson's theory (1947) is based on three fundamental principles: (1) All protein syntheses require the presence of nucleic acids. (2) Quantitatively the most important nucleic acids in the chromosomes are of the deoxyribose type. (3) The nucleus itself is a cell organelle especially organized as the main center for the formation of proteins.

Starting from these premises, Caspersson (1941, 1950) suggests that the euchromatin, genetically active and rich in DNA, controls the synthesis of the more complex and specific proteins: these would thus be products of the genes. Heterochromatin, especially the nucleolus-associated chromatin, controls the synthesis of histone-like proteins, which are rich in diamino acids. These substances accumulate to form the main bulk of the nucleolus. From the nucleolus, the basic proteins would diffuse toward the nuclear membrane, cross it, and induce in the perinuclear cytoplasm an intensive production of pentose nucleoproteins, the basic amino acids (arginine and histidine) being precursors of the RNA purines. These cytoplasmic pentose nucleic acids somehow would then induce the synthesis of cytoplasmic proteins. As indicated in Fig. 24, nucleoli and cytoplasmic RNA would be intermediaries between nucleolus-associated chromatin and cytoplasmic proteins.

Certain aspects of Caspersson's initial theory, which is now twenty years old, can hardly be retained at present. This is especially true for the role that the histones were supposed to play. Work by Mirsky and Pollister (1942) and by Mirsky (1943) has shown that the absorption spectra of purified histones cannot be distinguished from those of more complex proteins, e.g., albumins. Furthermore, Vincent (1952a, b), working with nucleoli isolated from starfish oöcytes, did not succeed in obtaining any histone-like protein from this material. Finally, recent work dealing with the biosynthesis of purines has shown that the purine ring is preferentially synthesized from simple precursors such as carbon dioxide, formate, and glycine; it does not arise directly from arginine and histidine, as was believed when Caspersson worked out his theory in 1941.

It should also be recalled that, in those days, very little was known about the existence of various cytoplasmic fractions, all of them containing RNA. It is therefore not surprising that no reference is to be found, in Caspersson's theory, to microsomes, mitochondria, or cell sap RNA.

We shall now concentrate on intracellular nucleic acids and protein metabolism. Caspersson's (1941, 1947, 1950) scheme provided a basis for much speculation among geneticists and biochemists, who have repeatedly suggested that RNA is synthesized under the influence of DNA, which would be the primary genetic substance. Proteins would, in turn, be synthesized under the influence of RNA. This idea has been developed by Rich and Watson (1954), by Gale (1955), by Crick (1958), and by Gale and Folkes (1954) among others. The latter, for instance, have proposed that DNA, while not itself capable of synthesizing proteins, might

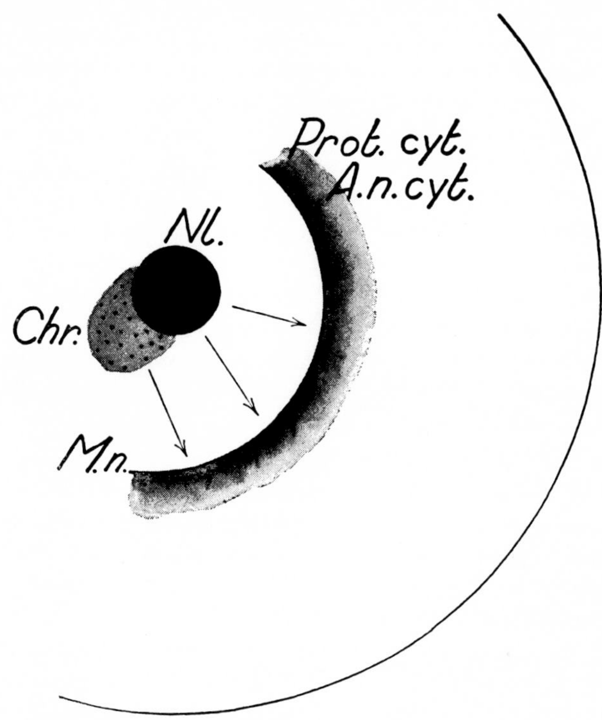

FIG. 24. Diagram of Caspersson's theory of cellular protein synthesis. *Nl.*, nucleolus; *Chr.*, nucleolus-associated chromatin; *M.n.*, nuclear membrane; *A.n.cyt.*, cytoplasmic RNA; *Prot.cyt.*, cytoplasmic proteins. The arrows indicate the migration of proteins from the heterochromatin, directly or via the nucleolus, to the nuclear membrane.

bring about exchange reactions in the proteins; DNA would act as an organizer for RNA synthesis, which in turn would catalyze protein synthesis. One cannot exclude, however, the possibility that DNA may play a direct part in the synthesis of the proteins of the chromosomes themselves, that is in the reproduction of the protein part of the genes. A detailed model for intracellular transfer of DNA, i.e., gene specificity, has been also proposed by Lockingen and De Busk (1955): they suggest that "the principal pathway for the determination of gene-controlled enzyme and protein specificity is by the synthesis of specific RNA moieties containing linear arrays of purine and pyrimidine bases determined by the hydrogen bonding of ribose nucleosides or nucleotides to the purine and pyrimidine bases of chromosomal DNA."

Similar views have been expressed with considerable talent by Crick (1958): according to him, the genetic information is contained in the nucleotide sequence of DNA and is transferred, through a coding mechanism, to the specific proteins; a precise amino acid sequence in the latter would thus correspond to a specific nucleotide sequence of DNA. RNA, as well as DNA (but not the protein), would also contain the genetic information and would be able to transfer it to the proteins. In other words, a specific RNA would be built on the DNA template, and this RNA, by another template mechanism, would control the synthesis of a specific protein: in this way, the secret "code" which is hidden in the architecture of the DNA molecule would be translated in the protein (as an amino acid sequence) with the help of RNA as an intermediate "messenger." All this can be (and has been) described in a much simpler form: "DNA makes RNA, and RNA makes protein." This slogan has become the motto of molecular biology.

But all slogans are simplifications, and we have seen, in this chapter, that difficulties can arise when we have to deal with all the complexities of the cell: nuclear and cytoplasmic proteins or RNA's are obviously different in many respects. Scheme I will perhaps be helpful for a better understanding of the proposed relationship between DNA, RNA, and proteins in the different parts of the cell:

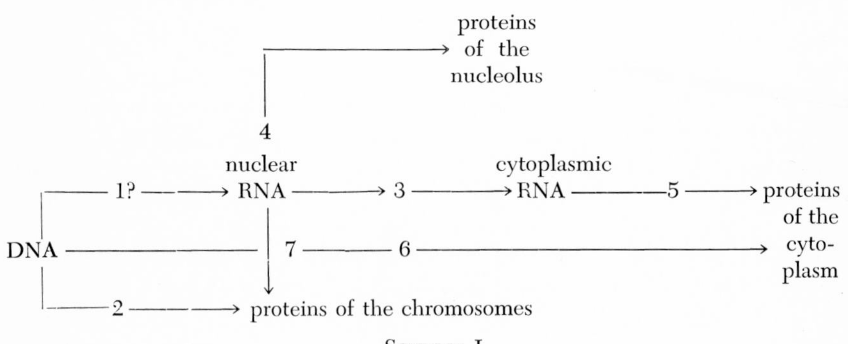

SCHEME I

The various parts of this scheme will now be critically discussed.

Step 5, that is the intervention of cytoplasmic RNA in protein synthesis, does not require much discussion, since this aspect of the role of RNA has been examined in great detail elsewhere (Brachet, 1955b). All that one could add is that the work on anucleate fragments of unicellular organisms, as we have seen, confirms the existence of a relationship between RNA and protein synthesis; the fact that, in *Acetabularia,* regeneration is inhibited by ribonuclease, trypaflavine, and UV irradiation, for which RNA is a favorite target, strongly suggests that RNA is involved in protein synthesis in this alga as elsewhere. The existence of a net protein synthesis in anucleate pieces of *Acetabularia* further shows that the presence of *cytoplasmic* RNA is sufficient to induce synthesis of cytoplasmic proteins, including specific enzymes.

We now come to the role of DNA in protein synthesis: as pointed out several times, present evidence is still conflicting. Definite indications in favor of a direct role of DNA in protein anabolism are found in Gale and Folkes' (1954) work, since they observed that DNA is more effective than RNA in promoting amino acid incorporation into the proteins of nucleic acid-depleted cells (disrupted staphylococci), regarding both the rate and the final amount of the incorporation attained.

The experiments of Allfrey *et al.* (1955b, 1957) on the isolated nuclei of thymocytes also speak strongly in favor of a direct intervention of DNA in the synthesis of *nuclear* proteins (step 2 of our scheme): they found that the incorporation of amino acids into nuclear proteins is markedly inhibited when DNA is partially broken down by deoxyribonuclease. Addition of thymus DNA restores to some extent the activity of the nuclei treated with deoxyribonuclease. The fact that, according to Allfrey *et al.* (1955b, 1957), the incorporation of the labeled amino acids is maximal in a protein fraction which is closely associated with DNA strongly suggests that DNA is directly involved in the synthesis of its protein counterpart in the gene.

However, the more recent experiments of Allfrey and Mirsky (1957, 1958) show that incorporation of amino acids into proteins of isolated nuclei partially depleted of their DNA content can be restored by unspecific DNA, RNA, and even synthetic polynucleotides. These puzzling results found their explanation when Allfrey and Mirsky (1957) showed that the incorporation reaction requires the existence of phosphorylations in the isolated nuclei; these phosphorylation reactions are inhibited when DNA is removed from the nuclei and are restored when polynucleotides are added to the system. These newer experiments of Allfrey and Mirsky, without casting any doubt on the necessity of DNA for the synthesis of nuclear proteins, show that the mechanism may be more complex than could be anticipated.

In fact, the work of Allfrey and Mirsky (1957, 1959) and that of their co-worker Hopkins (1959) shows that protein synthesis, in isolated nuclei, requires an initial synthesis of RNA; according to Hopkins (1959), nuclear RNA would play a *direct* role in the synthesis of proteins in isolated thymus nuclei, just as it does in the case of whole cells.

Mention should also be made of very recent observations of our colleague A. Ficq (1960) on DNA, RNA, and protein synthesis in the various bands of the giant chromosomes of larvae salivary glands in the insect *Rynchosciara:* most of these bands contain no, or very little, RNA; but a few bands are characterized by a high RNA content. Autoradiography observations on the incorporation of radioactive amino acids into the proteins of the individual bands lead to the general conclusion that only the RNA-containing bands incorporate measurable amounts of the label. If protein synthesis is *directly* controlled by DNA, as suggested by arrow 2 of Scheme I, it is probably a relatively slow process as compared to that (nuclear RNA ⟶ proteins of the chromosomes) indicated by arrow 7 in the scheme.

There is also very little evidence for the view that DNA might directly control the synthesis of *cytoplasmic* proteins (arrow 6 of Scheme I): our own observations on *Acetabularia* (Brachet *et al.*, 1955) have clearly shown that protein synthesis is possible in the absence of the nucleus, i.e., of DNA. It is therefore impossible that, in *Acetabularia,* DNA directly controls cytoplasmic protein synthesis. It has been suggested by Mazia and Prescott (1955), following an idea first expressed to them by Monod, that the high degree of genetic autonomy of the plastids might be the reason why regeneration and protein synthesis are possible in the absence of the nucleus in *Acetabularia,* but not in *Amoeba.* The idea is certainly interesting and deserves consideration; but if the plastids are responsible for the high regeneration capacity of *Acetabularia,* it is certainly not by virtue of a content in DNA. Attempts to estimate the DNA content in anucleate pieces of *Acetabularia,* by a thymine isotope dilution method, have so far failed to detect any measurable trace of DNA. We certainly believe that chloroplasts are essential for regeneration and protein synthesis in *Acetabularia*; but their primary role must be to provide energy through photosynthesis, since there is no regeneration in the dark and since the amount and the rate of initial protein synthesis, in anucleate halves, largely depends on illumination conditions.

It should be added, however, that chloroplasts, although devoid of any DNA content can, in *Acetabularia,* incorporate a certain amount of labeled thymidine, but the chemical form in which this incorporation occurs is not yet known (Brachet, 1959a). Furthermore, one should not forget that, according to Naora *et al.* (1960), a net synthesis of RNA occurs, in the absence of the nucleus, in the chloroplasts of *Acetabularia*.

The presence of chloroplasts might thus throw some doubt on the validity of generalizing the conclusions which have been drawn in the case of *Acetabularia* to other cells: but as we have seen, synthesis of hemoglobin is still possible in reticulocytes, and protein synthesis also occurs in enucleated sea urchin or newt eggs. There is thus no doubt that synthesis of *specific* proteins (enzymes, hemoglobin) is possible in the complete absence of the nucleus, thus of the whole genetic material: the genetic information, which is no longer present in the form of DNA, must therefore be stored elsewhere, in some intermediary substance. The best candidate for such an intermediary role is, as we shall see, RNA.

There is, therefore, no reason to draw a sharp line between the organisms that contain chloroplasts and those that do not, so far as the role of the nucleus in protein synthesis is concerned. In any event, it is easy to understand why various organisms or cells behave differently with respect to protein synthesis in the absence of the nucleus: photosynthesis remains normal in anucleate fragments of *Acetabularia* and these fragments are therefore well provided with ATP produced by photosynthetic phosphorylations. On the other hand, it is not surprising that no net protein synthesis can occur in anucleate ameba halves, which are unable to feed. Appreciable protein synthesis or turnover continues, as we know, in reticulocytes and in eggs after loss or removal of the nucleus: but the former are living in a nutrient medium and the eggs have their own large reserves of yolk. If we consider the energy supply as one of the important controlling factors, we see that the variety of the results obtained in enucleation experiments in a series of organisms or cells becomes perfectly logical and that the differences which have been observed are of a *quantitative* nature only.

Coming back to the possible intervention of DNA in over-all protein synthesis, we should like to mention Landman and Spiegelman's (1955) important experiments with bacterial protoplasts: they show that ribonuclease, but not deoxyribonuclease, strongly inhibits induced enzyme synthesis in this system. According to a more recent paper of Spiegelman (1957), removal of *all* the DNA present in the protoplasts does not suppress induced enzyme synthesis; but, surprisingly enough, a DNA synthesis occurs, during the experiments, in protoplasts that have been lysed by an osmotic shock.

Mention should also be made of an isolated, but very interesting, finding of Pardee (1959), who showed that the introduction, by "fertilization" of the chromosomal material containing the gene which controls β-galactosidase synthesis in *Escherichia coli,* can produce the *immediate* synthesis of the enzyme. The absence of any detectable lag in enzyme synthesis suggests that the DNA which is present in the injected chromosome might act as a direct template for protein synthesis. However, the presence of small amounts of specific RNA's in the bacterial chromosome cannot be ruled out; this RNA might contain genetic information received from adjacent DNA.

On the whole, the impression gathered from this survey of the present, still rather meager, evidence is that DNA, in contrast to RNA, is *not* directly involved in the over-all protein synthesis of complex cells; but the situation may be somewhat different in simpler systems, bacteria for instance, and the whole question deserves more study.

One thing is certain, however: there is a large body of evidence showing that, in microorganisms, simultaneous protein and RNA synthesis is possible in the absence of any DNA *synthesis.* This evidence will now be presented because of its obvious importance for the present discussion.

H. and R. Jeener were the first to show, in 1952, that, in *Thermobacterium acidophilus,* suppression of external DNA—which is required for normal growth in this organism—stops nuclear multiplication without affecting growth: the result is the formation of filamentous bacteria, which contain only a small number of nuclei. The cytochemical evidence in this case suggested that DNA synthesis had been effectively inhibited, without suppression of growth, which involves RNA and protein synthesis.

A much more complete biochemical analysis of a similar phenomenon has been made by Cohen and Barner (1954, 1955). They studied the effects of thymine deficiency on a thymine-requiring mutant of *E. coli* and found that these organisms can still increase in length and double their RNA content in the almost complete absence of any DNA synthesis. A very interesting contribution of Cohen and Barner (1954, 1955), which agrees well with the aforementioned findings of Landman and Spiegelman (1955), is that these bacteria are still capable of the induced synthesis of an enzyme, xylose isomerase. Cohen and Barner (1954, 1955) conclude that cytoplasmic and induced enzyme synthesis are still possible in the absence of appreciable DNA synthesis.

The same situation can be found in bacteria irradiated with UV. At low doses, UV light induces, in *E. coli,* the formation of filamentous forms; DNA synthesis is completely inhibited, while RNA and protein syntheses proceed almost normally (Kelner, 1953; Kanazir and Errera, 1954).

Recent observations made in Spiegelman's laboratory by Halvorson and Jackson (1956) on the effects of UV on the induced synthesis of α-glucosidase have further shown that the doses which inhibit DNA synthesis have no marked effect on the enzyme's synthesis; on the other hand, a 22% inhibition of RNA metabolism suppresses enzyme formation to the extent of 95%. We can conclude with Spiegelman (1956) and Spiegelman *et al.* (1955) that suppression of DNA synthesis does not inhibit protein synthesis which, on the other hand, is strongly dependent on the integrity of the RNA molecules.

It might finally be added that agents which are more active on DNA than on RNA synthesis (X-rays, sulfur, and nitrogen mustards) are ineffective in inhibiting induced enzyme synthesis (Baron *et al.*, 1953; Sher and Mallette, 1954; Pardee, 1954; Gros-Doulcet *et al.*, 1955). Conversely, as shown by Pardee and Prestidge (1955), simultaneous inhibition of protein and RNA synthesis, without any effect on DNA synthesis, can be obtained when bacteria are treated with a chemical analog of phenylalanine.

The conclusion of all this work is clear: while synthesis of new RNA molecules seems to be a compulsory concomitant of induced enzyme synthesis,[4] DNA synthesis is not necessary for cytoplasmic or induced enzyme synthesis.

Let us now go back to Scheme I. We have accepted as probable steps 5 and 2 and considered as unlikely a direct relation between DNA and cytoplasmic proteins (step 6), which would have bypassed steps 3 and 5. We now come to the next question: is nuclear RNA produced under the influence of DNA, as indicated in step 1 of the scheme?

This is a difficult question to answer: if we know a little about the effects of chromosomes (especially heterochromosomes) on the RNA content of the nucleolus and the cytoplasm, we have absolutely no proof that DNA is the only active constituent of the chromosomes. Present evidence for a control by heterochromatin of the RNA content of the cell has already been presented. It seems that, according to Caspersson and Schultz's (1939) cytochemical estimations, the amount of RNA in the nucleolus and in the cytoplasm depends on the DNA content and the chromosomal composition of the nuclei. When they compared the RNA content, in the nucleoli and in the cytoplasm, of *Drosophila* oöcytes having a chromosomal formula XX and XXY, they found that the introduction of an extra Y produces an increase. But these early results should be confirmed with the much more accurate methods now in existence.

More precise data exist about the influence of the chromosomal composition on the RNA content of the whole organisms or of special organs such as the larval salivary glands. As shown by Altorfer (1953), there are

[4] However, according to Peabody and Hurwitz (1960), RNA synthesis and protein synthesis are not necessarily linked together, even in bacterial systems.

no significant differences, in *Drosophila,* between normal males and males without a Y chromosome. In the same species, Patterson *et al.* (1954) showed that DNA and RNA contents of larval salivary glands are higher in females than in males, presumably owing to the additional X of the female. Addition or deficiency of a Y chromosome has no measurable effect on the content of either RNA or DNA.

More recent experiments by Levenbook *et al.* (1958) have shown that the situation is more complex than had been anticipated: the presence of a Y chromosome does not modify the actual RNA content of the oöcytes, but it changes the base composition of the pool of acid-soluble nucleotides, nucleosides, and free bases.

There are thus definite indications in favor of the view that heterochromatin, as suggested by Caspersson and Schultz (1939), affects RNA metabolism. But there is no evidence available for or against the view that the production of nuclear RNA is placed directly under control of DNA. A question mark should thus be left in our scheme for step 1.

We also have little information about the possible role of RNA in the synthesis of nucleolar proteins (step 4, Scheme I), but the little we have is all in favor of such a mechanism. For instance, Vincent (1952a, b) has shown that, in starfish oöcytes, a decrease in the RNA concentration occurs during the growth of the nucleolus. A high RNA content thus, as usual, precedes the synthesis of the nucleolar proteins. The facts that, in the same material, incorporation of adenine into RNA is more active than that of amino acids into proteins and that the first of these processes becomes much less active when the oöcyte stops growing, also favor of the same interpretation (Ficq, 1955a). Finally, according to Ficq and Errera (1955a), treatment of the living oöcytes with ribonuclease strongly inhibits the incorporation of amino acids into the nucleolar proteins. This effect of ribonuclease becomes less important in full-grown oöcytes, where the RNA concentration in the nucleolus has decreased. None of these observations constitute absolute proof that the proteins of the nucleolus are being built under the influence of nuclear RNA; but the general evidence indicates that such a conclusion is probably correct and it is thus likely that step 4 of the scheme really proceeds in the cell nucleus.

There is one last point to be discussed: the nuclear origin of cytoplasmic RNA, that is, step 3 of our scheme. We know, from the work of Marshak and Calvet (1949), Jeener and Szafarz (1950), Barnum and Huseby (1950), Hurlbert and Potter (1952), and Smellie *et al.* (1953), that the nuclear RNA always has a higher specific activity than cytoplasmic RNA in isotope experiments with different precursors (P^{32}, orotic acid, glycine, formate). The experimental results are compatible with the possibility that nuclear RNA is a precursor of cytoplasmic RNA, as first proposed by Marshak and Calvet (1949) and by Jeener and Szafarz (1950). There is, however, an alternative explanation, which has been suggested by Barnum and Huseby (1950) and by Hurlbert and Potter (1952): both nuclear and cytoplasmic RNA are synthesized independently at different rates from some unidentified precursor which could be of either nuclear or cytoplasmic origin. In a later analysis of the same problem, based on new experimental data and on extensive calculations, Barnum *et al.* (1953) have finally come to the conclusion, which is shared by Sacks and Samarth (1956), that their results are not consistent with the assumption that nuclear RNA is the precursor of any fraction of cytoplasmic RNA. The complexity of the interpretation of the experimental results remains apparent in the more recent analysis of the problem by Jardetzky and Barnum (1957).

Another difficulty lies in the fact that nuclear RNA and cytoplasmic RNA do not have the same molar composition (Smellie *et al.*, 1953; Elson *et al.*, 1955; Moldave and Heidelberger, 1954; Olmsted and Villee, 1955; de Lamirande *et al.* 1955): all the cytoplasmic RNA can therefore not originate from nuclear RNA by simple diffusion.

But the work done on base composition or radioactivity of nuclear and cytoplasmic RNA does not preclude the possibility that a small, particularly active, RNA fraction migrates from the nucleus to the cytoplasm. In fact, Vincent (1957a, b, 1958) found that starfish nucleoli contain at least two different RNA fractions when incorporation of P^{32} into RNA and solubility are studied; one of them might well be the precursor of cytoplasmic RNA. Similarly, Scholtissek (1960) has found that the RNA of the small microsomes behaves as a mixture of nuclear RNA and that of the mitochondria and larger microsomes; he could also detect, in liver cell nuclei labeled with P^{32}, a small, but very active fraction, which quickly undergoes degradation. Similar observations made by Allfrey and Mirsky (1959) on isolated thymus nuclei strongly suggest that very active RNA fractions of that type come out of the nucleoli. Further studies of these active nucleolar RNA fractions should lead to important conclusions regarding their biochemical role in protein synthesis and, possibly, in transfer of genetic information from the nucleus to the cytoplasm.

Finally, quantitative measurements of the RNA content of the cytoplasm and the nucleoli in various types of cells have also led Swift *et al.* (1956) to the conclusion that the nucleus is not directly involved in cytoplasmic RNA synthesis.

As discussed earlier in this chapter, the evidence obtained with anucleate fragments of unicellular organisms is also conflicting: in amebae, many experiments strongly suggest that a very large proportion, if not all, of the cytoplasmic RNA is of nuclear origin. On the other hand, in *Acetabularia,* net synthesis of chloroplastic RNA probably occurs in the absence of the nucleus.

It seems difficult, under these conditions, to avoid the conclusion that, generally speaking, two different mechanisms coexist for RNA synthesis: intensive RNA synthesis certainly occurs in the nucleus; part of this nuclear RNA very probably goes into the cytoplasm. But independent cytoplasmic RNA synthesis or turnover is also possible, in *Acetabularia,* reticulocytes, and eggs, at any rate.

In order to establish the relative importance of the two synthetic mechanisms, further work on anucleate fragments of unicellular organisms is required. A study of the molar composition of the RNA in these fragments is obviously called for.

What one would especially like to know, in *Acetabularia,* is the chemical composition of the RNA which is being synthesized in the chloroplasts in the absence of the nucleus: is it "nuclear" or "cytoplasmic" RNA that is being formed by the cytoplasm after removal of the nucleus? The question has more than academic interest, since it has important genetic implications: as shown very clearly by Hämmerling's (1934, 1939, 1943, 1953) fundamental experiments, the caps which form during regeneration of an anucleate part possess the genetic characters of the parent organism. The morphogenetic substances that are produced by the nucleus—and which look like gene products—are species specific, as we have seen. It is usually believed that the genetic information contained in DNA is transferred to the cytoplasmic protein-synthesizing systems through RNA. One would, of course, expect nuclear RNA to be the intermediate between DNA and these protein-synthesizing systems of the cytoplasm. Nevertheless, in *Acetabularia,* typical regeneration begins in the absence of either DNA or nuclear RNA. If nuclear RNA is really such an intermediate between the gene and the cytoplasmic protein-synthesizing systems, one would expect the RNA synthesized by the chloroplasts in the absence of the nucleus to retain the same molar base composition as that of nuclear RNA; the latter, in other words, should still be capable of reproducing itself in anucleate cytoplasm, where it would behave as what was once called a "gene-initiated plasmagene." If, on the other hand, it is cytoplasmic RNA which is synthesized in the absence of the nucleus, chloroplast multiplication would represent the reproduction of an independent, non-gene-initiated type of plasmagene. Such a possibility is by no means excluded, since we know that, in Hämmerling's (1934 to 1953) experiments with interspecific grafts, the grafted rhizoid contains cytoplasm besides the nucleus.

If one can venture a prediction, it is that the RNA that is present in the chloroplasts is different from that present in the microsomes and the cell sap; it is likely that soluble and microsomal RNA are more directly placed under nuclear control than chloroplastic RNA (Naora *et al.*, 1960); it is also likely that they play a more direct role in the synthesis of morphogenetic substances than does chloroplastic RNA, and it is a distinct possibility that regeneration stops when the RNA of nuclear origin has been completely utilized.

New experiments are obviously needed in order to provide more direct answers to these questions, which have far-reaching importance for molecular biology as well as biochemical cytology.

References

Abd-el-Wahab, A., and Pantelouris, E. M. (1957). *Exptl. Cell Research* **13**, 78.
Allen, E. H., and Schweet, R. S. (1960). *Biochim. et Biophys. Acta* **39**, 185.
Allfrey, V. G., and Mirsky, A. E. (1957). *Proc. Natl. Acad. Sci. U.S.* **43**, 589.
Allfrey, V. G., and Mirsky, A. E. (1958). *Proc. Natl. Acad. Sci. U.S.* **44**, 981.
Allfrey, V. G., and Mirsky, A. E. (1959). *Proc. Natl. Acad. Sci. U.S.* **45**, 1325.
Allfrey, V. G., Daly, M. M., and Mirsky, A. E. (1955a). *J. Gen. Physiol.* **38**, 415.
Allfrey, V. G., Mirsky, A. E., and Osawa, S. (1955b). *Nature* **176**, 1042.
Allfrey, V. G., Mirsky, A. E., and Stern, H. (1955c). *Advances in Enzymol.* **16**, 411.
Allfrey, V. G., Mirsky, A. E., and Osawa, S. (1957). *J. Gen. Physiol.* **40**, 451.
Altorfer, N. (1953). *Experientia* **9**, 563.
Arnon, D. I., Allen, M. B., Whatley, F. R., Capindale, J. B., and Rosenberg, L. L. (1956). *Proc. Intern. Congr. Biochem. 3rd Congr. Brussels, 1955,* p. 227.
Bacq, Z. M., Damblon, J., and Herve, A. (1955). *Compt. rend. soc. biol.* **149**, 1512.
Bacq, Z. M., Vanderhaeghe, F., Damblon, J., Errera, M., and Herve, A. (1957). *Exptl. Cell Research* **12**, 639.
Balamuth, W. (1940). *Quart. Rev. Biol.* **15**, 290.
Balbiani, E. G. (1888). *Rev. suisse zool.* **5**, 1.
Ballentine, R. (1939). *Biol. Bull.* **77**, 328.
Baltus, E. (1954). *Biochim. et Biophys. Acta* **15**, 263.
Baltus, E. (1955). *Proc. Intern. Congr. Biochem. 3rd Congr., Brussels,* p. 76.
Baltus, E. (1956). *Arch. intern. physiol. et biochim.* **64**, 124.
Baltus, E. (1959a). *Biochim. et Biophys. Acta* **33**, 340.
Baltus, E. (1959b). *Biochim. et Biophys. Acta* **33**, 337.
Barnum, C. P., and Huseby, R. A. (1950). *Arch. Biochem.* **29**, 7.
Barnum, C. P., Huseby, R. A., and Vermund, H. (1953). *Cancer Research* **13**, 880.
Baron, L. S., Spiegelman, S., and Quastler, H. J. (1953). *J. Gen. Physiol.* **36**, 631.
Berg, W. E., and Kutsky, P. B. (1951). *Biol. Bull.* **101**, 47.
Berg, W. E., and Prescott, D. M. (1958). *Exptl. Cell Research* **14**, 402.
Beth, K. (1943). *Z. Induktive Abstammungs- u. Vererbungslehre* **81**, 252, 271.
Beth, K. (1953a). *Z. Naturforsch.* **8b**, 334.
Beth, K. (1953b). *Z. Naturforsch.* **8b**, 771.
Beth, K. (1955). *Z. Naturforsch.* **10b**, 267, 276.
Borner, K., and Mattenheimer, H. (1959). *Biochim. et Biophys. Acta* **34**, 592.

Borsook, H., Deasy, C. L., Haagen-Smit, A. J., Keighley, O., and Lowy, P. H. (1952). *J. Biol. Chem.* **196**, 669.
Brachet, A. (1922). *Arch. biol. (Liège)* **32**, 205.
Brachet, J. (1939). *Arch. exptl. Zellforsch. Gewebezücht.* **22**, 541.
Brachet, J. (1941). *Arch. biol. (Liège)* **53**, 207.
Brachet, J. (1951). *Nature* **168**, 205.
Brachet, J. (1952). *Experientia* **8**, 347.
Brachet, J. (1955a). *Biochim. et Biophys. Acta* **18**, 247.
Brachet, J. (1955b). *In* "Nucleic Acids" (J. N. Davidson and E. Chargaff, eds.), Vol. 2. Academic Press, New York, 1955.
Brachet, J. (1957). "Biochemical Cytology," 516 pp. Academic Press, New York.
Brachet, J. (1959a). *Exptl. Cell Research Suppl.* **6**, 78.
Brachet, J. (1959b). *Ann. N.Y. Acad. Sci.* **78**, 688.
Brachet, J. (1959c). *J. Exptl. Zool.* **142**, 115
Brachet, J., and Ficq, A. (1956). *Arch. biol. (Liège)* **67**, 431.
Brachet, J., and Ledoux, L. (1955). *Exptl. Cell Research, Suppl.* **3**, 27.
Brachet, J., Chantrenne, H., and Vanderhaeghe, F. (1955). *Biochim. et Biophys. Acta* **18**, 544.
Briggs, R., and King, T. J. (1953). *J. Exptl. Zool.* **122**, 485.
Burch, H. B., Storvick, C. A., Bicknell, R. L., Kung, H. C., Alejo, L. G., Everhart, W. A., Lowy, O. H., King, C. G., and Bessey, C. A. (1955). *J. Biol. Chem.* **212**, 897.
Carneiro, J., and Leblond, C. P. (1959). *Science* **129**, 391.
Caspersson, T. (1941). *Naturwissenschaften* **29**, 33.
Caspersson, T. (1947). *Symposia Soc. Exptl. Biol. No.* **1**, 127.
Caspersson, T. (1950). "Cell Growth and Cell Function." Norton, New York.
Caspersson, T., and Schultz, J. (1939). *Nature* **143**, 602, 609.
Chalfin, D. (1956). *J. Cellular Comp. Physiol.* **47**, 215.
Clark, A. M. (1942). *Australian J. Exptl. Biol. Med. Sci.* **20**, 241.
Clauss, H. (1959). *Planta* **52**, 334-534.
Cohen, A. I. (1956). *J. Biophys. Biochem. Cytol.* **2**, 15.
Cohen, S. S., and Barner, H. D. (1954). *Proc. Natl. Acad. Sci. U.S.* **40**, 885.
Cohen, S. S., and Barner, H. D. (1955). *J. Bacteriol.* **69**, 59.
Comandon, J., and de Fonbrune, P. (1939). *Compt. rend. soc. biol.* **130**, 740.
Crick, F. H. S. (1958). *Symposia Soc. Exptl. Biol. No.* **12**, 138.
Crocker, T. T., Goldstein, L., and Cailleau, R. (1956). *Science* **124**, 935.
Cutter, V. M., Jr., Wilson, K. S., and Freeman, B. (1955). *Am. J. Botany* **42**, 109.
Daly, M. M., Allfrey, V. G., and Mirsky, A. E. (1952). *J. Gen. Physiol.* **36**, 173.
Danielli, J. F. (1955). *Exptl. Cell Research, Suppl.* **3**, 98.
Danielli, J. F. (1959a). *Ann. N.Y. Acad. Sci.* **78**, 675.
Danielli, J. F. (1959b). *Exptl. Cell Research, Suppl.* **6**, 252.
Danielli, J. F., Lorch, I. J., Lord, M. J., and Wilson, E. G. (1955). *Nature* **176**, 1114.
De Duve, C., Pressman, B. C., Gianetto, R. J., Wattiaux, R., and Appelmans, F. (1955). *Biochem. J.* **60**, 604.
Defendi, V., and Pearson, B. (1955). *Experientia* **11**, 355.
De Lamirande, G., Allard, C., and Cantero, A. (1955). *J. Biol. Chem.* **214**, 519.
Ellis, D., Sewell, C. E., and Skinner, L. G. (1956). *Nature* **177**, 190.
Elson, D., Trent, I. W., and Chargaff, E. (1955). *Biochim. et Biophys. Acta* **17**, 362.
Errera, M., and Vanderhaeghe, F. (1957). *Exptl. Cell Research* **13**, 10.

Errera, M., Ficq, A., Logan, R., Skreb, Y., and Vanderhaeghe, F. (1959). *Exptl. Cell Research, Suppl.* **6**, 268.

Ficq, A. (1955a). *Arch. biol. (Liège)* **66**, 509.

Ficq, A. (1955b). *Exptl. Cell Research* **9**, 286.

Ficq, A. (1956). *Arch. intern. physiol. et biochim* **64**, 129.

Ficq, A. (1960). *Pathol. et biol.* (in press).

Ficq, A., and Brachet, J. (1956). *Exptl. Cell Research* **11**, 135.

Ficq, A., and Errera, M. (1955a). *Arch. intern. physiol. et biochim.* **63**, 259.

Ficq, A., and Errera, M. (1955b). *Biochim. et Biophys. Acta* **16**, 45.

Frederic, J. (1954). *Ann. N.Y. Acad. Sci.* **58**, 1246.

Gale, E. F. (1955). *In* "Amino Acid Metabolism" (W. McElroy and B. Glass, eds.), p. 171. Johns Hopkins Univ. Press, Baltimore, Maryland.

Gale, E. F., and Folkes, J. P. (1954). *Nature* **173**, 1223.

Gavosto, F., and Rechenmann, R. (1954). *Biochim. et Biophys. Acta* **13**, 583.

Gay, H. (1955). *Proc. Natl. Acad. Sci. U.S.* **41**, 370.

Gay, H. (1956). *J. Biophys. Biochem. Cytol. Suppl.* **2**, 419.

Giardina, G. (1954). *Experientia* **10**, 215.

Goldstein, L. (1958). *Exptl. Cell Research* **15**, 635.

Goldstein, L., and Micou, J. (1959). *J. Biophys. Biochem. Cytol.* **6**, 301.

Goldstein, L., and Plaut, W. (1955). *Proc. Natl. Acad. Sci. U.S.* **41**, 874.

Gros-Doulcet, F., Gros, F., and Spiegelman, S. (1956). *Proc. Intern. Cong. Biochem. 3rd Congr., Brussels, 1955*, p. 74.

Halvorson, H. O., and Jackson, L. (1956). *J. Gen. Microbiol.* **14**, 26.

Hämmerling, J. (1934). *Wilhelm Roux' Arch. Entwicklungsmech. Organ.* **131**, 1.

Hämmerling, J. (1939). *Biol. Zentr.* **59**, 158.

Hämmerling, J. (1943). *Z. Induktive Abstammungs- u. Vererbungslehre* **81**, 114.

Hämmerling, J. (1946). *Z. Naturforsch.* **1b**, 337.

Hämmerling, J. (1953). *Intern. Rev. Cytol.* **2**, 475.

Hämmerling, J. (1956). *Z. Naturforsch.* **11b**, 217.

Hämmerling, J. (1959). *Biol. Zentr.* **78**, 703.

Hämmerling, J., and Hämmerling, Ch. (1959). *Planta* **53**, 522.

Hämmerling, J., and Stich, H. (1954). *Z. Naturforsch.* **9b**, 149.

Hämmerling, J., and Stich, H. (1956). *Z. Naturforsch.* **11b**, 158, 162.

Hämmerling, J., Clauss, H., Keck, K., Richter, G., and Werz, G. (1959). *Exptl. Cell Research, Suppl.* **6**, 210.

Harris, H. (1959). *Biochem. J.* **73**, 362.

Harvey, E. B. (1932). *Biol. Bull.* **62**, 155.

Hogeboom, G. H., and Schneider, W. C. (1952). *J. Biol. Chem.* **197**, 611.

Holloway, B. W., and Ripley, S. H. (1952). *J. Biol. Chem.* **196**, 695.

Holter, H. (1954). *Proc. Roy. Soc.* **B142**, 140.

Holter, H. (1955). *In* "Fine Structure of Cells." Symposium held at the 8th Congress of Cell Biology, Leyden, 1954, p. 71, Interscience, New York.

Holter, H., and Doyle, W. L. (1938). *Compt. rend. trav. Lab. Carlsberg Sér. chim.* **22**, 219.

Holter, H., and Lowy, B. A. (1959). *Compt. rend. trav. lab. Carlsberg* **31**, 105.

Hopkins, J. W. (1959). *Proc. Natl. Acad. Sci. U.S.* **45**, 1461.

Hurlbert, R. B., and Potter, V. R. (1952). *J. Biol. Chem.* **195**, 257.

James, T. W. (1954). *Biochim. et Biophys. Acta* **15**, 367.

Jardetzky, C. D., and Barnum, C. P. (1957). *Arch. Biochem. Biophys.* **67**, 350.

Jeener, H., and Jeener, R. (1952). *Exptl. Cell Research* **3**, 675.
Jeener, R., and Szafarz, D. (1950). *Arch. Biochem.* **26**, 54.
Johnson, R. B., and Ackermann, W. W. (1953). *J. Biol. Chem.* **200**, 263.
Kanazir, D., and Errera, M. (1954). *Biochim. et Biophys. Acta* **14**, 62.
Kay, E. R. M. (1956). *Federation Proc.* **15**, 107.
Keck, K., and Clauss, H. (1958). *Botan. Gaz.* **120**, 43.
Kelner, A. (1953). *J. Bacteriol.* **62**, 252.
Klebs, G. (1889). *Biol. Zentr.* **7**.
Koritz, S. B., and Chantrenne, H. (1954). *Biochim. et Biophys. Acta* **13**, 209.
Kruh, J., and Borsook, H. (1955). *Nature* **175**, 386.
Kruh, J., Schapira, G., and Dreyfus, J. C. (1960). *Biochim. et Biophys. Acta* **39**, 157.
Landman, O. E., and Spiegelman, S. (1955). *Proc. Natl. Acad. Sci. U.S.* **41**, 698.
Levenbook, L., Travaglini, E. K., and Schultz, J. (1956). *Proc. Intern. Cong. Biochem. 3rd Congr., Brussels, 1955*, p. 70.
Levenbook, L., Travaglini, E. C., and Schultz, J. (1958). *Exptl. Cell Research* **15**, 43.
Lockingen, L. S., and De Busk, A. G. (1955). *Proc. Natl. Acad. Sci. U.S.* **41**, 925.
Loeb, J. (1899). *Wilhelm Roux' Arch. Entwicklungsmech. Organ.* **8**, 689.
Lorch, I. J., and Danielli, J. F. (1950). *Nature* **166**, 329.
Malkin, A., and Denstedt, O. P. (1956). *Can. J. Biochem. and Physiol.* **34**, 130.
Malkin, H. M. (1954). *J. Cellular Comp. Physiol.* **44**, 105.
Marshak, A. (1948). *J. Cellular Comp. Physiol.* **32**, 481.
Marshak, A., and Calvet, F. (1949). *J. Cellular Comp. Physiol.* **34**, 451.
Maschlanka, H. (1946). *Biol. Zentr.* **65**, 157.
Mazia, D. (1952). *In* "Modern Trends in Physiology and Biochemistry" (E. S. G. Barron, ed.), p. 77. Academic Press, New York.
Mazia, D., and Hirshfield, H. (1950). *Science* **112**, 297.
Mazia, D., and Prescott, D. M. (1955). *Biochim. et Biophys. Acta* **17**, 23.
Mirsky, A. E. (1943). *Advances in Enzymol.* **3**, 1.
Mirsky, A. E., and Pollister, A. W. (1942). *Proc. Natl. Acad. Sci. U.S.* **28**, 344.
Moldave, K., and Heidelberger, C. (1954). *J. Am. Chem. Soc.* **76**, 679.
Moyson, F. (1955). *Arch. biol. (Liège)* **64**, 247.
Naora, H., Richter, G., and Naora, H. (1958). *Exptl. Cell Research* **16**, 434.
Naora, H., Naora, H., and Brachet, J. (1960). *J. Gen. Physiol.* **43**, 1083.
Nizet, A., and Lambert, S. (1953). *Bull. soc. chim. biol.* **35**, 771.
Odeblad, E., and Magnusson, G. (1954). *Acta Endocrinol.* **17**, 290.
Olmsted, P. S., and Villee, C. A. (1955). *J. Biol. Chem.* **212**, 179.
Olszewska, M., and Brachet, J. (1961). *Exptl. Cell Research* **22**, 370.
Pardee, A. B. (1954). *Proc. Natl. Acad. Sci. U.S.* **40**, 263.
Pardee, A. B. (1959). *Exptl. Cell Research, Suppl.* **6**, 142.
Pardee, A. B., and Prestidge, L. S. (1955). *Federation Proc.* **14**, 262.
Patterson, E. K., Lang, H. M., Dackerman, M. E., and Schultz, J. (1954). *Exptl. Cell Research* **6**, 181.
Peabody, R. A., and Hurwitz, C. (1960). *Biochim. et Biophys. Acta* **39**, 184.
Pease, D. C. (1956). *Blood* **11**, 501.
Pelc, S. R. (1956). *Nature* **178**, 359.
Perry, R., and Errera, M. (1960). *In* "The Cell Nucleus," pp. 24-29. Butterworth and Co., Ltd., London.
Plaut, W. (1959a). *Exptl. Cell Research, Suppl.* **6**, 69.
Plaut, W. (1959b). *Ann. N.Y. Acad. Sci.* **78**, 688.
Plaut, W., and Rustad, R. C. (1956). *Nature* **177**, 89.

Plaut, W., and Rustad, R. C. (1957). *J. Biophys. Biochem. Cytol.* **3**, 625.
Plaut, W., and Rustad, R. C. (1959). *Biochim. et Biophys. Acta* **33**, 59.
Plaut, W., and Sagan, L. A. (1958). *J. Biophys. Biochem. Cytol.* **4**, 483.
Porter, K. R. (1955). *Federation Proc.* **14**, 673.
Potter, V. R., Lyle, G. C., and Schneider, W. C. (1951). *J. Biol. Chem.* **190**, 293.
Prescott, D. M. (1957). *Exptl. Cell Research* **12**, 196.
Prescott, D. M. (1960) *Exptl. Cell Research* **19**, 29.
Prescott, D. M., and Mazia, D. (1954). *Exptl. Cell Research* **6**, 117.
Puiseux-Dao, S. (1958). *Compt. rend. acad. sci.* **246**, 1079, 2286.
Quertier, J., and Brachet, J. (1959). *Arch. biol. (Liège)* **70**, 153.
Rabinovitz, M., and Olson, M. E. (1959). *J. Biol. Chem.* **234**, 2085.
Rapoport, S., and Hofmann, P. C. G. (1955). *Biochem. Z.* **326**, 493.
Rebhun, L. I. (1956). *J. Biophys. Biochem. Cytol.* **2**, 93.
Rich, A., and Watson, J. D. (1954). *Proc. Natl. Acad. Sci. U.S.* **40**, 759.
Richter, G. (1957). *Naturwissenschaften* **44**, 515.
Richter, G. (1959a). *Biochim. et Biophys. Acta* **34**, 407.
Richter, G. (1959b). *Z. Naturforsch.* **14b**, 100.
Richter, G. (1959c). *Planta* **52**, 554.
Rojas, P., and De Robertis, E. (1936). *Rev. soc. arg. biol.* **12**, 325.
Rubinstein, D., and Denstedt, O. P. (1955). *J. Biol. Chem.* **204**, 623.
Rubinstein, D., Ottolenghi, P., and Denstedt, O. P. (1956). *Can. J. Biochem. and Physiol.* **34**, 222.
Sacks, J., and Samarth, E. D. (1956). *J. Biol. Chem.* **223**, 423.
Scholtissek, C. (1960). *Biochem. Z.* **332**, 458, 467.
Schulze, K. L. (1939). *Arch. Protistenk.* **92**, 179.
Schweet, R. S., Lamfrom, H., and Allen, E. (1958). *Proc. Natl. Acad. Sci. U.S.* **44**, 1029.
Shapiro, H. (1935). *J. Cellular Comp. Physiol.* **6**, 101.
Sher, H. I., and Mallette, M. P. (1954). *Arch. Biochem. Biophys.* **52**, 331.
Siekevitz, P. (1952). *J. Biol. Chem.* **195**, 549.
Sirlin, J. L. (1955). *Experientia* **11**, 112.
Sirlin, J. L. (1960). *Exptl. Cell Research* **19**, 177.
Sirlin, J. L., and Elsdale, T. R. (1959). *Exptl. Cell Research* **18**, 268.
Sirlin, J. L., and Waddington, C. H. (1956). *Exptl. Cell Research* **11**, 197.
Six, E. (1956). *Z. Naturforsch.* **11b**, 463.
Six, E. (1958). *Z. Naturforsch.* **13b**, 6.
Skreb-Guilcher, Y. (1955). *Biochim. et Biophys. Acta* **17**, 599.
Skreb-Guilcher, Y. (1956). Personal communication.
Skreb-Guilcher, Y., and Errera, M. (1957). *Exptl. Cell Research* **12**, 649.
Smellie, R. M. S., McIndoe, W. M., and Davidson, J. N. (1953). *Biochim. et Biophys. Acta* **11**, 559.
Spiegelman, S. (1956). *Proc. Intern. Cong. Biochem. 3rd Congr., Brussels, 1955,* p. 185.
Spiegelman, S. (1957). *In* "Symposium on the Chemical Basis of Heredity" (W. D. McElroy and B. Glass, eds.), p. 232. Johns Hopkins Univ. Press, Baltimore, Maryland.
Spiegelman, S., Halvorson, H. O., and Ben-Ishai, R. (1955). *In* "Amino Acid Metabolism" (W. McElroy and B. Glass, eds.), p. 124. Johns Hopkins Univ. Press, Baltimore, Maryland.

Stern, H. (1955). *Science* **121**, 144.
Stern, H., and Mirsky, A. E. (1952). *J. Gen. Physiol.* **36**, 181.
Stern, H., and Timonen, S. (1954). *J. Gen. Physiol.* **38**, 41.
Stern, H., Allfrey, V., Mirsky, A. E., and Saetren, H. (1952). *J. Gen. Physiol.* **35**, 559.
Stich, H. (1951a). *Naturwissenschaften* **38**, 435.
Stich, H. (1951b). *Z. Naturforsch.* **6b**, 36.
Stich, H. (1953). *Z. Naturforsch.* **8b**, 36.
Stich, H. (1955). *Z. Naturforsch.* **10b**, 281.
Stich, H. (1956a). *Experientia* **12**, 7.
Stich, H. (1956b). *Chromosoma* **7**, 693.
Stich, H., and Hämmerling, J. (1953). *Z. Naturforsch.* **8b**, 329.
Stich, H., and Plaut, W. (1958). *J. Biophys. Biochem. Cytol.* **4**, 119.
Swift, H., Rebhun, L., Rasch, E., and Woodard, J. (1956). *In* "Cellular Mechanisms in Differentiation and Growth" (D. Rudnick, ed.), p. 45. Princeton Univ. Press, Princeton, New Jersey.
Tartar, V. (1953). *J. Exptl. Zool.* **124**, 63.
Tartar, V. (1956). *In* "Cellular Mechanisms in Differentiation and Growth" (D. Rudnick, ed.), p. 73. Princeton Univ. Press, Princeton, New Jersey.
Tartar, V., and Chen, T. T. (1941). *Biol. Bull.* **80**, 130.
Taylor, J. H. (1953). *Science* **118**, 555.
Taylor, J. H. (1954). *Genetics* **39**, 998.
Taylor, J. H., and McMaster, R. D. (1955). *Genetics* **40**, 600.
Thimann, K. V., and Beth, K. (1959). *Nature* **183**, 946.
Tiedemann, H., and Tiedemann, H. (1954). *Naturwissenschaften* **41**, 535.
Townsend, C. O. (1897). *Jahrb. wiss. Botan.* **30**.
Urbani, E. (1952). *Biochim. et Biophys. Acta* **9**, 108.
Vanderhaeghe, F. (1954). *Biochim. et Biophys. Acta* **15**, 281.
Vanderhaeghe, F. (1957). Doctorate Thesis. Faculty of Sciences. University of Brussels.
Van Wysselingh, C. (1908). *Beitr. botan. Zentrbl.* **24**, 133.
Verworn, M. (1892). *Arch. ges. Physiol. Pflüger's* **51**, 1.
Vincent, W. S. (1952a). The Isolation and Chemistry of the Nucleoli of Starfish Oocytes. Dissertation, Univ. Pennsylvania.
Vincent, W. S. (1952b). *Proc. Natl. Acad. Sci. U.S.* **38**, 139.
Vincent, W. S. (1954). *Biol. Bull.* **107**, 326.
Vincent, W. S. (1957a). *Publ. Am. Assoc. Advance. Sci. No.* **48**, 1.
Vincent, W. S. (1957b). *Science* **126**, 306.
Vincent, W. S. (1958). *In* "Symposium on the Chemical Basis of Development" (W. D. McElroy and B. Glass, eds.), p. 153. Johns Hopkins Univ. Press, Baltimore, Maryland.
Vishniac, W., and Ochoa, S. (1952). *J. Biol. Chem.* **198**, 501.
Waddington, C. H., and Sirlin, J. L. (1954). *J. Embryol. Exptl. Morphol.* **2**, 340.
Wagner, R. P., and Mitchell, H. K. (1955). "Genetics and Metabolism." Wiley, New York.
Warburg, O. (1909). *Z. physiol. Chem.* **59**, 112.
Waris, H. (1951). *Physiol. Plantarum* **4**, 387.
Weiss, M. (1953). *J. Exptl. Med.* **98**, 607.
Weisz, P. B. (1948). *J. Exptl. Zool.* **107**, 269.
Weisz, P. B. (1949). *J. Exptl. Zool.* **111**, 141.

Weisz, P. B. (1950). *J. Morphol.* **87**, 275.
Weisz, P. B. (1954). *Quart. Rev. Biol.* **29**, 207.
Weisz, P. B. (1955). *J. Cellular Comp. Physiol.* **46**, 517.
Weisz, P. B. (1956). *J. Exptl. Zool.* **131**, 137.
Werz, G. (1957a). *Z. Naturforsch.* **12b**, 559.
Werz, G. (1957b). *Experientia* **13**, 79.
Werz, G. (1960). *Z. Naturforsch.* **15b**, 85.
Werz, G., and Hämmerling, J. (1959) *Planta* **53**, 145.
Whiteley, A. H. (1956). *J. Cellular Comp. Physiol.* **48**, 344.
Wilson, E. (1959). *Exptl. Cell Research, Suppl.* **6**, 132.
Wilson, E. B. (1925). "The Cell in Development and Heredity." Macmillan, New York.
Woodard, J. W. (1958). *J. Biophys. Biochem. Cytol.* **4**, 383.
Woods, P. S. (1959). *Brookhaven Symposium in Biol. No.* **12**, 153.
Wright, S. (1945). *Am. Naturalist* **79**, 289.
Yoshida, W. (1956). *J. Fac. Sci. Niigata Univ. Ser. II,* **2**, 73.
Zalokar, M. (1959). *Nature* **183**, 1330.
Zalokar, M. (1960). *Exptl. Cell Research* **19**, 184.

AUTHOR INDEX

Numbers in italic indicate the pages on which the references are listed.

A

C

E

H

I

J

K

S

SUBJECT INDEX

E

I

P

S